Student Solutions Manual

for

Stewart, Redlin, and Watson's

Algebra and Trigonometry

Second Edition

Andrew Bulman-Fleming

THOMSON
BROOKS/COLE

Australia • Brazil • Canada • Mexico • Singapore • Spain • United Kingdom • United States

Printer: Thomson/West

0-495-01359-5
Cover image: Bill Ralph

Thomson Higher Education
10 Davis Drive
Belmont, CA 94002-3098
USA

For more information about our products, contact us at:
Thomson Learning Academic Resource Center
1-800-423-0563

For permission to use material from this text or product, submit a request online at
http://www.thomsonrights.com.
Any additional questions about permissions can be submitted by email to **thomsonrights@thomson.com.**

Contents

11 Analytic Geometry 339

12 Sequences and Series 375

13 Counting and Probability   399

P Prerequisites

P.1 Modeling the Real World

1. (a) $M = \dfrac{N}{G} = \dfrac{230}{5.4} = 42.6$ miles/gallon

(b) $25 = \dfrac{185}{G} \quad \Leftrightarrow \quad G = \dfrac{185}{25} = 7.4$ gallons

3. (a) $P = 0.45d + 14.7 = 0.45\,(180) + 14.7 = 95.7$ lb/in^2

(b) $P = 0.45d + 14.7 = 0.45\,(330) + 14.7 = 163.2$ lb/in^2

5. (a) $P = 0.06s^3 = 0.06\,(12^3) = 103.7$ hp

(b) $7.5 = 0.06s^3 \quad \Leftrightarrow \quad s^3 = 125$ so $s = 5$ knots

7. $P = 0.8\,(1000) - 500 = \$300.$

9. The number N of days in w weeks is $N = 7w$.

11. The average A of two numbers, a and b, is $A = \dfrac{a+b}{2}$.

13. The sum S of two consecutive integers is $S = n + (n + 1) = 2n + 1$, where n is the first integer.

15. The sum S of a number n and its square is $S = n + n^2$.

17. The product of two consecutive integers is $P = n\,(n + 1) = n^2 + n$, where n is the smaller integer.

19. The sum S of an integer n and twice the integer $S = n + 2n = 3n$.

21. The time it takes an airplane to travel d miles at r miles per hour is $t = \dfrac{d}{r}$.

23. The area A of a square of side x is $A = x^2$.

25. The length is $4 + x$, so the perimeter P is $P = 2\,(4 + x) + 2x = 8 + 2x + 2x = 8 + 4x$.

27. The volume V of a cube of side x is $V = x^3$.

29. The surface area A of a cube of side x is the sum of the areas of the 6 sides, each with area x^2 ; so $A = 6 \cdot x^2 = 6x^2$.

31. The length of one side of the square is $2r$, so its area is $(2r)^2 = 4r^2$. Since the area of the circle is πr^2, the area of what remains is $A = 4r^2 - \pi r^2 = (4 - \pi)\,r^2$.

33. The race track consists of a rectangle of length x and width $2r$ and two semicircles of radius r. The area of the rectangle is $2rx$ and the area of a circle of radius r is πr^2. So the area enclosed is $A = 2rx + \pi r^2$.

35. This is the volume of the large ball minus the volume of the inside ball. The volume of a ball (sphere) is $\frac{4}{3}\pi$ (radius)^3. Thus the volume is $V = \frac{4}{3}\pi R^3 - \frac{4}{3}\pi r^3 = \frac{4}{3}\pi\,(R^3 - r^3)$.

37. (a) $\$7.95 + 2\,(\$1.25) = \$7.95 + \$3.75 = \$11.70$

(b) The cost C, in dollars, of a pizza with n toppings is $C = 7.95 + 1.25n$.

(c) Using the model $C = 7.95 + 1.25$ with $C = 14.20$, we get $14.20 = 7.95 + 1.25n \quad \Leftrightarrow \quad 1.25n = 6.25 \quad \Leftrightarrow \quad n = 5$. So the pizza has 5 toppings.

39. (a) 4 ft/s $\cdot$ 20 s $= 80$ ft

(b) $d = 4$ ft/s $\cdot\, t$ s $= 4t$ ft

(c) Since 1 mile $= 5280$ ft, solve $d = 4t = 5280$ for t. We have $4t = 5280 \quad \Leftrightarrow \quad t = \frac{5280}{4} = 1320$ s $= 22$ minutes.

(d) If R is the rate of mowing, then $60R = 9600$, so $R = \frac{9600}{60} = 160$ ft^2/min.

41. (a) If the width is 20, then the length is 40, so the volume is $20 \cdot 20 \cdot 40 = 16{,}000$ in^3.

(b) In terms of width, $V = x \cdot x \cdot 2x = 2x^3$.

(c) Solve $V = 2x^3 = 6750$ for x. We have $2x^3 = 6750 \quad \Leftrightarrow \quad x^3 = 3375 \quad \Leftrightarrow \quad x = 15$. So the width is 15 in. and the length is $2\,(15) = 30$ in. Thus, the dimensions are 15 in. $\times$ 15 in. $\times$ 30 in.

43. (a) The GPA is $\dfrac{4a + 3b + 2c + 1d + 0f}{a + b + c + d + f} = \dfrac{4a + 3b + 2c + d}{a + b + c + d + f}$.

   (b) Using $a = 2 \cdot 3 = 6$, $b = 4$, $c = 3 \cdot 3 = 9$, and $d = f = 0$ in the formula from part (a), we find the GPA to be
   $\dfrac{4 \cdot 6 + 3 \cdot 4 + 2 \cdot 9}{6 + 4 + 9} = \dfrac{54}{19} \approx 2.84$.

P.2   Real Numbers

1. (a) Natural number: 50

   (b) Integers: $0, -10, 50$

   (c) Rational numbers: $0, -10, 50, \frac{22}{7}, 0.538, 1.2\overline{3}, -\frac{1}{3}$

   (d) Irrational numbers: $\sqrt{7}, \sqrt[3]{2}$

3. Commutative Property for addition

5. Associative Property for addition

7. Distributive Property

9. Commutative Property for multiplication

11. $x + 3 = 3 + x$

13. $4(A + B) = 4A + 4B$

15. $3(x + y) = 3x + 3y$

17. $4(2m) = (4 \cdot 2)m = 8m$

19. $-\frac{5}{2}(2x - 4y) = -\frac{5}{2}(2x) + \frac{5}{2}(4y) = -5x + 10y$

21. (a) $\frac{3}{10} + \frac{4}{15} = \frac{9}{30} + \frac{8}{30} = \frac{17}{30}$

   (b) $\frac{1}{4} + \frac{1}{5} = \frac{5}{20} + \frac{4}{20} = \frac{9}{20}$

23. (a) $\frac{2}{3}\left(6 - \frac{3}{2}\right) = \frac{2}{3} \cdot 6 - \frac{2}{3} \cdot \frac{3}{2} = 4 - 1 = 3$

   (b) $0.25\left(\frac{8}{9} + \frac{1}{2}\right) = \frac{1}{4}\left(\frac{16}{18} + \frac{9}{18}\right) = \frac{1}{4} \cdot \frac{25}{18} = \frac{25}{72}$

25. (a) $\dfrac{\frac{2}{3}}{\frac{2}{3}} - \dfrac{\frac{2}{3}}{2} = 2 \cdot \frac{3}{2} - \frac{2}{3} \cdot \frac{1}{2} = 3 - \frac{1}{3} = \frac{9}{3} - \frac{1}{3} = \frac{8}{3}$

   (b) $\dfrac{\frac{1}{12}}{\frac{1}{8} - \frac{1}{9}} = \dfrac{\frac{1}{12}}{\frac{1}{8} - \frac{1}{9}} \cdot \frac{72}{72} = \frac{6}{9 - 8} = \frac{6}{1} = 6$

27. (a) $2 \cdot 3 = 6$ and $2 \cdot \frac{7}{2} = 7$, so $3 < \frac{7}{2}$

   (b) $-6 > -7$

   (c) $3.5 = \frac{7}{2}$

29. (a) False

   (b) True

31. (a) False

   (b) True

33. (a) $x > 0$

   (b) $t < 4$

   (c) $a \geq \pi$

   (d) $-5 < x < \frac{1}{3}$

   (e) $|p - 3| \leq 5$

35. (a) $A \cup B = \{1, 2, 3, 4, 5, 6, 7, 8\}$

   (b) $A \cap B = \{2, 4, 6\}$

37. (a) $A \cup C = \{1, 2, 3, 4, 5, 6, 7, 8, 9, 10\}$

   (b) $A \cap C = \{7\}$

39. (a) $B \cup C = \{x \mid x \leq 5\}$

   (b) $B \cap C = \{x \mid -1 < x < 4\}$

41. $(-3, 0) = \{x \mid -3 < x < 0\}$

43. $[2, 8) = \{x \mid 2 \leq x < 8\}$

45. $[2, \infty) = \{x \mid x \geq 2\}$

47. $x \leq 1 \quad \Leftrightarrow \quad x \in (-\infty, 1]$

49. $-2 < x \le 1 \quad \Leftrightarrow \quad x \in (-2, 1]$

51. $x > -1 \quad \Leftrightarrow \quad x \in (-1, \infty)$

53. (a) $[-3, 5]$

   (b) $(-3, 5]$

55. $(-2, 0) \cup (-1, 1) = (-2, 1)$

57. $[-4, 6] \cap [0, 8) = [0, 6]$

59. $(-\infty, -4) \cup (4, \infty)$

61. (a) $|100| = 100$

   (b) $|-73| = 73$

63. (a) $||-6| - |-4|| = |6 - 4| = |2| = 2$

   (b) $\frac{-1}{|-1|} = \frac{-1}{1} = -1$

65. (a) $|(-2) \cdot 6| = |-12| = 12$

   (b) $\left|\left(-\frac{1}{3}\right)(-15)\right| = |5| = 5$

67. $|(-2) - 3| = |-5| = 5$

69. (a) $|17 - 2| = 15$

   (b) $|21 - (-3)| = |21 + 3| = |24| = 24$

   (c) $\left|-\frac{3}{10} - \frac{11}{8}\right| = \left|-\frac{12}{40} - \frac{55}{40}\right| = \left|-\frac{67}{40}\right| = \frac{67}{40}$

71. (a) Let $x = 0.777\ldots$. Then $10x = 7.7777\ldots$ and $x = 0.7777\ldots$. Subtracting, we get $9x = 7$. Thus, $x = \dfrac{7}{9}$.

   (b) Let $x = 0.2888\ldots$. Then $100x = 28.8888\ldots$ and $10x = 2.8888\ldots$. Subtracting, we get $90x = 26$. Thus, $x = \dfrac{26}{90} = \dfrac{13}{45}$.

   (c) Let $x = 0.575757\ldots$. Then $100x = 57.5757\ldots$ and $x = 0.5757\ldots$. Subtracting, we get $99x = 57$. Thus, $x = \dfrac{57}{99} = \dfrac{19}{33}$.

73. Distributive Property

75. (a) Is $\frac{1}{28}x + \frac{1}{34}y \le 15$ when $x = 165$ and $y = 230$? We have $\frac{1}{28}(165) + \frac{1}{34}(230) = 5.89 + 6.76 = 12.65 \le 15$, so indeed the car can travel 165 city miles and 230 highway miles without running out of gas.

   (b) Here we must solve for y when $x = 280$. So $\frac{1}{28}(280) + \frac{1}{34}y = 15 \quad \Leftrightarrow \quad 10 + \frac{1}{34}y = 15 \quad \Leftrightarrow \quad \frac{1}{34}y = 5 \quad \Leftrightarrow \quad y = 170$. Thus, the car can travel up to 170 highway miles without running out of gas.

77. (a) Negative since $a > 0 \quad \Leftrightarrow \quad -a < 0$.

   (b) Positive since $b < 0 \quad \Leftrightarrow \quad -b > 0$.

   (c) Positive since the product of two negative numbers is positive.

   (d) Positive since $a - b = a + (-b)$ is the sum of two positive numbers.

   (e) Negative $c - a = c + (-a)$ is the sum of two negative numbers.

   (f) Positive since a is positive and bc is positive by part (c).

   (g) Negative. ab is negative (the product of a positive and a negative) and ac is negative for the same reason. Finally, the sum of two negative numbers is negative.

   (h) Negative, since bc is positive by Part (c) and $a(bc)$ is the product of two positive number, therefore positive, hence and $-abc$ is negative.

   (i) Positive since a is known to be positive and b^2 is also positive, being a square. The product of two positive numbers is positive.

79. $\frac{1}{2} + \sqrt{2}$ is irrational. Since the sum of two rational numbers is rational (see Exercise 78), if $\frac{1}{2} + \sqrt{2}$ were rational, then sum $\left(\frac{1}{2} + \sqrt{2}\right) + \left(-\frac{1}{2}\right) = \sqrt{2}$ is rational. But this is a contradiction. In general the sum of a rational number and an irrational number is irrational. Zero is rational and the product of zero and any real number is zero. If the rational number is not zero, $\frac{a}{b} \neq 0$, then the product of it and any irrational number w is irrational. Again, if we assume the product is rational, say $\frac{a}{b} \cdot w = \frac{c}{d}$, then since $\frac{a}{b} \neq 0$, we have $\frac{b}{a} \neq 0$ so $\frac{b}{a}\left(\frac{a}{b} \cdot w\right) = \frac{b}{a} \cdot \frac{c}{d}$ $\Leftrightarrow$ $w = \frac{bc}{ad}$ which is rational. But w is irrational.

81. (a) Construct the number $\sqrt{2}$ on the number line by transferring the length of the hypotenuse of a right triangle with legs of length 1 and 1.

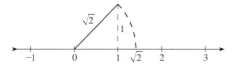

(b) Construct a right triangle with legs of length 1 and 2. By the Pythagorean Theorem, the length of the hypotenuse is $\sqrt{1^2 + 2^2} = \sqrt{5}$. Then transfer the length of the hypotenuse to the number line.

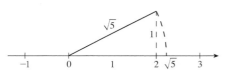

(c) Construct a right triangle with legs of length $\sqrt{2}$ and 2 [construct $\sqrt{2}$ as in part (a)]. By the Pythagorean Theorem, the length of the hypotenuse is $\sqrt{\left(\sqrt{2}\right)^2 + 2^2} = \sqrt{6}$. Then transfer the length of the hypotenuse to the number line.

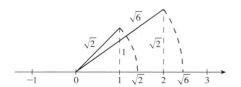

P.3   Integer Exponents

1. $5^2 \cdot 5 = 5^3 = 125$

3. $\left(2^3\right)^2 = 2^6 = 64$

5. $(-6)^0 = 1$

7. $-3^2 = -9$

9. $\left(\frac{1}{3}\right)^4 \cdot 3^6 = \frac{1}{3^4} \cdot 3^6 = 3^{-4} \cdot 3^6 = 3^2 = 9$

11. $\frac{10^7}{10^4} = 10^{7-4} = 10^3 = 1{,}000$

13. $\frac{4^{-3}}{2^{-8}} = \frac{\left(2^2\right)^{-3}}{2^{-8}} = \frac{2^{-6}}{2^{-8}} = 2^{-6-(-8)} = 2^{-6+8} = 2^2 = 4$

15. $\left(\frac{1}{4}\right)^{-2} = (4)^2 = 16$

17. $\left(\frac{3}{2}\right)^{-2} \cdot \frac{9}{16} = \left(\frac{2}{3}\right)^2 \cdot \frac{9}{16} = \frac{2^2}{3^2} \cdot \frac{9}{16} = \frac{4}{9} \cdot \frac{9}{16} = \frac{1}{4}$

19. $\left(\frac{1}{13}\right)^0 \left(\frac{2}{3}\right)^6 \left(\frac{4}{9}\right)^{-3} = 1 \cdot \left(\frac{2^6}{3^6}\right) \left(\frac{9}{4}\right)^3 = \left(\frac{2^6}{3^6}\right) \left(\frac{3^2}{2^2}\right)^3 = \left(\frac{2^6}{3^6}\right) \left(\frac{3^6}{2^6}\right) = 1$

21. $2^{-2} + 2^{-3} = \frac{1}{2^2} + \frac{1}{2^3} = \frac{1}{4} + \frac{1}{8} = \frac{2}{8} + \frac{1}{8} = \frac{3}{8}$

23. $(2x)^4 x^3 = 2^4 x^4 x^3 = 16x^{4+3} = 16x^7$

25. $(-3y)^4 = (-3)^4 y^4 = 3^4 y^4 = 81y^4$

27. $(3z)^2 \left(6z^2\right)^{-3} = \left(3^2 z^2\right) \left(6^{-3} z^{2(-3)}\right) = \left(3^2 z^2\right) \left(6^{-3} z^{-6}\right) = 3^2 (2 \cdot 3)^{-3} z^{2-6} = 3^2 \left(2^{-3} \cdot 3^{-3}\right) z^{2-6}$
$= 3^{2-3} 2^{-3} z^{-4} = 3^{-1} 2^{-3} z^{-4} = \frac{1}{3 \cdot 2^3 z^4} = \frac{1}{24z^4}$

29. $\frac{5x^2}{25x^5} = \frac{5}{25} \cdot \frac{x^2}{x^5} = \frac{1}{5} \cdot \frac{1}{x^{5-2}} = \frac{1}{5} \cdot \frac{1}{x^3} = \frac{1}{5x^3}$

31. $\dfrac{10\,(x+y)^4}{5\,(x+y)^3} = \dfrac{10}{5}\cdot\dfrac{(x+y)^4}{(x+y)^3} = 2\cdot(x+y)^{4-3} = 2\,(x+y) = 2x+2y$

33. $a^9 a^{-5} = a^{9-5} = a^4$

35. $\left(12x^2 y^4\right)\left(\tfrac{1}{2}x^5 y\right) = \left(12\cdot\tfrac{1}{2}\right)x^{2+5}y^{4+1} = 6x^7 y^5$

37. $\dfrac{x^9\,(2x)^4}{x^3} = 2^4\cdot x^{9+4-3} = 16x^{10}$

39. $b^4\left(\tfrac{1}{3}b^2\right)\left(12b^{-8}\right) = \dfrac{12}{3}b^{4+2-8} = 4b^{-2} = \dfrac{4}{b^2}$

41. $(rs)^3\,(2s)^{-2}\,(4r)^4 = r^3 s^3 2^{-2}s^{-2}4^4 r^4 = r^3 s^3 2^{-2}s^{-2}2^{2\cdot4}r^4 = 2^{-2+8}r^{3+4}s^{3-2} = 2^6 r^7 s = 64r^7 s$

43. $\dfrac{\left(6y^3\right)^4}{2y^5} = \dfrac{6^4 y^{3\cdot4}}{2y^5} = \dfrac{6^4}{2}\,y^{12-5} = 648y^7$

45. $\dfrac{\left(x^2 y^3\right)^4\left(xy^4\right)^{-3}}{x^2 y} = \dfrac{x^8 y^{12}x^{-3}y^{-12}}{x^2 y} = x^{8-3-2}y^{12-12-1} = x^3 y^{-1} = \dfrac{x^3}{y}$

47. $\dfrac{\left(xy^2 z^3\right)^4}{\left(x^3 y^2 z\right)^3} = \dfrac{x^4 y^8 z^{12}}{x^9 y^6 z^3} = x^{4-9}y^{8-6}z^{12-3} = x^{-5}y^2 z^9 = \dfrac{y^2 z^9}{x^5}$

49. $\left(\dfrac{q^{-1}rs^{-2}}{r^{-5}sq^{-8}}\right)^{-1} = \dfrac{qr^{-1}s^2}{r^5 s^{-1}q^8} = q^{1-8}r^{-1-5}s^{2-(-1)} = q^{-7}r^{-6}s^3 = \dfrac{s^3}{q^7 r^6}$

51. $69{,}300{,}000 = 6.93\times10^7$

53. $0.000028536 = 2.8536\times10^{-5}$

55. $129{,}540{,}000 = 1.2954\times10^8$

57. $0.0000000014 = 1.4\times10^{-9}$

59. $3.19\times10^5 = 319{,}000$

61. $2.670\times10^{-8} = 0.00000002670$

63. $7.1\times10^{14} = 710{,}000{,}000{,}000{,}000$

65. $8.55\times10^{-3} = 0.00855$

67. **(a)** $5{,}900{,}000{,}000{,}000$ mi $= 5.9\times10^{12}$ mi

 (b) 0.0000000000004 cm $= 4\times10^{-13}$ cm

 (c) 33 billion billion molecules $= 33\times10^9\times10^9 = 3.3\times10^{19}$ molecules

69. $\left(7.2\times10^{-9}\right)\left(1.806\times10^{-12}\right) = 7.2\times1.806\times10^{-9}\times10^{-12} \approx 13.0\times10^{-21} = 1.3\times10^{-20}$

71. $\dfrac{1.295643\times10^9}{\left(3.610\times10^{-17}\right)\left(2.511\times10^6\right)} = \dfrac{1.295643}{3.610\times2.511}\times10^{9+17-6} \approx 0.1429\times10^{19} = 1.429\times10^{19}$

73. $\dfrac{(0.0000162)(0.01582)}{(594621000)(0.0058)} = \dfrac{\left(1.62\times10^{-5}\right)\left(1.582\times10^{-2}\right)}{\left(5.94621\times10^8\right)\left(5.8\times10^{-3}\right)} = \dfrac{1.62\times1.582}{5.94621\times5.8}\times10^{-5-2-8+3} = 0.074\times10^{-12}$
$$= 7.4\times10^{-14}$$

75. **(a)** b^5 is negative since a negative number raised to an odd power is negative.

 (b) b^{10} is positive since a negative number raised to an even power is positive.

 (c) $ab^2 c^3$ we have (positive) (negative)^2 (negative)^3 = (positive) (positive) (negative) which is negative.

 (d) Since $b-a$ is negative, $(b-a)^3$ = (negative)^3 which is negative.

 (e) Since $b-a$ is negative, $(b-a)^4$ = (negative)^4 which is positive.

 (f) $\dfrac{a^3 c^3}{b^6 c^6} = \dfrac{(\text{positive})^3\,(\text{negative})^3}{(\text{negative})^6\,(\text{negative})^6} = \dfrac{(\text{positive})\,(\text{negative})}{(\text{positive})\,(\text{positive})} = \dfrac{\text{negative}}{\text{positive}}$ which is negative.

77. Since one light year is 5.9×10^{12} miles, Centauri is about $4.3\times5.9\times10^{12} \approx 2.54\times10^{13}$ miles away or $25{,}400{,}000{,}000{,}000$ miles away.

79. Volume $=$ (average depth) (area) $= \left(3.7\times10^3\text{ m}\right)\left(3.6\times10^{14}\text{ m}^2\right)\left(\dfrac{10^3\text{ liters}}{\text{m}^3}\right) \approx 1.33\times10^{21}$ liters

81. The number of molecules is equal to

$$(\text{volume})\cdot\left(\dfrac{\text{liters}}{\text{m}^3}\right)\cdot\left(\dfrac{\text{molecules}}{22.4\text{ liters}}\right) = (5\cdot10\cdot3)\cdot\left(10^3\right)\cdot\left(\dfrac{6.02\times10^{23}}{22.4}\right) \approx 4.03\times10^{27}$$

83.

Year	Total interest
1	$152.08
2	308.79
3	470.26
4	636.64
5	808.08

85. (a) $\dfrac{18^5}{9^5} = \left(\dfrac{18}{9}\right)^5 = 2^5 = 32$

    (b) $20^6 \cdot (0.5)^6 = (20 \cdot 0.5)^6 = 10^6 = 1{,}000{,}000$

P.4   Rational Exponents and Radicals

1. $\dfrac{1}{\sqrt{5}} = 5^{-1/2}$

3. $4^{2/3} = \sqrt[3]{4^2} = \sqrt[3]{16}$

5. $\sqrt[5]{5^3} = 5^{3/5}$

7. $a^{2/5} = \sqrt[5]{a^2}$

9. (a) $\sqrt{16} = \sqrt{4^2} = 4$

    (b) $\sqrt[4]{16} = \sqrt[4]{2^4} = 2$

    (c) $\sqrt[4]{\dfrac{1}{16}} = \sqrt[4]{\left(\dfrac{1}{2}\right)^4} = \dfrac{1}{2}$

11. (a) $\sqrt{\dfrac{4}{9}} = \sqrt{\left(\dfrac{2}{3}\right)^2} = \dfrac{2}{3}$

    (b) $\sqrt[4]{256} = \sqrt[4]{4^4} = 4$

    (c) $\sqrt[6]{\dfrac{1}{64}} = \sqrt[6]{\dfrac{1}{2^6}} = \dfrac{\sqrt[6]{1}}{\sqrt[6]{2^6}} = \dfrac{1}{2}$

13. (a) $\left(\dfrac{4}{9}\right)^{-1/2} = \left(\dfrac{2^2}{3^2}\right)^{-1/2} = \dfrac{2^{-1}}{3^{-1}} = \dfrac{3}{2}$

    (b) $(-32)^{2/5} = \left[(-2)^5\right]^{2/5} = (-2)^2 = 4$

    (c) $(-125)^{-1/3} = \left[(-5)^3\right]^{-1/3} = (-5)^{-1} = \dfrac{1}{-5} = -\dfrac{1}{5}$

15. (a) $\left(\dfrac{1}{32}\right)^{2/5} = \left(\dfrac{1}{2^5}\right)^{2/5} = \left(\dfrac{1}{2}\right)^{5\cdot(2/5)} = \left(\dfrac{1}{2}\right)^2 = \dfrac{1}{4}$

    (b) $(27)^{-4/3} = \left(3^3\right)^{-4/3} = (3)^{3\cdot(-4/3)} = 3^{-4} = \dfrac{1}{3^4} = \dfrac{1}{81}$

    (c) $\left(\dfrac{1}{8}\right)^{-2/3} = \left(\dfrac{1}{2^3}\right)^{-2/3} = \left(2^{-3}\right)^{-2/3} = 2^{(-3)\cdot(-2/3)} = 2^2 = 4$

17. (a) $100^{-1.5} = \left(10^2\right)^{-1.5} = 10^{2\cdot(-1.5)} = 10^{-3} = \dfrac{1}{10^3} = \dfrac{1}{1000}$

    (b) $4^{2/3} \cdot 6^{2/3} \cdot 9^{2/3} = \left(2^2\right)^{2/3} \cdot (2\cdot 3)^{2/3} \cdot \left(3^2\right)^{2/3} = 2^{4/3} \cdot 2^{2/3} \cdot 3^{2/3} \cdot 3^{4/3} = 2^2 \cdot 3^2 = 4 \cdot 9 = 36$

    (c) $0.001^{-2/3} = \left(10^{-3}\right)^{-2/3} = 10^2 = 100$

19. When $x = 3$, $y = 4$, $z = -1$ we have $\sqrt{x^2 + y^2} = \sqrt{3^2 + 4^2} = \sqrt{9 + 16} = \sqrt{25} = 5$.

21. When $x = 3$, $y = 4$, $z = -1$ we have

$$(9x)^{2/3} + (2y)^{2/3} + z^{2/3} = (9 \cdot 3)^{2/3} + (2 \cdot 4)^{2/3} + (-1)^{2/3} = \left(3^3\right)^{2/3} + \left(2^3\right)^{2/3} + (1)^{1/3}$$

$$= 3^2 + 2^2 + 1 = 9 + 4 + 1 = 14.$$

23. $\sqrt{32} + \sqrt{18} = \sqrt{16 \cdot 2} + \sqrt{9 \cdot 2} = \sqrt{4^2 \cdot 2} + \sqrt{3^2 \cdot 2} = 4\sqrt{2} + 3\sqrt{2} = 7\sqrt{2}$

25. $\sqrt{125} - \sqrt{45} = \sqrt{25 \cdot 5} - \sqrt{9 \cdot 5} = \sqrt{5^2 \cdot 5} - \sqrt{3^2 \cdot 5} = 5\sqrt{5} - 3\sqrt{5} = 2\sqrt{5}$

27. $\sqrt[3]{108} - \sqrt[3]{32} = 3\sqrt[3]{4} - 2\sqrt[3]{4} = \sqrt[3]{4}$

29. $\sqrt{245} - \sqrt{125} = 7\sqrt{5} - 5\sqrt{5} = 2\sqrt{5}$

31. $\sqrt[5]{96} + \sqrt[5]{3} = \sqrt[5]{32 \cdot 3} + \sqrt[5]{3} = \sqrt[5]{2^5 \cdot 3} + \sqrt[5]{3} = 2\sqrt[5]{3} + \sqrt[5]{3} = 3\sqrt[5]{3}$

33. $\sqrt[4]{x^4} = |x|$

35. $\sqrt[4]{16x} = \sqrt[4]{2^4 x^8} = 2x^2$

37. $\sqrt[3]{x^3 y} = \left(x^3\right)^{1/3} y^{1/3} = x\sqrt[3]{y}$

39. $\sqrt[5]{a^6 b^7} = a^{6/5} b^{7/5} = a \cdot a^{1/5} b \cdot b^{2/5} = ab\sqrt[5]{ab^2}$

41. $\sqrt[3]{\sqrt{64x^6}} = \left(8\left|x^3\right|\right)^{1/3} = 2\left|x\right|$

43. $x^{2/3} x^{1/5} = x^{(10/15 + 3/15)} = x^{13/15}$

45. $\left(-3a^{1/4}\right)(9a)^{-3/2} = \left(-1 \cdot 3a^{1/4}\right)\left(3^2 a\right)^{-3/2} = \left(-1 \cdot 3a^{1/4}\right)\left(3^{-3} a^{-3/2}\right) = -1 \cdot 3^{1-3} \cdot a^{1/4 - 3/2}$

$= -1 \cdot 3^{-3} a^{-5/4} = -\dfrac{1}{3^3 \cdot a^{5/4}} = -\dfrac{1}{9a^{5/4}}$

47. $(4b)^{1/2}\left(8b^{2/5}\right) = \sqrt{4} \cdot 8b^{1/2} b^{2/5} = 16b^{(5/10 + 4/10)} = 16b^{9/10}$

49. $\left(c^2 d^3\right)^{-1/3} = c^{-2/3} d^{-1} = \dfrac{1}{c^{2/3} d}$

51. $\left(y^{3/4}\right)^{2/3} = y^{(3/4)\cdot(2/3)} = y^{1/2}$

53. $\left(2x^4 y^{-4/5}\right)^3 \left(8y^2\right)^{2/3} = 2^3 x^{12} y^{-12/5} 8^{2/3} y^{4/3} = 2^{3+2} x^{12} y^{(-12/5 + 4/3)} = \dfrac{32x^{12}}{y^{16/15}}$. Note that $8^{2/3} = \left(8^{1/3}\right)^2 = 2^2$.

55. $\left(\dfrac{x^6 y}{y^4}\right)^{5/2} = \dfrac{x^{15} y^{5/2}}{y^{10}} = x^{15} y^{5/2 - 10} = x^{15} y^{-15/2} = \dfrac{x^{15}}{y^{15/2}}$

57. $\left(\dfrac{3a^{-2}}{4b^{-1/3}}\right)^{-1} = \dfrac{3^{-1} a^2}{4^{-1} b^{1/3}} = \dfrac{4a^2}{3b^{1/3}}$

59. $\dfrac{(9st)^{3/2}}{\left(27s^3 t^{-4}\right)^{2/3}} = \dfrac{27s^{3/2} t^{3/2}}{9s^2 t^{-8/3}} = 3s^{3/2 - 2} t^{3/2 + 8/3} = 3s^{-1/2} t^{25/6} = \dfrac{3t^{25/6}}{s^{1/2}}$

61. (a) $\dfrac{1}{\sqrt{6}} = \dfrac{1}{\sqrt{6}} \cdot \dfrac{\sqrt{6}}{\sqrt{6}} = \dfrac{\sqrt{6}}{6}$

(b) $\dfrac{3}{\sqrt{2}} = \dfrac{3}{\sqrt{2}} \cdot \dfrac{\sqrt{2}}{\sqrt{2}} = \dfrac{3\sqrt{2}}{2}$

(c) $\dfrac{9}{\sqrt{3}} = \dfrac{9}{\sqrt{3}} \cdot \dfrac{\sqrt{3}}{\sqrt{3}} = \dfrac{9\sqrt{3}}{3} = 3\sqrt{3}$

63. (a) $\dfrac{1}{\sqrt[3]{4}} = \dfrac{1}{\sqrt[3]{2^2}} \cdot \dfrac{\sqrt[3]{2}}{\sqrt[3]{2}} = \dfrac{\sqrt[3]{2}}{2}$

(b) $\dfrac{1}{\sqrt[4]{3}} = \dfrac{1}{\sqrt[4]{3}} \cdot \dfrac{\sqrt[4]{3^3}}{\sqrt[4]{3^3}} = \dfrac{\sqrt[4]{3^3}}{3} = \dfrac{\sqrt[4]{27}}{3}$

(c) $\dfrac{8}{\sqrt[5]{2}} = \dfrac{8}{\sqrt[5]{2}} \cdot \dfrac{\sqrt[5]{2^4}}{\sqrt[5]{2^4}} = \dfrac{8\sqrt[5]{2^4}}{2} = 4\sqrt[5]{2^4} = 4\sqrt[5]{16}$

65. (a) $\dfrac{1}{\sqrt[3]{x}} = \dfrac{1}{\sqrt[3]{x}} \cdot \dfrac{\sqrt[3]{x^2}}{\sqrt[3]{x^2}} = \dfrac{\sqrt[3]{x^2}}{x}$

(b) $\dfrac{1}{\sqrt[5]{x^2}} = \dfrac{1}{\sqrt[5]{x^2}} \cdot \dfrac{\sqrt[5]{x^3}}{\sqrt[5]{x^3}} = \dfrac{\sqrt[5]{x^3}}{x}$

(c) $\dfrac{1}{\sqrt[7]{x^3}} = \dfrac{1}{\sqrt[7]{x^3}} \cdot \dfrac{\sqrt[7]{x^4}}{\sqrt[7]{x^4}} = \dfrac{\sqrt[7]{x^4}}{x}$

67. First convert 1135 feet to miles. This gives

$1135 \text{ ft} = 1135 \cdot \dfrac{1 \text{ mile}}{5280 \text{ feet}} = 0.215 \text{ mi}$. Thus the distance you can see is given by

$$D = \sqrt{2rh + h^2} = \sqrt{2\,(3960)\,(0.215) + (0.215)^2}$$

$$\approx \sqrt{1702.8} \approx 41.3 \text{ miles}$$

69. (a) Substituting, we get $0.30\,(60) + 0.38\,(3400)^{1/2} - 3\,(650)^{1/3} \approx 18 + 0.38\,(58.31) - 3\,(8.66) \approx 18 + 22.16 - 25.98 \approx 14.18$. Since this value is less than 16, the sailboat qualifies for the race.

(b) Solve for A when $L = 65$ and $V = 600$. Substituting, we get $0.30\,(65) + 0.38A^{1/2} - 3\,(600)^{1/3} \leq 16 \Leftrightarrow$ $19.5 + 0.38A^{1/2} - 25.30 \leq 16 \Leftrightarrow 0.38A^{1/2} - 5.80 \leq 16 \Leftrightarrow 0.38A^{1/2} \leq 21.80 \Leftrightarrow A^{1/2} \leq 57.38 \Leftrightarrow A \leq 3292.0$. Thus, the largest possible sail is 3292 ft^2.

71. Since 1 day $= 86,400$ s, 365.25 days $= 31,557,600$ s. Substituting, we obtain $d = \left(\dfrac{6.67 \times 10^{-11} \times 1.99 \times 10^{30}}{4\pi^2} \right)^{1/3}$.

$\left(3.15576 \times 10^7 \right)^{2/3} \approx 1.5 \times 10^{11}$ m $= 1.5 \times 10^8$ km.

73. (a)

n	1	2	5	10	100
$2^{1/n}$	$2^{1/1} = 2$	$2^{1/2} = 1.414$	$2^{1/5} = 1.149$	$2^{1/10} = 1.072$	$2^{1/100} = 1.007$

So when n gets large, $2^{1/n}$ decreases to 1.

(b)

n	1	2	5	10	100
$\left(\frac{1}{2}\right)^{1/n}$	$\left(\frac{1}{2}\right)^{1/1} = 0.5$	$\left(\frac{1}{2}\right)^{1/2} = 0.707$	$\left(\frac{1}{2}\right)^{1/5} = 0.871$	$\left(\frac{1}{2}\right)^{1/10} = 0.933$	$\left(\frac{1}{2}\right)^{1/100} = 0.993$

So when n gets large, $\left(\frac{1}{2}\right)^{1/n}$ increases to 1.

75. Using $v = c/10$ in the formula $m = \dfrac{m_0}{\sqrt{1 - v^2/c^2}}$, we get $m = \dfrac{m_0}{\sqrt{1 - (c/10)^2/c^2}} = \dfrac{m_0}{\sqrt{1 - \frac{1}{100}}}$. Thus

$m = \dfrac{1}{\sqrt{\frac{99}{100}}} m_0 = \dfrac{10\sqrt{11}}{33} m_0$. So the rest mass of the spaceship is multiplied by $\dfrac{10\sqrt{11}}{33} \approx 1.005$.

Using $v = c/2$ in the formula, we get $m = \dfrac{m_0}{\sqrt{1 - (c/2)^2/c^2}} = \dfrac{m_0}{\sqrt{1 - \frac{1}{4}}}$. Thus $m = \dfrac{1}{\sqrt{\frac{3}{4}}} m_0 = \dfrac{2\sqrt{3}}{3} m_0$. So the rest

mass of the spaceship is multiplied by $\frac{2\sqrt{3}}{3} \approx 1.15$.

Using $v = 0.9c$ in the formula, we get $m = \dfrac{m_0}{\sqrt{1 - (0.9c)^2/c^2}} = \dfrac{m_0}{\sqrt{1 - 0.81}}$. Thus $m = \dfrac{1}{\sqrt{0.19}} m_0 \approx 2.29 m_0$. So the

rest mass of the spaceship is multiplied by $\frac{1}{\sqrt{0.19}} \approx 2.29$.

As the spaceship travels closer to the speed of light, $v \to c$, so the term $\dfrac{v^2}{c^2}$ approaches 1. So $\sqrt{1 - \dfrac{v^2}{c^2}}$ approaches 0,

and we obtain $m = \dfrac{m_0}{\text{very small number}}$, which means $m = (\text{very large number})\, m_0$. Hence the mass of the space ship

becomes arbitrarily large as its speed approaches the speed of light. The actual value of the speed of light does not affect the calculations.

P.5 Algebraic Expressions

1. Trinomial, terms x^2, $-3x$, and 7, degree 2

3. Monomial, term -8, degree 0

5. Polynomial, terms x, $-x^2$, x^3, and $-x^4$, degree 4

7. Not a polynomial

9. Polynomial, degree 3

11. Not a polynomial

13. $(12x - 7) - (5x - 12) = 12x - 7 - 5x + 12 = 7x + 5$

15. $\left(3x^2 + x + 1\right) + \left(2x^2 - 3x - 5\right) = 5x^2 - 2x - 4$

17. $\left(x^3 + 6x^2 - 4x + 7\right) - \left(3x^2 + 2x - 4\right) = x^3 + 6x^2 - 4x + 7 - 3x^2 - 2x + 4 = x^3 + 3x^2 - 6x + 11$

19. $8(2x + 5) - 7(x - 9) = 16x + 40 - 7x + 63 = 9x + 103$

21. $2(2 - 5t) + t^2(t - 1) - \left(t^4 - 1\right) = 4 - 10t + t^3 - t^2 - t^4 + 1 = -t^4 + t^3 - t^2 - 10t + 5$

23. $x^2\left(2x^2 - x + 1\right) = x^2\left(2x^2\right) - x^2(x) + x^2(1) = 2x^4 - x^3 + x^2$

25. $\sqrt{x}\left(x - \sqrt{x}\right) = x^{1/2}\left(x - x^{1/2}\right) = x^{1/2}x - x^{1/2}x^{1/2} = x^{3/2} - x$

27. $y^{1/3}\left(y^2-1\right)=y^{1/3}y^2-y^{1/3}=y^{7/3}-y^{1/3}$

29. $(3t-2)\,(7t-5)=21t^2-15t-14t+10=21t^2-29t+10$

31. $(x+2y)\,(3x-y)=3x^2-xy+6xy-2y^2=3x^2+5xy-2y^2$

33. $(1-2y)^2=1-4y+4y^2$

35. $\left(2x^2+3y^2\right)^2=\left(2x^2\right)^2+2\left(2x^2\right)\left(3y^2\right)+\left(3y^2\right)^2=4x^4+12x^2y^2+9y^4$

37. $(2x-5)\,\left(x^2-x+1\right)=2x^3-2x^2+2x-5x^2+5x-5=2x^3-7x^2+7x-5$

39. $\left(x^2-a^2\right)\left(x^2+a^2\right)=\left(x^2\right)^2-\left(a^2\right)^2=x^4-a^4$ (difference of squares)

41. $\left(\sqrt{a}-\dfrac{1}{b}\right)\left(\sqrt{a}+\dfrac{1}{b}\right)=(\sqrt{a})^2-\left(\dfrac{1}{b}\right)^2=a-\dfrac{1}{b^2}$ (difference of squares)

43. $\left(1+a^3\right)^3=1+3\left(a^3\right)+3\left(a^3\right)^2+\left(a^3\right)^3=1+3a^3+3a^6+a^9$ (perfect cube)

45. $\left(x^2+x-1\right)\left(2x^2-x+2\right)=x^2\left(2x^2-x+2\right)+x\left(2x^2-x+2\right)-\left(2x^2-x+2\right)$
$$=2x^4-x^3+2x^2+2x^3-x^2+2x-2x^2+x-2=2x^4+x^3-x^2+3x-2$$

47. $\left(x^2+x-2\right)\left(x^3-x+1\right)=x^5-x^3+x^2+x^4-x^2+x-2x^3+2x-2=x^5+x^4-3x^3+3x-2$

49. $\left(1+x^{4/3}\right)\left(1-x^{2/3}\right)=1-x^{2/3}+x^{4/3}-x^{6/3}=1-x^{2/3}+x^{4/3}-x^2$

51. $\left(3x^2y+7xy^2\right)\left(x^2y^3-2y^2\right)=3x^4y^4-6x^2y^3+7x^3y^5-14xy^4=3x^4y^4+7x^3y^5-6x^2y^3-14xy^4$ (arranging in decreasing powers of x)

53. $(2x+y-3)\,(2x+y+3)=[(2x+y)-3]\,[(2x+y)+3]=(2x+y)^2-3^2=4x^2+4xy+y^2-9$

55. $(x+y+z)\,(x-y-z)=[x+(y+z)]\,[x-(y+z)]=x^2-(y+z)^2=x^2-\left(y^2+2yz+z^2\right)=x^2-y^2-2yz-z^2$

57. (a) The height of the box is x, its width is $6-2x$, and its length is $10-2x$. Since Volume = height × width × length, we have $V=x\,(6-2x)\,(10-2x)$.

(b) $V=x\left(60-32x+4x^2\right)=60x-32x^2+4x^3$, degree 3.

(c) When $x=1$, the volume is $V=60\,(1)-32\left(1^2\right)+4\left(1^3\right)=32$, and when $x=2$, the volume is
$V=60\,(2)-32\left(2^2\right)+4\left(2^3\right)=24$.

59. (a) $A=2000\,(1+r)^3=2000\left(1+3r+3r^2+r^3\right)=2000+6000r+6000r^2+2000r^3$, degree 3.

(b) Remember that % means divide by 100, so 2% = 0.02.

Interest rate r	2%	3%	4.5%	6%	10%
Amount A	$2122.42	$2185.45	$2282.33	$2382.03	$2662.00

61. (a) When $x=1$, $(x+5)^2=(1+5)^2=36$ and $x^2+25=1^2+25=26$.

(b) $(x+5)^2=x^2+10x+25$

P.6   Factoring

1. $5a-20=5\,(a-4)$

3. $-2x^3+16x=-2x\left(x^2-8\right)$

5. $y\,(y-6)+9\,(y-6)=(y-6)\,(y+9)$

7. $2x^2y-6xy^2+3xy=xy\,(2x-6y+3)$

9. $x^2+2x-3=(x-1)\,(x+3)$

11. $6+y-y^2=(3-y)\,(2+y)$

13. $8x^2-14x-15=(2x-5)\,(4x+3)$

15. $(3x+2)^2+8\,(3x+2)+12=[(3x+2)+2]\,[(3x+2)+6]=(3x+4)\,(3x+8)$

17. $9a^2 - 16 = (3a)^2 - 4^2 = (3a - 4)(3a + 4)$

19. $27x^3 + y^3 = (3x)^3 + y^3 = (3x + y)\left[(3x)^2 + 3xy + y^2\right] = (3x + y)\left(9x^2 - 3xy + y^2\right)$

21. $8s^3 - 125t^3 = (2s)^3 - (5t)^3 = (2s - 5t)\left[(2s)^2 + (2s)(5t) + (5t)^2\right] = (2s - 5t)\left(4s^2 + 10st + 25t^2\right)$

23. $x^2 + 12x + 36 = x^2 + 2(6x) + 6^2 = (x + 6)^2$

25. $x^3 + 4x^2 + x + 4 = x^2(x + 4) + 1(x + 4) = (x + 4)\left(x^2 + 1\right)$

27. $2x^3 + x^2 - 6x - 3 = x^2(2x + 1) - 3(2x + 1) = (2x + 1)\left(x^2 - 3\right)$. If irrational coefficients are permitted, then this can be further factored as $(2x + 1)\left(x - \sqrt{3}\right)\left(x - \sqrt{3}\right)$.

29. $x^3 + x^2 + x + 1 = x^2(x + 1) + 1(x + 1) = (x + 1)\left(x^2 + 1\right)$

31. $12x^3 + 18x = 6x\left(2x^2 + 3\right)$

33. $6y^4 - 15y^3 = 3y^3(2y - 5)$

35. $x^2 - 2x - 8 = (x - 4)(x + 2)$

37. $y^2 - 8y + 15 = (y - 3)(y - 5)$

39. $2x^2 + 5x + 3 = (2x + 3)(x + 1)$

41. $9x^2 - 36x - 45 = 9\left(x^2 - 4x - 5\right) = 9(x - 5)(x + 1)$

43. $6x^2 - 5x - 6 = (3x + 2)(2x - 3)$

45. $4t^2 - 12t + 9 = (2t - 3)^2$

47. $r^2 - 6rs + 9s^2 = (r - 3s)^2$

49. $x^2 - 36 = (x - 6)(x + 6)$

51. $49 - 4y^2 = (7 - 2y)(7 + 2y)$

53. $(a + b)^2 - (a - b)^2 = [(a + b) - (a - b)][(a + b) + (a - b)] = (2b)(2a) = 4ab$

55. $x^2\left(x^2 - 1\right) - 9\left(x^2 - 1\right) = \left(x^2 - 1\right)\left(x^2 - 9\right) = (x - 1)(x + 1)(x - 3)(x + 3)$

57. $t^3 + 1 = (t + 1)\left(t^2 - t + 1\right)$

59. $8x^3 - 125 = (2x)^3 - 5^3 = (2x - 5)\left[(2x)^2 + (2x)(5) + 5^2\right] = (2x - 5)\left(4x^2 + 10x + 25\right)$

61. $x^6 - 8y^3 = \left(x^2\right)^3 - (2y)^3 = \left(x^2 - 2y\right)\left[\left(x^2\right)^2 + \left(x^2\right)(2y) + (2y)^2\right] = \left(x^2 - 2y\right)\left(x^4 + 2x^2y + 4y^2\right)$

63. $x^3 + 2x^2 + x = x\left(x^2 + 2x + 1\right) = x(x + 1)^2$

65. $x^4 + 2x^3 - 3x^2 = x^2\left(x^2 + 2x - 3\right) = x^2(x - 1)(x + 3)$

67. $y^3 - 3y^2 - 4y + 12 = \left(y^3 - 3y^2\right) + (-4y + 12) = y^2(y - 3) + (-4)(y - 3) = (y - 3)\left(y^2 - 4\right)$
$= (y - 3)(y - 2)(y + 2)$ (factor by grouping)

69. $2x^3 + 4x^2 + x + 2 = \left(2x^3 + 4x^2\right) + (x + 2) = 2x^2(x + 2) + (1)(x + 2) = (x + 2)\left(2x^2 + 1\right)$ (factor by grouping)

71. $(x - 1)(x + 2)^2 - (x - 1)^2(x + 2) = (x - 1)(x + 2)[(x + 2) - (x - 1)] = 3(x - 1)(x + 2)$

73. $y^4(y + 2)^3 + y^5(y + 2)^4 = y^4(y + 2)^3[(1) + y(y + 2)] = y^4(y + 2)^3\left(y^2 + 2y + 1\right) = y^4(y + 2)^3(y + 1)^2$

75. Start by factoring $y^2 - 7y + 10$, and then substitute $a^2 + 1$ for y. This gives
$$\left(a^2 + 1\right)^2 - 7\left(a^2 + 1\right) + 10 = \left[\left(a^2 + 1\right) - 2\right]\left[\left(a^2 + 1\right) - 5\right] = \left(a^2 - 1\right)\left(a^2 - 4\right) = (a - 1)(a + 1)(a - 2)(a + 2)$$

77. $x^{5/2} - x^{1/2} = x^{1/2}\left(x^2 - 1\right) = \sqrt{x}\,(x - 1)(x + 1)$

79. Start by factoring out the power of x with the smallest exponent, that is, $x^{-3/2}$. So
$$x^{-3/2} + 2x^{-1/2} + x^{1/2} = x^{-3/2}\left(1 + 2x + x^2\right) = \frac{(1 + x)^2}{x^{3/2}}.$$

81. Start by factoring out the power of $\left(x^2 + 1\right)$ with the smallest exponent, that is, $\left(x^2 + 1\right)^{-1/2}$. So
$$\left(x^2 + 1\right)^{1/2} + 2\left(x^2 + 1\right)^{-1/2} = \left(x^2 + 1\right)^{-1/2}\left[\left(x^2 + 1\right) + 2\right] = \frac{x^2 + 3}{\sqrt{x^2 + 1}}.$$

83. $2x^{1/3}(x - 2)^{2/3} - 5x^{4/3}(x - 2)^{-1/3} = x^{1/3}(x - 2)^{-1/3}[2(x - 2) - 5x] = x^{1/3}(x - 2)^{-1/3}(2x - 4 - 5x)$
$$= x^{1/3}(x - 2)^{-1/3}(-3x - 4) = \frac{(-3x - 4)\sqrt[3]{x}}{\sqrt[3]{x - 2}}$$

85. $3x^2 (4x - 12)^2 + x^3 (2) (4x - 12) (4) = x^2 (4x - 12) [3 (4x - 12) + x (2) (4)] = 4x^2 (x - 3) (12x - 36 + 8x)$
$$= 4x^2 (x - 3) (20x - 36) = 16x^2 (x - 3) (5x - 9)$$

87. $3 (2x - 1)^2 (2) (x + 3)^{1/2} + (2x - 1)^3 \left(\frac{1}{2}\right) (x + 3)^{-1/2} = (2x - 1)^2 (x + 3)^{-1/2} \left[6 (x + 3) + (2x - 1) \left(\frac{1}{2}\right)\right]$
$$= (2x - 1)^2 (x + 3)^{-1/2} \left(6x + 18 + x - \tfrac{1}{2}\right) = (2x - 1)^2 (x + 3)^{-1/2} \left(7x + \tfrac{35}{2}\right)$$

89. $(x^2 + 3)^{-1/3} - \frac{2}{3}x^2 (x^2 + 3)^{-4/3} = (x^2 + 3)^{-4/3} \left[(x^2 + 3) - \frac{2}{3}x^2\right] = (x^2 + 3)^{-4/3} \left(\frac{1}{3}x^2 + 3\right) = \dfrac{\frac{1}{3}x^2 + 3}{(x^2 + 3)^{4/3}}$

91. (a) $\frac{1}{2} \left[(a + b)^2 - (a^2 + b^2)\right] = \frac{1}{2} \left[a^2 + 2ab + b^2 - a^2 - b^2\right] = \frac{1}{2} (2ab) = ab.$

(b) $(a^2 + b^2)^2 - (a^2 - b^2)^2 = \left[(a^2 + b^2) - (a^2 - b^2)\right] \left[(a^2 + b^2) + (a^2 - b^2)\right]$
$$= (a^2 + b^2 - a^2 + b^2) (a^2 + b^2 + a^2 - b^2) = (2b^2) (2a^2) = 4a^2 b^2$$

(c) LHS $= (a^2 + b^2) (c^2 + d^2) = a^2 c^2 + a^2 d^2 + b^2 c^2 + b^2 d^2.$

RHS $= (ac + bd)^2 + (ad - bc)^2 = a^2 c^2 + 2abcd + b^2 d^2 + a^2 d^2 - 2abcd + b^2 c^2 = a^2 c^2 + a^2 d^2 + b^2 c^2 + b^2 d^2.$

So LHS = RHS, that is, $(a^2 + b^2) (c^2 + d^2) = (ac + bd)^2 + (ad - bc)^2.$

(d) $4a^2 c^2 - (c^2 - b^2 + a^2)^2 = (2ac)^2 - (c^2 - b^2 + a^2)$
$$= \left[(2ac) - (c^2 - b^2 + a^2)\right] \left[(2ac) + (c^2 - b^2 + a^2)\right] \text{ (difference of squares)}$$
$$= (2ac - c^2 + b^2 - a^2) (2ac + c^2 - b^2 + a^2)$$
$$= \left[b^2 - (c^2 - 2ac + a^2)\right] \left[(c^2 + 2ac + a^2) - b^2\right] \text{ (regrouping)}$$
$$= \left[b^2 - (c - a)^2\right] \left[(c + a)^2 - b^2\right] \text{ (perfect squares)}$$
$$= [b - (c - a)] [b + (c - a)] [(c + a) - b] [(c + a) + b] \text{ (each factor is a difference of squares)}$$
$$= (b - c + a) (b + c - a) (c + a - b) (c + a + b)$$
$$= (a + b - c) (-a + b + c) (a - b + c) (a + b + c)$$

(e) $x^4 + 3x^2 + 4 = (x^4 + 4x^2 + 4) - x^2 = (x^2 + 2)^2 - x^2 = \left[(x^2 + 2) - x\right] \left[(x^2 + 2) + x\right]$
$$= (x^2 - x + 2) (x^2 + x + 2)$$

93. The volume of the shell is the difference between the volumes of the outside cylinder (with radius R) and the inside cylinder

(with radius r). Thus $V = \pi R^2 h - \pi r^2 h = \pi (R^2 - r^2) h = \pi (R - r) (R + r) h = 2\pi \cdot \dfrac{R + r}{2} \cdot h \cdot (R - r)$. The

average radius is $\dfrac{R + r}{2}$ and $2\pi \cdot \dfrac{R + r}{2}$ is the average circumference (length of the rectangular box), h is the height, and

$R - r$ is the thickness of the rectangular box. Thus $V = \pi R^2 h - \pi r^2 h = 2\pi \cdot \dfrac{R + r}{2} \cdot h \cdot (R - r) = 2\pi \cdot$ (average radius) $\cdot$

(height) $\cdot$ (thickness)

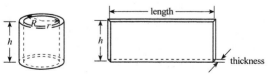

95. (a) $528^2 - 527^2 = (528 - 527) (528 + 527) = 1 (1055) = 1055$

(b) $122^2 - 120^2 = (122 - 120) (122 + 120) = 2 (242) = 484$

(c) $1020^2 - 1010^2 = (1020 - 1010) (1020 + 1010) = 10 (2030) = 20{,}300$

(d) $49 \cdot 51 = (50 - 1) (50 + 1) = 50^2 - 1 = 2500 - 1 = 2499$

(e) $998 \cdot 1002 = (1000 - 2) (1000 + 2) = 1000^2 - 2^2 = 1{,}000{,}000 - 4 = 999{,}996$

97.

$$
\begin{array}{r}
A + 1 \\
\times \quad A - 1 \\
\hline
-A - 1 \\
A^2 + A \\
\hline
A^2 \qquad - 1
\end{array}
\qquad
\begin{array}{r}
A^2 + A + 1 \\
\times \quad A - 1 \\
\hline
-A^2 - A - 1 \\
A^3 + A^2 + A \\
\hline
A^3 \qquad\qquad - 1
\end{array}
\qquad
\begin{array}{r}
A^3 + A^2 + A + 1 \\
\times \quad A - 1 \\
\hline
-A^3 - A^2 - A - 1 \\
A^4 + A^3 + A^2 + A \\
\hline
A^4 \qquad\qquad\qquad - 1
\end{array}
$$

Based on the pattern, we suspect that $A^5 - 1 = (A - 1)\left(A^4 + A^3 + A^2 + A + 1\right)$. Check:

$$
\begin{array}{r}
A^4 + A^3 + A^2 + A + 1 \\
\times \qquad\qquad\qquad A - 1 \\
\hline
-A^4 - A^3 - A^2 - A - 1 \\
A^5 + A^4 + A^3 + A^2 + A \\
\hline
A^5 \qquad\qquad\qquad\qquad - 1
\end{array}
$$

The general pattern is $A^n - 1 = (A - 1)\left(A^{n-1} + A^{n-2} + \cdots + A^2 + A + 1\right)$, where n is a positive integer.

P.7 Rational Expressions

1. (a) When $x = 5$, we get $4\left(5^2\right) - 10\left(5\right) + 3 = 53$.

   (b) The domain is all real numbers.

3. (a) When $x = 7$, we get $\dfrac{2\left(7\right) + 1}{7 - 4} = \dfrac{15}{3} = 5$.

   (b) Since $x - 4 \neq 0$, we have $x \neq 4$, so the domain is $\{x \mid x \neq 4\}$.

5. (a) When $x = 6$, we get $\sqrt{6 + 3} = \sqrt{9} = 3$.

   (b) $x + 3 \geq 0 \iff x \geq -3$, so the domain is $\{x \mid x \geq -3\}$.

7. $\dfrac{12x}{6x^2} = \dfrac{6x \cdot 2}{6x \cdot x} = \dfrac{2}{x}$

9. $\dfrac{5y^2}{10y + y^2} = \dfrac{y \cdot 5y}{y \cdot (10 + y)} = \dfrac{5y}{10 + y}$

11. $\dfrac{3\left(x + 2\right)\left(x - 1\right)}{6\left(x - 1\right)^2} = \dfrac{3\left(x - 1\right) \cdot \left(x + 2\right)}{3\left(x - 1\right) \cdot 2\left(x - 1\right)} = \dfrac{x + 2}{2\left(x - 1\right)}$

13. $\dfrac{x - 2}{x^2 - 4} = \dfrac{x - 2}{\left(x - 2\right)\left(x + 2\right)} = \dfrac{1}{x + 2}$

15. $\dfrac{x^2 + 6x + 8}{x^2 + 5x + 4} = \dfrac{\left(x + 2\right)\left(x + 4\right)}{\left(x + 1\right)\left(x + 4\right)} = \dfrac{x + 2}{x + 1}$

17. $\dfrac{y^2 + y}{y^2 - 1} = \dfrac{y\left(y + 1\right)}{\left(y - 1\right)\left(y + 1\right)} = \dfrac{y}{y - 1}$

19. $\dfrac{2x^3 - x^2 - 6x}{2x^2 - 7x + 6} = \dfrac{x\left(2x^2 - x - 6\right)}{\left(2x - 3\right)\left(x - 2\right)} = \dfrac{x\left(2x + 3\right)\left(x - 2\right)}{\left(2x - 3\right)\left(x - 2\right)} = \dfrac{x\left(2x + 3\right)}{2x - 3}$

21. $\dfrac{4x}{x^2 - 4} \cdot \dfrac{x + 2}{16x} = \dfrac{4x}{\left(x - 2\right)\left(x + 2\right)} \cdot \dfrac{x + 2}{16x} = \dfrac{1}{4\left(x - 2\right)}$

23. $\dfrac{x^2 - x - 12}{x^2 - 9} \cdot \dfrac{3 + x}{4 - x} = \dfrac{\left(x - 4\right)\left(x + 3\right)}{\left(x - 3\right)\left(x + 3\right)} \cdot \dfrac{x + 3}{-\left(x - 4\right)} = \dfrac{x + 3}{-\left(x - 3\right)} = \dfrac{3 + x}{3 - x}$

25. $\dfrac{t - 3}{t^2 + 9} \cdot \dfrac{t + 3}{t^2 - 9} = \dfrac{\left(t - 3\right)\left(t + 3\right)}{\left(t^2 + 9\right)\left(t - 3\right)\left(t + 3\right)} = \dfrac{1}{t^2 + 9}$

27. $\dfrac{x^2 + 7x + 12}{x^2 + 3x + 2} \cdot \dfrac{x^2 + 5x + 6}{x^2 + 6x + 9} = \dfrac{\left(x + 3\right)\left(x + 4\right)}{\left(x + 1\right)\left(x + 2\right)} \cdot \dfrac{\left(x + 2\right)\left(x + 3\right)}{\left(x + 3\right)\left(x + 3\right)} = \dfrac{x + 4}{x + 1}$

29. $\dfrac{2x^2 + 3x + 1}{x^2 + 2x - 15} \div \dfrac{x^2 + 6x + 5}{2x^2 - 7x + 3} = \dfrac{2x^2 + 3x + 1}{x^2 + 2x - 15} \cdot \dfrac{2x^2 - 7x + 3}{x^2 + 6x + 5} = \dfrac{\left(2x + 1\right)\left(x + 1\right)}{\left(x - 3\right)\left(x + 5\right)} \cdot \dfrac{\left(2x - 1\right)\left(x - 3\right)}{\left(x + 1\right)\left(x + 5\right)}$

$$= \dfrac{\left(2x + 1\right)\left(2x - 1\right)}{\left(x + 5\right)\left(x + 5\right)} = \dfrac{\left(2x + 1\right)\left(2x - 1\right)}{\left(x + 5\right)^2}$$

31. $\dfrac{\dfrac{x^3}{x+1}}{\dfrac{x}{x^2+2x+1}} = \dfrac{x^3}{x+1} \cdot \dfrac{x^2+2x+1}{x} = \dfrac{x^3(x+1)(x+1)}{(x+1)x} = x^2(x+1)$

33. $\dfrac{x/y}{z} = \dfrac{x}{y} \cdot \dfrac{1}{z} = \dfrac{x}{yz}$

35. $2 + \dfrac{x}{x+3} = \dfrac{2(x+3)}{x+3} + \dfrac{x}{x+3} = \dfrac{2x+6+x}{x+3} = \dfrac{3x+6}{x+3} = \dfrac{3(x+2)}{(x+3)}$

37. $\dfrac{1}{x+5} + \dfrac{2}{x-3} = \dfrac{x-3}{(x+5)(x-3)} + \dfrac{2(x+5)}{(x+5)(x-3)} = \dfrac{x-3+2x+10}{(x+5)(x-3)} = \dfrac{3x+7}{(x+5)(x-3)}$

39. $\dfrac{1}{x+1} - \dfrac{1}{x+2} = \dfrac{x+2}{(x+1)(x+2)} + \dfrac{-(x+1)}{(x+1)(x+2)} = \dfrac{x+2-x-1}{(x+1)(x+2)} = \dfrac{1}{(x+1)(x+2)}$

41. $\dfrac{x}{(x+1)^2} + \dfrac{2}{x+1} = \dfrac{x}{(x+1)^2} + \dfrac{2(x+1)}{(x+1)(x+1)} = \dfrac{x+2x+2}{(x+1)^2} = \dfrac{3x+2}{(x+1)^2}$

43. $u + 1 + \dfrac{u}{u+1} = \dfrac{(u+1)(u+1)}{u+1} + \dfrac{u}{u+1} = \dfrac{u^2+2u+1+u}{u+1} = \dfrac{u^2+3u+1}{u+1}$

45. $\dfrac{1}{x^2} + \dfrac{1}{x^2+x} = \dfrac{1}{x^2} + \dfrac{1}{x(x+1)} = \dfrac{x+1}{x^2(x+1)} + \dfrac{x}{x^2(x+1)} = \dfrac{2x+1}{x^2(x+1)}$

47. $\dfrac{2}{x+3} - \dfrac{1}{x^2+7x+12} = \dfrac{2}{x+3} - \dfrac{1}{(x+3)(x+4)} = \dfrac{2(x+4)}{(x+3)(x+4)} + \dfrac{-1}{(x+3)(x+4)}$

$= \dfrac{2x+8-1}{(x+3)(x+4)} = \dfrac{2x+7}{(x+3)(x+4)}$

49. $\dfrac{1}{x+3} + \dfrac{1}{x^2-9} = \dfrac{1}{x+3} + \dfrac{1}{(x-3)(x+3)} = \dfrac{x-3}{(x-3)(x+3)} + \dfrac{1}{(x-3)(x+3)} = \dfrac{x-2}{(x-3)(x+3)}$

51. $\dfrac{2}{x} + \dfrac{3}{x-1} - \dfrac{4}{x^2-x} = \dfrac{2}{x} + \dfrac{3}{x-1} - \dfrac{4}{x(x-1)} = \dfrac{2(x-1)}{x(x-1)} + \dfrac{3x}{x(x-1)} + \dfrac{-4}{x(x-1)} = \dfrac{2x-2+3x-4}{x(x-1)} = \dfrac{5x-6}{x(x-1)}$

53. $\dfrac{1}{x^2+3x+2} - \dfrac{1}{x^2-2x-3} = \dfrac{1}{(x+2)(x+1)} - \dfrac{1}{(x-3)(x+1)}$

$= \dfrac{x-3}{(x-3)(x+2)(x+1)} + \dfrac{-(x+2)}{(x-3)(x+2)(x+1)} = \dfrac{x-3-x-2}{(x-3)(x+2)(x+1)} = \dfrac{-5}{(x-3)(x+2)(x+1)}$

55. $\dfrac{\dfrac{x}{y} - \dfrac{y}{x}}{\dfrac{1}{x^2} - \dfrac{1}{y^2}} = \dfrac{\dfrac{x^2-y^2}{xy}}{\dfrac{y^2-x^2}{x^2y^2}} = \dfrac{x^2-y^2}{xy} \cdot \dfrac{x^2y^2}{y^2-x^2} = \dfrac{xy}{-1} = -xy$. An alternative method is to multiply the

numerator and denominator by the common denominator of both the numerator and denominator, in this case x^2y^2:

$\dfrac{\dfrac{x}{y} - \dfrac{y}{x}}{\dfrac{1}{x^2} - \dfrac{1}{y^2}} = \dfrac{\left(\dfrac{x}{y} - \dfrac{y}{x}\right)}{\left(\dfrac{1}{x^2} - \dfrac{1}{y^2}\right)} \cdot \dfrac{x^2y^2}{x^2y^2} = \dfrac{x^3y - xy^3}{y^2-x^2} = \dfrac{xy(x^2-y^2)}{y^2-x^2} = -xy$.

57. $\dfrac{1 + \dfrac{1}{c-1}}{1 - \dfrac{1}{c-1}} = \dfrac{\dfrac{c-1}{c-1} + \dfrac{1}{c-1}}{\dfrac{c-1}{c-1} + \dfrac{-1}{c-1}} = \dfrac{\dfrac{c}{c-1}}{\dfrac{c-2}{c-1}} = \dfrac{c}{c-1} \cdot \dfrac{c-1}{c-2} = \dfrac{c}{c-2}$. Using the alternative method, we obtain

$\dfrac{1 + \dfrac{1}{c-1}}{1 - \dfrac{1}{c-1}} = \dfrac{\left(1 + \dfrac{1}{c-1}\right)}{\left(1 - \dfrac{1}{c-1}\right)} \cdot \dfrac{c-1}{c-1} = \dfrac{c-1+1}{c-1-1} = \dfrac{c}{c-2}$.

59. $\dfrac{\dfrac{5}{x-1}-\dfrac{2}{x+1}}{\dfrac{x}{x-1}+\dfrac{1}{x+1}}=\dfrac{\dfrac{5(x+1)}{(x-1)(x+1)}+\dfrac{-2(x-1)}{(x-1)(x+1)}}{\dfrac{x(x+1)}{(x-1)(x+1)}+\dfrac{x-1}{(x-1)(x+1)}}=\dfrac{\dfrac{5x+5-2x+2}{(x-1)(x+1)}}{\dfrac{x^2+x+x-1}{(x-1)(x+1)}}$

$$=\dfrac{3x+7}{(x-1)(x+1)}\cdot\dfrac{(x-1)(x+1)}{x^2+2x-1}=\dfrac{3x+7}{x^2+2x-1}$$

Alternatively,

$$\dfrac{\dfrac{5}{x-1}-\dfrac{2}{x+1}}{\dfrac{x}{x-1}+\dfrac{1}{x+1}}=\dfrac{\left(\dfrac{5}{x-1}-\dfrac{2}{x+1}\right)}{\left(\dfrac{x}{x-1}+\dfrac{1}{x+1}\right)}\cdot\dfrac{(x-1)(x+1)}{(x-1)(x+1)}=\dfrac{5(x+1)-2(x-1)}{x(x+1)+(x-1)}$$

$$=\dfrac{5x+5-2x+2}{x^2+x+x-1}=\dfrac{3x+7}{x^2+2x-1}$$

61. $\dfrac{x^{-2}-y^{-2}}{x^{-1}+y^{-1}}=\dfrac{\dfrac{1}{x^2}-\dfrac{1}{y^2}}{\dfrac{1}{x}+\dfrac{1}{y}}=\dfrac{\dfrac{y^2}{x^2y^2}-\dfrac{x^2}{x^2y^2}}{\dfrac{y}{xy}+\dfrac{x}{xy}}=\dfrac{y^2-x^2}{x^2y^2}\cdot\dfrac{xy}{y+x}=\dfrac{(y-x)(y+x)\,xy}{x^2y^2(y+x)}=\dfrac{y-x}{xy}$

Alternatively, $\dfrac{x^{-2}-y^{-2}}{x^{-1}+y^{-1}}=\dfrac{\left(\dfrac{1}{x^2}-\dfrac{1}{y^2}\right)}{\left(\dfrac{1}{x}+\dfrac{1}{y}\right)}\cdot\dfrac{x^2y^2}{x^2y^2}=\dfrac{y^2-x^2}{xy^2+x^2y}=\dfrac{(y-x)(y+x)}{xy(y+x)}=\dfrac{y-x}{xy}.$

63. $\dfrac{1}{1+a^n}+\dfrac{1}{1+a^{-n}}=\dfrac{1}{1+a^n}+\dfrac{1}{1+a^{-n}}\cdot\dfrac{a^n}{a^n}=\dfrac{1}{1+a^n}+\dfrac{a^n}{a^n+1}=\dfrac{1+a^n}{1+a^n}=1$

65. $\dfrac{\dfrac{1}{a+h}-\dfrac{1}{a}}{h}=\dfrac{\dfrac{a}{a(a+h)}-\dfrac{a+h}{a(a+h)}}{h}=\dfrac{\dfrac{a-a-h}{a(a+h)}}{h}=\dfrac{-h}{a(a+h)}\cdot\dfrac{1}{h}=\dfrac{-1}{a(a+h)}$

67. $\dfrac{\dfrac{1-(x+h)}{2+(x+h)}-\dfrac{1-x}{2+x}}{h}=\dfrac{\dfrac{(2+x)(1-x-h)}{(2+x)(2+x+h)}-\dfrac{(1-x)(2+x+h)}{(2+x)(2+x+h)}}{h}$

$$=\dfrac{\dfrac{2-x-x^2-2h-xh}{(2+x)(2+x+h)}-\dfrac{2-x-x^2+h-xh}{(2+x)(2+x+h)}}{h}=\dfrac{-3h}{(2+x)(2+x+h)}\cdot\dfrac{1}{h}=\dfrac{-3}{(2+x)(2+x+h)}$$

69. $\sqrt{1+\left(\dfrac{x}{\sqrt{1-x^2}}\right)^2}=\sqrt{1+\dfrac{x^2}{1-x^2}}=\sqrt{\dfrac{1-x^2}{1-x^2}+\dfrac{x^2}{1-x^2}}=\sqrt{\dfrac{1}{1-x^2}}=\dfrac{1}{\sqrt{1-x^2}}$

71. $\dfrac{3(x+2)^2(x-3)^2-(x+2)^3(2)(x-3)}{(x-3)^4}=\dfrac{(x+2)^2(x-3)[3(x-3)-(x+2)(2)]}{(x-3)^4}$

$$=\dfrac{(x+2)^2(3x-9-2x-4)}{(x-3)^3}=\dfrac{(x+2)^2(x-13)}{(x-3)^3}$$

73. $\dfrac{2(1+x)^{1/2}-x(1+x)^{-1/2}}{1+x}=\dfrac{(1+x)^{-1/2}[2(1+x)-x]}{1+x}=\dfrac{x+2}{(1+x)^{3/2}}$

75. $\dfrac{3(1+x)^{1/3}-x(1+x)^{-2/3}}{(1+x)^{2/3}}=\dfrac{(1+x)^{-2/3}[3(1+x)-x]}{(1+x)^{2/3}}=\dfrac{2x+3}{(1+x)^{4/3}}$

77. $\dfrac{1}{2-\sqrt{3}}=\dfrac{1}{2-\sqrt{3}}\cdot\dfrac{2+\sqrt{3}}{2+\sqrt{3}}=\dfrac{2+\sqrt{3}}{4-3}=\dfrac{2+\sqrt{3}}{1}=2+\sqrt{3}$

79. $\dfrac{2}{\sqrt{2}+\sqrt{7}}=\dfrac{2}{\sqrt{2}+\sqrt{7}}\cdot\dfrac{\sqrt{2}-\sqrt{7}}{\sqrt{2}-\sqrt{7}}=\dfrac{2\left(\sqrt{2}-\sqrt{7}\right)}{2-7}=\dfrac{2\left(\sqrt{2}-\sqrt{7}\right)}{-5}=\dfrac{2\left(\sqrt{7}-\sqrt{2}\right)}{5}$

81. $\dfrac{y}{\sqrt{3}+\sqrt{y}}=\dfrac{y}{\sqrt{3}+\sqrt{y}}\cdot\dfrac{\sqrt{3}-\sqrt{y}}{\sqrt{3}-\sqrt{y}}=\dfrac{y\left(\sqrt{3}-\sqrt{y}\right)}{3-y}=\dfrac{y\sqrt{3}-y\sqrt{y}}{3-y}$

83. $\dfrac{1-\sqrt{5}}{3} = \dfrac{1-\sqrt{5}}{3} \cdot \dfrac{1+\sqrt{5}}{1+\sqrt{5}} = \dfrac{1-5}{3\left(1+\sqrt{5}\right)} = \dfrac{-4}{3\left(1+\sqrt{5}\right)}$

85. $\dfrac{\sqrt{r}+\sqrt{2}}{5} = \dfrac{\sqrt{r}+\sqrt{2}}{5} \cdot \dfrac{\sqrt{r}-\sqrt{2}}{\sqrt{r}-\sqrt{2}} = \dfrac{r-2}{5\left(\sqrt{r}-\sqrt{2}\right)}$

87. $\sqrt{x^2+1}-x = \dfrac{\sqrt{x^2+1}-x}{1} \cdot \dfrac{\sqrt{x^2+1}+x}{\sqrt{x^2+1}+x} = \dfrac{x^2+1-x^2}{\sqrt{x^2+1}+x} = \dfrac{1}{\sqrt{x^2+1}+x}$

89. $\dfrac{16+a}{16} = \dfrac{16}{16} + \dfrac{a}{16} = 1 + \dfrac{a}{16}$, so the statement is true.

91. This statement is false. For example, take $x = 2$. Then LHS $= \dfrac{2}{4+x} = \dfrac{2}{4+2} = \dfrac{2}{6} = \dfrac{1}{3}$, while

RHS $= \dfrac{1}{2} + \dfrac{2}{x} = \dfrac{1}{2} + \dfrac{2}{2} = \dfrac{3}{2}$, and $\dfrac{1}{3} \neq \dfrac{3}{2}$.

93. This statement is false. For example, take $x = 0$ and $y = 1$. Then LHS $= \dfrac{x}{x+y} = \dfrac{0}{0+1} = 0$, while

RHS $= \dfrac{1}{1+y} = \dfrac{1}{1+1} = \dfrac{1}{2}$, and $0 \neq \dfrac{1}{2}$.

95. This statement is true: $\dfrac{-a}{b} = (-a)\left(\dfrac{1}{b}\right) = (-1)(a)\left(\dfrac{1}{b}\right) = (-1)\left(\dfrac{a}{b}\right) = -\dfrac{a}{b}$.

97. (a) $R = \dfrac{1}{\dfrac{1}{R_1} + \dfrac{1}{R_2}} = \dfrac{1}{\dfrac{1}{R_1} + \dfrac{1}{R_2}} \cdot \dfrac{R_1 R_2}{R_1 R_2} = \dfrac{R_1 R_2}{R_2 + R_1}$

(b) Substituting $R_1 = 10$ ohms and $R_2 = 20$ ohms gives $R = \dfrac{(10)(20)}{(20)+(10)} = \dfrac{200}{30} \approx 6.7$ ohms.

99.

x	2.80	2.90	2.95	2.99	2.999	3	3.001	3.01	3.05	3.10	3.20
$\dfrac{x^2-9}{x-3}$	5.80	5.90	5.95	5.99	5.999	?	6.001	6.01	6.05	6.10	6.20

From the table, we see that the expression $\dfrac{x^2-9}{x-3}$ approaches 6 as x approaches 3. We simplify the expression:

$\dfrac{x^2-9}{x-3} = \dfrac{(x-3)(x+3)}{x-3} = x+3$, $x \neq 3$. Clearly as x approaches 3, $x+3$ approaches 6. This explains the result in the table.

101. Answers will vary.

Algebraic Error	Counterexample
$\dfrac{1}{a} + \dfrac{1}{b} \neq \dfrac{1}{a+b}$	$\dfrac{1}{2} + \dfrac{1}{2} \neq \dfrac{1}{2+2}$
$(a+b)^2 \neq a^2 + b^2$	$(1+3)^2 \neq 1^2 + 3^2$
$\sqrt{a^2+b^2} \neq a+b$	$\sqrt{5^2+12^2} \neq 5+12$
$\dfrac{a+b}{a} \neq b$	$\dfrac{2+6}{2} \neq 6$
$\left(a^3+b^3\right)^{1/3} \neq a+b$	$\left(2^3+2^3\right)^{1/3} \neq 2+2$
$\dfrac{a^m}{a^n} \neq a^{m/n}$	$\dfrac{3^5}{3^2} \neq 3^{5/2}$
$a^{-1/n} \neq \dfrac{1}{a^n}$	$64^{-1/3} \neq \dfrac{1}{64^3}$

Chapter P Review

1. Commutative Property for addition.

3. Distributive Property

5. $x \in [-2, 6) \iff -2 \le x < 6$

7. $x \in (-\infty, 4] \iff x \le 4$

9. $x \ge 5 \iff x \in [5, \infty)$

11. $-1 < x \le 5 \iff x \in (-1, 5]$

13. $|3 - |-9|| = |3 - 9| = |-6| = 6$

15. $2^{-3} - 3^{-2} = \frac{1}{8} - \frac{1}{9} = \frac{9}{72} - \frac{8}{72} = \frac{1}{72}$

17. $216^{-1/3} = \frac{1}{216^{1/3}} = \frac{1}{\sqrt[3]{216}} = \frac{1}{6}$

19. $\frac{\sqrt{242}}{\sqrt{2}} = \sqrt{\frac{242}{2}} = \sqrt{121} = 11$

21. $2^{1/2}8^{1/2} = \sqrt{2} \cdot \sqrt{8} = \sqrt{16} = 4$

23. $\frac{1}{x^2} = x^{-2}$

25. $x^2 x^m \left(x^3\right)^m = x^{2+m+3m} = x^{4m+2}$

27. $x^a x^b x^c = x^{a+b+c}$

29. $x^{c+1}\left(x^{2c-1}\right)^2 = x^{(c+1)+2(2c-1)} = x^{c+1+4c-2} = x^{5c-1}$

31. $\left(2x^3 y\right)^2 \left(3x^{-1}y^2\right) = 4x^6 y^2 \cdot 3x^{-1}y^2 = 4 \cdot 3x^{6-1}y^{2+2}$
$$= 12x^5 y^4$$

33. $\frac{x^4 \left(3x\right)^2}{x^3} = \frac{x^4 \cdot 9x^2}{x^3} = 9x^{4+2-3} = 9x^3$

35. $\sqrt[3]{\left(x^3 y\right)^2 y^4} = \sqrt[3]{x^6 y^4 y^2} = \sqrt[3]{x^6 y^6} = x^2 y^2$

37. $\frac{x}{2 + \sqrt{x}} = \frac{x}{2 + \sqrt{x}} \cdot \frac{2 - \sqrt{x}}{2 - \sqrt{x}} = \frac{x\left(2 - \sqrt{x}\right)}{4 - x}$. Here simplify means to rationalize the denominator.

39. $\frac{8r^{1/2}s^{-3}}{2r^{-2}s^4} = 4r^{(1/2)-(-2)}s^{-3-4} = 4r^{5/2}s^{-7} = \frac{4r^{5/2}}{s^7}$

41. $78{,}250{,}000{,}000 = 7.825 \times 10^{10}$

43. $\frac{ab}{c} \approx \frac{(0.00000293)\left(1.582 \times 10^{-14}\right)}{2.8064 \times 10^{12}} = \frac{\left(2.93 \times 10^{-6}\right)\left(1.582 \times 10^{-14}\right)}{2.8064 \times 10^{12}} = \frac{2.93 \cdot 1.582}{2.8064} \times 10^{-6-14-12}$
$$\approx 1.65 \times 10^{-32}$$

45. $2x^2 y - 6xy^2 = 2xy\left(x - 3y\right)$

47. $x^2 - 9x + 18 = \left(x - 6\right)\left(x - 3\right)$

49. $3x^2 - 2x - 1 = \left(3x + 1\right)\left(x - 1\right)$

51. $4t^2 - 13t - 12 = \left(4t + 3\right)\left(t - 4\right)$

53. $25 - 16t^2 = \left(5 - 4t\right)\left(5 + 4t\right)$

55. $x^6 - 1 = \left(x^3 - 1\right)\left(x^3 + 1\right) = \left(x - 1\right)\left(x^2 + x + 1\right)\left(x + 1\right)\left(x^2 - x + 1\right)$

57. $x^{-1/2} - 2x^{1/2} + x^{3/2} = x^{-1/2}\left(1 - 2x + x^2\right) = x^{-1/2}\left(1 - x\right)^2$

59. $4x^3 - 8x^2 + 3x - 6 = 4x^2\left(x - 2\right) + 3\left(x - 2\right) = \left(4x^2 + 3\right)\left(x - 2\right)$

61. $\left(x^2 + 2\right)^{5/2} + 2x\left(x^2 + 2\right)^{3/2} + x^2\sqrt{x^2 + 2} = \left(x^2 + 2\right)^{1/2}\left(\left(x^2 + 2\right)^2 + 2x\left(x^2 + 2\right) + x^2\right)$
$$= \sqrt{x^2 + 2}\left(x^4 + 4x^2 + 4 + 2x^3 + 4x + x^2\right) = \sqrt{x^2 + 2}\left(x^4 + 2x^3 + 5x^2 + 4x + 4\right) = \sqrt{x^2 + 2}\left(x^2 + x + 2\right)^2$$

63. $a^2 y - b^2 y = y\left(a^2 - b^2\right) = y\left(a - b\right)\left(a + b\right)$

65. $\left(x + 1\right)^2 - 2\left(x + 1\right) + 1 = \left[\left(x + 1\right) - 1\right]^2 = x^2$. You can also obtain this result by expanding each term and then simplifying.

67. $(2x+1)(3x-2) - 5(4x-1) = 6x^2 - 4x + 3x - 2 - 20x + 5 = 6x^2 - 21x + 3$

69. $(2a^2 - b)^2 = (2a^2)^2 - 2(2a^2)(b) + (b)^2 = 4a^4 - 4a^2b + b^2$

71. $(x-1)(x-2)(x-3) = (x-1)(x^2 - 5x + 6) = x^3 - 5x^2 + 6x - x^2 + 5x - 6 = x^3 - 6x^2 + 11x - 6$

73. $\sqrt{x}(\sqrt{x}+1)(2\sqrt{x}-1) = (x + \sqrt{x})(2\sqrt{x}-1) = 2x\sqrt{x} - x + 2x - \sqrt{x} = 2x^{3/2} + x - x^{1/2}$

75. $x^2(x-2) + x(x-2)^2 = x^3 - 2x^2 + x(x^2 - 4x + 4) = x^3 - 2x^2 + x^3 - 4x^2 + 4x = 2x^3 - 6x^2 + 4x$

77. $\dfrac{x^2 - 2x - 3}{2x^2 + 5x + 3} = \dfrac{(x-3)(x+1)}{(2x+3)(x+1)} = \dfrac{x-3}{2x+3}$

79. $\dfrac{x^2 + 2x - 3}{x^2 + 8x + 16} \cdot \dfrac{3x + 12}{x - 1} = \dfrac{(x+3)(x-1)}{(x+4)(x+4)} \cdot \dfrac{3(x+4)}{(x-1)} = \dfrac{3(x+3)}{x+4}$

81. $\dfrac{x^2 - 2x - 15}{x^2 - 6x + 5} \div \dfrac{x^2 - x - 12}{x^2 - 1} = \dfrac{(x-5)(x+3)}{(x-5)(x-1)} \cdot \dfrac{(x-1)(x+1)}{(x-4)(x+3)} = \dfrac{x+1}{x-4}$

83. $\dfrac{1}{x-1} - \dfrac{x}{x^2+1} = \dfrac{x^2+1}{(x-1)(x^2+1)} - \dfrac{x(x-1)}{(x-1)(x^2+1)} = \dfrac{x^2+1-x^2+x}{(x-1)(x^2+1)} = \dfrac{x+1}{(x-1)(x^2+1)}$

85. $\dfrac{1}{x-1} - \dfrac{2}{x^2-1} = \dfrac{1}{x-1} - \dfrac{2}{(x-1)(x+1)} = \dfrac{x+1}{(x-1)(x+1)} - \dfrac{2}{(x-1)(x+1)}$

$= \dfrac{x+1-2}{(x-1)(x+1)} = \dfrac{x-1}{(x-1)(x+1)} = \dfrac{1}{x+1}$

87. $\dfrac{\frac{1}{x} - \frac{1}{2}}{x-2} = \dfrac{\frac{2}{2x} - \frac{x}{2x}}{x-2} = \dfrac{2-x}{2x} \cdot \dfrac{1}{x-2} = \dfrac{-1(x-2)}{2x} \cdot \dfrac{1}{x-2} = \dfrac{-1}{2x}$

89. $\dfrac{3(x+h)^2 - 5(x+h) - (3x^2 - 5x)}{h} = \dfrac{3x^2 + 6xh + 3h^2 - 5x - 5h - 3x^2 + 5x}{h} = \dfrac{6xh + 3h^2 - 5h}{h}$

$= \dfrac{h(6x + 3h - 5)}{h} = 6x + 3h - 5$

91. This statement is false. For example, take $x = 1$ and $y = 1$. Then LHS $= (x+y)^3 = (1+1)^3 = 2^3 = 8$, while RHS $= x^3 + y^3 = 1^3 + 1^3 = 1 + 1 = 2$, and $8 \neq 2$.

93. This statement is true: $\dfrac{12+y}{y} = \dfrac{12}{y} + \dfrac{y}{y} = \dfrac{12}{y} + 1$.

95. This statement is false. For example, take $a = -1$. Then LHS $= \sqrt{a^2} = \sqrt{(-1)^2} = \sqrt{1} = 1$, which does not equal $a = -1$. The true statement is $\sqrt{a^2} = |a|$.

97. This statement is false. For example, take $x = 1$ and $y = 1$. Then LHS $= x^3 + y^3 = 1^3 + 1^3 = 1 + 1 = 2$, while RHS $= (x+y)(x^2 + xy + y^2) = (1+1)[1^2 + (1)(1) + 1^2] = 2(3) = 6$.

99. Substituting for a and b, we obtain

$$a^2 + b^2 = (2mn)^2 + (m^2 - n^2)^2 = 4m^2n^2 + m^4 - 2m^2n^2 + n^4 = m^4 + 2m^2n^2 + n^4 = (m^2 + n^2)^2$$

Since this last expression is c^2, we have $a^2 + b^2 = c^2$ for these values.

Chapter P Test

1. (a)

$[-2, 5)$ $(7, \infty)$

(b) $x \leq 3 \iff x \in (-\infty, 3]; \, 0 < x \leq 8 \iff x \in (0, 8]$

(c) The distance is $|-12 - 39| = |-51| = 51$.

2. (a) $(-3)^4 = 81$

(b) $-3^4 = -81$

(c) $3^{-4} = \dfrac{1}{3^4} = \dfrac{1}{81}$

(d) $\dfrac{5^{23}}{5^{21}} = 5^{23-21} = 5^2 = 25$

(e) $\left(\dfrac{2}{3}\right)^{-2} = \dfrac{3^2}{2^2} = \dfrac{9}{4}$

(f) $\dfrac{\sqrt[6]{64}}{\sqrt{32}} = \dfrac{\sqrt[6]{2^6}}{\sqrt{16\cdot 2}} = \dfrac{2}{4\sqrt{2}} = \dfrac{1}{2\sqrt{2}}$

(g) $16^{-3/4} = \left(2^4\right)^{-3/4} = 2^{-3} = \frac{1}{8}$

3. (a) $186{,}000{,}000{,}000 = 1.86 \times 10^{11}$

(b) $0.0000003965 = 3.965 \times 10^{-7}$

4. (a) $\sqrt{200} - \sqrt{32} = 10\sqrt{2} - 4\sqrt{2} = 6\sqrt{2}$

(b) $\left(3a^3 b^3\right)\left(4ab^2\right)^2 = 3a^3 b^3 \cdot 4^2 a^2 b^4 = 48a^5 b^7$

(c) $\sqrt[3]{\dfrac{125}{x^{-9}}} = \sqrt[3]{5^3 \cdot x^9} = 5x^3$

(d) $\left(\dfrac{3x^{3/2}y^3}{x^2 y^{-1/2}}\right)^{-2} = \dfrac{3^{-2}x^{-3}y^{-6}}{x^{-4}y^1} = \frac{1}{9}x^{-3-(-4)}y^{-6-1} = \frac{1}{9}xy^{-7} = \dfrac{x}{9y^7}$

5. (a) $3(x+6) + 4(2x-5) = 3x + 18 + 8x - 20 = 11x - 2$

(b) $(x+3)(4x-5) = 4x^2 - 5x + 12x - 15 = 4x^2 + 7x - 15$

(c) $\left(\sqrt{a} + \sqrt{b}\right)\left(\sqrt{a} - \sqrt{b}\right) = \left(\sqrt{a}\right)^2 - \left(\sqrt{b}\right)^2 = a - b$

(d) $(2x+3)^2 = (2x)^2 + 2(2x)(3) + (3)^2 = 4x^2 + 12x + 9$

(e) $(x+2)^3 = (x)^3 + 3(x)^2(2) + 3(x)(2)^2 + (2)^3 = x^3 + 6x^2 + 12x + 8$

6. (a) $4x^2 - 25 = (2x-5)(2x+5)$

(b) $2x^2 + 5x - 12 = (2x-3)(x+4)$

(c) $x^3 - 3x^2 - 4x + 12 = x^2(x-3) - 4(x-3) = (x-3)\left(x^2-4\right) = (x-3)(x-2)(x+2)$

(d) $x^4 + 27x = x\left(x^3+27\right) = x(x+3)\left(x^2-3x+9\right)$

(e) $3x^{3/2} - 9x^{1/2} + 6x^{-1/2} = 3x^{-1/2}\left(x^2 - 3x + 2\right) = 3x^{-1/2}(x-2)(x-1)$

(f) $x^3 y - 4xy = xy\left(x^2-4\right) = xy(x-2)(x+2)$

7. (a) $\dfrac{x^2+3x+2}{x^2-x-2} = \dfrac{(x+1)(x+2)}{(x+1)(x-2)} = \dfrac{x+2}{x-2}$

(b) $\dfrac{2x^2-x-1}{x^2-9} \cdot \dfrac{x+3}{2x+1} = \dfrac{(2x+1)(x-1)}{(x-3)(x+3)} \cdot \dfrac{x+3}{2x+1} = \dfrac{x-1}{x-3}$

(c) $\dfrac{x^2}{x^2-4} - \dfrac{x+1}{x+2} = \dfrac{x^2}{(x-2)(x+2)} - \dfrac{x+1}{x+2} = \dfrac{x^2}{(x-2)(x+2)} + \dfrac{-(x+1)(x-2)}{(x-2)(x+2)}$

$$= \dfrac{x^2 - \left(x^2-x-2\right)}{(x-2)(x+2)} = \dfrac{x+2}{(x-2)(x+2)} = \dfrac{1}{x-2}$$

(d) $\dfrac{\dfrac{y}{x} - \dfrac{x}{y}}{\dfrac{1}{y} - \dfrac{1}{x}} = \dfrac{\dfrac{y}{x} - \dfrac{x}{y}}{\dfrac{1}{y} - \dfrac{1}{x}} \cdot \dfrac{xy}{xy} = \dfrac{y^2-x^2}{x-y} = \dfrac{(y-x)(y+x)}{x-y} = \dfrac{-(x-y)(y+x)}{x-y} = -(y+x)$

8. (a) $\dfrac{6}{\sqrt[3]{4}} = \dfrac{6}{\sqrt[3]{2^2}} = \dfrac{6}{\sqrt[3]{2^2}} \cdot \dfrac{\sqrt[3]{2}}{\sqrt[3]{2}} = \dfrac{6\sqrt[3]{2}}{2} = 3\sqrt[3]{2}$

(b) $\dfrac{\sqrt{6}}{2+\sqrt{3}} = \dfrac{\sqrt{6}}{2+\sqrt{3}} \cdot \dfrac{2-\sqrt{3}}{2-\sqrt{3}} = \dfrac{2\sqrt{6}-\sqrt{18}}{4-3} = \dfrac{2\sqrt{6}-\sqrt{9\cdot 2}}{1} = 2\sqrt{6}-3\sqrt{2}$

Focus on Problem Solving: General Principles

1. Let d be the distance traveled to and from work. Let t_1 and t_2 be the times for the trip from home to work and the trip from work to home, respectively. Using time$= \dfrac{\text{distance}}{\text{rate}}$, we get $t_1 = \dfrac{d}{50}$ and $t_2 = \dfrac{d}{30}$. Since average speed $= \dfrac{\text{distance traveled}}{\text{total time}}$, we have average speed $= \dfrac{2d}{t_1 + t_2} = \dfrac{2d}{\dfrac{d}{50} + \dfrac{d}{30}} \quad \Leftrightarrow$

average speed $= \dfrac{150\,(2d)}{150\left(\dfrac{d}{50} + \dfrac{d}{30}\right)} = \dfrac{300d}{3d + 5d} = \dfrac{300}{8} = 37.5$ mi/h.

3. We use the formula $d = rt$. Since the car and the van each travel at a speed of 40 mi/h, they approach each other at a combined speed of 80 mi/h. (The distance between them decreases at a rate of 80 mi/h.) So the time spent driving till they meet is $t = \dfrac{d}{r} = \dfrac{120}{80} = 1.5$ hours. Thus, the fly flies at a speed of 100 mi/h for 1.5 hours, and therefore travels a distance of $d = rt = (100)\,(1.5) = 150$ miles.

5. We continue the pattern. Three parallel cuts produce 10 pieces. Thus, each new cut produces an additional 3 pieces. Since the first cut produces 4 pieces, we get the formula $f(n) = 4 + 3(n-1)$, $n \geq 1$. Since $f(142) = 4 + 3(141) = 427$, we see that 142 parallel cuts produce 427 pieces.

7. George's speed is $\frac{1}{50}$ lap/s, and Sue's is $\frac{1}{30}$ lap/s. So after t seconds, George has run $\dfrac{t}{50}$ laps, and Sue has run $\dfrac{t}{30}$ laps. They will next be side by side when Sue has run exactly one more lap than George, that is, when $\dfrac{t}{30} = \dfrac{t}{50} + 1$. We solve for t by first multiplying by 150, so $50t = 30t + 150 \quad \Leftrightarrow \quad 20t = 150 \quad \Leftrightarrow \quad t = 75$. Therefore, they will be even after 75 s.

9. *Method 1:* After the exchanges, the volume of liquid in the pitcher and in the cup is the same as it was to begin with. Thus, any coffee in the pitcher of cream must be replacing an equal amount of cream that has ended up in the coffee cup.

Method 2: Alternatively, look at the drawing of the spoonful of coffee and cream mixture being returned to the pitcher of cream. Suppose it is possible to separate the cream and the coffee, as shown. Then you can see that the coffee going into the cream occupies the same volume as the cream that was left in the coffee.

Method 3 (an algebraic approach): Suppose the cup of coffee has y spoonfuls of coffee. When one spoonful of cream is added to the coffee cup, the resulting mixture has the following ratios: $\dfrac{\text{cream}}{\text{mixture}} = \dfrac{1}{y+1}$ and $\dfrac{\text{coffee}}{\text{mixture}} = \dfrac{y}{y+1}$.

So, when we remove a spoonful of the mixture and put it into the pitcher of cream, we are really removing $\dfrac{1}{y+1}$ of a spoonful of cream and $\dfrac{y}{y+1}$ spoonful of coffee. Thus the amount of cream left in the mixture (cream in the coffee) is $1 - \dfrac{1}{y+1} = \dfrac{y}{y+1}$ of a spoonful. This is the same as the amount of coffee we added to the cream.

11. Let r be the radius of the earth in feet. Then the circumference (length of the ribbon) is $2\pi r$. When we increase the radius by 1 foot, the new radius is $r + 1$, so the new circumference is $2\pi(r + 1)$. Thus you need $2\pi(r + 1) - 2\pi r = 2\pi$ extra feet of ribbon.

13. Let $r_1 = 8$. $r_2 = \frac{1}{2}\left(8 + \frac{72}{8}\right) = \frac{1}{2}(8 + 9) = 8.5$, $r_3 = \frac{1}{2}\left(8.5 + \dfrac{72}{8.5}\right) = \frac{1}{2}(8.5 + 8.471) = 8.485$, and

$r_4 = \frac{1}{2}\left(8.485 + \dfrac{72}{8.485}\right) = \frac{1}{2}(8.485 + 8.486) = 8.485$. Thus $\sqrt{72} \approx 8.49$.

15. The first few powers of 3 are $3^1 = 3$, $3^2 = 9$, $3^3 = 27$, $3^4 = 81$, $3^5 = 243$, $3^6 = 729$. It appears that the final digit cycles in a pattern, namely $3 \rightarrow 9 \rightarrow 7 \rightarrow 1 \rightarrow 3 \rightarrow 9 \rightarrow 7 \rightarrow 1$, of length 4. Since $459 = 4 \times 114 + 3$, the final digit is the third in the cycle, namely 7.

17. Let p be an odd prime and label the other integer sides of the triangle as shown.
Then $p^2 + b^2 = c^2$ $\Leftrightarrow$ $p^2 = c^2 - b^2$ $\Leftrightarrow$ $p^2 = (c - b)(c + b)$. Now
$c - b$ and $c + b$ are integer factors of p^2 where p is a prime.

Case 1: $c - b = p$ and $c + b = p$ Subtracting gives $b = 0$ which contradicts that b is the length of a side of a triangle.

Case 2: $c - b = 1$ and $c + b = p^2$ Adding gives $2c = 1 + p^2$ $\Leftrightarrow$ $c = \dfrac{1 + p^2}{2}$ and $b = \dfrac{p^2 - 1}{2}$. Since p is odd, p^2

is odd. Thus $\dfrac{1 + p^2}{2}$ and $\dfrac{p^2 - 1}{2}$ both define integers and are the only solutions. (For example, if we choose $p = 3$, then $c = 5$ and $b = 4$.)

19. The north pole is such a point. And there are others: Consider a point a_1 near the south pole such that the parallel passing through a_1 forms a circle C_1 with circumference exactly one mile. Any point P_1 exactly one mile north of the circle C_1 along a meridian is a point satisfying the conditions in the problem: starting at P_1 she walks one mile south to the point a_1 on the circle C_1, then one mile east along C_1 returning to the point a_1, then north for one mile to P_1. That's not all. If a point a_2 (or a_3, a_4, a_5, . . .) is chosen near the south pole so that the parallel passing through it forms a circle C_2 (C_3, C_4, C_5, . . .) with a circumference of exactly $\frac{1}{2}$ mile ($\frac{1}{3}$ mi, $\frac{1}{4}$ mi, $\frac{1}{5}$ mi, . . .), then the point P_2 (P_3, P_4, P_5, . . .) one mile north of a_2 (a_3, a_4, a_5, . . .) along a meridian satisfies the conditions of the problem: she walks one mile south from P_2 (P_3, P_4, P_5, . . .) arriving at a_2 (a_3, a_4, a_5, . . .) along the circle C_2 (C_3, C_4, C_5, . . .), walks east along the circle for one mile thus traversing the circle twice (three times, four times, five times, . . .) returning to a_2 (a_3, a_4, a_5, . . .), and then walks north one mile to P_2 (P_3, P_4, P_5, . . .).

21. Let r and R be the radius of the smaller circle and larger circle, respectively. Since the line is tangent to the smaller circle, using the Pythagorean Theorem, we get the following relationship between the radii:
$R^2 = 1^2 + r^2$. Thus the area of the shaded region is
$\pi R^2 - \pi r^2 = \pi \left(1 + r^2\right) - \pi r^2 = \pi$.

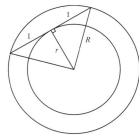

23. Since $\sqrt[3]{1729} \approx 12.0023$, we start with $n = 12$ and find the other perfect cube.

n	$1729 - n^3$
12	$1729 - (12)^3 = 1$
11	$1729 - (11)^3 = 398$
10	$1729 - (10)^3 = 729$

Since 1 and 729 are perfect cubes, the two representations we seek are $1^3 + 12^3 = 1729$ and $9^3 + 10^3 = 1729$.

25. (a) Imagine that the rooms are colored alternately black and white (like a checkerboard) with the entrance being a white room. So there are 18 black rooms and 18 white rooms. If a tourist is in a white room, he must next go to a black room, and if he is in a black room, he must next go to a white room.

(b) Since the entrance is white, any path that the tourist takes must be of the form W, B, W, B, W, B, ... Since the museum contains an even number of rooms and the tourist starts in a white room, he must end his tour in a black room. But the exit is in a white room, so the proposed tour is impossible.

1 Equations and Inequalities

1.1 Basic Equations

1. **(a)** When $x = -2$, LHS $= 4(-2) + 7 = -8 + 7 = -1$ and RHS $= 9(-2) - 3 = -18 - 3 = -21$. Since LHS $\neq$ RHS, $x = -2$ is not a solution.

(b) When $x = 2$, LHS $= 4(-2) + 7 = 8 + 7 = 15$ and RHS $= 9(2) - 3 = 18 - 3 = 15$. Since LHS $=$ RHS, $x = 2$ is a solution.

3. **(a)** When $x = 2$, LHS $= 1 - [2 - (3 - (2))] = 1 - [2 - 1] = 1 - 1 = 0$ and RHS $= 4(2) - (6 + (2)) = 8 - 8 = 0$. Since LHS $=$ RHS, $x = 2$ is a solution.

(b) When $x = 4$ LHS $= 1 - [2 - (3 - (4))] = 1 - [2 - (-1)] = 1 - 3 = -2$ and RHS $= 4(4) - (6 + (4)) = 16 - 10 = 6$. Since LHS $\neq$ RHS, $x = 4$ is not a solution.

5. **(a)** When $x = -1$, LHS $= 2(-1)^{1/3} - 3 = 2(-1) - 3 = -2 - 3 = -5$. Since LHS $\neq 1$, $x = -1$ is not a solution.

(b) When $x = 8$ LHS $= 2(8)^{1/3} - 3 = 2(2) - 3 = 4 - 3 = 1 =$ RHS. So $x = 8$ is a solution.

7. **(a)** When $x = 0$, LHS $= \dfrac{0 - a}{0 - b} = \dfrac{-a}{-b} = \dfrac{a}{b} =$ RHS. So $x = 0$ is a solution.

(b) When $x = b$, LHS $= \dfrac{b - a}{b - b} = \dfrac{b - a}{0}$ is not defined, so $x = b$ is not a solution.

9. $2x + 7 = 31 \iff 2x = 24 \iff x = 12$

11. $\frac{1}{2}x - 8 = 1 \iff \frac{1}{2}x = 9 \iff x = 18$

13. $x - 3 = 2x + 6 \iff -9 = x$

15. $-7w = 15 - 2w \iff -5w = 15 \iff w = -3$

17. $\frac{1}{2}y - 2 = \frac{1}{3}y \iff 3y - 12 = 2y$ (multiply both sides by the LCD, 6) $\iff y = 12$

19. $2(1 - x) = 3(1 + 2x) + 5 \iff 2 - 2x = 3 + 6x + 5 \iff 2 - 2x = 8 + 6x \iff -6 = 8x \iff x = -\frac{3}{4}$

21. $4\left(y - \frac{1}{2}\right) - y = 6(5 - y) \iff 4y - 2 - y = 30 - 6y \iff 3y - 2 = 30 - 6y \iff 9y = 32 \iff y = \frac{32}{9}$

23. $\dfrac{1}{x} = \dfrac{4}{3x} + 1 \implies 3 = 4 + 3x$ (multiply both sides by the LCD, $3x$) $\iff -1 = 3x \iff x = -\frac{1}{3}$

25. $\dfrac{2}{t + 6} = \dfrac{3}{t - 1} \implies 2(t - 1) = 3(t + 6)$ [multiply both sides by the LCD, $(t - 1)(t + 6)$] $\iff 2t - 2 = 3t + 18$ $\iff -20 = t$

27. $r - 2[1 - 3(2r + 4)] = 61 \iff r - 2(1 - 6r - 12) = 61 \iff r - 2(-6r - 11) = 61 \iff r + 12r + 22 = 61$ $\iff 13r = 39 \iff r = 3$

29. $\sqrt{3}x + \sqrt{12} = \dfrac{x + 5}{\sqrt{3}} \iff 3x + 6 = x + 5$ (multiply both sides by $\sqrt{3}$) $\iff 2x = -1 \iff x = -\frac{1}{2}$

31. $\dfrac{2}{x} - 5 = \dfrac{6}{x} + 4 \implies 2 - 5x = 6 + 4x \iff -4 = 9x \iff -\frac{4}{9} = x$

33. $\dfrac{3}{x + 1} - \dfrac{1}{2} = \dfrac{1}{3x + 3} \implies 3(6) - (3x + 3) = 2$ [multiply both sides by $6(x + 1)$] $\iff 18 - 3x - 3 = 2 \iff$ $-3x + 15 = 2 \iff -3x = -13 \iff x = \frac{13}{3}$

35. $\dfrac{2x - 7}{2x + 4} = \frac{2}{3} \implies (2x - 7)3 = 2(2x + 4)$ (cross multiply) $\iff 6x - 21 = 4x + 8 \iff 2x = 29 \iff x = \frac{29}{2}$

37. $x - \frac{1}{3}x - \frac{1}{2}x - 5 = 0 \iff 6x - 2x - 3x - 30 = 0$ (multiply both sides by 6) $\iff x = 30$

39. $\dfrac{1}{z} - \dfrac{1}{2z} - \dfrac{1}{5z} = \dfrac{10}{z + 1} \implies 10(z + 1) - 5(z + 1) - 2(z + 1) = 10(10z)$ [multiply both sides by $10z(z + 1)$] $\iff$ $3(z + 1) = 100z \iff 3z + 3 = 100z \iff 3 = 97z \iff \frac{3}{97} = z$

41. $\dfrac{u}{u - \dfrac{u+1}{2}} = 4 \;\Rightarrow\; u = 4\left(u - \dfrac{u+1}{2}\right)$ (cross multiply) $\Leftrightarrow\; u = 4u - 2u - 2 \;\Leftrightarrow\; u = 2u - 2 \;\Leftrightarrow\; 2 = u$

43. $\dfrac{x}{2x-4} - 2 = \dfrac{1}{x-2} \;\Rightarrow\; x - 2(2x-4) = 2$ [multiply both sides by $2(x-2)$] $\Leftrightarrow\; x - 4x + 8 = 2 \;\Leftrightarrow$
$-3x = -6 \;\Leftrightarrow\; x = 2$. But substituting $x = 2$ into the original equation does not work, since we cannot divide by 0.
Thus there is no solution.

45. $\dfrac{3}{x+4} = \dfrac{1}{x} + \dfrac{6x+12}{x^2+4x} \;\Rightarrow\; 3(x) = (x+4) + 6x + 12$ (multiply both sides by $x(x+4)$] $\Leftrightarrow\; 3x = 7x + 16 \;\Leftrightarrow$
$-4x = 16 \;\Leftrightarrow\; x = -4$. But substituting $x = -4$ into the original equation does not work, since we cannot divide by 0.
Thus, there is no solution.

47. $x^2 = 49 \;\Rightarrow\; x = \pm 7$

49. $x^2 - 24 = 0 \;\Leftrightarrow\; x^2 = 24 \;\Rightarrow\; x = \pm\sqrt{24} = \pm 2\sqrt{6}$

51. $8x^2 - 64 = 0 \;\Leftrightarrow\; x^2 - 8 = 0 \;\Leftrightarrow\; x^2 = 8 \;\Rightarrow\; x = \pm\sqrt{8} = \pm 2\sqrt{2}$

53. $x^2 + 16 = 0 \;\Leftrightarrow\; x^2 = -16$ which has no real solution.

55. $(x+2)^2 = 4 \;\Leftrightarrow\; (x+2)^2 = 4 \;\Rightarrow\; x + 2 = \pm 2$. If $x + 2 = 2$, then $x = 0$. If $x + 2 = -2$, then $x = -4$. The
solutions are -4 and 0.

57. $x^3 = 27 \;\Leftrightarrow\; x = 27^{1/3} = 3$

59. $0 = x^4 - 16 = (x^2+4)(x^2-4) = (x^2+4)(x-2)(x+2) . x^2 + 4 = 0$ has no real solution. If $x - 2 = 0$, then
$x = 2$. If $x + 2 = 0$, then $x = -2$. The solutions are ± 2.

61. $x^4 + 64 = 0 \;\Leftrightarrow\; x^4 = -64$ which has no real solution.

63. $(x+2)^4 - 81 = 0 \;\Leftrightarrow\; (x+2)^4 = 81 \;\Leftrightarrow\; \left[(x+2)^4\right]^{1/4} = \pm 81^{1/4} \;\Leftrightarrow\; x + 2 = \pm 3$. So $x + 2 = 3$, then $x = 1$.
If $x + 2 = -3$, then $x = -5$. The solutions are -5 and 1.

65. $3(x-3)^3 = 375 \;\Leftrightarrow\; (x-3)^3 = 125 \;\Leftrightarrow\; (x-3) = 125^{1/3} = 5 \;\Leftrightarrow\; x = 3 + 5 = 8$

67. $\sqrt[3]{x} = 5 \;\Leftrightarrow\; x = 5^3 = 125$

69. $2x^{5/3} + 64 = 0 \;\Leftrightarrow\; 2x^{5/3} = -64 \;\Leftrightarrow\; x^{5/3} = -32 \;\Leftrightarrow\; x = (-32)^{3/5} = \left(-2^5\right)^{1/5} = (-2)^3 = -8$

71. $3.02x + 1.48 = 10.92 \;\Leftrightarrow\; 3.02x = 9.44 \;\Leftrightarrow\; x = \dfrac{9.44}{3.02} \approx 3.13$

73. $2.15x - 4.63 = x + 1.19 \;\Leftrightarrow\; 1.15x = 5.82 \;\Leftrightarrow\; x = \dfrac{5.82}{1.19} \approx 5.06$

75. $3.16(x + 4.63) = 4.19(x - 7.24) \;\Leftrightarrow\; 3.16x + 14.63 = 4.19x - 30.34 \;\Leftrightarrow$
$44.97 = 1.03x \Leftrightarrow\; x = \dfrac{44.97}{1.03} \approx 43.66$

77. $\dfrac{0.26x - 1.94}{3.03 - 2.44x} = 1.76 \;\Rightarrow\; 0.26x - 1.94 = 1.76(3.03 - 2.44x) \;\Leftrightarrow\; 0.26x - 1.94 = 5.33 - 4.29x \;\Leftrightarrow$
$4.55x = 7.27 \;\Leftrightarrow\; x = \dfrac{7.27}{4.55} \approx 1.60$

79. $PV = nRT \;\Leftrightarrow\; R = \dfrac{PV}{nT}$

81. $\dfrac{1}{R} = \dfrac{1}{R_1} + \dfrac{1}{R_2} \;\Leftrightarrow\; R_1 R_2 = RR_2 + RR_1$ (multiply both sides by the LCD, $RR_1 R_2$). Thus $R_1 R_2 - RR_1 = RR_2$
$\Leftrightarrow\; R_1(R_2 - R) = RR_2 \;\Leftrightarrow\; R_1 = \dfrac{RR_2}{R_2 - R}$.

83. $\dfrac{ax+b}{cx+d} = 2 \;\Leftrightarrow\; ax + b = 2(cx+d) \;\Leftrightarrow\; ax + b = 2cx + 2d \;\Leftrightarrow\; ax - 2cx = 2d - b \;\Leftrightarrow\; (a - 2c)x = 2d - b$
$\Leftrightarrow\; x = \dfrac{2d-b}{a-2c}$

85. $a^2 x + (a-1) = (a+1)x \;\Leftrightarrow\; a^2 x - (a+1)x = -(a-1) \;\Leftrightarrow\; \left(a^2 - (a+1)\right)x = -a + 1 \;\Leftrightarrow$
$\left(a^2 - a - 1\right)x = -a + 1 \;\Leftrightarrow\; x = \dfrac{-a+1}{a^2 - a - 1}$

87. $V = \frac{1}{3}\pi r^2 h \;\; \Leftrightarrow \;\; r^2 = \dfrac{3V}{\pi h} \;\; \Rightarrow \;\; r = \pm\sqrt{\dfrac{3V}{\pi h}}$

89. $a^2 + b^2 = c^2 \;\; \Leftrightarrow \;\; b^2 = c^2 - a^2 \;\; \Rightarrow \;\; b = \pm\sqrt{c^2 - a^2}$

91. $V = \frac{4}{3}\pi r^3 \;\; \Leftrightarrow \;\; r^3 = \dfrac{3V}{4\pi} \;\; \Leftrightarrow \;\; r = \sqrt[3]{\dfrac{3V}{4\pi}}$

93. (a) The shrinkage factor when $w = 250$ is $S = \dfrac{0.032\,(250) - 2.5}{10{,}000} = \dfrac{8 - 2.5}{10{,}000} = 0.00055$. So the beam shrinks $0.00055 \times 12.025 \approx 0.007$ m, so when it dries it will be $12.025 - 0.007 = 12.018$ m long.

(b) Substituting $S = 0.00050$ we get $0.00050 = \dfrac{0.032w - 2.5}{10{,}000} \;\; \Leftrightarrow \;\; 5 = 0.032w - 2.5 \;\; \Leftrightarrow \;\; 7.5 = 0.032w \;\; \Leftrightarrow$

$w = \dfrac{7.5}{0.032} \approx 234.375$. So the water content should be 234.375 kg/m^3.

95. (a) Solving for v when $P = 10{,}000$ we get $10{,}000 = 15.6v^3 \;\; \Leftrightarrow \;\; v^3 \approx 641.02 \;\; \Leftrightarrow \;\; v \approx 8.6$ km/h.

(b) Solving for v when $P = 50{,}000$ we get $50{,}000 = 15.6v^3 \;\; \Leftrightarrow \;\; v^3 \approx 3205.13 \;\; \Leftrightarrow \;\; v \approx 14.7$ km/h.

97. (a) $3\,(0) + k - 5 = k\,(0) - k + 1 \;\; \Leftrightarrow \;\; k - 5 = -k + 1 \;\; \Leftrightarrow \;\; 2k = 6 \;\; \Leftrightarrow \;\; k = 3$

(b) $3\,(1) + k - 5 = k\,(1) - k + 1 \;\; \Leftrightarrow \;\; 3 + k - 5 = k - k + 1 \;\; \Leftrightarrow \;\; k - 2 = 1 \;\; \Leftrightarrow \;\; k = 3$

(c) $3\,(2) + k - 5 = k\,(2) - k + 1 \;\; \Leftrightarrow \;\; 6 + k - 5 = 2k - k + 1 \;\; \Leftrightarrow \;\; k + 1 = k + 1$. $x = 2$ is a solution for every value of k. That is, $x = 2$ is a solution to every member of this family of equations.

99. (a) Using the formulas for volume we have: for the sphere, $V = \frac{4}{3}\pi r^3$; for the cylinder, $V = \pi r^2 h_1$; and for the cone, $V = \frac{1}{3}\pi r^2 h_2$. Since these have the same volume we must have $\frac{4}{3}\pi r^3 = \pi r^2 h_1$ and $\frac{4}{3}\pi r^3 = \frac{1}{3}\pi r^2 h_2$.

(b) Setting the volume of the sphere equal to the volume of the cylinder, we have $\frac{4}{3}\pi r^3 = \pi r^2 h_1 \;\; \Leftrightarrow \;\; \frac{4}{3}r = h_1$. So the height of the cylinder is $h_1 = \frac{4}{3}r$.

Likewise, setting the volume of the sphere equal to the volume of the cone, we have $\frac{4}{3}\pi r^3 = \frac{1}{3}\pi r^2 h_2 \;\; \Leftrightarrow \;\; 4r = h_2$. So the height of the cone is $h_2 = 4r$.

1.2   Modeling with Equations

1. If n is the first integer, then $n + 1$ is the middle integer, and $n + 2$ is the third integer. So the sum of the three consecutive integers is $n + (n + 1) + (n + 2) = 3n + 3$.

3. If s is the third test score, then since the other test scores are 78 and 82, the average of the three test scores is $\dfrac{78 + 82 + s}{3} = \dfrac{160 + s}{3}$.

5. If x dollars are invested at $2\frac{1}{2}\%$ simple interest, then the first year you will receive $0.025x$ dollars in interest.

7. Since w is the width of the rectangle, the length is three times the width, or $3w$. Then area $=$ length $\times$ width $= 3w \times w = 3w^2$ ft^2.

9. Since distance $=$ rate$\times$ time we have distance $= s \times (45 \text{ min}) \dfrac{1 \text{ h}}{60 \text{ min}} = \frac{3}{4}s$ mi.

11. If x is the quantity of pure water added, the mixture will contain 25 oz of salt and $3 + x$ gallons of water. Thus the concentration is $\dfrac{25}{3 + x}$.

13. Let x be the first integer. Then $x + 1$ and $x + 2$ are the next consecutive integers. So $x + (x + 1) + (x + 2) = 156 \;\; \Leftrightarrow \;\; 3x + 3 = 156 \;\; \Leftrightarrow \;\; 3x = 153 \;\; \Leftrightarrow \;\; x = 51$. Thus the consecutive integers are 51, 52, and 53.

15. Let m be the amount invested at $4\frac{1}{2}\%$. Then $12{,}000 - m$ is the amount invested at 4%.

Since the total interest is equal to the interest earned at $4\frac{1}{2}\%$ plus the interest earned at 4%, we have

$525 = 0.045m + 0.04\,(12{,}000 - m)$ ⟺ $525 = 0.045m + 480 - 0.04m$ ⟺ $45 = 0.005m$ ⟺

$m = \dfrac{45}{0.005} = 9000$. Thus 9000 is invested at $4\frac{1}{2}\%$, and $12{,}000 - 9000 = \$3000$ is invested at 4%.

17. Let x be her monthly salary. Since her annual salary $= 12 \times$ (monthly salary) $+$ (Christmas bonus) we have
$97{,}300 = 12x + 8{,}500$ ⟺ $88{,}800 = 12x$ ⟺ $x \approx 7{,}400$. Her monthly salary is $\$7{,}400$.

19. Let h be the amount that Craig inherits. So $(x + 22{,}000)$ is the amount that he invests and doubles. Thus
$2\,(x + 22{,}000) = 134{,}000$ ⟺ $2x + 44{,}000 = 134{,}000$ ⟺ $2x = 90{,}000$ ⟺ $x = 45{,}000$. So Craig inherits
$\$45{,}000$.

21. Let x be the hours the assistant worked. Then $2x$ is the hours the plumber worked. Since the labor charge is equal
to the plumber's labor plus the assistant's labor, we have $4025 = 45\,(2x) + 25x$ ⟺ $4025 = 90x + 25x$ ⟺
$4025 = 115x$ ⟺ $x = \frac{4025}{115} = 35$. Thus the assistant works for 35 hours, and the plumber works for $2 \times 35 = 70$ hours.

23. All ages are in terms of the daughter's age 7 years ago. Let y be age of the daughter 7 years ago. Then $11y$ is the age of
the movie star 7 years ago. Today, the daughter is $y + 7$, and the movie star is $11y + 7$. But the movie star is also 4 times
his daughter's age today. So $4\,(y + 7) = 11y + 7$ ⟺ $4y + 28 = 11y + 7$ ⟺ $21 = 7y$ ⟺ $y = 3$. Thus the
movie star's age today is $11\,(3) + 7 = 40$ years.

25. Let p be the number of pennies. Then p is the number of nickels and p is the number of dimes. So the value of
the coins in the purse is the value of the pennies plus the value of the nickels plus the value of the dimes. Thus
$1.44 = 0.01p + 0.05p + 0.10p$ ⟺ $1.44 = 0.16p$ ⟺ $p = \frac{1.44}{0.16} = 9$. So the purse contains 9 pennies, 9 nickels,
and 9 dimes.

27. Let l be the length of the garden. Since area $=$ width $\cdot$ length, we obtain the equation $1125 = 25l$ ⟺ $l = \frac{1125}{25} = 45$ ft.
So the garden is 45 feet long.

29. Let x be the length of a side of the square plot. As shown in the figure,
area of the plot $=$ area of the building $+$ area of the parking lot. Thus,
$x^2 = 60\,(40) + 12{,}000 = 2{,}400 + 12{,}000 = 14{,}400$ ⟹ $x = \pm 120$. So the
plot of land measures 120 feet by 120 feet.

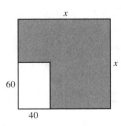

31. The figure is a trapezoid, so its area is $\dfrac{\text{base}_1 + \text{base}_2}{2}$ (height). Putting in the known quantities, we have

$120 = \dfrac{y + 2y}{2}\,(y) = \frac{3}{2}y^2$ ⟺ $y^2 = 80$ ⟹ $y = \pm\sqrt{80} = \pm 4\sqrt{5}$. Since length is positive, $y = 4\sqrt{5} \approx 8.94$.

33. Let x be the width of the strip. Then the length of the mat is $20 + 2x$, and the width of the mat is $15 + 2x$. Now the
perimeter is twice the length plus twice the width, so $102 = 2\,(20 + 2x) + 2\,(15 + 2x)$ ⟺ $102 = 40 + 4x + 30 + 4x$
 ⟺ $102 = 70 + 8x$ ⟺ $32 = 8x$ ⟺ $x = 4$. Thus the strip of mat is 4 inches wide.

35. Let x be the grams of silver added. The weight of the rings is 5×18 g $= 90$ g.

	5 rings	Pure silver	Mixture
Grams	90	x	$90 + x$
Rate (% gold)	0.90	0	0.75
Value	$0.90\,(90)$	$0x$	$0.75\,(90 + x)$

So $0.90\,(90) + 0x = 0.75\,(90 + x)$ $\Leftrightarrow$ $81 = 67.5 + 0.75x$ $\Leftrightarrow$ $0.75x = 13.5$ $\Leftrightarrow$ $x = \frac{13.5}{0.75} = 18$. Thus 18 grams of silver must be added to get the required mixture.

37. Let x be the liters of coolant removed and replaced by water.

	60% antifreeze	60% antifreeze (removed)	Water	Mixture
Liters	3.6	x	x	3.6
Rate (% antifreeze)	0.60	0.60	0	0.50
Value	$0.60\,(3.6)$	$-0.60x$	$0x$	$0.50\,(3.6)$

so $0.60\,(3.6) - 0.60x + 0x = 0.50\,(3.6)$ $\Leftrightarrow$ $2.16 - 0.6x = 1.8$ $\Leftrightarrow$ $-0.6x = -0.36$ $\Leftrightarrow$ $x = \frac{-0.36}{-0.6} = 0.6$. Thus 0.6 liters must be removed and replaced by water.

39. Let c be the concentration of fruit juice in the cheaper brand. The new mixture that Jill makes will consist of 650 mL of the original fruit punch and 100 ml of the cheaper fruit punch.

	Original Fruit Punch	Cheaper Fruit Punch	Mixture
mL	650	100	750
Concentration	0.50	c	0.48
Juice	$0.50 \cdot 650$	$100c$	$0.48 \cdot 750$

So $0.50 \cdot 650 + 100c = 0.48 \cdot 750$ $\Leftrightarrow$ $325 + 100c = 360$ $\Leftrightarrow$ $100c = 35$ $\Leftrightarrow$ $c = 0.35$. Thus the cheaper brand is only 35% fruit juice.

41. Let x be the length of the man's shadow, in meters. Using similar triangles, $\frac{10 + x}{6} = \frac{x}{2}$ $\Leftrightarrow$ $20 + 2x = 6x$ $\Leftrightarrow$ $4x = 20$ $\Leftrightarrow$ $x = 5$. Thus the man's shadow is 5 meters long.

43. Let x be the distance from the fulcrum to where the mother sits. Then substituting the known values into the formula given, we have $100\,(8) = 125x$ $\Leftrightarrow$ $800 = 125x$ $\Leftrightarrow$ $x = 6.4$. So the mother should sit 6.4 feet from the fulcrum.

45. Let t be the time in minutes it would take Candy and Tim if they work together. Candy delivers the papers at a rate of $\frac{1}{70}$ of the job per minute, while Tim delivers the paper at a rate of $\frac{1}{80}$ of the job per minute. The sum of the fractions of the job that each can do individually in one minute equals the fraction of the job they can do working together. So we have $\frac{1}{t} = \frac{1}{70} + \frac{1}{80}$ $\Leftrightarrow$ $560 = 8t + 7t$ $\Leftrightarrow$ $560 = 15t$ $\Leftrightarrow$ $t = 37\frac{1}{3}$ minutes. Since $\frac{1}{3}$ of a minute is 20 seconds, it would take them 37 minutes 20 seconds if they worked together.

47. Let t be the time, in hours, it takes Karen to paint a house alone. Then working together, Karen and Betty can paint a house in $\frac{2}{3}t$ hours. The sum of their individual rates equals their rate working together, so $\frac{1}{t} + \frac{1}{6} = \frac{1}{\frac{2}{3}t}$ $\Leftrightarrow$ $\frac{1}{t} + \frac{1}{6} = \frac{3}{2t}$ $\Leftrightarrow$ $6 + t = 9$ $\Leftrightarrow$ $t = 3$. Thus it would take Karen 3 hours to paint a house alone.

49. Let t be the time in hours that Wendy spent on the train. Then $\frac{11}{2} - t$ is the time in hours that Wendy spent on the bus. We construct a table:

	Rate	Time	Distance
By train	40	t	$40t$
By bus	60	$\frac{11}{2} - t$	$60\left(\frac{11}{2} - t\right)$

The total distance traveled is the sum of the distances traveled by bus and by train, so $300 = 40t + 60\left(\frac{11}{2} - t\right)$ $\Leftrightarrow$ $300 = 40t + 330 - 60t$ $\Leftrightarrow$ $-30 = -20t$ $\Leftrightarrow$ $t = \frac{-30}{-20} = 1.5$ hours. So the time spent on the train is $5.5 - 1.5 = 4$ hours.

51. Let r be the speed of the plane from Montreal to Los Angeles. Then $r + 0.20r = 1.20r$ is the speed of the plane from Los Angeles to Montreal.

	Rate	Time	Distance
Montreal to L.A.	r	$\dfrac{2500}{r}$	2500
L.A. to Montreal	$1.2r$	$\dfrac{2500}{1.2r}$	2500

The total time is the sum of the times each way, so $9\frac{1}{6} = \frac{2500}{r} + \frac{2500}{1.2r}$ $\Leftrightarrow$ $\frac{55}{6} = \frac{2500}{r} + \frac{2500}{1.2r}$ $\Leftrightarrow$ $55 \cdot 1.2r = 2500 \cdot 6 \cdot 1.2 + 2500 \cdot 6$ $\Leftrightarrow$ $66r = 18{,}000 + 15{,}000$ $\Leftrightarrow$ $66r = 33{,}000$ $\Leftrightarrow$ $r = \frac{33{,}000}{66} = 500$. Thus the plane flew at a speed of 500 mi/h on the trip from Montreal to Los Angeles.

53. Let l be the length of the lot in feet. Then the length of the diagonal is $l + 10$. We apply the Pythagorean Theorem with the hypotenuse as the diagonal. So $l^2 + 50^2 = (l + 10)^2$ $\Leftrightarrow$ $l^2 + 2500 = l^2 + 20l + 100$ $\Leftrightarrow$ $20l = 2400$ $\Leftrightarrow$ $l = 120$. Thus the length of the lot is 120 feet.

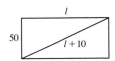

55. Let h be the height in feet of the structure. The structure is composed of a right cylinder with radius 10 and height $\frac{2}{3}h$ and a cone with base radius 10 and height $\frac{1}{3}h$. Using the formulas for the volume of a cylinder and that of a cone, we obtain the equation $1400\pi = \pi (10)^2 \left(\frac{2}{3}h\right) + \frac{1}{3}\pi (10)^2 \left(\frac{1}{3}h\right)$ $\Leftrightarrow$ $1400\pi = \frac{200\pi}{3}h + \frac{100\pi}{9}h$ $\Leftrightarrow$ $126 = 6h + h$ (multiply both sides by $\frac{9}{100\pi}$) $\Leftrightarrow$ $126 = 7h$ $\Leftrightarrow$ $h = 18$. Thus the height of the structure is 18 feet.

57. Pythagoras was born about 569 BC in Samos, Ionia and died about 475 BC.
Euclid was born about 325 BC and died about 265 BC in Alexandria, Egypt.
Archimedes was born in 287 BC in Syracuse, Sicily and died in 212 BC in Syracuse.

1.3   Quadratic Equations

1. $x^2 + x - 12 = 0$ $\Leftrightarrow$ $(x - 3)(x + 4) = 0$ $\Leftrightarrow$ $x - 3 = 0$ or $x + 4 = 0$. Thus, $x = 3$ or $x = -4$.

3. $x^2 - 7x + 12 = 0$ $\Leftrightarrow$ $(x - 4)(x - 3) = 0$ $\Leftrightarrow$ $x - 4 = 0$ or $x - 3 = 0$. Thus, $x = 4$ or $x = 3$.

5. $3x^2 - 5x - 2 = 0$ $\Leftrightarrow$ $(3x + 1)(x - 2) = 0$ $\Leftrightarrow$ $3x + 1 = 0$ or $x - 2 = 0$. Thus, $x = -\frac{1}{3}$ or $x = 2$.

7. $2y^2 + 7y + 3 = 0$ $\Leftrightarrow$ $(2y + 1)(y + 3) = 0$ $\Leftrightarrow$ $2y + 1 = 0$ or $y + 3 = 0$. Thus, $y = -\frac{1}{2}$ or $y = -3$.

9. $6x^2 + 5x = 4$ $\Leftrightarrow$ $6x^2 + 5x - 4 = 0$ $\Leftrightarrow$ $(2x - 1)(3x + 4) = 0$ $\Leftrightarrow$ $2x - 1 = 0$ or $3x + 4 = 0$. If $2x - 1 = 0$, then $x = \frac{1}{2}$; if $3x + 4 = 0$, then $x = -\frac{4}{3}$.

11. $x^2 = 5(x + 100)$ $\Leftrightarrow$ $x^2 = 5x + 500$ $\Leftrightarrow$ $x^2 - 5x - 500 = 0$ $\Leftrightarrow$ $(x - 25)(x + 20) = 0$ $\Leftrightarrow$ $x - 25 = 0$ or $x + 20 = 0$. Thus, $x = 25$ or $x = -20$.

13. $x^2 + 2x - 5 = 0$ $\Leftrightarrow$ $x^2 + 2x = 5$ $\Leftrightarrow$ $x^2 + 2x + 1 = 5 + 1$ $\Leftrightarrow$ $(x + 1)^2 = 6$ $\Rightarrow$ $x + 1 = \pm\sqrt{6} \Leftrightarrow$ $x = -1 \pm \sqrt{6}$.

15. $x^2 - 6x - 11 = 0$ $\Leftrightarrow$ $x^2 - 6x = 11$ $\Leftrightarrow$ $x^2 - 6x + 9 = 11 + 9$ $\Leftrightarrow$ $(x - 3)^2 = 20$ $\Rightarrow$ $x - 3 = \pm 2\sqrt{5}$ $\Leftrightarrow$ $x = 3 \pm 2\sqrt{5}$.

17. $x^2 + x - \frac{3}{4} = 0$ $\Leftrightarrow$ $x^2 + x = \frac{3}{4}$ $\Leftrightarrow$ $x^2 + x + \frac{1}{4} = \frac{3}{4} + \frac{1}{4}$ $\Leftrightarrow$ $\left(x + \frac{1}{2}\right)^2 = 1$ $\Rightarrow$ $x + \frac{1}{2} = \pm 1 \Leftrightarrow$ $x = -\frac{1}{2} \pm 1$.
So $x = -\frac{1}{2} - 1 = -\frac{3}{2}$ and $x = -\frac{1}{2} + 1 = \frac{1}{2}$.

19. $x^2 + 22x + 21 = 0$ $\Leftrightarrow$ $x^2 + 22x = -21$ $\Leftrightarrow$ $x^2 + 22x + 11^2 = -21 + 11^2 = -21 + 121$ $\Leftrightarrow$ $(x + 11)^2 = 100$ $\Rightarrow$ $x + 11 = \pm 10$ $\Leftrightarrow$ $x = -11 \pm 10$. Thus, $x = -1$ or $x = -21$.

21. $2x^2 + 8x + 1 = 0$ $\Leftrightarrow$ $x^2 + 4x + \frac{1}{2} = 0$ $\Leftrightarrow$ $x^2 + 4x = -\frac{1}{2}$ $\Leftrightarrow$ $x^2 + 4x + 4 = -\frac{1}{2} + 4$ $\Leftrightarrow$ $(x + 2)^2 = \frac{7}{2}$ $\Rightarrow$ $x + 2 = \pm\sqrt{\frac{7}{2}} \Leftrightarrow$ $x = -2 \pm \frac{\sqrt{14}}{2}$.

23. $4x^2 - x = 0$ $\Leftrightarrow$ $x^2 - \frac{1}{4}x = 0$ $\Leftrightarrow$ $x^2 - \frac{1}{4}x + \frac{1}{64} = \frac{1}{64}$ $\Leftrightarrow$ $\left(x - \frac{1}{8}\right)^2 = \frac{1}{64}$ $\Rightarrow$ $x - \frac{1}{8} = \pm\frac{1}{8}$ $\Leftrightarrow$ $x = \frac{1}{8} \pm \frac{1}{8}$, so $x = \frac{1}{8} - \frac{1}{8} = 0$ or $x = \frac{1}{8} + \frac{1}{8} = \frac{1}{4}$.

25. $x^2 - 2x - 15 = 0$ $\Leftrightarrow$ $(x + 3)(x - 5) = 0$ $\Leftrightarrow$ $x + 3 = 0$ or $x - 5 = 0$. Thus, $x = -3$ or $x = 5$.

27. $x^2 - 7x + 10 = 0$ $\Leftrightarrow$ $(x - 5)(x - 2) = 0$ $\Leftrightarrow$ $x - 5 = 0$ or $x - 2 = 0$. Thus, $x = 5$ or $x = 2$.

29. $2x^2 + x - 3 = 0$ $\Leftrightarrow$ $(x - 1)(2x + 3) = 0$ $\Leftrightarrow$ $x - 1 = 0$ or $2x + 3 = 0$. If $x - 1 = 0$, then $x = 1$; if $2x + 3 = 0$, then $x = -\frac{3}{2}$.

31. $x^2 + 3x + 1 = 0$ $\Leftrightarrow$ $x^2 + 3x = -1$ $\Leftrightarrow$ $x^2 + 3x + \frac{9}{4} = -1 + \frac{9}{4} = \frac{5}{4}$ $\Leftrightarrow$ $\left(x + \frac{3}{2}\right)^2 = \frac{5}{4}$ $\Rightarrow$ $x + \frac{3}{2} = \pm\sqrt{\frac{5}{4}} = \pm\frac{\sqrt{5}}{2}$ $\Leftrightarrow$ $x = -\frac{3}{2} \pm \frac{\sqrt{5}}{2}$.

33. $x^2 + 12x - 27 = 0$ $\Leftrightarrow$ $x^2 + 12x = 27$ $\Leftrightarrow$ $x^2 + 12x + 36 = 27 + 36$ $\Leftrightarrow$ $(x + 6)^2 = 63$ $\Rightarrow$ $x + 6 = \pm 3\sqrt{7} \Leftrightarrow$ $x = -6 \pm 3\sqrt{7}$.

35. $3x^2 + 6x - 5 = 0$ $\Leftrightarrow$ $x^2 + 2x - \frac{5}{3} = 0$ $\Leftrightarrow$ $x^2 + 2x = \frac{5}{3}$ $\Leftrightarrow$ $x^2 + 2x + 1 = \frac{5}{3} + 1$ $\Leftrightarrow$ $(x + 1)^2 = \frac{8}{3}$ $\Rightarrow$ $x + 1 = \pm\sqrt{\frac{8}{3}} \Leftrightarrow$ $x = -1 \pm \frac{2\sqrt{6}}{3}$.

37. $2y^2 - y - \frac{1}{2} = 0$ $\Rightarrow$ $y = \dfrac{-b \pm \sqrt{b^2 - 4ac}}{2a} = \dfrac{-(-1) \pm \sqrt{(-1)^2 - 4(2)\left(-\frac{1}{2}\right)}}{2(2)} = \dfrac{1 \pm \sqrt{1 + 4}}{4} = \dfrac{1 \pm \sqrt{5}}{4}$.

39. $4x^2 + 16x - 9 = 0$ $\Leftrightarrow$ $(2x - 1)(2x + 9) = 0$ $\Leftrightarrow$ $2x - 1 = 0$ or $2x + 9 = 0$. If $2x - 1 = 0$, then $x = \frac{1}{2}$; if $2x + 9 = 0$, then $x = -\frac{9}{2}$.

41. $3 + 5z + z^2 = 0$ $\Rightarrow$ $z = \dfrac{-b \pm \sqrt{b^2 - 4ac}}{2a} = \dfrac{-(5) \pm \sqrt{(5)^2 - 4(1)(3)}}{2(1)} = \dfrac{-5 \pm \sqrt{25 - 12}}{2} = \dfrac{-5 \pm \sqrt{13}}{2}$.

43. $x^2 - \sqrt{5}x + 1 = 0$ $\Rightarrow$ $x = \dfrac{-b \pm \sqrt{b^2 - 4ac}}{2a} = \dfrac{-(-\sqrt{5}) \pm \sqrt{\left(-\sqrt{5}\right)^2 - 4(1)(1)}}{2(1)} = \dfrac{\sqrt{5} \pm \sqrt{5 - 4}}{2} = \dfrac{\sqrt{5} \pm 1}{2}$.

45. $10y^2 - 16y + 5 = 0$ $\Rightarrow$
$$x = \dfrac{-b \pm \sqrt{b^2 - 4ac}}{2a} = \dfrac{-(-16) \pm \sqrt{(-16)^2 - 4(10)(5)}}{2(10)} = \dfrac{16 \pm \sqrt{256 - 200}}{20} = \dfrac{16 \pm \sqrt{56}}{20} = \dfrac{8 \pm \sqrt{14}}{10}.$$

47. $3x^2 + 2x + 2 = 0$ $\Rightarrow$ $x = \dfrac{-b \pm \sqrt{b^2 - 4ac}}{2a} = \dfrac{-(2) \pm \sqrt{(2)^2 - 4(3)(2)}}{2(3)} = \dfrac{-2 \pm \sqrt{4 - 24}}{6} = \dfrac{-2 \pm \sqrt{-20}}{6}$.
Since the discriminant is less than 0, the equation has no real solution.

49. $x^2 - 0.011x - 0.064 = 0 \Rightarrow$

$$x = \frac{-(-0.011) \pm \sqrt{(-0.011)^2 - 4(1)(-0.064)}}{2(1)} = \frac{0.011 \pm \sqrt{0.000121 + 0.256}}{2} \approx \frac{0.011 \pm 0.506}{2}.$$

Thus, $x \approx \dfrac{0.011 + 0.506}{2} = 0.259$ or $x \approx \dfrac{0.011 - 0.506}{2} = -0.248.$

51. $x^2 - 2.450x + 1.501 = 0 \Rightarrow$

$$x = \frac{-(-2.450) \pm \sqrt{(-2.450)^2 - 4(1)(1.501)}}{2(1)} = \frac{2.450 \pm \sqrt{6.0025 - 6.004}}{2} = \frac{2.450 \pm \sqrt{-0.0015}}{2}.$$

Thus, there is no real solution.

53. $2.232x^2 - 4.112x = 6.219 \Leftrightarrow 2.232x^2 - 4.112x - 6.219 = 0 \Rightarrow$

$$x = \frac{-(-4.112) \pm \sqrt{(-4.112)^2 - 4(2.232)(-6.219)}}{2(2.232)} = \frac{4.112 \pm \sqrt{16.9085 + 55.5232}}{4.464} \approx \frac{4.112 \pm 8.511}{4.464}. \text{ Thus,}$$

$$x = \frac{4.112 - 8.511}{4.464} = -0.985 \text{ or } x = \frac{4.112 + 8.511}{4.464} = 2.828.$$

55. $h = \frac{1}{2}gt^2 + v_0 t \Leftrightarrow \frac{1}{2}gt^2 + v_0 t - h = 0.$ Using the quadratic formula,

$$t = \frac{-(v_0) \pm \sqrt{(v_0)^2 - 4\left(\frac{1}{2}g\right)(-h)}}{2\left(\frac{1}{2}g\right)} = \frac{-v_0 \pm \sqrt{v_0^2 + 2gh}}{g}.$$

57. $A = 2x^2 + 4xh \Leftrightarrow 2x^2 + 4xh - A = 0.$ Using the quadratic formula,

$$x = \frac{-(4h) \pm \sqrt{(4h)^2 - 4(2)(-A)}}{2(2)} = \frac{-4h \pm \sqrt{16h^2 + 8A}}{4} = \frac{-4h \pm \sqrt{4(4h^2 + 2A)}}{4} = \frac{-4h \pm 2\sqrt{4h^2 + 2A}}{4}$$

$$= \frac{2\left(-2h \pm \sqrt{4h^2 + 2A}\right)}{4} = \frac{-2h \pm \sqrt{4h^2 + 2A}}{2}$$

59. $\dfrac{1}{s+a} + \dfrac{1}{s+b} = \dfrac{1}{c} \Leftrightarrow c(s+b) + c(s+a) = (s+a)(s+b) \Leftrightarrow cs + bc + cs + ac = s^2 + as + bs + ab \Leftrightarrow$

$s^2 + (a + b - 2c)s + (ab - ac - bc) = 0.$ Using the quadratic formula,

$$s = \frac{-(a+b-2c) \pm \sqrt{(a+b-2c)^2 - 4(1)(ab - ac - bc)}}{2(1)}$$

$$= \frac{-(a+b-2c) \pm \sqrt{a^2 + b^2 + 4c^2 + 2ab - 4ac - 4bc - 4ab + 4ac + 4bc}}{2}$$

$$= \frac{-(a+b-2c) \pm \sqrt{a^2 + b^2 + 4c^2 - 2ab}}{2}$$

61. $D = b^2 - 4ac = (-6)^2 - 4(1)(1) = 32.$ Since D is positive, this equation has two real solutions.

63. $D = b^2 - 4ac = (2.20)^2 - 4(1)(1.21) = 4.84 - 4.84 = 0.$ Since $D = 0$, this equation has one real solution.

65. $D = b^2 - 4ac = (5)^2 - 4(4)\left(\frac{13}{8}\right) = 25 - 26 = -1.$ Since D is negative, this equation has no real solution.

67. $D = b^2 - 4ac = (r)^2 - 4(1)(-s) = r^2 + 4s.$ Since D is positive, this equation has two real solutions.

69. $a^2 x^2 + 2ax + 1 = 0 \Leftrightarrow (ax + 1)^2 = 0 \Leftrightarrow ax + 1 = 0.$ So $ax + 1 = 0$ then $ax = -1 \Leftrightarrow x = -\dfrac{1}{a}.$

71. $ax^2 - (2a + 1)x + (a + 1) = 0 \Leftrightarrow [ax - (a + 1)](x - 1) = 0 \Leftrightarrow ax - (a + 1) = 0$ or $x - 1 = 0.$ If

$ax - (a + 1) = 0,$ then $x = \dfrac{a + 1}{a};$ if $x - 1 = 0,$ then $x = 1.$

73. We want to find the values of k that make the discriminant 0. Thus $k^2 - 4(4)(25) = 0 \Leftrightarrow k^2 = 400 \Leftrightarrow k = \pm 20.$

75. Let n be one number. Then the other number must be $55 - n$, since $n + (55 - n) = 55$. Because the product is 684, we have $(n)(55 - n) = 684 \quad\Leftrightarrow\quad 55n - n^2 = 684 \quad\Leftrightarrow\quad n^2 - 55n + 684 = 0 \quad\Rightarrow$ $n = \frac{-(-55)\pm\sqrt{(-55)^2 - 4(1)(684)}}{2(1)} = \frac{55\pm\sqrt{3025 - 2736}}{2} = \frac{55\pm\sqrt{289}}{2} = \frac{55\pm 17}{2}$. So $n = \frac{55+17}{2} = \frac{72}{2} = 36$ or $n = \frac{55-17}{2} = \frac{38}{2} = 19$. In either case, the two numbers are 19 and 36.

77. Let w be the width of the garden in feet. Then the length is $w + 10$. Thus $875 = w(w + 10) \quad\Leftrightarrow\quad w^2 + 10w - 875 = 0$ $\Leftrightarrow\quad (w + 35)(w - 25) = 0$. So $w + 35 = 0$ in which case $w = -35$, which is not possible, or $w - 25 = 0$ and so $w = 25$. Thus the width is 25 feet and the length is 35 feet.

79. Let w be the width of the garden in feet. We use the perimeter to express the length l of the garden in terms of width. Since the perimeter is twice the width plus twice the length, we have $200 = 2w + 2l \quad\Leftrightarrow\quad 2l = 200 - 2w \quad\Leftrightarrow$ $l = 100 - w$. Using the formula for area, we have $2400 = w(100 - w) = 100w - w^2 \quad\Leftrightarrow\quad w^2 - 100w + 2400 = 0$ $\Leftrightarrow\quad (w - 40)(w - 60) = 0$. So $w - 40 = 0 \quad\Leftrightarrow\quad w = 40$, or $w - 60 = 0 \quad\Leftrightarrow\quad w = 60$. If $w = 40$, then $l = 100 - 40 = 60$. And if $w = 60$, then $l = 100 - 60 = 40$. So the length is 60 feet and the width is 40 feet.

81. The shaded area is the sum of the area of a rectangle and the area of a triangle. So $A = y(1) + \frac{1}{2}(y)(y) = \frac{1}{2}y^2 + y$. We are given that the area is 1200 cm^2, so $1200 = \frac{1}{2}y^2 + y \quad\Leftrightarrow\quad y^2 + 2y - 2400 = 0 \quad\Leftrightarrow\quad (y + 50)(y - 48) = 0$. y is positive, so $y = 48$ cm.

83. Let x be the length of one side of the cardboard, so we start with a piece of cardboard x by x. When 4 inches are removed from each side, the base of the box is $x - 8$ by $x - 8$. Since the volume is 100 in^3, we get $4(x - 8)^2 = 100 \quad\Leftrightarrow$ $x^2 - 16x + 64 = 25 \quad\Leftrightarrow\quad x^2 - 16x + 39 = 0 \quad\Leftrightarrow\quad (x - 3)(x - 13) = 0$. So $x = 3$ or $x = 13$. But $x = 3$ is not possible, since then the length of the base would be $3 - 8 = -5$, and all lengths must be positive. Thus $x = 13$, and the piece of cardboard is 13 inches by 13 inches.

85. Let w be the width of the lot in feet. Then the length is $w + 6$. Using the Pythagorean Theorem, we have $w^2 + (w + 6)^2 = (174)^2 \quad\Leftrightarrow\quad w^2 + w^2 + 12w + 36 = 30{,}276 \quad\Leftrightarrow\quad 2w^2 + 12w - 30240 = 0 \quad\Leftrightarrow$ $w^2 + 6w - 15120 = 0 \quad\Leftrightarrow\quad (w + 126)(w - 120) = 0$. So either $w + 126 = 0$ in which case $w = -126$, which is not possible, or $w - 120 = 0$ in which case $w = 120$. Thus the width is 120 feet and the length is 126 feet.

87. Using $h_0 = 288$, we solve $0 = -16t^2 + 288$, for $t \geq 0$. So $0 = -16t^2 + 288 \quad\Leftrightarrow\quad 16t^2 = 288 \quad\Leftrightarrow\quad t^2 = 18 \quad\Rightarrow$ $t = \pm\sqrt{18} = \pm 3\sqrt{2}$. Thus it takes $3\sqrt{2} \approx 4.24$ seconds for the ball the hit the ground.

89. We are given $v_o = 40$ ft/s.

(a) Setting $h = 24$, we have $24 = -16t^2 + 40t \quad\Leftrightarrow\quad 16t^2 - 40t + 24 = 0 \quad\Leftrightarrow\quad 8(2t^2 - 5t + 3) = 0 \quad\Leftrightarrow$ $8(2t - 3)(t - 1) = 0 \quad\Leftrightarrow\quad t = 1$ or $t = 1\frac{1}{2}$. Therefore, the ball reaches 24 feet in 1 second (ascending) and again after $1\frac{1}{2}$ seconds (descending).

(b) Setting $h = 48$, we have $48 = -16t^2 + 40t \quad\Leftrightarrow\quad 16t^2 - 40t + 48 = 0 \quad\Leftrightarrow\quad 2t^2 - 5t + 6 = 0 \quad\Leftrightarrow$ $t = \frac{5\pm\sqrt{25 - 48}}{4} = \frac{5\pm\sqrt{-23}}{4}$. However, since the discriminant $D < 0$, there is no real solution, and hence the ball never reaches a height of 48 feet.

(c) The greatest height h is reached only once. So $h = -16t^2 + 40t \quad\Leftrightarrow\quad 16t^2 - 40t + h = 0$ has only one solution. Thus $D = (-40)^2 - 4(16)(h) = 0 \quad\Leftrightarrow\quad 1600 - 64h = 0 \quad\Leftrightarrow\quad h = 25$. So the greatest height reached by the ball is 25 feet.

(d) Setting $h = 25$, we have $25 = -16t^2 + 40t \quad\Leftrightarrow\quad 16t^2 - 40t + 25 = 0 \quad\Leftrightarrow\quad (4t - 5)^2 = 0 \quad\Leftrightarrow\quad t = 1\frac{1}{4}$. Thus the ball reaches the highest point of its path after $1\frac{1}{4}$ seconds.

(e) Setting $h = 0$ (ground level), we have $0 = -16t^2 + 40t \quad\Leftrightarrow\quad 2t^2 - 5t = 0 \quad\Leftrightarrow\quad t(2t - 5) = 0 \quad\Leftrightarrow\quad t = 0$ (start) or $t = 2\frac{1}{2}$. So the ball hits the ground in $2\frac{1}{2}$ s.

91. (a) The fish population on January 1, 2002 corresponds to $t = 0$, so $F = 1000\left(30 + 17\left(0\right) - \left(0\right)^2\right) = 30,000$. To find when the population will again reach this value, we set $F = 30,000$, giving

$$30000 = 1000\left(30 + 17t - t^2\right) = 30000 + 17000t - 1000t^2 \quad \Leftrightarrow \quad 0 = 17000t - 1000t^2 = 1000t\left(17 - t\right) \quad \Leftrightarrow$$

$t = 0$ or $t = 17$. Thus the fish population will again be the same 17 years later, that is, on January 1, 2019.

(b) Setting $F = 0$, we have $0 = 1000\left(30 + 17t - t^2\right) \quad \Leftrightarrow \quad t^2 - 17t - 30 = 0 \quad \Leftrightarrow$

$t = \dfrac{17 \pm \sqrt{289 + 120}}{-2} = \dfrac{17 \pm \sqrt{409}}{-2} = \dfrac{17 \pm 20.22}{2}$. Thus $t \approx -1.612$ or $t \approx 18.612$. Since $t < 0$ is inadmissible, it follows that the fish in the lake will have died out 18.612 years after January 1, 2002, that is on August 12, 2020.

93. Let x be the rate, in mi/h, at which the salesman drove between Ajax and Barrington.

Cities	Distance	Rate	Time
Ajax $\to$ Barrington	120	x	$\dfrac{120}{x}$
Barrington $\to$ Collins	150	$x + 10$	$\dfrac{150}{x + 10}$

We have used the equation time $= \dfrac{\text{distance}}{\text{rate}}$ to fill in the "Time" column of the table. Since the second part of the trip took 6 minutes (or $\frac{1}{10}$ hour) more than the first, we can use the time column to get the equation $\dfrac{120}{x} + \dfrac{1}{10} = \dfrac{150}{x + 10} \quad \Rightarrow$

$120\left(10\right)\left(x + 10\right) + x\left(x + 10\right) = 150\left(10x\right) \quad \Leftrightarrow \quad 1200x + 12,000 + x^2 + 10x = 1500x \quad \Leftrightarrow \quad x^2 - 290x + 12,000 = 0$

$\Leftrightarrow \quad x = \dfrac{-\left(-290\right) \pm \sqrt{\left(-290\right)^2 - 4\left(1\right)\left(12,000\right)}}{2} = \dfrac{290 \pm \sqrt{84,100 - 48,000}}{2} = \dfrac{290 \pm \sqrt{36,100}}{2} = \dfrac{290 \pm 190}{2} = 145 \pm 95$. Hence, the salesman drove either 50 mi/h or 240 mi/h between Ajax and Barrington. (The first choice seems more likely!)

95. Let r be the rowing rate in km/h of the crew in still water. Then their rate upstream was $r - 3$ km/h, and their rate downstream was $r + 3$ km/h.

	Distance	Rate	Time
Upstream	6	$r - 3$	$\dfrac{6}{r - 3}$
Downstream	6	$r + 3$	$\dfrac{6}{r + 3}$

Since the time to row upstream plus the time to row downstream was 2 hours 40 minutes $= \frac{8}{3}$ hour, we get the equation

$\dfrac{6}{r - 3} + \dfrac{6}{r + 3} = \dfrac{8}{3} \quad \Leftrightarrow \quad 6\left(3\right)\left(r + 3\right) + 6\left(3\right)\left(r - 3\right) = 8\left(r - 3\right)\left(r + 3\right) \quad \Leftrightarrow \quad 18r + 54 + 18r - 54 = 8r^2 - 72$

$\Leftrightarrow \quad 0 = 8r^2 - 36r - 72 = 4\left(2r^2 - 9r - 18\right) = 4\left(2r + 3\right)\left(r - 6\right)$. Since $2r + 3 = 0 \quad \Leftrightarrow \quad r = -\frac{3}{2}$ is impossible, the solution is $r - 6 = 0 \quad \Leftrightarrow \quad r = 6$. So the rate of the rowing crew in still water is 6 km/h.

97. Let w be the uniform width of the lawn. With w cut off each end, the area of the factory is $\left(240 - 2w\right)\left(180 - 2w\right)$. Since the lawn and the factory are equal in size this area, is $\frac{1}{2} \cdot 240 \cdot 180$. So $21,600 = 43,200 - 480w - 360w + 4w^2 \quad \Leftrightarrow$

$0 = 4w^2 - 840w + 21,600 = 4\left(w^2 - 210w + 5400\right) = 4\left(w - 30\right)\left(w - 180\right) \quad \Rightarrow \quad w = 30$ or $w = 180$. Since 180 ft is too wide, the width of the lawn is 30 ft, and the factory is 120 ft by 180 ft.

99. Let t be the time, in hours it takes Irene to wash all the windows. Then it takes Henry $t + \frac{3}{2}$ hours to wash all the windows, and the sum of the fraction of the job per hour they can do individually equals the fraction of the job they can do together. Since 1 hour 48 minutes $= 1 + \frac{48}{60} = 1 + \frac{4}{5} = \frac{9}{5}$, we have $\dfrac{1}{t} + \dfrac{1}{t + \frac{3}{2}} = \dfrac{1}{\frac{9}{5}}$ $\Leftrightarrow$ $\dfrac{1}{t} + \dfrac{2}{2t + 3} = \dfrac{5}{9}$

$\Rightarrow$ $9(2t + 3) + 2(9t) = 5t(2t + 3)$ $\Leftrightarrow$ $18t + 27 + 18t = 10t^2 + 15t$ $\Leftrightarrow$ $10t^2 - 21t - 27 = 0$

$\Leftrightarrow$ $t = \dfrac{-(-21) \pm \sqrt{(-21)^2 - 4(10)(-27)}}{2(10)} = \dfrac{21 \pm \sqrt{441 + 1080}}{20} = \dfrac{21 \pm 39}{20}$. So $t = \dfrac{21 - 39}{20} = -\dfrac{9}{10}$

or $t = \dfrac{21 + 39}{20} = 3$. Since $t < 0$ is impossible, all the windows are washed by Irene alone in 3 hours and by Henry alone in $3 + \frac{3}{2} = 4\frac{1}{2}$ hours.

101. Let x be the distance from the center of the earth to the dead spot (in thousands of miles). Now setting $F = 0$,

we have $0 = -\dfrac{K}{x^2} + \dfrac{0.012K}{(239 - x)^2}$ $\Leftrightarrow$ $\dfrac{K}{x^2} = \dfrac{0.012K}{(239 - x)^2}$ $\Leftrightarrow$ $K(239 - x)^2 = 0.012Kx^2$ $\Leftrightarrow$

$57121 - 478x + x^2 = 0.012x^2$ $\Leftrightarrow$ $0.988x^2 - 478x + 57121 = 0$. Using the quadratic formula, we obtain

$x = \dfrac{-(-478) \pm \sqrt{(-478)^2 - 4(0.988)(57121)}}{2(0.988)} = \dfrac{478 \pm \sqrt{228484 - 225742.192}}{1.976} = \dfrac{478 \pm \sqrt{2741.808}}{1.976} \approx \dfrac{478 \pm 52.362}{1.976} \approx 241.903 \pm 26.499$.

So either $x \approx 241.903 + 26.499 \approx 268$ or $x \approx 241.903 - 26.499 \approx 215$. Since 268 is greater than the distance from the earth to the moon, we reject it; thus $x \approx 215{,}000$ miles.

103. Let x equal the original length of the reed in cubits. Then $x - 1$ is the piece that fits 60 times along the length of the field, that is, the length is $60(x - 1)$. The width is $30x$. Then converting cubits to ninda, we have

$375 = 60(x - 1) \cdot 30x \cdot \dfrac{1}{12^2} = \dfrac{25}{2}x(x - 1)$ $\Leftrightarrow$ $30 = x^2 - x$ $\Leftrightarrow$ $x^2 - x - 30 = 0$ $\Leftrightarrow$ $(x - 6)(x + 5) = 0$.

So $x = 6$ or $x = -5$. Since x must be positive, the original length of the reed is 6 cubits.

1.4   **Complex Numbers**

1. $5 - 7i$: real part 5, imaginary part -7.

3. $\dfrac{-2 - 5i}{3} = -\dfrac{2}{3} - \dfrac{5}{3}i$: real part $-\dfrac{2}{3}$, imaginary part $-\dfrac{5}{3}$.

5. 3: real part 3, imaginary part 0.

7. $-\dfrac{2}{3}i$: real part 0, imaginary part $-\dfrac{2}{3}$.

9. $\sqrt{3} + \sqrt{-4} = \sqrt{3} + 2i$: real part $\sqrt{3}$, imaginary part 2.

11. $(2 - 5i) + (3 + 4i) = (2 + 3) + (-5 + 4)i = 5 - i$

13. $(-6 + 6i) + (9 - i) = (-6 + 9) + (6 - 1)i = 3 + 5i$

15. $\left(7 - \frac{1}{2}i\right) - \left(5 + \frac{3}{2}i\right) = (7 - 5) + \left(-\frac{1}{2} - \frac{3}{2}\right)i = 2 - 2i$

17. $(-12 + 8i) - (7 + 4i) = -12 + 8i - 7 - 4i = (-12 - 7) + (8 - 4)i = -19 + 4i$

19. $4(-1 + 2i) = -4 + 8i$

21. $(7 - i)(4 + 2i) = 28 + 14i - 4i - 2i^2 = (28 + 2) + (14 - 4)i = 30 + 10i$

23. $(3 - 4i)(5 - 12i) = 15 - 36i - 20i + 48i^2 = (15 - 48) + (-36 - 20)i = -33 - 56i$

25. $(6 + 5i)(2 - 3i) = 12 - 18i + 10i - 15i^2 = (12 + 15) + (-18 + 10)i = 27 - 8i$

27. $\dfrac{1}{i} = \dfrac{1}{i} \cdot \dfrac{i}{i} = \dfrac{i}{i^2} = \dfrac{i}{-1} = -i$

29. $\dfrac{2 - 3i}{1 - 2i} = \dfrac{2 - 3i}{1 - 2i} \cdot \dfrac{1 + 2i}{1 + 2i} = \dfrac{2 + 4i - 3i - 6i^2}{1 - 4i^2} = \dfrac{(2 + 6) + (4 - 3)i}{1 + 4} = \dfrac{8 + i}{5}$ or $\dfrac{8}{5} + \dfrac{1}{5}i$

31. $\dfrac{26 + 39i}{2 - 3i} = \dfrac{26 + 39i}{2 - 3i} \cdot \dfrac{2 + 3i}{2 + 3i} = \dfrac{52 + 78i + 78i + 117i^2}{4 - 9i^2} = \dfrac{(52 - 117) + (78 + 78)i}{4 + 9} = \dfrac{-65 + 156i}{13}$

$= \dfrac{13(-5 + 12i)}{13} = -5 + 12i$

33. $\dfrac{10i}{1 - 2i} = \dfrac{10i}{1 - 2i} \cdot \dfrac{1 + 2i}{1 + 2i} = \dfrac{10i + 20i^2}{1 - 4i^2} = \dfrac{-20 + 10i}{1 + 4} = \dfrac{5(-4 + 2i)}{5} = -4 + 2i$

35. $\dfrac{4+6i}{3i} = \dfrac{4+6i}{3i} \cdot \dfrac{i}{i} = \dfrac{4i+6i^2}{3i^2} = \dfrac{-6+4i}{-3} = \dfrac{-6}{-3} + \dfrac{4}{-3}i = 2 - \tfrac{4}{3}i$

37. $\dfrac{1}{1+i} - \dfrac{1}{1-i} = \dfrac{1}{1+i} \cdot \dfrac{1-i}{1-i} - \dfrac{1}{1-i} \cdot \dfrac{1+i}{1+i} = \dfrac{1-i}{1-i^2} - \dfrac{1+i}{1-i^2} = \dfrac{1-i}{2} + \dfrac{-1-i}{2} = -i$

39. $i^3 = i^2 \cdot i = -1 \cdot i = -i$

41. $i^{100} = \left(i^4\right)^{25} = (1)^{25} = 1$

43. $\sqrt{-25} = 5i$

45. $\sqrt{-3}\sqrt{-12} = i\sqrt{3} \cdot 2i\sqrt{3} = 6i^2 = -6$

47. $\left(3 - \sqrt{-5}\right)\left(1 + \sqrt{-1}\right) = \left(3 - i\sqrt{5}\right)(1+i) = 3 + 3i - i\sqrt{5} - i^2\sqrt{5} = \left(3 + \sqrt{5}\right) + \left(3 - \sqrt{5}\right)i$

49. $\dfrac{2+\sqrt{-8}}{1+\sqrt{-2}} = \dfrac{2+2i\sqrt{2}}{1+i\sqrt{2}} = \dfrac{2+2i\sqrt{2}}{1+i\sqrt{2}} \cdot \dfrac{1-i\sqrt{2}}{1-i\sqrt{2}} = \dfrac{2-2i\sqrt{2}+2i\sqrt{2}-4i^2}{1-2i^2} = \dfrac{(2+4)+\left(-2\sqrt{2}+2\sqrt{2}\right)i}{1+2} = \dfrac{6}{3} = 2$

51. $\dfrac{\sqrt{-36}}{\sqrt{-2}\sqrt{-9}} = \dfrac{6i}{i\sqrt{2}\cdot 3i} = \dfrac{2}{i\sqrt{2}} \cdot \dfrac{i\sqrt{2}}{i\sqrt{2}} = \dfrac{2i\sqrt{2}}{2i^2} = \dfrac{i\sqrt{2}}{-1} = -i\sqrt{2}$

53. $x^2 + 9 = 0 \;\Leftrightarrow\; x^2 = -9 \;\Rightarrow\; x = \pm 3i$

55. $x^2 - 4x + 5 = 0 \;\Rightarrow\; x = \dfrac{-(-4) \pm \sqrt{(-4)^2 - 4(1)(5)}}{2(1)} = \dfrac{4 \pm \sqrt{16-20}}{2} = \dfrac{4 \pm \sqrt{-4}}{2} = \dfrac{4 \pm 2i}{2} = 2 \pm i$

57. $x^2 + x + 1 = 0 \;\Rightarrow\; x = \dfrac{-(1) \pm \sqrt{(1)^2 - 4(1)(1)}}{2(1)} = \dfrac{-1 \pm \sqrt{1-4}}{2} = \dfrac{-1 \pm \sqrt{-3}}{2} = \dfrac{-1 \pm i\sqrt{3}}{2} = -\tfrac{1}{2} \pm \dfrac{i\sqrt{3}}{2}$

59. $2x^2 - 2x + 1 = 0 \;\Rightarrow\; x = \dfrac{-(-2) \pm \sqrt{(-2)^2 - 4(2)(1)}}{2(2)} = \dfrac{2 \pm \sqrt{4-8}}{4} = \dfrac{2 \pm \sqrt{-4}}{4} = \dfrac{2 \pm 2i}{4} = \tfrac{1}{2} \pm \tfrac{1}{2}i$

61. $t + 3 + \dfrac{3}{t} = 0 \;\Leftrightarrow\; t^2 + 3t + 3 = 0 \;\Rightarrow\; t = \dfrac{-(3) \pm \sqrt{(3)^2 - 4(1)(3)}}{2(1)} = \dfrac{-3 \pm \sqrt{9-12}}{2} = \dfrac{-3 \pm \sqrt{-3}}{2} = \dfrac{-3 \pm i\sqrt{3}}{2} = -\tfrac{3}{2} \pm i\dfrac{\sqrt{3}}{2}$

63. $6x^2 + 12x + 7 = 0 \;\Rightarrow\;$

$x = \dfrac{-(12) \pm \sqrt{(12)^2 - 4(6)(7)}}{2(6)} = \dfrac{-12 \pm \sqrt{144-168}}{12} = \dfrac{-12 \pm \sqrt{-24}}{12} = \dfrac{-12 \pm 2i\sqrt{6}}{12} = \dfrac{-12}{12} \pm \dfrac{2i\sqrt{6}}{12} = -1 \pm \dfrac{\sqrt{6}}{6}i$

65. $\tfrac{1}{2}x^2 - x + 5 = 0 \;\Rightarrow\; x = \dfrac{-(-1) \pm \sqrt{(-1)^2 - 4\left(\tfrac{1}{2}\right)(5)}}{2\left(\tfrac{1}{2}\right)} = \dfrac{1 \pm \sqrt{1-10}}{1} = 1 \pm \sqrt{-9} = 1 \pm 3i$

67. $\text{LHS} = \overline{z} + \overline{w} = \overline{(a+bi)} + \overline{(c+di)} = a - bi + c - di = (a+c) + (-b-d)i = (a+c) - (b+d)i.$

$\text{RHS} = \overline{z+w} = \overline{(a+bi) + (c+di)} = \overline{(a+c) + (b+d)i} = (a+c) - (b+d)i.$

Since LHS = RHS, this proves the statement.

69. $\text{LHS} = (\overline{z})^2 = \left(\overline{(a+bi)}\right)^2 = (a-bi)^2 = a^2 - 2abi + b^2 i^2 = \left(a^2 - b^2\right) - 2abi.$

$\text{RHS} = \overline{z^2} = \overline{(a+bi)^2} = \overline{a^2 + 2abi + b^2 i^2} = \overline{\left(a^2 - b^2\right) + 2abi} = \left(a^2 - b^2\right) - 2abi.$

Since LHS = RHS, this proves the statement.

71. $z + \overline{z} = (a+bi) + \overline{(a+bi)} = a + bi + a - bi = 2a$, which is a real number.

73. $z \cdot \overline{z} = (a+bi) \cdot \overline{(a+bi)} = (a+bi) \cdot (a-bi) = a^2 - b^2 i^2 = a^2 + b^2$, which is a real number.

75. Using the quadratic formula, the solutions to the equation are $x = \dfrac{-b \pm \sqrt{b^2 - 4ac}}{2a}$. Since both solutions are imaginary,

we have $b^2 - 4ac < 0 \;\Leftrightarrow\; 4ac - b^2 > 0$, so the solutions are $x = \dfrac{-b}{2a} \pm \dfrac{\sqrt{4ac - b^2}}{2a}i$, where $\sqrt{4ac - b^2}$ is a real

number. Thus the solutions are complex conjugates of each other.

1.5   Other Types of Equations

1. $x^3 = 16x \Leftrightarrow 0 = x^3 - 16x = x\left(x^2 - 16\right) = x\left(x - 4\right)\left(x + 4\right)$. So $x = 0$, $x - 4 = 0 \Leftrightarrow x = 4$, or $x + 4 = 0$
$\Leftrightarrow x = -4$. The solutions are 0 and ± 4.

3. $0 = x^6 - 81x^2 = x^2\left(x^4 - 81\right) = x^2\left(x^2 - 9\right)\left(x^2 + 9\right) = x^2\left(x - 3\right)\left(x + 3\right)\left(x^2 + 9\right)$. So $x^2 = 0 \Leftrightarrow x = 0$, or
$x - 3 = 0 \Leftrightarrow x = 3$, or $x + 3 = 0 \Leftrightarrow x = -3$. $x^2 + 9 = 0 \Leftrightarrow x^2 = -9$ which has no real solution. The
solutions are 0 and ± 3.

5. $0 = x^5 + 8x^2 = x^2\left(x^3 + 8\right) = x^2\left(x + 2\right)\left(x^2 - 2x + 4\right) \Leftrightarrow x^2 = 0$, $x + 2 = 0$, or $x^2 - 2x + 4 = 0$. If $x^2 = 0$, then
$x = 0$; if $x + 2 = 0$, then $x = -2$, and $x^2 - 2x + 4 = 0$ has no real solution. Thus the solutions are $x = 0$ and $x = -2$.

7. $0 = x^3 - 5x^2 + 6x = x\left(x^2 - 5x + 6\right) = x\left(x - 2\right)\left(x - 3\right) \Leftrightarrow x = 0$, $x - 2 = 0$, or $x - 3 = 0$. Thus $x = 0$, or
$x = 2$, or $x = 3$. The solutions are $x = 0$, $x = 2$, and $x = 3$.

9. $0 = x^4 + 4x^3 + 2x^2 = x^2\left(x^2 + 4x + 2\right)$. So either $x^2 = 0 \Leftrightarrow x = 0$, or using the quadratic formula on
$x^2 + 4x + 2 = 0$, we have $x = \dfrac{-4 \pm \sqrt{4^2 - 4(1)(2)}}{2(1)} = \dfrac{-4 \pm \sqrt{16 - 8}}{2} = \dfrac{-4 \pm \sqrt{8}}{2} = \dfrac{-4 \pm 2\sqrt{2}}{2} = -2 \pm \sqrt{2}$. The solutions are 0,
$-2 - \sqrt{2}$, and $-2 + \sqrt{2}$.

11. $0 = x^3 - 5x^2 - 2x + 10 = x^2\left(x - 5\right) - 2\left(x - 5\right) = \left(x - 5\right)\left(x^2 - 2\right)$. If $x - 5 = 0$, then $x = 5$. If $x^2 - 2 = 0$, then
$x^2 = 2 \Leftrightarrow x = \pm\sqrt{2}$. The solutions are 5 and $\pm\sqrt{2}$.

13. $x^3 - x^2 + x - 1 = x^2 + 1 \Leftrightarrow 0 = x^3 - 2x^2 + x - 2 = x^2\left(x - 2\right) + \left(x - 2\right) = \left(x - 2\right)\left(x^2 + 1\right)$. Since $x^2 + 1 = 0$
has no real solution, the only solution comes from $x - 2 = 0 \Leftrightarrow x = 2$.

15. $\dfrac{1}{x - 1} + \dfrac{1}{x + 2} = \dfrac{5}{4} \Leftrightarrow 4\left(x - 1\right)\left(x + 2\right)\left(\dfrac{1}{x - 1} + \dfrac{1}{x + 2}\right) = 4\left(x - 1\right)\left(x + 2\right)\left(\dfrac{5}{4}\right) \Leftrightarrow$
$4\left(x + 2\right) + 4\left(x - 1\right) = 5\left(x - 1\right)\left(x + 2\right) \Leftrightarrow 4x + 8 + 4x - 4 = 5x^2 + 5x - 10 \Leftrightarrow 5x^2 - 3x - 14 = 0 \Leftrightarrow$
$\left(5x + 7\right)\left(x - 2\right) = 0$. If $5x + 7 = 0$, then $x = -\frac{7}{5}$; if $x - 2 = 0$, then $x = 2$. The solutions are $-\dfrac{7}{5}$ and 2.

17. $\dfrac{x^2}{x + 100} = 50 \Rightarrow x^2 = 50\left(x + 100\right) = 50x + 5000 \Leftrightarrow x^2 - 50x - 5000 = 0 \Leftrightarrow \left(x - 100\right)\left(x + 50\right) = 0$
$\Leftrightarrow x - 100 = 0$ or $x + 50 = 0$. Thus $x = 100$ or $x = -50$. The solutions are 100 and -50.

19. $\dfrac{x + 5}{x - 2} = \dfrac{5}{x + 2} + \dfrac{28}{x^2 - 4} \Rightarrow \left(x + 2\right)\left(x + 5\right) = 5\left(x - 2\right) + 28 \Leftrightarrow x^2 + 7x + 10 = 5x - 10 + 28 \Leftrightarrow$
$x^2 + 2x - 8 = 0 \Leftrightarrow \left(x - 2\right)\left(x + 4\right) = 0 \Leftrightarrow x - 2 = 0$ or $x + 4 = 0 \Leftrightarrow x = 2$ or $x = -4$. However, $x = 2$ is
inadmissible since we can't divide by 0 in the original equation, so the only solution is -4.

21. $\dfrac{1}{x - 1} - \dfrac{2}{x^2} = 0 \Leftrightarrow x^2 - 2\left(x - 1\right) = 0 \Leftrightarrow x^2 - 2x + 2 = 0 \Rightarrow$
$x = \dfrac{-\left(-2\right) \pm \sqrt{\left(-2\right)^2 - 4\left(1\right)\left(2\right)}}{2\left(1\right)} = \dfrac{2 \pm \sqrt{4 - 8}}{2} = \dfrac{2 \pm \sqrt{-4}}{2}$. Since the radicand is negative, there is no real solution.

23. $0 = \left(x + 5\right)^2 - 3\left(x + 5\right) - 10 = \left[\left(x + 5\right) - 5\right]\left[\left(x + 5\right) + 2\right] = x\left(x + 7\right) \Leftrightarrow x = 0$ or $x = -7$. The solutions are 0
and -7.

25. Let $w = \dfrac{1}{x + 1}$. Then $\left(\dfrac{1}{x + 1}\right)^2 - 2\left(\dfrac{1}{x + 1}\right) - 8 = 0$ becomes $w^2 - 2w - 8 = 0 \Leftrightarrow \left(w - 4\right)\left(w + 2\right) = 0$. So
$w - 4 = 0 \Leftrightarrow w = 4$, and $w + 2 = 0 \Leftrightarrow w = -2$. When $w = 4$, we have $\dfrac{1}{x + 1} = 4 \Leftrightarrow 1 = 4x + 4 \Leftrightarrow$
$-3 = 4x \Leftrightarrow x = -\frac{3}{4}$. When $w = -2$, we have $\dfrac{1}{x + 1} = -2 \Leftrightarrow 1 = -2x - 2 \Leftrightarrow 3 = -2x \Leftrightarrow x = -\frac{3}{2}$.
Solutions are $-\frac{3}{4}$ and $-\frac{3}{2}$.

27. Let $w = x^2$. Then $x^4 - 13x^2 + 40 = (x^2)^2 - 13x^2 + 40 = 0$ becomes $w^2 - 13w + 40 = 0$ $\Leftrightarrow$ $(w - 5)(w - 8) = 0$. So $w - 5 = 0$ $\Leftrightarrow$ $w = 5$, and $w - 8 = 0$ $\Leftrightarrow$ $w = 8$. When $w = 5$, we have $x^2 = 5$ $\Rightarrow$ $x = \pm\sqrt{5}$. When $w = 8$, we have $x^2 = 8$ $\Rightarrow$ $x = \pm\sqrt{8} = \pm 2\sqrt{2}$. The solutions are $\pm\sqrt{5}$ and $\pm 2\sqrt{2}$.

29. $2x^4 + 4x^2 + 1 = 0$. The LHS is the sum of two nonnegative numbers and a positive number, so $2x^4 + 4x^2 + 1 \geq 1 \neq 0$. This equation has no real solution.

31. $0 = x^6 - 26x^3 - 27 = (x^3 - 27)(x^3 + 1)$. If $x^3 - 27 = 0$ $\Leftrightarrow$ $x^3 = 27$, so $x = 3$. If $x^3 + 1 = 0$ $\Leftrightarrow$ $x^3 = -1$, so $x = -1$. The solutions are 3 and -1.

33. Let $u = x^{2/3}$. Then $0 = x^{4/3} - 5x^{2/3} + 6$ becomes $u^2 - 5u + 6 = 0$ $\Leftrightarrow$ $(u - 3)(u - 2) = 0$ $\Leftrightarrow$ $u - 3 = 0$ or $u - 2 = 0$. If $u - 3 = 0$, then $x^{2/3} - 3 = 0$ $\Leftrightarrow$ $x^{2/3} = 3$ $\Rightarrow$ $x = \pm 3^{3/2} = \pm 3\sqrt{3}$. If $u - 2 = 0$, then $x^{2/3} - 2 = 0$ $\Leftrightarrow$ $x^{2/3} = 2$ $\Rightarrow$ $x = \pm 2^{3/2} = 2\sqrt{2}$. The solutions are $\pm 3\sqrt{3}$ and $\pm 2\sqrt{2}$.

35. $4(x + 1)^{1/2} - 5(x + 1)^{3/2} + (x + 1)^{5/2} = 0$ $\Leftrightarrow$ $\sqrt{x + 1}\left[4 - 5(x + 1) + (x + 1)^2\right] = 0$ $\Leftrightarrow$ $\sqrt{x + 1}\left(4 - 5x - 5 + x^2 + 2x + 1\right) = 0$ $\Leftrightarrow$ $\sqrt{x + 1}\left(x^2 - 3x\right) = 0$ $\Leftrightarrow$ $\sqrt{x + 1} \cdot x(x - 3) = 0$ $\Leftrightarrow$ $x = -1$ or $x = 0$ or $x = 3$. The solutions are -1, 0, and 3.

37. $0 = x^{3/2} + 8x^{1/2} + 16x^{-1/2} = x^{-1/2}(x^2 + 8x + 16) = x^{-1/2}(x + 4)^2$. Now $x^{-1/2} = \dfrac{1}{\sqrt{x}}$ so $x \neq 0$. So $x + 4 = 0$ $\Leftrightarrow$ $x = -4$. But $\sqrt{-4}$ is not a real number, so this equation has no real solution. Alternatively, we see that this is the sum of three positive real numbers (remember $x \neq 0$), so it never equals zero.

39. Let $u = x^{1/6}$. (We choose the exponent $\frac{1}{6}$ because the LCD of 2, 3, and 6 is 6.) Then $x^{1/2} - 3x^{1/3} = 3x^{1/6} - 9$ $\Leftrightarrow$ $x^{3/6} - 3x^{2/6} = 3x^{1/6} - 9$ $\Leftrightarrow$ $u^3 - 3u^2 = 3u - 9$ $\Leftrightarrow$ $0 = u^3 - 3u^2 - 3u + 9 = u^2(u - 3) - 3(u - 3) = (u - 3)(u^2 - 3)$. So $u - 3 = 0$ or $u^2 - 3 = 0$. If $u - 3 = 0$, then $x^{1/6} - 3 = 0$ $\Leftrightarrow$ $x^{1/6} = 3$ $\Leftrightarrow$ $x = 3^6 = 729$. If $u^2 - 3 = 0$, then $x^{1/3} - 3 = 0$ $\Leftrightarrow$ $x^{1/3} = 3$ $\Leftrightarrow$ $x = 3^3 = 27$. The solutions are 729 and 27.

41. $\dfrac{1}{x^3} + \dfrac{4}{x^2} + \dfrac{4}{x} = 0$ $\Rightarrow$ $1 + 4x + 4x^2 = 0$ $\Leftrightarrow$ $(1 + 2x)^2 = 0$ $\Leftrightarrow$ $1 + 2x = 0$ $\Leftrightarrow$ $2x = -1$ $\Leftrightarrow$ $x = -\frac{1}{2}$. The solution is $-\frac{1}{2}$.

43. $\sqrt{2x + 1} + 1 = x$ $\Leftrightarrow$ $\sqrt{2x + 1} = x - 1$ $\Rightarrow$ $2x + 1 = (x - 1)^2$ $\Leftrightarrow$ $2x + 1 = x^2 - 2x + 1$ $\Leftrightarrow$ $0 = x^2 - 4x = x(x - 4)$. Potential solutions are $x = 0$ and $x - 4$ $\Leftrightarrow$ $x = 4$. These are only potential solutions since squaring is not a reversible operation. We must check each potential solution in the original equation.
Checking $x = 0$: $\sqrt{2(0) + 1} + 1 = (0)$ $\Leftrightarrow$ $\sqrt{1} + 1 = 0$ is false.
Checking $x = 4$: $\sqrt{2(4) + 1} + 1 = (4)$ $\Leftrightarrow$ $\sqrt{9} + 1 = 4$ $\Leftrightarrow$ $3 + 1 = 4$ is true. The only solution is $x = 4$.

45. $\sqrt{5 - x} + 1 = x - 2$ $\Leftrightarrow$ $\sqrt{5 - x} = x - 3$ $\Rightarrow$ $5 - x = (x - 3)^2$ $\Leftrightarrow$ $5 - x = x^2 - 6x + 9$ $\Leftrightarrow$ $0 = x^2 - 5x + 4 = (x - 4)(x - 1)$. Potential solutions are $x = 4$ and $x = 1$. We must check each potential solution in the original equation.
Checking $x = 4$: $\sqrt{5 - (4)} + 1 = (4) - 2$ $\Leftrightarrow$ $\sqrt{1} + 1 = 4 - 2$ $\Leftrightarrow$ $1 + 1 = 2$ is true.
Checking $x = 1$: $\sqrt{5 - (1)} + 1 = (1) - 2$ $\Leftrightarrow$ $\sqrt{4} + 1 = -1$ $\Leftrightarrow$ $2 + 1 = -1$ is false. The only solution is $x = 4$.

47. $x - \sqrt{x + 3} = \dfrac{x}{2}$ $\Leftrightarrow$ $\dfrac{x}{2} = \sqrt{x + 3}$ $\Rightarrow$ $\left(\frac{1}{2}x\right)^2 = x + 3$ $\Leftrightarrow$ $\frac{1}{4}x^2 = x + 3$ $\Leftrightarrow$ $x^2 = 4x + 12$ $\Leftrightarrow$ $0 = x^2 - 4x - 12 = (x - 6)(x + 2)$. Potential solutions are $x = 6$ and $x = -2$. We must check each potential solution in the original equation. Checking $x = 6$: $6 - \sqrt{6 + 3} = \frac{6}{2}$ $\Leftrightarrow$ $6 - \sqrt{6 + 3} = 6 - \sqrt{9} = 6 - 3$, which is true. Checking $x = -2$: $-2 - \sqrt{-2 + 3} = -\frac{2}{2}$ $\Leftrightarrow$ $-2 - \sqrt{-2 + 3} = -2 - \sqrt{1} = -2 - 1 = -3 = -\frac{2}{2}$ $\Leftrightarrow$ $-3 = -1$, hence this not a solution. The only solution is $x = 6$.

49. $\sqrt{\sqrt{x-5}+x} = 5$. Squaring both sides, we get $\sqrt{x-5}+x = 25$ $\Leftrightarrow$ $\sqrt{x-5} = 25 - x$. Squaring both sides again, we get $x - 5 = (25 - x)^2$ $\Leftrightarrow$ $x - 5 = 625 - 50x + x^2$ $\Leftrightarrow$ $0 = x^2 - 51x + 630 = (x - 30)(x - 21)$. Potential solutions are $x = 30$ and $x = 21$. We must check each potential solution in the original equation.

Checking $x = 30$: $\sqrt{\sqrt{(30) - 5} + (30)} = 5$ $\Leftrightarrow$ $\sqrt{\sqrt{(30) - 5} + (30)} = \sqrt{\sqrt{25} + 30} = \sqrt{35} = 5$, hence $x = 30$ is not a solution.

Checking $x = 21$: $\sqrt{\sqrt{(21) - 5} + 21} = 5$ $\Leftrightarrow$ $\sqrt{\sqrt{(21) - 5} + 21} = \sqrt{\sqrt{16} + 21} = \sqrt{25} = 5$, hence $x = 21$ is the only solution.

51. $x^2\sqrt{x+3} = (x+3)^{3/2}$ $\Leftrightarrow$ $0 = x^2\sqrt{x+3} - (x+3)^{3/2}$ $\Leftrightarrow$ $0 = \sqrt{x+3}\left[(x^2) - (x+3)\right]$ $\Leftrightarrow$ $0 = \sqrt{x+3}\left(x^2 - x - 3\right)$. If $(x+3)^{1/2} = 0$, then $x + 3 = 0$ $\Leftrightarrow$ $x = -3$. If $x^2 - x - 3 = 0$, then using the quadratic formula $x = \frac{1 \pm \sqrt{13}}{2}$. The solutions are -3 and $\frac{1 \pm \sqrt{13}}{2}$.

53. $\sqrt{x + \sqrt{x+2}} = 2$. Squaring both sides, we get $x + \sqrt{x+2} = 4$ $\Leftrightarrow$ $\sqrt{x+2} = 4 - x$. Squaring both sides again, we get $x + 2 = (4 - x)^2 = 16 - 8x + x^2$ $\Leftrightarrow$ $0 = x^2 - 9x + 14$ $\Leftrightarrow$ $0 = (x - 7)(x - 2)$. If $x - 7 = 0$, then $x = 7$. If $x - 2 = 0$, then $x = 2$. So $x = 2$ is a solution but $x = 7$ is not, since it does not satisfy the original equation.

55. $x^3 = 1$ $\Leftrightarrow$ $x^3 - 1 = 0$ $\Leftrightarrow$ $(x - 1)\left(x^2 + x + 1\right) = 0$ $\Leftrightarrow$ $x - 1 = 0$ or $x^2 + x + 1 = 0$. If $x - 1 = 0$, then $x = 1$. If $x^2 + x + 1 = 0$, then using the quadratic formula $x = \frac{-1 \pm i\sqrt{3}}{2}$. The solutions are 1 and $\frac{-1 \pm i\sqrt{3}}{2}$.

57. $x^3 + x^2 + x = 0$ $\Leftrightarrow$ $x\left(x^2 + x + 1\right) = 0$ $\Leftrightarrow$ $x = 0$ or $x = \frac{-1 \pm i\sqrt{3}}{2}$. The solutions are 0 and $\frac{-1 \pm i\sqrt{3}}{2}$.

59. $x^4 - 6x^2 + 8 = 0$ $\Leftrightarrow$ $\left(x^2 - 4\right)\left(x^2 - 2\right) = 0$ $\Leftrightarrow$ $x = \pm 2$ or $x = \pm\sqrt{2}$. The solutions are ± 2 and $\pm\sqrt{2}$.

61. $x^6 - 9x^3 + 8 = 0$ $\Leftrightarrow$ $\left(x^3 - 8\right)\left(x^3 - 1\right) = 0$ $\Leftrightarrow$ $(x - 2)\left(x^2 + 2x + 4\right)(x - 1)\left(x^2 + x + 1\right) = 0$ $\Leftrightarrow$ $x = 2$ or $x = \frac{-2 \pm 2i\sqrt{3}}{2} = -1 \pm i\sqrt{3}$ or $x = 1$ or $x = \frac{-1 \pm i\sqrt{3}}{2}$. The solutions are 2, $-1 \pm i\sqrt{3}$, 1, and $\frac{-1 \pm i\sqrt{3}}{2}$.

63. $\sqrt{x^2 + 1} + \frac{8}{\sqrt{x^2 + 1}} = \sqrt{x^2 + 9}$. Squaring both sides, we have $\left(x^2 + 1\right) + 16 + \frac{64}{x^2 + 1} = x^2 + 9$ $\Leftrightarrow$ $\frac{64}{x^2 + 1} = -8$

$\Leftrightarrow$ $\frac{8}{x^2 + 1} = -1$ $\Leftrightarrow$ $x^2 + 1 = -8$ $\Leftrightarrow$ $x^2 = -9$ $\Leftrightarrow$ $x = \pm 3i$. We must check each potential

solution in the original equation. Checking $x = \pm 3i$: $\sqrt{(\pm 3i)^2 + 1} + \frac{8}{\sqrt{(\pm 3i)^2 + 1}} \overset{?}{=} \sqrt{(\pm 3i)^2 + 9}$,

LHS $= \sqrt{-8} + \frac{8}{\sqrt{-8}} = 2\sqrt{2}i + \frac{8}{2\sqrt{2}i} = 2\sqrt{2}i - 2\sqrt{2}i = 0$. RHS $= \sqrt{(\pm 3i)^2 + 9} = \sqrt{-9 + 9} = 0$. Since

LHS $=$ RHS, $-3i$ and $3i$ are solutions.

65. $0 = x^4 + 5ax^2 + 4a^2 = \left(x^2 + a\right)\left(x^2 + 4a\right)$. Since a is positive, $x^2 + a = 0$ $\Leftrightarrow$ $x^2 = -a$ $\Leftrightarrow$ $x = \pm i\sqrt{a}$. Again, since a is positive, $x^2 + 4a = 0$ $\Leftrightarrow$ $x^2 = -4a$ $\Leftrightarrow$ $x = \pm 2i\sqrt{a}$. Thus the four solutions are: $\pm i\sqrt{a}$, $\pm 2i\sqrt{a}$.

67. $\sqrt{x + a} + \sqrt{x - a} = \sqrt{2}\sqrt{x + 6}$. Squaring both sides, we have
$x + a + 2\left(\sqrt{x + a}\right)\left(\sqrt{x - a}\right) + x - a = 2(x + 6)$ $\Leftrightarrow$ $2x + 2\left(\sqrt{x + a}\right)\left(\sqrt{x - a}\right) = 2x + 12$ $\Leftrightarrow$
$2\left(\sqrt{x + a}\right)\left(\sqrt{x - a}\right) = 12$ $\Leftrightarrow$ $\left(\sqrt{x + a}\right)\left(\sqrt{x - a}\right) = 6$. Squaring both sides again we have $(x + a)(x - a) = 36$
$\Leftrightarrow$ $x^2 - a^2 = 36$ $\Leftrightarrow$ $x^2 = a^2 + 36$ $\Leftrightarrow$ $x = \pm\sqrt{a^2 + 36}$. Checking these answers, we see that $x = -\sqrt{a^2 + 36}$ is not a solution (for example, try substituting $a = 8$), but $x = \sqrt{a^2 + 36}$ is a solution.

69. Let x be the number of people originally intended to take the trip. Then originally, the cost of the trip is $\frac{900}{x}$.

After 5 people cancel, there are now $x - 5$ people, each paying $\frac{900}{x} + 2$. Thus $900 = (x - 5)\left(\frac{900}{x} + 2\right)$ $\Leftrightarrow$

$900 = 900 + 2x - \frac{4500}{x} - 10$ $\Leftrightarrow$ $0 = 2x - 10 - \frac{4500}{x}$ $\Leftrightarrow$ $0 = 2x^2 - 10x - 4500 = (2x - 100)(x + 45)$.

Thus either $2x - 100 = 0$, so $x = 50$, or $x + 45 = 0$, $x = -45$. Since the number of people on the trip must be positive, originally 50 people intended to take the trip.

71. We want to solve for t when $P = 500$. Letting $u = \sqrt{t}$ and substituting, we have $500 = 3t + 10\sqrt{t} + 140$ $\Leftrightarrow$ $500 = 3u^2 + 10u + 140$ $\Leftrightarrow$ $0 = 3u^2 + 10u - 360$ $\Rightarrow$ $u = \dfrac{-5 \pm \sqrt{1105}}{3}$. Since $u = \sqrt{t}$, we must have $u \geq 0$. So $\sqrt{t} = u = \dfrac{-5 + \sqrt{1105}}{3} \approx 9.414$ $\Rightarrow$ $t = \approx 88.62$. So it will take 89 days for the fish population to reach 500.

73. We have that the volume is 180 ft^3, so $x(x-4)(x+9) = 180$ $\Leftrightarrow$ $x^3 + 5x^2 - 36x = 180$ $\Leftrightarrow$ $x^3 + 5x^2 - 36x - 180 = 0$ $\Leftrightarrow$ $x^2(x+5) - 36(x+5) = 0$ $\Leftrightarrow$ $(x+5)(x^2 - 36) = 0$ $\Leftrightarrow$ $(x+5)(x+6)(x-6) = 0$ $\Rightarrow$ $x = 6$ is the only positive solution. So the box is 2 feet by 6 feet by 15 feet.

75. Let x be the length, in miles, of the abandoned road to be used. Then the length of the abandoned road not used is $40 - x$, and the length of the new road is $\sqrt{10^2 + (40-x)^2}$ miles, by the Pythagorean Theorem. Since the cost of the road is cost per mile $\times$ number of miles, we have $100{,}000x + 200{,}000\sqrt{x^2 - 80x + 1700} = 6{,}800{,}000$ $\Leftrightarrow$ $2\sqrt{x^2 - 80x + 1700} = 68 - x$. Squaring both sides, we get $4x^2 - 320x + 6800 = 4624 - 136x + x^2$ $\Leftrightarrow$ $3x^2 - 184x + 2176 = 0$ $\Leftrightarrow$ $x = \dfrac{184 \pm \sqrt{33856 - 26112}}{6} = \dfrac{184 \pm 88}{6}$ $\Leftrightarrow$ $x = \dfrac{136}{3}$ or $x = 16$. Since $45\frac{1}{3}$ is longer than the existing road, 16 miles of the abandoned road should be used. A completely new road would have length $\sqrt{10^2 + 40^2}$ (let $x = 0$) and would cost $\sqrt{1700} \times 200{,}000 \approx 8.3$ million dollars. So no, it would not be cheaper.

77. Let x be the height of the pile in feet. Then the diameter is $3x$ and the radius is $\frac{3}{2}x$ feet. Since the volume of the cone is 1000 ft^3, we have $\dfrac{\pi}{3}\left(\dfrac{3x}{2}\right)^2 x = 1000$ $\Leftrightarrow$ $\dfrac{3\pi x^3}{4} = 1000$ $\Leftrightarrow$ $x^3 = \dfrac{4000}{3\pi}$ $\Leftrightarrow$ $x = \sqrt[3]{\dfrac{4000}{3\pi}} \approx 7.52$ feet.

79. Let x be the length of the hypotenuse of the triangle, in feet. Then one of the other sides has length $x - 7$ feet, and since the perimeter is 392 feet, the remaining side must have length $392 - x - (x - 7) = 399 - 2x$. From the Pythagorean Theorem, we get $(x-7)^2 + (399 - 2x)^2 = x^2$ $\Leftrightarrow$ $4x^2 - 1610x + 159250 = 0$. Using the quadratic formula, we get

$x = \dfrac{1610 \pm \sqrt{1610^2 - 4(4)(159250)}}{2(4)} = \dfrac{1610 \pm \sqrt{44100}}{8} = \dfrac{1610 \pm 210}{8}$, and so $x = 227.5$ or $x = 175$. But if $x = 227.5$, then the side of length $x - 7$ combined with the hypotenuse already exceeds the perimeter of 392 feet, and so we must have $x = 175$. Thus the other sides have length $175 - 7 = 168$ and $399 - 2(175) = 49$. The lot has sides of length 49 feet, 168 feet, and 175 feet.

81. Since the total time is 3 s, we have $3 = \dfrac{\sqrt{d}}{4} + \dfrac{d}{1090}$. Letting $w = \sqrt{d}$, we have $3 = \frac{1}{4}w + \frac{1}{1090}w^2$ $\Leftrightarrow$ $\frac{1}{1090}w^2 + \frac{1}{4}w - 3 = 0$ $\Leftrightarrow$ $2w^2 + 545w - 6540 = 0$ $\Rightarrow$ $w = \dfrac{-545 \pm 591.054}{4}$. Since $w \geq 0$, we have $\sqrt{d} = w \approx 11.51$, so $d = 132.56$. The well is 132.6 ft deep.

1.6 Inequalities

1. $x = -2$: $-2 - 3 \overset{?}{>} 0$. No, $-5 \not> 0$. $\quad x = -1$: $-1 - 3 \overset{?}{>} 0$. No, $-4 \not> 0$. $\quad x = 0$: $0 - 3 \overset{?}{>} 0$. No, $-3 \not> 0$. $x = \frac{1}{2}$: $\frac{1}{2} - 3 \overset{?}{>} 0$. No, $-\frac{5}{2} \not> 0$. $\quad x = 1$: $1 - 3 \overset{?}{>} 0$. No, $-2 \not> 0$. $\quad x = \sqrt{2}$: $\sqrt{2} - 3 \overset{?}{>} 0$. No, $\sqrt{2} - 3 \not> 0$. $x = 2$: $2 - 3 \overset{?}{>} 0$. No, $-1 \not> 0$. $\quad x = 4$: $4 - 3 \overset{?}{>} 0$. Yes, $1 > 0$. $\quad$ Only 4 satisfies the inequality.

3. $x = -2$: $3 - 2(-2) \overset{?}{\leq} \frac{1}{2}$. No, $7 \not\leq \frac{1}{2}$. $x = -1$: $3 - 2(-1) \overset{?}{\leq} \frac{1}{2}$. No, $6 \not\leq \frac{1}{2}$. $x = 0$: $3 - 2(0) \overset{?}{\leq} \frac{1}{2}$. No, $3 \not\leq \frac{1}{2}$.

$x = \frac{1}{2}$: $3 - 2\left(\frac{1}{2}\right) \overset{?}{\leq} \frac{1}{2}$. No, $2 \not\leq \frac{1}{2}$. $x = 1$: $3 - 2(1) \overset{?}{\leq} \frac{1}{2}$. No, $1 \not\leq \frac{1}{2}$.

$x = \sqrt{2}$: $3 - 2\left(\sqrt{2}\right) \overset{?}{\leq} \frac{1}{2}$. Yes, $3 - 2\sqrt{2} \leq \frac{1}{2}$. $x = 2$: $3 - 2(2) \overset{?}{\leq} \frac{1}{2}$. Yes, $-1 \leq \frac{1}{2}$.

$x = 4$: $3 - 2(4) \overset{?}{\leq} \frac{1}{2}$. Yes, $5 \leq \frac{1}{2}$.   The elements $\sqrt{2}$, 2, and 4 all satisfy the inequality.

5. $x = -2$: $1 \overset{?}{<} 2(-2) - 4 \overset{?}{\leq} 7$. No, $2(-2) - 4 = -8$ and $1 \not< -8$.

$x = -1$: $1 \overset{?}{<} 2(-1) - 4 \overset{?}{\leq} 7$. No, $2(-1) - 4 = -6$ and $1 \not< -6$.

$x = 0$: $1 \overset{?}{<} 2(0) - 4 \overset{?}{\leq} 7$. No, $2(0) - 4 = -4$ and $1 \not< -4$.

$x = \frac{1}{2}$: $1 \overset{?}{<} 2\left(\frac{1}{2}\right) - 4 \overset{?}{\leq} 7$. No, $2\left(\frac{1}{2}\right) - 4 = -3$ and $1 \not< -3$.

$x = 1$: $1 \overset{?}{<} 2(1) - 4 \overset{?}{\leq} 7$. No, $2(1) - 4 = -2$ and $1 \not< -2$.

$x = \sqrt{2}$: $1 \overset{?}{<} 2\left(\sqrt{2}\right) - 4 \overset{?}{\leq} 7$. No, $2\sqrt{2} - 4 < 0$ and $1 \not< 0$.

$x = 2$: $1 \overset{?}{<} 2(2) - 4 \overset{?}{\leq} 7$. No, $2(1) - 4 = -2$ and $1 \not< -2$.

$x = 4$: $1 \overset{?}{<} 2(4) - 4 \overset{?}{\leq} 7$. Yes, $2(4) - 4 = 4$ and $1 < 4 \leq 7$.   Only 4 satisfies the inequality.

7. $x = -2$: $\frac{1}{(-2)} \overset{?}{\leq} \frac{1}{2}$. Yes, $-\frac{1}{2} \leq \frac{1}{2}$. $x = -1$: $\frac{1}{(-1)} \overset{?}{\leq} \frac{1}{2}$. Yes, $-1 \leq \frac{1}{2}$. $x = 0$: $\frac{1}{0} \overset{?}{\leq} \frac{1}{2}$. No, $\frac{1}{0}$ is not defined.

$x = \frac{1}{2}$: $\frac{1}{1/2} \overset{?}{\leq} \frac{1}{2}$. No, $2 \not\leq \frac{1}{2}$. $x = 1$: $\frac{1}{1} \overset{?}{\leq} \frac{1}{2}$. No, $1 \not\leq \frac{1}{2}$. $x = \sqrt{2}$: $\frac{1}{\sqrt{2}} \overset{?}{\leq} \frac{1}{2}$. No, $2 \not\leq \sqrt{2}$.

$x = 2$: $\frac{1}{2} \overset{?}{\leq} \frac{1}{2}$. Yes, $\frac{1}{2} \leq \frac{1}{2}$. $x = 4$: $\frac{1}{4} \overset{?}{\leq} \frac{1}{2}$. Yes. The elements $-2, -1, 2$, and 4 all satisfy the inequality.

9. $2x \leq 7 \iff x \leq \frac{7}{2}$. Interval: $\left(-\infty, \frac{7}{2}\right]$

Graph: $\frac{7}{2}$

11. $2x - 5 > 3 \iff 2x > 8 \iff x > 4$
Interval: $(4, \infty)$

Graph: 4

13. $7 - x \geq 5 \iff -x \geq -2 \iff x \leq 2$
Interval: $(-\infty, 2]$

Graph: 2

15. $2x + 1 < 0 \iff 2x < -1 \iff x < -\frac{1}{2}$
Interval: $\left(-\infty, -\frac{1}{2}\right)$

Graph: $-\frac{1}{2}$

17. $3x + 11 \leq 6x + 8 \iff 3 \leq 3x \iff 1 \leq x$
Interval: $[1, \infty)$

Graph: 1

19. $\frac{1}{2}x - \frac{2}{3} > 2 \iff \frac{1}{2}x > \frac{8}{3} \iff x > \frac{16}{3}$
Interval: $\left(\frac{16}{3}, \infty\right)$

Graph: $\frac{16}{3}$

21. $\frac{1}{3}x + 2 < \frac{1}{6}x - 1 \iff \frac{1}{6}x < -3 \iff x < -18$
Interval: $(-\infty, -18)$

Graph: -18

23. $4 - 3x \leq -(1 + 8x) \iff 4 - 3x \leq -1 - 8x \iff$
$5x \leq -5 \iff x \leq -1$
Interval: $(-\infty, -1]$

Graph: -1

25. $2 \le x + 5 < 4$ $\Leftrightarrow$ $-3 \le x < -1$

Interval: $[-3, -1)$

Graph:

$$\begin{array}{c} \bullet \!\!-\!\!-\!\!-\!\!-\!\!-\!\! \circ \longrightarrow \\ {\scriptstyle -3} \qquad\quad {\scriptstyle -1} \end{array}$$

27. $-1 < 2x - 5 < 7$ $\Leftrightarrow$ $4 < 2x < 12$ $\Leftrightarrow$

$2 < x < 6$

Interval: $(2, 6)$

Graph:

$$\begin{array}{c} \circ \!\!-\!\!-\!\!-\!\!-\!\!-\!\! \circ \longrightarrow \\ {\scriptstyle 2} \qquad\quad {\scriptstyle 6} \end{array}$$

29. $-2 < 8 - 2x \le -1$ $\Leftrightarrow$ $-10 < -2x \le -9$ $\Leftrightarrow$

$5 > x \ge \frac{9}{2}$ $\Leftrightarrow$ $\frac{9}{2} \le x < 5$

Interval: $\left[\frac{9}{2}, 5\right)$

Graph:

$$\begin{array}{c} \bullet \!\!-\!\!-\!\!-\!\!-\!\!-\!\! \circ \longrightarrow \\ {\scriptstyle \frac{9}{2}} \qquad\quad {\scriptstyle 5} \end{array}$$

31. $\dfrac{2}{3} \ge \dfrac{2x - 3}{12} > \dfrac{1}{6}$ $\Leftrightarrow$ $8 \ge 2x - 3 > 2$ (multiply

each expression by 12) $\Leftrightarrow$ $11 \ge 2x > 5$ $\Leftrightarrow$

$\frac{11}{2} \ge x > \frac{5}{2}$. Interval: $\left(\frac{5}{2}, \frac{11}{2}\right]$.

Graph:

$$\begin{array}{c} \circ \!\!-\!\!-\!\!-\!\!-\!\!-\!\! \bullet \longrightarrow \\ {\scriptstyle \frac{5}{2}} \qquad\quad {\scriptstyle \frac{11}{2}} \end{array}$$

33. $(x + 2)(x - 3) < 0$. The expression on the left of the inequality changes sign where $x = -2$ and where $x = 3$. Thus we must check the intervals in the following table.

Interval	$(-\infty, -2)$	$(-2, 3)$	$(3, \infty)$
Sign of $x + 2$	$-$	$+$	$+$
Sign of $x - 3$	$-$	$-$	$+$
Sign of $(x + 2)(x - 3)$	$+$	$-$	$+$

From the table, the solution set is
$\{x \mid -2 < x < 3\}$. Interval: $(-2, 3)$.

Graph:

$$\begin{array}{c} \circ \!\!-\!\!-\!\!-\!\!-\!\!-\!\! \circ \longrightarrow \\ {\scriptstyle -2} \qquad\quad {\scriptstyle 3} \end{array}$$

35. $x(2x + 7) \ge 0$. The expression on the left of the inequality changes sign where $x = 0$ and where $x = -\frac{7}{2}$. Thus we must check the intervals in the following table.

Interval	$\left(-\infty, -\frac{7}{2}\right)$	$\left(-\frac{7}{2}, 0\right)$	$(0, \infty)$
Sign of x	$-$	$-$	$+$
Sign of $2x + 7$	$-$	$+$	$+$
Sign of $x(2x + 7)$	$+$	$-$	$+$

From the table, the solution set is
$\left\{x \mid x \le -\frac{7}{2} \text{ or } 0 \le x\right\}$.
Interval: $\left(-\infty, -\frac{7}{2}\right] \cup [0, \infty)$.

Graph:

$$\begin{array}{c} \bullet \!\!-\!\!-\!\!-\!\!-\!\!-\!\! \bullet \longrightarrow \\ {\scriptstyle -\frac{7}{2}} \qquad\quad {\scriptstyle 0} \end{array}$$

37. $x^2 - 3x - 18 \le 0$ $\Leftrightarrow$ $(x + 3)(x - 6) \le 0$. The expression on the left of the inequality changes sign where $x = 6$ and where $x = -3$. Thus we must check the intervals in the following table.

Interval	$(-\infty, -3)$	$(-3, 6)$	$(6, \infty)$
Sign of $x + 3$	$-$	$+$	$+$
Sign of $x - 6$	$-$	$-$	$+$
Sign of $(x + 3)(x - 6)$	$+$	$-$	$+$

From the table, the solution set is
$\{x \mid -3 \le x \le 6\}$. Interval: $[-3, 6]$.

Graph:

$$\begin{array}{c} \bullet \!\!-\!\!-\!\!-\!\!-\!\!-\!\! \bullet \longrightarrow \\ {\scriptstyle -3} \qquad\quad {\scriptstyle 6} \end{array}$$

39. $2x^2 + x \geq 1$ $\Leftrightarrow$ $2x^2 + x - 1 \geq 0$ $\Leftrightarrow$ $(x + 1)(2x - 1) \geq 0$. The expression on the left of the inequality changes sign where $x = -1$ and where $x = \frac{1}{2}$. Thus we must check the intervals in the following table.

Interval	$(-\infty, -1)$	$(-1, \frac{1}{2})$	$(\frac{1}{2}, \infty)$
Sign of $x + 1$	$-$	$+$	$+$
Sign of $2x - 1$	$-$	$-$	$+$
Sign of $(x + 1)(2x - 1)$	$+$	$-$	$+$

From the table, the solution set is $\left\{ x \mid x \leq -1 \text{ or } \frac{1}{2} \leq x \right\}$.

Interval: $(-\infty, -1] \cup \left[\frac{1}{2}, \infty \right)$.

Graph:

41. $3x^2 - 3x < 2x^2 + 4$ $\Leftrightarrow$ $x^2 - 3x - 4 < 0$ $\Leftrightarrow$ $(x + 1)(x - 4) < 0$. The expression on the left of the inequality changes sign where $x = -1$ and where $x = 4$. Thus we must check the intervals in the following table.

Interval	$(-\infty, -1)$	$(-1, 4)$	$(4, \infty)$
Sign of $x + 1$	$-$	$+$	$+$
Sign of $x - 4$	$-$	$-$	$+$
Sign of $(x + 1)(x - 4)$	$+$	$-$	$+$

From the table, the solution set is $\{ x \mid -1 < x < 4 \}$. Interval: $(-1, 4)$.

Graph:

43. $x^2 > 3(x + 6)$ $\Leftrightarrow$ $x^2 - 3x - 18 > 0$ $\Leftrightarrow$ $(x + 3)(x - 6) > 0$. The expression on the left of the inequality changes sign where $x = 6$ and where $x = -3$. Thus we must check the intervals in the following table.

Interval	$(-\infty, -3)$	$(-3, 6)$	$(6, \infty)$
Sign of $x + 3$	$-$	$+$	$+$
Sign of $x - 6$	$-$	$-$	$+$
Sign of $(x + 3)(x - 6)$	$+$	$-$	$+$

From the table, the solution set is $\{ x \mid x < -3 \text{ or } 6 < x \}$.

Interval: $(-\infty, -3) \cup (6, \infty)$.

Graph:

45. $x^2 < 4$ $\Leftrightarrow$ $x^2 - 4 < 0$ $\Leftrightarrow$ $(x + 2)(x - 2) < 0$. The expression on the left of the inequality changes sign where $x = -2$ and where $x = 2$. Thus we must check the intervals in the following table.

Interval	$(-\infty, -2)$	$(-2, 2)$	$(2, \infty)$
Sign of $x + 2$	$-$	$+$	$+$
Sign of $x - 2$	$-$	$-$	$+$
Sign of $(x + 2)(x - 2)$	$+$	$-$	$+$

From the table, the solution set is $\{ x \mid -2 < x < 2 \}$. Interval: $(-2, 2)$.

Graph:

47. $-2x^2 \leq 4$ $\Leftrightarrow$ $-2x^2 - 4 \leq 0$ $\Leftrightarrow$ $-2(x^2 + 1) \leq 0$. Since $x^2 + 1 > 0$, $-2(x^2 + 1) \leq 0$ for all x.

Interval: $(-\infty, \infty)$. Graph:

49. $x^3 - 4x > 0 \iff x(x^2 - 4) > 0 \iff x(x+2)(x-2) > 0$. The expression on the left of the inequality changes sign where $x = 0$, $x = -2$ and where $x = 4$. Thus we must check the intervals in the following table.

Interval	$(-\infty, -2)$	$(-2, 0)$	$(0, 2)$	$(2, \infty)$
Sign of x	$-$	$-$	$+$	$+$
Sign of $x + 2$	$-$	$+$	$+$	$+$
Sign of $x - 2$	$-$	$-$	$-$	$+$
Sign of $x(x+2)(x-2)$	$-$	$+$	$-$	$+$

From the table, the solution set is $\{x \mid -2 < x < 0 \text{ or } x > 2\}$. Interval: $(-2, 0) \cup (2, \infty)$.

Graph:

51. $\dfrac{x-3}{x+1} \geq 0$. The expression on the left of the inequality changes sign where $x = -1$ and where $x = 3$. Thus we must check the intervals in the following table.

Interval	$(-\infty, -1)$	$(-1, 3)$	$(3, \infty)$
Sign of $x + 1$	$-$	$+$	$+$
Sign of $x - 3$	$-$	$-$	$+$
Sign of $\dfrac{x-3}{x+1}$	$+$	$-$	$+$

From the table, the solution set is $\{x \mid x < -1 \text{ or } x \leq 3\}$. Since the denominator cannot equal 0 we must have $x \neq -1$. Interval: $(-\infty, -1) \cup [3, \infty)$.

Graph:

53. $\dfrac{4x}{2x+3} > 2 \iff \dfrac{4x}{2x+3} - 2 > 0 \iff \dfrac{4x}{2x+3} - \dfrac{2(2x+3)}{2x+3} > 0 \iff \dfrac{-6}{2x+3} > 0$. The expression on the left of the inequality changes sign where $x = -\frac{3}{2}$. Thus we must check the intervals in the following table.

Interval	$\left(-\infty, -\frac{3}{2}\right)$	$\left(-\frac{3}{2}, \infty\right)$
Sign of -6	$-$	$-$
Sign of $2x + 3$	$-$	$+$
Sign of $\dfrac{-6}{2x+3}$	$+$	$-$

From the table, the solution set is $\left\{x \mid x < -\frac{3}{2}\right\}$. Interval: $\left(-\infty, -\frac{3}{2}\right)$.

Graph:

55. $\dfrac{2x+1}{x-5} \leq 3 \iff \dfrac{2x+1}{x-5} - 3 \leq 0 \iff \dfrac{2x+1}{x-5} - \dfrac{3(x-5)}{x-5} \leq 0 \iff \dfrac{-x+16}{x-5} \leq 0$. The expression on the left of the inequality changes sign where $x = 16$ and where $x = 5$. Thus we must check the intervals in the following table.

Interval	$(-\infty, 5)$	$(5, 16)$	$(16, \infty)$
Sign of $-x + 16$	$+$	$+$	$-$
Sign of $x - 5$	$-$	$+$	$+$
Sign of $\dfrac{-x+16}{x-5}$	$-$	$+$	$-$

From the table, the solution set is $\{x \mid x < 5 \text{ or } x \geq 16\}$. Since the denominator cannot equal 0, we must have $x \neq 5$. Interval: $(-\infty, 5) \cup [16, \infty)$.

Graph:

57. $\dfrac{4}{x} < x \quad \Leftrightarrow \quad \dfrac{4}{x} - x < 0 \quad \Leftrightarrow \quad \dfrac{4}{x} - \dfrac{x \cdot x}{x} < 0 \quad \Leftrightarrow \quad \dfrac{4 - x^2}{x} < 0 \quad \Leftrightarrow \quad \dfrac{(2-x)(2+x)}{x} < 0$. The expression

on the left of the inequality changes sign where $x = 0$, where $x = -2$, and where $x = 2$. Thus we must check the intervals
in the following table.

Interval	$(-\infty, -2)$	$(-2, 0)$	$(0, 2)$	$(2, \infty)$
Sign of $2 + x$	$-$	$+$	$+$	$+$
Sign of x	$-$	$-$	$+$	$+$
Sign of $2 - x$	$+$	$+$	$+$	$-$
Sign of $\dfrac{(2-x)(2+x)}{x}$	$+$	$-$	$+$	$-$

From the table, the solution set is $\{x \mid -2 < x < 0 \text{ or } 2 < x\}$.

Interval: $(-2, 0) \cup (2, \infty)$.

Graph:

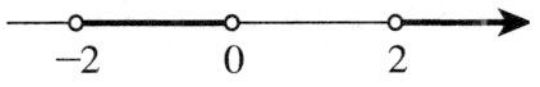

59. $1 + \dfrac{2}{x+1} \leq \dfrac{2}{x} \quad \Leftrightarrow \quad 1 + \dfrac{2}{x+1} - \dfrac{2}{x} \leq 0 \quad \Leftrightarrow \quad \dfrac{x(x+1)}{x(x+1)} + \dfrac{2x}{x(x+1)} - \dfrac{2(x+1)}{x(x+1)} \leq 0 \quad \Leftrightarrow$

$\dfrac{x^2 + x + 2x - 2x - 2}{x(x+1)} \leq 0 \quad \Leftrightarrow \quad \dfrac{x^2 + x - 2}{x(x+1)} \leq 0 \quad \Leftrightarrow \quad \dfrac{(x+2)(x-1)}{x(x+1)} \leq 0$. The expression on the left of the

inequality changes sign where $x = -2$, where $x = -1$, where $x = 0$, and where $x = 1$. Thus we must check the intervals
in the following table.

Interval	$(-\infty, -2)$	$(-2, -1)$	$(-1, 0)$	$(0, 1)$	$(1, \infty)$
Sign of $x + 2$	$-$	$+$	$+$	$+$	$+$
Sign of $x - 1$	$-$	$-$	$-$	$-$	$+$
Sign of x	$-$	$-$	$-$	$+$	$+$
Sign of $x + 1$	$-$	$-$	$+$	$+$	$+$
Sign of $\dfrac{(x+2)(x-1)}{x(x+1)}$	$+$	$-$	$+$	$-$	$+$

Since $x = -1$ and $x = 0$ yield undefined expressions, we cannot include them in the solution. From the table, the solution
set is $\{x \mid -2 \leq x < -1 \text{ or } 0 < x \leq 1\}$.

Interval: $[-2, -1) \cup (0, 1]$.

Graph:

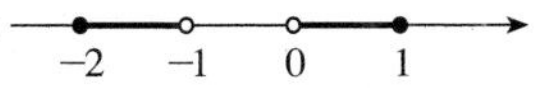

61. $\dfrac{6}{x-1} - \dfrac{6}{x} \geq 1$ $\Leftrightarrow$ $\dfrac{6}{x-1} - \dfrac{6}{x} - 1 \geq 0$ $\Leftrightarrow$ $\dfrac{6x}{x(x-1)} - \dfrac{6(x-1)}{x(x-1)} - \dfrac{x(x-1)}{x(x-1)} \geq 0$ $\Leftrightarrow$

$\dfrac{6x - 6x + 6 - x^2 + x}{x(x-1)} \geq 0$ $\Leftrightarrow$ $\dfrac{-x^2 + x + 6}{x(x-1)} \geq 0$ $\Leftrightarrow$ $\dfrac{(-x+3)(x+2)}{x(x-1)} \geq 0$. The

expression on the left of the inequality changes sign where $x = 3$, where $x = -2$, where $x = 0$, and where $x = 1$. Thus we

must check the intervals in the following table.

Interval	$(-\infty, -2)$	$(-2, 0)$	$(0, 1)$	$(1, 3)$	$(3, \infty)$
Sign of $-x + 3$	$+$	$+$	$+$	$+$	$-$
Sign of $x + 2$	$-$	$+$	$+$	$+$	$+$
Sign of x	$-$	$-$	$+$	$+$	$+$
Sign of $x - 1$	$-$	$-$	$-$	$+$	$+$
Sign of $\dfrac{(-x+3)(x+2)}{x(x-1)}$	$-$	$+$	$-$	$+$	$-$

From the table, the solution set is $\{x \mid -2 \leq x < 0 \text{ or } 1 < x \leq 3\}$. The points $x = 0$ and $x = 1$ are excluded from the

solution set because they make the denominator zero.

Interval: $[-2, 0) \cup (1, 3]$.

Graph:

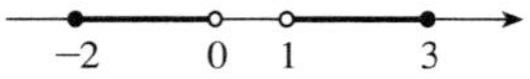

63. $\dfrac{x+2}{x+3} < \dfrac{x-1}{x-2}$ $\Leftrightarrow$ $\dfrac{x+2}{x+3} - \dfrac{x-1}{x-2} < 0$ $\Leftrightarrow$ $\dfrac{(x+2)(x-2)}{(x+3)(x-2)} - \dfrac{(x-1)(x+3)}{(x-2)(x+3)} < 0$ $\Leftrightarrow$

$\dfrac{x^2 - 4 - x^2 - 2x + 3}{(x+3)(x-2)} < 0$ $\Leftrightarrow$ $\dfrac{-2x - 1}{(x+3)(x-2)} < 0$. The expression on the left of the inequality

changes sign where $x = -\frac{1}{2}$, where $x = -3$, and where $x = 2$. Thus we must check the intervals in the following table.

Interval	$(-\infty, -3)$	$\left(-3, -\frac{1}{2}\right)$	$\left(-\frac{1}{2}, 2\right)$	$(2, \infty)$
Sign of $-2x - 1$	$+$	$+$	$-$	$-$
Sign of $x + 3$	$-$	$+$	$+$	$+$
Sign of $x - 2$	$-$	$-$	$-$	$+$
Sign of $\dfrac{-2x - 1}{(x+3)(x-2)}$	$+$	$-$	$+$	$-$

From the table, the solution set is $\left\{x \mid -3 < x < -\frac{1}{2} \text{ or } 2 < x\right\}$.

Interval: $\left(-3, -\frac{1}{2}\right) \cup (2, \infty)$.

Graph:

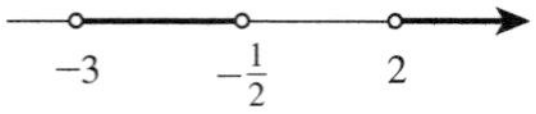

65. $x^4 > x^2 \quad \Leftrightarrow \quad x^4 - x^2 > 0 \quad \Leftrightarrow \quad x^2 \left(x^2 - 1 \right) > 0 \quad \Leftrightarrow \quad x^2 \left(x - 1 \right) \left(x + 1 \right) > 0$. The expression on the left of the inequality changes sign where $x = 0$, where $x = 1$, and where $x = -1$. Thus we must check the intervals in the following table.

Interval	$(-\infty, -1)$	$(-1, 0)$	$(0, 1)$	$(1, \infty)$
Sign of x^2	+	+	+	+
Sign of $x - 1$	−	−	−	+
Sign of $x + 1$	−	+	+	+
Sign of $x^2 \left(x - 1 \right) \left(x + 1 \right)$	+	−	−	+

From the table, the solution set is $\{x \mid x < -1 \text{ or } 1 < x\}$. Interval: $(-\infty, -1) \cup (1, \infty)$. Graph:

67. For $\sqrt{16 - 9x^2}$ to be defined as a real number we must have $16 - 9x^2 \geq 0 \quad \Leftrightarrow \quad (4 - 3x)(4 + 3x) \geq 0$. The expression in the inequality changes sign at $x = \frac{4}{3}$ and $x = -\frac{4}{3}$.

Interval	$\left(-\infty, -\frac{4}{3}\right)$	$\left(-\frac{4}{3}, \frac{4}{3}\right)$	$\left(\frac{4}{3}, \infty\right)$
Sign of $4 - 3x$	+	+	−
Sign of $4 + 3x$	−	+	+
Sign of $(4 - 3x)(4 + 3x)$	−	+	−

Thus $-\frac{4}{3} \leq x \leq \frac{4}{3}$.

69. For $\left(\dfrac{1}{x^2 - 5x - 14} \right)^{1/2}$ to be defined as a real number we must have $x^2 - 5x - 14 > 0 \quad \Leftrightarrow \quad (x - 7)(x + 2) > 0$. The expression in the inequality changes sign at $x = 7$ and $x = -2$.

Interval	$(-\infty, -2)$	$(-2, 7)$	$(7, \infty)$
Sign of $x - 7$	−	−	+
Sign of $x + 2$	−	+	+
Sign of $(x - 7)(x + 2)$	+	−	+

Thus $x < -2$ or $7 < x$, and the solution set is $(-\infty, -2) \cup (7, \infty)$.

71. (a) $a \left(bx - c \right) \geq bc$ (where $a, b, c > 0$) $\quad \Leftrightarrow \quad bx - c \geq \dfrac{bc}{a} \quad \Leftrightarrow \quad bx \geq \dfrac{bc}{a} + c \quad \Leftrightarrow \quad x \geq \dfrac{1}{b} \left(\dfrac{bc}{a} + c \right) = \dfrac{c}{a} + \dfrac{c}{b}$

$\Leftrightarrow \quad x \geq \dfrac{c}{a} + \dfrac{c}{b}$.

(b) We have $a \leq bx + c < 2a$, where $a, b, c > 0 \quad \Leftrightarrow \quad a - c \leq bx < 2a - c \quad \Leftrightarrow \quad \dfrac{a - c}{b} \leq x < \dfrac{2a - c}{b}$.

73. Inserting the relationship $C = \frac{5}{9} \left(F - 32 \right)$, we have $20 \leq C \leq 30 \quad \Leftrightarrow \quad 20 \leq \frac{5}{9} \left(F - 32 \right) \leq 30 \quad \Leftrightarrow$
$36 \leq F - 32 \leq 54 \quad \Leftrightarrow \quad 68 \leq F \leq 86$.

75. Let x be the average number of miles driven per day. Each day the cost of Plan A is $30 + 0.10x$, and the cost of Plan B is 50. Plan B saves money when $50 < 30 + 0.10x \quad \Leftrightarrow \quad 20 < 0.1x \quad \Leftrightarrow \quad 200 < x$. So Plan B saves money when you average more than 200 miles a day.

77. We need to solve $6400 \leq 0.35m + 2200 \leq 7100$ for m. So $6400 \leq 0.35m + 2200 \leq 7100 \quad \Leftrightarrow \quad 4200 \leq 0.35m \leq 4900$
$\Leftrightarrow \quad 12{,}000 \leq m \leq 14{,}000$. She plans on driving between 12,000 and 14,000 miles.

79. (a) Let x be the number of $3 increases. Then the number of seats sold is $120 - x$. So $P = 200 + 3x$ $\Leftrightarrow$ $3x = P - 200$ $\Leftrightarrow$ $x = \frac{1}{3}(P - 200)$. Substituting for x we have that the number of seats sold is $120 - x = 120 - \frac{1}{3}(P - 200) = -\frac{1}{3}P + \frac{560}{3}$.

(b) $90 \leq -\frac{1}{3}P + \frac{560}{3} \leq 115$ $\Leftrightarrow$ $270 \leq 360 - P + 200 \leq 345$ $\Leftrightarrow$ $270 \leq -P + 560 \leq 345$ $\Leftrightarrow$ $-290 \leq -P \leq -215$ $\Leftrightarrow$ $290 \geq P \geq 215$. Putting this into standard order, we have $215 \leq P \leq 290$. So the ticket prices are between $215 and $290.

81. $0.0004 \leq \dfrac{4{,}000{,}000}{d^2} \leq 0.01$. Since $d^2 \geq 0$ and $d \neq 0$, we can multiply each expression by d^2 to obtain $0.0004d^2 \leq 4{,}000{,}000 \leq 0.01d^2$. Solving each pair, we have $0.0004d^2 \leq 4{,}000{,}000$ $\Leftrightarrow$ $d^2 \leq 10{,}000{,}000{,}000$ $\Rightarrow$ $d \leq 100{,}000$ (recall that d represents distance, so it is always nonnegative). Solving $4{,}000{,}000 \leq 0.01d^2$ $\Leftrightarrow$ $400{,}000{,}000 \leq d^2 \Rightarrow 20{,}000 \leq d$. Putting these together, we have $20{,}000 \leq d \leq 100{,}000$.

83. $128 + 16t - 16t^2 \geq 32$ $\Leftrightarrow$ $-16t^2 + 16t + 96 \geq 0$ $\Leftrightarrow$ $-16(t^2 - t - 6) \geq 0$ $\Leftrightarrow$ $-16(t - 3)(t + 2) \geq 0$. The expression on the left of the inequality changes sign at $x = -2$, at $t = 3$, and at $t = -2$. However, $t \geq 0$, so the only endpoint is $t = 3$.

Interval	$(0, 3)$	$(3, \infty)$
Sign of -16	$-$	$-$
Sign of $t - 3$	$-$	$+$
Sign of $t + 2$	$+$	$+$
Sign of $-16(t - 3)(t + 2)$	$+$	$-$

So $0 \leq t \leq 3$.

85. $240 > v + \dfrac{v^2}{20}$ $\Leftrightarrow$ $\frac{1}{20}v^2 + v - 240 < 0$ $\Leftrightarrow$ $\left(\frac{1}{20}v - 3\right)(v + 80) < 0$. The expression in the inequality changes sign at $v = 60$ and $v = -80$. However, since v represents the speed, we must have $v \geq 0$.

Interval	$(0, 60)$	$(60, \infty)$
Sign of $\frac{1}{20}v - 3$	$-$	$+$
Sign of $v + 80$	$+$	$+$
Sign of $\left(\frac{1}{20}v - 3\right)(v + 80)$	$-$	$+$

So Kerry must drive less than 60 mi/h.

87. Let n be the number of people in the group. Then the bus fare is $\dfrac{360}{n}$, and the cost of the theater tickets is $30 - 0.25n$.

We want the total cost to be less than \$39 per person, that is, $\dfrac{360}{n} + (30 - 0.25n) < 39$. If we multiply this inequality

by n, we will not change the direction of the inequality; n is positive since it represents the number of people. So we

get $360 + n(30 - 0.25n) < 39n$ $\Leftrightarrow$ $-0.25n^2 + 30n + 360 < 39n$ $\Leftrightarrow$ $-0.25n^2 - 9n + 360 < 0$ $\Leftrightarrow$

$(0.25n + 15)(-n + 24)$. The expression on the left of the inequality changes sign when $n = -60$ and $n = 24$. Since

$n > 0$, we check the intervals in the following table.

Interval	$(0, 24)$	$(24, \infty)$
Sign of $0.25n + 15$	$+$	$+$
Sign of $-n + 24$	$+$	$-$
Sign of $(0.25n + 15)(-n + 24)$	$+$	$-$

So the group must have more than 24 people in order that the cost of the theater tour is less than \$39.

89. *Case 1: $a < b < 0$*   We have $a \cdot a > a \cdot b$, since $a < 0$, and $b \cdot a > b \cdot b$, since $b < 0$. So $a^2 > a \cdot b > b^2$, that is $a < b < 0$

$\Rightarrow$ $a^2 > b^2$. Continuing, we have $a \cdot a^2 < a \cdot b^2$, since $a < 0$ and $b^2 \cdot a < b^2 \cdot b$, since $b^2 > 0$. So $a^3 < ab^2 < b^3$. Thus

$a < b < 0$ $\Rightarrow$ $a^3 > b^3$. So $a < b < 0$ $\Rightarrow$ $a^n > b^n$, if n is even, and $a^n < b$, if n is odd.

Case 2: $0 < a < b$ We have $a \cdot a < a \cdot b$, since $a > 0$, and $b \cdot a < b \cdot b$, since $b < 0$. So $a^2 < a \cdot b < b^2$. Thus $0 <$

$a < b$ $\Rightarrow$ $a^2 < b^2$. Likewise, $a^2 \cdot a < a^2 \cdot b$ and $b \cdot a^2 < b \cdot b^2$, thus $a^3 < b^3$. So $0 < a < b$ $\Rightarrow$ $a^n < b^n$, for all

positive integers n.

Case 3: $a < 0 < b$ If n is odd, then $a^n < b^n$, because a^n is negative and b^n is positive. If n is even, then we could have

either $a^n < b^n$ or $a^n > b^n$. For example, $-1 < 2$ and $(-1)^2 < 2^2$, but $-3 < 2$ and $(-3)^2 > 2^2$.

1.7   Absolute Value Equations and Inequalities

1. $|4x| = 24$ $\Leftrightarrow$ $4x = \pm 24$ $\Leftrightarrow$ $x = \pm 6$.

3. $5|x| + 3 = 28$ $\Leftrightarrow$ $5|x| = 25$ $\Leftrightarrow$ $|x| = 5$ $\Leftrightarrow$ $x = \pm 5$.

5. $|x - 3| = 2$ is equivalent to $x - 3 = \pm 2$ $\Leftrightarrow$ $x = 3 \pm 2$ $\Leftrightarrow$ $x = 1$ or $x = 5$.

7. $|x + 4| = 0.5$ is equivalent to $x + 4 = \pm 0.5$ $\Leftrightarrow$ $x = -4 \pm 0.5$ $\Leftrightarrow$ $x = -4.5$ or $x = -3.5$.

9. $|4x + 7| = 9$ is equivalent to either $4x + 7 = 9$ $\Leftrightarrow$ $4x = 2$ $\Leftrightarrow$ $x = \dfrac{1}{2}$; or $4x + 7 = -9$ $\Leftrightarrow$ $4x = -16$ $\Leftrightarrow$

$x = -4$. The two solutions are $x = \frac{1}{2}$ and $x = -4$.

11. $4 - |3x + 6| = 1$ $\Leftrightarrow$ $-|3x + 6| = -3$ $\Leftrightarrow$ $|3x + 6| = 3$, which is equivalent to either $3x + 6 = 3$ $\Leftrightarrow$ $3x = -3$

$\Leftrightarrow$ $x = -1$; or $3x + 6 = -3$ $\Leftrightarrow$ $3x = -9$ $\Leftrightarrow$ $x = -3$. The two solutions are $x = -1$ and $x = -3$.

13. $3|x + 5| + 6 = 15$ $\Leftrightarrow$ $3|x + 5| = 9$ $\Leftrightarrow$ $|x + 5| = 3$, which is equivalent to either $x + 5 = 3$ $\Leftrightarrow$ $x = -2$; or

$x + 5 = -3$ $\Leftrightarrow$ $x = -8$. The two solutions are $x = -2$ and $x = -8$.

15. $8 + 5\left|\frac{1}{3}x - \frac{5}{6}\right| = 33$ $\Leftrightarrow$ $5\left|\frac{1}{3}x - \frac{5}{6}\right| = 25$ $\Leftrightarrow$ $\left|\frac{1}{3}x - \frac{5}{6}\right| = 5$, which is equivalent to either $\frac{1}{3}x - \frac{5}{6} = 5$ $\Leftrightarrow$

$\frac{1}{3}x = \frac{35}{6}$ $\Leftrightarrow$ $x = \frac{35}{2}$; or $\frac{1}{3}x - \frac{5}{6} = -5$ $\Leftrightarrow$ $\frac{1}{3}x = -\frac{25}{6}$ $\Leftrightarrow$ $x = -\frac{25}{2}$. The two solutions are $x = -\frac{25}{2}$ and

$x = \frac{35}{2}$.

17. $|x - 1| = |3x + 2|$, which is equivalent to either $x - 1 = 3x + 2$ $\Leftrightarrow$ $-2x = 3$ $\Leftrightarrow$ $x = -\frac{3}{2}$; or $x - 1 = -(3x + 2)$

$\Leftrightarrow$ $x - 1 = -3x - 2$ $\Leftrightarrow$ $4x = -1$ $\Leftrightarrow$ $x = -\frac{1}{4}$. The two solutions are $x = -\frac{3}{2}$ and $x = -\frac{1}{4}$.

19. $|x| \leq 4$ $\Leftrightarrow$ $-4 \leq x \leq 4$. Interval: $[-4, 4]$.

21. $|2x| > 7$ is equivalent to $2x > 7$ $\Leftrightarrow$ $x > \frac{7}{2}$; or $2x < 7$ $\Leftrightarrow$ $x < -\frac{7}{2}$. Interval: $\left(-\infty, -\frac{7}{2}\right) \cup \left(\frac{7}{2}, \infty\right)$.

23. $|x - 5| \leq 3$ $\Leftrightarrow$ $-3 \leq x - 5 \leq 3$ $\Leftrightarrow$ $2 \leq x \leq 8$. Interval: $[2, 8]$.

25. $|x + 1| \geq 1$ is equivalent to $x + 1 \geq 1$ $\Leftrightarrow$ $x \geq 0$; or $x + 1 \leq -1$ $\Leftrightarrow$ $x \leq -2$. Interval: $(-\infty, -2] \cup [0, \infty)$.

27. $|x + 5| \geq 2$ is equivalent to $x + 5 \geq 2$ $\Leftrightarrow$ $x \geq -3$; or $x + 5 \leq -2$ $\Leftrightarrow$ $x \leq -7$. Interval: $(-\infty, -7] \cup [-3, \infty)$.

29. $|2x - 3| \leq 0.4$ $\Leftrightarrow$ $-0.4 \leq 2x - 3 \leq 0.4$ $\Leftrightarrow$ $2.6 \leq 2x \leq 3.4$ $\Leftrightarrow$ $1.3 \leq x \leq 1.7$. Interval: $[1.3, 1.7]$.

31. $\left|\dfrac{x - 2}{3}\right| < 2$ $\Leftrightarrow$ $-2 < \dfrac{x - 2}{3} < 2$ $\Leftrightarrow$ $-6 < x - 2 < 6$ $\Leftrightarrow$ $-4 < x < 8$.
Interval: $(-4, 8)$.

33. $|x + 6| < 0.001$ $\Leftrightarrow$ $-0.001 < x + 6 < 0.001$ $\Leftrightarrow$ $-6.001 < x < -5.999$.
Interval: $(-6.001, -5.999)$.

35. $4|x + 2| - 3 < 13$ $\Leftrightarrow$ $4|x + 2| < 16$ $\Leftrightarrow$ $|x + 2| < 4$ $\Leftrightarrow$ $-4 < x + 2 < 4$ $\Leftrightarrow$ $-6 < x < 2$.
Interval: $(-6, 2)$.

37. $8 - |2x - 1| \geq 6$ $\Leftrightarrow$ $-|2x - 1| \geq -2$ $\Leftrightarrow$ $|2x - 1| \leq 2$ $\Leftrightarrow$ $-2 \leq 2x - 1 \leq 2$ $\Leftrightarrow$ $-1 \leq 2x \leq 3$ $\Leftrightarrow$ $-\frac{1}{2} \leq x \leq \frac{3}{2}$. Interval: $\left[-\frac{1}{2}, \frac{3}{2}\right]$.

39. $\frac{1}{2}\left|4x + \frac{1}{3}\right| > \frac{5}{6}$ $\Leftrightarrow$ $\left|4x + \frac{1}{3}\right| > \frac{5}{3}$, which is equivalent to $4x + \frac{1}{3} > \frac{5}{3}$ $\Leftrightarrow$ $4x > \frac{4}{3}$ $\Leftrightarrow$ $x > \frac{1}{3}$, or $4x + \frac{1}{3} < -\frac{5}{3}$ $\Leftrightarrow$ $4x < -2$ $\Leftrightarrow$ $x < -\frac{1}{2}$. Interval: $\left(-\infty, -\frac{1}{2}\right) \cup \left(\frac{1}{3}, \infty\right)$.

41. $1 \leq |x| \leq 4$. If $x \geq 0$, then this is equivalent to $1 \leq x \leq 4$. If $x < 0$, then this is equivalent to $1 \leq -x \leq 4$ $\Leftrightarrow$ $-1 \geq x \geq -4$ $\Leftrightarrow$ $-4 \leq x \leq -1$. Interval: $[-4, -1] \cup [1, 4]$.

43. $\dfrac{1}{|x + 7|} > 2$ $\Leftrightarrow$ $1 > 2|x + 7|(x \neq -7)$ $\Leftrightarrow$ $|x + 7| < \frac{1}{2}$ $\Leftrightarrow$ $-\frac{1}{2} < x + 7 < \frac{1}{2}$ $\Leftrightarrow$ $-\frac{15}{2} < x < -\frac{13}{2}$ and $x \neq -7$. Interval: $\left(-\frac{15}{2}, -7\right) \cup \left(-7, -\frac{13}{2}\right)$.

45. $|x| < 3$ **47.** $|x - 7| \geq 5$ **49.** $|x| \leq 2$ **51.** $|x| > 3$

53. (a) Let x be the thickness of the laminate. Then $|x - 0.020| \leq 0.003$.

(b) $|x - 0.020| \leq 0.003$ $\Leftrightarrow$ $-0.003 \leq x - 0.020 \leq 0.003$ $\Leftrightarrow$ $0.017 \leq x \leq 0.023$.

55. $|x - 1|$ is the distance between x and 1; $|x - 3|$ is the distance between x and 3. So $|x - 1| < |x - 3|$ represents those points closer to 1 than to 3, and the solution is $x < 2$, since 2 is the point halfway between 1 and 3. If $a < b$, then the solution to $|x - a| < |x - b|$ is $x < \dfrac{a + b}{2}$.

Chapter 1 Review

1. $5x + 11 = 36$ $\Leftrightarrow$ $5x = 25$ $\Leftrightarrow$ $x = 5$

3. $3x + 12 = 24$ $\Leftrightarrow$ $3x = 12$ $\Leftrightarrow$ $x = 4$

5. $7x - 6 = 4x + 9$ $\Leftrightarrow$ $3x = 15$ $\Leftrightarrow$ $x = 5$

7. $\frac{1}{3}x - \frac{1}{2} = 2$ $\Leftrightarrow$ $2x - 3 = 12$ $\Leftrightarrow$ $2x = 15$ $\Leftrightarrow$ $x = \frac{15}{2}$

9. $2(x + 3) - 4(x - 5) = 8 - 5x$ $\Leftrightarrow$ $2x + 6 - 4x + 20 = 8 - 5x$ $\Leftrightarrow$ $-2x + 26 = 8 - 5x$ $\Leftrightarrow$ $3x = -18$ $\Leftrightarrow$ $x = -6$

11. $\dfrac{x + 1}{x - 1} = \dfrac{2x - 1}{2x + 1}$ $\Leftrightarrow$ $(x + 1)(2x + 1) = (2x - 1)(x - 1)$ $\Leftrightarrow$ $2x^2 + 3x + 1 = 2x^2 - 3x + 1$ $\Leftrightarrow$ $6x = 0$ $\Leftrightarrow$ $x = 0$

13. $x^2 = 144$ $\Rightarrow$ $x = \pm 12$

15. $5x^4 - 16 = 0$ $\Leftrightarrow$ $5x^4 = 16$ $\Leftrightarrow$ $x^4 = \frac{16}{5}$ $\Rightarrow$ $x^2 = \pm\frac{4}{\sqrt{5}}$. Now $x^2 = -\frac{4}{\sqrt{5}}$ has no real solution, so we solve $x^2 = \frac{4}{\sqrt{5}}$ $\Rightarrow$ $x = \pm\frac{2}{\sqrt[4]{5}} = \pm\frac{2\sqrt[4]{5^3}}{5}$

17. $5x^3 - 15 = 0$ $\Leftrightarrow$ $5x^3 = 15$ $\Leftrightarrow$ $x^3 = 3$ $\Leftrightarrow$ $x = \sqrt[3]{3}$.

19. $(x + 1)^3 = -64$ $\Leftrightarrow$ $x + 1 = -4$ $\Leftrightarrow$ $x = -1 - 4 = -5$.

21. $\sqrt[3]{x} = -3$ $\Leftrightarrow$ $x = (-3)^3 = -27$.

23. $4x^{3/4} - 500 = 0$ $\Leftrightarrow$ $4x^{3/4} = 500$ $\Leftrightarrow$ $x^{3/4} = 125$ $\Leftrightarrow$ $x = 125^{4/3} = 5^4 = 625$.

25. $\dfrac{x+1}{x-1} = \dfrac{3x}{3x-6} = \dfrac{3x}{3(x-2)} = \dfrac{x}{x-2}$ $\Leftrightarrow$ $(x+1)(x-2) = x(x-1)$ $\Leftrightarrow$ $x^2 - x - 2 = x^2 - x$ $\Leftrightarrow$ $-2 = 0$.
Since this last equation is never true, there is no real solution to the original equation.

27. $x^2 - 9x + 14 = 0$ $\Leftrightarrow$ $(x - 7)(x - 2) = 0$ $\Leftrightarrow$ $x = 7$ or $x = 2$.

29. $2x^2 + x = 1$ $\Leftrightarrow$ $2x^2 + x - 1 = 0$ $\Leftrightarrow$ $(2x - 1)(x + 1) = 0$. So either $2x - 1 = 0$ $\Leftrightarrow$ $2x = 1$ $\Leftrightarrow$ $x = \frac{1}{2}$; or $x + 1 = 0$ $\Leftrightarrow$ $x = -1$.

31. $0 = 4x^3 - 25x = x(4x^2 - 25) = x(2x - 5)(2x + 5) = 0$. So either $x = 0$; or $2x - 5 = 0$ $\Leftrightarrow$ $2x = 5$ $\Leftrightarrow$ $x = \frac{5}{2}$; or $2x + 5 = 0$ $\Leftrightarrow$ $2x = -5$ $\Leftrightarrow$ $x = -\frac{5}{2}$.

33. $3x^2 + 4x - 1 = 0 \Rightarrow$
$x = \dfrac{-b \pm \sqrt{b^2 - 4ac}}{2a} = \dfrac{-(4) \pm \sqrt{(4)^2 - 4(3)(-1)}}{2(3)} = \dfrac{-4 \pm \sqrt{16+12}}{6} = \dfrac{-4 \pm \sqrt{28}}{6} = \dfrac{-4 \pm 2\sqrt{7}}{6} = \dfrac{2(-2 \pm \sqrt{7})}{6} = \dfrac{-2 \pm \sqrt{7}}{3}$.

35. $\dfrac{1}{x} + \dfrac{2}{x-1} = 3$ $\Leftrightarrow$ $(x-1) + 2(x) = 3(x)(x-1)$ $\Leftrightarrow$ $x - 1 + 2x = 3x^2 - 3x$ $\Leftrightarrow$ $0 = 3x^2 - 6x + 1$ $\Rightarrow$
$x = \dfrac{-b \pm \sqrt{b^2 - 4ac}}{2a} = \dfrac{-(-6) \pm \sqrt{(-6)^2 - 4(3)(1)}}{2(3)} = \dfrac{6 \pm \sqrt{36-12}}{6} = \dfrac{6 \pm \sqrt{24}}{6} = \dfrac{6 \pm 2\sqrt{6}}{6} = \dfrac{2(3 \pm \sqrt{6})}{6} = \dfrac{3 \pm \sqrt{6}}{3}$.

37. $x^4 - 8x^2 - 9 = 0$ $\Leftrightarrow$ $(x^2 - 9)(x^2 + 1) = 0$ $\Leftrightarrow$ $(x - 3)(x + 3)(x^2 + 1) = 0$ $\Rightarrow$ $x - 3 = 0$ $\Leftrightarrow$ $x = 3$, or $x + 3 = 0$ $\Leftrightarrow$ $x = -3$, however $x^2 + 1 = 0$ has no real solution. The solutions are $x = \pm 3$.

39. $x^{-1/2} - 2x^{1/2} + x^{3/2} = 0$ $\Leftrightarrow$ $x^{-1/2}(1 - 2x + x^2) = 0$ $\Leftrightarrow$ $x^{1/2}(1 - x)^2 = 0$. Since $x^{-1/2} \neq 0$, the only solution comes from $(1 - x)^2 = 0$ $\Leftrightarrow$ $1 - x = 0$ $\Leftrightarrow$ $x = 1$.

41. $|x - 7| = 4$ $\Leftrightarrow$ $x - 7 = \pm 4$ $\Leftrightarrow$ $x = 7 \pm 4$, so $x = 11$ or $x = 3$.

43. $|2x - 5| = 9$ is equivalent to $2x - 5 = \pm 9$ $\Leftrightarrow$ $2x = 5 \pm 9$ $\Leftrightarrow$ $x = \frac{5 \pm 9}{2}$. So $x = -2$ or $x = 7$.

45. Let x be the number of pounds of raisins. Then the number of pounds of nuts is $50 - x$.

	Raisins	Nuts	Mixture
Pounds	x	$50 - x$	50
Rate (cost per pound)	3.20	2.40	2.72

So $3.20x + 2.40(50 - x) = 2.72(50)$ $\Leftrightarrow$ $3.20x + 120 - 2.40x = 136$ $\Leftrightarrow$ $0.8x = 16$ $\Leftrightarrow$ $x = 20$. Thus the mixture uses 20 pounds of raisins and $50 - 20 = 30$ pounds of nuts.

47. Let r be the rate the woman runs in mi/h. Then she cycles at $r + 8$ mi/h.

	Rate	Time	Distance
Cycle	$r + 8$	$\dfrac{4}{r+8}$	4
Run	r	$\dfrac{2.5}{r}$	2.5

Since the total time of the workout is 1 hour, we have $\dfrac{4}{r+8} + \dfrac{2.5}{r} = 1$. Multiplying by $2r(r+8)$, we get
$4(2r) + 2.5(2)(r + 8) = 2r(r + 8)$ $\Leftrightarrow$ $8r + 5r + 40 = 2r^2 + 16r$ $\Leftrightarrow$ $0 = 2r^2 + 3r - 40$ $\Rightarrow$
$r = \dfrac{-3 \pm \sqrt{(3)^2 - 4(2)(-40)}}{2(2)} = \dfrac{-3 \pm \sqrt{9+320}}{4} = \dfrac{-3 \pm \sqrt{329}}{4}$. Since $r \geq 0$, we reject the negative value. She runs at $r = \dfrac{-3 + \sqrt{329}}{4} \approx 3.78$ mi/h.

49. Let x be the length of one side in cm. Then $28 - x$ is the length of the other side. Using the Pythagorean Theorem, we have $x^2 + (28 - x)^2 = 20^2$ $\Leftrightarrow$ $x^2 + 784 - 56x + x^2 = 400$ $\Leftrightarrow$ $2x^2 - 56x + 384 = 0$ $\Leftrightarrow$ $2(x^2 - 28x + 192) = 0$ $\Leftrightarrow$ $2(x - 12)(x - 16) = 0$. So $x = 12$ or $x = 16$. If $x = 12$, then the other side is $28 - 12 = 16$. Similarly, if $x = 16$, then the other side is 12. The sides are 12 cm and 16 cm.

51. Let w be width of the pool. Then the length of the pool is $2w$, and its volume is $8(w)(2w) = 8464$ $\Leftrightarrow$ $16w^2 = 8464$ $\Leftrightarrow$ $w^2 = 529$ $\Rightarrow$ $w = \pm 23$. Since $w > 0$, we reject the negative value. The pool is 23 feet wide, $2(23) = 46$ feet long, and 8 feet deep.

53. $(3 - 5i) - (6 + 4i) = 3 - 5i - 6 - 4i = -3 - 9i$

55. $(2 + 7i)(6 - i) = 12 - 2i + 42i - 7i^2 = 12 + 40i + 7 = 19 + 40i$

57. $\dfrac{2 - 3i}{2 + 3i} = \dfrac{2 - 3i}{2 + 3i} \cdot \dfrac{2 - 3i}{2 - 3i} = \dfrac{4 - 12i + 9i^2}{4 - 9i^2} = \dfrac{4 - 12i - 9}{4 + 9} = \dfrac{-5 - 12i}{13} = -\dfrac{5}{13} - \dfrac{12}{13}i$

59. $i^{45} = i^{44}i = \left(i^4\right)^{11} i = (1)^{11} i = i$

61. $\left(1 - \sqrt{-3}\right)\left(2 + \sqrt{-4}\right) = \left(1 - \sqrt{3}i\right)(2 + 2i) = 2 + 2i - 2\sqrt{3}i - 2\sqrt{3}i^2 = 2 + \left(2 - 2\sqrt{3}\right)i + 2\sqrt{3}$
$$= \left(2 + 2\sqrt{3}\right) + \left(2 - 2\sqrt{3}\right)i$$

63. $x^2 + 16 = 0$ $\Leftrightarrow$ $x^2 = -16$ $\Rightarrow$ $x = \pm\sqrt{-16} = \pm 4i$

65. $x^2 + 6x + 10 = 0$ $\Rightarrow$ $x = \dfrac{-6 \pm \sqrt{6^2 - 4(1)(10)}}{2(1)} = \dfrac{-6 \pm \sqrt{36 - 40}}{2} = \dfrac{-6 \pm \sqrt{-4}}{2} = \dfrac{-6 \pm 2i}{2} = -3 \pm i$

67. $x^4 - 256 = 0$ $\Leftrightarrow$ $\left(x^2 + 16\right)\left(x^2 - 16\right) = 0$. Thus either $x^2 + 16 = 0$ $\Leftrightarrow$ $x^2 = -16$ $\Rightarrow$ $x = \pm\sqrt{-16}$ $\Leftrightarrow$ $x = \pm 4i$; or $x^2 - 16 = 0$ $\Leftrightarrow$ $(x - 4)(x + 4) = 0$ $\Rightarrow$ $x - 4 = 0$ $\Leftrightarrow$ $x = 4$; or $x + 4 = 0$ $\Leftrightarrow$ $x = -4$. The solutions are $\pm 4i$ and ± 4.

69. $x^2 + 4x = (2x + 1)^2$ $\Leftrightarrow$ $x^2 + 4x = 4x^2 + 4x + 1$ $\Leftrightarrow$ $0 = 3x^2 + 1$ $\Leftrightarrow$ $3x^2 = -1$ $\Leftrightarrow$ $x^2 = -\dfrac{1}{3}$ $\Rightarrow$ $x = \pm\sqrt{-\dfrac{1}{3}} = \pm\dfrac{\sqrt{3}}{3}i$.

71. $3x - 2 > -11$ $\Leftrightarrow$ $3x > -9$ $\Leftrightarrow$ $x > -3$.
Interval: $(-3, \infty)$

Graph:

73. $-1 < 2x + 5 \le 3$ $\Leftrightarrow$ $-6 < 2x \le -2$ $\Leftrightarrow$ $-3 < x \le -1$
Interval: $(-3, -1]$

Graph:

75. $x^2 + 4x - 12 > 0$ $\Leftrightarrow$ $(x - 2)(x + 6) > 0$. The expression on the left of the inequality changes sign where $x = 2$ and where $x = -6$. Thus we must check the intervals in the following table.

Interval	$(-\infty, -6)$	$(-6, 2)$	$(2, \infty)$
Sign of $x - 2$	$-$	$-$	$+$
Sign of $x + 6$	$-$	$+$	$+$
Sign of $(x - 2)(x + 6)$	$+$	$-$	$+$

Interval: $(-\infty, -6) \cup (2, \infty)$.

Graph:

77. $\dfrac{2x + 5}{x + 1} \le 1$ $\Leftrightarrow$ $\dfrac{2x + 5}{x + 1} - 1 \le 0$ $\Leftrightarrow$ $\dfrac{2x + 5}{x + 1} - \dfrac{x + 1}{x + 1} \le 0$ $\Leftrightarrow$ $\dfrac{x + 4}{x + 1} \le 0$. The expression on the left of the inequality changes sign where $x = -1$ and where $x = -4$. Thus we must check the intervals in the following table.

Interval	$(-\infty, -4)$	$(-4, -1)$	$(-1, \infty)$
Sign of $x + 4$	$-$	$+$	$+$
Sign of $x + 1$	$-$	$-$	$+$
Sign of $\dfrac{x + 4}{x + 1}$	$+$	$-$	$+$

We exclude $x = -1$, since the expression is not defined at this value. Thus the solution is $[-4, -1)$.

Graph:

79. $\dfrac{x-4}{x^2-4} \le 0 \quad \Leftrightarrow \quad \dfrac{x-4}{(x-2)(x+2)} \le 0$. The expression on the left of the inequality changes sign where $x = -2$, where $x = 2$, and where $x = 4$. Thus we must check the intervals in the following table.

Interval	$(-\infty, -2)$	$(-2, 2)$	$(2, 4)$	$(4, \infty)$
Sign of $x - 4$	$-$	$-$	$-$	$+$
Sign of $x - 2$	$-$	$-$	$+$	$+$
Sign of $x + 2$	$-$	$+$	$+$	$+$
Sign of $\dfrac{x-4}{(x-2)(x+2)}$	$-$	$+$	$-$	$+$

Since the expression is not defined when $x = \pm 2$, we exclude these values and the solution is $(-\infty, -2) \cup (2, 4]$.

Graph:

81. $|x - 5| \le 3 \quad \Leftrightarrow \quad -3 \le x - 5 \le 3 \quad \Leftrightarrow \quad 2 \le x \le 8$. Interval: $[2, 8]$.

Graph:

83. $|2x + 1| \ge 1$ is equivalent to $2x + 1 \ge 1$ or $2x + 1 \le -1$.

Case 1: $2x + 1 \ge 1 \quad \Leftrightarrow \quad 2x \ge 0 \quad \Leftrightarrow \quad x \ge 0$

Case 2: $2x + 1 \le -1 \quad \Leftrightarrow \quad 2x \le -2 \quad \Leftrightarrow \quad x \le -1$

Interval: $(-\infty, -1] \cup [0, \infty)$.

Graph:

85. (a) For $\sqrt{24 - x - 3x^2}$ to define a real number, we must have $24 - x - 3x^2 \ge 0 \quad \Leftrightarrow \quad (8 - 3x)(3 + x) \ge 0$. The expression on the left of the inequality changes sign where $8 - 3x = 0 \quad \Leftrightarrow \quad -3x = -8 \quad \Leftrightarrow \quad x = \frac{8}{3}$; or where $x = -3$. Thus we must check the intervals in the following table.

Interval	$(-\infty, -3)$	$\left(-3, \frac{8}{3}\right)$	$\left(\frac{8}{3}, \infty\right)$
Sign of $8 - 3x$	$+$	$+$	$-$
Sign of $3 + x$	$-$	$+$	$+$
Sign of $(8 - 3x)(3 + x)$	$-$	$+$	$-$

Interval: $\left[-3, \frac{8}{3}\right]$.

Graph:

(b) For $\dfrac{1}{\sqrt[4]{x - x^4}}$ to define a real number we must have $x - x^4 > 0 \quad \Leftrightarrow \quad x\left(1 - x^3\right) > 0 \quad \Leftrightarrow \quad x(1 - x)\left(1 + x + x^2\right) > 0$. The expression on the left of the inequality changes sign where $x = 0$; or where $x = 1$; or where $1 + x + x^2 = 0 \quad \Rightarrow \quad x = \dfrac{-1 \pm \sqrt{1^2 - 4(1)(1)}}{2(1)} = \dfrac{1 \pm \sqrt{1 - 4}}{2}$ which is imaginary. We check the intervals in the following table.

Interval	$(-\infty, 0)$	$(0, 1)$	$(1, \infty)$
Sign of x	$-$	$+$	$+$
Sign of $1 - x$	$+$	$+$	$-$
Sign of $1 + x + x^2$	$+$	$+$	$+$
Sign of $x(1 - x)\left(1 + x + x^2\right)$	$-$	$+$	$-$

Interval: $(0, 1)$.

Graph:

Chapter 1 Test

1. **(a)** $3x - 7 = \frac{1}{2}x$ $\Leftrightarrow$ $-7 = -\frac{5}{2}x$ $\Leftrightarrow$ $x = \frac{14}{5}$

(b) $x^{2/3} - 9 = 0$ $\Leftrightarrow$ $x^{2/3} = 9$ $\Leftrightarrow$ $x^{1/3} = \pm 3$ $\Leftrightarrow$ $x = \pm 27$

(c) $\dfrac{2x}{2x-5} = \dfrac{x+2}{x-1}$ $\Leftrightarrow$ $2x(x-1) = (2x-5)(x+2)$ $\Leftrightarrow$ $2x^2 - 2x = 2x^2 - x - 10$ $\Leftrightarrow$ $-x = -10$ $\Leftrightarrow$
$x = 10$

2. Let d be the distance between Ajax and Bixby. Then we have the following table.

	Rate	Time	Distance
Ajax to Bixby	50	$\dfrac{d}{50}$	d
Bixby to Ajax	60	$\dfrac{d}{60}$	d

We use the fact that the total time is $4\frac{2}{5}$ hours to get the equation $\dfrac{d}{50} + \dfrac{d}{60} = \dfrac{22}{5}$ $\Leftrightarrow$ $6d + 5d = 1320$ $\Leftrightarrow$ $11d = 1320$
$\Leftrightarrow$ $d = 120$. Thus Ajax and Bixby are 120 miles apart.

3. **(a)** $(6 - 2i) - \left(7 - \frac{1}{2}i\right) = 6 - 2i - 7 + \frac{1}{2}i = -1 - \frac{3}{2}i$

(b) $(1 + i)(3 - 2i) = 3 - 2i + 3i - 2i^2 = 3 + i + 2 = 5 + i$

(c) $\dfrac{5 + 10i}{3 - 4i} = \dfrac{5 + 10i}{3 - 4i} \cdot \dfrac{3 + 4i}{3 + 4i} = \dfrac{15 + 20i + 30i + 40i^2}{9 - 16i^2} = \dfrac{15 + 50i - 40}{9 + 16} = \dfrac{-25 + 50i}{25} = -1 + 2i$

(d) $i^{50} = i^{48} \cdot i^2 = \left(i^4\right)^{12} \cdot i^2 = (1)^{12} \cdot (-1) = -1$

(e) $\left(2 - \sqrt{-2}\right)\left(\sqrt{8} + \sqrt{-4}\right) = \left(2 - \sqrt{2}i\right)\left(2\sqrt{2} + 2i\right) = 4\sqrt{2} + 4i - 4i - 2\sqrt{2}i^2 = 4\sqrt{2} + 2\sqrt{2} = 6\sqrt{2}$

4. **(a)** $x^2 - x - 12 = 0$ $\Leftrightarrow$ $(x - 4)(x + 3) = 0$. So $x = 4$ or $x = -3$.

(b) $2x^2 + 4x + 3 = 0$ $\Rightarrow$ $x = \dfrac{-4 \pm \sqrt{4^2 - 4(2)(3)}}{2(2)} = \dfrac{-4 \pm \sqrt{16 - 24}}{4} = \dfrac{-4 \pm \sqrt{-8}}{4} = \dfrac{-4 \pm 2\sqrt{2}i}{4} = -1 \pm \dfrac{\sqrt{2}}{2}i$

(c) $\sqrt{3 - \sqrt{x+5}} = 2$ $\Rightarrow$ $3 - \sqrt{x+5} = 4$ $\Leftrightarrow$ $-1 = \sqrt{x+5}$. (Note that this is impossible, so there can be no
solution.) Squaring both sides again, we get $1 = x + 5$ $\Leftrightarrow$ $x = -4$. But this does not satisfy the original equation,
so there is no solution. (You must always check your final answers if you have squared both sides when solving an
equation, since extraneous answers may be introduced, as here.)

(d) $x^{1/2} - 3x^{1/4} + 2 = 0$. Let $u = x^{1/4}$, then we have $u^2 - 3u + 2 = 0$ $\Leftrightarrow$ $(u - 2)(u - 1) = 0$. So
$u - 2 = x^{1/4} - 2 = 0$ $\Leftrightarrow$ $x^{1/4} = 2$ $\Rightarrow$ $x = 2^4 = 16$; or $u - 1 = x^{1/4} - 1 = 0$ $\Leftrightarrow$ $x^{1/4} = 1$ $\Rightarrow$
$x = 1^4 = 1$. Thus the solutions are $x = 16$ and $x = 1$.

(e) $x^4 - 16x^2 = 0$ $\Leftrightarrow$ $x^2\left(x^2 - 16\right) = 0$ $\Leftrightarrow$ $x^2(x - 4)(x + 4) = 0$. So $x = 0, -4$, or 4.

(f) $3|x - 4| - 10 = 0$ $\Leftrightarrow$ $3|x - 4| = 10$ $\Leftrightarrow$ $|x - 4| = \frac{10}{3}$ $\Leftrightarrow$ $x - 4 = \pm\frac{10}{3}$ $\Leftrightarrow$ $x = 4 \pm \frac{10}{3}$. So
$x = 4 - \frac{10}{3} = \frac{2}{3}$ or $x = 4 + \frac{10}{3} = \frac{22}{3}$. Thus the solutions are $x = \frac{2}{3}$ and $x = \frac{22}{3}$.

5. Let w be the width of the parcel of land. Then $w + 70$ is the length of the parcel of land. Then $w^2 + (w + 70)^2 = 130^2$
$\Leftrightarrow$ $w^2 + w^2 + 140w + 4900 = 16{,}900$ $\Leftrightarrow$ $2w^2 + 140w - 12{,}000 = 0$ $\Leftrightarrow$ $w^2 + 70w - 6000 = 0$
$\Leftrightarrow$ $(w - 50)(w + 120) = 0$. So $w = 50$ or $w = -120$. Since $w \geq 0$, the width is $w = 50$ ft and the length is
$w + 70 = 120$ ft.

6. (a) $-1 \le 5 - 2x < 10 \quad \Leftrightarrow \quad -6 \le -2x < 5 \quad \Leftrightarrow \quad 3 \ge x > -\frac{5}{2}$ or $-\frac{5}{2} < x \le 3$.

Interval: $\left(-\frac{5}{2}, 3\right]$. Graph:

(b) $x(x-1)(x-2) > 0$. The expression on the left of the inequality changes sign when $x = 0$, $x = 1$, and $x = 2$. Thus we must check the intervals in the following table.

Interval	$(-\infty, 0)$	$(0, 1)$	$(1, 2)$	$(2, \infty)$
Sign of x	$-$	$+$	$+$	$+$
Sign of $x - 1$	$-$	$-$	$+$	$+$
Sign of $x - 2$	$-$	$-$	$-$	$+$
Sign of $x(x-1)(x-2)$	$-$	$+$	$-$	$+$

From the table, the solution set is $\{x \mid 0 < x < 1 \text{ or } 2 < x\}$.

Interval: $(0, 1) \cup (2, \infty)$. Graph:

(c) $|x - 3| < 2$ is equivalent to $-2 < x - 3 < 2 \quad \Leftrightarrow \quad 1 < x < 5$.

Interval: $(1, 5)$. Graph:

(d) $\dfrac{2x+5}{x+1} \le 1 \quad \Leftrightarrow \quad \dfrac{2x+5}{x+1} - 1 \le 0 \quad \Leftrightarrow \quad \dfrac{2x+5}{x+1} - \dfrac{x+1}{x+1} \le 0 \quad \Leftrightarrow \quad \dfrac{x+4}{x+1} \le 0$. The expression on the left of the inequality changes sign where $x = -4$ and where $x = -1$. Thus we must check the intervals in the following table.

Interval	$(-\infty, -4)$	$(-4, -1)$	$(-1, \infty)$
Sign of $x + 4$	$-$	$+$	$+$
Sign of $x + 1$	$-$	$-$	$+$
Sign of $\dfrac{x+4}{x+1}$	$+$	$-$	$+$

Since $x = -1$ makes the expression in the inequality undefined, we exclude this value.

Interval: $[-4, -1)$. Graph:

7. $5 \le \frac{5}{9}(F - 32) \le 10 \quad \Leftrightarrow \quad 9 \le F - 32 \le 18 \quad \Leftrightarrow \quad 41 \le F \le 50$. Thus the medicine is to be stored at a temperature between $41°$ F and $50°$ F.

8. For $\sqrt{4x - x^2}$ to be defined as a real number $4x - x^2 \ge 0 \quad \Leftrightarrow \quad x(4 - x) \ge 0$. The expression on the left of the inequality changes sign when $x = 0$ and $x = 4$. Thus we must check the intervals in the following table.

Interval	$(-\infty, 0)$	$(0, 4)$	$(4, \infty)$
Sign of x	$-$	$+$	$+$
Sign of $4 - x$	$+$	$+$	$-$
Sign of $x(4 - x)$	$-$	$+$	$-$

From the table, we see that $\sqrt{4x - x^2}$ is defined when $0 \le x \le 4$.

Focus on Modeling: Making the Best Decisions

1. (a) The total cost is $\left(\begin{array}{c}\text{cost of}\\\text{copier}\end{array}\right) + \left(\begin{array}{c}\text{maintenance}\\\text{cost}\end{array}\right)\left(\begin{array}{c}\text{number}\\\text{of months}\end{array}\right) + \left(\begin{array}{c}\text{copy}\\\text{cost}\end{array}\right)\left(\begin{array}{c}\text{number}\\\text{of months}\end{array}\right)$. Each

month the copy cost is $8000 \cdot 0.03 = 240$. Thus we get $C_1(n) = 5800 + 25n + 240n = 5800 + 265n$.

(b) In this case the cost is $\left(\begin{array}{c}\text{rental}\\\text{cost}\end{array}\right)\left(\begin{array}{c}\text{number}\\\text{of months}\end{array}\right) + \left(\begin{array}{c}\text{copy}\\\text{cost}\end{array}\right)\left(\begin{array}{c}\text{number}\\\text{of months}\end{array}\right)$. Each month the copy cost is

$8000 \cdot 0.06 = 480$. Thus we get $C_2(n) = 95n + 480n = 575n$.

(c)

Years	n	Purchase	Rental
1	12	8,980	6,900
2	24	12,160	13,800
3	36	15,340	20,700
4	48	18,520	27,600
5	60	21,700	34,500
6	72	24,880	41,400

(d) The cost is the same when $C_1(n) = C_2(n)$. So $5800 + 265n = 575n \iff 5800 = 310n \iff n \approx 18.71$ months.

3. (a) The total cost is $\left(\begin{array}{c}\text{setup}\\\text{cost}\end{array}\right) + \left(\begin{array}{c}\text{cost per}\\\text{tire}\end{array}\right)\left(\begin{array}{c}\text{number}\\\text{of tires}\end{array}\right)$. So $C(x) = 8000 + 22x$.

(b) The revenue is $\left(\begin{array}{c}\text{price per}\\\text{tire}\end{array}\right)\left(\begin{array}{c}\text{number}\\\text{of tires}\end{array}\right)$. So $R(x) = 49x$.

(c) Profit = Revenue − Cost. So $P(x) = R(x) - C(x) = 49x - (8000 + 22x) = 27x - 8000$.

(d) Break even is when profit is zero. Thus $27x - 8000 = 0 \iff 27x = 8000 \iff x \approx 296.3$. So they need to sell at least 297 tires to break even.

5. (a) Design 1 is a square and the perimeter of a square is four times the length of a side. $24 = 4x$, so each side is $x = 6$ feet long. Thus the area is $6^2 = 36 \text{ ft}^2$.

Design 2 is a circle with perimeter $2\pi r$ and area πr^2. Thus we must solve $2\pi r = 24 \iff r = \dfrac{12}{\pi}$. Thus, the area is

$\pi\left(\dfrac{12}{\pi}\right)^2 = \dfrac{144}{\pi} \approx 45.8 \text{ ft}^2$. Design 2 gives the largest area.

(b) In Design 1, the cost is \$3 times the perimeter p, so $120 = 3p$ and the perimeter is 40 feet. By part (a), each side is then $\frac{40}{4} = 10$ feet long. So the area is $10^2 = 100 \text{ ft}^2$.

In Design 2, the cost is \$4 times the perimeter p. Because the perimeter is $2\pi r$, we get $120 = 4(2\pi r)$ so

$r = \dfrac{120}{8\pi} = \dfrac{15}{\pi}$. The area is $\pi r^2 = \pi\left(\dfrac{15}{\pi}\right)^2 = \dfrac{225}{\pi} \approx 71.6 \text{ ft}^2$. Design 1 gives the largest area.

7. (a)

Minutes	Plan A	Plan B	Plan C
500	$39 + 100\,(0.20) = \$59$	$49	$69
600	$39 + 200\,(0.20) = \$79$	$49	$69
700	$39 + 300\,(0.20) = \$99$	$49 + 100\,(0.15) = \$64$	$69
800	$39 + 400\,(0.20) = \$119$	$49 + 200\,(0.15) = \$79$	$69
900	$39 + 500\,(0.20) = \$139$	$49 + 300\,(0.15) = \$94$	$69
1000	$39 + 600\,(0.20) = \$159$	$49 + 400\,(0.15) = \$109$	$69
1100	$39 + 700\,(0.20) = \$179$	$49 + 500\,(0.15) = \$124$	$69 + 100\,(0.15) = \$84$

(b) If Jane uses either 500 or 700 minutes, Plan B is the cheapest, but if she uses 1100 minutes, Plan C is the cheapest.

2 Coordinates and Graphs

2.1 The Coordinate Plane

1.

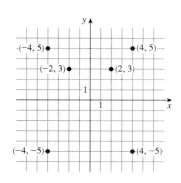

3. The two points are $(0, 2)$ and $(3, 0)$.

(a) $d = \sqrt{(3 - 0)^2 + (0 - (-2))^2}$
$= \sqrt{3^2 + 2^2} = \sqrt{9 + 4} = \sqrt{13}$

(b) midpoint: $\left(\dfrac{3 + 0}{2}, \dfrac{0 + 2}{2}\right) = \left(\dfrac{3}{2}, 1\right)$

5. The two points are $(-3, 3)$ and $(5, -3)$.

(a) $d = \sqrt{(-3 - 5)^2 + (3 - (-3))^2} = \sqrt{(-8)^2 + 6^2}$
$= \sqrt{64 + 36} = \sqrt{100} = 10$

(b) midpoint: $\left(\dfrac{-3 + 5}{2}, \dfrac{3 + (-3)}{2}\right) = (1, 0)$

7. (a)

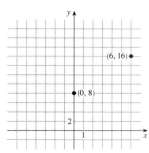

(b) $d = \sqrt{(0 - 6)^2 + (8 - 16)^2}$
$= \sqrt{(-6)^2 + (-8)^2} = \sqrt{100} = 10$

(c) Midpoint: $\left(\dfrac{0+6}{2}, \dfrac{8+16}{2}\right) = (3, 12)$

9. (a)

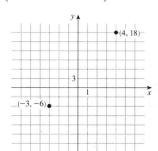

(b) $d = \sqrt{(-3 - 4)^2 + (-6 - 18)^2}$
$= \sqrt{(-7)^2 + (-24)^2} = \sqrt{49 + 576} = \sqrt{625} = 25$

(c) Midpoint: $\left(\dfrac{-3+4}{2}, \dfrac{-6+18}{2}\right) = \left(\dfrac{1}{2}, 6\right)$

11. (a)

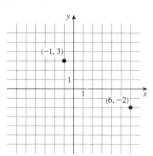

(b) $d = \sqrt{(6 - (-1))^2 + (-2 - 3)^2}$
$= \sqrt{7^2 + (-5)^2} = \sqrt{49 + 25} = \sqrt{74}$

(c) Midpoint: $\left(\dfrac{6-1}{2}, \dfrac{-2+3}{2}\right) = \left(\dfrac{5}{2}, \dfrac{1}{2}\right)$

13. (a)

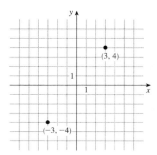

(b) $d = \sqrt{(3 - (-3))^2 + (4 - (-4))^2}$
$= \sqrt{6^2 + 8^2} = \sqrt{36 + 64} = \sqrt{100} = 10$

(c) Midpoint: $\left(\dfrac{3+(-3)}{2}, \dfrac{4+(-4)}{2}\right) = (0, 0)$

15. $(1,3)(5,3)$ $\frac{y_2 - x_1}{}$ $d(A, B) = \sqrt{(1-5)^2 + (3-3)^2} = \sqrt{(-4)^2} = 4.$

$d(A, C) = \sqrt{(1-1)^2 + (3-(-3))^2} = \sqrt{(6)^2} = 6.$ So

the area is $4 \cdot 6 = 24$.

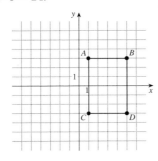

17. From the graph, the quadrilateral $ABCD$ has a pair of parallel sides, so $ABCD$ is a trapezoid. The area is $\frac{1}{2}(b_1 + b_2) h$. From the graph we see that

$b_1 = d(A, B) = \sqrt{(1-5)^2 + (0-0)^2} = \sqrt{4^2} = 4;$

$b_2 = d(C, D) = \sqrt{(4-2)^2 + (3-3)^2} = \sqrt{2^2} = 2;$

and h is the difference in y-coordinates is $|3 - 0| = 3$.

Thus the area of the trapezoid is $\frac{1}{2}(4 + 2) 3 = 9$.

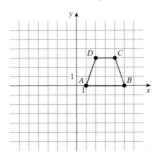

19. $\{(x, y) \mid x \leq 0\}$

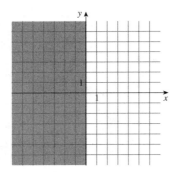

21. $\{(x, y) \mid x = 3\}$

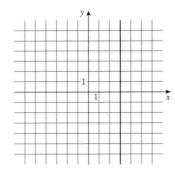

23. $\{(x, y) \mid 1 < x < 2\}$

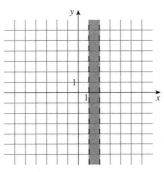

25. $\{(x, y) \mid xy < 0\}$

$= \{(x, y) \mid x < 0 \text{ and } y > 0 \text{ or } x > 0 \text{ and } y < 0\}$

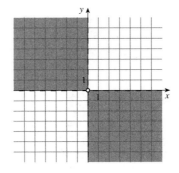

27. $\{(x, y) \mid x \geq 1 \text{ and } y < 3\}$

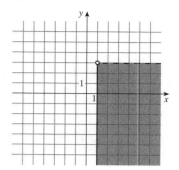

29. $\{(x, y) \mid |x| > 4\}$

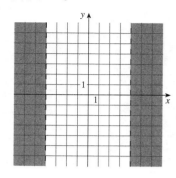

31. $\{(x, y) \mid |x| \leq 2 \text{ and } |y| \leq 3\}$

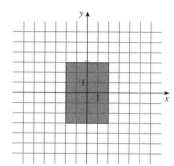

33. $d(0, A) = \sqrt{(6 - 0)^2 + (7 - 0)^2}$

$$= \sqrt{6^2 + 7^2} = \sqrt{36 + 49} = \sqrt{85}$$

$d(0, B) = \sqrt{(-5 - 0)^2 + (8 - 0)^2}$

$$= \sqrt{(-5)^2 + 8^2} = \sqrt{25 + 64} = \sqrt{89}$$

Thus point $A(6, 7)$ is closer to the origin.

35. $d(P, R) = \sqrt{(-1 - 3)^2 + (-1 - 1)^2} = \sqrt{(-4)^2 + (-2)^2} = \sqrt{16 + 4} = \sqrt{20} = 2\sqrt{5}$.

$d(Q, R) = \sqrt{(-1 - (-1))^2 + (-1 - 3)^2} = \sqrt{0 + (-4)^2} = \sqrt{16} = 4$. Thus point $Q(-1, 3)$ is closer to point R.

37. Since we do not know which pair are isosceles, we find the length of all three sides.

$d(A, B) = \sqrt{(-3 - 0)^2 + (-1 - 2)^2} = \sqrt{(-3)^2 + (-3)^2} = \sqrt{9 + 9} = \sqrt{18} = 3\sqrt{2}$.

$d(C, B) = \sqrt{(-3 - (-4))^2 + (-1 - 3)^2} = \sqrt{1^2 + (-4)^2} = \sqrt{1 + 16} = \sqrt{17}$.

$d(A, C) = \sqrt{(0 - (-4))^2 + (2 - 3)^2} = \sqrt{4^2 + (-1)^2} = \sqrt{16 + 1} = \sqrt{17}$. So sides AC and CB have the same length.

39. (a) Here we have $A = (2, 2)$, $B = (3, -1)$, and $C = (-3, -3)$. So

$$d(A, B) = \sqrt{(3 - 2)^2 + (-1 - 2)^2} = \sqrt{1^2 + (-3)^2} = \sqrt{1 + 9} = \sqrt{10};$$

$$d(C, B) = \sqrt{(3 - (-3))^2 + (-1 - (-3))^2} = \sqrt{6^2 + 2^2} = \sqrt{36 + 4} = \sqrt{40} = 2\sqrt{10};$$

$$d(A, C) = \sqrt{(-3 - 2)^2 + (-3 - 2)^2} = \sqrt{(-5)^2 + (-5)^2} = \sqrt{25 + 25} = \sqrt{50} = 5\sqrt{2}.$$

Since $[d(A, B)]^2 + [d(C, B)]^2 = [d(A, C)]^2$, we conclude that the triangle is a right triangle.

(b) The area of the triangle is $\frac{1}{2} \cdot d(C, B) \cdot d(A, B) = \frac{1}{2} \cdot \sqrt{10} \cdot 2\sqrt{10} = 10$.

41. We show that all sides are the same length (its a rhombus) and then show that the diagonals are equal. Here we have $A = (-2, 9)$, $B = (4, 6)$, $C = (1, 0)$, and $D = (-5, 3)$. So

$$d(A, B) = \sqrt{(4 - (-2))^2 + (6 - 9)^2} = \sqrt{6^2 + (-3)^2} = \sqrt{36 + 9} = \sqrt{45};$$

$$d(B, C) = \sqrt{(1 - 4)^2 + (0 - 6)^2} = \sqrt{(-3)^2 + (-6)^2} = \sqrt{9 + 36} = \sqrt{45};$$

$$d(C, D) = \sqrt{(-5 - 1)^2 + (3 - 0)^2} = \sqrt{(-6)^2 + (-3)^2} = \sqrt{36 + 9} = \sqrt{45};$$

$$d(D, A) = \sqrt{(-2 - (-5))^2 + (9 - 3)^2} = \sqrt{3^2 + 6^2} = \sqrt{9 + 36} = \sqrt{45}.$$ So the points form a rhombus. Also $d(A, C) = \sqrt{(1 - (-2))^2 + (0 - 9)^2} = \sqrt{3^2 + (-9)^2} = \sqrt{9 + 81} = \sqrt{90} = 3\sqrt{10}$,

and $d(B, D) = \sqrt{(-5 - 4)^2 + (3 - 6)^2} = \sqrt{(-9)^2 + (-3)^2} = \sqrt{81 + 9} = \sqrt{90} = 3\sqrt{10}$. Since the diagonals are equal, the rhombus is a square.

43. Let $P = (0, y)$ be such a point. Setting the distances equal we get

$$\sqrt{(0 - 5)^2 + (y - (-5))^2} = \sqrt{(0 - 1)^2 + (y - 1)^2} \quad \Leftrightarrow$$

$$\sqrt{25 + y^2 + 10y + 25} = \sqrt{1 + y^2 - 2y + 1} \quad \Rightarrow \quad y^2 + 10y + 50 = y^2 - 2y + 2 \quad \Leftrightarrow \quad 12y = -48 \quad \Leftrightarrow$$

$y = -4$. Thus, the point is $P = (0, -4)$. Check:

$$\sqrt{(0 - 5)^2 + (-4 - (-5))^2} = \sqrt{(-5)^2 + 1^2} = \sqrt{25 + 1} = \sqrt{26};$$

$$\sqrt{(0 - 1)^2 + (-4 - 1)^2} = \sqrt{(-1)^2 + (-5)^2} = \sqrt{25 + 1} = \sqrt{26}.$$

45. We find the midpoint M of PQ, and then the midpoint of PM. Now $M = \left(\frac{-1 + 7}{2}, \frac{3 + 5}{2}\right) = (3, 4)$, and the midpoint of PM is thus $\left(\frac{-1 + 3}{2}, \frac{3 + 4}{2}\right) = \left(1, \frac{7}{2}\right)$.

47. As indicated by Example 5, we must find a point $S(x_1, y_1)$ such that the midpoints of PR

and QS are the same. Thus $\left(\frac{4 + (-1)}{2}, \frac{2 + (-4)}{2}\right) = \left(\frac{x_1 + 1}{2}, \frac{y_1 + 1}{2}\right)$. Setting the

x-coordinates equal, we get $\frac{4 + (-1)}{2} = \frac{x_1 + 1}{2} \quad \Leftrightarrow \quad 4 - 1 = x_1 + 1 \quad \Leftrightarrow \quad x_1 = 2$.

Setting the y-coordinates equal, we get $\frac{2 + (-4)}{2} = \frac{y_1 + 1}{2} \quad \Leftrightarrow \quad 2 - 4 = y_1 + 1 \quad \Leftrightarrow$

$y_1 = -3$. Thus $S = (2, -3)$.

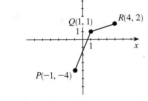

49. (a)

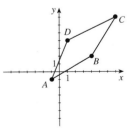

(b) The midpoint of AC is $\left(\frac{-2 + 7}{2}, \frac{-1 + 7}{2}\right) = \left(\frac{5}{2}, 3\right)$, the

midpoint of BD is $\left(\frac{4 + 1}{2}, \frac{2 + 4}{2}\right) = \left(\frac{5}{2}, 3\right)$.

(c) Since the they have the same midpoint, we conclude that the diagonals bisect each other.

51. (a) $d(A, B) = \sqrt{3^2 + 4^2} = \sqrt{25} = 5$.

(b) We want the distances from $C = (4, 2)$ to $D = (11, 26)$. The walking distance is $|4 - 11| + |2 - 26| = 7 + 24 = 31$ blocks. Straight-line distance is

$$\sqrt{(4 - 11)^2 + (2 - 26)^2} = \sqrt{7^2 + 24^2} = \sqrt{625} = 25 \text{ blocks}.$$

(c) The two points are on the same avenue or the same street.

53. The midpoint of the line segment is $(66, 45)$. The pressure experienced by an ocean diver at a depth of 66 feet is 45 lb/in^2.

55. (a) The point $(3, 7)$ is reflected to the point $(-3, 7)$.

(b) The point (a, b) is reflected to the point $(-a, b)$.

(c) Since the point $(-a, b)$ is the reflection of (a, b), the point $(-4, -1)$ is the reflection of $(4, -1)$.

(d) $A = (3, 3)$, so $A' = (-3, 3)$; $B = (6, 1)$, so $B' = (-6, 1)$; and $C = (1, -4)$, so $C' = (-1, -4)$.

57. We need to find a point $S(x_1, y_1)$ such that $PQRS$ is a parallelogram. As indicated by Example 3, this will be the case if the diagonals PR and QS bisect each other. So the midpoints of PR and QS are the same. Thus

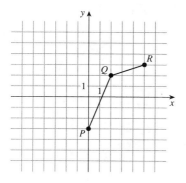

$$\left(\frac{0+5}{2}, \frac{-3+3}{2}\right) = \left(\frac{x_1+2}{2}, \frac{y_1+2}{2}\right).$$ Setting the x-coordinates equal, we

get $\dfrac{0+5}{2} = \dfrac{x_1+2}{2} \quad \Leftrightarrow \quad 0+5 = x_1+2 \quad \Leftrightarrow \quad x_1 = 3$.

Setting the y-coordinates equal, we get $\dfrac{-3+3}{2} = \dfrac{y_1+2}{2} \quad \Leftrightarrow$

$-3+3 = y_1+2 \quad \Leftrightarrow \quad y_1 = -2$. Thus $S = (3, -2)$.

2.2 **Graphs of Equations in Two Variables**

1. $(0, 2)$: $2 \overset{?}{=} 3(0) - 2 \quad \Leftrightarrow \quad 2 \overset{?}{=} -2$. No. $\left(\frac{1}{3}, 1\right)$: $1 \overset{?}{=} 3\left(\frac{1}{3}\right) - 2 \quad \Leftrightarrow \quad 1 \overset{?}{=} 1 - 2$. No.

$(1, 1)$: $1 \overset{?}{=} 3(1) - 2 \quad \Leftrightarrow \quad 1 \overset{?}{=} 3 - 2$. Yes. So $(1, 1)$ is on the graph of this equation.

3. $(0, 0)$: $0 - 2(0) - 1 \overset{?}{=} 0 \quad \Leftrightarrow \quad -1 \overset{?}{=} 0$. No. $(1, 0)$: $1 - 2(0) - 1 \overset{?}{=} 0 \quad \Leftrightarrow \quad -1 + 1 \overset{?}{=} 0$. Yes.

$(-1, -1)$: $(-1) - 2(-1) - 1 \overset{?}{=} 0 \quad \Leftrightarrow \quad -1 + 2 - 1 \overset{?}{=} 0$. Yes.

So $(1, 0)$ and $(-1, -1)$ are points on the graph of this equation.

5. $(0, -2)$: $(0)^2 + (0)(-2) + (-2)^2 \overset{?}{=} 4 \quad \Leftrightarrow \quad 0 + 0 + 4 \overset{?}{=} 4$. Yes.

$(1, -2)$: $(1)^2 + (1)(-2) + (-2)^2 \overset{?}{=} 4 \quad \Leftrightarrow \quad 1 - 2 + 4 \overset{?}{=} 4$. No.

$(2, -2)$: $(2)^2 + (2)(-2) + (-2)^2 \overset{?}{=} 4 \quad \Leftrightarrow \quad 4 - 4 + 4 \overset{?}{=} 4$. Yes.

So $(0, -2)$ and $(2, -2)$ are points on the graph of this equation.

7. To find x-intercepts, set $y = 0$. This gives $0 = 4x - x^2 \quad \Leftrightarrow \quad 0 = x(4 - x) \quad \Leftrightarrow \quad 0 = x$ or $x = 4$, so the x-intercept are 0 and 4. To find y-intercepts, set $x = 0$. This gives $y = 4(0) - 0^2 \quad \Leftrightarrow \quad y = 0$, so the y-intercept is 0.

9. To find x-intercepts, set $y = 0$. This gives $x^4 + 0^2 - x(0) = 16 \quad \Leftrightarrow \quad x^4 = 16 \quad \Leftrightarrow \quad x = \pm 2$. So the x-intercept are -2 and 2.

To find y-intercepts, set $x = 0$. This gives $0^4 + y^2 - (0)y = 16 \quad \Leftrightarrow \quad y^2 = 16 \quad \Leftrightarrow \quad y = \pm 4$. So the y-intercept are -4 and 4.

11. To find x-intercepts, set $y = 0$. This gives $0 = x - 3 \quad \Leftrightarrow \quad x = 3$, so the x-intercept is 3.

To find y-intercepts, set $x = 0$. This gives $y = 0 - 3 \quad \Leftrightarrow \quad y = -3$, so the y-intercept is -3.

13. To find x-intercepts, set $y = 0$. This gives $0 = x^2 - 9 \quad \Leftrightarrow \quad x^2 = 9 \quad \Rightarrow \quad x = \pm 3$, so the x-intercept are ± 3.

To find y-intercepts, set $x = 0$. This gives $y = (0)^2 - 9 \quad \Leftrightarrow \quad y = -9$, so the y-intercept is -9.

15. To find x-intercepts, set $y = 0$. This gives $x^2 + (0)^2 = 4 \quad \Leftrightarrow \quad x^2 = 4 \quad \Rightarrow \quad x = \pm 2$, so the x-intercept are ± 2.

To find y-intercepts, set $x = 0$. This gives $(0)^2 + y^2 = 4 \quad \Leftrightarrow \quad y^2 = 4 \quad \Rightarrow \quad y = \pm 2$, so the y-intercept are ± 2.

17. To find x-intercepts, set $y = 0$. This gives $x(0) = 5 \quad \Leftrightarrow \quad 0 = 5$, which is impossible, so there is no x-intercept.

To find y-intercepts, set $x = 0$. This gives $(0)y = 5 \quad \Leftrightarrow \quad 0 = 5$, which is again impossible, so there is no y-intercept.

19. $y = -x$

x	y
-4	4
-2	2
0	0
1	-1
2	-2
3	-3
4	-4

When $y = 0$, $x = 0$. So the x-intercept is 0, and the y-intercept is also 0.

x-axis symmetry: $(-y) = -x$ $\Leftrightarrow$ $y = x$, which is not the same as $y = -x$, so the graph is not symmetric with respect to the x-axis.

y-axis symmetry: $y = -(-x)$ $\Leftrightarrow$ $y = x$, which is not the same as $y = -x$, so the graph is not symmetric with respect to the y-axis.

Origin symmetry: $(-y) = -(-x)$ is the same as $y = -x$, so the graph is symmetric with respect to the origin.

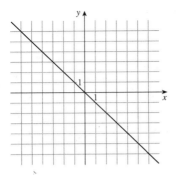

21. $y = -x + 4$

x	y
-4	8
-2	6
0	4
1	3
2	2
3	1
4	0

When $y = 0$ we get $x = 4$. So the x-intercept is 4, and $x = 0$ $\Rightarrow$ $y = 4$, so the y-intercept is 4.

x-axis symmetry: $(-y) = -x + 4$ $\Leftrightarrow$ $y = x - 4$, which is not the same as $y = -x + 4$, so the graph is not symmetric with respect to the x-axis.

y-axis symmetry: $y = -(-x) + 4$ $\Leftrightarrow$ $y = x + 4$, which is not the same as $y = -x + 4$, so the graph is not symmetric with respect to the y-axis.

Origin symmetry: $(-y) = -(-x) + 4$ $\Leftrightarrow$ $y = -x - 4$, which is not the same as $y = -x + 4$, so the graph is not symmetric with respect to the origin.

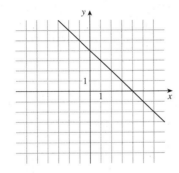

23. $2x - y = 6$

x	y
-1	-8
0	-6
1	-4
2	-2
3	0
4	2
5	4

When $y = 0$ we get $2x = 6$ So the x-intercept is 3. When $x = 0$ we get $-y = 6$ so the y-intercept is -6.

x-axis symmetry: $2x - (-y) = 6$ $\Leftrightarrow$ $2x + y = 6$, which is not the same, so the graph is not symmetric with respect to the x-axis.

y-axis symmetry: $2(-x) - y = 6$ $\Leftrightarrow$ $2x + y = -6$, so the graph is not symmetric with respect to the y-axis.

Origin symmetry: $2(-x) - (-y) = 6$ $\Leftrightarrow$ $-2x + y = 6$, which not he same, so the graph is not symmetric with respect to the origin.

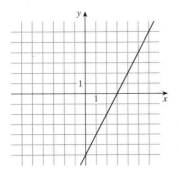

25. $y = 1 - x^2$

x	y
-3	-8
-2	-3
-1	0
0	1
1	0
2	-3
3	-8

$y = 0 \Rightarrow 0 = 1 - x^2 \Leftrightarrow x^2 = 1 \Rightarrow x = \pm 1$, so the x-intercepts are 1 and -1, and $x = 0 \Rightarrow y = 1 - (0)^2 = 1$, so the y-intercept is 1.

x-axis symmetry: $(-y) = 1 - x^2 \Leftrightarrow -y = 1 - x^2$, which is not the same as $y = 1 - x^2$, so the graph is not symmetric with respect to the x-axis.

y-axis symmetry: $y = 1 - (-x)^2 \Leftrightarrow y = 1 - x^2$, so the graph is symmetric with respect to the y-axis.

Origin symmetry: $(-y) = 1 - (-x)^2 \Leftrightarrow -y = 1 - x^2$ which is not the same as $y = 1 - x^2$. The graph is not symmetric with respect to the origin.

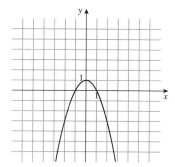

27. $4y = x^2 \Leftrightarrow y = \frac{1}{4}x^2$

x	y
-6	9
-4	4
-2	1
0	0
2	1
4	4
3	2

$y = 0 \Rightarrow 0 = \frac{1}{4}x^2 \Leftrightarrow x^2 = 0 \Rightarrow x = 0$, so the x-intercept is 0, and $x = 0 \Rightarrow y = \frac{1}{4}(0)^2 = 0$, so the y-intercept is 0.

x-axis symmetry: $(-y) = \frac{1}{4}x^2$, which is not the same as $y = \frac{1}{4}x^2$, so the graph is not symmetric with respect to the x-axis.

y-axis symmetry: $y = \frac{1}{4}(-x)^2 \Leftrightarrow y = \frac{1}{4}x^2$, so the graph is symmetric with respect to the y-axis.

Origin symmetry: $(-y) = \frac{1}{4}(-x)^2 \Leftrightarrow -y = \frac{1}{4}x^2$, which is not the same as $y = \frac{1}{4}x^2$, so the graph is not symmetric with respect to the origin.

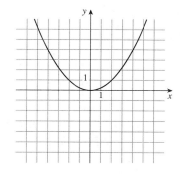

29. $y = x^2 - 9$

x	y
-4	7
-3	0
-2	-5
-1	-8
0	-9
1	-8
2	-5
3	0
4	7

$y = 0 \Rightarrow 0 = x^2 - 9 \Leftrightarrow x^2 = 9 \Rightarrow x = \pm 3$, so the x-intercepts are 3 and -3, and $x = 0 \Rightarrow y = (0)^2 - 9 = -9$, so the y-intercept is -9.

x-axis symmetry: $(-y) = x^2 - 9$, which is not the same as $y = x^2 - 9$, so the graph is not symmetric with respect to the x-axis.

y-axis symmetry: $y = (-x)^2 - 9 \Leftrightarrow y = x^2 - 9$, so the graph is symmetric with respect to the y-axis.

Origin symmetry: $(-y) = (-x)^2 - 9 \Leftrightarrow -y = x^2 - 9$, which is not the same as $y = x^2 - 9$, so the graph is not symmetric with respect to the origin.

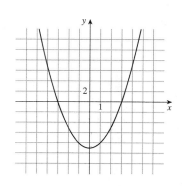

31. $xy = 2$ $\Leftrightarrow$ $y = \dfrac{2}{x}$

x	y
-4	$-\frac{1}{2}$
-2	-1
-1	-2
$-\frac{1}{2}$	-4
$-\frac{1}{4}$	-8
$\frac{1}{4}$	8
$\frac{1}{2}$	4
1	2
2	1
4	$\frac{1}{2}$

$y = 0$ or $x = 0$ $\Rightarrow$ $0 = 2$, which is impossible, so this equation has no x-intercept and no y-intercept.

x-axis symmetry: $x(-y) = 2$ $\Leftrightarrow$ $-xy = 2$, which is not the same as $xy = 2$, so the graph is not symmetric with respect to the x-axis.

y-axis symmetry: $(-x)y = 2$ $\Leftrightarrow$ $-xy = 2$, which is not the same as $xy = 2$, so the graph is not symmetric with respect to the y-axis.

Origin symmetry: $(-x)(-y) = 2$ $\Leftrightarrow$ $xy = 2$, so the graph is symmetric with respect to the origin.

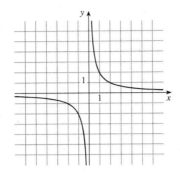

33. $y = \sqrt{x}$

x	y
0	0
$\frac{1}{4}$	$\frac{1}{2}$
1	1
2	$\sqrt{2}$
4	2
9	3
16	4

$y = 0$ $\Rightarrow$ $0 = \sqrt{x}$ $\Leftrightarrow$ $x = 0$. so the x-intercept is 0, and $x = 0$ $\Rightarrow$ $y = \sqrt{0} = 0$, so the y-intercept is 0. Since we are graphing real numbers and $\sqrt{x}$ is defined to be a nonnegative number, the equation is not symmetric with respect to the x-axis nor with respect to the y-axis. Also, the equation is not symmetric with respect to the origin.

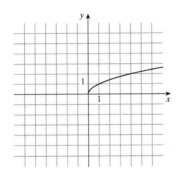

35. $y = \sqrt{4 - x^2}$. Since the radicand (the inside of the square root) cannot be negative, we must have $4 - x^2 \geq 0$ $\Leftrightarrow$ $x^2 \leq 4$ $\Leftrightarrow$ $|x| \leq 2$.

x	y
-2	0
-1	$\sqrt{3}$
0	4
1	$\sqrt{3}$
2	0

$y = 0$ $\Rightarrow$ $0 = \sqrt{4 - x^2}$ $\Leftrightarrow$ $4 - x^2 = 0$ $\Leftrightarrow$ $x^2 = 4$ $\Rightarrow$ $x = \pm 2$, so the x-intercept are -2 and 2, and $x = 0$ $\Rightarrow$ $y = \sqrt{4 - (0)^2} = \sqrt{4} = 2$, so the y-intercept is 2. Since $y \geq 0$, the graph is not symmetric with respect to the x-axis.

y-axis symmetry: $y = \sqrt{4 - (-x)^2} = \sqrt{4 - x^2}$, so the graph is symmetric with respect to the y-axis. Also, since $y \geq 0$ the graph is not symmetric with respect to the origin.

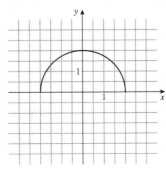

37. $y = |x|$

x	y
-3	3
-2	2
-1	1
0	0
1	1
2	2
3	3

$y = 0 \Rightarrow 0 = |x| \Leftrightarrow x = 0$, so the x-intercept is 0, and $x = 0$ $\Rightarrow y = |0| = 0$, so the y-intercept is 0.

Since $y \geq 0$, the graph is not symmetric with respect to the x-axis.

y-axis symmetry: $y = |-x| = |x|$, so the graph is symmetric with respect to the y-axis.

Since $y \geq 0$, the graph is not symmetric with respect to the origin.

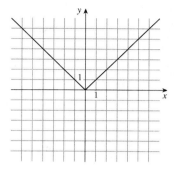

39. $y = 4 - |x|$

x	y
-6	-2
-4	0
-2	2
0	4
2	2
4	0
6	-2

$y = 0 \Rightarrow 0 = 4 - |x| \Leftrightarrow |x| = 4 \Rightarrow x = \pm 4$, so the x-intercepts are -4 and 4, and $x = 0 \Rightarrow y = 4 - |0| = 4$, so the y-intercept is 4.

x-axis symmetry: $(-y) = 4 - |x| \Leftrightarrow y = -4 + |x|$, which is not the same as $y = 4 - |x|$, so the graph is not symmetric with respect to the x-axis.

y-axis symmetry: $y = 4 - |-x| = 4 - |x|$, so the graph is symmetric with respect to the y-axis.

Origin symmetry: $(-y) = 4 - |-x| \Leftrightarrow y = -4 + |x|$, which is not the same as $y = 4 - |x|$, so the graph is not symmetric with respect to the origin.

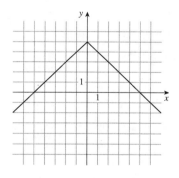

41. Since $x = y^3$ is solved for x in terms of y, we insert values for y and find the corresponding values of x.

x	y
-27	-3
-8	-2
-1	-1
0	0
1	1
8	2
27	3

$y = 0 \Rightarrow x = (0)^3 = 0$, so the x-intercept is 0, and $x = 0 \Rightarrow 0 = y^3 \Rightarrow y = 0$, so the y-intercept is 0.

x-axis symmetry: $x = (-y)^3 = -y^3$, which is not the same as $x = y^3$, so the graph is not symmetric with respect to the x-axis.

y-axis symmetry: $(-x) = y^3 \Leftrightarrow x = -y^3$, which is not the same as $x = y^3$, so the graph is not symmetric with respect to the y-axis.

Origin symmetry: $(-x) = (-y)^3 \Leftrightarrow -x = -y^3 \Leftrightarrow x = y^3$, so the graph is symmetric with respect to the origin.

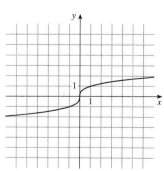

43. $y = x^4$

x	y
-3	81
-2	16
-1	1
0	0
1	1
2	16
3	81

$y = 0 \;\Rightarrow\; 0 = x^4 \;\Rightarrow\; x = 0$, so the x-intercept is 0, and $x = 0$ $\Rightarrow\; y = 0^4 = 0$, so the y-intercept is 0.

x-axis symmetry: $(-y) = x^4 \;\Leftrightarrow\; y = -x^4$, which is not the same as $y = x^4$, so the graph is not symmetric with respect to the x-axis.

y-axis symmetry: $y = (-x)^4 = x^4$, so the graph is symmetric with respect to the y-axis.

Origin symmetry: $(-y) = (-x)^4 \;\Leftrightarrow\; -y = x^4$, which is not the same as $y = x^4$, so the graph is not symmetric with respect to the origin.

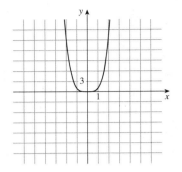

45. x-axis symmetry: $(-y) = x^4 + x^2 \;\Leftrightarrow\; y = -x^4 - x^2$, which is not the same as $y = x^4 + x^2$, so the graph is not symmetric with respect to the x-axis.

y-axis symmetry: $y = (-x)^4 + (-x)^2 = x^4 + x^2$, so the graph is symmetric with respect to the y-axis.

Origin symmetry: $(-y) = (-x)^4 + (-x)^2 \;\Leftrightarrow\; -y = x^4 + x^2$, which is not the same as $y = x^4 + x^2$, so the graph is not symmetric with respect to the origin.

47. x-axis symmetry: $x^2 (-y)^2 + x (-y) = 1 \;\Leftrightarrow\; x^2 y^2 - xy = 1$, which is not the same as $x^2 y^2 + xy = 1$, so the graph is not symmetric with respect to the x-axis.

y-axis symmetry: $(-x)^2 y^2 + (-x) y = 1 \;\Leftrightarrow\; x^2 y^2 - xy = 1$, which is not the same as $x^2 y^2 + xy = 1$, so the graph is not symmetric with respect to the y-axis.

Origin symmetry: $(-x)^2 (-y)^2 + (-x)(-y) = 1 \;\Leftrightarrow\; x^2 y^2 + xy = 1$, so the graph is symmetric with respect to the origin.

49. x-axis symmetry: $(-y) = x^3 + 10x \;\Leftrightarrow\; y = -x^3 - 10x$, which is not the same as $y = x^3 + 10x$, so the graph is not symmetric with respect to the x-axis.

y-axis symmetry: $y = (-x)^3 + 10(-x) \;\Leftrightarrow\; y = -x^3 - 10x$, which is not the same as $y = x^3 + 10x$, so the graph is not symmetric with respect to the y-axis.

Origin symmetry: $(-y) = (-x)^3 + 10(-x) \;\Leftrightarrow\; -y = -x^3 - 10x \;\Leftrightarrow\; y = x^3 + 10x$, so the graph is symmetric with respect to the origin.

51. Symmetric with respect to the y-axis.

53. Symmetric with respect to the origin.

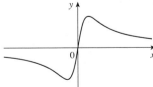

55. Using $h = 2$, $k = -1$, and $r = 3$, we get $(x - 2)^2 + (y - (-1))^2 = 3^2 \;\Leftrightarrow\; (x - 2)^2 + (y + 1)^2 = 9$.

57. The equation of a circle centered at the origin is $x^2 + y^2 = r^2$. Using the point $(4, 7)$ we solve for r^2. This gives $(4)^2 + (7)^2 = r^2 \;\Leftrightarrow\; 16 + 49 = 65 = r^2$. Thus, the equation of the circle is $x^2 + y^2 = 65$.

59. The center is at the midpoint of the line segment, which is $\left(\frac{-1+5}{2}, \frac{1+5}{2}\right) = (2, 3)$. The radius is one half the diameter, so $r = \frac{1}{2}\sqrt{(-1-5)^2 + (1-5)^2} = \frac{1}{2}\sqrt{36 + 16} = \frac{1}{2}\sqrt{52} = \sqrt{13}$. Thus, an equation of the circle is $(x - 2)^2 + (y - 3)^2 = \left(\sqrt{13}\right)^2$ or $(x - 2)^2 + (y - 3)^2 = 13$.

61. Since the circle is tangent to the x-axis, it must contain the point $(7, 0)$, so the radius is the change in the y-coordinates. That is, $r = |-3 - 0| = 3$. So the equation of the circle is $(x - 7)^2 + (y - (-3))^2 = 3^2$, which is $(x - 7)^2 + (y + 3)^2 = 9$.

63. From the figure, the center of the circle is at $(-2, 2)$. The radius is the change in the y-coordinates, so $r = |2 - 0| = 2$. Thus the equation of the circle is $(x - (-2))^2 + (y - 2)^2 = 2^2$, which is $(x + 2)^2 + (y - 2)^2 = 4$.

65. Completing the square gives $x^2 + y^2 - 2x + 4y + 1 = 0 \Leftrightarrow x^2 - 2x + \left(\frac{-2}{2}\right)^2 + y^2 + 4y + \left(\frac{4}{2}\right)^2 = -1 + \left(\frac{-2}{2}\right)^2 + \left(\frac{4}{2}\right)^2$
$\Leftrightarrow x^2 - 2x + 1 + y^2 + 4y + 4 = -1 + 1 + 4 \Leftrightarrow (x - 1)^2 + (y + 2)^2 = 4$. Thus, the center is $(1, -2)$, and the radius is 2.

67. Completing the square gives $x^2 + y^2 - 4x + 10y + 13 = 0 \quad \Leftrightarrow \quad x^2 - 4x + \left(\frac{-4}{2}\right)^2 + y^2 + 10y + \left(\frac{10}{2}\right)^2 = -13 + \left(\frac{4}{2}\right)^2 + \left(\frac{10}{2}\right)^2$
$\Leftrightarrow \quad x^2 - 4x + 4 + y^2 + 10y + 25 = -13 + 4 + 25 \quad \Leftrightarrow \quad (x - 2)^2 + (y + 5)^2 = 16$.
Thus, the center is $(2, -5)$, and the radius is 4.

69. Completing the square gives $x^2 + y^2 + x = 0 \quad \Leftrightarrow \quad x^2 + x + \left(\frac{1}{2}\right)^2 + y^2 = \left(\frac{1}{2}\right)^2 \quad \Leftrightarrow \quad x^2 + x + \frac{1}{4} + y^2 = \frac{1}{4} \quad \Leftrightarrow$
$\left(x + \frac{1}{2}\right)^2 + y^2 = \frac{1}{4}$. Thus, the circle has center $\left(-\frac{1}{2}, 0\right)$ and radius $\frac{1}{2}$.

71. Completing the square gives $x^2 + y^2 - \frac{1}{2}x + \frac{1}{2}y = \frac{1}{8} \quad \Leftrightarrow \quad x^2 - \frac{1}{2}x + \left(\frac{-1/2}{2}\right)^2 + y^2 + \frac{1}{2}y + \left(\frac{1/2}{2}\right)^2 = \frac{1}{8} +$
$\left(\frac{-1/2}{2}\right)^2 + \left(\frac{1/2}{2}\right)^2 \quad \Leftrightarrow \quad x^2 - \frac{1}{2}x + \frac{1}{16} + y^2 + \frac{1}{2}y + \frac{1}{16} = \frac{1}{8} + \frac{1}{16} + \frac{1}{16} = \frac{2}{8} = \frac{1}{4} \quad \Leftrightarrow \quad \left(x - \frac{1}{4}\right)^2 + \left(y + \frac{1}{4}\right)^2 = \frac{1}{4}$.
Thus, the circle has center $\left(\frac{1}{4}, -\frac{1}{4}\right)$ and radius $\frac{1}{2}$.

73. Completing the square gives $x^2 + y^2 + 4x - 10y = 21$
$\Leftrightarrow \quad x^2 + 4x + \left(\frac{4}{2}\right)^2 + y^2 - 10y + \left(\frac{-10}{2}\right)^2 = 21 +$
$\left(\frac{4}{2}\right)^2 + \left(\frac{-10}{2}\right)^2 \quad \Leftrightarrow$
$(x + 2)^2 + (y - 5)^2 = 21 + 4 + 25 = 50$. Thus, the
circle has center $(-2, 5)$ and radius $\sqrt{50} = 5\sqrt{2}$.

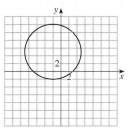

75. Completing the square gives
$x^2 + y^2 + 6x - 12y + 45 = 0 \quad \Leftrightarrow$
$(x + 3)^2 + (y - 6)^2 = -45 + 9 + 36 = 0$. Thus, the
center is $(-3, 6)$, and the radius is 0. This is a degenerate
circle whose graph consists only of the point $(-3, 6)$.

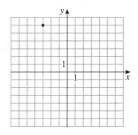

77. $\{(x, y) \mid x^2 + y^2 \leq 1\}$. This is the set of points inside
(and on) the circle $x^2 + y^2 = 1$.

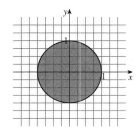

79. $\{(x, y) \mid 1 \leq x^2 + y^2 < 9\}$. This is the set of points
outside the circle $x^2 + y^2 = 1$ and inside (but not on) the
circle $x^2 + y^2 = 9$.

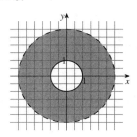

81. Completing the square gives $x^2 + y^2 - 4y - 12 = 0 \quad \Leftrightarrow \quad x^2 + y^2 - 4y + \left(\frac{-4}{2}\right)^2 = 12 + \left(\frac{-4}{2}\right)^2 \quad \Leftrightarrow$
$x^2 + (y - 2)^2 = 16$. Thus, the center is $(0, 2)$, and the radius is 4. So the circle $x^2 + y^2 = 4$, with center $(0, 0)$ and radius
2, sits completely inside the larger circle. Thus, the area is $\pi 4^2 - \pi 2^2 = 16\pi - 4\pi = 12\pi$.

83. (a) In 1980 inflation was 14%; in 1990, it was 6%; in 1999, it was 2%.

(b) Inflation exceeded 6% from 1975 to 1976 and from 1978 to 1982.

(c) Between 1980 and 1985 the inflation rate generally decreased. Between 1987 and 1992 the inflation rate generally increased.

(d) The highest rate was about 14% in 1980. The lowest was about 1% in 2002.

85. Completing the square gives $x^2 + y^2 + ax + by + c = 0 \quad \Leftrightarrow \quad x^2 + ax + \left(\dfrac{a}{2}\right)^2 + y^2 + by + \left(\dfrac{b}{2}\right)^2 = -c + \left(\dfrac{a}{2}\right)^2 + \left(\dfrac{b}{2}\right)^2$

$\Leftrightarrow \quad \left(x + \dfrac{a}{2}\right)^2 + \left(y + \dfrac{b}{2}\right)^2 = -c + \dfrac{a^2 + b^2}{4}$. This equation represents a circle only when $-c + \dfrac{a^2 + b^2}{4} > 0$. This

equation represents a point when $-c + \dfrac{a^2 + b^2}{4} = 0$, and this equation represents the empty set when $-c + \dfrac{a^2 + b^2}{4} < 0$.

When the equation represents a circle, the center is $\left(-\dfrac{a}{2}, -\dfrac{b}{2}\right)$, and the radius is $\sqrt{-c + \dfrac{a^2 + b^2}{4}} = \frac{1}{2}\sqrt{a^2 + b^2 - 4ac}$.

87. (a) Symmetric about the x-axis. **(b)** Symmetric about the y-axis. **(c)** Symmetric about the origin.

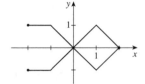 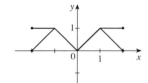 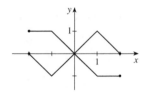

2.3 Graphing Calculators; Solving Equations and Inequalities Graphically

1. $y = x^4 + 2$

(a) $[-2, 2]$ by $[-2, 2]$ **(b)** $[0, 4]$ by $[0, 4]$

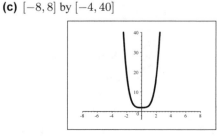 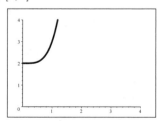

(c) $[-8, 8]$ by $[-4, 40]$ **(d)** $[-40, 40]$ by $[-80, 800]$

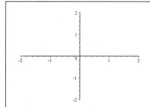 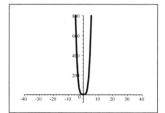

The viewing rectangle in part (c) produces the most appropriate graph of the equation.

3. $y = 100 - x^2$

   (a) $[-4, 4]$ by $[-4, 4]$

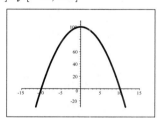

   (b) $[-10, 10]$ by $[-10, 10]$

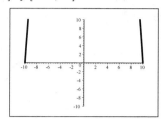

   (c) $[-15, 15]$ by $[-30, 110]$

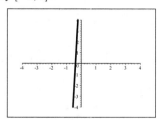

   (d) $[-4, 4]$ by $[-30, 110]$

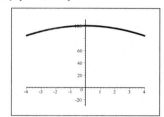

The viewing rectangle in part (c) produces the most appropriate graph of the equation.

5. $y = 10 + 25x - x^3$

   (a) $[-4, 4]$ by $[-4, 4]$

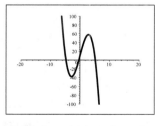

   (b) $[-10, 10]$ by $[-10, 10]$

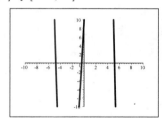

   (c) $[-20, 20]$ by $[-100, 100]$

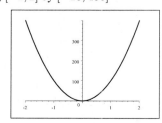

   (d) $[-100, 100]$ by $[-200, 200]$

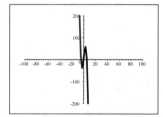

The viewing rectangle in part (c) produces the most appropriate graph of the equation.

7. $y = 100x^2$, $[-2, 2]$ by $[-10, 400]$

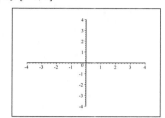

9. $y = 4 + 6x - x^2$, $[-4, 10]$ by $[-10, 20]$

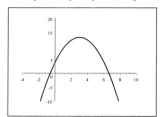

11. $y = \sqrt[4]{256 - x^2}$. We require that $256 - x^2 \geq 0$ $\Rightarrow$ $-16 \leq x \leq 16$, so we graph $y = \sqrt[4]{256 - x^2}$ in the viewing rectangle $[-20, 20]$ by $[-1, 5]$.

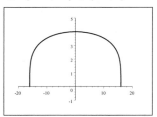

13. $y = 0.01x^3 - x^2 + 5$, $[-50, 150]$ by $[-2000, 2000]$

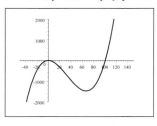

15. $y = x^4 - 4x^3$, $[-4, 6]$ by $[-50, 100]$

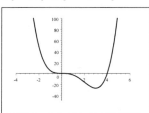

17. $y = 1 + |x - 1|$, $[-3, 5]$ by $[-1, 5]$

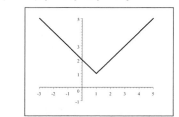

19. $x^2 + y^2 = 9$ $\Leftrightarrow$ $y^2 = 9 - x^2$ $\Rightarrow$ $y = \pm\sqrt{9 - x^2}$. So we graph the functions $y_1 = \sqrt{9 - x^2}$ and $y_2 = -\sqrt{9 - x^2}$ in the viewing rectangle $[-8, 8]$ by $[-6, 6]$.

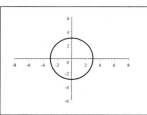

21. $4x^2 + 2y^2 = 1$ $\Leftrightarrow$ $2y^2 = 1 - 4x^2$ $\Leftrightarrow$ $y^2 = \dfrac{1 - 4x^2}{2}$ $\Rightarrow$ $y = \pm\sqrt{\dfrac{1 - 4x^2}{2}}$. So we graph the functions $y_1 = \sqrt{\dfrac{1 - 4x^2}{2}}$ and $y_2 = -\sqrt{\dfrac{1 - 4x^2}{2}}$ in the viewing rectangle $[-1.2, 1.2]$ by $[-0.8, 0.8]$.

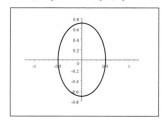

23. Although the graphs of $y = -3x^2 + 6x - \frac{1}{2}$ and $y = \sqrt{7 - \frac{7}{12}x^2}$ appear to intersect in the viewing rectangle $[-4, 4]$ by $[-1, 3]$, there is no point of intersection. You can verify this by zooming in.

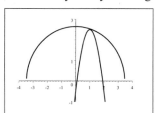

25. The graphs of $y = 6 - 4x - x^2$ and $y = 3x + 18$ appear to have two points of intersection in the viewing rectangle $[-6, 2]$ by $[-5, 20]$. You can verify that $x = -4$ and $x = -3$ are exact solutions.

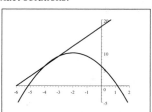

27. Algebraically: $x - 4 = 5x + 12 \quad \Leftrightarrow \quad -16 = 4x \quad \Leftrightarrow$
$x = -4$.

Graphically: We graph the two equations $y_1 = x - 4$ and $y_2 = 5x + 12$ in the viewing rectangle $[-6, 4]$ by $[-10, 2]$. Zooming in, we see that the solution is $x = -4$.

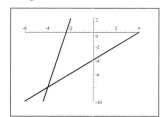

29. Algebraically: $\dfrac{2}{x} + \dfrac{1}{2x} = 7 \quad \Leftrightarrow$

$2x\left(\dfrac{2}{x} + \dfrac{1}{2x}\right) = 2x\,(7) \quad \Leftrightarrow \quad 4 + 1 = 14x \quad \Leftrightarrow$

$x = \frac{5}{14}$.

Graphically: We graph the two equations $y_1 = \dfrac{2}{x} + \dfrac{1}{2x}$ and $y_2 = 7$ in the viewing rectangle $[-2, 2]$ by $[-2, 8]$. Zooming in, we see that the solution is $x \approx 0.36$.

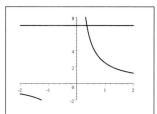

31. Algebraically: $x^2 - 32 = 0 \quad \Leftrightarrow \quad x^2 = 32 \quad \Rightarrow$
$x = \pm\sqrt{32} = \pm 4\sqrt{2}$.

Graphically: We graph the equation $y_1 = x^2 - 32$ and determine where this curve intersects the x-axis. We use the viewing rectangle $[-10, 10]$ by $[-5, 5]$. Zooming in, we see that solutions are $x \approx 5.66$ and $x \approx -5.66$.

33. Algebraically: $16x^4 = 625 \quad \Leftrightarrow \quad x^4 = \frac{625}{16} \quad \Rightarrow$
$x = \pm\frac{5}{2} = \pm 2.5$.

Graphically: We graph the two equations $y_1 = 16x^4$ and $y_2 = 625$ in the viewing rectangle $[-5, 5]$ by $[610, 640]$. Zooming in, we see that solutions are $x = \pm 2.5$.

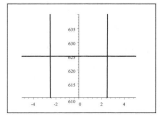

35. Algebraically: $(x - 5)^4 - 80 = 0 \quad \Leftrightarrow \quad (x - 5)^4 = 80$
$\Rightarrow \quad x - 5 = \pm\sqrt[4]{80} = \pm 2\sqrt[4]{5} \quad \Leftrightarrow \quad x = 5 \pm 2\sqrt[4]{5}$.

Graphically: We graph the equation $y_1 = (x - 5)^4 - 80$ and determine where this curve intersects the x-axis. We use the viewing rectangle $[-1, 9]$ by $[-5, 5]$. Zooming in, we see that solutions are $x \approx 2.01$ and $x \approx 7.99$.

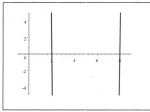

37. We graph $y = x^2 - 7x + 12$ in the viewing rectangle $[0, 6]$ by $[-0.1, 0.1]$. The solutions appear to be exactly $x = 3$ and $x = 4$. [In fact $x^2 - 7x + 12 = (x - 3)(x - 4)$.]

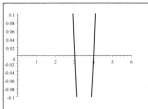

39. We graph $y = x^3 - 6x^2 + 11x - 6$ in the viewing rectangle $[-1, 4]$ by $[-0.1, 0.1]$. The solutions are $x = 1.00$, $x = 2.00$, and $x = 3.00$.

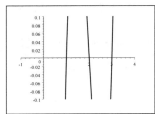

41. We first graph $y = x - \sqrt{x + 1}$ in the viewing rectangle $[-1, 5]$ by $[-0.1, 0.1]$ and find that the solution is near 1.6. Zooming in, we see that solutions is $x \approx 1.62$.

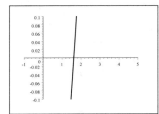

43. We graph $y = x^{1/3} - x$ in the viewing rectangle $[-3, 3]$ by $[-1, 1]$. The solutions are $x = -1$, $x = 0$, and $x = 1$, as can be verified by substitution.

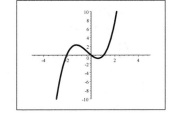

45. $x^3 - 2x^2 - x - 1 = 0$, so we start by graphing the function $y = x^3 - 2x^2 - x - 1$ in the viewing rectangle $[-10, 10]$ by $[-100, 100]$. There appear to be two solutions, one near $x = 0$ and another one between $x = 2$ and $x = 3$. We then use the viewing rectangle $[-1, 5]$ by $[-1, 1]$ and zoom in on the only solution, $x \approx 2.55$.

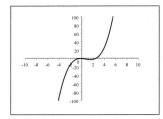
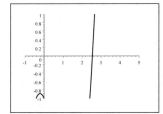

47. $x(x - 1)(x + 2) = \frac{1}{6}x \iff$
$x(x - 1)(x + 2) - \frac{1}{6}x = 0$. We start by graphing the function
$y = x(x - 1)(x + 2) - \frac{1}{6}x$ in the viewing rectangle $[-5, 5]$ by $[-10, 10]$. There appear to be three solutions. We then use the viewing rectangle $[-2.5, 2.5]$ by $[-1, 1]$ and zoom into the solutions at $x \approx -2.05$, $x = 0.00$, and $x \approx 1.05$.

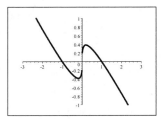
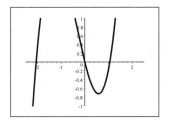

49. We graph $y = x^2 - 3x - 10$ in the viewing rectangle $[-5, 6]$ by $[-14, 2]$. The the solution to the inequality is $[-2, 5]$.

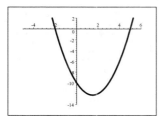

51. Since $x^3 + 11x \leq 6x^2 + 6$ $\Leftrightarrow$ $x^3 - 6x^2 + 11x - 6 \leq 0$, we graph $y = x^3 - 6x^2 + 11x - 6$ in the viewing rectangle $[0, 5]$ by $[-5, 5]$. The solution set is $(-\infty, 1.0] \cup [2.0, 3.0]$.

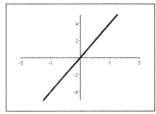

53. Since $x^{1/3} \leq x$ $\Leftrightarrow$ $x^{1/3} - x < 0$, we graph $y = x^{1/3} - x$ in the viewing rectangle $[-3, 3]$ by $[-1, 1]$. From this, we find that the solution set is $(-1, 0) \cup (1, \infty)$.

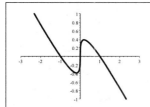

55. Since $(x + 1)^2 < (x - 1)^2$ $\Leftrightarrow$ $(x + 1)^2 - (x - 1)^2 < 0$, we graph $y = (x + 1)^2 - (x - 1)^2$ in the viewing rectangle $[-2, 2]$ by $[-5, 5]$. The solution set is $(-\infty, 0)$.

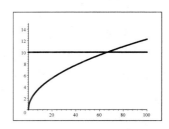

57. As in Example 6, we graph the equation $y = x^3 - 6x^2 + 9x - \sqrt{x}$ in the viewing rectangle $[0, 10]$ by $[-2, 15]$. We see the two solutions found in Example 6 and what appears to be an additional solution at $x = 0$. In the viewing rectangle $[0, 0.05]$ by $[-0.25, 0.25]$, we find yet another solution at $x \approx 0.01$. We can verify that $x = 0$ is an exact solution by substitution.

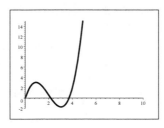

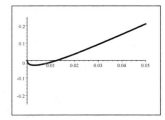

59. Using a zoom or trace function, we find that $y \geq 10$ for $x \geq 66.7$. We could estimate this since if $x < 100$, then $\left(\frac{x}{5280}\right)^2 \leq 0.00036$. So for $x < 100$ we have $\sqrt{1.5x + \left(\frac{x}{5280}\right)^2} \approx \sqrt{1.5x}$. Solving $\sqrt{1.5x} > 10$ we get $1.5 > 100$ or $x > \frac{100}{1.5} = 66.7$ mi.

61. Calculators perform operations in the following order: exponents are applied before division and division is applied before addition. Therefore, `Y_1=x^1/3` is interpreted as $y = \frac{x^1}{3} = \frac{x}{3}$, which is the equation of a line. Likewise, `Y_2=x/x+4` is interpreted as $y = \frac{x}{x} + 4 = 1 + 4 = 5$. Instead, enter the following: `Y_1=x^(1/3)`, `Y_2=x/(x+4)`.

63. (a) We graph $y_1 = x^3 - 3x$ and $y_2 = k$ for $k = -4, -2, 0, 2,$ and 4 in the viewing rectangle $[-5, 5]$ by $[-10, 10]$. The number of solutions and the solutions are shown in the table below.

k	Number of solutions	Solutions
-4	1	$x \approx -2.20$
-2	2	$x = -2, x = 1$
0	3	$x \approx \pm 1.73, x = 0$
2	2	$x = -1, x = 2$
4	1	$x \approx 2.20$

(b) The equation $x^3 - 3x = k$ will have one solution for all $k < -2$ or $k > 2$, it will have exactly two solutions when $k = \pm 2$, and it will have three solutions for $-2 < k < 2$.

2.4 Lines

1. $m = \dfrac{y_2 - y_1}{x_2 - x_1} = \dfrac{2 - 0}{4 - 0} = \dfrac{2}{4} = \dfrac{1}{2}$

3. $m = \dfrac{y_2 - y_1}{x_2 - x_1} = \dfrac{0 - 2}{-10 - 2} = \dfrac{-2}{-12} = \dfrac{1}{6}$

5. $m = \dfrac{y_2 - y_1}{x_2 - x_1} = \dfrac{4 - 3}{2 - 4} = \dfrac{1}{-2} = -\dfrac{1}{2}$

7. $m = \dfrac{y_2 - y_1}{x_2 - x_1} = \dfrac{6 - (-3)}{-1 - 1} = \dfrac{9}{-2} = -\dfrac{9}{2}$

9. For ℓ_1, we find two points, $(-1, 2)$ and $(0, 0)$ that lie on the line. Thus the slope of ℓ_1 is $m = \dfrac{y_2 - y_1}{x_2 - x_1} = \dfrac{2 - 0}{-1 - 0} = -2$.

For ℓ_2, we find two points $(0, 2)$ and $(2, 3)$. Thus, the slope of ℓ_2 is $m = \dfrac{y_2 - y_1}{x_2 - x_1} = \dfrac{3 - 2}{2 - 0} = \dfrac{1}{2}$. For ℓ_3 we find the points

$(2, -2)$ and $(3, 1)$. Thus, the slope of ℓ_3 is $m = \dfrac{y_2 - y_1}{x_2 - x_1} = \dfrac{1 - (-2)}{3 - 2} = 3$. For ℓ_4, we find the points $(-2, -1)$ and

$(2, -2)$. Thus, the slope of ℓ_4 is $m = \dfrac{y_2 - y_1}{x_2 - x_1} = \dfrac{-2 - (-1)}{2 - (-2)} = \dfrac{-1}{4} = -\dfrac{1}{4}$.

11. First we find two points $(0, 4)$ and $(4, 0)$ that lie on the line. So the slope is $m = \dfrac{0 - 4}{4 - 0} = -1$. Since the y-intercept is 4, the equation of the line is $y = mx + b = -1x + 4$. So $y = -x + 4$, or $x + y - 4 = 0$.

13. We choose the two intercepts as points, $(0, -3)$ and $(2, 0)$. So the slope is $m = \dfrac{0 - (-3)}{2 - 0} = \dfrac{3}{2}$. Since the y-intercept is -3, the equation of the line is $y = mx + b = \dfrac{3}{2}x - 3$, or $3x - 2y - 6 = 0$.

15. Using the equation $y - y_1 = m(x - x_1)$, we get $y - 3 = 1(x - 2)$ $\Leftrightarrow$ $-x + y = 1$ $\Leftrightarrow$ $x - y + 1 = 0$.

17. Using the equation $y - y_1 = m(x - x_1)$, we get $y - 7 = \dfrac{2}{3}(x - 1)$ $\Leftrightarrow$ $3y - 21 = 2x - 2$ $\Leftrightarrow$ $-2x + 3y = 19$ $\Leftrightarrow$ $2x - 3y + 19 = 0$.

19. First we find the slope, which is $m = \dfrac{y_2 - y_1}{x_2 - x_1} = \dfrac{6 - 1}{1 - 2} = \dfrac{5}{-1} = -5$. Substituting into $y - y_1 = m(x - x_1)$, we get

$y - 6 = -5(x - 1)$ $\Leftrightarrow$ $y - 6 = -5x + 5$ $\Leftrightarrow$ $5x + y - 11 = 0$.

21. Using $y = mx + b$, we have $y = 3x + (-2)$ or $3x - y - 2 = 0$.

23. We are given two points, $(1, 0)$ and $(0, -3)$. Thus, the slope is $m = \dfrac{y_2 - y_1}{x_2 - x_1} = \dfrac{-3 - 0}{0 - 1} = \dfrac{-3}{-1} = 3$. Using the

y-intercept, we have $y = 3x + (-3)$ or $y = 3x - 3$ or $3x - y - 3 = 0$.

25. Since the equation of a horizontal line passing through (a, b) is $y = b$, the equation of the horizontal line passing through $(4, 5)$ is $y = 5$.

27. Since $x + 2y = 6$ $\Leftrightarrow$ $2y = -x + 6$ $\Leftrightarrow$ $y = -\dfrac{1}{2}x + 3$, the slope of this line is $-\dfrac{1}{2}$. Thus, the line we seek is given

by $y - (-6) = -\dfrac{1}{2}(x - 1)$ $\Leftrightarrow$ $2y + 12 = -x + 1$ $\Leftrightarrow$ $x + 2y + 11 = 0$.

29. Any line parallel to $x = 5$ will have undefined slope and be of the form $x = a$. Thus the equation of the line is $x = -1$.

31. First find the slope of $2x + 5y + 8 = 0$. This gives $2x + 5y + 8 = 0 \quad \Leftrightarrow \quad 5y = -2x - 8 \quad \Leftrightarrow \quad y = -\frac{2}{5}x - \frac{8}{5}$. So the slope of the line that is perpendicular to $2x + 5y + 8 = 0$ is $m = -\dfrac{1}{-2/5} = \frac{5}{2}$. The equation of the line we seek is $y - (-2) = \frac{5}{2}(x - (-1)) \quad \Leftrightarrow \quad 2y + 4 = 5x + 5 \quad \Leftrightarrow \quad 5x - 2y + 1 = 0$.

33. First find the slope of the line passing through $(2, 5)$ and $(-2, 1)$. This gives $m = \dfrac{1 - 5}{-2 - 2} = \dfrac{-4}{-4} = 1$, and so the equation of the line we seek is $y - 7 = 1(x - 1) \quad \Leftrightarrow \quad x - y + 6 = 0$.

35. (a)

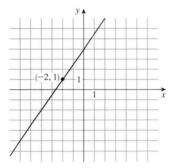

(b) $y - 1 = \frac{3}{2}(x - (-2)) \quad \Leftrightarrow \quad 2y - 2 = 3(x + 2)$
$\Leftrightarrow \quad 2y - 2 = 3x + 6 \quad \Leftrightarrow \quad 3x - 2y + 8 = 0$.

37.

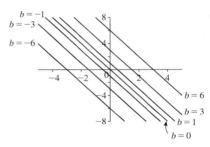

$y = -2x + b$, $b = 0, \pm 1, \pm 3, \pm 6$. They have the same slope, so they are parallel.

39.

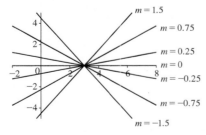

$y = m(x - 3)$, $m = 0, \pm 0.25, \pm 0.75, \pm 1.5$. Each of the lines contains the point $(3, 0)$ because the point $(3, 0)$ satisfies each equation $y = m(x - 3)$. Since $(3, 0)$ is on the x-axis, we could also say that they all have the same x-intercept.

41. $x + y = 3 \quad \Leftrightarrow \quad y = -x + 3$. So the slope is -1, and the y-intercept is 3.

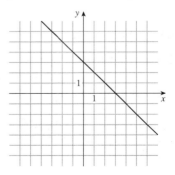

43. $x + 3y = 0 \iff 3y = -x \iff y = -\frac{1}{3}x$. So the slope is $-\frac{1}{3}$, and the y-intercept is 0.

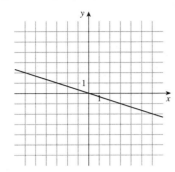

45. $\frac{1}{2}x - \frac{1}{3}y + 1 = 0 \iff -\frac{1}{3}y = -\frac{1}{2}x - 1 \iff y = \frac{3}{2}x + 3$. So the slope is $\frac{3}{2}$, and the y-intercept is 3.

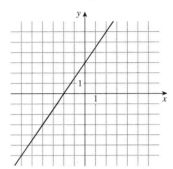

47. $y = 4$ can also be expressed as $y = 0x + 4$. So the slope is 0, and the y-intercept is 4.

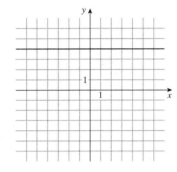

49. $3x - 4y = 12 \iff -4y = -3x + 12 \iff y = \frac{3}{4}x - 3$. So the slope is $\frac{3}{4}$, and the y-intercept is -3.

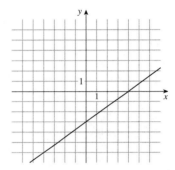

51. $3x + 4y - 1 = 0 \iff 4y = -3x + 1 \iff y = -\frac{3}{4}x + \frac{1}{4}$. So the slope is $-\frac{3}{4}$, and the y-intercept is $\frac{1}{4}$.

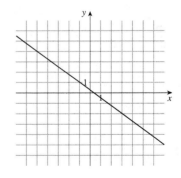

53. We first plot the points to find the pairs of points that determine each side. Next we find the slopes of opposite sides. The slope of AB is $\dfrac{4 - 1}{7 - 1} = \dfrac{3}{6} = \dfrac{1}{2}$, and the slope of DC is $\dfrac{10 - 7}{5 - (-1)} = \dfrac{3}{6} = \dfrac{1}{2}$. Since these slope are equal, these two sides are parallel. The slope of AD is $\dfrac{7 - 1}{-1 - 1} = \dfrac{6}{-2} = -3$, and the slope of BC is $\dfrac{10 - 4}{5 - 7} = \dfrac{6}{-2} = -3$. Since these slope are equal, these two sides are parallel. Hence $ABCD$ is a parallelogram.

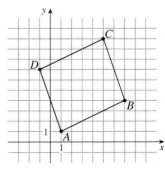

55. We first plot the points to find the pairs of points that determine each side. Next we

find the slopes of opposite sides. The slope of AB is $\dfrac{3-1}{11-1}=\dfrac{2}{10}=\dfrac{1}{5}$ and the

slope of DC is $\dfrac{6-8}{0-10}=\dfrac{-2}{-10}=\dfrac{1}{5}$. Since these slope are equal, these two sides

are parallel. Slope of AD is $\dfrac{6-1}{0-1}=\dfrac{5}{-1}=-5$, and the slope of BC is

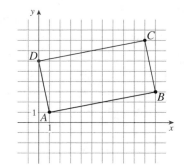

$\dfrac{3-8}{11-10}=\dfrac{-5}{1}=-5$. Since these slope are equal, these two sides are parallel.

Since (slope of AB) $\times$ (slope of AD) $=\frac{1}{5}\times(-5)=-1$, the first two sides are

each perpendicular to the second two sides. So the sides form a rectangle.

57. We need the slope and the midpoint of the line AB. The midpoint of AB is $\left(\dfrac{1+7}{2},\dfrac{4-2}{2}\right)=(4,1)$, and the slope of

AB is $m=\dfrac{-2-4}{7-1}=\dfrac{-6}{6}=-1$. The slope of the perpendicular bisector will have slope $\dfrac{-1}{m}=\dfrac{-1}{-1}=1$. Using the

point-slope form, the equation of the perpendicular bisector is $y-1=1\,(x-4)$ or $x-y-3=0$.

59. (a) We start with the two points $(a,0)$ and $(0,b)$. The slope of the line that contains them is $\dfrac{b-0}{0-a}=-\dfrac{b}{a}$. So the equation

of the line containing them is $y=-\dfrac{b}{a}x+b$ (using the slope-intercept form). Dividing by b (since $b\neq0$) gives

$\dfrac{y}{b}=-\dfrac{x}{a}+1\quad\Leftrightarrow\quad\dfrac{x}{a}+\dfrac{y}{b}=1$.

(b) Setting $a=6$ and $b=-8$, we get $\dfrac{x}{6}+\dfrac{y}{-8}=1\quad\Leftrightarrow\quad 4x-3y=24\quad\Leftrightarrow\quad 4x-3y-24=0$.

61. Let h be the change in your horizontal distance, in feet. Then $-\frac{6}{100}=\dfrac{-1000}{h}\quad\Leftrightarrow\quad h=\dfrac{100{,}000}{6}\approx16{,}667$. So the

change in your horizontal distance is about 16,667 feet.

63. (a) The slope is $0.0417D=0.0417\,(200)=8.34$. It represents the increase in dosage for each one-year increase in the

child's age.

(b) When $a=0$, $c=8.34\,(0+1)=8.34$ mg.

65. (a)

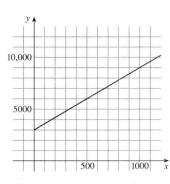

(b) The slope is the cost per toaster oven, \$6. The y-intercept, \$3000, is

the monthly fixed cost — the cost that is incurred no matter how

many toaster ovens are produced.

67. (a) Using n in place of x and t in place of y, we find that the slope is $\dfrac{t_2-t_1}{n_2-n_1}=\dfrac{80-70}{168-120}=\dfrac{10}{48}=\dfrac{5}{24}$. So the linear

equation is $t-80=\frac{5}{24}\,(n-168)\quad\Leftrightarrow\quad t-80=\frac{5}{24}n-35\quad\Leftrightarrow\quad t=\frac{5}{24}n+45$.

(b) When $n=150$, the temperature is approximately given by $t=\frac{5}{24}\,(150)+45=76.25°$ F $\approx76°$ F.

69. (a) We are given $\dfrac{\text{change in pressure}}{10\text{ feet change in depth}} = \dfrac{4.34}{10} = 0.434$. Using P for pressure and d for depth, and using the point $P = 15$ when $d = 0$, we have $P - 15 = 0.434\,(d - 0)$ $\Leftrightarrow$ $P = 0.434d + 15$.

(c) The slope represents the increase in pressure per foot of descent. The y-intercept represents the pressure at the surface.

(d) When $P = 100$, then $100 = 0.434d + 15$ $\Leftrightarrow$ $0.434d = 85$ $\Leftrightarrow$ $d = 195.9$ ft. Thus the pressure is 100 lb/in^3 at a depth of approximately 196 ft.

(b)

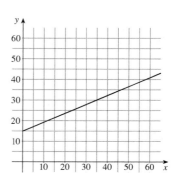

71. (a) Using d in place of x and C in place of y, we find the slope to be $\dfrac{C_2 - C_1}{d_2 - d_1} = \dfrac{460 - 380}{800 - 480} = \dfrac{80}{320} = \dfrac{1}{4}$. So the linear equation is $C - 460 = \frac{1}{4}\,(d - 800)$ $\Leftrightarrow$ $C - 460 = \frac{1}{4}d - 200$ $\Leftrightarrow$ $C = \frac{1}{4}d + 260$.

(b) Substituting $d = 1500$ we get $C = \frac{1}{4}\,(1500) + 260 = 635$. Thus, the cost of driving 1500 miles is \$635.

(d) The y-intercept represents the fixed cost, \$260.

(e) It is a suitable model because you have fixed monthly costs such as insurance and car payments, as well as costs that occur as you drive, such as gasoline, oil, tires, etc., and the cost of these for each additional mile driven is a constant.

(c)

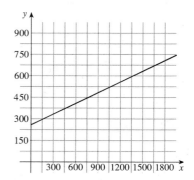

The slope of the line represents the cost per mile, \$0.25.

73. Slope is the rate of change of one variable per unit change in another variable. So if the slope is positive, then the temperature is rising. Likewise, if the slope is negative then the temperature is decreasing. If the slope is 0, then the temperature is not changing.

2.5   Modeling Variation

1. $T = kx$, where k is constant.

3. $v = \dfrac{k}{z}$, where k is constant.

5. $y = \dfrac{ks}{t}$, where k is constant.

7. $z = k\sqrt{y}$, where k is constant.

9. $V = klwh$, where k is constant.

11. $R = \dfrac{ki}{Pt}$, where k is constant.

13. Since y is directly proportional to x, $y = kx$. Since $y = 42$ when $x = 6$, we have $42 = k\,(6)$ $\Leftrightarrow$ $k = 7$. So $y = 7x$.

15. Since M varies directly as x and inversely as y, $M = \dfrac{kx}{y}$. Since $M = 5$ when $x = 2$ and $y = 6$, we have $5 = \dfrac{k\,(2)}{6}$ $\Leftrightarrow$ $k = 15$. Therefore $M = \dfrac{15x}{y}$.

17. Since W is inversely proportional to the square of r, $W = \dfrac{k}{r^2}$. Since $W = 10$ when $r = 6$, we have $10 = \dfrac{k}{(6)^2}$ $\Leftrightarrow$ $k = 360$. So $W = \dfrac{360}{r^2}$.

19. Since C is jointly proportional to l, w, and h, we have $C = klwh$. Since $C = 128$ when $l = w = h = 2$, we have $128 = k\,(2)\,(2)\,(2)$ $\Leftrightarrow$ $128 = 8k$ $\Leftrightarrow$ $k = 16$. Therefore, $C = 16lwh$.

21. Since s is inversely proportional to the square root of t, we have $s = \dfrac{k}{\sqrt{t}}$. Since $s = 100$ when $t = 25$, we have

$$100 = \frac{k}{\sqrt{25}} \quad \Leftrightarrow \quad 100 = \frac{k}{5} \quad \Leftrightarrow \quad k = 500. \text{ So } s = \frac{500}{\sqrt{t}}.$$

23. (a) The force F needed is $F = kx$.

   (b) Since $F = 40$ when $x = 5$, we have $40 = k\,(5) \quad \Leftrightarrow \quad k = 8$.

   (c) From part (b), we have $F = 8x$. Substituting $x = 4$ into $F = 8x$ gives $F = 8\,(4) = 32$ N.

25. (a) $C = kpm$

   (b) Since $C = 60{,}000$ when $p = 120$ and $m = 4000$, we get $60{,}000 = k\,(120)\,(4000) \quad \Leftrightarrow \quad k = \frac{1}{8}$. So $C = \frac{1}{8}pm$.

   (c) Substituting $p = 92$ and $m = 5{,}000$, we get $C = \frac{1}{8}\,(92)\,(5{,}000) = \$57{,}500$ and $k = \frac{1}{8}$.

27. (a) $P = ks^3$.

   (b) Since $P = 96$ when $s = 20$, we get $96 = k \cdot 20^3 \quad \Leftrightarrow \quad k = 0.012$. So $P = 0.012s^3$.

   (c) Substituting $x = 30$, we get $P = 0.012 \cdot 30^3 = 324$ watts.

29. $L = \dfrac{k}{d^2}$. Since $L = 70$ when $d = 10$, we have $70 = \dfrac{k}{10^2}$ so $k = 7{,}000$. Thus $L = \dfrac{7{,}000}{d^2}$. When $d = 100$ we get
$L = \dfrac{7{,}000}{100^2} = 0.7$ dB.

31. $P = kAv^3$. If $A = \frac{1}{2}A_0$ and $v = 2v_0$, then $P = k\left(\frac{1}{2}A_0\right)(2v_0)^3 = \frac{1}{2}kA_0\left(8v_0^3\right) = 4kA_0v_0^3$. The power is increased by a factor of 4.

33. $F = kAs^2$. Since $F = 220$ when $A = 40$ and $s = 5$. Solving for k we have $220 = k\,(40)\,(5)^2 \quad \Leftrightarrow \quad 220 = 1000k$
$\Leftrightarrow \quad k = 0.22$. Now when $A = 28$ and $F = 175$ we get $175 = 0.220\,(28)\,s^2 \quad \Leftrightarrow \quad 28.4090 = s^2$ so
$s = \sqrt{28.4090} = 5.33$ mi/h.

35. (a) $R = \dfrac{kL}{d^2}$

   (b) Since $R = 140$ when $L = 1.2$ and $d = 0.005$, we get $140 = \dfrac{k\,(1.2)}{(0.005)^2} \quad \Leftrightarrow \quad k = \frac{7}{2400} = 0.0029\overline{16}$.

   (c) Substituting $L = 3$ and $d = 0.008$, we have $R = \dfrac{7}{2400} \cdot \dfrac{3}{(0.008)^2} = \dfrac{4375}{32} \approx 137\ \Omega$.

37. (a) For the sun, $E_S = k6000^4$ and for earth $E_E = k300^4$. Thus $\dfrac{E_S}{E_E} = \dfrac{k6000^4}{k300^4} = \left(\frac{6000}{300}\right)^4 = 20^4 = 160{,}000$. So the sun produces 160,000 times the radiation energy per unit area than the Earth.

   (b) The surface area of the sun is $4\pi\,(435{,}000)^2$ and the surface area of the Earth is $4\pi\,(3{,}960)^2$. So the sun has
   $\dfrac{4\pi\,(435{,}000)^2}{4\pi\,(3{,}960)^2} = \left(\dfrac{435{,}000}{3{,}960}\right)^2$ times the surface area of the Earth. Thus the total radiation emitted by the sun is

   $$210{,}000 \times \left(\frac{435{,}000}{3{,}960}\right)^2 = 1{,}930{,}670{,}340 \text{ times the total radiation emitted by the Earth.}$$

39. Let S be the final size of the cabbage, in pounds, let N be the amount of nutrients it receives, in ounces, and let c be the number of other cabbages around it. Then $S = k\dfrac{N}{c}$. When $N = 20$ and $c = 12$, we have $S = 30$, so substituting, we have
$30 = k\frac{20}{12} \quad \Leftrightarrow \quad k = 18$. Thus $S = 18\dfrac{N}{c}$. When $N = 10$ and $c = 5$, the final size is $S = 18\left(\frac{10}{5}\right) = 36$ lb.

41. (a) Since f is inversely proportional to L, we have $f = \dfrac{k}{L}$, where k is a positive constant.

   (b) If we replace L by $2L$ we have $\dfrac{k}{2L} = \frac{1}{2} \cdot \dfrac{k}{L} = \frac{1}{2}f$. So the frequency of the vibration is cut in half.

43. Examples include radioactive decay and exponential growth in biology.

Chapter 2 Review

1. (a)

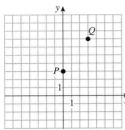

(b) The distance from P to Q is

$$d(P,Q) = \sqrt{(0-3)^2 + (3-7)^2}$$
$$= \sqrt{9+16} = \sqrt{25} = 5$$

(c) The midpoint is $\left(\dfrac{0+3}{2}, \dfrac{3+7}{2} \right) = \left(\dfrac{3}{2}, 5 \right)$.

(d) The line has slope $m = \dfrac{3-7}{0-3} = \dfrac{4}{3}$ and equation

$y - 3 = \frac{4}{3}(x-0) \quad \Leftrightarrow \quad y - 3 = \frac{4}{3}x \quad \Leftrightarrow$

$y = \frac{4}{3}x + 3.$

(e) The radius of this circle was found in part (b). It is $r = d(P,Q) = 5$. So an equation is

$$(x-0)^2 + (y-3)^2 = (5)^2 \quad \Leftrightarrow \quad x^2 + (y-3)^2 = 25.$$

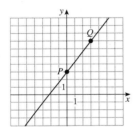

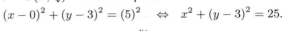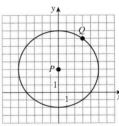

3. (a)

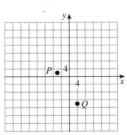

(b) The distance from P to Q is

$$d(P,Q) = \sqrt{(-6-4)^2 + [2-(-14)]^2}$$
$$= \sqrt{100 + 256} = \sqrt{356} = 2\sqrt{89}$$

(c) The midpoint is $\left(\dfrac{-6+4}{2}, \dfrac{2+(-14)}{2} \right) = (-1, -6)$.

(d) The line has slope

$m = \dfrac{2-(-14)}{-6-4} = \dfrac{16}{-10} = -\dfrac{8}{5}$ and equation

$y - 2 = -\frac{8}{5}(x+6) \quad \Leftrightarrow \quad y - 2 = -\frac{8}{5}x - \frac{48}{5}$

$\Leftrightarrow \quad y = -\frac{8}{5}x - \frac{38}{5}.$

(e) The radius of this circle was found in part (b). It is $r = d(P,Q) = 2\sqrt{89}$. So an equation is

$[x-(-6)]^2 + (y-2)^2 = \left(2\sqrt{89}\right)^2 \quad \Leftrightarrow$

$(x+6)^2 + (y-2)^2 = 356.$

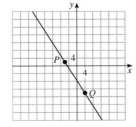

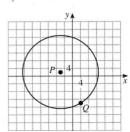

5.

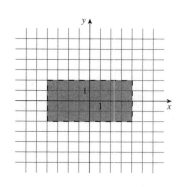

7. $d(A,C) = \sqrt{(4-(-1))^2 + (4-(-3))^2}$

$\qquad\qquad = \sqrt{(4+1)^2 + (4+3)^2}$

$\qquad\qquad = \sqrt{74}$

and

$d(B,C) = \sqrt{(5-(-1))^2 + (3-(-3))^2}$

$\qquad\qquad = \sqrt{(5+1)^2 + (3+3)^2}$

$\qquad\qquad = \sqrt{72}$

Therefore, B is closer to C.

9. The center is $C = (-5, -1)$, and the point $P = (0,0)$ is on the circle. The radius of the circle is

$r = d(P,C) = \sqrt{(0-(-5))^2 + (0-(-1))^2} = \sqrt{(0+5)^2 + (0+1)^2} = \sqrt{26}$. Thus, an equation of the circle is

$(x+5)^2 + (y+1)^2 = 26$.

11. $x^2 + y^2 + 2x - 6y + 9 = 0 \quad\Leftrightarrow\quad (x^2 + 2x) + (y^2 - 6y) = -9 \quad\Leftrightarrow\quad (x^2 + 2x + 1) + (y^2 - 6y + 9) = -9 + 1 + 9$

$\Leftrightarrow \quad (x+1)^2 + (y-3)^2 = 1$. This equation represents a circle with center at $(-1, 3)$ and radius 1.

13. $x^2 + y^2 + 72 = 12x \quad\Leftrightarrow\quad (x^2 - 12x) + y^2 = -72 \quad\Leftrightarrow\quad (x^2 - 12x + 36) + y^2 = -72 + 36 \quad\Leftrightarrow$

$(x-6)^2 + y^2 = -36$. Since the left side of this equation must be greater than or equal to zero, this equation has no graph.

15. $y = 2 - 3x$

x	y
-2	8
0	2
$\frac{2}{3}$	0

x-axis symmetry: $(-y) = 2 - 3x \quad\Leftrightarrow\quad y = -2 + 3x$, which is
not the same as the original equation, so the graph is not symmetric
with respect to the x-axis.

y-axis symmetry: $y = 2 - 3(-x) \quad\Leftrightarrow\quad y = 2 + 3x$, which is not
the same as the original equation, so the graph is not symmetric with
respect to the y-axis.

Origin symmetry: $(-y) = 2 - 3(-x) \quad\Leftrightarrow\quad -y = 2 + 3x \quad\Leftrightarrow$
$y = -2 - 3x$, which is not the
same as the original equation, so the graph is not symmetric with
respect to the origin.

Hence the graph has no symmetry.

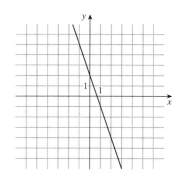

17. $x + 3y = 21 \quad\Leftrightarrow\quad y = -\frac{1}{3}x + 7$

x	y
-3	8
0	7
21	0

x-axis symmetry: $x + 3(-y) = 21 \quad\Leftrightarrow\quad x - 3y = 21$, which is
not the same as the original equation, so the graph is not symmetric
with respect to the x-axis.

y-axis symmetry: $(-x) + 3y = 21 \quad\Leftrightarrow\quad x - 3y = -21$, which is
not the same as the original equation, so the graph is not symmetric
with respect to the y-axis.

Origin symmetry: $(-x) + 3(-y) = 21 \quad\Leftrightarrow\quad x + 3y = -21$,
which is not the same as the original equation, so the graph is not
symmetric with respect to the origin. Hence the graph has no
symmetry.

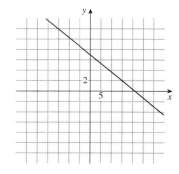

19. $\dfrac{x}{2} - \dfrac{y}{7} = 1 \iff y = \dfrac{7}{2}x - 7$

x	y
-2	-14
0	-7
2	0

x-axis symmetry: $\dfrac{x}{2} - \dfrac{(-y)}{7} = 1 \iff \dfrac{x}{2} + \dfrac{y}{7} = 1$, which is not the same as the original equation, so the graph is not symmetric with respect to the x-axis.

y-axis symmetry: $\dfrac{(-x)}{2} - \dfrac{y}{7} = 1 \iff \dfrac{x}{2} + \dfrac{y}{7} = -1$, which is not the same as the original equation, so the graph is not symmetric with respect to the y-axis.

Origin symmetry: $\dfrac{(-x)}{2} - \dfrac{(-y)}{7} = 1 \iff \dfrac{x}{2} - \dfrac{y}{7} = -1$, which is not the same as the original equation, so the graph is not symmetric with respect to the origin. Hence, the graph has no symmetry.

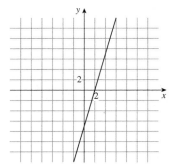

21. $y = 16 - x^2$

x	y
-3	7
-1	15
0	16
1	15
3	7

x-axis symmetry: $(-y) = 16 - x^2 \iff y = -16 + x^2$, which is not the same as the original equation, so the graph is not symmetric with respect to the x-axis.

y-axis symmetry: $y = 16 - (-x)^2 \iff y = 16 - x^2$, which is the same as the original equation, so its is symmetric with respect to the y-axis.

Origin symmetry: $(-y) = 16 - (-x)^2 \iff y = -16 + x^2$, which is not the same as the original equation, so the graph is not symmetric with respect to the origin. Hence, the graph is symmetric with respect to the y-axis.

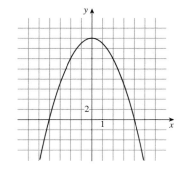

23. $x = \sqrt{y}$

x	y
0	0
1	1
2	4
3	9

x-axis symmetry: $x = \sqrt{-y}$, which is not the same as the original equation, so the graph is not symmetric with respect to the x-axis.

y-axis symmetry: $(-x) = \sqrt{y} \iff x = -\sqrt{y}$, which is not the same as the original equation, so the graph is not symmetric with respect to the y-axis.

Origin symmetry: $(-x) = \sqrt{-y}$, which is not the same as the original equation, so the graph is not symmetric with respect to the origin. Hence, the graph has no symmetry.

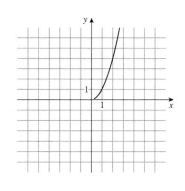

25. $y = x^2 - 6x$. Viewing rectangle $[-10, 10]$ by $[-10, 10]$.

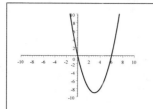

27. $y = x^3 - 4x^2 - 5$. Viewing rectangle $[-4, 10]$ by $[-30, 20]$.

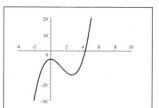

29. $x^2 - 4x = 2x + 7$. We graph the equations $y_1 = x^2 - 4x$ and $y_2 = 2x + 7$ in the viewing rectangle rectangle $[-10, 10]$ by $[-5, 25]$. Using a zoom or trace function, we get the solutions $x = -1$ and $x = 7$.

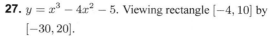

31. $x^4 - 9x^2 = x - 9$. We graph the equations $y_1 = x^4 - 9x^2$ and $y_2 = x - 9$ in the viewing rectangle $[-5, 5]$ by $[-25, 10]$. Using a zoom or trace function, we get the solutions $x \approx -2.72$, $x \approx -1.15$, $x = 1.00$, and $x \approx 2.87$.

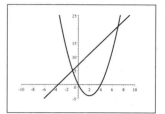

33. $4x - 3 \geq x^2$. We graph the equations $y_1 = 4x - 3$ and $y_2 = x^2$ in the viewing rectangle $[-5, 5]$ by $[0, 15]$. Using a zoom or trace function, we find the points of intersection are at $x = 1$ and $x = 3$. Since we want $4x - 3 \geq x^2$, the solution is the interval $[1, 3]$.

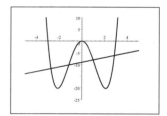

35. $x^4 - 4x^2 < \frac{1}{2}x - 1$. We graph the equations $y_1 = x^4 - 4x^2$ and $y_2 = \frac{1}{2}x - 1$ in the viewing rectangle $[-5, 5]$ by $[-5, 5]$. We find the points of intersection are at $x \approx -1.85$, $x \approx -0.60$, $x \approx 0.45$, and $x = 2.00$. Since we want $x^4 - 4x^2 < \frac{1}{2}x - 1$, the solution is $(-1.85, -0.60) \cup (0.45, 2.00)$.

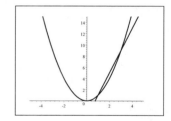

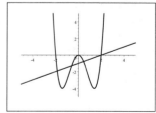

37. The line has slope $m = \dfrac{-4 + 6}{2 + 1} = \dfrac{2}{3}$, and so, by the point-slope formula, the equation is $y + 4 = \frac{2}{3}(x - 2)$ $\Leftrightarrow$ $y = \frac{2}{3}x - \frac{16}{3}$ $\Leftrightarrow$ $2x - 3y - 16 = 0$.

39. The x-intercept is 4, and the y-intercept is 12, so the slope is $m = \dfrac{12 - 0}{0 - 4} = -3$. Therefore, by the slope-intercept formula, the equation of the line is $y = -3x + 12$ $\Leftrightarrow$ $3x + y - 12 = 0$.

41. We first find the slope of the line $3x + 15y = 22$. This gives $3x + 15y = 22$ $\Leftrightarrow$ $15y = -3x + 22$ $\Leftrightarrow$ $y = -\frac{1}{5}x + \frac{22}{15}$. So this line has slope $m = -\frac{1}{5}$, as does any line parallel to it. Then the parallel line passing through the origin has equation $y - 0 = -\frac{1}{5}(x - 0)$ $\Leftrightarrow$ $x + 5y = 0$.

43. (a) The slope, 0.3, represents the increase in length of the spring for each unit increase in weight w. The S-intercept is the resting or natural length of the spring.

(b) When $w = 5$, $S = 0.3\,(5) + 2.5 = 1.5 + 2.5 = 4.0$ inches.

45. Since M varies directly as z we have $M = kz$. Substituting $M = 120$ when $z = 15$, we find $120 = k\,(15) \quad \Leftrightarrow \quad k = 8$. Therefore, $M = 8z$.

47. (a) The intensity I varies inversely as the square of the distance d, so $I = \dfrac{k}{d^2}$.

(b) Substituting $I = 1000$ when $d = 8$, we get $1000 = \dfrac{k}{(8)^2} \quad \Leftrightarrow \quad k = 64{,}000$.

(c) From parts (a) and (b), we have $I = \dfrac{64{,}000}{d^2}$. Substituting $d = 20$, we get $I = \dfrac{64{,}000}{(20)^2} = 160$ candles.

49. Let v be the terminal velocity of the parachutist in mi/h and w be his weight in pounds. Since the terminal velocity is directly proportional to the square root of the weight, we have $v = k\sqrt{w}$. Substituting $v = 9$ when $w = 160$, we solve for k. This gives $9 = k\sqrt{160} \quad \Leftrightarrow \quad k = \dfrac{9}{\sqrt{160}} \approx 0.712$. Thus $v = 0.712\sqrt{w}$. When $w = 240$, the terminal velocity is $v = 0.712\sqrt{240} \approx 11$ mi/h.

51. Here the center is at $(0,0)$, and the circle passes through the point $(-5, 12)$, so the radius is $r = \sqrt{(-5-0)^2 + (12-0)^2} = \sqrt{25 + 144} = \sqrt{169} = 13$. The equation of the circle is $x^2 + y^2 = 13^2 \quad \Leftrightarrow \quad x^2 + y^2 = 169$. The line shown is the tangent that passes through the point $(-5, 12)$, so it is perpendicular to the line through the points $(0,0)$ and $(-5, 12)$. This line has slope $m_1 = \dfrac{12 - 0}{-5 - 0} = -\dfrac{12}{5}$. The slope of the line we seek is $m_2 = -\dfrac{1}{m_1} = -\dfrac{1}{-12/5} = \dfrac{5}{12}$. Thus, an equation of the tangent line is $y - 12 = \frac{5}{12}\,(x + 5) \quad \Leftrightarrow \quad y - 12 = \frac{5}{12}x + \frac{25}{12} \quad \Leftrightarrow \quad y = \frac{5}{12}x + \frac{169}{12} \quad \Leftrightarrow \quad 5x - 12y + 169 = 0$.

Chapter 2 Test

1. (a)

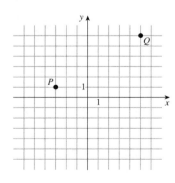

(b) The distance between P and Q is
$$d\,(P, Q) = \sqrt{(-3 - 5)^2 + (1 - 6)^2} = \sqrt{64 + 25} = \sqrt{89}.$$

(c) The midpoint is $\left(\dfrac{-3 + 5}{2}, \dfrac{1 + 6}{2}\right) = \left(1, \frac{7}{2}\right)$.

(d) The slope of the line is $\dfrac{1 - 6}{-3 - 5} = \dfrac{-5}{-8} = \dfrac{5}{8}$.

(e) The perpendicular bisector of PQ contains the midpoint, $\left(1, \frac{7}{2}\right)$, and it slope is the negative reciprocal of $\frac{5}{8}$. Thus the slope is $-\dfrac{1}{5/8} = -\dfrac{8}{5}$. Hence the equation is $y - \frac{7}{2} = -\frac{8}{5}\,(x - 1) \quad \Leftrightarrow \quad y = -\frac{8}{5}x + \frac{8}{5} + \frac{7}{2} = -\frac{8}{5}x + \frac{51}{10}$. That is, $y = -\frac{8}{5}x + \frac{51}{10}$.

(f) The center of the circle is the midpoint, $\left(1, \frac{7}{2}\right)$, and the length of the radius is $\frac{1}{2}\sqrt{89}$. Thus the equation of the circle whose diameter is PQ is $(x - 1)^2 + \left(y - \frac{7}{2}\right)^2 = \left(\frac{1}{2}\sqrt{89}\right)^2 \quad \Leftrightarrow \quad (x - 1)^2 + \left(y - \frac{7}{2}\right)^2 = \frac{89}{4}$.

2. (a) $x^2 + y^2 = 25 = 5^2$ has center $(0,0)$ and radius 5.

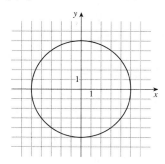

(b) $(x-2)^2 + (y+1)^2 = 9 = 3^2$ has center $(2,-1)$ and radius 3.

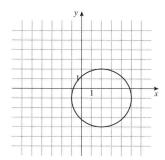

(c) $x^2 + 6x + y^2 - 2y + 6 = 0 \Leftrightarrow$ $x^2 + 6x + 9 + y^2 - 2y + 1 = 4 \Leftrightarrow$ $(x+3)^2 + (y-1)^2 = 4 = 2^2$ has center $(-3, 1)$ and radius 2.

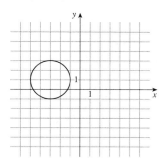

3. $2x - 3y = 15 \quad \Leftrightarrow$ $-3y = -2x + 15 \quad \Leftrightarrow \quad y = \frac{2}{3}x - 5.$

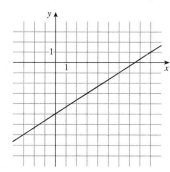

4. (a) $3x + y - 10 = 0 \quad \Leftrightarrow \quad y = -3x + 10$, so the slope of the line we seek is -3. Using the point-slope, $y - (-6) = -3(x-3) \quad \Leftrightarrow$ $y + 6 = -3x + 9 \quad \Leftrightarrow \quad 3x + y - 3 = 0.$

(b) Using the intercept form we get $\frac{x}{6} + \frac{y}{4} = 1 \quad \Leftrightarrow \quad 2x + 3y = 12$ $\Leftrightarrow \quad 2x + 3y - 12 = 0.$

5. (a) When $x = 100$ we have $T = 0.08(100) - 4 = 8 - 4 = 4$, so the temperature at one meter is $4°$ C.

(c) The slope represents the raise in temperature as the depth increase. The T-intercept is the surface temperature of the soil and the x-intercept represents the depth of the "frost line", where the soil below is not frozen.

(b)

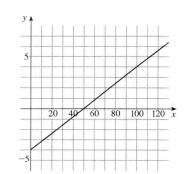

6. (a) $x^3 - 9x - 1 = 0$. We graph the equation $y = x^3 - 9x - 1$ in the viewing rectangle $[-5, 5]$ by $[-10, 10]$. We find that the points of intersection occur at $x \approx -2.94, -0.11, 3.05$.

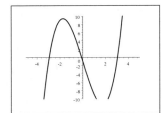

(b) $\frac{1}{2}x + 2 \geq \sqrt{x^2 + 1}$. We graph the equations $y_1 = \frac{1}{2}x + 2$ and $y_2 = \sqrt{x^2 + 1}$ in the viewing rectangle $[-5, 5]$ by $[0, 5]$. We find that the points of intersection occur at $x \approx -0.78$ and $x \approx 1.28$. Since we want $\frac{1}{2}x + 2 \geq \sqrt{x^2 + 1}$, the solution is approximately the interval $[-0.78, 1.28]$.

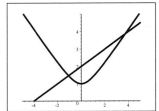

7. (a) $M = k\dfrac{wh^2}{L}$

(b) Substituting $w = 4$, $h = 6$, $L = 12$, and $M = 4800$, we have $4800 = k\dfrac{(4)\left(6^2\right)}{12}$ $\quad \Leftrightarrow \quad k = 400$. Thus $M = 400\dfrac{wh^2}{L}$.

(c) Now if $L = 10$, $w = 3$, and $h = 10$, then $M = 400\dfrac{(3)\left(10^2\right)}{10} = 12{,}000$. So the beam can support 12,000 pounds.

Focus on Modeling: Fitting Lines to Data

1. (a)

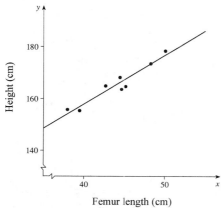

Femur length (cm)

(b) Using a graphing calculator, we obtain the regression line $y = 1.8807x + 82.65$.

(c) Using $x = 58$ in the equation $y = 1.8807x + 82.65$, we get $y = 1.8807(58) + 82.65 \approx 191.7$ cm.

3. (a)

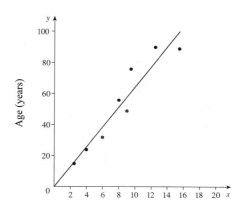

Diameter (in.)

(b) Using a graphing calculator, we obtain the regression line $y = 6.451x - 0.1523$.

(c) Using $x = 18$ in the equation $y = 6.451x - 0.1523$, we get $y = 6.451(18) - 0.1523 \approx 116$ years.

5. (a)

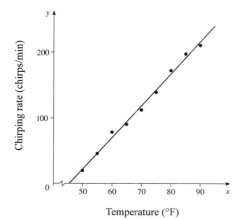

Temperature (°F)

(b) Using a graphing calculator, we obtain the regression line $y = 4.857x - 220.97$.

(c) Using $x = 100°$ F in the equation $y = 4.857x - 220.97$, we get $y \approx 265$ chirps per minute.

7. (a)

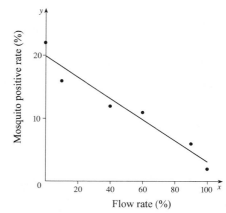

Flow rate (%)

(b) Using a graphing calculator, we obtain the regression line $y = -0.168x + 19.89$.

(c) Using the regression line equation $y = -0.168x + 19.89$, we get $y \approx 8.13\%$ when $x = 70\%$.

9. (a)

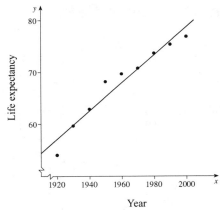

Year

(b) Using a graphing calculator, we obtain
$y = 0.27083x - 462.9$.

(c) We substitute $x = 2004$ in the model
$y = 0.27083x - 462.9$ to get $y = 79.8$, that is, a life
expectancy of 79.8 years.

(d) As of this writing, data for 2004 are not yet available.
The life expectancy of a child born in the US in 2003
is 77.6 years.

13. The students should find a fairly strong correlation between shoe size and height.

11. (a) If we take $x = 0$ in 1900 for both men and women,
then the regression equation for the men's data is
$y = -0.173x + 64.72$ and the regression equation for
the women's data is $y = -0.269x + 78.67$.

(b)

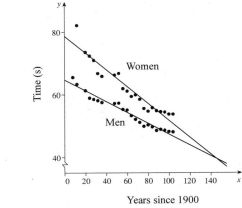

Years since 1900

These lines predict that the women will overtake the
men in this event when
$-0.173x + 64.72 = -0.269x + 78.67$ $\Leftrightarrow$
$0.096x = 13.95$ $\Leftrightarrow$ $x = 145.31$, or in 2045.
This seems unlikely, but who knows?

3 Functions

3.1 What is a Function?

1. $f(x) = 2(x+3)$

3. $f(x) = (x-5)^2$

5. Subtract 4, then divide by 3.

7. Square, then add 2.

9. Machine diagram for $f(x) = \sqrt{x-1}$.

x	$f(x)$
-1	$2(-1-1)^2 = 8$
0	$2(-1)^2 = 2$
1	$2(1-1)^2 = 0$
2	$2(2-1)^2 = 2$
3	$2(3-1)^2 = 8$

11. $f(x) = 2(x-1)^2$

13. $f(1) = 2(1)+1 = 3$; $f(-2) = 2(-2)+1 = -3$; $f\left(\frac{1}{2}\right) = 2\left(\frac{1}{2}\right)+1 = 2$; $f(a) = 2(a)+1 = 2a+1$;
$f(-a) = 2(-a)+1 = -2a+1$; $f(a+b) = 2(a+b)+1 = 2a+2b+1$.

15. $g(2) = \dfrac{1-(2)}{1+(2)} = \dfrac{-1}{3} = -\dfrac{1}{3}$; $g(-2) = \dfrac{1-(-2)}{1+(-2)} = \dfrac{3}{-1} = -3$; $g\left(\dfrac{1}{2}\right) = \dfrac{1-\left(\frac{1}{2}\right)}{1+\left(\frac{1}{2}\right)} = \dfrac{\frac{1}{2}}{\frac{3}{2}} = \dfrac{1}{3}$;
$g(a) = \dfrac{1-(a)}{1+(a)} = \dfrac{1-a}{1+a}$; $g(a-1) = \dfrac{1-(a-1)}{1+(a-1)} = \dfrac{1-a+1}{1+a-1} = \dfrac{2-a}{a}$; $g(-1) = \dfrac{1-(-1)}{1+(-1)} = \dfrac{2}{0}$, so $g(-1)$ is not
defined.

17. $f(0) = 2(0)^2 + 3(0) - 4 = -4$; $f(2) = 2(2)^2 + 3(2) - 4 = 8 + 6 - 4 = 10$;
$f(-2) = 2(-2)^2 + 3(-2) - 4 = 8 - 6 - 4 = -2$; $f(\sqrt{2}) = 2(\sqrt{2})^2 + 3(\sqrt{2}) - 4 = 4 + 3\sqrt{2} - 4 = 3\sqrt{2}$;
$f(x+1) = 2(x+1)^2 + 3(x+1) - 4 = 2x^2 + 4x + 2 + 3x + 3 - 4 = 2x^2 + 7x + 1$;
$f(-x) = 2(-x)^2 + 3(-x) - 4 = 2x^2 - 3x - 4$.

19. $f(-2) = 2|-2-1| = 2(3) = 6$; $f(0) = 2|0-1| = 2(1) = 2$;
$f\left(\frac{1}{2}\right) = 2\left|\frac{1}{2}-1\right| = 2\left(\frac{1}{2}\right) = 1$; $f(2) = 2|2-1| = 2(1) = 2$; $f(x+1) = 2|(x+1)-1| = 2|x|$;
$f(x^2+2) = 2|(x^2+2)-1| = 2|x^2+1| = 2x^2 + 2$ (since $x^2 + 1 > 0$).

21. Since $-2 < 0$, we have $f(-2) = (-2)^2 = 4$. Since $-1 < 0$, we have $f(-1) = (-1)^2 = 1$. Since $0 \geq 0$, we have
$f(0) = 0 + 1 = 1$. Since $1 \geq 0$, we have $f(1) = 1 + 1 = 2$. Since $2 \geq 0$, we have $f(2) = 2 + 1 = 3$.

23. Since $-4 \leq -1$, we have $f(-4) = (-4)^2 + 2(-4) = 16 - 8 = 8$. Since $-\frac{3}{2} \leq -1$, we have
$f\left(-\frac{3}{2}\right) = \left(-\frac{3}{2}\right)^2 + 2\left(-\frac{3}{2}\right) = \frac{9}{4} - 3 = -\frac{3}{4}$. Since $-1 \leq -1$, we have $f(-1) = (-1)^2 + 2(-1) = 1 - 2 = -1$. Since
$-1 < 0 \leq 1$, we have $f(0) = 0$. Since $25 > 1$, we have $f(25) = -1$.

25. $f(x+2) = (x+2)^2 + 1 = x^2 + 4x + 4 + 1 = x^2 + 4x + 5$; $f(x) + f(2) = x^2 + 1 + (2)^2 + 1 = x^2 + 1 + 4 + 1 = x^2 + 6$.

27. $f(x^2) = x^2 + 4$; $[f(x)]^2 = [x+4]^2 = x^2 + 8x + 16$.

29. $f(a) = 3(a) + 2 = 3a + 2$; $f(a+h) = 3(a+h) + 2 = 3a + 3h + 2$;
$\dfrac{f(a+h) - f(a)}{h} = \dfrac{(3a+3h+2) - (3a+2)}{h} = \dfrac{3a + 3h + 2 - 3a - 2}{h} = \dfrac{3h}{h} = 3$.

31. $f(a) = 5$; $f(a+h) = 5$; $\dfrac{f(a+h) - f(a)}{h} = \dfrac{5-5}{h} = 0$.

33. $f(a) = \dfrac{a}{a+1}$; $f(a+h) = \dfrac{a+h}{a+h+1}$;

$$\frac{f(a+h) - f(a)}{h} = \frac{\dfrac{a+h}{a+h+1} - \dfrac{a}{a+1}}{h} = \frac{\dfrac{(a+h)(a+1)}{(a+h+1)(a+1)} - \dfrac{a(a+h+1)}{(a+h+1)(a+1)}}{h}$$

$$= \frac{\dfrac{(a+h)(a+1) - a(a+h+1)}{(a+h+1)(a+1)}}{h} = \frac{a^2 + a + ah + h - (a^2 + ah + a)}{h(a+h+1)(a+1)}$$

$$= \frac{1}{(a+h+1)(a+1)}$$

35. $f(a) = 3 - 5a + 4a^2$;

$$f(a+h) = 3 - 5(a+h) + 4(a+h)^2 = 3 - 5a - 5h + 4\left(a^2 + 2ah + h^2\right)$$
$$= 3 - 5a - 5h + 4a^2 + 8ah + 4h^2;$$

$$\frac{f(a+h) - f(a)}{h} = \frac{\left(3 - 5a - 5h + 4a^2 + 8ah + 4h^2\right) - \left(3 - 5a + 4a^2\right)}{h}$$

$$= \frac{3 - 5a - 5h + 4a^2 + 8ah + 4h^2 - 3 + 5a - 4a^2}{h} = \frac{-5h + 8ah + 4h^2}{h}$$

$$= \frac{h(-5 + 8a + 4h)}{h} = -5 + 8a + 4h.$$

37. $f(x) = 2x$. Since there is no restrictions, the domain is the set of real numbers, $(-\infty, \infty)$.

39. $f(x) = 2x$. The domain is restricted by the exercise to $[-1, 5]$.

41. $f(x) = \dfrac{1}{x-3}$. Since the denominator cannot equal 0 we have $x - 3 \neq 0 \quad \Leftrightarrow \quad x \neq 3$. Thus the domain is $\{x \mid x \neq 3\}$. In interval notation, the domain is $(-\infty, 3) \cup (3, \infty)$.

43. $f(x) = \dfrac{x+2}{x^2 - 1}$. Since the denominator cannot equal 0 we have $x^2 - 1 \neq 0 \quad \Leftrightarrow \quad x^2 \neq 1 \quad \Rightarrow \quad x \neq \pm 1$. Thus the domain is $\{x \mid x \neq \pm 1\}$. In interval notation, the domain is $(-\infty, -1) \cup (-1, 1) \cup (1, \infty)$.

45. $f(x) = \sqrt{x - 5}$. We require $x - 5 \geq 0 \quad \Leftrightarrow \quad x \geq 5$. Thus the domain is $\{x \mid x \geq 5\}$. The domain can also be expressed in interval notation as $[5, \infty)$.

47. $f(t) = \sqrt[3]{t - 1}$. Since the odd root is defined for all real numbers, the domain is the set of real numbers, $(-\infty, \infty)$.

49. $h(x) = \sqrt{2x - 5}$. Since the square root is defined as a real number only for nonnegative numbers, we require that $2x - 5 \geq 0 \quad \Leftrightarrow \quad 2x \geq 5 \quad \Leftrightarrow \quad x \geq \frac{5}{2}$. So the domain is $\{x \mid x \geq \frac{5}{2}\}$. In interval notation, the domain is $\left[\frac{5}{2}, \infty\right)$.

51. $g(x) = \dfrac{\sqrt{2 + x}}{3 - x}$. We require $2 + x \geq 0$, and the denominator cannot equal 0. Now $2 + x \geq 0 \quad \Leftrightarrow \quad x \geq -2$, and $3 - x \neq 0 \quad \Leftrightarrow \quad x \neq 3$. Thus the domain is $\{x \mid x \geq -2 \text{ and } x \neq 3\}$, which can be expressed in interval notation as $[-2, 3) \cup (3, \infty)$.

53. $g(x) = \sqrt[4]{x^2 - 6x}$. Since the input to an even root must be nonnegative, we have $x^2 - 6x \geq 0 \quad \Leftrightarrow \quad x(x - 6) \geq 0$. We make a table:

	$(-\infty, 0)$	$(0, 6)$	$(6, \infty)$
Sign of x	$-$	$+$	$+$
Sign of $x - 6$	$-$	$-$	$+$
Sign of $x(x - 6)$	$+$	$-$	$+$

Thus the domain is $(-\infty, 0] \cup [6, \infty)$.

55. $f(x) = \dfrac{3}{\sqrt{x-4}}$. Since the input to an even root must be nonnegative and the denominator cannot equal 0, we have

$x - 4 > 0 \quad \Leftrightarrow \quad x > 4$. Thus the domain is $(4, \infty)$.

57. $f(x) = \dfrac{(x+1)^2}{\sqrt{2x-1}}$. Since the input to an even root must be nonnegative and the denominator cannot equal 0, we have

$2x - 1 > 0 \quad \Leftrightarrow \quad x > \frac{1}{2}$. Thus the domain is $\left(\frac{1}{2}, \infty\right)$.

59. **(a)** $C(10) = 1500 + 3(10) + 0.02(10)^2 + 0.0001(10)^3 = 1500 + 30 + 2 + 0.1 = 1532.1$

$C(100) = 1500 + 3(100) + 0.02(100)^2 + 0.0001(100)^3 = 1500 + 300 + 200 + 100 = 2100$

(b) $C(10)$ represents the cost of producing 10 yards of fabric and $C(100)$ represents the cost of producing 100 yards of fabric.

(c) $C(0) = 1500 + 3(0) + 0.02(0)^2 + 0.0001(0)^3 = 1500$

61. **(a)** $D(0.1) = \sqrt{2(3960)(0.1) + (0.1)^2} = \sqrt{792.01} \approx 28.1$ miles

$D(0.2) = \sqrt{2(3960)(0.2) + (0.2)^2} = \sqrt{1584.04} \approx 39.8$ miles

(b) 1135 feet $= \frac{1135}{5280}$ miles ≈ 0.215 miles. $D(0.215) = \sqrt{2(3960)(0.215) + (0.215)^2} = \sqrt{1702.846} \approx 41.3$ miles

(c) $D(7) = \sqrt{2(3960)(7) + (7)^2} = \sqrt{55489} \approx 235.6$ miles

63. **(a)** $v(0.1) = 18500\left(0.25 - 0.1^2\right) = 4440$,

$v(0.4) = 18500\left(0.25 - 0.4^2\right) = 1665$.

(b) They tell us that the blood flows much faster (about 2.75 times faster) 0.1 cm from the center than 0.1 cm from the edge.

(c)

r	$v(r)$
0	4625
0.1	4440
0.2	3885
0.3	2960
0.4	1665
0.5	0

65. **(a)** $L(0.5c) = 10\sqrt{1 - \dfrac{(0.5c)^2}{c^2}} \approx 8.66$ m, $L(0.75c) = 10\sqrt{1 - \dfrac{(0.75c)^2}{c^2}} \approx 6.61$ m, and

$L(0.9c) = 10\sqrt{1 - \dfrac{(0.9c)^2}{c^2}} \approx 4.36$ m.

(b) It will appear to get shorter.

67. **(a)** $C(75) = 75 + 15 = \$90$; $C(90) = 90 + 15 = \$105$; $C(100) = \$100$; and $C(105) = \$105$.

(b) The total price of the books purchased, including shipping.

69. **(a)** $F(x) = \begin{cases} 15(40 - x) & \text{if } 0 < x < 40 \\ 0 & \text{if } 40 \le x \le 65 \\ 15(x - 65) & \text{if } x > 65 \end{cases}$.

(b) $F(30) = 15(40 - 10) = 15 \cdot 10 = \$150$; $F(50) = \$0$; and $F(75) = 15(75 - 65)\, 15 \cdot 10 = \$150$.

(c) The fines for violating the speed limits on the freeway.

71.

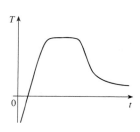

73.

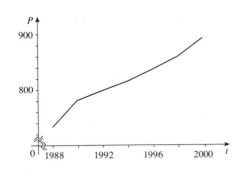

75. Answers will vary.

3.2 Graphs of Functions

1.

x	$f(x) = 2$
-9	2
-6	2
-3	2
0	2
3	2
6	2

3.

x	$f(x) = 2x - 4$
-1	-6
0	-4
1	-2
2	0
3	2
4	4
5	6

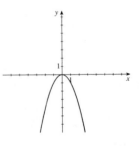

5.

x	$f(x) = -x + 3,$ $-3 \le x \le 3$
-3	6
-2	5
0	3
1	2
2	1
3	0

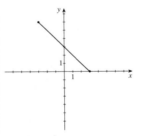

7.

x	$f(x) = -x^2$
± 4	-16
± 3	-9
± 2	-4
± 1	-1
0	0

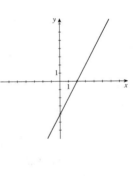

9.

x	$g(x) = x^3 - 8$
-2	-16
-1	-9
0	-8
1	-7
2	0
3	19

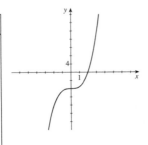

11.

x	$g(x) = \sqrt{x + 4}$
-4	0
-3	1
-2	1.414
-1	1.732
0	2
1	2.236

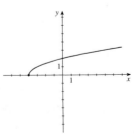

13.

x	$F(x) = \dfrac{1}{x}$
-5	-0.2
-1	-1
-0.1	-10
0.1	10
1	1
5	0.2

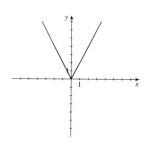

15.

x	$H(x) = |2x|$
± 5	10
± 4	8
± 3	6
± 2	4
± 1	2
0	0

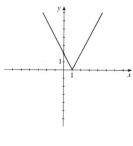

17.

x	$G(x) = |x| + x$
-5	0
-2	0
0	0
1	2
2	4
5	10

19.

x	$f(x) = |2x - 2|$
-5	12
-2	8
0	2
1	0
2	2
5	8

21.

x	$g(x) = \dfrac{2}{x^2}$
± 5	0.08
± 4	0.125
± 3	0.2
± 2	0.5
± 1	2
± 0.5	8

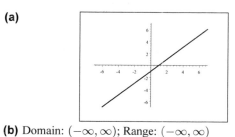

23. (a) $h(-2) = 1$; $h(0) = -1$; $h(2) = 3$; $h(3) = 4$.

 (b) Domain: $[-3, 4]$. Range: $[-1, 4]$.

25. (a) $f(0) = 3 > \frac{1}{2} = g(0)$. So $f(0)$ is larger.

 (b) $f(-3) \approx -\frac{3}{2} < 2 = g(-3)$. So $g(-3)$ is larger.

 (c) For $x = -2$ and $x = 2$.

27. (a)

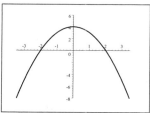

 (b) Domain: $(-\infty, \infty)$; Range: $(-\infty, \infty)$

29. (a)

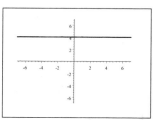

 (b) Domain: $(-\infty, \infty)$; Range: $\{4\}$

31. (a)

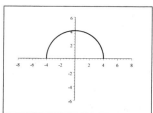

 (b) Domain: $(-\infty, \infty)$; Range: $(-\infty, 4]$

33. (a)

 (b) Domain: $[-4, 4]$; Range: $[0, 4]$

35. (a)

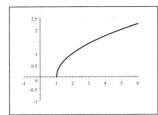

(b) Domain: $[1, \infty)$; Range: $[0, \infty)$

37. $f(x) = \begin{cases} 0 & \text{if } x < 2 \\ 1 & \text{if } x \geq 2 \end{cases}$

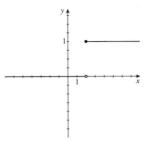

39. $f(x) = \begin{cases} 3 & \text{if } x < 2 \\ x - 1 & \text{if } x \geq 2 \end{cases}$

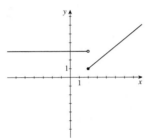

41. $f(x) = \begin{cases} x & \text{if } x \leq 0 \\ x + 1 & \text{if } x > 0 \end{cases}$

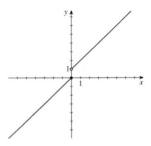

43. $f(x) = \begin{cases} -1 & \text{if } x < -1 \\ 1 & \text{if } -1 \leq x \leq 1 \\ -1 & \text{if } x > 1 \end{cases}$

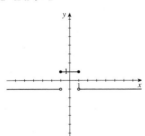

45. $f(x) = \begin{cases} 2 & \text{if } x \leq -1 \\ x^2 & \text{if } x > -1 \end{cases}$

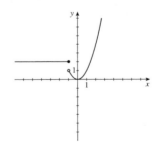

47. $f(x) = \begin{cases} 0 & \text{if } |x| \leq 2 \\ 3 & \text{if } |x| > 2 \end{cases}$

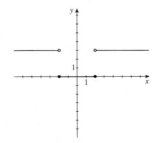

49. $f(x) = \begin{cases} 4 & \text{if } x < -2 \\ x^2 & \text{if } -2 \leq x \leq 2 \\ -x + 6 & \text{if } x > 2 \end{cases}$

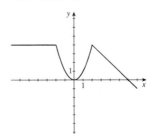

51. $f(x) = \begin{cases} x + 2 & \text{if } x \le -1 \\ x^2 & \text{if } x > -1 \end{cases}$

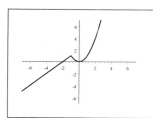

53. $f(x) = \begin{cases} -2 & \text{if } x < -2 \\ x & \text{if } -2 \le x \le 2 \\ 2 & \text{if } x > 2 \end{cases}$

55. The curves in parts (a) and (c) are graphs of a function of x, by the Vertical Line Test.

57. The given curve is the graph of a function of x. Domain: $[-3, 2]$. Range: $[-2, 2]$.

59. No, the given curve is not the graph of a function of x, by the Vertical Line Test.

61. Solving for y in terms of x gives $x^2 + 2y = 4$ $\Leftrightarrow$ $2y = 4 - x^2$ $\Leftrightarrow$ $y = 2 - \frac{1}{2}x^2$. This defines y as a function of x.

63. Solving for y in terms of x gives $x = y^2$ $\Leftrightarrow$ $y = \pm\sqrt{x}$. The last equation gives two values of y for a given value of x. Thus, this equation does not define y as a function of x.

65. Solving for y in terms of x gives $x + y^2 = 9$ $\Leftrightarrow$ $y^2 = 9 - x$ $\Leftrightarrow$ $y = \pm\sqrt{9 - x}$. The last equation gives two values of y for a given value of x. Thus, this equation does not define y as a function of x.

67. Solving for y in terms of x gives $x^2 y + y = 1 \Leftrightarrow y(x^2 + 1) = 1 \Leftrightarrow y = \dfrac{1}{x^2 + 1}$. This defines y as a function of x.

69. Solving for y in terms of x gives $2|x| + y = 0$ $\Leftrightarrow$ $y = -2|x|$. This defines y as a function of x.

71. Solving for y in terms of x gives $x = y^3$ $\Leftrightarrow$ $y = \sqrt[3]{x}$. This defines y as a function of x.

73. (a) $f(x) = x^2 + c$, for $c = 0, 2, 4,$ and 6.

(b) $f(x) = x^2 + c$, for $c = 0, -2, -4$, and -6.

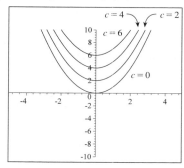

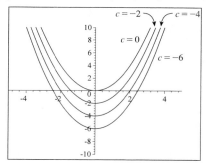

(c) The graphs in part (a) are obtained by shifting the graph of $f(x) = x^2$ upward c units, $c > 0$. The graphs in part (b) are obtained by shifting the graph of $f(x) = x^2$ downward c units.

75. (a) $f(x) = (x - c)^3$, for $c = 0, 2, 4,$ and 6.

(b) $f(x) = (x - c)^3$, for $c = 0, -2, -4$, and -6.

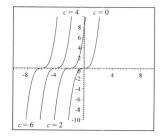

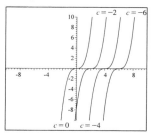

(c) The graphs in part (a) are obtained by shifting the graph of $f(x) = x^3$ to the right c units, $c > 0$. The graphs in part (b) are obtained by shifting the graph of $f(x) = x^3$ to the left $|c|$ units, $c < 0$.

77. (a) $f(x) = x^c$, for $c = \frac{1}{2}, \frac{1}{4}$, and $\frac{1}{6}$.

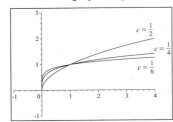

(b) $f(x) = x^c$, for $c = 1, \frac{1}{3}$, and $\frac{1}{5}$.

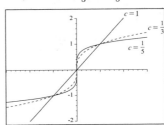

(c) Graphs of even roots are similar to $y = \sqrt{x}$, graphs of odd roots are similar to $y = \sqrt[3]{x}$. As c increases, the graph of $y = \sqrt[c]{x}$ becomes steeper near $x = 0$ and flatter when $x > 1$.

79. The slope of the line segment joining the points $(-2, 1)$ and $(4, -6)$ is $m = \dfrac{-6 - 1}{4 - (-2)} = -\frac{7}{6}$. Using the point-slope form,

we have $y - 1 = -\frac{7}{6}(x + 2)$ $\Leftrightarrow$ $y = -\frac{7}{6}x - \frac{7}{3} + 1$ $\Leftrightarrow$ $y = -\frac{7}{6}x - \frac{4}{3}$. Thus the function is $f(x) = -\frac{7}{6}x - \frac{4}{3}$ for $-2 \le x \le 4$.

81. First solve the circle for y: $x^2 + y^2 = 9$ $\Leftrightarrow$ $y^2 = 9 - x^2$ $\Rightarrow$ $y = \pm\sqrt{9 - x^2}$. Since we seek the top half of the circle, we choose $y = \sqrt{9 - x^2}$. So the function is $f(x) = \sqrt{9 - x^2}$, $-3 \le x \le 3$.

83. This person appears to gain weight steadily until the age of 21 when this person's weight gain slows down. At age 30, this person experiences a sudden weight loss, but recovers, and by about age 40, this person's weight seems to level off at around 200 pounds. It then appears that the person dies at about age 68. The sudden weight loss could be due to a number of reasons, among them major illness, a weight loss program, etc.

85. Runner A won the race. All runners finished the race. Runner B fell, but got up and finished the race.

87. (a) The first noticeable movements occurred at time $t = 5$ seconds.

(b) It seemed to end at time $t = 30$ seconds.

(c) Maximum intensity was reached at $t = 17$ seconds.

89. $C(x) = \begin{cases} 2.00 & \text{if } 0 < x \le 1 \\ 2.20 & \text{if } 1 < x \le 1.1 \\ 2.40 & \text{if } 1.1 < x \le 1.2 \\ \ \vdots \\ 4.00 & \text{if } 1.9 < x < 2 \end{cases}$

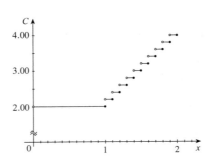

91. The graph of $x = y^2$ is not the graph of a function because both $(1, 1)$ and $(-1, 1)$ satisfy the equation $x = y^2$. The graph of $x = y^3$ is the graph of a function because $x = y^3$ $\Leftrightarrow$ $x^{1/3} = y$. If n is even, then both $(1, 1)$ and $(-1, 1)$ satisfies the equation $x = y^n$, so the graph of $x = y^n$ is not the graph of a function. When n is odd, $y = x^{1/n}$ is defined for all real numbers, and since $y = x^{1/n}$ $\Leftrightarrow$ $x = y^n$, the graph of $x = y^n$ is the graph of a function.

93.

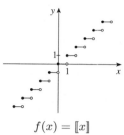

$$f(x) = [\![x]\!]$$

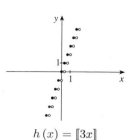

$$g(x) = [\![2x]\!]$$

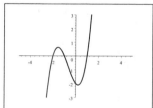

$$h(x) = [\![3x]\!]$$

The graph of $k(x) = [\![nx]\!]$ is a step function whose steps are each $\dfrac{1}{n}$ wide.

3.3   Increasing and Decreasing Functions; Average Rate of Change

1. The function is increasing on $[-1, 1]$ and $[2, 4]$.

(b) The function is decreasing on $[1, 2]$.

3. The function is increasing on $[-2, -1]$ and $[1, 2]$.

(b) The function is decreasing on $[-3, -2]$, $[-1, 1]$, and $[2, 3]$.

5. (a) $f(x) = x^{2/5}$ is graphed in the viewing rectangle $[-10, 10]$ by $[-5, 5]$.

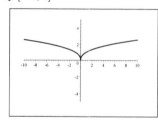

(b) The function is increasing on $[0, \infty)$. It is decreasing on $(-\infty, 0]$.

7. (a) $f(x) = x^2 - 5x$ is graphed in the viewing rectangle $[-2, 7]$ by $[-10, 10]$.

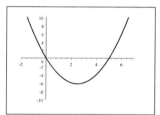

(b) The function is increasing on $[2.5, \infty)$. It is decreasing on $(-\infty, 2.5]$.

9. (a) $f(x) = 2x^3 - 3x^2 - 12x$ is graphed in the viewing rectangle $[-3, 5]$ by $[-25, 20]$.

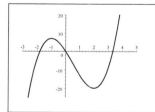

(b) The function is increasing on $(-\infty, -1]$ and $[2, \infty)$. It is decreasing on $[-1, 2]$.

11. (a) $f(x) = x^3 + 2x^2 - x - 2$ is graphed in the viewing rectangle $[-5, 5]$ by $[-3, 3]$.

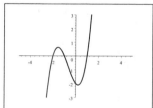

(b) The function is increasing on $(-\infty, -1.55]$ and $[0.22, \infty)$. It is decreasing on $[-1.55, 0.22]$.

13. We use the points $(1, 3)$ and $(4, 5)$, so the average rate of change is $\dfrac{5-3}{4-1} = \dfrac{2}{3}$.

15. We use the points $(0, 6)$ and $(5, 2)$, so the average rate of change is $\dfrac{2-6}{5-0} = -\dfrac{4}{5}$.

17. The average rate of change is $\dfrac{f(3) - f(2)}{3 - 2} = \dfrac{[3(3) - 2] - [3(2) - 2]}{1} = 7 - 4 = 3$.

19. The average rate of change is $\dfrac{h\left(4\right) - h\left(-1\right)}{4 - \left(-1\right)} = \dfrac{\left[4^2 + 2\left(4\right)\right] - \left[\left(-1\right)^2 + 2\left(-1\right)\right]}{5} = \dfrac{24 - \left(-1\right)}{5} = 5.$

21. The average rate of change is $\dfrac{f\left(10\right) - f\left(0\right)}{10 - 0} = \dfrac{\left[10^3 - 4\left(10^2\right)\right] - \left[0^3 - 4\left(0^2\right)\right]}{10 - 0} = \dfrac{600 - 0}{10} = 60.$

23. The average rate of change is

$\dfrac{f\left(2 + h\right) - f\left(2\right)}{\left(2 + h\right) - 2} = \dfrac{\left[3\left(2 + h\right)^2\right] - \left[3\left(2^2\right)\right]}{h} = \dfrac{12 + 12h + 3h^2 - 12}{h} = \dfrac{12h + 3h^2}{h} = \dfrac{h\left(12 + 3h\right)}{h} = 12 + 3h.$

25. The average rate of change is $\dfrac{g\left(1\right) - g\left(a\right)}{1 - a} = \dfrac{\frac{1}{1} - \frac{1}{a}}{1 - a} \cdot \dfrac{a}{a} = \dfrac{a - 1}{a\left(1 - a\right)} = \dfrac{-1\left(1 - a\right)}{a\left(1 - a\right)} = \dfrac{-1}{a}.$

27. The average rate of change is

$\dfrac{f\left(a + h\right) - f\left(a\right)}{\left(a + h\right) - a} = \dfrac{\frac{2}{a + h} - \frac{2}{a}}{h} \cdot \dfrac{a\left(a + h\right)}{a\left(a + h\right)} = \dfrac{2a - 2\left(a + h\right)}{ah\left(a + h\right)} = \dfrac{-2h}{ah\left(a + h\right)} = \dfrac{-2}{a\left(a + h\right)}.$

29. (a) The average rate of change is

$\dfrac{f\left(a + h\right) - f\left(a\right)}{\left(a + h\right) - a} = \dfrac{\left[\frac{1}{2}\left(a + h\right) + 3\right] - \left[\frac{1}{2}a + 3\right]}{h} = \dfrac{\frac{1}{2}a + \frac{1}{2}h + 3 - \frac{1}{2}a - 3}{h} = \dfrac{\frac{1}{2}h}{h} = \dfrac{1}{2}.$

(b) The slope of the line $f\left(x\right) = \frac{1}{2}x + 3$ is $\frac{1}{2}$, which is also the average rate of change.

31. (a) The function P is increasing on $[0, 150]$ and $[300, 365]$ and decreasing on $[150, 300]$.

(b) The average rate of change is $\dfrac{W\left(200\right) - W\left(100\right)}{200 - 100} = \dfrac{50 - 75}{200 - 100} = \dfrac{-25}{100} = -\dfrac{1}{4}$ ft/day.

33. (a) The average rate of change of population is $\dfrac{1,591 - 856}{2001 - 1998} = \dfrac{735}{3} = 245$ persons/yr.

(b) The average rate of change of population is $\dfrac{826 - 1,483}{2004 - 2002} = \dfrac{-657}{2} = -328.5$ persons/yr.

(c) The population was increasing from 1997 to 2001.

(d) The population was decreasing from 2001 to 2006.

35. (a) The average rate of change of sales is $\dfrac{584 - 512}{2003 - 1993} = \dfrac{72}{10} = 7.2$ units/yr.

(b) The average rate of change of sales is $\dfrac{520 - 512}{1994 - 1993} = \dfrac{8}{1} = 8$ units/yr.

(c) The average rate of change of sales is $\dfrac{410 - 520}{1996 - 1994} = \dfrac{-110}{2} = -55$ units/yr.

(d)

Year	CD players sold	Change in sales from previous year		Year	CD players sold	Change in sales from previous year
1993	512	—		1999	590	80
1994	520	8		2000	607	17
1995	413	−107		2001	732	125
1996	410	−3		2002	612	−120
1997	468	58		2003	584	−28
1998	510	42				

Sales increased most quickly between 2000 and 2001. Sales decreased most quickly between 2001 and 2002.

37. (a) For all three runners, the average rate of change is $\dfrac{d\left(10\right) - d\left(0\right)}{10 - 0} = \dfrac{100}{10} = 10.$

(b) Runner A gets a great jump out of the blocks but tires at the end of the race. Runner B runs a steady race. Runner C is slow at the beginning but accelerates down the track.

39. (a) $f(x)$ is always increasing, and $f(x) > 0$ for all x.

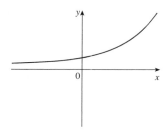

(b) $f(x)$ is always decreasing, and $f(x) > 0$ for all x.

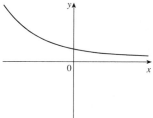

(c) $f(x)$ is always increasing, and $f(x) < 0$ for all x.

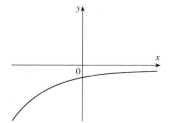

(d) $f(x)$ is always decreasing, and $f(x) < 0$ for all x.

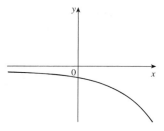

3.4 Transformations of Functions

1. (a) Shift the graph of $y = f(x)$ downward 5 units.
   (b) Shift the graph of $y = f(x)$ to the right 5 units.

3. (a) Shift the graph of f to the left $\frac{1}{2}$ unit.
   (b) Shift the graph of f upward $\frac{1}{2}$ unit.

5. (a) Reflect the graph of $y = f(x)$ about the x-axis, and stretch vertically by a factor of 2.

   (b) Reflect the graph of $y = f(x)$ about the x-axis, and shrink vertically by a factor of $\frac{1}{2}$.

7. (a) Shift the graph of $y = f(x)$ to the right 4 units, and then upward $\frac{3}{4}$ of a unit.

   (b) Shift the graph of $y = f(x)$ to the left 4 units, and then downward $\frac{3}{4}$ of a unit.

9. (a) Shrink the graph of $y = f(x)$ horizontally by a factor of $\frac{1}{4}$.

   (b) Stretch the graph of $y = f(x)$ horizontally by a factor of 4.

11. $g(x) = f(x - 2) = (x - 2)^2 = x^2 - 4x + 4$

13. $g(x) = f(x + 1) + 2 = |x + 1| + 2$

15. $g(x) = -f(x + 2) = -\sqrt{x + 2}$

16. $g(x) = -f(x - 2) + 1 = -(x - 2)^2 + 1 = -x^2 + 4x - 3$

17. (a) $y = f(x - 4)$ is graph #3.
   (b) $y = f(x) + 3$ is graph #1.

(c) $y = 2f(x + 6)$ is graph #2.
(d) $y = -f(2x)$ is graph #4.

19. (a) $y = f(x-2)$

(b) $y = f(x) - 2$

(c) $y = 2f(x)$

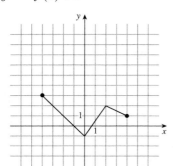

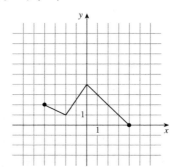

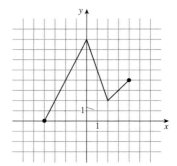

(d) $y = -f(x) + 3$

(e) $y = f(-x)$

(f) $y = \frac{1}{2}f(x-1)$

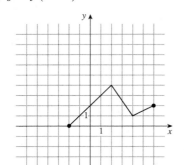

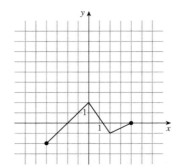

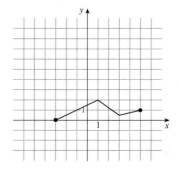

21. (a) $f(x) = \dfrac{1}{x}$

(b) (i) $y = -\dfrac{1}{x}$. Reflect the graph of f about the x-axis.

(ii) $y = \dfrac{1}{x-1}$. Shift the graph of f to the right 1 unit.

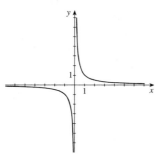

(iii) $y = \dfrac{2}{x+2}$. Shift the graph of f to the left 2 units and stretch vertically by a factor of 2.

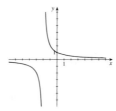

(iv) $y = 1 + \dfrac{1}{x-3}$. Shift the graph of f to the right 3 units and upward 1 unit.

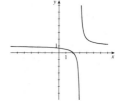

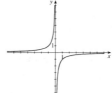

23. (a) The graph of $g(x) = (x+2)^2$ is obtained by shifting the graph of $f(x)$ to the left 2 units.

(b) The graph of $g(x) = x^2 + 2$ is obtained by shifting the graph of $f(x)$ upward 2 units.

25. (a) The graph of $g(x) = 2\sqrt{x}$ is obtained by stretching the graph of $f(x)$ vertically by a factor of 2.

(b) The graph of $g(x) = \frac{1}{2}\sqrt{x-2}$ is obtained by shifting the graph of $f(x)$ to the right 2 units, and then shrinking the graph vertically by a factor of $\frac{1}{2}$.

27. $y = f(x-2) + 3$. When $f(x) = x^2$, we get $y = (x-2)^2 + 3 = x^2 - 4x + 4 + 3 = x^2 - 4x + 7$.

29. $y = -5f(x+3)$. When $f(x) = \sqrt{x}$, we get $y = -5\sqrt{x+3}$

31. $y = 0.1f\left(x - \frac{1}{2}\right) - 2$. When $f(x) = |x|$, we get $y = 0.1\left|x - \frac{1}{2}\right| - 2$

33. $f(x) = (x-2)^2$. Shift the graph of $y = x^2$ to the right 2 units.

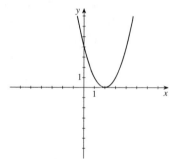

35. $f(x) = -(x+1)^2$. Shift the graph of $y = x^2$ to the left 1 unit, then reflect about the x-axis.

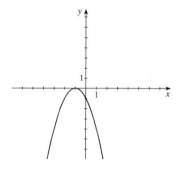

37. $f(x) = x^3 + 2$. Shift the graph of $y = x^3$ upward 2 units.

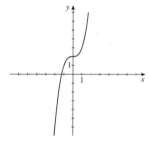

39. $y = 1 + \sqrt{x}$. Shift the graph of $y = \sqrt{x}$ upward 1 unit.

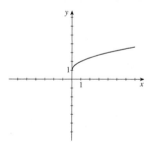

41. $y = \frac{1}{2}\sqrt{x+4} - 3$. Shift the graph of $y = \sqrt{x}$ to the left 4 units, shrink vertically by a factor of $\frac{1}{2}$, and then shift it downward 3 units.

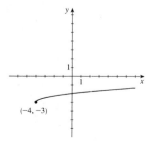

43. $y = 5 + (x+3)^2$. Shift the graph of $y = x^2$ to the left 3 units, then upward 5 units.

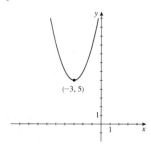

45. $y = |x| - 1$. Shift the graph of $y = |x|$ downward 1 unit.

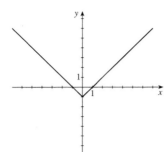

47. $y = |x + 2| + 2$. Shift the graph of $y = |x|$ to the left 2 units and upward 2 units.

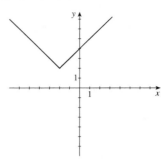

49.

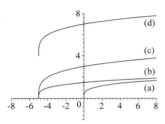
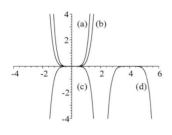

For part (b), shift the graph in (a) to the left 5 units; for part (c), shift the graph in (a) to the left 5 units, and stretch it vertically by a factor of 2; for part (d), shift the graph in (a) to the left 5 units, stretch it vertically by a factor of 2, and then shift it upward 4 units.

51.

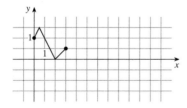

For part (b), shrink the graph in (a) vertically by a factor of $\frac{1}{3}$; for part (c), shrink the graph in (a) vertically by a factor of $\frac{1}{3}$, and reflect it about the x-axis; for part (d), shift the graph in (a) to the right 4 units, shrink vertically by a factor of $\frac{1}{3}$, and then reflect it about the x-axis.

53. (a) $y = g(2x)$

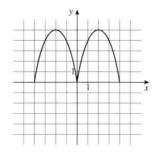

(b) $y = g\left(\frac{1}{2}x\right)$

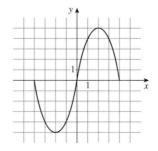

55. (a) Even

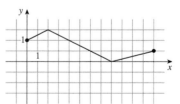

(b) Odd

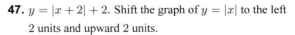

57. $y = [\![2x]\!]$

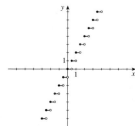

59. (a) $y = f(x) = \sqrt{2x - x^2}$ **(b)** $y = f(2x) = \sqrt{2(2x) - (2x)^2}$ **(c)** $y = f\left(\frac{1}{2}x\right) = \sqrt{2\left(\frac{1}{2}x\right) - \left(\frac{1}{2}x\right)^2}$

$$= \sqrt{4x - 4x^2} \qquad\qquad\qquad = \sqrt{x - \frac{1}{4}x^2}$$

The graph in part (b) is obtained by horizontally shrinking the graph in part (a) by a factor of 2 (so the graph is half as wide). The graph in part (c) is obtained by horizontally stretching the graph in part (a) by a factor of 2 (so the graph is twice as wide).

61. $f(x) = x^{-2}$. $f(-x) = (-x)^{-2} = x^{-2} = f(x)$. Thus $f(x)$ is even.

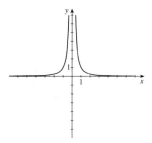

63. $f(x) = x^2 + x$. $f(-x) = (-x)^2 + (-x) = x^2 - x$. Thus $f(-x) \neq f(x)$. Also, $f(-x) \neq -f(x)$, so $f(x)$ is neither odd nor even.

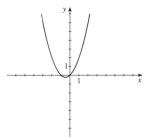

65. $f(x) = x^3 - x$.

$$f(-x) = (-x)^3 - (-x) = -x^3 + x$$
$$= -\left(x^3 - x\right) = -f(x).$$

Thus $f(x)$ is odd.

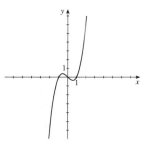

67. $f(x) = 1 - \sqrt[3]{x}$. $f(-x) = 1 - \sqrt[3]{(-x)} = 1 + \sqrt[3]{x}$. Thus $f(-x) \neq f(x)$. Also $f(-x) \neq -f(x)$, so $f(x)$ is neither odd nor even.

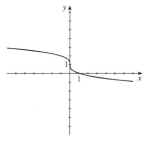

69. Since $f(x) = x^2 - 4 < 0$, for $-2 < x < 2$, the graph of $y = g(x)$ is found by sketching the graph of $y = f(x)$ for $x \leq -2$ and $x \geq 2$, then reflecting about the x-axis the part of the graph of $y = f(x)$ for $-2 < x < 2$.

71. (a) $f(x) = 4x - x^2$

(b) $f(x) = |4x - x^2|$

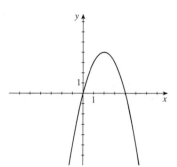

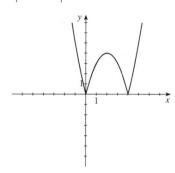

73. (a) The graph of $y = t^2$ must be shrunk vertically by a factor of 0.01 and shifted vertically 4 units up to obtain the graph of $y = f(t)$.

(b) The graph of $y = f(t)$ must be shifted horizontally 10 units to the right to obtain the graph of $y = g(t)$. So
$g(t) = f(t - 10) = 4 + 0.01(t - 10)^2 = 5 - 0.02t + 0.01t^2$.

75. f even implies $f(-x) = f(x)$; g even implies $g(-x) = g(x)$; f odd implies $f(-x) = -f(x)$; and g odd implies $g(-x) = -g(x)$

If f and g are both even, then $(f + g)(-x) = f(-x) + g(-x) = f(x) + g(x) = (f + g)(x)$ and $f + g$ is even.

If f and g are both odd, then $(f + g)(-x) = f(-x) + g(-x) = -f(x) - g(x) = -(f + g)(x)$ and $f + g$ is odd.

If f odd and g even, then $(f + g)(-x) = f(-x) + g(-x) = -f(x) + g(x)$, which is neither odd nor even.

77. $f(x) = x^n$ is even when n is an even integer and $f(x) = x^n$ is odd when n is an odd integer.

These names were chosen because polynomials with only terms with odd powers are odd functions, and polynomials with only terms with even powers are even functions.

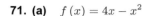 **3.5** **Quadratic Functions; Maxima and Minima**

1. (a) Vertex: $(3, 4)$

(b) Maximum value of f: 4.

3. (a) Vertex: $(1, -3)$

(b) Minimum value of f: -3.

5. (a) $f(x) = x^2 - 6x = (x - 3)^2 - 9$

(b) Vertex: $y = x^2 - 6x = x^2 - 6x + 9 - 9 = (x - 3)^2 - 9$. So the vertex is at $(3, -9)$.

x-intercepts: $y = 0 \implies 0 = x^2 - 6x = x(x - 6)$. So $x = 0$ or $x = 6$. The x-intercepts are $x = 0$ and $x = 6$.

y-intercept: $x = 0 \implies y = 0$. The y-intercept is $y = 0$.

(c)

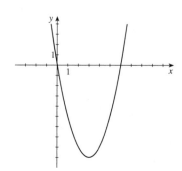

7. (a) $f(x) = 2x^2 + 6x = 2\left(x + \frac{3}{2}\right)^2 - \frac{9}{2}$

  (b) The vertex is $\left(-\frac{3}{2}, -\frac{9}{2}\right)$.

  x-intercepts: $y = 0 \Rightarrow 0 = 2x^2 + 6x = 2x(x + 3) \Rightarrow$
  $x = 0$ or $x = -3$. The x-intercepts are $x = 0$ and $x = -3$.
  y-intercept: $x = 0 \Rightarrow y = 0$. The y-intercept is $y = 0$.

  (c)

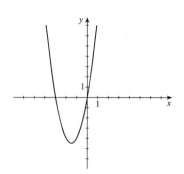

9. (a) $f(x) = x^2 + 4x + 3 = (x + 2)^2 - 1$

  (b) The vertex is $(-2, -1)$.

  x-intercepts: $y = 0 \Rightarrow 0 = x^2 + 4x + 3 = (x + 1)(x + 3)$. So
  $x = -1$ or $x = -3$. The x-intercepts are $x = -1$ and $x = -3$.
  y-intercept: $x = 0 \Rightarrow y = 3$. The y-intercept is $y = 3$.

  (c)

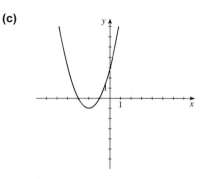

11. (a) $f(x) = -x^2 + 6x + 4 = -(x - 3)^2 + 13$

  (b) The vertex is $(3, 13)$.

  x-intercepts: $y = 0 \Rightarrow 0 = -(x - 3)^2 + 13 \Leftrightarrow$
  $(x - 3)^2 = 13 \Rightarrow x - 3 = \pm\sqrt{13} \Leftrightarrow x = 3 \pm \sqrt{13}$. The
  x-intercepts are $x = 3 - \sqrt{13}$ and $x = 3 + \sqrt{13}$.
  y-intercept: $x = 0 \Rightarrow y = 4$. The y-intercept is $y = 4$.

  (c)

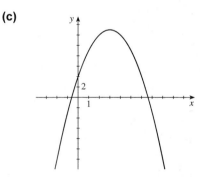

13. (a) $f(x) = 2x^2 + 4x + 3 = 2(x + 1)^2 + 1$

  (b) The vertex is $(-1, 1)$.

  x-intercepts: $y = 0 \Rightarrow 0 = 2x^2 + 4x + 3 = 2(x + 1)^2 + 1$
  $\Leftrightarrow 2(x + 1)^2 = -1$. Since this last equation has no real solution,
  there is no x-intercept.
  y-intercept: $x = 0 \Rightarrow y = 3$. The y-intercept is $y = 3$.

  (c)

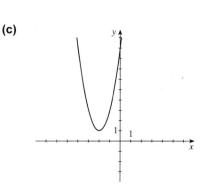

15. (a) $f(x) = 2x^2 - 20x + 57 = 2(x-5)^2 + 7$

(b) The vertex is $(5, 7)$.

    x-intercepts: $y = 0$ $\Rightarrow$

    $0 = 2x^2 - 20x + 57 = 2(x-5)^2 + 7 \Leftrightarrow 2(x-5)^2 = -7$. Since

    this last equation has no real solution, there is no x-intercept.

    y-intercept: $x = 0$ $\Rightarrow$ $y = 57$. The y-intercept is $y = 57$.

(c)

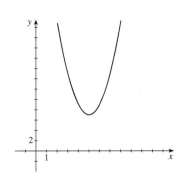

17. (a) $f(x) = -4x^2 - 16x + 3 = -4(x+2)^2 + 19$

(b) The vertex is $(-2, 19)$.

    x-intercepts: $y = 0$ $\Rightarrow$

    $0 = -4x^2 - 16x + 3 = -4(x+2)^2 + 19 \Leftrightarrow 4(x+2)^2 = 19$

    $\Leftrightarrow$ $(x+2)^2 = \frac{19}{4} \Rightarrow x + 2 = \pm\sqrt{\frac{19}{4}} = \pm\frac{\sqrt{19}}{2}$ $\Leftrightarrow$

    $x = -2 \pm \frac{\sqrt{19}}{2}$. The x-intercepts are $x = -2 - \frac{\sqrt{19}}{2}$ and

    $x = -2 + \frac{\sqrt{19}}{2}$.

    y-intercept: $x = 0$ $\Rightarrow$ $y = 3$. The y-intercept is $y = 3$.

(c)

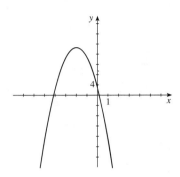

19. (a) $f(x) = 2x - x^2 = -(x^2 - 2x)$
$$= -(x^2 - 2x + 1) + 1 = -(x-1)^2 + 1$$

(b)

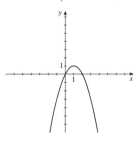

(c) The maximum value is $f(1) = 1$.

21. (a) $f(x) = x^2 + 2x - 1 = (x^2 + 2x) - 1$
$$= (x^2 + 2x + 1) - 1 - 1 = (x+1)^2 - 2$$

(b)

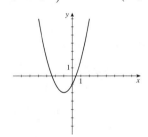

(c) The minimum value is $f(-1) = -2$.

23. (a) $f(x) = -x^2 - 3x + 3 = -(x^2 + 3x) + 3$
$$= -\left(x^2 + 3x + \frac{9}{4}\right) + 3 + \frac{9}{4}$$
$$= -\left(x + \frac{3}{2}\right)^2 + \frac{21}{4}$$

(b)

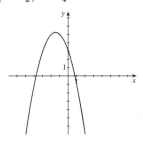

(c) The maximum value is $f\left(-\frac{3}{2}\right) = \frac{21}{4}$.

25. (a) $g(x) = 3x^2 - 12x + 13 = 3(x^2 - 4x) + 13$
$$= 3(x^2 - 4x + 4) + 13 - 12$$
$$= 3(x-2)^2 + 1$$

(b)

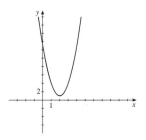

(c) The minimum value is $g(2) = 1$.

27. (a) $h(x) = 1 - x - x^2 = -(x^2 + x) + 1$

$$= -\left(x^2 + x + \tfrac{1}{4}\right) + 1 + \tfrac{1}{4}$$

$$= -\left(x + \tfrac{1}{2}\right)^2 + \tfrac{5}{4}$$

(b)

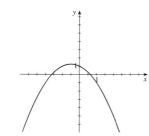

(c) The maximum value is $h\left(-\tfrac{1}{2}\right) = \tfrac{5}{4}$.

29. $f(x) = x^2 + x + 1 = (x^2 + x) + 1$

$$= \left(x^2 + x + \tfrac{1}{4}\right) + 1 + \tfrac{1}{4} = \left(x + \tfrac{1}{2}\right)^2 + \tfrac{3}{4}$$

Therefore, the minimum value is $f\left(-\tfrac{1}{2}\right) = \tfrac{3}{4}$.

31. $f(t) = 100 - 49t - 7t^2 = -7\left(t^2 + 7t\right) + 100$

$$= -7\left(t^2 + 7t + \tfrac{49}{4}\right) + 100 + \tfrac{343}{4}$$

$$= -7\left(t + \tfrac{7}{2}\right)^2 + \tfrac{743}{4}$$

Therefore, the maximum value is

$$f\left(-\tfrac{7}{2}\right) = \tfrac{743}{4} = 185.75.$$

33. $f(s) = s^2 - 1.2s + 16 = (s^2 - 1.2s) + 16 = (s^2 - 1.2s + 0.36) + 16 - 0.36 = (s - 0.6)^2 + 15.64$.
Therefore, the minimum value is $f(0.6) = 15.64$.

35. $h(x) = \tfrac{1}{2}x^2 + 2x - 6 = \tfrac{1}{2}\left(x^2 + 4x\right) - 6 = \tfrac{1}{2}\left(x^2 + 4x + 4\right) - 6 - 2 = \tfrac{1}{2}(x + 2)^2 - 8$.
Therefore, the minimum value is $h(-2) = -8$.

37. $f(x) = 3 - x - \tfrac{1}{2}x^2 = -\tfrac{1}{2}\left(x^2 + 2x\right) + 3 = -\tfrac{1}{2}\left(x^2 + 2x + 1\right) + 3 + \tfrac{1}{2} = -\tfrac{1}{2}(x + 1)^2 + \tfrac{7}{2}$. Therefore, the maximum
value is $f(-1) = \tfrac{7}{2}$.

39. Since the vertex is at $(1, -2)$, the function is of the form $f(x) = a(x - 1)^2 - 2$. Substituting the point
$(4, 16)$, we get $16 = a(4 - 1)^2 - 2 \quad \Leftrightarrow \quad 16 = 9a - 2 \quad \Leftrightarrow \quad 9a = 18 \quad \Leftrightarrow \quad a = 2$. So the function is
$f(x) = 2(x - 1)^2 - 2 = 2x^2 - 4x$.

41. $f(x) = -x^2 + 4x - 3 = -\left(x^2 - 4x\right) - 3 = -\left(x^2 - 4x + 4\right) - 3 + 4 = -(x - 2)^2 + 1$. So the domain of $f(x)$ is
$(-\infty, \infty)$. Since $f(x)$ has a maximum value of 1, the range is $(-\infty, 1]$.

43. $f(x) = 2x^2 + 6x - 7 = 2\left(x + \tfrac{3}{2}\right)^2 - 7 - \tfrac{9}{2} = 2\left(x + \tfrac{3}{2}\right)^2 - \tfrac{23}{2}$. The domain of the function is all real numbers, and
since the minimum value of the function is $f\left(-\tfrac{3}{2}\right) = -\tfrac{23}{2}$, the range of the function is $\left[-\tfrac{23}{2}, \infty\right)$.

45. (a) The graph of $f(x) = x^2 + 1.79x - 3.21$ is
shown. The minimum value is
$f(x) \approx -4.01$.

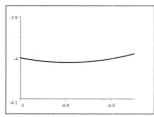

(b) $f(x) = x^2 + 1.79x - 3.21$

$$= \left[x^2 + 1.79x + \left(\tfrac{1.79}{2}\right)^2\right] - 3.21 - \left(\tfrac{1.79}{2}\right)^2$$

$$= (x + 0.895)^2 - 4.011025$$

Therefore, the exact minimum of $f(x)$ is -4.011025.

47. Local maximum: 2 at $x = 0$. Local minimum: -1 at $x = -2$ and 0 at $x = 2$.

49. Local maximum: 0 at $x = 0$ and 1 at $x = 3$. Local minimum: -2 at $x = -2$ and -1 at $x = 1$.

51. In the first graph, we see that $f(x) = x^3 - x$ has a local minimum and a local maximum. Smaller x- and y-ranges show that $f(x)$ has a local maximum of about 0.38 when $x \approx -0.58$ and a local minimum of about -0.38 when $x \approx 0.58$.

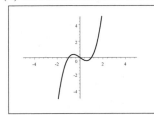 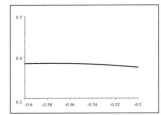 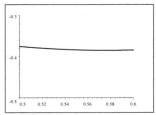

53. In the first graph, we see that $g(x) = x^4 - 2x^3 - 11x^2$ has two local minimums and a local maximum. The local maximum is $g(x) = 0$ when $x = 0$. Smaller x- and y-ranges show that local minima are $g(x) \approx -13.61$ when $x \approx -1.71$ and $g(x) \approx -73.32$ when $x \approx 3.21$.

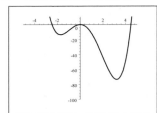 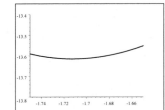 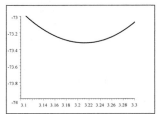

55. In the first graph, we see that $U(x) = x\sqrt{6-x}$ only has a local maximum. Smaller x- and y-ranges show that $U(x)$ has a local maximum of about 5.66 when $x \approx 4.00$.

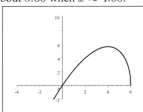 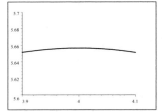

57. In the first graph, we see that $V(x) = \dfrac{1 - x^2}{x^3}$ has a local minimum and a local maximum. Smaller x- and y-ranges show that $V(x)$ has a local maximum of about 0.38 when $x \approx -1.73$ and a local minimum of about -0.38 when $x \approx 1.73$.

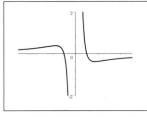 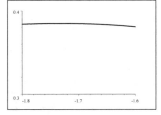 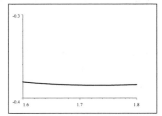

59. $y = f(t) = 40t - 16t^2 = -16\left(t^2 - \frac{5}{2}\right) = -16\left[t^2 - \frac{5}{2}t + \left(\frac{5}{4}\right)^2\right] + 16\left(\frac{5}{4}\right)^2 = -16\left(t - \frac{5}{4}\right)^2 + 25$. Thus the maximum height attained by the ball is $f\left(\frac{5}{4}\right) = 25$ feet.

61. $R(x) = 80x - 0.4x^2 = -0.4\left(x^2 - 200x\right) = -0.4\left(x^2 - 200x + 10{,}000\right) + 4{,}000 = -0.4\left(x - 100\right)^2 + 4{,}000$. So revenue is maximized at $\$4{,}000$ when 100 units are sold.

63. $E(n) = \frac{2}{3}n - \frac{1}{90}n^2 = -\frac{1}{90}\left(n^2 - 60n\right) = -\frac{1}{90}\left(n^2 - 60n + 900\right) + 10 = -\dfrac{1}{90}\left(n - 30\right)^2 + 10$. Since the maximum of the function occurs when $n = 30$, the viewer should watch the commercial 30 times for maximum effectiveness.

65. Graphing $A(n) = n(900 - 9n)$ in the viewing rectangle $[0, 100]$ by $[0, 25000]$, we see that maximum yield of apples occurs when there are 50 trees per acre.

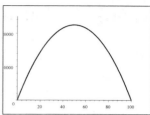

67. In the first graph, we see the general location of the maximum of $N(s) = \dfrac{88s}{17 + 17\left(\dfrac{s}{20}\right)^2}$. In the second graph we isolate

the maximum, and from this graph we see that at the speed of 20 mi/h the largest number of cars that can use the highway safely is 52.

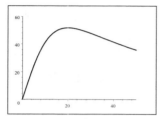
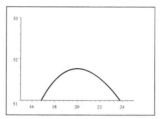

69. In the first graph, we see the general location of the maximum of $v(r) = 3.2(1 - r)r^2$ is around $r = 0.7$ cm. In the second graph, we isolate the maximum, and from this graph we see that at the maximum velocity is 0.47 when $r \approx 0.67$ cm.

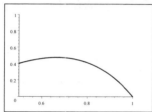

71. (a) If $x = a$ is a local maximum of $f(x)$ then
$f(a) \geq f(x) \geq 0$ for all x around $x = a$. So
$[g(a)]^2 \geq [g(x)]^2$ and thus $g(a) \geq g(x)$.
Similarly, if $x = b$ is a local minimum of $f(x)$,
then $f(x) \geq f(b) \geq 0$ for all x around $x = b$.
So $[g(x)]^2 \geq [g(b)]^2$ and thus $g(x) \geq g(b)$.

(b) Using the distance formula,
$$g(x) = \sqrt{(x - 3)^2 + (x^2 - 0)^2}$$
$$= \sqrt{x^4 + x^2 - 6x + 9}$$

(c) Let $f(x) = x^4 + x^2 - 6x + 9$. From the graph, we see that $f(x)$ has a minimum at $x = 1$. Thus $g(x)$ also has a minimum at $x = 1$ and this minimum value is
$$g(1) = \sqrt{1^4 + 1^2 - 6(1) + 9} = \sqrt{5}.$$

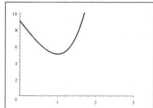

3.6   Combining Functions

1. $f(x) = x - 3$ has domain $(-\infty, \infty)$. $g(x) = x^2$ has domain $(-\infty, \infty)$. The intersection of the domains of f and g is $(-\infty, \infty)$.

$(f + g)(x) = (x - 3) + (x^2) = x^2 + x - 3$, and the domain is $(-\infty, \infty)$.

$(f - g)(x) = (x - 3) - (x^2) = -x^2 + x - 3$, and the domain is $(-\infty, \infty)$.

$(fg)(x) = (x - 3)(x^2) = x^3 - 3x^2$, and the domain is $(-\infty, \infty)$.

$\left(\dfrac{f}{g}\right)(x) = \dfrac{x - 3}{x^2}$, and the domain is $\{x \mid x \neq 0\}$.

3. $f(x) = \sqrt{4 - x^2}$, has domain $[-2, 2]$. $g(x) = \sqrt{1 + x}$, has domain $[-1, \infty)$. The intersection of the domains of f and g is $[-1, 2]$.

$(f + g)(x) = \sqrt{4 - x^2} + \sqrt{1 + x}$, and the domain is $[-1, 2]$.

$(f - g)(x) = \sqrt{4 - x^2} - \sqrt{1 + x}$, and the domain is $[-1, 2]$.

$(fg)(x) = \sqrt{4 - x^2}\sqrt{1 + x} = \sqrt{-x^3 - x^2 + 4x + 4}$, and the domain is $[-1, 2]$.

$\left(\dfrac{f}{g}\right)(x) = \dfrac{\sqrt{4 - x^2}}{\sqrt{1 + x}} = \sqrt{\dfrac{4 - x^2}{1 + x}}$, and the domain is $(-1, 2]$.

5. $f(x) = \dfrac{2}{x}$ has domain $x \neq 0$. $g(x) = \dfrac{4}{x + 4}$, has domain $x \neq -4$. The intersection of the domains of f and g is $\{x \mid x \neq 0, -4\}$; in interval notation, this is $(-\infty, -4) \cup (-4, 0) \cup (0, \infty)$.

$(f + g)(x) = \dfrac{2}{x} + \dfrac{4}{x + 4} = \dfrac{2}{x} + \dfrac{4}{x + 4} = \dfrac{2(3x + 4)}{x(x + 4)}$, and the domain is $(-\infty, -4) \cup (-4, 0) \cup (0, \infty)$.

$(f - g)(x) = \dfrac{2}{x} - \dfrac{4}{x + 4} = -\dfrac{2(x - 4)}{x(x + 4)}$, and the domain is $(-\infty, -4) \cup (-4, 0) \cup (0, \infty)$.

$(fg)(x) = \dfrac{2}{x} \cdot \dfrac{4}{x + 4} = \dfrac{8}{x(x + 4)}$, and the domain is $(-\infty, -4) \cup (-4, 0) \cup (0, \infty)$.

$\left(\dfrac{f}{g}\right)(x) = \dfrac{\dfrac{2}{x}}{\dfrac{4}{x + 4}} = \dfrac{x + 4}{2x}$, and the domain is $(-\infty, -4) \cup (-4, 0) \cup (0, \infty)$.

7. $f(x) = \sqrt{x} + \sqrt{1 - x}$. The domain of $\sqrt{x}$ is $[0, \infty)$, and the domain of $\sqrt{1 - x}$ is $(-\infty, 1]$. Thus the domain is $(-\infty, 1] \cap [0, \infty) = [0, 1]$.

9. $h(x) = (x - 3)^{-1/4} = \dfrac{1}{(x - 3)^{1/4}}$. Since $1/4$ is an even root and the denominator can not equal 0, $x - 3 > 0 \iff x > 3$. So the domain is $(3, \infty)$.

11.

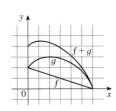

13.

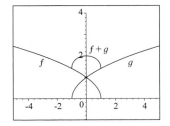

15.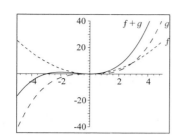

17. (a) $f(g(0)) = f(2 - (0)^2) = f(2) = 3(2) - 5 = 1$

(b) $g(f(0)) = g(3(0) - 5) = g(-5) = 2 - (-5)^2 = -23$

19. (a) $(f \circ g)(-2) = f(g(-2)) = f(2 - (-2)^2) = f(-2) = 3(-2) - 5 = -11$

(b) $(g \circ f)(-2) = g(f(-2)) = g(3(-2) - 5) = g(-11) = 2 - (-11)^2 = -119$

21. (a) $(f \circ g)(x) = f(g(x)) = f(2 - x^2) = 3(2 - x^2) - 5 = 6 - 3x^2 - 5 = 1 - 3x^2$

(b) $(g \circ f)(x) = g(f(x)) = g(3x - 5) = 2 - (3x - 5)^2 = 2 - (9x^2 - 30x + 25) = -9x^2 + 30x - 23$

23. $f(g(2)) = f(5) = 4$

25. $(g \circ f)(4) = g(f(4)) = g(2) = 5$

27. $(g \circ g)(-2) = g(g(-2)) = g(1) = 4$

29. $f(x) = 2x + 3$, has domain $(-\infty, \infty)$; $g(x) = 4x - 1$, has domain $(-\infty, \infty)$.
$(f \circ g)(x) = f(4x - 1) = 2(4x - 1) + 3 = 8x + 1$, and the domain is $(-\infty, \infty)$.
$(g \circ f)(x) = g(2x + 3) = 4(2x + 3) - 1 = 8x + 11$, and the domain is $(-\infty, \infty)$.
$(f \circ f)(x) = f(2x + 3) = 2(2x + 3) + 3 = 4x + 9$, and the domain is $(-\infty, \infty)$.
$(g \circ g)(x) = g(4x - 1) = 4(4x - 1) - 1 = 16x - 5$, and the domain is $(-\infty, \infty)$.

31. $f(x) = x^2$, has domain $(-\infty, \infty)$; $g(x) = x + 1$, has domain $(-\infty, \infty)$.
$(f \circ g)(x) = f(x + 1) = (x + 1)^2 = x^2 + 2x + 1$, and the domain is $(-\infty, \infty)$.
$(g \circ f)(x) = g(x^2) = (x^2) + 1 = x^2 + 1$, and the domain is $(-\infty, \infty)$.
$(f \circ f)(x) = f(x^2) = (x^2)^2 = x^4$, and the domain is $(-\infty, \infty)$.
$(g \circ g)(x) = g(x + 1) = (x + 1) + 1 = x + 2$, and the domain is $(-\infty, \infty)$.

33. $f(x) = \dfrac{1}{x}$, has domain $\{x \mid x \neq 0\}$; $g(x) = 2x + 4$, has domain $(-\infty, \infty)$.

$(f \circ g)(x) = f(2x + 4) = \dfrac{1}{2x + 4}$. $(f \circ g)(x)$ is defined for $2x + 4 \neq 0 \iff x \neq -2$. So the domain is $\{x \mid x \neq -2\} = (-\infty, -2) \cup (-2, \infty)$.

$(g \circ f)(x) = g\left(\dfrac{1}{x}\right) = 2\left(\dfrac{1}{x}\right) + 4 = \dfrac{2}{x} + 4$, the domain is $\{x \mid x \neq 0\} = (-\infty, 0) \cup (0, \infty)$.

$(f \circ f)(x) = f\left(\dfrac{1}{x}\right) = \dfrac{1}{\left(\dfrac{1}{x}\right)} = x$. $(f \circ f)(x)$ is defined whenever both $f(x)$ and $f(f(x))$ are defined; that is,

whenever $\{x \mid x \neq 0\} = (-\infty, 0) \cup (0, \infty)$.
$(g \circ g)(x) = g(2x + 4) = 2(2x + 4) + 4 = 4x + 8 + 4 = 4x + 12$, and the domain is $(-\infty, \infty)$.

35. $f(x) = |x|$, has domain $(-\infty, \infty)$; $g(x) = 2x + 3$, has domain $(-\infty, \infty)$
$(f \circ g)(x) = f(2x + 4) = |2x + 3|$, and the domain is $(-\infty, \infty)$.
$(g \circ f)(x) = g(|x|) = 2|x| + 3$, and the domain is $(-\infty, \infty)$.
$(f \circ f)(x) = f(|x|) = ||x|| = |x|$, and the domain is $(-\infty, \infty)$.
$(g \circ g)(x) = g(2x + 3) = 2(2x + 3) + 3 = 4x + 6 + 3 = 4x + 9$. Domain is $(-\infty, \infty)$.

37. $f(x) = \dfrac{x}{x+1}$, has domain $\{x \mid x \neq -1\}$; $g(x) = 2x - 1$, has domain $(-\infty, \infty)$

$(f \circ g)(x) = f(2x - 1) = \dfrac{2x - 1}{(2x - 1) + 1} = \dfrac{2x - 1}{2x}$, and the domain is $\{x \mid x \neq 0\} = (-\infty, 0) \cup (0, \infty)$.

$(g \circ f)(x) = g\left(\dfrac{x}{x+1}\right) = 2\left(\dfrac{x}{x+1}\right) - 1 = \dfrac{2x}{x+1} - 1$, and the domain is $\{x \mid x \neq -1\} = (-\infty, -1) \cup (-1, \infty)$

$(f \circ f)(x) = f\left(\dfrac{x}{x+1}\right) = \dfrac{\frac{x}{x+1}}{\frac{x}{x+1} + 1} \cdot \dfrac{x+1}{x+1} = \dfrac{x}{x + x + 1} = \dfrac{x}{2x+1}$. $(f \circ f)(x)$ is defined whenever

both $f(x)$ and $f(f(x))$ are defined; that is, whenever $x \neq -1$ and $2x + 1 \neq 0 \Rightarrow x \neq -\frac{1}{2}$, which is
$(-\infty, -1) \cup \left(-1, -\frac{1}{2}\right) \cup \left(-\frac{1}{2}, \infty\right)$.

$(g \circ g)(x) = g(2x - 1) = 2(2x - 1) - 1 = 4x - 2 - 1 = 4x - 3$, and the domain is $(-\infty, \infty)$.

39. $f(x) = \sqrt[3]{x}$, has domain $(-\infty, \infty)$; $g(x) = \sqrt[4]{x}$, has domain $[0, \infty)$.

$(f \circ g)(x) = f(\sqrt[4]{x}) = \sqrt[3]{\sqrt[4]{x}} = \sqrt[12]{x}$. $(f \circ g)(x)$ is defined whenever both $g(x)$ and $f(g(x))$ are defined. Since $f(x)$
has no restriction, the domain is $[0, \infty)$.

$(g \circ f)(x) = g(\sqrt[3]{x}) = \sqrt[4]{\sqrt[3]{x}} = \sqrt[12]{x}$. $(g \circ f)(x)$ is defined whenever both $f(x)$ and $g(f(x))$ are defined; that is,
whenever $x \geq 0$. So the domain is $[0, \infty)$.

$(f \circ f)(x) = f(\sqrt[3]{x}) = \sqrt[3]{\sqrt[3]{x}} = \sqrt[9]{x}$. $(f \circ f)(x)$ is defined whenever both $f(x)$ and $f(f(x))$ are defined. Since $f(x)$
is defined everywhere, the domain is $(-\infty, \infty)$.

$(g \circ g)(x) = g(\sqrt[4]{x}) = \sqrt[4]{\sqrt[4]{x}} = \sqrt[16]{x}$. $(g \circ g)(x)$ is defined whenever both $g(x)$ and $g(g(x))$ are defined; that is,
whenever $x \geq 0$. So the domain is $[0, \infty)$.

41. $(f \circ g \circ h)(x) = f(g(h(x))) = f(g(x - 1)) = f(\sqrt{x - 1}) = \sqrt{x - 1} - 1$

43. $(f \circ g \circ h)(x) = f(g(h(x))) = f(g(\sqrt{x})) = f(\sqrt{x} - 5) = (\sqrt{x} - 5)^4 + 1$

For Exercises 45–54, many answers are possible.

45. $F(x) = (x - 9)^5$. Let $f(x) = x^5$ and $g(x) = x - 9$, then $F(x) = (f \circ g)(x)$.

47. $G(x) = \dfrac{x^2}{x^2 + 4}$. Let $f(x) = \dfrac{x}{x + 4}$ and $g(x) = x^2$, then $G(x) = (f \circ g)(x)$.

49. $H(x) = |1 - x^3|$. Let $f(x) = |x|$ and $g(x) = 1 - x^3$, then $H(x) = (f \circ g)(x)$.

51. $F(x) = \dfrac{1}{x^2 + 1}$. Let $f(x) = \dfrac{1}{x}$, $g(x) = x + 1$, and $h(x) = x^2$, then $F(x) = (f \circ g \circ h)(x)$.

53. $G(x) = (4 + \sqrt[3]{x})^9$. Let $f(x) = x^9$, $g(x) = 4 + x$, and $h(x) = \sqrt[3]{x}$, then $G(x) = (f \circ g \circ h)(x)$.

55. The price per sticker is $0.15 - 0.000002x$ and the number sold is x, so the revenue is
$R(x) = (0.15 - 0.000002x)x = 0.15x - 0.000002x^2$.

57. (a) Since the ripple travels at a speed of 60 cm/s, the distance traveled in t seconds is the radius, so $g(t) = 60t$.

(b) The area of a circle is πr^2, so $f(r) = \pi r^2$.

(c) $f \circ g = \pi(g(t))^2 = \pi(60t)^2 = 3600\pi t^2$ cm². This function represents the area of the ripple as a function of time.

59. Let r be the radius of the spherical balloon in centimeters. Since the radius is increasing at a rate of 2 cm/s, the radius is $r = 2t$
after t seconds. Therefore, the surface area of the balloon can be written as $S = 4\pi r^2 = 4\pi(2t)^2 = 4\pi(4t^2) = 16\pi t^2$.

61. (a) $f(x) = 0.90x$

(b) $g(x) = x - 100$

(c) $f \circ g = f(x - 100) = 0.90(x - 100) = 0.90x - 90$. $f \circ g$ represents applying the $100 coupon, then the
10% discount. $g \circ f = g(0.90x) = 0.90x - 100$. $g \circ f$ represents applying the 10% discount, then the $100 coupon.
So applying the 10% discount, then the $100 coupon gives the lower price.

63. $A(x) = 1.05x$. $(A \circ A)(x) = A(A(x)) = A(1.05x) = 1.05(1.05x) = (1.05)^2 x$.

$(A \circ A \circ A)(x) = A(A \circ A(x)) = A((1.05)^2 x) = 1.05[(1.05)^2 x] = (1.05)^3 x$.

$(A \circ A \circ A \circ A)(x) = A(A \circ A \circ A(x)) = A((1.05)^3 x) = 1.05[(1.05)^3 x] = (1.05)^4 x$. A represents the amount in the account after 1 year; $A \circ A$ represents the amount in the account after 2 years; $A \circ A \circ A$ represents the amount in the account after 3 years; and $A \circ A \circ A \circ A$ represents the amount in the account after 4 years. We can see that if we compose n copies of A, we get $(1.05)^n x$.

65. $g(x) = 2x + 1$ and $h(x) = 4x^2 + 4x + 7$.

Method 1: Notice that $(2x + 1)^2 = 4x^2 + 4x + 1$. We see that adding 6 to this quantity gives $(2x + 1)^2 + 6 = 4x^2 + 4x + 1 + 6 = 4x^2 + 4x + 7$, which is $h(x)$. So let $f(x) = x^2 + 6$, and we have $(f \circ g)(x) = (2x + 1)^2 + 6 = h(x)$.

Method 2: Since $g(x)$ is linear and $h(x)$ is a second degree polynomial, $f(x)$ must be a second degree polynomial, that is, $f(x) = ax^2 + bx + c$ for some a, b, and c. Thus $f(g(x)) = f(2x + 1) = a(2x + 1)^2 + b(2x + 1) + c \Leftrightarrow 4ax^2 + 4ax + a + 2bx + b + c = 4ax^2 + (4a + 2b)x + (a + b + c) = 4x^2 + 4x + 7$. Comparing this with $f(g(x))$, we have $4a = 4$ (the x^2 coefficients), $4a + 2b = 4$ (the x coefficients), and $a + b + c = 7$ (the constant terms) $\Leftrightarrow a = 1$ and $2a + b = 2$ and $a + b + c = 7 \Leftrightarrow a = 1, b = 0, c = 6$. Thus $f(x) = x^2 + 6$.

$f(x) = 3x + 5$ and $h(x) = 3x^2 + 3x + 2$.

Note since $f(x)$ is linear and $h(x)$ is quadratic, $g(x)$ must also be quadratic. We can then use trial and error to find $g(x)$. Another method is the following: We wish to find g so that $(f \circ g)(x) = h(x)$. Thus $f(g(x)) = 3x^2 + 3x + 2 \Leftrightarrow 3(g(x)) + 5 = 3x^2 + 3x + 2 \Leftrightarrow 3(g(x)) = 3x^2 + 3x - 3 \Leftrightarrow g(x) = x^2 + x - 1$.

3.7 One-to-One Functions and Their Inverses

1. By the Horizontal Line Test, f is not one-to-one.

3. By the Horizontal Line Test, f is one-to-one.

5. By the Horizontal Line Test, f is not one-to-one.

7. $f(x) = -2x + 4$. If $x_1 \neq x_2$, then $-2x_1 \neq -2x_2$ and $-2x_1 + 4 \neq -2x_2 + 4$. So f is a one-to-one function.

9. $g(x) = \sqrt{x}$. If $x_1 \neq x_2$, then $\sqrt{x_1} \neq \sqrt{x_2}$ because two different numbers cannot have the same square root. Therefore, g is a one-to-one function.

11. $h(x) = x^2 - 2x$. Since $h(0) = 0$ and $h(2) = (2) - 2(2) = 0$ we have $h(0) = h(2)$. So f is not a one-to-one function.

13. $f(x) = x^4 + 5$. Every nonzero number and its negative have the same fourth power. For example, $(-1)^4 = 1 = (1)^4$, so $f(-1) = f(1)$. Thus f is not a one-to-one function.

15. $f(x) = \dfrac{1}{x^2}$. Every nonzero number and its negative have the same square. For example, $\dfrac{1}{(-1)^2} = 1 = \dfrac{1}{(1)^2}$, so $f(-1) = f(1)$. Thus f is not a one-to-one function.

17. (a) $f(2) = 7$. Since f is one-to-one, $f^{-1}(7) = 2$.

   (b) $f^{-1}(3) = -1$. Since f is one-to-one, $f(-1) = 3$.

19. $f(x) = 5 - 2x$. Since f is one-to-one and $f(1) = 5 - 2(1) = 3$, then $f^{-1}(3) = 1$. (Find 1 by solving the equation $5 - 2x = 3$.)

21. $f(g(x)) = f(x + 6) = (x + 6) - 6 = x$ for all x.

$g(f(x)) = g(x - 6) = (x - 6) + 6 = x$ for all x. Thus f and g are inverses of each other.

23. $f(g(x)) = f\left(\dfrac{x + 5}{2}\right) = 2\left(\dfrac{x + 5}{2}\right) - 5 = x + 5 - 5 = x$ for all x.

$g(f(x)) = g(2x - 5) = \dfrac{(2x - 5) + 5}{2} = x$ for all x. Thus f and g are inverses of each other.

25. $f(g(x)) = f\left(\dfrac{1}{x}\right) = \dfrac{1}{1/x} = x$ for all x. Since $f(x) = g(x)$, we also have $g(f(x)) = x$. Thus f and g are inverses of each other.

27. $f(g(x)) = f\left(\sqrt{x+4}\right) = \left(\sqrt{x+4}\right)^2 - 4 = x + 4 - 4 = x$ for all $x \geq -4$.

$g(f(x)) = g(x^2 - 4) = \sqrt{(x^2 - 4) + 4} = \sqrt{x^2} = x$ for all $x \geq 0$. Thus f and g are inverses of each other.

29. $f(g(x)) = f\left(\dfrac{1}{x} + 1\right) = \dfrac{1}{\left(\dfrac{1}{x} + 1\right) - 1} = x$ for all $x \neq 0$.

$g(f(x)) = g\left(\dfrac{1}{x-1}\right) = \dfrac{1}{\left(\dfrac{1}{x-1}\right)} + 1 = (x-1) + 1 = x$ for all $x \neq 1$. Thus f and g are inverses of each other.

31. $f(x) = 2x + 1$. $y = 2x + 1$ $\Leftrightarrow$ $2x = y - 1$ $\Leftrightarrow$ $x = \frac{1}{2}(y - 1)$. So $f^{-1}(x) = \frac{1}{2}(x - 1)$.

33. $f(x) = 4x + 7$. $y = 4x + 7$ $\Leftrightarrow$ $4x = y - 7$ $\Leftrightarrow$ $x = \frac{1}{4}(y - 7)$. So $f^{-1}(x) = \frac{1}{4}(x - 7)$.

35. $f(x) = \dfrac{x}{2}$. $y = \dfrac{x}{2}$ $\Leftrightarrow$ $x = 2y$. So $f^{-1}(x) = 2x$.

37. $f(x) = \dfrac{1}{x+2}$. $y = \dfrac{1}{x+2}$ $\Leftrightarrow$ $x + 2 = \dfrac{1}{y}$ $\Leftrightarrow$ $x = \dfrac{1}{y} - 2$. So $f^{-1}(x) = \dfrac{1}{x} - 2$.

39. $f(x) = \dfrac{1 + 3x}{5 - 2x}$. $y = \dfrac{1 + 3x}{5 - 2x}$ $\Leftrightarrow$ $y(5 - 2x) = 1 + 3x$ $\Leftrightarrow$ $5y - 2xy = 1 + 3x$ $\Leftrightarrow$ $3x + 2xy = 5y - 1$

$\Leftrightarrow$ $x(3 + 2y) = 5y - 1$ $\Leftrightarrow$ $x = \dfrac{5y - 1}{2y + 3}$. So $f^{-1}(x) = \dfrac{5x - 1}{2x + 3}$.

41. $f(x) = \sqrt{2 + 5x}$, $x \geq -\frac{2}{5}$. $y = \sqrt{2 + 5x}$, $y \geq 0$ $\Leftrightarrow$ $y^2 = 2 + 5x$ $\Leftrightarrow$ $5x = y^2 - 2$ $\Leftrightarrow$ $x = \frac{1}{5}(y^2 - 2)$ and $y \geq 0$. So $f^{-1}(x) = \frac{1}{5}(x^2 - 2)$, $x \geq 0$.

43. $f(x) = 4 - x^2$, $x \geq 0$. $y = 4 - x^2$ $\Leftrightarrow$ $x^2 = 4 - y$ $\Leftrightarrow$ $x = \sqrt{4 - y}$. So $f^{-1}(x) = \sqrt{4 - x}$. Note: $x \geq 0$ $\Rightarrow$ $f(x) \leq 4$.

45. $f(x) = 4 + \sqrt[3]{x}$. $y = 4 + \sqrt[3]{x}$ $\Leftrightarrow$ $\sqrt[3]{x} = y - 4$ $\Leftrightarrow$ $x = (y - 4)^3$. So $f^{-1}(x) = (x - 4)^3$.

47. $f(x) = 1 + \sqrt{1 + x}$. $y = 1 + \sqrt{1 + x}$, $y \geq 1$ $\Leftrightarrow$ $\sqrt{1 + x} = y - 1$ $\Leftrightarrow$ $1 + x = (y - 1)^2$
$\Leftrightarrow$ $x = (y - 1)^2 - 1 = y^2 - 2y$. So $f^{-1}(x) = x^2 - 2x$, $x \geq 1$.

49. $f(x) = x^4$, $x \geq 0$. $y = x^4$, $y \geq 0$ $\Leftrightarrow$ $x = \sqrt[4]{y}$. So $f^{-1}(x) = \sqrt[4]{x}$, $x \geq 0$.

51. (a), (b) $f(x) = 3x - 6$

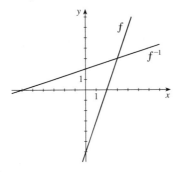

(c) $f(x) = 3x - 6$. $y = 3x - 6$ $\Leftrightarrow$
$3x = y + 6$ $\Leftrightarrow$ $x = \frac{1}{3}(y + 6)$. So
$f^{-1}(x) = \frac{1}{3}(x + 6)$.

53. (a), (b) $f(x) = \sqrt{x + 1}$

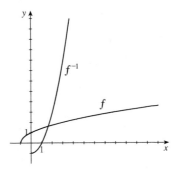

(c) $f(x) = \sqrt{x + 1}$, $x \geq -1$.
$y = \sqrt{x + 1}$, $y \geq 0$ $\Leftrightarrow$ $y^2 = x + 1$ $\Leftrightarrow$
$x = y^2 - 1$ and $y \geq 0$. So $f^{-1}(x) = x^2 - 1$,
$x \geq 0$.

55. $f(x) = x^3 - x$. Using a graphing device and the Horizontal Line Test, we see that f is not a one-to-one function. For example, $f(0) = 0 = f(-1)$.

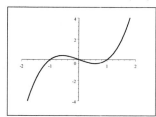

57. $f(x) = \dfrac{x + 12}{x - 6}$. Using a graphing device and the Horizontal Line Test, we see that f is a one-to-one function.

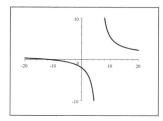

59. $f(x) = |x| - |x - 6|$. Using a graphing device and the Horizontal Line Test, we see that f is not a one-to-one function. For example $f(0) = -6 = f(-2)$.

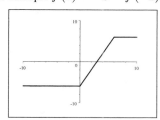

61. (a) $y = f(x) = 2 + x \quad \Leftrightarrow \quad x = y - 2$. So
$$f^{-1}(x) = x - 2.$$

(b)

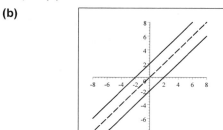

63. (a) $y = g(x) = \sqrt{x + 3}, \ y \geq 0 \quad \Leftrightarrow \quad x + 3 = y^2,$
$y \geq 0 \quad \Leftrightarrow \quad x = y^2 - 3, \ y \geq 0$. So
$g^{-1}(x) = x^2 - 3, \ x \geq 0$.

(b)

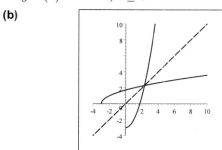

65. If we restrict the domain of $f(x)$ to $[0, \infty)$, then
$$y = 4 - x^2 \quad \Leftrightarrow \quad x^2 = 4 - y \quad \Rightarrow \quad x = \sqrt{4 - y}$$
(since $x \geq 0$, we take the positive square root). So
$f^{-1}(x) = \sqrt{4 - x}$.
If we restrict the domain of $f(x)$ to $(-\infty, 0]$, then
$$y = 4 - x^2 \quad \Leftrightarrow \quad x^2 = 4 - y \quad \Rightarrow \quad x = -\sqrt{4 - y}$$
(since $x \leq 0$, we take the negative square root). So
$f^{-1}(x) = -\sqrt{4 - x}$.

67. If we restrict the domain of $h(x)$ to $[-2, \infty)$, then $y = (x + 2)^2 \Rightarrow x + 2 = \sqrt{y}$ (since
$x \geq -2$, we take the positive square root)$\Leftrightarrow x = -2 + \sqrt{y}$. So $h^{-1}(x) = -2 + \sqrt{x}$.
If we restrict the domain of $h(x)$ to $(-\infty, -2]$, then $y = (x + 2)^2 \Rightarrow x + 2 = -\sqrt{y}$ (since
$x \leq -2$, we take the negative square root) $\Leftrightarrow x = -2 - \sqrt{y}$. So $h^{-1}(x) = -2 - \sqrt{x}$.

69.

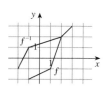

71. (a) $f(x) = 500 + 80x$.

(b) $f(x) = 500 + 80x$. $y = 500 + 80x \quad \Leftrightarrow \quad 80x = y - 500 \quad \Leftrightarrow \quad x = \dfrac{y - 500}{80}$. So $f^{-1}(x) = \dfrac{x - 500}{80}$. f^{-1}
represents the number of hours of investigation the investigate spends on a case for x dollars.

(c) $f^{-1}(1220) = \dfrac{1220 - 500}{80} = \dfrac{720}{80} = 9$. The investigator spent 9 hours investigating this case.

73. (a) $v(r) = 18{,}500 (0.25 - r^2)$. $t = 18{,}500 (0.25 - r^2)$ $\Leftrightarrow$ $t = 4625 - 18{,}500 r^2$ $\Leftrightarrow$ $18500 r^2 = 4625 - t$

$\Leftrightarrow$ $r^2 = \dfrac{4625 - t}{18{,}500}$ $\Rightarrow$ $r = \pm\sqrt{\dfrac{4625 - t}{18{,}500}}$. Since r represents a distance, $r \geq 0$, so $v^{-1}(t) = \sqrt{\dfrac{4625 - t}{18{,}500}}$.

v^{-1} represents the radius in the vein that has the velocity v.

(b) $v^{-1}(30) = \sqrt{\dfrac{4625 - 30}{18{,}500}} \approx 0.498$ cm. The velocity is 30 at 0.498 cm from the center of the artery or vein.

75. (a) $F(x) = \frac{9}{5} x + 32$. $y = \frac{9}{5} x + 32$ $\Leftrightarrow$ $\frac{9}{5} x = y - 32$ $\Leftrightarrow$ $x = \frac{5}{9}(y - 32)$. So $F^{-1}(x) = \frac{5}{9}(x - 32)$. F^{-1} represents the Celsius temperature that corresponds to the Fahrenheit temperature of F.

(b) $F^{-1}(86) = \frac{5}{9}(86 - 32) = \frac{5}{9}(54) = 30$. So $86°$ Fahrenheit is the same as $30°$ Celsius.

77. (a) $f(x) = \begin{cases} 0.1x, & \text{if } 0 \leq x \leq 20{,}000 \\ 2000 + 0.2(x - 20{,}000) & \text{if } x > 20{,}000 \end{cases}$

(b) We will find the inverse of each piece of the function f.

$f_1(x) = 0.1x$. $y = 0.1x$ $\Leftrightarrow$ $x = 10y$. So $f_1^{-1}(x) = 10x$.

$f_2(x) = 2000 + 0.2(x - 20{,}000) = 0.2x - 2000$. $y = 0.2x - 2000$ $\Leftrightarrow$ $0.2x = y + 2000$ $\Leftrightarrow$ $x = 5y + 10{,}000$. So $f_2^{-1}(x) = 5x + 10{,}000$.

Since $f(0) = 0$ and $f(20{,}000) = 2000$ we have $f^{-1}(x) = \begin{cases} 10x, & \text{if } 0 \leq x \leq 2000 \\ 5x + 10{,}000 & \text{if } x > 2000 \end{cases}$ It represents the taxpayer's income.

(b) $f^{-1}(10{,}000) = 5(10{,}000) + 10{,}000 = 60{,}000$. The required income is $60{,}000.

79. $f(x) = 7 + 2x$. $y = 7 + 2x$ $\Leftrightarrow$ $2x = y - 7$ $\Leftrightarrow$ $x = \dfrac{y - 7}{2}$. So $f^{-1}(x) = \dfrac{x - 7}{2}$. f^{-1} is the number of toppings on a pizza that costs x dollars.

81. (a) $f(x) = \dfrac{2x + 1}{5}$ is "multiply by 2, add 1, and then divide by 5 ". So the reverse is "multiply by 5, subtract 1, and then divide by 2 " or $f^{-1}(x) = \dfrac{5x - 1}{2}$. Check: $f \circ f^{-1}(x) = f\left(\dfrac{5x - 1}{2}\right) = \dfrac{2\left(\dfrac{5x - 1}{2}\right) + 1}{5} = \dfrac{5x - 1 + 1}{5} = \dfrac{5x}{5} = x$

and $f^{-1} \circ f(x) = f^{-1}\left(\dfrac{2x + 1}{5}\right) = \dfrac{5\left(\dfrac{2x + 1}{5}\right) - 1}{2} = \dfrac{2x + 1 - 1}{2} = \dfrac{2x}{2} = x$.

(b) $f(x) = 3 - \dfrac{1}{x} = \dfrac{-1}{x} + 3$ is "take the negative reciprocal and add 3 ". Since the reverse of "take the negative reciprocal" is "take the negative reciprocal ", $f^{-1}(x)$ is "subtract 3 and take the negative reciprocal ", that is, $f^{-1}(x) = \dfrac{-1}{x - 3}$. Check: $f \circ f^{-1}(x) = f\left(\dfrac{-1}{x - 3}\right) = 3 - \dfrac{1}{\dfrac{-1}{x - 3}} = 3 - \left(1 \cdot \dfrac{x - 3}{-1}\right) = 3 + x - 3 = x$ and

$f^{-1} \circ f(x) = f^{-1}\left(3 - \dfrac{1}{x}\right) = \dfrac{-1}{\left(3 - \dfrac{1}{x}\right) - 3} = \dfrac{-1}{-\dfrac{1}{x}} = -1 \cdot \dfrac{x}{-1} = x$.

(c) $f(x) = \sqrt{x^3 + 2}$ is "cube, add 2, and then take the square root". So the reverse is "square, subtract 2, then take the cube root " or $f^{-1}(x) = \sqrt[3]{x^2 - 2}$. Domain for $f(x)$ is $\left[-\sqrt[3]{2}, \infty\right)$; domain for $f^{-1}(x)$ is $[0, \infty)$. Check:

$f \circ f^{-1}(x) = f\left(\sqrt[3]{x^2 - 2}\right) = \sqrt{\left(\sqrt[3]{x^2 - 2}\right)^3 + 2} = \sqrt{x^2 - 2 + 2} = \sqrt{x^2} = x$ (on the appropriate domain) and

$f^{-1} \circ f(x) = f^{-1}\left(\sqrt{x^3 + 2}\right) = \sqrt[3]{\left(\sqrt{x^3 + 2}\right)^2 - 2} = \sqrt[3]{x^3 + 2 - 2} = \sqrt[3]{x^3} = x$ (on the appropriate domain).

(d) $f(x) = (2x - 5)^3$ is "double, subtract 5, and then cube". So the reverse is "take the cube root, add 5, and divide by 2" or $f^{-1}(x) = \dfrac{\sqrt[3]{x} + 5}{2}$ Domain for both $f(x)$ and $f^{-1}(x)$ is $(-\infty, \infty)$. Check:

$$f \circ f^{-1}(x) = f\left(\frac{\sqrt[3]{x} + 5}{2}\right) = \left[2\left(\frac{\sqrt[3]{x} + 5}{2}\right) - 5\right]^3 = \left(\sqrt[3]{x} + 5 - 5\right)^3 = \left(\sqrt[3]{x}\right)^3 = \sqrt[3]{x^3} = x \text{ and}$$

$$f^{-1} \circ f(x) = f^{-1}\left((2x - 5)^3\right) = \frac{\sqrt[x]{(2x - 5)^3} + 5}{2} = \frac{(2x - 5) + 5}{2} = \frac{2x}{2} = x.$$

In a function like $f(x) = 3x - 2$, the variable occurs only once and it easy to see how to reverse the operations step by step. But in $f(x) = x^3 + 2x + 6$, you apply two different operations to the variable x (cubing and multiplying by 2) and then add 6, so it is not possible to reverse the operations step by step.

83. (a) We find $g^{-1}(x)$: $y = 2x + 1 \quad \Leftrightarrow \quad 2x = y - 1 \quad \Leftrightarrow \quad x = \frac{1}{2}(y - 1)$. So $g^{-1}(x) = \frac{1}{2}(x - 1)$. Thus
$f(x) = h \circ g^{-1}(x) = h\left(\frac{1}{2}(x - 1)\right) = 4\left[\frac{1}{2}(x - 1)\right]^2 + 4\left[\frac{1}{2}(x - 1)\right] + 7 = x^2 - 2x + 1 + 2x - 2 + 7 = x^2 + 6.$

(b) $f \circ g = h \quad \Leftrightarrow \quad f^{-1} \circ f \circ g = f^{-1} \circ h \quad \Leftrightarrow \quad I \circ g = f^{-1} \circ h \quad \Leftrightarrow \quad g = f^{-1} \circ h.$
Note that we compose with f^{-1} on the left on each side of the equation. We find f^{-1}:

$y = 3x + 5 \quad \Leftrightarrow \quad 3x = y - 5 \quad \Leftrightarrow \quad x = \dfrac{1}{3}(y - 5)$. So $f^{-1}(x) = \frac{1}{3}(x - 5)$. Thus

$g(x) = f^{-1} \circ h(x) = f^{-1}(3x^2 + 3x + 2) = \frac{1}{3}\left[(3x^2 + 3x + 2) - 5\right] = \frac{1}{3}\left[3x^2 + 3x - 3\right] = x^2 + x - 1.$

Chapter 3 Review

1. $f(x) = x^2 - 4x + 6$; $f(0) = (0)^2 - 4(0) + 6 = 6$; $f(2) = (2)^2 - 4(2) + 6 = 2$;
$f(-2) = (-2)^2 - 4(-2) + 6 = 18$; $f(a) = (a)^2 - 4(a) + 6 = a^2 - 4a + 6$; $f(-a) = (-a)^2 - 4(-a) + 6 = a^2 + 4a + 6$;
$f(x + 1) = (x + 1)^2 - 4(x + 1) + 6 = x^2 + 2x + 1 - 4x - 4 + 6 = x^2 - 2x + 3$; $f(2x) = (2x)^2 - 4(2x) + 6 = 4x^2 - 8x + 6$;
$2f(x) - 2 = 2(x^2 - 4x + 6) - 2 = 2x^2 - 8x + 12 - 2 = 2x^2 - 8x + 10.$

3. (a) $f(-2) = -1$. $f(2) = 2$.

(b) The domain of f is $[-4, 5]$.

(b) The range of f is $[-4, 4]$.

(d) f is increasing on $[-4, -2]$ and $[-1, 4]$; f is decreasing on $[-2, -1]$ and $[4, 5]$.

(e) f is not a one-to-one, for example, $f(-2) = -1 = f(0)$. There are many more examples.

5. Domain: We must have $x + 3 \geq 0 \quad \Leftrightarrow \quad x \geq -3$. In interval notation, the domain is $[-3, \infty)$.
Range: For x in the domain of f, we have $x \geq -3 \quad \Leftrightarrow \quad x + 3 \geq 0 \quad \Leftrightarrow \quad \sqrt{x + 3} \geq 0 \quad \Leftrightarrow \quad f(x) \geq 0$. So the range is $[0, \infty)$.

7. $f(x) = 7x + 15$. The domain is all real numbers, $(-\infty, \infty)$.

9. $f(x) = \sqrt{x + 4}$. We require $x + 4 \geq 0 \quad \Leftrightarrow \quad x \geq -4$. Thus the domain is $[-4, \infty)$.

11. $f(x) = \dfrac{1}{x} + \dfrac{1}{x + 1} + \dfrac{1}{x + 2}$. The denominators cannot equal 0, therefore the domain is $\{x \mid x \neq 0, -1, -2\}$.
$\{x \mid 2x + 1 \neq 0 \text{ and } x - 3 \neq 0\} = \{x \mid x \neq -\frac{1}{2} \text{ and } x \neq 3\}.$

13. $h(x) = \sqrt{4-x} + \sqrt{x^2-1}$. We require the expression inside the radicals be nonnegative. So $4 - x \geq 0 \quad \Leftrightarrow \quad 4 \geq x$; also $x^2 - 1 \geq 0 \quad \Leftrightarrow \quad (x-1)(x+1) \geq 0$. We make a table:

Interval	$(-\infty, -1)$	$(-1, 1)$	$(1, \infty)$
Sign of $x - 1$	$-$	$-$	$+$
Sign of $x + 1$	$-$	$+$	$+$
Sign of $(x-1)(x+1)$	$+$	$-$	$+$

Thus the domain is $(-\infty, 4] \cap \{(-\infty, -1] \cup [1, \infty)\} = (-\infty, -1] \cup [1, 4]$.

15. $f(x) = 1 - 2x$

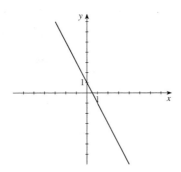

17. $f(t) = 1 - \frac{1}{2}t^2$

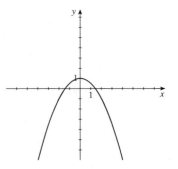

19. $f(x) = x^2 - 6x + 6$

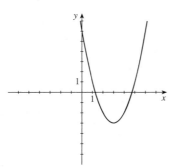

21. $g(x) = 1 - \sqrt{x}$

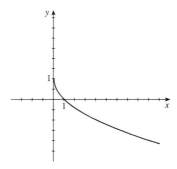

23. $h(x) = \frac{1}{2}x^3$

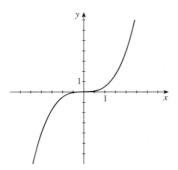

25. $h(x) = \sqrt[3]{x}$

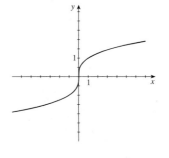

27. $g(x) = \dfrac{1}{x^2}$

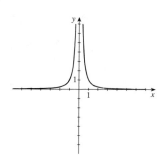

29. $f(x) = \begin{cases} 1 - x & \text{if } x < 0 \\ 1 & \text{if } x \geq 0 \end{cases}$

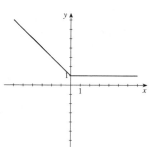

31. $f(x) = \begin{cases} x + 6 & \text{if } x < -2 \\ x^2 & \text{if } x \geq -2 \end{cases}$

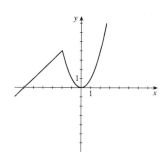

33. $f(x) = 6x^3 - 15x^2 + 4x - 1$

(i) $[-2, 2]$ by $[-2, 2]$

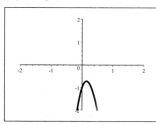

(ii) $[-8, 8]$ by $[-8, 8]$

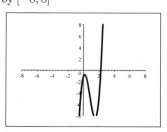

(iii) $[-4, 4]$ by $[-12, 12]$

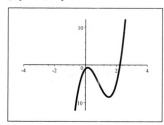

(iv) $[-100, 100]$ by $[-100, 100]$

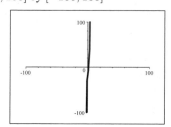

From the graphs, we see that the viewing rectangle in (iii) produces the most appropriate graph.

35. $f(x) = x^2 + 25x + 173$
$$= \left(x^2 + 25x + \tfrac{625}{4}\right) + 173 - \tfrac{625}{4}$$
$$= \left(x + \tfrac{25}{2}\right)^2 + \tfrac{67}{4}$$

We use the viewing rectangle $[-30, 5]$ by $[-20, 250]$.

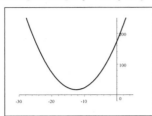

37. $f(x) = \dfrac{x}{\sqrt{x^2 + 16}}$. Since $\sqrt{x^2 + 16} \ge \sqrt{x^2} = |x|$, it

follows that y should behave like $\dfrac{x}{|x|}$. Thus we use the

viewing rectangle $[-20, 20]$ by $[-2, 2]$.

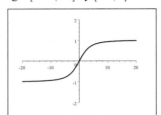

39. $f(x) = \sqrt{x^3 - 4x + 1}$. The domain consists of all x where $x^3 - 4x + 1 \ge 0$.
Using a graphing device, we see that the domain is approximately
$[-2.1, 0.2] \cup [1.9, \infty)$.

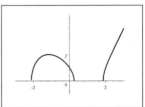

41. The average rate of change is $\dfrac{f(2) - f(0)}{2 - 0} = \dfrac{\left[(2)^2 + 3(2)\right] - \left[0^2 + 3(0)\right]}{2} = \dfrac{4 + 6 - 0}{2} = 5.$

43. The average rate of change is $\dfrac{f(3 + h) - f(3)}{(3 + h) - 3} = \dfrac{\dfrac{1}{3 + h} - \dfrac{1}{3}}{h} \cdot \dfrac{3(3 + h)}{3(3 + h)} = \dfrac{3 - (3 + h)}{3h(3 + h)} = \dfrac{-h}{3h(3 + h)} = -\dfrac{1}{3(3 + h)}.$

45. $f(x) = x^3 - 4x^2$ is graphed in the viewing rectangle $[-5, 5]$ by $[-20, 10]$. $f(x)$
is increasing on $(-\infty, 0]$ and $[2.67, \infty)$. It is decreasing on $[0, 2.67]$.

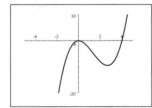

47. (a) $y = f(x) + 8$. Shift the graph of $f(x)$ upward 8 units.

   (b) $y = f(x + 8)$. Shift the graph of $f(x)$ to the left 8 units.

   (c) $y = 1 + 2f(x)$. Stretch the graph of $f(x)$ vertically by a factor of 2, then shift it upward 1 unit.

   (d) $y = f(x - 2) - 2$. Shift the graph of $f(x)$ to the right 2 units, then downward 2 units.

   (e) $y = f(-x)$. Reflect the graph of $f(x)$ about the y-axis.

   (f) $y = -f(-x)$. Reflect the graph of $f(x)$ first about the y-axis, then reflect about the x-axis.

   (g) $y = -f(x)$. Reflect the graph of $f(x)$ about the x-axis.

   (h) $y = f^{-1}(x)$. Reflect the graph of $f(x)$ about the line $y = x$.

49. (a) $f(x) = 2x^5 - 3x^2 + 2$. $f(-x) = 2(-x)^5 - 3(-x)^2 + 2 = -2x^5 - 3x^2 + 2$. Since $f(x) \ne f(-x)$, f is not even.
   $-f(x) = -2x^5 + 3x^2 - 2$. Since $-f(x) \ne f(-x)$, f is not odd.

   (b) $f(x) = x^3 - x^7$. $f(-x) = (-x)^3 - (-x)^7 = -\left(x^3 - x^7\right) = -f(x)$, hence f is odd.

   (c) $f(x) = \dfrac{1 - x^2}{1 + x^2}$. $f(-x) = \dfrac{1 - (-x)^2}{1 + (-x)^2} = \dfrac{1 - x^2}{1 + x^2} = f(x)$. Since $f(x) = f(-x)$, f is even.

   (d) $f(x) = \dfrac{1}{x + 2}$. $f(-x) = \dfrac{1}{(-x) + 2} = \dfrac{1}{2 - x}$. $-f(x) = -\dfrac{1}{x + 2}$. Since $f(x) \ne f(-x)$, f is not even, and since
   $f(-x) \ne -f(x)$, f is not odd.

51. $f(x) = x^2 + 4x + 1 = (x^2 + 4x + 4) + 1 - 4 = (x + 2)^2 - 3.$

53. $g(x) = 2x^2 + 4x - 5 = 2(x^2 + 2x) - 5 = 2(x^2 + 2x + 1) - 5 - 2 = 2(x + 1)^2 - 7.$ So the minimum value is $g(-1) = -7.$

55. $h(t) = -16t^2 + 48t + 32 = -16(t^2 - 3t) + 32 = -16\left(t^2 - 3t + \frac{9}{4}\right) + 32 + 36 = -16\left(t^2 - 3t + \frac{9}{4}\right) + 68$

$= -16\left(t - \frac{3}{2}\right)^2 + 68.$ The stone reaches a maximum height of 68 feet.

57. $f(x) = 3.3 + 1.6x - 2.5x^3.$ In the first viewing rectangle, $[-2, 2]$ by $[-4, 8]$, we see that $f(x)$ has a local maximum and a local minimum. In the next viewing rectangle, $[0.4, 0.5]$ by $[3.78, 3.80]$, we isolate the local maximum value as approximately 3.79 when $x \approx 0.46.$ In the last viewing rectangle, $[-0.5, -0.4]$ by $[2.80, 2.82]$, we isolate the local minimum value as 2.81 when $x \approx -0.46.$

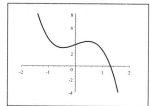 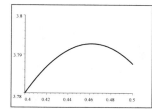 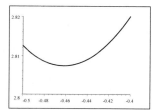

59. $f(x) = x^2 - 3x + 2$ and $g(x) = 4 - 3x.$

(a) $(f + g)(x) = (x^2 - 3x + 2) + (4 - 3x) = x^2 - 6x + 6$

(b) $(f - g)(x) = (x^2 - 3x + 2) - (4 - 3x) = x^2 - 2$

(c) $(fg)(x) = (x^2 - 3x + 2)(4 - 3x) = 4x^2 - 12x + 8 - 3x^3 + 9x^2 - 6x = -3x^3 + 13x^2 - 18x + 8$

(d) $\left(\dfrac{f}{g}\right)(x) = \dfrac{x^2 - 3x + 2}{4 - 3x}, x \neq \dfrac{4}{3}$

(e) $(f \circ g)(x) = f(4 - 3x) = (4 - 3x)^2 - 3(4 - 3x) + 2 = 16 - 24x + 9x^2 - 12 + 9x + 2 = 9x^2 - 15x + 6$

(f) $(g \circ f)(x) = g(x^2 - 3x + 2) = 4 - 3(x^2 - 3x + 2) = -3x^2 + 9x - 2$

61. $f(x) = 3x - 1$ and $g(x) = 2x - x^2.$

$(f \circ g)(x) = f(2x - x^2) = 3(2x - x^2) - 1 = -3x^2 + 6x - 1,$ and the domain is $(-\infty, \infty).$

$(g \circ f)(x) = g(3x - 1) = 2(3x - 1) - (3x - 1)^2 = 6x - 2 - 9x^2 + 6x - 1 = -9x^2 + 12x - 3,$ and the domain is $(-\infty, \infty)$

$(f \circ f)(x) = f(3x - 1) = 3(3x - 1) - 1 = 9x - 4,$ and the domain is $(-\infty, \infty).$

$(g \circ g)(x) = g(2x - x^2) = 2(2x - x^2) - (2x - x^2)^2 = 4x - 2x^2 - 4x^2 + 4x^3 - x^4 = -x^4 + 4x^3 - 6x^2 + 4x,$ and domain is $(-\infty, \infty).$

63. $f(x) = \sqrt{1 - x}, g(x) = 1 - x^2$ and $h(x) = 1 + \sqrt{x}.$

$(f \circ g \circ h)(x) = f(g(h(x))) = f(g(1 + \sqrt{x})) = f\left(1 - (1 + \sqrt{x})^2\right) = f(1 - (1 + 2\sqrt{x} + x))$

$= f(-x - 2\sqrt{x}) = \sqrt{1 - (-x - 2\sqrt{x})} = \sqrt{1 + 2\sqrt{x} + x} = \sqrt{(1 + \sqrt{x})^2} = 1 + \sqrt{x}$

65. $f(x) = 3 + x^3.$ If $x_1 \neq x_2,$ then $x_1^3 \neq x_2^3$ (unequal numbers have unequal cubes), and therefore $3 + x_1^3 \neq 3 + x_2^3.$ Thus f is a one-to-one function.

67. $h(x) = \dfrac{1}{x^4}.$ Since the fourth powers of a number and its negative are equal, h is not one-to-one. For example,

$h(-1) = \dfrac{1}{(-1)^4} = 1$ and $h(1) = \dfrac{1}{(1)^4} = 1,$ so $h(-1) = h(1).$

69. $p(x) = 3.3 + 1.6x - 2.5x^3$. Using a graphing device and the Horizontal Line Test, we see that p is not a one-to-one function.

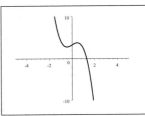

71. $f(x) = 3x - 2$ $\Leftrightarrow$ $y = 3x - 2$ $\Leftrightarrow$ $3x = y + 2$ $\Leftrightarrow$ $x = \frac{1}{3}(y + 2)$. So $f^{-1}(x) = \frac{1}{3}(x + 2)$.

73. $f(x) = (x + 1)^3$ $\Leftrightarrow$ $y = (x + 1)^3$ $\Leftrightarrow$ $x + 1 = \sqrt[3]{y}$ $\Leftrightarrow$ $x = \sqrt[3]{y} - 1$. So $f^{-1}(x) = \sqrt[3]{x} - 1$.

75. (a), (b) $f(x) = x^2 - 4$, $x \geq 0$

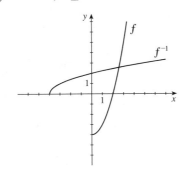

(c) $f(x) = x^2 - 4$, $x \geq 0$ $\Leftrightarrow$ $y = x^2 - 4$, $y \geq -4$
$\Leftrightarrow$ $x^2 = y + 4$ $\Leftrightarrow$ $x = \sqrt{y + 4}$. So
$f^{-1}(x) = \sqrt{x + 4}$, $x \geq -4$.

Chapter 3 Test

1. By the Vertical Line Test, figures (a) and (b) are graphs of functions. By the Horizontal Line Test, only figure (a) is the graph of a one-to-one function.

2. (a) $f(3) = \dfrac{\sqrt{3 + 1}}{3} = \dfrac{\sqrt{4}}{3} = \dfrac{2}{3}$; $f(5) = \dfrac{\sqrt{5 + 1}}{5} = \dfrac{\sqrt{6}}{5}$; $f(a - +1) = \dfrac{\sqrt{(a - 1) + 1}}{a - 1} = \dfrac{\sqrt{a}}{a - 1}$.

(b) $f(x) = \dfrac{\sqrt{x + 1}}{x}$. Our restrictions are that the input to the radical is nonnegative, and the denominator must not be equal to zero. Thus $x + 1 \geq 0$ $\Leftrightarrow$ $x \geq -1$ and $x \neq 0$. In interval notation, the domain is $[-1, 0) \cup (0, \infty)$.

3. The average rate of change is $\dfrac{f(2) - f(5)}{2 - 5} = \dfrac{[2^2 - 2(2)] - [5^2 - 2(5)]}{-3} = \dfrac{4 - 4 - (25 - 10)}{-3} = \dfrac{-15}{-3} = 5$.

4. (a) $f(x) = x^3$

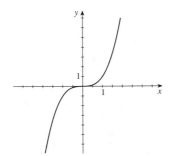

(b) $g(x) = (x - 1)^3 - 2$. To obtain the graph of g, shift the graph of f to the right 1 unit and downward 2 units.

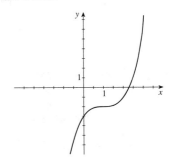

5. (a) $y = f(x-3) + 2$. Shift the graph of $f(x)$ to the right 3 units, then shift the graph upward 2 units.

   (b) $y = f(-x)$. Reflect the graph of $f(x)$ about the y-axis.

6. (a) $f(x) = 2x^2 - 8x + 13$

$$= 2\left(x^2 - 4x\right) + 13$$

$$= 2\left(x^2 - 4x + 4\right) + 13 - 8$$

$$= 2(x-2)^2 + 5$$

(b)

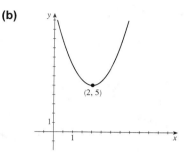

   (c) Since $f(x) = 2(x-2)^2 + 5$ is in standard form, the minimum value of f is $f(2) = 5$.

7. (a) $f(-2) = 1 - (-2)^2 = 1 - 4 = -3$ (since $-2 \le 0$).

$f(1) = 2(1) + 1 = 2 + 1 = 3$ (since $1 > 0$).

(b)

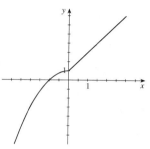

8. $f(x) = x^2 + 1$; $g(x) = x - 3$.

   (a) $(f \circ g)(x) = f(g(x)) = f(x-3) = (x-3)^2 + 1 = x^2 - 6x + 9 + 1 = x^2 - 6x + 10$

   (b) $(g \circ f)(x) = g(f(x)) = g(x^2 + 1) = (x^2 + 1) - 3 = x^2 - 2$

   (c) $f(g(2)) = f(-1) = (-1)^2 + 1 = 2$. (We have used the fact that $g(2) = (2) - 3 = -1$.)

   (d) $g(f(2)) = g(5) = 5 - 3 = 2$. (We have used the fact that $f(2) = 2^2 + 1 = 5$.)

   (e) $(g \circ g \circ g)(x) = g(g(g(x))) = g(g(x-3)) = g(x-6) = (x-6) - 3 = x - 9$. (We have used the fact that $g(x-3) = (x-3) - 3 = x - 6$.)

9. (a) $f(x) = \sqrt{3-x}$, $x \le 3$ $\quad\Leftrightarrow$

$$y = \sqrt{3-x} \quad\Leftrightarrow\quad y^2 = 3 - x \quad\Leftrightarrow$$

$x = 3 - y^2$. Thus $f^{-1}(x) = 3 - x^2$,

$x \ge 0$.

(b) $f(x) = \sqrt{3-x}$, $x \le 3$ and $f^{-1}(x) = 3 - x^2$, $x \ge 0$

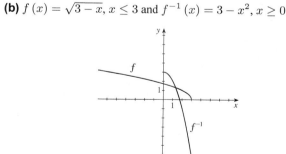

10. (a) The domain of f is $[0, 6]$, and the range of f is $[1, 7]$.

   (c) The average rate of change is $\dfrac{f(6) - f(2)}{6 - 2} = \dfrac{7 - 2}{4} = \dfrac{5}{4}$.

(b)

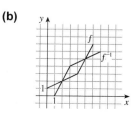

11. (a) $f(x) = 3x^4 - 14x^2 + 5x - 3$. The graph is shown in the viewing rectangle
   $[-10, 10]$ by $[-30, 10]$.

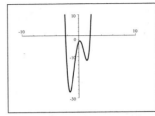

   (b) No, by the Horizontal Line Test.

   (c) The local maximum is approximately -2.55 when $x \approx 0.18$, as shown in the first viewing rectangle $[0.15, 0.25]$ by
   $[-2.6, -2.5]$. One local minimum is approximately -27.18 when $x \approx -1.61$, as shown in the second viewing
   rectangle $[-1.65, -1.55]$ by $[-27.5, -27]$. The other local minimum is approximately -11.93 when $x \approx 1.43$, as
   shown is the viewing rectangle $[1.4, 1.5]$ by $[-12, -11.9]$.

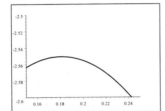 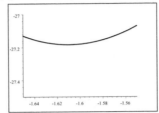 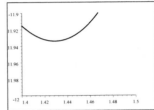

   (d) Using the graph in part (a) and the local minimum, -27.18, found in part (c), we see that the range is $[-27.18, \infty)$.

   (e) Using the information from part (c) and the graph in part (a), $f(x)$ is increasing on the intervals $[-1.61, 0.18]$ and
   $[1.43, \infty)$ and decreasing on the intervals $(-\infty, -1.61]$ and $[0.18, 1.43]$.

Focus on Modeling: Modeling with Functions

1. Let w be the width of the building lot. Then the length of the lot is $3w$. So the area of the building lot is $A(w) = 3w^2$, $w > 0$.

3. Let w be the width of the base of the rectangle. Then the height of the rectangle is $\frac{1}{2}w$. Thus the volume of the box is given by the function $V(w) = \frac{1}{2}w^3$, $w > 0$.

5. Let P be the perimeter of the rectangle and y be the length of the other side. Since $P = 2x + 2y$ and the perimeter is 20, we have $2x + 2y = 20 \quad \Leftrightarrow \quad x + y = 10 \quad \Leftrightarrow \quad y = 10 - x$. Since area is $A = xy$, substituting gives $A(x) = x(10 - x) = 10x - x^2$, and since A must be positive, the domain is $0 < x < 10$.

7.

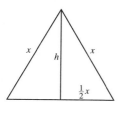

Let h be the height of an altitude of the equilateral triangle whose side has length x, as shown in the diagram. Thus the area is given by $A = \frac{1}{2}xh$. By the Pythagorean Theorem, $h^2 + \left(\frac{1}{2}x\right)^2 = x^2 \quad \Leftrightarrow \quad h^2 + \frac{1}{4}x^2 = x^2 \quad \Leftrightarrow$

$h^2 = \frac{3}{4}x^2 \Leftrightarrow h = \frac{\sqrt{3}}{2}x$. Substituting into the area of a triangle, we get

$$A(x) = \frac{1}{2}xh = \frac{1}{2}x\left(\frac{\sqrt{3}}{2}x\right) = \frac{\sqrt{3}}{4}x^2,\ x > 0.$$

9. We solve for r in the formula for the area of a circle. This gives $A = \pi r^2 \quad \Leftrightarrow \quad r^2 = \dfrac{A}{\pi} \quad \Rightarrow \quad r = \sqrt{\dfrac{A}{\pi}}$, so the model

is $r(A) = \sqrt{\dfrac{A}{\pi}},\ A > 0$.

11. Let h be the height of the box in feet. The volume of the box is $V = 60$. Then $x^2 h = 60 \quad \Leftrightarrow \quad h = \dfrac{60}{x^2}$.

The surface area, S, of the box is the sum of the area of the 4 sides and the area of the base and top. Thus

$S = 4xh + 2x^2 = 4x\left(\dfrac{60}{x^2}\right) + 2x^2 = \dfrac{240}{x} + 2x^2$, so the model is $S(x) = \dfrac{240}{x} + 2x^2,\ x > 0$.

13.

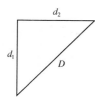

Let d_1 be the distance traveled south by the first ship and d_2 be the distance traveled east by the second ship. The first ship travels south for t hours at 5 mi/h, so $d_1 = 15t$ and, similarly, $d_2 = 20t$. Since the ships are traveling at right angles to each other, we can apply the Pythagorean Theorem to get

$$D^2 = d_1^2 + d_2^2 = (15t)^2 + (20t)^2 = 225t^2 + 400t^2 = 625t^2.$$

15.

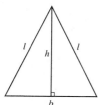

Let b be the length of the base, l be the length of the equal sides, and h be the height in centimeters. Since the perimeter is 8, $2l + b = 8 \quad \Leftrightarrow \quad 2l = 8 - b$

$\Leftrightarrow \quad l = \frac{1}{2}(8 - b)$. By the Pythagorean Theorem, $h^2 + \left(\frac{1}{2}b\right)^2 = l^2 \quad \Leftrightarrow$

$h = \sqrt{l^2 - \frac{1}{4}b^2}$. Therefore the area of the triangle is

$$A = \frac{1}{2} \cdot b \cdot h = \frac{1}{2} \cdot b\sqrt{l^2 - \frac{1}{4}b^2} = \frac{b}{2}\sqrt{\frac{1}{4}(8 - b)^2 - \frac{1}{4}b^2}$$
$$= \frac{b}{4}\sqrt{64 - 16b + b^2 - b^2} = \frac{b}{4}\sqrt{64 - 16b} = \frac{b}{4} \cdot 4\sqrt{4 - b} = b\sqrt{4 - b}$$

so the model is $A(b) = b\sqrt{4 - b},\ 0 < b < 4$.

17. Let w be the length of the rectangle. By the Pythagorean Theorem, $\left(\frac{1}{2}w\right)^2 + h^2 = 10^2 \quad \Leftrightarrow \quad \dfrac{w^2}{4} + h^2 = 10^2$

$\Leftrightarrow \quad w^2 = 4(100 - h^2) \quad \Leftrightarrow \quad w = 2\sqrt{100 - h^2}$ (since $w > 0$). Therefore, the area of the rectangle is $A = wh = 2h\sqrt{100 - h^2}$, so the model is $A(h) = 2h\sqrt{100 - h^2},\ 0 < h < 10$.

19. (a) We complete the table.

First number	Second number	Product
1	18	18
2	17	34
3	16	48
4	15	60
5	14	70
6	13	78
7	12	84
8	11	88
9	10	90
10	9	90
11	8	88

From the table we conclude that the numbers is still increasing, the numbers whose product is a maximum should both be 9.5.

(b) Let x be one number: then $19 - x$ is the other number, and so the product, p, is

$$p(x) = x(19 - x) = 19x - x^2.$$

(c) $p(x) = 19x - x^2 = -\left(x^2 - 19x\right)$
$$= -\left[x^2 - 19x + \left(\tfrac{19}{2}\right)^2\right] + \left(\tfrac{19}{2}\right)^2$$
$$= -\left(x - 9.5\right)^2 + 90.25$$

So the product is maximized when the numbers are both 9.5.

21. Let x and y be the two numbers. Since their sum is -24, we have $x + y = -24 \quad \Leftrightarrow \quad y = -x - 24$. The product of the two numbers is $P = xy = x(-x - 24) = -x^2 - 24x$, which we wish to maximize. So $P = -x^2 - 24x = -\left(x^2 + 24x\right) = -\left(x^2 + 24x + 144\right) + 144 = -\left(x + 12\right)^2 + 144$. Thus the maximum product is 144, and it occurs when $x = -12$ and $y = -(-12) - 24 = -12$. Thus the two numbers are -12 and -12.

23. (a) Let x be the width of the field (in feet) and l be the length of the field (in feet). Since the farmer has 2400 ft of fencing we must have $2x + l = 2400$.

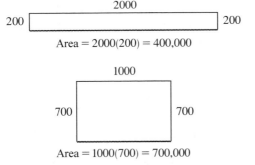

Width	Length	Area
200	2000	400,000
300	1800	540,000
400	1600	640,000
500	1400	700,000
600	1200	720,000
700	1000	700,000
800	800	640,000

It appears that the field of largest area is about 600 ft × 1200 ft.

(b) Let x be the width of the field (in feet) and l be the length of the field (in feet). Since the farmer has 2400 ft of fencing we must have $2x + l = 2400 \quad \Leftrightarrow \quad l = 2400 - 2x$. The area of the fenced-in field is given by
$$A(x) = l \cdot x = (2400 - 2x)x = -2x^2 + 2400x = -2\left(x^2 - 1200x\right).$$

(c) The area is $A(x) = -2\left(x^2 - 1200x + 600^2\right) + 2\left(600^2\right) = -2\left(x - 600\right)^2 + 720000$. So the maximum area occurs when $x = 600$ feet and $l = 2400 - 2(600) = 1200$ feet.

25. (a) Let x be the length of the fence along the road. If the area is 1200, we have $1200 = x \cdot$ width, so the width of the garden is $\dfrac{1200}{x}$. Then the cost of the fence is given by the function $C(x) = 5(x) + 3 \left[x + 2 \cdot \dfrac{1200}{x} \right] = 8x + \dfrac{7200}{x}$.

(b) We graph the function $y = C(x)$ in the viewing rectangle $[0, 75] \times [0, 800]$. From this we get the cost is minimized when $x = 30$ ft. Then the width is $\frac{1200}{30} = 40$ ft. So the length is 30 ft and the width is 40 ft.

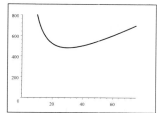

(c) We graph the function $y = C(x)$ and $y = 600$ in the viewing rectangle $[10, 65] \times [450, 650]$. From this we get that the cost is at most \$600 when $15 \le x \le 60$. So the range of lengths he can fence along the road is 15 feet to 60 feet.

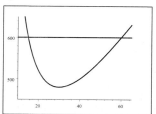

27. (a) Let p be the price of the ticket. So $10 - p$ is the difference in ticket price and therefore the number of tickets sold is $27{,}000 + 3000(10 - p) = 57{,}000 - 3000p$. Thus the revenue is $R(p) = p(57{,}000 - 3000p) = 57{,}000p - 3000p^2$.

(b) $R(p) = 0 = p(57{,}000 - 3000p)$. So $p = 0$ *or* $= \dfrac{57{,}000}{3000} = 19$. So at \$19 no one will come.

(c) We complete the square:

$$R(p) = 57{,}000p - 3000p^2 = -3000\left(p^2 - 19p\right) = -3000\left(p^2 - 19p + \tfrac{19^2}{4}\right) + 270{,}750$$

$$= -3000\left(p - \tfrac{19}{2}\right)^2 + 270{,}750$$

The revenue is maximized when $p = \frac{19}{2}$, and so the price should be set at \$9.50.

29. (a) Let h be the height in feet of the straight portion of the window. The circumference of the semicircle is $C = \frac{1}{2}\pi x$. Since the perimeter of the window is 30 feet, we have $x + 2h + \frac{1}{2}\pi x = 30$. Solving for h, we get $2h = 30 - x - \frac{1}{2}\pi x \quad \Leftrightarrow \quad h = 15 - \frac{1}{2}x - \frac{1}{4}\pi x$. The area of the window is $A(x) = xh + \frac{1}{2}\pi\left(\frac{1}{2}x\right)^2 = x\left(15 - \frac{1}{2}x - \frac{1}{4}\pi x\right) + \frac{1}{8}\pi x^2 = 15x - \frac{1}{2}x^2 - \frac{1}{8}\pi x^2$.

(b) $A(x) = 15x - \frac{1}{2}x^2 - \frac{1}{8}\pi x^2 = 15x - \frac{1}{8}(\pi + 4)x^2 = -\frac{1}{8}(\pi + 4)\left[x^2 - \dfrac{120}{\pi + 4}x\right]$

$= -\frac{1}{8}(\pi + 4)\left[x^2 - \dfrac{120}{\pi + 4}x + \left(\dfrac{60}{\pi + 4}\right)^2\right] + \dfrac{450}{\pi + 4} = -\frac{1}{8}(\pi + 4)\left(x - \dfrac{60}{\pi + 4}\right)^2 + \dfrac{450}{\pi + 4}$

The area is maximized when $x = \dfrac{60}{\pi + 4} \approx 8.40$, and hence $h \approx 15 - \frac{1}{2}(8.40) - \frac{1}{4}\pi(8.40) \approx 4.20$.

31. (a) Let x be the length of one side of the base and let h be the height of the box in feet. Since the volume of

the box is $V = x^2h = 12$, we have $x^2h = 12 \quad \Leftrightarrow \quad h = \dfrac{12}{x^2}$. The surface area, A, of the box is sum of

the area of the four sides and the area of the base. Thus the surface area of the box is given by the formula

$$A(x) = 4xh + x^2 = 4x\left(\dfrac{12}{x^2}\right) + x^2 = \dfrac{48}{x} + x^2, x > 0.$$

(b) The function $y = A(x)$ is shown in the first viewing rectangle below. In the second viewing rectangle, we isolate the minimum, and we see that the amount of material is minimized when x (the length and width) is 2.88 ft. Then the

height is $h = \dfrac{12}{x^2} \approx 1.44$ ft.

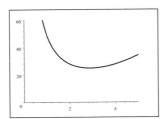
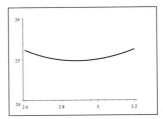

33. (a) Let w be the width of the pen and l be the length in meters. We use the area to establish a relationship between

w and l. Since the area is 100 m², we have $l \cdot w = 100 \quad \Leftrightarrow \quad l = \dfrac{100}{w}$. So the amount of fencing used is

$$F = 2l + 2w = 2\left(\dfrac{100}{w}\right) + 2w = \dfrac{200 + 2w^2}{w}.$$

(b) Using a graphing device, we first graph F in the viewing rectangle $[0, 40]$ by $[0, 100]$, and locate the approximate location of the minimum value. In the second viewing rectangle, $[8, 12]$ by $[39, 41]$, we see that the minimum value of F occurs when $w = 10$. Therefore the pen should be a square with side 10 m.

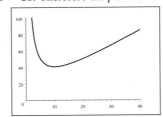
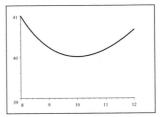

35. (a) Let x be the distance from point B to C, in miles. Then the distance from A to C is $\sqrt{x^2 + 25}$, and the energy used in flying from A to C then C to D is $f(x) = 14\sqrt{x^2 + 25} + 10(12 - x)$.

(b) By using a graphing device, the energy expenditure is minimized when the distance from B to C is about 5.1 miles.

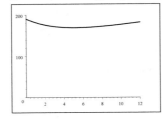
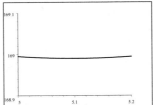

4 Polynomial and Rational Functions

4.1 Polynomial Functions and Their Graphs

1. (a) $P(x) = x^2 - 4$

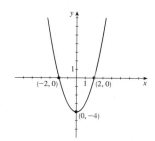

(b) $Q(x) = (x-4)^2$

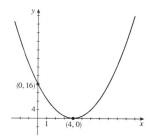

(c) $R(x) = 2x^2 - 2$

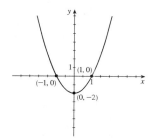

(d) $S(x) = 2(x-2)^2$

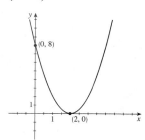

3. (a) $P(x) = x^3 - 8$

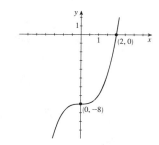

(b) $Q(x) = -x^3 + 27$

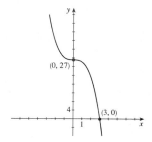

(c) $R(x) = -(x+2)^3$

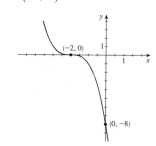

(d) $S(x) = \frac{1}{2}(x-1)^3 + 4$

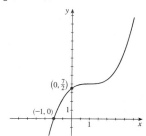

5. III

7. V

9. VI

11. $P(x) = (x-1)(x+2)$

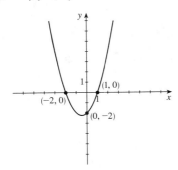

13. $P(x) = x(x-3)(x+2)$

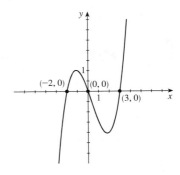

15. $P(x) = (x-3)(x+2)(3x-2)$

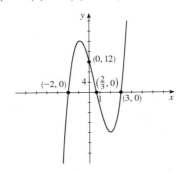

17. $P(x) = (x-1)^2(x-3)$

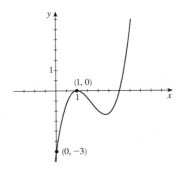

19. $P(x) = \frac{1}{12}(x+2)^2(x-3)^2$

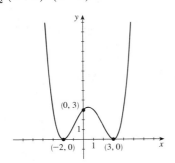

21. $P(x) = x^3(x+2)(x-3)^2$

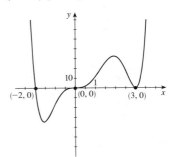

23. $P(x) = x^3 - x^2 - 6x = x(x+2)(x-3)$

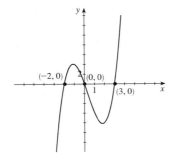

25. $P(x) = -x^3 + x^2 + 12x = -x(x+3)(x-4)$

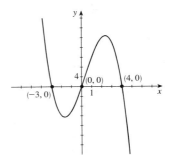

27. $P(x) = x^4 - 3x^3 + 2x^2 = x^2(x-1)(x-2)$

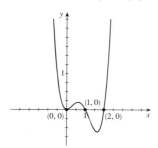

29. $P(x) = x^3 + x^2 - x - 1 = (x-1)(x+1)^2$

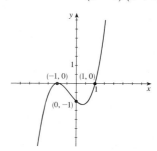

31. $P(x) = 2x^3 - x^2 - 18x + 9$
$= (x-3)(2x-1)(x+3)$

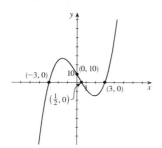

33. $P(x) = x^4 - 2x^3 - 8x + 16$
$= (x-2)^2(x^2+2x+4)$

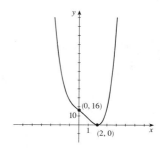

35. $P(x) = x^4 - 3x^2 - 4$
$= (x-2)(x+2)(x^2+1)$

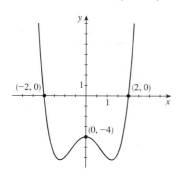

37. $P(x) = 3x^3 - x^2 + 5x + 1$; $Q(x) = 3x^3$. Since P has odd degree and positive leading coefficient, it has the following end behavior: $y \to \infty$ as $x \to \infty$ and $y \to -\infty$ as $x \to -\infty$.

On a large viewing rectangle, the graphs of P and Q look almost the same. On a small viewing rectangle, we see that the graphs of P and Q have different intercepts.

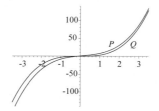
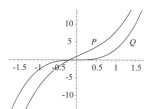

39. $P(x) = x^4 - 7x^2 + 5x + 5$; $Q(x) = x^4$. Since P has even degree and positive leading coefficient, it has the following end behavior: $y \to \infty$ as $x \to \infty$ and $y \to \infty$ as $x \to -\infty$.

On a large viewing rectangle, the graphs of P and Q look almost the same. On a small viewing rectangle, the graphs of P and Q look very different and we see that they have different intercepts.

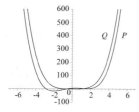
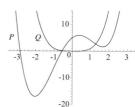

41. $P(x) = x^{11} - 9x^9$; $Q(x) = x^{11}$. Since P has odd degree and positive leading coefficient, it has the following end behavior: $y \to \infty$ as $x \to \infty$ and $y \to -\infty$ as $x \to -\infty$.

On a large viewing rectangle, the graphs of P and Q look like they have the same end behavior. On a small viewing rectangle, the graphs of P and Q look very different and seem (wrongly) to have different end behavior.

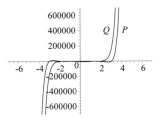   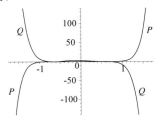

43. (a) x-intercepts at 0 and 4, y-intercept at 0.

  (b) Local maximum at $(2, 4)$, no local minimum.

45. (a) x-intercepts at -2 and 1, y-intercept at -1.

  (b) Local maximum at $(1, 0)$, local minimum at $(-1, -2)$.

47. $y = -x^2 + 8x$, $[-4, 12]$ by $[-50, 30]$

  No local minimum. Local maximum at $(4, 16)$.

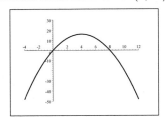

49. $y = x^3 - 12x + 9$, $[-5, 5]$ by $[-30, 30]$

  Local minimum at $(-2, 25)$, Local maximum at $(2, -7)$.

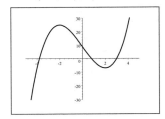

51. $y = x^4 + 4x^3$, $[-5, 5]$ by $[-30, 30]$

  Local minimum at $(-3, -27)$. No local maximum.

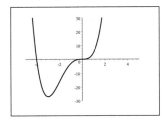

53. $y = 3x^5 - 5x^3 + 3$, $[-3, 3]$ by $[-5, 10]$

  Local maximum at $(-1, 5)$. Local minimum at $(1, 1)$.

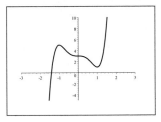

55. $y = -2x^2 + 3x + 5$ has one local maximum at $(0.75, 6.13)$.

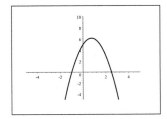

57. $y = x^3 - x^2 - x$ has one local maximum at $(-0.33, 0.19)$ and one local minimum at $(1.00, -1.00)$.

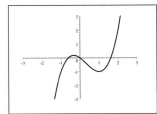

59. $y = x^4 - 5x^2 + 4$ has one local maximum at $(0, 4)$ and two local minima at $(-1.58, -2.25)$ and $(1.58, -2.25)$.

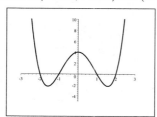

61. $y = (x - 2)^5 + 32$ has no maximum or minimum.

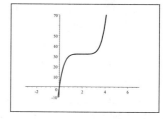

63. $y = x^8 - 3x^4 + x$ has one local maximum at $(0.44, 0.33)$ and two local minima at $(1.09, -1.15)$ and $(-1.12, -3.36)$.

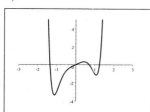

65. $y = cx^3$; $c = 1, 2, 5, \frac{1}{2}$. Increasing the value of c stretches the graph vertically.

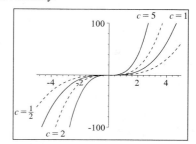

67. $P(x) = x^4 + c$; $c = -1, 0, 1$, and 2. Increasing the value of c moves the graph up.

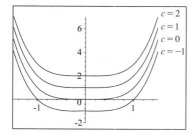

69. $P(x) = x^4 - cx$; $c = 0, 1, 8$, and 27. Increasing the value of c causes a deeper dip in the graph, in the fourth quadrant, and moves the positive x-intercept to the right.

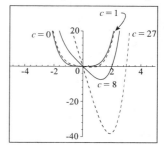

71. (a)

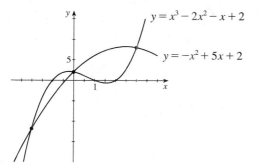

(b) The two graphs appear to intersect at 3 points.

(c) $x^3 - 2x^2 - x + 2 = -x^2 + 5x + 2$ $\Leftrightarrow$
$x^3 - x^2 - 6x = 0$ $\Leftrightarrow$ $x(x^2 - x - 6) = 0 \Leftrightarrow$
$x(x - 3)(x + 2) = 0$. Then either $x = 0$, $x = 3$, or $x = -2$. If $x = 0$, then $y = 2$; if $x = 3$ then $y = 8$; if $x = -2$, then $y = -12$. Hence the points where the two graphs intersect are $(0, 2)$, $(3, 8)$, and $(-2, -12)$.

73. (a) Let $P(x)$ be a polynomial containing only odd powers of x. Then each term of $P(x)$ can be written as Cx^{2n+1}, for some constant C and integer n. Since $C(-x)^{2n+1} = -Cx^{2n+1}$, each term of $P(x)$ is an odd function. Thus by part (a), $P(x)$ is an odd function.

(b) Let $P(x)$ be a polynomial containing only even powers of x. Then each term of $P(x)$ can be written as Cx^{2n}, for some constant C and integer n. Since $C(-x)^{2n} = Cx^{2n}$, each term of $P(x)$ is an even function. Thus by part (b), $P(x)$ is an even function.

(c) Since $P(x)$ contains both even and odd powers of x, we can write it in the form $P(x) = R(x) + Q(x)$, where $R(x)$ contains all the even-powered terms in $P(x)$ and $Q(x)$ contains all the odd-powered terms. By part (d), $Q(x)$ is an odd function, and by part (e), $R(x)$ is an even function. Thus, since neither $Q(x)$ nor $R(x)$ are constantly 0 (by assumption), by part (c), $P(x) = R(x) + Q(x)$ is neither even nor odd.

(d) $P(x) = x^5 + 6x^3 - x^2 - 2x + 5 = (x^5 + 6x^3 - 2x) + (-x^2 + 5) = P_O(x) + P_E(x)$ where $P_O(x) = x^5 + 6x^3 - 2x$ and $P_E(x) = -x^2 + 5$. Since $P_O(x)$ contains only odd powers of x, it is an odd function, and since $P_E(x)$ contains only even powers of x, it is an even function.

75. (a) $P(x) = (x-2)(x-4)(x-5)$ has one local maximum and one local minimum.

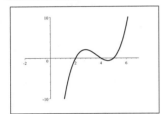

(b) Since $P(a) = P(b) = 0$, and $P(x) > 0$ for $a < x < b$ (see the table below), the graph of P must first rise and then fall on the interval (a, b), and so P must have at least one local maximum between a and b. Using similar reasoning, the fact that $P(b) = P(c) = 0$ and $P(x) < 0$ for $b < x < c$ shows that P must have at least one local minimum between b and c. Thus P has at least two local extrema.

Interval	$(-\infty, a)$	(a, b)	(b, c)	(c, ∞)
Sign of $x - a$	$-$	$+$	$+$	$+$
Sign of $x - b$	$-$	$-$	$+$	$+$
Sign of $x - c$	$-$	$-$	$-$	$+$
Sign of $(x-a)(x-b)(x-c)$	$-$	$+$	$-$	$+$

77. $P(x) = 8x + 0.3x^2 - 0.0013x^3 - 372$

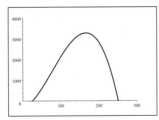

(a) For the firm to break even, $P(x) = 0$. From the graph, we see that $P(x) = 0$ when $x \approx 25.2$. Of course, the firm cannot produce fractions of a blender, so the manufacturer must produce at least 26 blenders a year.

(b) No, the profit does not increase indefinitely. The largest profit is approximately $\$3276.22$, which occurs when the firm produces 166 blenders per year.

79. (a) The length of the bottom is $40 - 2x$, the width of the bottom is $20 - 2x$, and the height is x, so the volume of the box is

$$V = x(20 - 2x)(40 - 2x)$$
$$= 4x^3 - 120x^2 + 800x$$

(b) Since the height and width must be positive, we must have $x > 0$ and $20 - 2x > 0$, and so the domain of V is $0 < x < 10$.

(c) Using the domain from part (b), we graph V in the viewing rectangle $[0, 10]$ by $[0, 1600]$. The maximum volume is $V \approx 1539.6$ when $x = 4.23$.

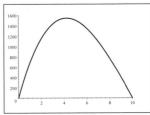

81.

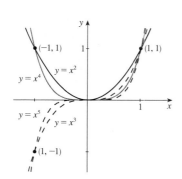

The graph of $y = x^{100}$ is close to the x-axis for $|x| < 1$, but passes through the points $(1, 1)$ and $(-1, 1)$. The graph of $y = x^{101}$ behaves similarly except that the y-values are negative for negative values of x, and it passes through $(-1, -1)$ instead of $(-1, 1)$.

83. No, it is impossible. The end behavior of a third degree polynomial is the same as that of $y = kx^3$, and for this function, the values of y go off in opposite directions as $x \to \infty$ and $x \to -\infty$. But for a function with just one extremum, the values of y would head off in the same direction (either both up or both down) on either side of the extremum. An nth-degree polynomial can have $n - 1$ extrema or $n - 3$ extrema or $n - 5$ extrema, and so on (decreasing by 2). A polynomial that has six local extrema must be of degree 7 or higher. For example, $P(x) = (x - 1)(x - 2)(x - 3)(x - 4)(x - 5)(x - 6)(x - 7)$ has six local extrema.

4.2   Dividing Polynomials

1.

$$
\begin{array}{r}
3x - 4 \\
x + 3 \overline{\smash{\big)}\ 3x^2 + 5x - 4} \\
\underline{3x^2 + 9x} \\
-4x - 4 \\
\underline{-4x - 12} \\
8
\end{array}
$$

Thus the quotient is $3x - 4$ and the remainder is 8, so $P(x) = 3x^2 + 5x - 4 = (x + 3) \cdot (3x - 4) + 8$.

3.

$$
\begin{array}{r}
x^2 \qquad - 1 \\
2x - 3 \overline{\smash{\big)}\ 2x^3 - 3x^2 - 2x} \\
\underline{2x^3 - 3x^2} \\
-2x \\
-2x + 3 \\
-3
\end{array}
$$

Thus the quotient is $x^2 - 1$ and the remainder is -3, and $P(x) = 2x^3 - 3x^2 - 2x = (x^2 - 1)(2x - 3) - 3$.

5.

$$
\begin{array}{r}
x^2 - x - 3 \\
x^2 + 3 \overline{\smash{\big)}\ x^4 - x^3 + 0x^2 + 4x + 2} \\
\underline{x^4 \qquad + 3x^2} \\
-x^3 - 3x^2 + 4x \\
\underline{-x^2 \qquad - 3x} \\
-3x^2 + 7x + 2 \\
\underline{-3x^2 \qquad - 9} \\
7x + 11
\end{array}
$$

Thus the quotient is $x^2 - x - 3$ and the remainder is $7x + 11$, and

$$
\begin{aligned}
P(x) &= x^3 + 4x^2 - 6x + 1 \\
&= (x^2 + 3) \cdot (x^2 - x - 3) + (7x + 11)
\end{aligned}
$$

7.

$$
\begin{array}{r|rrr}
-3 & 1 & 4 & -8 \\
   &   & -3 & -3 \\
\hline
   & 1 & 1 & -11
\end{array}
$$

Thus the quotient is $x + 1$ and the remainder is -11, and

$$
\frac{P(x)}{D(x)} = \frac{x^2 + 4x - 8}{x + 3} = (x + 1) + \frac{-11}{x + 3}.
$$

9.

$$
\begin{array}{r}
2x - \frac{1}{2} \\
2x - 1 \enclose{longdiv}{4x^2 - 3x - 7} \\
\underline{4x^2 - 2x} \\
-x - 7 \\
\underline{-x + \frac{1}{2}} \\
-\frac{15}{2}
\end{array}
$$

Thus the quotient is $2x - \frac{1}{2}$ and the remainder is $-\frac{15}{2}$, and

$$\frac{P(x)}{D(x)} = \frac{4x^2 - 3x - 7}{2x - 1} = \left(2x - \frac{1}{2}\right) + \frac{-\frac{15}{2}}{2x - 1}.$$

11.

$$
\begin{array}{r}
2x^2 - x + 1 \\
x^2 + 4 \enclose{longdiv}{2x^4 - x^3 + 9x^2} \\
\underline{2x^4 \quad\ + 8x^2} \\
-x^3 + x^2 \\
\underline{-x^3 \quad\quad - 4x} \\
x^2 + 4x \\
\underline{x^2 \quad\quad + 4} \\
4x - 4
\end{array}
$$

Thus the quotient is $2x^2 - x + 1$ and the remainder is $4x - 4$, and

$$\frac{P(x)}{D(x)} = \frac{2x^4 - x^3 + 9x^2}{x^2 + 4} = \left(2x^2 - x + 1\right) + \frac{4x - 4}{x^2 + 4}.$$

13.

$$
\begin{array}{r}
x - 2 \\
x - 4 \enclose{longdiv}{x^2 - 6x - 8} \\
\underline{x^2 - 4x} \\
-2x - 8 \\
\underline{-2x + 8} \\
-16
\end{array}
$$

Thus the quotient is $x - 2$ and the remainder is -16.

15.

$$
\begin{array}{r}
2x^2 \quad\ - 1 \\
2x + 1 \enclose{longdiv}{4x^3 + 2x^2 - 2x - 3} \\
\underline{4x^3 + 2x^2} \\
-2x - 3 \\
\underline{-2x - 1} \\
-2
\end{array}
$$

Thus the quotient is $2x^2 - 1$ and the remainder is -2.

17.

$$
\begin{array}{r}
x + 2 \\
x^2 - 2x + 2 \enclose{longdiv}{x^3 + 0x^2 + 6x + 3} \\
\underline{x^3 - 2x^2 + 2x} \\
2x^2 + 4x + 3 \\
\underline{2x^2 - 4x + 4} \\
8x - 1
\end{array}
$$

Thus the quotient is $x + 2$, and the remainder is $8x - 1$.

19.

$$
\begin{array}{r}
3x + 1 \\
2x^2 + 0x + 5 \enclose{longdiv}{6x^3 + 2x^2 + 22x + 0} \\
\underline{6x^3 \quad\quad + 15x} \\
2x^2 + 7x + 0 \\
\underline{2x^2 \quad\quad + 5} \\
7x - 5
\end{array}
$$

Thus the quotient is $3x + 1$, and the remainder is $7x - 5$.

21.

$$
\begin{array}{r}
x^4 \quad\quad\quad\quad\quad + 1 \\
x^2 + 1 \enclose{longdiv}{x^6 + 0x^5 + x^4 + 0x^3 + x^2 + 0x + 1} \\
\underline{x^6 \quad\quad + x^4} \\
0 \quad\quad + x^2 \quad\quad + 1 \\
\underline{x^2 \quad\quad + 1} \\
0
\end{array}
$$

Thus the quotient is $x^4 + 1$, and the remainder is 0.

23. The synthetic division table for this problem takes the following form.

$$
\begin{array}{r|rrr}
3 & 1 & -5 & 4 \\
& & 3 & -6 \\
\hline
& 1 & -2 & -2
\end{array}
$$

Thus the quotient is $x - 2$, and the remainder is -2.

25. The synthetic division table for this problem takes the following form.

$$
\begin{array}{r|rrr}
6 & 3 & 5 & 0 \\
  &   & 18 & 138 \\
\hline
  & 3 & 23 & 138
\end{array}
$$

Thus the quotient is $3x + 23$, and the remainder is 138.

27. Since $x + 2 = x - (-2)$, the synthetic division table for this problem takes the following form.

$$
\begin{array}{r|rrrr}
-2 & 1 & 2 & 2 & 1 \\
   &   & -2 & 0 & -4 \\
\hline
   & 1 & 0 & 2 & -3
\end{array}
$$

Thus the quotient is $x^2 + 2$, and the remainder is -3.

29. Since $x + 3 = x - (-3)$ and $x^3 - 8x + 2 = x^3 + 0x^2 - 8x + 2$, the synthetic division table for this problem takes the following form.

$$
\begin{array}{r|rrrr}
-3 & 1 & 0 & -8 & 2 \\
   &   & -3 & 9 & -3 \\
\hline
   & 1 & -3 & 1 & -1
\end{array}
$$

Thus the quotient is $x^2 - 3x + 1$, and the remainder is -1.

31. Since $x^5 + 3x^3 - 6 = x^5 + 0x^4 + 3x^3 + 0x^2 + 0x - 6$, the synthetic division table for this problem takes the following form.

$$
\begin{array}{r|rrrrrr}
1 & 1 & 0 & 3 & 0 & 0 & -6 \\
  &   & 1 & 1 & 4 & 4 & 4 \\
\hline
  & 1 & 1 & 4 & 4 & 4 & -2
\end{array}
$$

Thus the quotient is $x^4 + x^3 + 4x^2 + 4x + 4$, and the remainder is -2.

33. The synthetic division table for this problem takes the following form.

$$
\begin{array}{r|rrrr}
\frac{1}{2} & 2 & 3 & -2 & 1 \\
            &   & 1 & 2 & 0 \\
\hline
            & 2 & 4 & 0 & 1
\end{array}
$$

Thus the quotient is $2x^2 + 4x$, and the remainder is 1.

35. Since $x^3 - 27 = x^3 + 0x^2 + 0x - 27$, the synthetic division table for this problem takes the following form.

$$
\begin{array}{r|rrrr}
3 & 1 & 0 & 0 & -27 \\
  &   & 3 & 9 & 27 \\
\hline
  & 1 & 3 & 9 & 0
\end{array}
$$

Thus the quotient is $x^2 + 3x + 9$, and the remainder is 0.

37. $P(x) = 4x^2 + 12x + 5$, $c = -1$

$$
\begin{array}{r|rrr}
-1 & 4 & 12 & 5 \\
   &   & -4 & -8 \\
\hline
   & 4 & 8 & -3
\end{array}
$$

Therefore, by the Remainder Theorem, $P(-1) = -3$.

39. $P(x) = x^3 + 3x^2 - 7x + 6$, $c = 2$

$$
\begin{array}{r|rrrr}
2 & 1 & 3 & -7 & 6 \\
  &   & 2 & 10 & 6 \\
\hline
  & 1 & 5 & 3 & 12
\end{array}
$$

Therefore, by the Remainder Theorem, $P(2) = 12$.

41. $P(x) = x^3 + 2x^2 - 7$, $c = -2$

$$
\begin{array}{r|rrrr}
-2 & 1 & 2 & 0 & -7 \\
   &   & -2 & 0 & 0 \\
\hline
   & 1 & 0 & 0 & -7
\end{array}
$$

Therefore, by the Remainder Theorem, $P(-2) = -7$.

43. $P(x) = 5x^4 + 30x^3 - 40x^2 + 36x + 14$, $c = -7$

$$
\begin{array}{r|rrrrr}
-7 & 5 & 30 & -40 & 36 & 14 \\
   &   & -35 & 35 & 35 & -497 \\
\hline
   & 5 & -5 & -5 & 71 & -483
\end{array}
$$

Therefore, by the Remainder Theorem, $P(-7) = -483$.

45. $P(x) = x^7 - 3x^2 - 1$
$$= x^7 + 0x^6 + 0x^5 + 0x^4 + 0x^3 - 3x^2 + 0x - 1$$
$c = 3$

$$
\begin{array}{r|rrrrrrrr}
3 & 1 & 0 & 0 & 0 & 0 & -3 & 0 & -1 \\
  &   & 3 & 9 & 27 & 81 & 243 & 720 & 2160 \\
\hline
  & 1 & 3 & 9 & 27 & 81 & 240 & 720 & 2159
\end{array}
$$

Therefore by the Remainder Theorem, $P(3) = 2159$.

47. $P(x) = 3x^3 + 4x^2 - 2x + 1$, $c = \frac{2}{3}$

$$
\begin{array}{r|rrrr}
\frac{2}{3} & 3 & 4 & -2 & 1 \\
            &   & 2 & 4 & \frac{4}{3} \\
\hline
            & 3 & 6 & 2 & \frac{7}{3}
\end{array}
$$

Therefore, by the Remainder Theorem, $P\left(\frac{2}{3}\right) = \frac{7}{3}$.

49. $P(x) = x^3 + 2x^2 - 3x - 8$, $c = 0.1$

$$
\begin{array}{r|rrrr}
0.1 & 1 & 2 & -3 & -8 \\
    &   & 0.1 & 0.21 & -0.279 \\
\hline
    & 1 & 2.1 & -2.79 & -8.279
\end{array}
$$

Therefore, by the Remainder Theorem, $P(0.1) = -8.279$.

51. $P(x) = x^3 - 3x^2 + 3x - 1$, $c = 1$

$$
\begin{array}{r|rrrr}
1 & 1 & -3 & 3 & -1 \\
  &   & 1 & -2 & 1 \\
\hline
  & 1 & -2 & 1 & 0
\end{array}
$$

Since the remainder is 0, $x - 1$ is a factor.

53. $P(x) = 2x^3 + 7x^2 + 6x - 5$, $c = \frac{1}{2}$

$$
\begin{array}{r|rrrr}
\frac{1}{2} & 2 & 7 & 6 & -5 \\
            &   & 1 & 4 & 5 \\
\hline
            & 2 & 8 & 10 & 0
\end{array}
$$

Since the remainder is 0, $x - \frac{1}{2}$ is a factor.

55. $P(x) = x^3 - x^2 - 11x + 15$, $c = 3$

$$
\begin{array}{r|rrrr}
3 & 1 & -1 & -11 & 15 \\
  &   & 3 & 6 & -15 \\
\hline
  & 1 & 2 & -5 & 0
\end{array}
$$

Since the remainder is 0, we know that 3 is a zero $x^3 - x^2 - 11x + 15 = (x - 3)(x^2 + 2x - 5)$. Now $x^2 + 2x - 5 = 0$ when $x = \frac{-2 \pm \sqrt{2^2 + 4(1)(5)}}{2} = -1 \pm \sqrt{6}$. Hence, the zeros are $-1 - \sqrt{6}$, $-1 + \sqrt{6}$, and 3.

57. Since the zeros are $x = -1$, $x = 1$, and $x = 3$, the factors are $x + 1$, $x - 1$, and $x - 3$. Thus $P(x) = (x + 1)(x - 1)(x - 3) = x^3 - 3x^2 - x + 3$.

59. Since the zeros are $x = -1$, $x = 1$, $x = 3$, and $x = 5$, the factors are $x + 1$, $x - 1$, $x - 3$, and $x - 5$. Thus $P(x) = (x + 1)(x - 1)(x - 3)(x - 5) = x^4 - 8x^3 + 14x^2 + 8x - 15$.

61. Since the zeros of the polynomial are 1, -2, and 3, it follows that $P(x) = C(x - 1)(x + 2)(x - 3) = C(x^3 - 2x^2 - 5x + 6) = Cx^3 - 2Cx^2 - 5Cx + 6C$. Since the coefficient of x^2 is to be 3, $-2C = 3$ so $C = -\frac{3}{2}$. Therefore, $P(x) = -\frac{3}{2}(x^3 - 2x^2 - 5x + 6) = -\frac{3}{2}x^3 + 3x^2 + \frac{15}{2}x - 9$ is the polynomial.

63. The y-intercept is 2 and the zeros of the polynomial are -1, 1, and 2. It follows that $P(x) = C(x + 1)(x - 1)(x - 2) = C(x^3 - 2x^2 - x + 2)$. Since $P(0) = 2$ we have $2 = C[(0)^3 - 2(0)^2 - (0) + 2]$ $\Leftrightarrow$ $2 = 2C \Leftrightarrow C = 1$ and $P(x) = (x + 1)(x - 1)(x - 2) = x^3 - 2x^2 - x + 2$.

65. The y-intercept is 4 and the zeros of the polynomial are -2 and 1 both being degree two. It follows that $P(x) = C(x + 2)^2(x - 1)^2 = C(x^4 + 2x^3 - 3x^2 - 4x + 4)$. Since $P(0) = 4$ we have $4 = C[(0)^4 + 2(0)^3 - 3(0)^2 - 4(0) + 4]$ $\Leftrightarrow$ $4 = 4C \Leftrightarrow C = 1$. Thus $P(x) = (x + 2)^2(x - 1)^2 = x^4 + 2x^3 - 3x^2 - 4x + 4$.

67. A. By the Remainder Theorem, the remainder when $P(x) = 6x^{1000} - 17x^{562} + 12x + 26$ is divided by $x + 1$ is $P(-1) = 6(-1)^{1000} - 17(-1)^{562} + 12(-1) + 26 = 6 - 17 - 12 + 26 = 3$.

B. If $x - 1$ is a factor of $Q(x) = x^{567} - 3x^{400} + x^9 + 2$, then $Q(1)$ must equal 0. $Q(1) = (1)^{567} - 3(1)^{400} + (1)^9 + 2 = 1 - 3 + 1 + 2 = 1 \neq 0$, so $x - 1$ is not a factor.

4.3   Real Zeros of Polynomials

1. $P(x) = x^3 - 4x^2 + 3$ has possible rational zeros ± 1 and ± 3.

3. $R(x) = 2x^5 + 3x^3 + 4x^2 - 8$ has possible rational zeros ± 1, ± 2, ± 4, ± 8, $\pm \frac{1}{2}$.

5. $T(x) = 4x^4 - 2x^2 - 7$ has possible rational zeros ± 1, ± 7, $\pm \frac{1}{2}$, $\pm \frac{7}{2}$, $\pm \frac{1}{4}$, $\pm \frac{7}{4}$.

7. (a) $P(x) = 5x^3 - x^2 - 5x + 1$ has possible rational zeros ± 1, $\pm \frac{1}{5}$.

(b) From the graph, the actual zeroes are -1, $\frac{1}{5}$, and 1.

9. (a) $P(x) = 2x^4 - 9x^3 + 9x^2 + x - 3$ has possible rational zeros ± 1, ± 3, $\pm \frac{1}{2}$, $\pm \frac{3}{2}$.

(b) From the graph, the actual zeroes are $-\frac{1}{2}$, 1, and 3.

11. $P(x) = x^3 + 3x^2 - 4$. The possible rational zeros are ± 1, ± 2, ± 4. $P(x)$ has 1 variation in sign and hence 1 positive real zero. $P(-x) = -x^3 + 3x^2 - 4$ has 2 variations in sign and hence 0 or 2 negative real zeros.

$$
\begin{array}{r|rrrr}
1 & 1 & 3 & 0 & -4 \\
  &   & 1 & 4 & 4 \\
\hline
  & 1 & 4 & 4 & 0
\end{array}
\Rightarrow x = 1 \text{ is a zero.}
$$

$P(x) = x^3 + 3x^2 - 4 = (x - 1)(x^2 + 4x + 4)$
$\qquad = (x - 1)(x + 2)^2$

Therefore, the zeros are $x = -2, 1$.

13. $P(x) = x^3 - 3x - 2$. The possible rational zeros are ± 1, ± 2. $P(x)$ has 1 variation in sign and hence 1 positive real zero. $P(-x) = -x^3 + 3x - 2$ has 2 variations in sign and hence 0 or 2 negative real zeros.

$$
\begin{array}{r|rrrr}
1 & 1 & 0 & -3 & -2 \\
  &   & 1 & 1 & -2 \\
\hline
  & 1 & 1 & -2 & -4
\end{array}
\Rightarrow x = 1 \text{ is not a zero.}
$$

$$
\begin{array}{r|rrrr}
2 & 1 & 0 & -3 & -2 \\
  &   & 2 & 4 & 2 \\
\hline
  & 1 & 2 & 1 & 0
\end{array}
\Rightarrow x = 2 \text{ is a zero.}
$$

$P(x) = x^3 - 3x -$
$2 = (x - 2)(x^2 + 2x + 1) = (x - 2)(x + 1)^2$.

Therefore, the zeros are $x = 2, -1$.

15. $P(x) = x^3 - 6x^2 + 12x - 8$. The possible rational zeros are ± 1, ± 2, ± 4, ± 8. $P(x)$ has 3 variations in sign and hence 1 or 3 positive real zeros. $P(-x) = -x^3 - 6x^2 - 12x - 8$ has no variations in sign and hence 0 negative real zeros.

$$
\begin{array}{r|rrrr}
1 & 1 & -6 & 12 & -8 \\
  &   & 1 & -5 & 7 \\
\hline
  & 1 & -5 & 7 & -1
\end{array}
\Rightarrow x = 1 \text{ is not a zero.}
\qquad
\begin{array}{r|rrrr}
2 & 1 & -6 & 12 & -8 \\
  &   & 2 & -8 & 8 \\
\hline
  & 1 & -4 & 4 & 0
\end{array}
\Rightarrow x = 2 \text{ is a zero.}
$$

$P(x) = x^3 - 6x^2 + 12x - 8 = (x - 2)(x^2 - 4x + 4) = (x - 2)^3$. Therefore, the zero is $x = 2$.

17. $P(x) = x^3 - 4x^2 + x + 6$. The possible rational zeros are ± 1, ± 2, ± 3, ± 6. $P(x)$ has 2 variations in sign and hence 0 or 2 positive real zeros. $P(-x) = -x^3 - 4x^2 - x + 6$ has 1 variation in sign and hence 1 negative real zeros.

$$
\begin{array}{r|rrrr}
-1 & 1 & -4 & 1 & 6 \\
   &   & -1 & 5 & -6 \\
\hline
   & 1 & -5 & 6 & 0
\end{array}
\Rightarrow x + 1 \text{ is a factor.}
$$

So

$P(x) = x^3 - 4x^2 + x + 6 = (x + 1)(x^2 - 5x + 6)$
$\qquad = (x + 1)(x - 3)(x - 2)$

Therefore, the zeros are $x = -1, 2, 3$.

19. $P(x) = x^3 + 3x^2 + 6x + 4$. The possible rational zeros are ± 1, ± 2, ± 4. $P(x)$ has no variation in sign and hence no positive real zeros. $P(-x) = -x^3 + 3x^2 - 6x + 4$ has 3 variations in sign and hence 1 or 3 negative real zeros.

$$
\begin{array}{r|rrrr}
-1 & 1 & 3 & 6 & 4 \\
   &   & -1 & -2 & -4 \\
\hline
   & 1 & 2 & 4 & 0
\end{array}
\Rightarrow x + 1 \text{ is a factor.}
$$

So

$P(x) = x^3 + 3x^2 + 6x + 4 = (x + 1)(x^2 + 2x + 4)$.

Now, $Q(x) = x^2 + 2x + 4$ has no real zeros, since the discriminant of this quadratic is

$b^2 - 4ac = (2)^2 - 4(1)(4) = -12 < 0$. Thus, the only real zero is $x = -1$.

21. *Method 1:* $P(x) = x^4 - 5x^2 + 4$. The possible rational zeros are $\pm 1, \pm 2, \pm 4$. $P(x)$ has 1 variation in sign and hence 1 positive real zero. $P(-x) = x^4 - 5x^2 + 4$ has 2 variations in sign and hence 0 or 2 negative real zeros.

$$
\begin{array}{r|rrrrr}
1 & 1 & 0 & -5 & 0 & 4 \\
  &   & 1 & 1 & -4 & -4 \\
\hline
  & 1 & 1 & -4 & -4 & 0 \\
\end{array} \Rightarrow x = 1 \text{ is a zero.}
$$

Thus $P(x) = x^4 - 5x^2 + 4 = (x-1)(x^3 + x^2 - 4x - 4)$. Continuing with the quotient we have:

$$
\begin{array}{r|rrrr}
-1 & 1 & 1 & -4 & -4 \\
   &   & -1 & 0 & 4 \\
\hline
   & 1 & 0 & -4 & 0 \\
\end{array} \Rightarrow x = -1 \text{ is a zero.}
$$

$P(x) = x^4 - 5x^2 + 4 = (x-1)(x+1)(x^2 - 4) = (x-1)(x+1)(x-2)(x+2)$. The zeros are $x = \pm 1, \pm 2$.
Method 2: Substituting $u = x^2$, the polynomial becomes $P(u) = u^2 - 5u + 4$, which factors:
$u^2 - 5u + 4 = (u-1)(u-4) = (x^2 - 1)(x^2 - 4)$, so either $x^2 = 1$ or $x^2 = 4$. If $x^2 = 1$, then $x = \pm 1$; if $x^2 = 4$, then $x = \pm 2$. Therefore, the zeros are $x = \pm 1, \pm 2$.

23. $P(x) = x^4 + 6x^3 + 7x^2 - 6x - 8$. The possible rational zeros are $\pm 1, \pm 2, \pm 4, \pm 8$. $P(x)$ has 1 variation in sign and hence 1 positive real zero. $P(-x) = x^4 - 6x^3 + 7x^2 + 6x - 8$ has 3 variations in sign and hence 1 or 3 negative real zeros.

$$
\begin{array}{r|rrrrr}
1 & 1 & 6 & 7 & -6 & -8 \\
  &   & 1 & 7 & 14 & 8 \\
\hline
  & 1 & 7 & 14 & 8 & 0 \\
\end{array} \Rightarrow x = 1 \text{ is a zero}
$$

and there are no other positive zeros. Thus $P(x) = x^4 + 6x^3 + 7x^2 - 6x - 8 = (x-1)(x^3 + 7x^2 + 14x + 8)$. Continuing by factoring the quotient, we have:

$$
\begin{array}{r|rrrr}
-1 & 1 & 7 & 14 & 8 \\
   &   & -1 & -6 & -8 \\
\hline
   & 1 & 6 & 8 & 0 \\
\end{array} \Rightarrow x = -1 \text{ is a zero.}
$$

So $P(x) = x^4 + 6x^3 + 7x^2 - 6x - 8 = (x-1)(x+1)(x^2 + 6x + 8) = (x-1)(x+1)(x+2)(x+4)$. Therefore, the zeros are $x = -4, -2, \pm 1$.

25. $P(x) = 4x^4 - 25x^2 + 36$ has possible rational zeros $\pm 1, \pm 2, \pm 3, \pm 4, \pm 6, \pm 9, \pm 12, \pm 18, \pm 36, \pm \frac{1}{2}, \pm \frac{1}{4}, \pm \frac{3}{2}, \pm \frac{3}{4}, \pm \frac{9}{2}, \pm \frac{9}{4}$. Since $P(x)$ has 2 variations in sign, there are 0 or 2 positive real zeros. Since $P(-x) = 4x^4 - 25x^2 + 36$ has 2 variations in sign, there are 0 or 2 negative real zeros.

$$
\begin{array}{r|rrrrr}
1 & 4 & 0 & -25 & 0 & 36 \\
  &   & 4 & 4 & -21 & -21 \\
\hline
  & 4 & 4 & -21 & -21 & 15 \\
\end{array}
\qquad
\begin{array}{r|rrrrr}
2 & 4 & 0 & -25 & 0 & 36 \\
  &   & 8 & 16 & -18 & -36 \\
\hline
  & 4 & 8 & -9 & -18 & 0 \\
\end{array} \Rightarrow x = 2 \text{ is a zero.}
$$

$P(x) = (x-2)(4x^3 + 8x^2 - 9x - 18)$

$$
\begin{array}{r|rrrr}
2 & 4 & 8 & -9 & -18 \\
  &   & 8 & 32 & 46 \\
\hline
  & 4 & 16 & 23 & 28 \\
\end{array} \Rightarrow \text{all positive, } x = 2 \text{ is an upper bound.}
\qquad
\begin{array}{r|rrrr}
\frac{1}{2} & 4 & 8 & -9 & -18 \\
  &   & 2 & 5 & -2 \\
\hline
  & 4 & 10 & -4 & -20 \\
\end{array}
$$

$$
\begin{array}{r|rrrr}
\frac{1}{4} & 4 & 8 & -9 & -18 \\
  &   & 1 & \frac{9}{4} & -\frac{27}{16} \\
\hline
  & 4 & 9 & -\frac{27}{4} & -\frac{315}{16} \\
\end{array}
\qquad
\begin{array}{r|rrrr}
\frac{3}{2} & 4 & 8 & -9 & -18 \\
  &   & 6 & 21 & 18 \\
\hline
  & 4 & 14 & 12 & 0 \\
\end{array} \Rightarrow x = \frac{3}{2} \text{ is a zero.}
$$

$P(x) = (x-2)(2x-3)(2x^2 + 7x + 6) = (x-2)(2x-3)(2x+3)(x+2)$. Therefore, the zeros are $x = \pm 2, \pm \frac{3}{2}$.
Note: Since $P(x)$ has only even terms, factoring by substitution also works. Let $x^2 = u$; then
$P(u) = 4u^2 - 25u + 36 = (u-4)(4u-9) = (x^2 - 4)(4x^2 - 9)$, which gives the same results.

27. $P(x) = x^4 + 8x^3 + 24x^2 + 32x + 16$. The possible rational zeros are $\pm 1, \pm 2, \pm 4, \pm 8, \pm 16$. $P(x)$ has no variations in sign and hence no positive real zero. $P(-x) = x^4 - 8x^3 + 24x^2 - 32x + 16$ has 4 variations in sign and hence 0 or 2 or 4 negative real zeros.

$$
\begin{array}{r|rrrrr}
-1 & 1 & 8 & 24 & 32 & 16 \\
   &   & -1 & -7 & -17 & -15 \\
\hline
   & 1 & 7 & 17 & 15 & 1
\end{array}
\Rightarrow x = -1 \text{ is not a zero.}
\qquad
\begin{array}{r|rrrrr}
-2 & 1 & 8 & 24 & 32 & 16 \\
   &   & -2 & -12 & -24 & -16 \\
\hline
   & 1 & 6 & 12 & 8 & 0
\end{array}
\Rightarrow x = -2 \text{ is a zero.}
$$

Thus $P(x) = x^4 + 8x^3 + 24x^2 + 32x + 16 = (x+2)\left(x^3 + 6x^2 + 12x + 8\right)$. Continuing by factoring the quotient, we have

$$
\begin{array}{r|rrrr}
-2 & 1 & 6 & 12 & 8 \\
   &   & -2 & -8 & -8 \\
\hline
   & 1 & 4 & 4 & 0
\end{array}
\Rightarrow x = -2 \text{ is a zero.}
$$

Thus $P(x) = (x+2)^2 \left(x^2 + 4x + 4\right) = (x+2)^4$. Therefore, the zero is $x = -2$

29. Factoring by grouping can be applied to this exercise. $4x^3 + 4x^2 - x - 1 = 4x^2(x+1) - (x+1) = (x+1)\left(4x^2 - 1\right) = (x+1)(2x+1)(2x-1)$. Therefore, the zeros are $x = -1, \pm\frac{1}{2}$.

31. $P(x) = 4x^3 - 7x + 3$. The possible rational zeros are $\pm 1, \pm 3, \pm\frac{1}{2}, \pm\frac{3}{2}, \pm\frac{1}{4}, \pm\frac{3}{4}$. Since $P(x)$ has 2 variations in sign, there are 0 or 2 positive zeros. Since $P(-x) = -4x^3 + 7x + 3$ has 1 variation in sign, there is 1 negative zero.

$$
\begin{array}{r|rrrr}
\frac{1}{2} & 4 & 0 & -7 & 3 \\
   &   & 2 & 1 & -3 \\
\hline
   & 4 & 2 & -6 & 0
\end{array}
\Rightarrow x = \tfrac{1}{2} \text{ is a zero.}
$$

$P(x) = \left(x - \frac{1}{2}\right)\left(4x^2 + 2x - 6\right) = (2x-1)\left(2x^2 + x - 3\right) = (2x-1)(x-1)(2x+3) = 0$. Thus, the zeros are $x = -\frac{3}{2}, \frac{1}{2}, 1$.

33. $P(x) = 4x^3 + 8x^2 - 11x - 15$. The possible rational zeros are $\pm 1, \pm 3, \pm 5, \pm\frac{1}{2}, \pm\frac{1}{4}, \pm\frac{3}{2}, \pm\frac{3}{4}, \pm\frac{5}{2}, \pm\frac{5}{4}$. $P(x)$ has 1 variation in sign and hence 1 positive real zero. $P(-x) = -4x^3 + 8x^2 + 11x - 15$ has 2 variations in sign, so P has 0 or 2 negative real zeros.

$$
\begin{array}{r|rrrr}
1 & 4 & 8 & -11 & -15 \\
  &   & 4 & 12 & 1 \\
\hline
  & 4 & 12 & 1 & -14
\end{array}
\Rightarrow x = 1 \text{ is not a zero.}
\qquad
\begin{array}{r|rrrr}
3 & 4 & 8 & -11 & -15 \\
  &   & 12 & 60 & 147 \\
\hline
  & 4 & 20 & 49 & 132
\end{array}
\Rightarrow x = 3 \text{ is not a zero.}
$$

$$
\begin{array}{r|rrrr}
5 & 4 & 8 & -11 & -15 \\
  &   & 20 & 140 & 645 \\
\hline
  & 4 & 28 & 129 & 630
\end{array}
\Rightarrow x = 5 \text{ is not a zero.}
\qquad
\begin{array}{r|rrrr}
3 & 4 & 8 & -11 & -15 \\
  &   & 12 & 60 & 147 \\
\hline
  & 4 & 20 & 49 & 132
\end{array}
\Rightarrow x = 3 \text{ is not a zero.}
$$

$$
\begin{array}{r|rrrr}
\frac{1}{2} & 4 & 8 & -11 & -15 \\
  &   & 2 & 5 & -3 \\
\hline
  & 4 & 10 & -6 & -18
\end{array}
\Rightarrow x = \tfrac{1}{2} \text{ is not a zero.}
\qquad
\begin{array}{r|rrrr}
\frac{1}{4} & 4 & 8 & -11 & -15 \\
  &   & 1 & \frac{9}{4} & -\frac{35}{16} \\
\hline
  & 4 & 9 & -\frac{35}{4} & -\frac{275}{16}
\end{array}
\Rightarrow x = \tfrac{1}{4} \text{ is not a zero.}
$$

$$
\begin{array}{r|rrrr}
\frac{3}{2} & 4 & 8 & -11 & -15 \\
  &   & 6 & 21 & 15 \\
\hline
  & 4 & 14 & 10 & 0
\end{array}
\Rightarrow x = \tfrac{3}{2} \text{ is a zero.}
$$

Thus $P(x) = 4x^3 + 8x^2 - 11x - 15 = \left(x - \frac{3}{2}\right)\left(4x^2 + 14x + 10\right)$. Continuing by factoring the quotient, whose possible rational zeros are $-1, -5, -\frac{1}{2}, -\frac{1}{4}, -\frac{5}{2}$, and $-\frac{5}{4}$, we have

$$
\begin{array}{r|rrr}
-1 & 4 & 14 & 10 \\
   &   & -4 & -10 \\
\hline
   & 4 & 10 & 0
\end{array} \quad \Rightarrow x = -1 \text{ is a zero.}
$$

Thus $P(x) = \left(x - \frac{3}{2}\right)(x + 1)(4x + 10)$ has zeros $\frac{3}{2}, -1$, and $-\frac{5}{2}$.

35. $P(x) = 2x^4 - 7x^3 + 3x^2 + 8x - 4$. The possible rational zeros are $\pm 1, \pm 2, \pm 4, \pm \frac{1}{2}$. $P(x)$ has 3 variations in sign and hence 1 or 3 positive real zeros. $P(-x) = 2x^4 + 7x^3 + 3x^2 - 8x - 4$ has 1 variation in sign and hence 1 negative real zero.

$$
\begin{array}{r|rrrrr}
1 & 2 & -7 & 3 & 8 & -4 \\
  &   & 2 & -5 & -2 & 6 \\
\hline
  & 2 & -5 & -2 & 6 & 2
\end{array} \quad \Rightarrow x = 1 \text{ is not a zero.}
\qquad
\begin{array}{r|rrrrr}
\frac{1}{2} & 2 & -7 & 3 & 8 & -4 \\
            &   & 1 & -3 & 0 & 4 \\
\hline
            & 2 & -6 & 0 & 8 & 0
\end{array} \quad \Rightarrow x = \frac{1}{2} \text{ is a zero.}
$$

Thus $P(x) = 2x^4 - 7x^3 + 3x^2 + 8x - 4 = \left(x - \frac{1}{2}\right)\left(2x^3 - 6x^2 + 8\right)$. Continuing by factoring the quotient, we have:

$$
\begin{array}{r|rrrr}
2 & 2 & -6 & 0 & 8 \\
  &   & 4 & -4 & -8 \\
\hline
  & 2 & -2 & -4 & 0
\end{array} \quad \Rightarrow x = 2 \text{ is a zero.}
$$

$P(x) = \left(x - \frac{1}{2}\right)(x - 2)\left(2x^2 - 2x - 4\right) = 2\left(x - \frac{1}{2}\right)(x - 2)\left(x^2 - x - 2\right) = 2\left(x - \frac{1}{2}\right)(x - 2)^2(x + 1)$. Thus, the zeros are $x = \frac{1}{2}, 2, -1$.

37. $P(x) = x^5 + 3x^4 - 9x^3 - 31x^2 + 36$. The possible rational zeros are $\pm 1, \pm 2, \pm 3, \pm 4, \pm 6, \pm 8, \pm 9, \pm 12, \pm 18$. $P(x)$ has 2 variations in sign and hence 0 or 2 positive real zeros. $P(-x) = -x^5 + 3x^4 + 9x^3 - 31x^2 + 36$ has 3 variations in sign and hence 1 or 3 negative real zeros.

$$
\begin{array}{r|rrrrrr}
1 & 1 & 3 & -9 & -31 & 0 & 36 \\
  &   & 1 & 4 & -5 & -36 & -36 \\
\hline
  & 1 & 4 & -5 & -36 & -36 & 0
\end{array} \quad \Rightarrow x = 1 \text{ is a zero.}
$$

So $P(x) = x^5 + 3x^4 - 9x^3 - 31x^2 + 36 = (x - 1)\left(x^4 + 4x^3 - 5x^2 - 36x - 36\right)$. Continuing by factoring the quotient, we have:

$$
\begin{array}{r|rrrrr}
1 & 1 & 4 & -5 & -36 & -36 \\
  &   & 1 & 5 & 0   & -36 \\
\hline
  & 1 & 1 & 0 & -36 & -72
\end{array}
\qquad
\begin{array}{r|rrrrr}
2 & 1 & 4 & -5 & -36 & -36 \\
  &   & 2 & 12 & 14 & -44 \\
\hline
  & 1 & 6 & 7 & -22 & -80
\end{array}
$$

$$
\begin{array}{r|rrrrr}
3 & 1 & 4 & -5 & -36 & -36 \\
  &   & 3 & 21 & 48 & 36 \\
\hline
  & 1 & 7 & 16 & 12 & 0
\end{array} \quad \Rightarrow x = 3 \text{ is a zero.}
$$

So $P(x) = (x - 1)(x - 3)\left(x^3 + 7x^2 + 16x + 12\right)$. Since we have 2 positive zeros, there are no more positive zeros, so we continue by factoring the quotient with possible negative zeros.

$$
\begin{array}{r|rrrr}
-1 & 1 & 7 & 16 & 12 \\
   &   & -1 & -6 & -10 \\
\hline
   & 1 & 6 & 10 & 2
\end{array}
\qquad
\begin{array}{r|rrrr}
-2 & 1 & 7 & 16 & 12 \\
   &   & -2 & -10 & -12 \\
\hline
   & 1 & 5 & 6 & 0
\end{array} \quad \Rightarrow x = -2 \text{ is a zero.}
$$

Then $P(x) = (x - 1)(x - 3)(x + 2)\left(x^2 + 5x + 6\right) = (x - 1)(x - 3)(x + 2)^2(x + 3)$. Thus, the zeros are $x = 1, 3, -2, -3$.

39. $P(x) = 3x^5 - 14x^4 - 14x^3 + 36x^2 + 43x + 10$ has possible rational zeros $\pm 1, \pm 2, \pm 5, \pm 10, \pm\frac{1}{3}, \pm\frac{2}{3}, \pm\frac{5}{3}, \pm\frac{10}{3}$. Since $P(x)$ has 2 variations in sign, there are 0 or 2 positive real zeros. Since $P(-x) = -3x^5 - 14x^4 + 14x^3 + 36x^2 - 43x + 10$ has 3 variations in sign, there are 1 or 3 negative real zeros.

1	3	-14	-14	36	43	10
		3	-11	-25	11	54
	3	-11	-25	11	54	64

2	3	-14	-14	36	43	10
		6	-16	-60	-48	-10
	3	-8	-30	-24	-5	0

$\Rightarrow x = 2$ is a zero.

$P(x) = (x - 2)\left(3x^4 - 8x^3 - 30x^2 - 24x - 5\right)$

2	3	-8	-30	-24	-5
		6	-4	-68	-184
	3	-2	-34	-92	-189

5	3	-8	-30	-24	-5
		15	35	25	5
	3	7	5	1	0

$\Rightarrow x = 5$ is a zero.

$P(x) = (x - 2)(x - 5)\left(3x^3 + 7x^2 + 5x + 1\right)$. Since $3x^3 + 7x^2 + 5x + 1$ has no variation in sign, there are no more positive zeros.

-1	3	7	5	1
		-3	-4	-1
	3	4	1	0

$\Rightarrow x = -1$ is a zero.

$P(x) = (x - 2)(x - 5)(x + 1)\left(3x^2 + 4x + 1\right) = (x - 2)(x - 5)(x + 1)(x + 1)(3x + 1)$. Therefore, the zeros are $x = -1, -\frac{1}{3}, 2, 5$.

41. $P(x) = x^3 + 4x^2 + 3x - 2$. The possible rational zeros are $\pm 1, \pm 2$. $P(x)$ has 1 variation in sign and hence 1 positive real zero. $P(-x) = -x^3 + 4x^2 - 3x - 2$ has 2 variations in sign and hence 0 or 2 negative real zeros.

1	1	4	3	-2
		1	5	8
	1	5	8	6

$\Rightarrow x = 1$ is an upper bound.

-1	1	4	3	-2
		-1	-3	0
	1	3	0	-2

-2	1	4	3	-2
		-2	-4	2
	1	2	-1	0

$\Rightarrow x = -2$ is a zero.

So $P(x) = (x + 2)\left(x^2 + 2x - 1\right)$. Using the quadratic formula on the second factor, we have:

$x = \frac{-2 \pm \sqrt{2^2 - 4(1)(-1)}}{2(1)} = \frac{-2 \pm \sqrt{8}}{2} = \frac{-2 \pm 2\sqrt{2}}{2} = -1 \pm \sqrt{2}$. Therefore, the zeros are $x = -2, -1 + \sqrt{2}, -1 - \sqrt{2}$.

43. $P(x) = x^4 - 6x^3 + 4x^2 + 15x + 4$. The possible rational zeros are $\pm 1, \pm 2, \pm 4$. $P(x)$ has 2 variations in sign and hence 0 or 2 positive real zeros. $P(-x) = x^4 + 6x^3 + 4x^2 - 15x + 4$ has 2 variations in sign and hence 0 or 2 negative real zeros.

1	1	-6	4	15	4
		1	-5	-1	14
	1	-5	-1	14	18

2	1	-6	4	15	4
		2	-8	-8	14
	1	-4	-4	7	18

4	1	-6	4	15	4
		4	-8	-16	-4
	1	-2	-4	-1	0

$\Rightarrow x = 4$ is a zero.

So $P(x) = (x - 4)\left(x^3 - 2x^2 - 4x - 1\right)$. Continuing by factoring the quotient, we have:

4	1	-2	-4	-1
		4	8	16
	1	2	4	15

$\Rightarrow x = 4$ is an upper bound.

-1	1	-2	-4	-1
		-1	3	1
	1	-3	-1	0

$\Rightarrow x = -1$ is a zero.

So $P(x) = (x - 4)(x + 1)\left(x^2 - 3x - 1\right)$. Using the quadratic formula on the third factor, we have:

$x = \frac{-(-3) \pm \sqrt{(-3)^2 - 4(1)(-1)}}{2(1)} = \frac{3 \pm \sqrt{13}}{2}$. Therefore, the zeros are $x = 4, -1, \frac{3 \pm \sqrt{13}}{2}$.

45. $P(x) = x^4 - 7x^3 + 14x^2 - 3x - 9$. The possible rational zeros are $\pm 1, \pm 3, \pm 9$. $P(x)$ has 3 variations in sign and hence 1 or 3 positive real zeros. $P(-x) = x^4 + 7x^3 + 14x^2 + 3x - 4$ has 1 variation in sign and hence 1 negative real zero.

$$
\begin{array}{r|rrrrr}
1 & 1 & -7 & 14 & -3 & -9 \\
  &   & 1 & -6 & 8 & 5 \\
\hline
  & 1 & -6 & 8 & 5 & 4
\end{array}
\qquad
\begin{array}{r|rrrrr}
3 & 1 & -7 & 14 & -3 & -9 \\
  &   & 3 & -12 & 6 & 9 \\
\hline
  & 1 & -4 & 2 & 3 & 0
\end{array}
\Rightarrow x = 3 \text{ is a zero.}
$$

So $P(x) = (x-3)(x^3 - 4x^2 + 2x + 3)$. Since the constant term of the second term is 3, ± 9 are no longer possible

zeros. Continuing by factoring the quotient, we have:
$$
\begin{array}{r|rrrr}
3 & 1 & -4 & 2 & 3 \\
  &   & 3 & -3 & -3 \\
\hline
  & 1 & -1 & -1 & 0
\end{array}
\Rightarrow x = 3 \text{ is a zero again.}
$$

So $P(x) = (x-3)^2 (x^2 - x - 1)$. Using the quadratic formula on the second factor, we have:

$x = \frac{-(-1)\pm\sqrt{(-1)^2 - 4(1)(-1)}}{2(1)} = \frac{1\pm\sqrt{5}}{2}$. Therefore, the zeros are $x = 3, \frac{1\pm\sqrt{5}}{2}$.

47. $P(x) = 4x^3 - 6x^2 + 1$. The possible rational zeros are $\pm 1, \pm\frac{1}{2}, \pm\frac{1}{4}$. $P(x)$ has 2 variations in sign and hence 0 or 2 positive real zeros. $P(-x) = -4x^3 - 6x^2 + 1$ has 1 variation in sign and hence 1 negative real zero.

$$
\begin{array}{r|rrrr}
1 & 4 & -6 & 0 & 1 \\
  &   & 4 & -2 & -2 \\
\hline
  & 4 & -2 & -2 & -1
\end{array}
\qquad
\begin{array}{r|rrrr}
\frac{1}{2} & 4 & -6 & 0 & 1 \\
  &   & 2 & -2 & -1 \\
\hline
  & 4 & -4 & -2 & 0
\end{array}
\Rightarrow x = \frac{1}{2} \text{ is a zero.}
$$

So $P(x) = \left(x - \frac{1}{2}\right)\left(4x^2 - 4x - 2\right)$. Using the quadratic formula on the second factor, we have:

$x = \frac{-(-4)\pm\sqrt{(-4)^2 - 4(4)(-2)}}{2(4)} = \frac{4\pm\sqrt{48}}{8} = \frac{4\pm 4\sqrt{3}}{8} = \frac{1\pm\sqrt{3}}{2}$. Therefore, the zeros are $x = \frac{1}{2}, \frac{1\pm\sqrt{3}}{2}$.

49. $P(x) = 2x^4 + 15x^3 + 17x^2 + 3x - 1$. The possible rational zeros are $\pm 1, \pm\frac{1}{2}$. $P(x)$ has 1 variation in sign and hence 1 positive real zero. $P(-x) = 2x^4 - 15x^3 + 17x^2 - 3x - 1$ has 3 variations in sign and hence 1 or 3 negative real zeros.

$$
\begin{array}{r|rrrrr}
\frac{1}{2} & 2 & 15 & 17 & 3 & -1 \\
  &   & 1 & 8 & \frac{25}{2} & \frac{31}{4} \\
\hline
  & 2 & 16 & 25 & \frac{31}{2} & \frac{27}{4}
\end{array}
\Rightarrow x = \frac{1}{2} \text{ is an upper bound.}
$$

$$
\begin{array}{r|rrrrr}
-\frac{1}{2} & 2 & 15 & 17 & 3 & -1 \\
  &   & -1 & -7 & -5 & 1 \\
\hline
  & 2 & 14 & 10 & -2 & 0
\end{array}
\Rightarrow x = -\frac{1}{2} \text{ is a zero.}
$$

So $P(x) = \left(x + \frac{1}{2}\right)\left(2x^3 + 14x^2 + 10x - 2\right) = 2\left(x + \frac{1}{2}\right)\left(x^3 + 7x^2 + 5x - 1\right)$.

$$
\begin{array}{r|rrrr}
-1 & 1 & 7 & 5 & -1 \\
   &   & -1 & -6 & 1 \\
\hline
   & 1 & 6 & -1 & 0
\end{array}
\Rightarrow x = -1 \text{ is a zero.}
$$

So $P(x) = \left(x + \frac{1}{2}\right)\left(2x^3 + 14x^2 + 10x - 2\right) = 2\left(x + \frac{1}{2}\right)(x+1)\left(x^2 + 6x - 1\right)$ Using the quadratic formula on the

third factor, we have $x = \frac{-(6)\pm\sqrt{(6)^2 - 4(1)(-1)}}{2(1)} = \frac{-6\pm\sqrt{40}}{2} = \frac{-6\pm 2\sqrt{10}}{2} = -3\pm\sqrt{10}$. Therefore, the zeros are $x = -1$,

$-\frac{1}{2}, -3\pm\sqrt{10}$.

51. (a) $P(x) = x^3 - 3x^2 - 4x + 12$ has possible rational zeros $\pm 1, \pm 2,$ **(b)**
$\pm 3, \pm 4, \pm 6, \pm 12.$

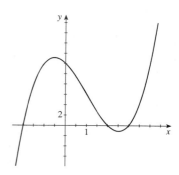

$$
\begin{array}{r|rrrr}
1 & 1 & -3 & -4 & 12 \\
  &   & 1 & -2 & -6 \\
\hline
  & 1 & -2 & -6 & 6 \\
\end{array}
$$

$$
\begin{array}{r|rrrr}
2 & 1 & -3 & -4 & 12 \\
  &   & 2 & -2 & -12 \\
\hline
  & 1 & -1 & -6 & 0 \\
\end{array}
$$
$\Rightarrow x = 2$ is a zero.

So $P(x) = (x-2)(x^2 - x - 6) = (x-2)(x+2)(x-3)$. The real zeros of P are $-2, 2, 3$.

53. (a) $P(x) = 2x^3 - 7x^2 + 4x + 4$ has possible rational zeros $\pm 1, \pm 2,$ **(b)**
$\pm 4, \pm \frac{1}{2}.$

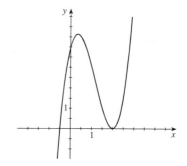

$$
\begin{array}{r|rrrr}
1 & 2 & -7 & 4 & 4 \\
  &   & 2 & -5 & -1 \\
\hline
  & 2 & -5 & -1 & -3 \\
\end{array}
\qquad
\begin{array}{r|rrrr}
2 & 2 & -7 & 4 & 4 \\
  &   & 4 & -6 & -4 \\
\hline
  & 2 & -3 & -2 & 0 \\
\end{array}
\Rightarrow x = 2 \text{ is a zero.}
$$

So $P(x) = (x-2)(2x^2 - 3x - 2)$. Continuing:

$$
\begin{array}{r|rrr}
2 & 2 & -3 & -2 \\
  &   & 4 & 2 \\
\hline
  & 2 & 1 & 0 \\
\end{array}
\Rightarrow x = 2 \text{ is a zero again.}
$$

Thus $P(x) = (x-2)^2(2x+1)$. The real zeros of P are 2 and $-\frac{1}{2}$.

55. (a) $P(x) = x^4 - 5x^3 + 6x^2 + 4x - 8$ has possible rational zeros $\pm 1,$ **(b)**
$\pm 2, \pm 4, \pm 8.$

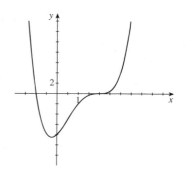

$$
\begin{array}{r|rrrrr}
1 & 1 & -5 & 6 & 4 & -8 \\
  &   & 1 & -4 & 2 & 6 \\
\hline
  & 1 & -4 & 2 & 6 & -2 \\
\end{array}
$$

$$
\begin{array}{r|rrrrr}
2 & 1 & -5 & 6 & 4 & -8 \\
  &   & 2 & -6 & 0 & 8 \\
\hline
  & 1 & -3 & 0 & 4 & 0 \\
\end{array}
\Rightarrow x = 2 \text{ is a zero.}
$$

So $P(x) = (x-2)(x^3 - 3x^2 + 4)$ and the possible rational zeros are restricted to $-1, \pm 2, \pm 4$.

$$
\begin{array}{r|rrrr}
2 & 1 & -3 & 0 & 4 \\
  &   & 2 & -2 & -4 \\
\hline
  & 1 & -1 & -2 & 0 \\
\end{array}
\Rightarrow x = 2 \text{ is a zero again.}
$$

$P(x) = (x-2)^2(x^2 - x - 2) = (x-2)^2(x-2)(x+1) = (x-2)^3(x+1)$. The real zeros of P are -1 and 2.

57. (a) $P(x) = x^5 - x^4 - 5x^3 + x^2 + 8x + 4$ has possible rational zeros $\pm 1, \pm 2, \pm 4$.

(b)

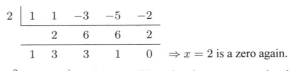

$$
\begin{array}{r|rrrrrr}
1 & 1 & -1 & -5 & 1 & 8 & 4 \\
  &   & 1 & 0 & -5 & -4 & 4 \\
\hline
  & 1 & 0 & -5 & -4 & 4 & 8
\end{array}
$$

$$
\begin{array}{r|rrrrrr}
2 & 1 & -1 & -5 & 1 & 8 & 4 \\
  &   & 2 & 2 & -6 & -10 & -4 \\
\hline
  & 1 & 1 & -3 & -5 & -2 & 0
\end{array} \Rightarrow x = 2 \text{ is a zero.}
$$

So $P(x) = (x - 2)(x^4 + x^3 - 3x^2 - 5x - 2)$, and the possible rational zeros are restricted to $-1, \pm 2$.

$$
\begin{array}{r|rrrrr}
2 & 1 & 1 & -3 & -5 & -2 \\
  &   & 2 & 6 & 6 & 2 \\
\hline
  & 1 & 3 & 3 & 1 & 0
\end{array} \Rightarrow x = 2 \text{ is a zero again.}
$$

So $P(x) = (x - 2)^2(x^3 + 3x^2 + 3x + 1)$, and the possible rational zeros are restricted to -1.

$$
\begin{array}{r|rrrr}
-1 & 1 & 3 & 3 & 1 \\
   &   & -1 & -2 & -1 \\
\hline
   & 1 & 2 & 1 & 0
\end{array} \Rightarrow x = -1 \text{ is a zero.}
$$

So $P(x) = (x - 2)^2(x + 1)(x^2 + 2x + 1) = (x - 2)^2(x + 1)^3$., and the real zeros of P are -1 and 2.

59. $P(x) = x^3 - x^2 - x - 3$. Since $P(x)$ has 1 variation in sign, P has 1 positive real zero. Since $P(-x) = -x^3 - x^2 + x - 3$ has 2 variations in sign, P has 2 or 0 negative real zeros. Thus, P has 1 or 3 real zeros.

61. $P(x) = 2x^6 + 5x^4 - x^3 - 5x - 1$. Since $P(x)$ has 1 variation in sign, P has 1 positive real zero. Since $P(-x) = 2x^6 + 5x^4 + x^3 + 5x - 1$ has 1 variation in sign, P has 1 negative real zero. Therefore, P has 2 real zeros.

63. $P(x) = x^5 + 4x^3 - x^2 + 6x$. Since $P(x)$ has 2 variations in sign, P has 2 or 0 positive real zeros. Since $P(-x) = -x^5 - 4x^3 - x^2 - 6x$ has no variation in sign, P has no negative real zero. Therefore, P has a total of 1 or 3 real zeros (since $x = 0$ is a zero, but is neither positive nor negative).

65. $P(x) = 2x^3 + 5x^2 + x - 2$; $a = -3, b = 1$

$$
\begin{array}{r|rrrr}
-3 & 2 & 5 & 1 & -2 \\
   &   & -6 & 3 & -12 \\
\hline
   & 2 & -1 & 4 & -14
\end{array} \text{ alternating signs } \Rightarrow \text{ lower bound.}
$$

$$
\begin{array}{r|rrrr}
1 & 2 & 5 & 1 & -2 \\
  &   & 2 & 7 & 8 \\
\hline
  & 2 & 7 & 8 & 6
\end{array} \text{ all nonnegative } \Rightarrow \text{ upper bound.}
$$

Therefore $a = -3$ and $b = 1$ are lower and upper bounds.

67. $P(x) = 8x^3 + 10x^2 - 39x + 9$; $a = -3, b = 2$

$$
\begin{array}{r|rrrr}
-3 & 8 & 10 & -39 & 9 \\
 & & -24 & 42 & -9 \\
\hline
 & 8 & -14 & 3 & 0
\end{array}
$$
alternating signs $\Rightarrow$ lower bound.

$$
\begin{array}{r|rrrr}
2 & 8 & 10 & -39 & 9 \\
 & & 16 & 52 & 26 \\
\hline
 & 8 & 26 & 13 & 35
\end{array}
$$
all nonnegative $\Rightarrow$ upper bound.

Therefore $a = -3$ and $b = 2$ are lower and upper bounds. Note that $x = -3$ is also a zero.

69. $P(x) = x^3 - 3x^2 + 4$ and use the Upper and Lower Bounds Theorem:

$$
\begin{array}{r|rrrr}
-1 & 1 & -3 & 0 & 4 \\
 & & -1 & 4 & -4 \\
\hline
 & 1 & -4 & 4 & 0
\end{array}
$$
alternating signs $\Rightarrow$ lower bound.

$$
\begin{array}{r|rrrr}
3 & 1 & -3 & 0 & 4 \\
 & & 3 & 0 & 0 \\
\hline
 & 1 & 0 & 0 & 4
\end{array}
$$
all nonnegative $\Rightarrow$ upper bound.

Therefore -1 is a lower bound (and a zero) and 3 is an upper bound. (There are many possible solutions.)

71. $P(x) = x^4 - 2x^3 + x^2 - 9x + 2$.

$$
\begin{array}{r|rrrrr}
1 & 1 & -2 & 1 & -9 & 2 \\
 & & 1 & -1 & 0 & -9 \\
\hline
 & 1 & -1 & 0 & -9 & -7
\end{array}
\qquad
\begin{array}{r|rrrrr}
2 & 1 & -2 & 1 & -9 & 2 \\
 & & 2 & 0 & 2 & -14 \\
\hline
 & 1 & 0 & 1 & -7 & -12
\end{array}
$$

$$
\begin{array}{r|rrrrr}
3 & 1 & -2 & 1 & -9 & 2 \\
 & & 3 & 3 & 12 & 9 \\
\hline
 & 1 & 1 & 4 & 3 & 11
\end{array}
$$
all positive $\Rightarrow$ upper bound.

$$
\begin{array}{r|rrrrr}
-1 & 1 & -2 & 1 & -9 & 2 \\
 & & -1 & 3 & -4 & 13 \\
\hline
 & 1 & -3 & 4 & -13 & 15
\end{array}
$$
alternating signs $\Rightarrow$ lower bound.

Therefore -1 is a lower bound and 3 is an upper bound. (There are many possible solutions.)

73. $P(x) = 2x^4 + 3x^3 - 4x^2 - 3x + 2$.

$$
\begin{array}{r|rrrrr}
1 & 2 & 3 & -4 & -3 & 2 \\
 & & 2 & 5 & 1 & -2 \\
\hline
 & 2 & 5 & 1 & -2 & 0
\end{array}
$$
$\Rightarrow x = 1$ is a zero.

$P(x) = (x - 1)(2x^3 + 5x^2 + x - 2)$

$$
\begin{array}{r|rrrr}
-1 & 2 & 5 & 1 & -2 \\
 & & -2 & -3 & 2 \\
\hline
 & 2 & 3 & -2 & 0
\end{array}
$$
$\Rightarrow x = -1$ is a zero.

$P(x) = (x - 1)(x + 1)(2x^2 + 3x - 2) = (x - 1)(x + 1)(2x - 1)(x + 2)$. Therefore, the zeros are $x = -2, \frac{1}{2}, \pm 1$.

75. *Method 1:* $P(x) = 4x^4 - 21x^2 + 5$ has 2 variations in sign, so by Descartes' rule of signs there are either 2 or 0 positive zeros. If we replace x with $(-x)$, the function does not change, so there are either 2 or 0 negative zeros. Possible rational zeros are $\pm 1, \pm\frac{1}{2}, \pm\frac{1}{4}, \pm 5, \pm\frac{5}{2}, \pm\frac{5}{4}$. By inspection, ± 1 and ± 5 are not zeros, so we must look for non-integer solutions:

$$
\begin{array}{r|rrrrr}
\frac{1}{2} & 4 & 0 & -21 & 0 & 5 \\
 & & 2 & 1 & -10 & -5 \\
\hline
 & 4 & 2 & -20 & -10 & 0
\end{array}
\quad \Rightarrow x = \tfrac{1}{2} \text{ is a zero.}
$$

$P(x) = \left(x - \frac{1}{2}\right)\left(4x^3 + 2x^2 - 20x - 10\right)$, continuing with the quotient, we have:

$$
\begin{array}{r|rrrr}
-\frac{1}{2} & 4 & 2 & -20 & -10 \\
 & & -2 & 0 & 10 \\
\hline
 & 4 & 0 & -20 & 0
\end{array}
\quad \Rightarrow x = -\tfrac{1}{2} \text{ is a zero.}
$$

$P(x) = \left(x - \frac{1}{2}\right)\left(x + \frac{1}{2}\right)\left(4x^2 - 20\right) = 0$. If $4x^2 - 20 = 0$, then $x = \pm\sqrt{5}$. Thus the zeros are $x = \pm\frac{1}{2}, \pm\sqrt{5}$.

Method 2: Substituting $u = x^2$, the equation becomes $4u^2 - 21u + 5 = 0$, which factors: $4u^2 - 21u + 5 = (4u - 1)(u - 5) = (4x^2 - 1)(x^2 - 5)$. Then either we have $x^2 = 5$, so that $x = \pm\sqrt{5}$, or we have $x^2 = \frac{1}{4}$, so that $x = \pm\sqrt{\frac{1}{4}} = \pm\frac{1}{2}$. Thus the zeros are $x = \pm\frac{1}{2}, \pm\sqrt{5}$.

77. $P(x) = x^5 - 7x^4 + 9x^3 + 23x^2 - 50x + 24$. The possible rational zeros are $\pm 1, \pm 2, \pm 3, \pm 4, \pm 6, \pm 8, \pm 12, \pm 24$. $P(x)$ has 4 variations in sign and hence 0, 2, or 4 positive real zeros. $P(-x) = -x^5 - 7x^4 - 9x^3 + 23x^2 + 50x + 24$ has 1 variation in sign, and hence 1 negative real zero.

$$
\begin{array}{r|rrrrrr}
1 & 1 & -7 & 9 & 23 & -50 & 24 \\
 & & 1 & -6 & 3 & 26 & -24 \\
\hline
 & 1 & -6 & 3 & 26 & -24 & 0
\end{array}
\quad \Rightarrow x = 1 \text{ is a zero.}
$$

$P(x) = (x - 1)\left(x^4 - 6x^3 + 3x^2 + 26x - 24\right)$; continuing with the quotient, we try 1 again.

$$
\begin{array}{r|rrrrr}
1 & 1 & -6 & 3 & 26 & -24 \\
 & & 1 & -5 & -2 & 24 \\
\hline
 & 1 & -5 & -2 & 24 & 0
\end{array}
\quad \Rightarrow x = 1 \text{ is a zero again.}
$$

$P(x) = (x - 1)^2\left(x^3 - 5x^2 - 2x + 24\right)$; continuing with the quotient, we start by trying 1 again.

$$
\begin{array}{r|rrrr}
1 & 1 & -5 & -2 & 24 \\
 & & 1 & -4 & -6 \\
\hline
 & 1 & -4 & -6 & 18
\end{array}
\qquad
\begin{array}{r|rrrr}
2 & 1 & -5 & -2 & 24 \\
 & & 2 & -6 & -16 \\
\hline
 & 1 & -3 & -8 & 8
\end{array}
\qquad
\begin{array}{r|rrrr}
3 & 1 & -5 & -2 & 24 \\
 & & 3 & -6 & -24 \\
\hline
 & 1 & -2 & -8 & 0
\end{array}
\quad \Rightarrow x = 3 \text{ is a zero.}
$$

$P(x) = (x - 1)^2(x - 3)\left(x^2 - 2x - 8\right) = (x - 1)^2(x - 3)(x - 4)(x + 2)$. Therefore, the zeros are $x = -2, 1, 3, 4$.

79. $P(x) = x^3 - x - 2$. The only possible rational zeros of $P(x)$ are ± 1 and ± 2.

$$
\begin{array}{r|rrrr}
1 & 1 & 0 & -1 & -2 \\
 & & 1 & 1 & 0 \\
\hline
 & 1 & 1 & 0 & -2
\end{array}
\qquad
\begin{array}{r|rrrr}
2 & 1 & 0 & -1 & -2 \\
 & & 2 & 4 & 6 \\
\hline
 & 1 & 2 & 3 & 4
\end{array}
\qquad
\begin{array}{r|rrrr}
-1 & 1 & 0 & -1 & -2 \\
 & & -1 & 1 & 0 \\
\hline
 & 1 & -1 & 0 & -2
\end{array}
$$

Since the row that contains -1 alternates between nonnegative and nonpositive, -1 is a lower bound and there is no need to try -2. Therefore, $P(x)$ does not have any rational zeros.

81. $P(x) = 3x^3 - x^2 - 6x + 12$ has possible rational zeros $\pm 1, \pm 2, \pm 3, \pm 4, \pm 6, \pm 12, \pm\frac{1}{3}, \pm\frac{2}{3}, \pm\frac{4}{3}$.

	3	−1	−6	12
1	3	2	−4	8
2	3	5	4	20
−1	3	−4	−2	14
−2	3	−7	8	−4

all positive $\Rightarrow$ $x = 2$ is an upper bound

alternating signs $\Rightarrow$ $x = -2$ is a lower bound

	3	−1	−6	12
$\frac{1}{3}$	3	0	−6	10
$\frac{2}{3}$	3	1	$-\frac{16}{3}$	$\frac{76}{9}$
$\frac{4}{3}$	3	3	−2	$\frac{28}{3}$
$-\frac{1}{3}$	3	−2	$-\frac{16}{3}$	$\frac{124}{9}$
$-\frac{2}{3}$	3	−3	−4	$\frac{44}{3}$
$-\frac{4}{3}$	3	−5	$\frac{2}{3}$	$\frac{100}{9}$

Therefore, there is no rational zero.

83. $P(x) = x^3 - 3x^2 - 4x + 12$, $[-4, 4]$ by $[-15, 15]$. The possible rational zeros are $\pm 1, \pm 2, \pm 3, \pm 4, \pm 6, \pm 12$. By observing the graph of P, the rational zeros are $x = -2$, 2, 3.

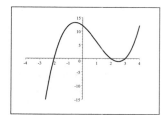

85. $P(x) = 2x^4 - 5x^3 - 14x^2 + 5x + 12$, $[-2, 5]$ by $[-40, 40]$. The possible rational zeros are $\pm 1, \pm 2, \pm 3, \pm 4, \pm 6, \pm 12, \pm\frac{1}{2}, \pm\frac{3}{2}$. By observing the graph of P, the zeros are $x = -\frac{3}{2}, -1, 1, 4$.

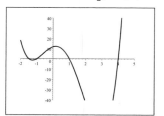

87. $x^4 - x - 4 = 0$. Possible rational solutions are $\pm 1, \pm 2, \pm 4$.

	1	0	0	−1	−4
1		1	1	1	0
	1	1	1	0	−4

	1	0	0	−1	−4
2		2	4	8	14
	1	2	4	7	10

$\Rightarrow x = 2$ is an upper bound.

	1	0	0	−1	−4
−1		−1	1	−1	2
	1	−1	1	−2	−2

	1	0	0	−1	−4
−2		−2	4	−8	18
	1	−2	4	−9	14

$\Rightarrow x = -2$ is a lower bound.

Therefore, we graph the function $P(x) = x^4 - x - 4$ in the viewing rectangle $[-2, 2]$ by $[-5, 20]$ and see there are two solutions. In the viewing rectangle $[-1.3, -1.25]$ by $[-0.1, 0.1]$, we find the solution $x \approx -1.28$. In the viewing rectangle $[1.5, 1.6]$ by $[-0.1, 0.1]$, we find the solution $x \approx 1.53$. Thus the solutions are $x \approx -1.28, 1.53$.

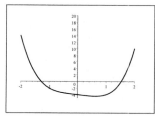

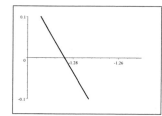

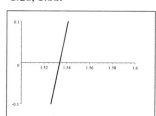

89. $4.00x^4 + 4.00x^3 - 10.96x^2 - 5.88x + 9.09 = 0$.

1	4	4	-10.96	-5.88	9.09
		4	8	-2.96	-8.84
	4	8	-2.96	-8.84	0.25

2	4	4	-10.96	-5.88	9.09
		8	24	26.08	40.40
	4	12	13.04	20.2	49.49

$\Rightarrow x = 2$ is an upper bound.

-2	4	4	-10.96	-5.88	9.09
		-8	8	5.92	-0.08
	4	-4	-2.96	0.04	9.01

-3	4	4	-10.96	-5.88	9.09
		-12	24	-39.12	135
	4	-8	13.04	-45	144.09

$\Rightarrow x = -3$ is a lower bound.

Therefore, we graph the function $P(x) = 4.00x^4 + 4.00x^3 - 10.96x^2 - 5.88x + 9.09$ in the viewing rectangle $[-3, 2]$ by $[-10, 40]$. There appear to be two solutions. In the viewing rectangle $[-1.6, -1.4]$ by $[-0.1, 0.1]$, we find the solution $x \approx -1.50$. In the viewing rectangle $[0.8, 1.2]$ by $[0, 1]$, we see that the graph comes close but does not go through the x-axis. Thus there is no solution here. Therefore, the only solution is $x \approx -1.50$.

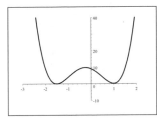 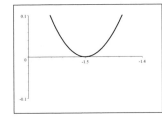 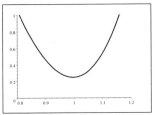

91. (a) Since $z > b$, we have $z - b > 0$. Since all the coefficients of $Q(x)$ are nonnegative, and since $z > 0$, we have $Q(z) > 0$ (being a sum of positive terms). Thus, $P(z) = (z - b) \cdot Q(z) + r > 0$, since the sum of a positive number and a nonnegative number.

(b) In part (a), we showed that if b satisfies the conditions of the first part of the Upper and Lower Bounds Theorem and $z > b$, then $P(z) > 0$. This means that no real zero of P can be larger than b, so b is an upper bound for the real zeros.

(c) Suppose $-b$ is a negative lower bound for the real zeros of $P(x)$. Then clearly b is an upper bound for $P_1(x) = P(-x)$. Thus, as in Part (a), we can write $P_1(x) = (x - b) \cdot Q(x) + r$, where $r > 0$ and the coefficients of Q are all nonnegative, and $P(x) = P_1(-x) = (-x - b) \cdot Q(-x) + r = (x + b) \cdot [-Q(-x)] + r$. Since the coefficients of $Q(x)$ are all nonnegative, the coefficients of $-Q(-x)$ will be alternately nonpositive and nonnegative, which proves the second part of the Upper and Lower Bounds Theorem.

93. Let r be the radius of the silo. The volume of the hemispherical roof is $\frac{1}{2}\left(\frac{4}{3}\pi r^3\right) = \frac{2}{3}\pi r^3$. The volume of the cylindrical section is $\pi\left(r^2\right)(30) = 30\pi r^2$. Because the total volume of the silo is 15,000 ft³, we get the following equation: $\frac{2}{3}\pi r^3 + 30\pi r^2 = 15000 \Leftrightarrow \frac{2}{3}\pi r^3 + 30\pi r^2 - 15000 = 0 \Leftrightarrow \pi r^3 + 45\pi r^2 - 22500 = 0$. Using a graphing device, we first graph the polynomial in the viewing rectangle $[0, 15]$ by $[-10000, 10000]$. The solution, $r \approx 11.28$ ft., is shown in the viewing rectangle $[11.2, 11.4]$ by $[-1, 1]$.

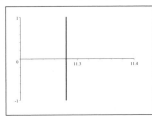 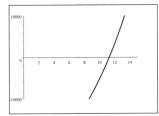

95. $h(t) = 11.60t - 12.41t^2 + 6.20t^3$
$$- 1.58t^4 + 0.20t^5 - 0.01t^6$$
is shown in the viewing rectangle
$[0, 10]$ by $[0, 6]$.

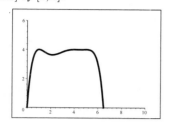

(a) It started to snow again.

(b) No, $h(t) \leq 4$.

(c) The function $h(t)$ is shown in the viewing rectangle $[6, 6.5]$ by $[0, 0.5]$. The x-intercept of the function is a little less than 6.5, which means that the snow melted just before midnight on Saturday night.

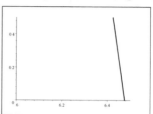

97. Let r be the radius of the cone and cylinder and let h be the height of the cone. Since the height and diameter are equal, we get $h = 2r$. So the volume of the cylinder is $V_1 = \pi r^2 \cdot$ (cylinder height) $= 20\pi r^2$, and the volume of the cone is $V_2 = \frac{1}{3}\pi r^2 h = \frac{1}{3}\pi r^2 (2r) = \frac{2}{3}\pi r^3$. Since the total volume is $\dfrac{500\pi}{3}$, it follows that $\frac{2}{3}\pi r^3 + 20\pi r^2 = \dfrac{500\pi}{3} \quad \Leftrightarrow \quad r^3 + 30r^2 - 250 = 0$. By Descartes' Rule of Signs, there is 1 positive zero. Since r is between 2.76 and 2.765 (see the table), the radius should be 2.76 m (correct to two decimals).

r	$r^3 + 30r^2 - 250$
1	-219
2	-122
3	47
2.7	-11.62
2.76	-2.33
2.77	1.44
2.765	1.44
2.8	7.15

99. Let b be the width of the base, and let l be the length of the box. Then the length plus girth is $l + 4b = 108$, and the volume is $V = lb^2 = 2200$. Solving the first equation for l and substituting this value into the second equation yields $l = 108 - 4b \Rightarrow V = (108 - 4b)\,b^2 = 2200 \quad \Leftrightarrow \quad 4b^3 - 108b^2 + 2200 = 0 \Leftrightarrow 4\left(b^3 - 27b^2 + 550\right) = 0$. Now $P(b) = b^3 - 27b^2 + 550$ has two variations in sign, so there are 0 or 2 positive real zeros. We also observe that since $l > 0, b < 27$, so $b = 27$ is an upper bound. Thus the possible positive rational real zeros are $1, 2, 3, 10, 11, 22, 25$.

$$
\begin{array}{r|rrrr}
1 & 1 & -27 & 0 & 550 \\
  &   & 1 & -26 & -26 \\
\hline
  & 1 & -26 & -26 & 524
\end{array}
\qquad
\begin{array}{r|rrrr}
2 & 1 & -27 & 0 & 550 \\
  &   & 2 & -50 & -100 \\
\hline
  & 1 & -25 & -50 & 450
\end{array}
$$

$$
\begin{array}{r|rrrr}
5 & 1 & -27 & 0 & 550 \\
  &   & 5 & -110 & -550 \\
\hline
  & 1 & -22 & -110 & 0
\end{array}
\quad \Rightarrow b = 5 \text{ is a zero.}
$$

$P(b) = (b - 5)\left(b^2 - 22b - 110\right)$. The other zeros are $b = \dfrac{22 \pm \sqrt{484 - 4(1)(-110)}}{2} = \dfrac{22 \pm \sqrt{924}}{2} = \dfrac{22 \pm 30.397}{2}$. The positive answer from this factor is $b \approx 26.20$. Thus we have two possible solutions, $b = 5$ or $b \approx 26.20$. If $b = 5$, then $l = 108 - 4(5) = 88$; if $b \approx 26.20$, then $l = 108 - 4(26.20) = 3.20$. Thus the length of the box is either 88 in. or 3.20 in.

101. (a) Substituting $X - \frac{a}{3}$ for x we have

$$
\begin{aligned}
x^3 + ax^2 + bx + c &= \left(X - \tfrac{1}{3}a\right)^3 + a\left(X - \tfrac{1}{3}a\right)^2 + b\left(X - \tfrac{1}{3}a\right) + c \\
&= X^3 - aX^2 + \tfrac{1}{3}a^2 X + \tfrac{1}{27}a^3 + a\left(X^2 - \tfrac{2}{3}aX + \tfrac{1}{9}a^2\right) + bX - \tfrac{1}{3}ab + c \\
&= X^3 - aX^2 + \tfrac{1}{3}a^2 X + \tfrac{1}{27}a^3 + aX^2 - \tfrac{2}{3}a^2 X + \tfrac{1}{9}a^3 + bX - \tfrac{1}{3}ab + c \\
&= X^3 + (-a + a) X^2 + \left(-\tfrac{1}{3}a^2 - \tfrac{2}{3}a^2 + b\right) X + \left(\tfrac{1}{27}a^3 + \tfrac{1}{9}a^3 - \tfrac{1}{3}ab + c\right) \\
&= X^3 + \left(b - a^2\right) X + \left(\tfrac{4}{27}a^3 - \tfrac{1}{3}ab + c\right)
\end{aligned}
$$

(b) $x^3 + 6x^2 + 9x + 4 = 0$. Setting $a = 6$, $b = 9$, and $c = 4$, we have: $X^3 + \left(9 - 6^2\right) X + (32 - 18 + 4) = X^3 - 27X + 18$.

4.4   Complex Zeros and the Fundamental Theorem of Algebra

1. (a) $x^4 + 4x^2 = 0$ $\Leftrightarrow$ $x^2\left(x^2 + 4\right) = 0$. So $x = 0$ or $x^2 + 4 = 0$. If $x^2 + 4 = 0$ then $x^2 = -4 \Leftrightarrow x = \pm 2i$. Therefore, the solutions are $x = 0$ and $\pm 2i$.

(b) To get the complete factorization, we factor the remaining quadratic factor $P(x) = x^2(x + 4) = x^2(x - 2i)(x + 2i)$.

3. (a) $x^3 - 2x^2 + 2x = 0$ $\Leftrightarrow$ $x\left(x^2 - 2x + 2\right) = 0$. So $x = 0$ or $x^2 - 2x + 2 = 0$. If $x^2 - 2x + 2 = 0$ then $x = \frac{-(-2) \pm \sqrt{(-2)^2 - 4(1)(2)}}{2} = \frac{2 \pm \sqrt{-4}}{2} = \frac{2 \pm 2i}{2} = 1 \pm i$. Therefore, the solutions are $x = 0$, $1 \pm i$.

(b) Since $1 - i$ and $1 + i$ are zeros, $x - (1 - i) = x - 1 + i$ and $x - (1 + i) = x - 1 - i$ are the factors of $x^2 - 2x + 2$. Thus the complete factorization is $P(x) = x\left(x^2 - 2x + 2\right) = x(x - 1 + i)(x - 1 - i)$.

5. (a) $x^4 + 2x^2 + 1 = 0$ $\Leftrightarrow$ $\left(x^2 + 1\right)^2 = 0$ $\Leftrightarrow$ $x^2 + 1 = 0$ $\Leftrightarrow$ $x^2 = -1 \Leftrightarrow x = \pm i$. Therefore the zeros of P are $x = \pm i$.

(b) Since $-i$ and i are zeros, $x + i$ and $x - i$ are the factors of $x^2 + 1$. Thus the complete factorization is $P(x) = \left(x^2 + 1\right)^2 = [(x + i)(x - i)]^2 = (x + i)^2(x - i)^2$.

7. (a) $x^4 - 16 = 0$ $\Leftrightarrow$ $0 = \left(x^2 - 4\right)\left(x^2 + 4\right) = (x - 2)(x + 2)\left(x^2 + 4\right)$. So $x = \pm 2$ or $x^2 + 4 = 0$. If $x^2 + 4 = 0$ then $x^2 = -4$ $\Rightarrow$ $x = \pm 2i$. Therefore the zeros of P are $x = \pm 2$, $\pm 2i$.

(b) Since $-i$ and i are zeros, $x + i$ and $x - i$ are the factors of $x^2 + 1$. Thus the complete factorization is $P(x) = (x - 2)(x + 2)\left(x^2 + 4\right) = (x - 2)(x + 2)(x - 2i)(x + 2i)$.

9. (a) $x^3 + 8 = 0$ $\Leftrightarrow$ $(x + 2)\left(x^2 - 2x + 4\right) = 0$. So $x = -2$ or $x^2 - 2x + 4 = 0$. If $x^2 - 2x + 4 = 0$ then $x = \frac{-(-2) \pm \sqrt{(-2)^2 - 4(1)(4)}}{2} = \frac{2 \pm \sqrt{-12}}{2} = \frac{2 \pm 2i\sqrt{3}}{2} = 1 \pm i\sqrt{3}$. Therefore, the zeros of P are $x = -2$, $1 \pm i\sqrt{3}$.

(b) Since $1 - i\sqrt{3}$ and $1 + i\sqrt{3}$ are the zeros from the $x^2 - 2x + 4 = 0$, $x - \left(1 - i\sqrt{3}\right)$ and $x - \left(1 + i\sqrt{3}\right)$ are the factors of $x^2 - 2x + 4$. Thus the complete factorization is

$$
\begin{aligned}
P(x) &= (x + 2)\left(x^2 - 2x + 4\right) = (x + 2)\left[x - \left(1 - i\sqrt{3}\right)\right]\left[x - \left(1 + i\sqrt{3}\right)\right] \\
&= (x + 2)\left(x - 1 + i\sqrt{3}\right)\left(x - 1 - i\sqrt{3}\right)
\end{aligned}
$$

11. (a) $x^6 - 1 = 0$ $\Leftrightarrow$ $0 = \left(x^3 - 1\right)\left(x^3 + 1\right) = (x - 1)\left(x^2 + x + 1\right)(x + 1)\left(x^2 - x + 1\right)$. Clearly, $x = \pm 1$ are solutions. If $x^2 + x + 1 = 0$, then $x = \frac{-1 \pm \sqrt{1 - 4(1)(1)}}{2} = \frac{-1 \pm \sqrt{-3}}{2} = -\frac{1}{2} \pm \frac{\sqrt{-3}}{2}$ so $x = -\frac{1}{2} \pm i\frac{\sqrt{3}}{2}$. And if $x^2 - x + 1 = 0$, then $x = \frac{1 \pm \sqrt{1 - 4(1)(1)}}{2} = \frac{1 \pm \sqrt{-3}}{2} = \frac{1}{2} \pm \frac{\sqrt{-3}}{2} = \frac{1}{2} \pm i\frac{\sqrt{3}}{2}$. Therefore, the zeros of P are $x = \pm 1$, $-\frac{1}{2} \pm i\frac{\sqrt{3}}{2}$, $\frac{1}{2} \pm i\frac{\sqrt{3}}{2}$.

(b) The zeros of $x^2 + x + 1 = 0$ are $-\frac{1}{2} - i\frac{\sqrt{3}}{2}$ and $-\frac{1}{2} + i\frac{\sqrt{3}}{2}$, so $x^2 + x + 1$ factors

as $\left[x - \left(-\frac{1}{2} - i\frac{\sqrt{3}}{2}\right)\right]\left[x - \left(-\frac{1}{2} + i\frac{\sqrt{3}}{2}\right)\right] = \left(x + \frac{1}{2} + i\frac{\sqrt{3}}{2}\right)\left(x + \frac{1}{2} - i\frac{\sqrt{3}}{2}\right)$. Simi-

larly, since the zeros of $x^2 - x + 1 = 0$ are $\frac{1}{2} - i\frac{\sqrt{3}}{2}$ and $\frac{1}{2} + i\frac{\sqrt{3}}{2}$, so $x^2 - x + 1$ factors as

$\left[x - \left(\frac{1}{2} - i\frac{\sqrt{3}}{2}\right)\right]\left[x - \left(\frac{1}{2} + i\frac{\sqrt{3}}{2}\right)\right] = \left(x - \frac{1}{2} + i\frac{\sqrt{3}}{2}\right)\left(x - \frac{1}{2} - i\frac{\sqrt{3}}{2}\right)$. The complete factorization is

$$P(x) = (x - 1)\left(x^2 + x + 1\right)(x + 1)\left(x^2 - x + 1\right)$$
$$= (x - 1)(x + 1)\left(x + \frac{1}{2} + i\frac{\sqrt{3}}{2}\right)\left(x + \frac{1}{2} - i\frac{\sqrt{3}}{2}\right)\left(x - \frac{1}{2} + i\frac{\sqrt{3}}{2}\right)\left(x - \frac{1}{2} - i\frac{\sqrt{3}}{2}\right)$$

13. $P(x) = x^2 + 25 = (x - 5i)(x + 5i)$. The zeros of P are $5i$ and $-5i$, both multiplicity 1.

15. $Q(x) = x^2 + 2x + 2$. Using the quadratic formula $x = \frac{-(2) \pm \sqrt{(2)^2 - 4(1)(2)}}{2(1)} = \frac{-2 \pm \sqrt{-4}}{2} = \frac{-2 \pm 2i}{2} = -1 \pm i$. So $Q(x) = (x + 1 - i)(x + 1 + i)$. The zeros of Q are $-1 - i$ (multiplicity 1) and $-1 + i$ (multiplicity 1).

17. $P(x) = x^3 + 4x = x\left(x^2 + 4\right) = x(x - 2i)(x + 2i)$. The zeros of P are 0, $2i$, and $-2i$ (all multiplicity 1).

19. $Q(x) = x^4 - 1 = \left(x^2 - 1\right)\left(x^2 + 1\right) = (x - 1)(x + 1)\left(x^2 + 1\right) = (x - 1)(x + 1)(x - i)(x + i)$. The zeros of Q are 1, -1, i, and $-i$ (all of multiplicity 1).

21. $P(x) = 16x^4 - 81 = \left(4x^2 - 9\right)\left(4x^2 + 9\right) = (2x - 3)(2x + 3)(2x - 3i)(2x + 3i)$. The zeros of P are $\frac{3}{2}$, $-\frac{3}{2}$, $\frac{3}{2}i$, and $-\frac{3}{2}i$ (all of multiplicity 1).

23. $P(x) = x^3 + x^2 + 9x + 9 = x^2(x + 1) + 9(x + 1) = (x + 1)\left(x^2 + 9\right) = (x + 1)(x - 3i)(x + 3i)$. The zeros of P are -1, $3i$, and $-3i$ (all of multiplicity 1).

25. $Q(x) = x^4 + 2x^2 + 1 = \left(x^2 + 1\right)^2 = (x - i)^2(x + i)^2$. The zeros of Q are i and $-i$ (both of multiplicity 2).

27. $P(x) = x^4 + 3x^2 - 4 = \left(x^2 - 1\right)\left(x^2 + 4\right) = (x - 1)(x + 1)(x - 2i)(x + 2i)$. The zeros of P are 1, -1, $2i$, and $-2i$ (all of multiplicity 1).

29. $P(x) = x^5 + 6x^3 + 9x = x\left(x^4 + 6x^2 + 9\right) = x\left(x^2 + 3\right)^2 = x\left(x - \sqrt{3}i\right)^2\left(x + \sqrt{3}i\right)^2$. The zeros of P are 0 (multiplicity 1), $\sqrt{3}i$ (multiplicity 2), and $-\sqrt{3}i$ (multiplicity 2).

31. Since $1 + i$ and $1 - i$ are conjugates, the factorization of the polynomial must be $P(x) = a(x - [1 + i])(x - [1 - i]) = a\left(x^2 - 2x + 2\right)$. If we let $a = 1$, we get $P(x) = x^2 - 2x + 2$.

33. Since $2i$ and $-2i$ are conjugates, the factorization of the polynomial must be $Q(x) = b(x - 3)(x - 2i)(x + 2i) = b(x - 3)\left(x^2 + 4\right) = b\left(x^3 - 3x^2 + 4x - 12\right)$. If we let $b = 1$, we get $Q(x) = x^3 - 3x^2 + 4x - 12$.

35. Since i is a zero, by the Conjugate Roots Theorem, $-i$ is also a zero. So the factorization of the polynomial must be $P(x) = a(x - 2)(x - i)(x + i) = a\left(x^3 - 2x^2 + x - 2\right)$. If we let $a = 1$, we get $P(x) = x^3 - 2x^2 + x - 2$.

37. Since the zeros are $1 - 2i$ and 1 (with multiplicity 2), by the Conjugate Roots Theorem, the other zero is $1 + 2i$. So a factorization is

$$\begin{aligned} R(x) &= c(x - [1 - 2i])(x - [1 + 2i])(x - 1)^2 = c([x - 1] + 2i)([x - 1] - 2i)(x - 1)^2 \\ &= c\left([x - 1]^2 - [2i]^2\right)\left(x^2 - 2x + 1\right) = c\left(x^2 - 2x + 1 + 4\right)\left(x^2 - 2x + 1\right) = c\left(x^2 - 2x + 5\right)\left(x^2 - 2x + 1\right) \\ &= c\left(x^4 - 2x^3 + x^2 - 2x^3 + 4x^2 - 2x + 5x^2 - 10x + 5\right) = c\left(x^4 - 4x^3 + 10x^2 - 12x + 5\right) \end{aligned}$$

If we let $c = 1$ we get $R(x) = x^4 - 4x^3 + 10x^2 - 12x + 5$.

39. Since the zeros are i and $1 + i$, by the Conjugate Roots Theorem, the other zeros are $-i$ and $1 - i$. So a factorization is

$$\begin{aligned} T(x) &= C(x - i)(x + i)(x - [1 + i])(x - [1 - i]) \\ &= C\left(x^2 - i^2\right)([x - 1] - i)([x - 1] + i) = C\left(x^2 + 1\right)\left(x^2 - 2x + 1 - i^2\right) = C\left(x^2 + 1\right)\left(x^2 - 2x + 2\right) \\ &= C\left(x^4 - 2x^3 + 2x^2 + x^2 - 2x + 2\right) = C\left(x^4 - 2x^3 + 3x^2 - 2x + 2\right) = Cx^4 - 2Cx^3 + 3Cx^2 - 2Cx + 2C \end{aligned}$$

Since the constant coefficient is 12, it follows that $2C = 12 \Leftrightarrow C = 6$, and so $T(x) = 6\left(x^4 - 2x^3 + 3x^2 - 2x + 2\right) = 6x^4 - 12x^3 + 18x^2 - 12x + 12$.

41. $P(x) = x^3 + 2x^2 + 4x + 8 = x^2(x+2) + 4(x+2) = (x+2)(x^2+4) = (x+2)(x-2i)(x+2i)$. Thus the zeros are -2 and $\pm 2i$.

43. $P(x) = x^3 - 2x^2 + 2x - 1$. By inspection, $P(1) = 1 - 2 + 2 - 1 = 0$, and hence $x = 1$ is a zero.

$$
\begin{array}{r|rrrr}
1 & 1 & -2 & 2 & -1 \\
  &   & 1 & -1 & 1 \\
\hline
  & 1 & -1 & 1 & 0
\end{array}
$$

Thus $P(x) = (x-1)(x^2 - x + 1)$. So $x = 1$ or $x^2 - x + 1 = 0$.

Using the quadratic formula, we have $x = \frac{1 \pm \sqrt{1 - 4(1)(1)}}{2} = \frac{1 \pm i\sqrt{3}}{2}$. Hence, the zeros are 1 and $\frac{1 \pm i\sqrt{3}}{2}$.

45. $P(x) = x^3 - 3x^2 + 3x - 2$.

$$
\begin{array}{r|rrrr}
2 & 1 & -3 & 3 & -2 \\
  &   & 2 & -2 & 2 \\
\hline
  & 1 & -1 & 1 & 0
\end{array}
$$

Thus $P(x) = (x-2)(x^2 - x + 1)$. So $x = 2$ or $x^2 - x + 1 = 0$

Using the quadratic formula we have $x = \frac{1 \pm \sqrt{1 - 4(1)(1)}}{2} = \frac{1 \pm i\sqrt{3}}{2}$. Hence, the zeros are 2, and $\frac{1 \pm i\sqrt{3}}{2}$.

47. $P(x) = 2x^3 + 7x^2 + 12x + 9$ has possible rational zeros ± 1, ± 3, ± 9, $\pm \frac{1}{2}$, $\pm \frac{3}{2}$, $\pm \frac{9}{2}$. Since all coefficients are positive, there are no positive real zeros.

$$
\begin{array}{r|rrrr}
-1 & 2 & 7 & 12 & 9 \\
   &   & -2 & -5 & -7 \\
\hline
   & 2 & 5 & 7 & 2
\end{array}
\qquad
\begin{array}{r|rrrr}
-2 & 2 & 7 & 12 & 9 \\
   &   & -4 & -6 & -12 \\
\hline
   & 2 & 3 & 6 & -3
\end{array}
$$

There is a zero between -1 and -2.

$$
\begin{array}{r|rrrr}
-\frac{3}{2} & 2 & 7 & 12 & 9 \\
             &   & -3 & -6 & -9 \\
\hline
             & 2 & 4 & 6 & 0
\end{array}
\Rightarrow x = -\frac{3}{2} \text{ is a zero.}
$$

$P(x) = \left(x + \frac{3}{2}\right)\left(2x^2 + 4x + 6\right) = 2\left(x + \frac{3}{2}\right)\left(x^2 + 2x + 3\right)$. Now $x^2 + 2x + 3$ has zeros

$x = \frac{-2 \pm \sqrt{4 - 4(3)(1)}}{2} = \frac{-2 \pm 2\sqrt{-2}}{2} = -1 \pm i\sqrt{2}$. Hence, the zeros are $-\frac{3}{2}$ and $-1 \pm i\sqrt{2}$.

49. $P(x) = x^4 + x^3 + 7x^2 + 9x - 18$. Since $P(x)$ has one change in sign, we are guaranteed a positive zero, and since $P(-x) = x^4 - x^3 + 7x^2 - 9x - 18$, there are 1 or 3 negative zeros.

$$
\begin{array}{r|rrrrr}
1 & 1 & 1 & 7 & 9 & -18 \\
  &   & 1 & 2 & 9 & 18 \\
\hline
  & 1 & 2 & 9 & 18 & 0
\end{array}
$$

Therefore, $P(x) = (x-1)(x^3 + 2x^2 + 9x + 18)$. Continuing with the quotient, we try negative zeros.

$$
\begin{array}{r|rrrr}
-1 & 1 & 2 & 9 & 18 \\
   &   & -1 & -1 & -8 \\
\hline
   & 1 & 1 & 8 & 10
\end{array}
\qquad
\begin{array}{r|rrrr}
-2 & 1 & 2 & 9 & 18 \\
   &   & -2 & 0 & -18 \\
\hline
   & 1 & 0 & 9 & 0
\end{array}
$$

$P(x) = (x-1)(x+2)(x^2+9) = (x-1)(x+2)(x-3i)(x+3i)$. Therefore, the zeros are 1, -2, and $\pm 3i$.

51. We see a pattern and use it to factor by grouping. This gives

$$P(x) = x^5 - x^4 + 7x^3 - 7x^2 + 12x - 12 = x^4(x-1) + 7x^2(x-1) + 12(x-1) = (x-1)(x^4 + 7x^2 + 12)$$
$$= (x-1)(x^2+3)(x^2+4) = (x-1)(x-i\sqrt{3})(x+i\sqrt{3})(x-2i)(x+2i)$$

Therefore, the zeros are 1, $\pm i\sqrt{3}$, and $\pm 2i$.

53. $P(x) = x^4 - 6x^3 + 13x^2 - 24x + 36$ has possible rational zeros $\pm 1, \pm 2, \pm 3, \pm 4, \pm 6, \pm 9, \pm 12, \pm 18$. $P(x)$ has 4 variations in sign and $P(-x)$ has no variation in sign.

$$
\begin{array}{r|rrrrr}
1 & 1 & -6 & 13 & -24 & 36 \\
  &   & 1 & -5 & 8 & -16 \\
\hline
  & 1 & -5 & 8 & -16 & 20
\end{array}
\qquad
\begin{array}{r|rrrrr}
2 & 1 & -6 & 13 & -24 & 36 \\
  &   & 2 & -8 & 10 & -28 \\
\hline
  & 1 & -4 & 5 & -14 & 8
\end{array}
\qquad
\begin{array}{r|rrrrr}
3 & 1 & -6 & 13 & -24 & 36 \\
  &   & 3 & -9 & 12 & -36 \\
\hline
  & 1 & -3 & 4 & -12 & 0
\end{array}
$$

$\Rightarrow x = 3$ is a zero.

Continuing:

$$
\begin{array}{r|rrrr}
3 & 1 & -3 & 4 & -12 \\
  &   & 3 & 0 & 12 \\
\hline
  & 1 & 0 & 4 & 0
\end{array}
$$

$\Rightarrow x = 3$ is a zero.

$P(x) = (x-3)^2(x^2+4) = (x-3)^2(x-2i)(x+2i)$. Therefore, the zeros are 3 (multiplicity 2) and $\pm 2i$.

55. $P(x) = 4x^4 + 4x^3 + 5x^2 + 4x + 1$ has possible rational zeros $\pm 1, \pm\frac{1}{2}, \pm\frac{1}{4}$. Since there is no variation in sign, all real zeros (if there are any) are negative.

$$
\begin{array}{r|rrrrr}
-1 & 4 & 4 & 5 & 4 & 1 \\
   &   & -4 & 0 & -5 & 1 \\
\hline
   & 4 & 0 & 5 & -1 & 2
\end{array}
\qquad
\begin{array}{r|rrrrr}
-\frac{1}{2} & 4 & 4 & 5 & 4 & 1 \\
   &   & -2 & -1 & -2 & -1 \\
\hline
   & 4 & 2 & 4 & 2 & 0
\end{array}
$$

$\Rightarrow x = -\frac{1}{2}$ is a zero.

$P(x) = \left(x+\frac{1}{2}\right)(4x^3 + 2x^2 + 4x + 2)$. Continuing:

$$
\begin{array}{r|rrrr}
-\frac{1}{2} & 4 & 2 & 4 & 2 \\
   &   & -2 & 0 & -2 \\
\hline
   & 4 & 0 & 4 & 0
\end{array}
$$

$\Rightarrow x = -\frac{1}{2}$ is a zero again.

$P(x) = \left(x+\frac{1}{2}\right)^2(4x^2+4)$. Thus, the zeros of $P(x)$ are $-\frac{1}{2}$ and $\pm i$.

57. $P(x) = x^5 - 3x^4 + 12x^3 - 28x^2 + 27x - 9$ has possible rational zeros $\pm 1, \pm 3, \pm 9$. $P(x)$ has 4 variations in sign and $P(-x)$ has 1 variation in sign.

$$
\begin{array}{r|rrrrrr}
1 & 1 & -3 & 12 & -28 & 27 & -9 \\
  &   & 1 & -2 & 10 & -18 & 9 \\
\hline
  & 1 & -2 & 10 & -18 & 9 & 0
\end{array}
$$

$\Rightarrow x = 1$ is a zero.

$$
\begin{array}{r|rrrrr}
1 & 1 & -2 & 10 & -18 & 9 \\
  &   & 1 & -1 & 9 & -9 \\
\hline
  & 1 & -1 & 9 & -9 & 0
\end{array}
\qquad
\begin{array}{r|rrrr}
1 & 1 & -1 & 9 & -9 \\
  &   & 1 & 0 & 9 \\
\hline
  & 1 & 0 & 9 & 0
\end{array}
$$

$\Rightarrow x = 1$ is a zero. $\qquad \Rightarrow x = 1$ is a zero.

$P(x) = (x-1)^3(x^2+9) = (x-1)^3(x-3i)(x+3i)$. Therefore, the zeros are 1 (multiplicity 3) and $\pm 3i$.

59. (a) $P(x) = x^3 - 5x^2 + 4x - 20 = x^2(x-5) + 4(x-5) = (x-5)(x^2+4)$

 (b) $P(x) = (x-5)(x-2i)(x+2i)$

61. (a) $P(x) = x^4 + 8x^2 - 9 = (x^2-1)(x^2+9) = (x-1)(x+1)(x^2+9)$

 (b) $P(x) = (x-1)(x+1)(x-3i)(x+3i)$

63. (a) $P(x) = x^6 - 64 = (x^3-8)(x^3+8) = (x-2)(x^2+2x+4)(x+2)(x^2-2x+4)$

 (b) $P(x) = (x-2)(x+2)(x+1-i\sqrt{3})(x+1+i\sqrt{3})(x-1-i\sqrt{3})(x-1+i\sqrt{3})$

65. (a) $x^4 - 2x^3 - 11x^2 + 12x = x\left(x^3 - 2x^2 - 11x + 12\right) = 0$. We first find the

bounds for our viewing rectangle.

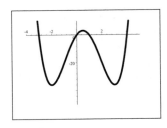

$$\begin{array}{r|rrrr} & 1 & -2 & -11 & 12 \\ \hline 5 & 1 & 3 & 4 & 32 \\ -4 & 1 & -6 & 13 & -50 \end{array}$$

$\Rightarrow x = 5$ is an upper bound.

$\Rightarrow x = -4$ is a lower bound.

We graph $P\left(x\right) = x^4 - 2x^3 - 11x^2 + 12x$ in the viewing rectangle $[-4, 5]$ by $[-50, 10]$ and see that it has 4 real solutions. Since this matches the degree of $P\left(x\right)$, $P\left(x\right)$ has no imaginary solution.

(b) $x^4 - 2x^3 - 11x^2 + 12x - 5 = 0$. We use the same bounds for our viewing rectangle, $[-4, 5]$ by $[-50, 10]$, and see that $R\left(x\right) = x^4 - 2x^3 - 11x^2 + 12x - 5$ has 2 real solutions. Since the degree of $R\left(x\right)$ is 4, $R\left(x\right)$ must have 2 imaginary solutions.

(c) $x^4 - 2x^3 - 11x^2 + 12x + 40 = 0$. We graph $T\left(x\right) = x^4 - 2x^3 - 11x^2 + 12x + 40$ in the viewing rectangle $[-4, 5]$ by $[-10, 50]$, and see that T has no real solution. Since the degree of T is 4, T must have 4 imaginary solutions.

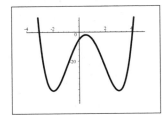

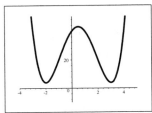

67. (a) $P\left(x\right) = x^2 - \left(1 + i\right)x + \left(2 + 2i\right)$. So $P\left(2i\right) = \left(2i\right)^2 - \left(1 + i\right)\left(2i\right) + 2 + 2i = -4 - 2i + 2 + 2 + 2i = 0$, and $P\left(1 - i\right) = \left(1 - i\right)^2 - \left(1 + i\right)\left(1 - i\right) + \left(2 + 2i\right) = 1 - 2i - 1 - 1 - 1 + 2 + 2i = 0$. Therefore, $2i$ and $1 - i$ are solutions of the equation $x^2 - \left(1 + i\right)x + \left(2 + 2i\right) = 0$. However, $P\left(-2i\right) = \left(-2i\right)^2 - \left(1 + i\right)\left(-2i\right) + 2 + 2i = -4 + 2i - 2 + 2 + 2i = -4 + 4i$, and $P\left(1 + i\right) = \left(1 + i\right)^2 - \left(1 + i\right)\left(1 + i\right) + 2 + 2i = 2 + 2i$. Since, $P\left(-2i\right) \neq 0$ and $P\left(1 + i\right) \neq 0$, $-2i$ and $1 + i$ are not solutions.

(b) This does not violate the Conjugate Roots Theorem because the coefficients of the polynomial $P\left(x\right)$ are not all real.

69. Since P has real coefficients, the imaginary zeros come in pairs: $a \pm bi$ (by the Conjugate Roots Theorem), where $b \neq 0$. Thus there must be an even number of imaginary zeros. Since P is of odd degree, it has an odd number of zeros (counting multiplicity). It follows that P has at least one real zero.

4.5   Rational Functions

1. $r\left(x\right) = \dfrac{x}{x - 2}$

(a)

x	$r\left(x\right)$
1.5	-3
1.9	-19
1.99	-199
1.999	-1999

x	$r\left(x\right)$
2.5	5
2.1	21
2.01	201
2.001	2001

x	$r\left(x\right)$
10	1.25
50	1.042
100	1.020
1000	1.002

x	$r\left(x\right)$
-10	0.833
-50	0.962
-100	0.980
-1000	0.998

(b) $r\left(x\right) \to -\infty$ as $x \to 2^-$ and $r\left(x\right) \to \infty$ as $x \to 2^+$.   **(c)** r has horizontal asymptote $y = 1$.

3. $r(x) = \dfrac{3x - 10}{(x - 2)^2}$

(a)

x	$r(x)$
1.5	-22
1.9	-430
1.99	$-40,300$
1.999	$-4,003,000$

x	$r(x)$
2.5	-10
2.1	-370
2.01	$-39,700$
2.001	$-3,997,000$

x	$r(x)$
10	0.3125
50	0.0608
100	0.0302
1000	0.0030

x	$r(x)$
-10	-0.2778
-50	-0.0592
-100	-0.0298
-1000	-0.0030

(b) $r(x) \to -\infty$ as $x \to 2$.

(c) r has horizontal asymptote $y = 0$.

5. $r(x) = \dfrac{x - 1}{x + 4}$. When $x = 0$, we have $r(0) = -\frac{1}{4}$, so the y-intercept is $-\frac{1}{4}$. The numerator is 0 when $x = 1$, so the x-intercept is 1.

7. $t(x) = \dfrac{x^2 - x - 2}{x - 6}$. When $x = 0$, we have $t(0) = \dfrac{-2}{-6} = \frac{1}{3}$, so the y-intercept is $\frac{1}{3}$. The numerator is 0 when $x^2 - x - 2 = (x - 2)(x + 1) = 0$ or when $x = 2$ or $x = -1$, so the x-intercepts are 2 and -1.

9. $r(x) = \dfrac{x^2 - 9}{x^2}$. Since 0 is not in the domain of $r(x)$, there is no y-intercept. The numerator is 0 when $x^2 - 9 = (x - 3)(x + 3) = 0$ or when $x = \pm 3$, so the x-intercepts are ± 3.

11. From the graph, the x-intercept is 3, the y-intercept is 3, the vertical asymptote is $x = 2$, and the horizontal asymptote is $y = 2$.

13. From the graph, the x-intercepts are -1 and 1, the y-intercept is about $\frac{1}{4}$, the vertical asymptotes are $x = -2$ and $x = 2$, and the horizontal asymptote is $y = 1$.

15. $r(x) = \dfrac{3}{x + 2}$. There is a vertical asymptote where $x + 2 = 0 \Leftrightarrow x = -2$. We have $r(x) = \dfrac{3}{x + 2} = \dfrac{3/x}{1 + (2/x)} \to 0$ as $x \to \pm\infty$, so the horizontal asymptote is $y = 0$.

17. $t(x) = \dfrac{x^2}{x^2 - x - 6} = \dfrac{x^2}{(x - 3)(x + 2)} = \dfrac{1}{1 - \dfrac{1}{x} - \dfrac{6}{x^2}} \to 1$ as $x \to \pm\infty$. Hence, the horizontal asymptote is $y = 1$.

The vertical asymptotes occur when $(x - 3)(x + 2) = 0 \Leftrightarrow x = 3$ or $x = -2$, and so the vertical asymptotes are $x = 3$ and $x = -2$.

19. $s(x) = \dfrac{6}{x^2 + 2}$. There is no vertical asymptote since $x^2 + 2$ is never 0. Since $s(x) = \dfrac{6}{x^2 + 2} = \dfrac{\dfrac{6}{x^2}}{1 + \dfrac{2}{x^2}} \to 0$ as $x \to \pm\infty$, the horizontal asymptote is $y = 0$.

21. $r(x) = \dfrac{6x - 2}{x^2 + 5x - 6}$. A vertical asymptote occurs when $x^2 + 5x - 6 = (x + 6)(x - 1) = 0 \Leftrightarrow x = 1$ or $x = -6$. Because the degree of the denominator is greater than the degree of the numerator, the horizontal asymptote is $y = 0$.

23. $y = \dfrac{x^2 + 2}{x - 1}$. A vertical asymptote occurs when $x - 1 = 0 \Leftrightarrow x = 1$. There are no horizontal asymptotes because the degree of the numerator is greater than the degree of the denominator.

In Exercises 25–31, let $f(x) = \dfrac{1}{x}$.

25. $r(x) = \dfrac{1}{x-1} = f(x-1)$. From this form we see that the graph of r is obtained

from the graph of f by shifting 1 unit to the right. Thus r has vertical asymptote

$x = 1$ and horizontal asymptote $y = 0$.

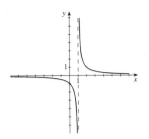

27. $s(x) = \dfrac{3}{x+1} = 3\left(\dfrac{1}{x+1}\right) = 3f(x+1)$. From this form we see that the graph

of s is obtained from the graph of f by shifting 1 unit to the left and stretching

vertically by a factor of 3. Thus s has vertical asymptote $x = -1$ and horizontal

asymptote $y = 0$.

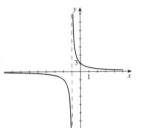

29. $t(x) = \dfrac{2x-3}{x-2} = 2 + \dfrac{1}{x-2} = f(x-2) + 2$ (see long

division below). From this form we see that the graph of t is

obtained from the graph of f by shifting 2 units to the right

and 2 units vertically. Thus t has vertical asymptote $x = 2$

and horizontal asymptote $y = 2$.

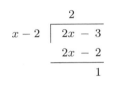

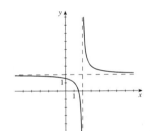

31. $r(x) = \dfrac{x+2}{x+3} = 1 - \dfrac{1}{x+3} = -f(x+3) + 1$ (see long

division below). From this form we see that the graph of r is

obtained from the graph of f by shifting 3 units to the left,

reflect about the x-axis, and then shifting vertically 1 unit.

Thus r has vertical asymptote $x = -3$ and horizontal

asymptote $y = 1$.

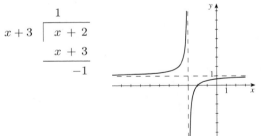

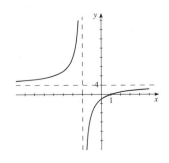

33. $y = \dfrac{4x-4}{x+2}$. When $x = 0$, $y = -2$, so the y-intercept is -2. When $y = 0$,

$4x - 4 = 0 \quad \Leftrightarrow \quad x = 1$, so the x-intercept is 1. Since the degree of the

numerator and denominator are the same the horizontal asymptote is $y = \frac{4}{1} = 4$. A

vertical asymptote occurs when $x = -2$. As $x \to -2^+$, $y = \dfrac{4x-4}{x+2} \to -\infty$, and

as $x \to -2^-$, $y = \dfrac{4x-4}{x+2} \to \infty$.

35. $s(x) = \dfrac{4 - 3x}{x + 7}$. When $x = 0$, $y = \frac{4}{7}$, so the y-intercept is $\frac{4}{7}$. The x-intercepts

occur when $y = 0 \quad \Leftrightarrow \quad 4 - 3x = 0 \quad \Leftrightarrow \quad x = \frac{4}{3}$. A vertical asymptote

occurs when $x = -7$. Since the degree of the numerator and denominator are the

same the horizontal asymptote is $y = \dfrac{-3}{1} = -3$.

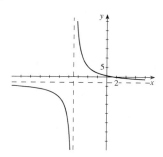

37. $r(x) = \dfrac{18}{(x - 3)^2}$. When $x = 0$, $y = \frac{18}{9} = 2$, and so the y-intercept is 2. Since the

numerator can never be zero, there is no x-intercept. There is a vertical asymptote

when $x - 3 = 0 \quad \Leftrightarrow \quad x = 3$, and because the degree of the asymptote is $y = 0$.

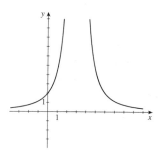

39. $s(x) = \dfrac{4x - 8}{(x - 4)(x + 1)}$. When $x = 0$, $y = \dfrac{-8}{(-4)(1)} = 2$, so the y-intercept is 2.

When $y = 0$, $4x - 8 = 0 \quad \Leftrightarrow \quad x = 2$, so the x-intercept is 2. The vertical

asymptotes are $x = -1$ and $x = 4$, and because the degree of the numerator is less

than the degree of the denominator, the horizontal asymptote is $y = 0$.

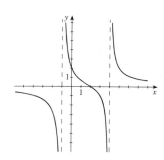

41. $s(x) = \dfrac{6}{x^2 - 5x - 6}$. When $x = 0$, $y = \dfrac{6}{-6} = -1$, so the y-intercept is -1.

Since the numerator is never zero, there is no x-intercept. The vertical asymptotes

occur when $x^2 - 5x - 6 = (x + 1)(x - 6) \Leftrightarrow x = -1$ and $x = 6$, and because

the degree of the numerator is less less than the degree of the denominator, the

horizontal asymptote is $y = 0$.

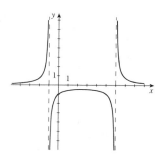

43. $t(x) = \dfrac{3x + 6}{x^2 + 2x - 8}$. When $x = 0$, $y = \dfrac{6}{-8} = -\dfrac{3}{4}$, so the y-intercept is $-\frac{3}{4}$.

When $y = 0$, $3x + 6 = 0 \quad \Leftrightarrow \quad x = -2$, so the x-intercept is -2. The vertical

asymptotes occur when $x^2 + 2x - 8 = (x - 2)(x + 4) = 0 \quad \Leftrightarrow \quad x = 2$ and

$x = -4$. Since the degree of the numerator is less than the degree of the

denominator, the horizontal asymptote is $y = 0$.

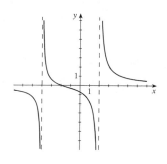

45. $r(x) = \dfrac{(x-1)(x+2)}{(x+1)(x-3)}$. When $x = 0$, $y = \frac{2}{3}$, so the y-intercept is $\frac{2}{3}$. When

$y = 0$, $(x-1)(x+2) = 0$ $\Rightarrow$ $x = -2, 1$, so, the x-intercepts are -2 and 1.
The vertical asymptotes are $x = -1$ and $x = 3$, and because the degree of the
numerator and denominator are the same the horizontal asymptote is $y = \frac{1}{1} = 1$.

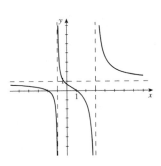

47. $r(x) = \dfrac{x^2 - 2x + 1}{x^2 + 2x + 1} = \dfrac{(x-1)^2}{(x+1)^2} = \left(\dfrac{x-1}{x+1}\right)^2$. When $x = 0$, $y = 1$, so the

y-intercept is 1. When $y = 0$, $x = 1$, so the x-intercept is 1 . A vertical asymptote
occurs at $x + 1 = 0 \Leftrightarrow x = -1$. Because the degree of the numerator and
denominator are the same the horizontal asymptote is $y = \frac{1}{1} = 1$.

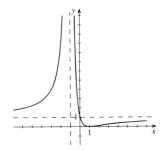

49. $r(x) = \dfrac{2x^2 + 10x - 12}{x^2 + x - 6} = \dfrac{2(x-1)(x+6)}{(x-2)(x+3)}$. When $x = 0$,

$y = \dfrac{2(-1)(6)}{(-2)(3)} = 2$, so the y-intercept is 2. When $y = 0$, $2(x-1)(x+6) = 0$

$\Rightarrow$ $x = -6, 1$, so the x-intercepts are -6 and 1. Vertical asymptotes occur when
$(x-2)(x+3) = 0 \Leftrightarrow x = -3$ or $x = 2$. Because the degree of the numerator
and denominator are the same the horizontal asymptote is $y = \frac{2}{1} = 2$.

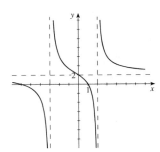

51. $y = \dfrac{x^2 - x - 6}{x^2 + 3x} = \dfrac{(x-3)(x+2)}{x(x+3)}$. The x-intercept occurs when $y = 0 \Leftrightarrow$

$(x-3)(x+2) = 0$ $\Rightarrow$ $x = -2, 3$, so the x-intercepts are -2 and 3. There
is no y-intercept because y is undefined when $x = 0$. The vertical asymptotes are
$x = 0$ and $x = -3$. Because the degree of the numerator and denominator are the
same, the horizontal asymptotes is $y = \frac{1}{1} = 1$.

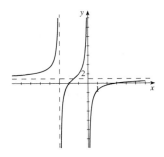

53. $r(x) = \dfrac{3x^2 + 6}{x^2 - 2x - 3} = \dfrac{3(x^2 + 2)}{(x-3)(x+1)}$. When $x = 0$, $y = -2$, so the

y-intercept is -2. Since the numerator can never equal zero, there is no
x-intercept. Vertical asymptotes occur when $x = -1, 3$. Because the degree of the
numerator and denominator are the same, the horizontal asymptote is.$y = \frac{3}{1} = 3$.

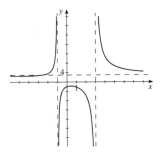

55. $s(x) = \dfrac{x^2 - 2x + 1}{x^3 - 3x^2} = \dfrac{(x-1)^2}{x^2(x-3)}$. Since $x = 0$ is not in the domain of $s(x)$,

there is no y-intercept. The x-intercept occurs when $y = 0 \Leftrightarrow$

$x^2 - 2x + 1 = (x-1)^2 = 0 \quad\Rightarrow\quad x = 1$, so the x-intercept is 1. Vertical

asymptotes occur when $x = 0, 3$. Since the degree of the numerator is less than the

degree of the denominator, the horizontal asymptote is $y = 0$.

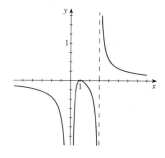

57. $r(x) = \dfrac{x^2}{x-2}$. When $x = 0$, $y = 0$, so the graph passes through the origin. There

is a vertical asymptote when $x - 2 = 0 \Leftrightarrow x = 2$, with $y \to \infty$ as $x \to 2^+$, and

$y \to -\infty$ as $x \to 2^-$. Because the degree of the numerator is greater than the

degree of the denominator, there is no horizontal asymptotes. By using long

division, we see that $y = x + 2 + \dfrac{4}{x-2}$, so $y = x + 2$ is a slant asymptote.

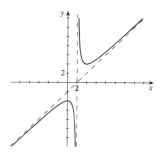

59. $r(x) = \dfrac{x^2 - 2x - 8}{x} = \dfrac{(x-4)(x+2)}{x}$. The vertical asymptote is $x = 0$, thus,

there is no y-intercept. If $y = 0$, then $(x-4)(x+2) = 0 \Rightarrow x = -2, 4$, so the

x-intercepts are -2 and 4. Because the degree of the numerator is greater than the

degree of the denominator, there are no horizontal asymptotes. By using long

division, we see that $y = x - 2 - \dfrac{8}{x}$, so $y = x - 2$ is a slant asymptote.

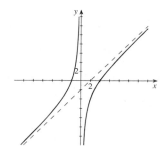

61. $r(x) = \dfrac{x^2 + 5x + 4}{x-3} = \dfrac{(x+4)(x+1)}{x-3}$. When $x = 0$, $y = -\dfrac{4}{3}$, so the

y-intercept is $-\dfrac{4}{3}$. When $y = 0$, $(x+4)(x+1) = 0 \quad\Leftrightarrow\quad x = -4, -1$, so the

two x-intercepts are -4 and -1. A vertical asymptote occurs when $x = 3$, with

$y \to \infty$ as $x \to 3^+$, and $y \to -\infty$ as $x \to 3^-$. Using long division, we see that

$y = x + 8 + \dfrac{28}{x-3}$, so $y = x + 8$ is a slant asymptote.

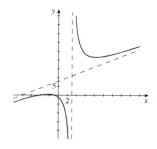

63. $r(x) = \dfrac{x^3 + x^2}{x^2 - 4} = \dfrac{x^2(x+1)}{(x-2)(x+2)}$. When $x = 0$, $y = 0$, so the graph passes

through the origin. Moreover, when $y = 0$, we have $x^2(x+1) = 0 \quad\Rightarrow$

$x = 0, -1$, so the x-intercepts are 0 and -1. Vertical asymptotes occur when

$x = \pm 2$; as $x \to \pm 2^-$, $y = -\infty$ and as $x \to \pm 2^+$, $y \to \infty$. Because the degree

of the numerator is greater than the degree of the denominator, there is no

horizontal asymptote. Using long division, we see that $y = x + 1 + \dfrac{4x + 4}{x^2 - 4}$, so

$y = x + 1$ is a slant asymptote.

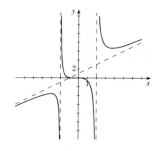

65. $f(x) = \dfrac{2x^2 + 6x + 6}{x + 3}$, $g(x) = 2x$. f has vertical asymptote $x = -3$.

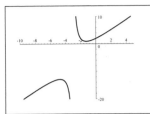

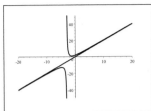

67. $f(x) = \dfrac{x^3 - 2x^2 + 16}{x - 2}$, $g(x) = x^2$. f has vertical asymptote $x = 2$.

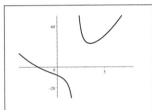

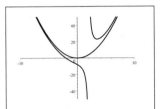

69. $f(x) = \dfrac{2x^2 - 5x}{2x + 3}$ has vertical asymptote $x = -1.5$, x-intercepts 0 and 2.5, y-intercept 0, local maximum $(-3.9, -10.4)$,

and local minimum $(0.9, -0.6)$. Using long division, we get $f(x) = x - 4 + \dfrac{12}{2x + 3}$. From the graph, we see that the end

behavior of $f(x)$ is like the end behavior of $g(x) = x - 4$.

$$
\begin{array}{r}
x \;-\; 4 \\
2x + 3 \;\overline{\big)\; 2x^2 \;-\; 5x } \\
2x^2 \;+\; 3x \\
\hline
-\;8x \\
-\;8x \;-\; 12 \\
\hline
12
\end{array}
$$

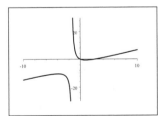

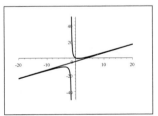

71. $f(x) = \dfrac{x^5}{x^3 - 1}$ has vertical asymptote $x = 1$, x-intercept 0, y-intercept 0, and local minimum $(1.4, 3.1)$.

Thus $y = x^2 + \dfrac{x^2}{x^3 - 1}$. From the graph we see that the end behavior of $f(x)$ is like the end behavior of $g(x) = x^2$.

$$
\begin{array}{r}
x^2 \\
x^3 - 1 \;\overline{\big)\; x^5 } \\
x^5 \;-\; x^2 \\
\hline
x^2
\end{array}
$$

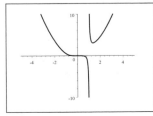

Graph of f

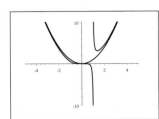

Graph of f and g

73. $f(x) = \dfrac{x^4 - 3x^3 + 6}{x - 3}$ has vertical asymptote $x = 3$, x-intercepts 1.6 and 2.7, y-intercept -2, local maxima $(-0.4, -1.8)$

and $(2.4, 3.8)$, and local minima $(0.6, -2.3)$ and $(3.4, 54.3)$. Thus $y = x^3 + \dfrac{6}{x - 3}$. From the graphs, we see that the end

behavior of $f(x)$ is like the end behavior of $g(x) = x^3$.

$$
\begin{array}{r}
x^3 \\
x - 3 \overline{\smash{\big)}\, x^4 - 3x^3 + 6} \\
\underline{x^4 - 3x^3 } \\
6
\end{array}
$$

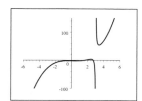 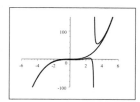

75. (a)

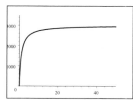

(b) $p(t) = \dfrac{3000t}{t + 1} = 3000 - \dfrac{3000}{t + 1}$. So as $t \to \infty$, we have $p(t) \to 3000$.

77. $c(t) = \dfrac{5t}{t^2 + 1}$

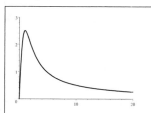

(a) The highest concentration of drug is 2.50 mg/L, and it is reached 1 hour after the drug is administered.

(b) The concentration of the drug in the bloodstream goes to 0.

(c) From the first viewing rectangle, we see that an approximate solution is near $t = 15$. Thus we graph $y = \dfrac{5t}{t^2 + 1}$ and $y = 0.3$ in the

viewing rectangle $[14, 18]$ by $[0, 0.5]$. So it takes about 16.61 hours for the concentration to drop below 0.3 mg/L.

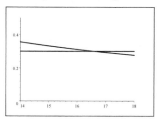

79. $P(v) = P_0 \left(\dfrac{s_0}{s_0 - v} \right) \quad \Rightarrow \quad P(v) = 440 \left(\dfrac{332}{332 - v} \right)$

If the speed of the train approaches the speed of sound, the pitch of the whistle becomes very loud. This would be experienced as a "sonic boom"— an effect seldom heard with trains.

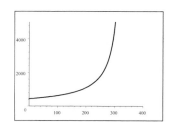

81. Vertical asymptote $x = 3$: $p(x) = \dfrac{1}{x - 3}$. Vertical asymptote $x = 3$ and horizontal asymptote $y = 2$: $r(x) = \dfrac{2x}{x - 3}$.

Vertical asymptotes $x = 1$ and $x = -1$, horizontal asymptote 0, and x-intercept 4: $q(x) = \dfrac{x - 4}{(x - 1)(x + 1)}$. Of course,

other answers are possible.

83. (a) $r(x) = \dfrac{3x^2 - 3x - 6}{x - 2} = \dfrac{3(x - 2)(x + 1)}{x - 2} = 3(x + 1)$, for $x \neq 2$. Therefore,

$r(x) = 3x + 3$, $x \neq 2$. Since $3(2) + 3 = 9$, the graph is the line $y = 3x + 3$ with

the point $(2, 9)$ removed.

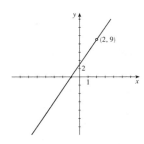

(b) $s(x) = \dfrac{x^2 + x - 20}{x + 5} = \dfrac{(x - 4)(x + 5)}{x + 5} = x - 4$, for $x \neq -5$. Therefore, $s(x) = x - 4$, $x \neq -5$. Since

$(-5) - 4 = -9$, the graph is the line $y = x - 4$ with the point $(-5, -9)$ removed.

$t(x) = \dfrac{2x^2 - x - 1}{x - 1} = \dfrac{(2x + 1)(x - 1)}{x - 1} = 2x + 1$, for $x \neq 1$. Therefore, $t(x) = 2x + 1$, $x \neq 1$. Since

$2(1) + 1 = 3$, the graph is the line $y = 2x + 1$ with the point $(1, 3)$ removed.

$u(x) = \dfrac{x - 2}{x^2 - 2x} = \dfrac{x - 2}{x(x - 2)} = \dfrac{1}{x}$, for $x \neq 2$. Therefore, $u(x) = \dfrac{1}{x}$, $x \neq 2$. When $x = 2$, $\dfrac{1}{x} = \dfrac{1}{2}$, so the graph is

the curve $y = \dfrac{1}{x}$ with the point $\left(2, \frac{1}{2}\right)$ removed.

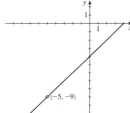

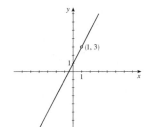

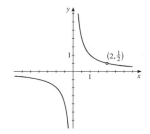

Chapter 4 Review

1. $P(x) = -x^3 + 64$

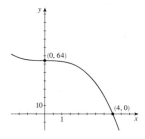

3. $P(x) = 2(x - 1)^4 - 32$

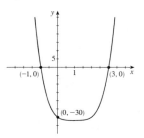

5. $P(x) = 32 + (x - 1)^5$

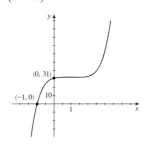

7. $P(x) = x^3 - 4x + 1$. x-intercepts: -2.1, 0.3, and 1.9.
y-intercept: 1. Local maximum is $(-1.2, 4.1)$. Local
minimum is $(1.2, -2.1)$. $y \to \infty$ as $x \to \infty$; $y \to -\infty$ as
$x \to -\infty$.

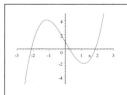

9. $P(x) = 3x^4 - 4x^3 - 10x - 1$. x-intercepts: -0.1 and 2.1. y-intercept: -1. Local maximum is $(1.4, -14.5)$. There is no local maximum. $y \to \infty$ as $x \to \pm\infty$.

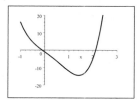

11. (a) Use the Pythagorean Theorem and solving for y^2 we have, $x^2 + y^2 = 10^2 \quad \Leftrightarrow \quad y^2 = 100 - x^2$. Substituting we get
$$S = 13.8x\left(100 - x^2\right) = 1380x - 13.8x^3.$$

(b) Domain is $[0, 10]$.

(c)

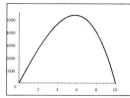

(d) The strongest beam has width 5.8 inches.

13. $\dfrac{x^2 - 3x + 5}{x - 2}$

$$
\begin{array}{r|rrr}
2 & 1 & -3 & 5 \\
  &   & 2 & -2 \\
\hline
  & 1 & -1 & 3 \\
\end{array}
$$

Using synthetic division, we see that $Q(x) = x - 1$ and $R(x) = 3$.

15. $\dfrac{x^3 - x^2 + 11x + 2}{x - 4}$

$$
\begin{array}{r|rrrr}
4 & 1 & -1 & 11 & 2 \\
  &   & 4 & 12 & 92 \\
\hline
  & 1 & 3 & 23 & 94 \\
\end{array}
$$

Using synthetic division, we see that $Q(x) = x^2 + 3x + 23$ and $R(x) = 94$.

17. $\dfrac{x^4 - 8x^2 + 2x + 7}{x + 5}$

$$
\begin{array}{r|rrrrr}
-5 & 1 & 0 & -8 & 2 & 7 \\
   &   & -5 & 25 & -85 & 415 \\
\hline
   & 1 & -5 & 17 & -83 & 422 \\
\end{array}
$$

Using synthetic division, we see that $Q(x) = x^3 - 5x^2 + 17x - 83$ and $R(x) = 422$.

19. $\dfrac{2x^3 + x^2 - 8x + 15}{x^2 + 2x - 1}$

$$
\require{enclose}
\begin{array}{r}
2x - 3 \\
x^2 + 2x - 1 \enclose{longdiv}{2x^3 + x^2 - 8x + 15} \\
\end{array}
$$

$$
\begin{array}{r}
2x^3 + 4x^2 - 2x \\
\hline
-3x^2 - 6x + 15 \\
-3x^2 - 6x + 3 \\
\hline
12
\end{array}
$$

Therefore, $Q(x) = 2x - 3$, and $R(x) = 12$.

21. $P(x) = 2x^3 - 9x^2 - 7x + 13$; find $P(5)$.

$$
\begin{array}{r|rrrr}
5 & 2 & -9 & -7 & 13 \\
  &   & 10 & 5 & -10 \\
\hline
  & 2 & 1 & -2 & 3 \\
\end{array}
$$

Therefore, $P(5) = 3$.

23. $\frac{1}{2}$ is a zero of $P(x) = 2x^4 + x^3 - 5x^2 + 10x - 4$ if $P\left(\frac{1}{2}\right) = 0$.

$$
\begin{array}{r|rrrrr}
\frac{1}{2} & 2 & 1 & -5 & 10 & -4 \\
  &   & 1 & 1 & -2 & 4 \\
\hline
  & 2 & 2 & -4 & 8 & 0 \\
\end{array}
$$

Since $P\left(\frac{1}{2}\right) = 0$, $\frac{1}{2}$ is a zero of the polynomial.

25. $P(x) = x^{500} + 6x^{201} - x^2 - 2x + 4$. The remainder from dividing $P(x)$ by $x - 1$ is $P(1) = (1)^{500} + 6(1)^{201} - (1)^2 - 2(1) + 4 = 8$.

27. (a) $P(x) = x^5 - 6x^3 - x^2 + 2x + 18$ has possible rational zeros $\pm 1, \pm 2, \pm 3, \pm 6, \pm 9, \pm 18$.

(b) Since $P(x)$ has 2 variations in sign, there are either 0 or 2 positive real zeros. Since $P(-x) = -x^5 + 6x^3 - x^2 - 2x + 18$ has 3 variations in sign, there are 1 or 3 negative real zeros.

29. (a) $P(x) = x^3 - 16x = x(x^2 - 16)$
$$= x(x-4)(x+4)$$

has zeros $-4, 0, 4$ (all of multiplicity 1).

(b)

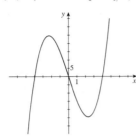

31. (a) $P(x) = x^4 + x^3 - 2x^2 = x^2(x^2 + x - 2)$
$$= x^2(x+2)(x-1)$$

The zeros are 0 (multiplicity 2), -2 (multiplicity 1), and 1 (multiplicity 1)

(b)

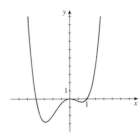

33. (a) $P(x) = x^4 - 2x^3 - 7x^2 + 8x + 12$. The possible rational zeros are $\pm 1, \pm 2, \pm 3, \pm 4, \pm 6, \pm 12$. P has 2 variations in sign, so it has either 2 or 0 positive real zeros.

$$
\begin{array}{r|rrrrr}
1 & 1 & -2 & -7 & 8 & 12 \\
  &   & 1 & -1 & -8 & 0 \\
\hline
  & 1 & -1 & -8 & 0 & 12
\end{array}
\qquad
\begin{array}{r|rrrrr}
2 & 1 & -2 & -7 & 8 & 12 \\
  &   & 2 & 0 & -14 & -12 \\
\hline
  & 2 & 0 & -7 & -6 & 0
\end{array}
\Rightarrow x = 2 \text{ is a root.}
$$

$P(x) = x^4 - 2x^3 - 7x^2 + 8x + 12 = (x-2)(x^3 - 7x - 6)$. Continuing:

$$
\begin{array}{r|rrrr}
2 & 1 & 0 & -6 & -6 \\
  &   & 2 & 4 & -4 \\
\hline
  & 1 & 2 & -2 & -10
\end{array}
\qquad
\begin{array}{r|rrrr}
3 & 1 & 0 & -7 & -6 \\
  &   & 3 & 9 & 6 \\
\hline
  & 1 & 3 & 2 & 0
\end{array}
$$

so $x = 3$ is a root and
$$P(x) = (x-2)(x-3)(x^2 + 3x + 2)$$
$$= (x-2)(x-3)(x+1)(x+2)$$
Therefore the real roots are $-2, -1, 2$, and 3 (all of multiplicity 1).

(b)

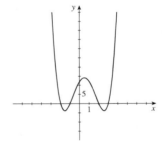

35. (a) $P(x) = 2x^4 + x^3 + 2x^2 - 3x - 2$. The possible rational roots are $\pm 1, \pm 2, \pm \frac{1}{2}$. P has one variation in sign, and hence 1 positive real root. $P(-x)$ has 3 variations in sign and hence either 3 or 1 negative real roots.

$$
\begin{array}{r|rrrrr}
1 & 2 & 1 & 2 & -3 & -2 \\
  &   & 2 & 3 & 5 & 2 \\
\hline
  & 2 & 3 & 5 & 2 & 0
\end{array}
\Rightarrow x = 1 \text{ is a zero.}
$$

$P(x) = 2x^4 + x^3 + 2x^2 - 3x - 2 = (x-1)(2x^3 + 3x^2 + 5x + 2)$. Continuing:

$$
\begin{array}{r|rrrr}
-1 & 2 & 3 & 5 & 2 \\
   &   & -2 & -1 & -4 \\
\hline
   & 2 & 1 & 4 & -2
\end{array}
\qquad
\begin{array}{r|rrrr}
-2 & 2 & 3 & 5 & 2 \\
   &   & -4 & 2 & -14 \\
\hline
   & 2 & -1 & 7 & -12
\end{array}
$$

$$
\begin{array}{r|rrrr}
-\frac{1}{2} & 2 & 3 & 5 & 2 \\
   &   & -1 & -1 & -2 \\
\hline
   & 2 & 2 & 4 & 0
\end{array}
\Rightarrow x = -\frac{1}{2} \text{ is a zero.}
$$

$P(x) = (x-1)\left(x+\frac{1}{2}\right)(2x^2 + 2x + 4)$. The quadratic is irreducible, so the real zeros are 1 and $-\frac{1}{2}$ (each of multiplicity 1).

(b)

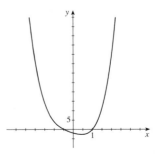

37. Since the zeros are $-\frac{1}{2}$, 2, and 3, a factorization is

$$
\begin{aligned}
P(x) &= C\left(x + \tfrac{1}{2}\right)(x - 2)(x - 3) = \tfrac{1}{2}C(2x + 1)\left(x^2 - 5x + 6\right) \\
&= \tfrac{1}{2}C\left(2x^3 - 10x^2 + 12x + x^2 - 5x + 6\right) = \tfrac{1}{2}C\left(2x^3 - 9x^2 + 7x + 6\right)
\end{aligned}
$$

Since the constant coefficient is 12, $\tfrac{1}{2}C(6) = 12 \quad \Leftrightarrow \quad C = 4$, and so the polynomial is $P(x) = 4x^3 - 18x^2 + 14x + 12$.

39. No, there is no polynomial of degree 4 with integer coefficients that has zeros i, $2i$, $3i$ and $4i$. Since the imaginary zeros of polynomial equations with real coefficients come in complex conjugate pairs, there would have to be 8 zeros, which is impossible for a polynomial of degree 4.

41. $P(x) = x^3 - 3x^2 - 13x + 15$ has possible rational zeros $\pm 1, \pm 3, \pm 5, \pm 15$.

$$
\begin{array}{r|rrrr}
1 & 1 & -3 & -13 & 15 \\
  &   & 1 & -2 & -15 \\
\hline
  & 1 & -2 & -15 & 0
\end{array} \quad \Rightarrow x = 1 \text{ is a zero.}
$$

So $P(x) = x^3 - 3x^2 - 13x + 15 = (x - 1)\left(x^2 - 2x - 15\right) = (x - 1)(x - 5)(x + 3)$. Therefore, the zeros are -3, 1, and 5.

43. $P(x) = x^4 + 6x^3 + 17x^2 + 28x + 20$ has possible rational zeros $\pm 1, \pm 2, \pm 4, \pm 5, \pm 10, \pm 20$. Since all of the coefficients are positive, there are no positive real zeros.

$$
\begin{array}{r|rrrrr}
-1 & 1 & 6 & 17 & 28 & 20 \\
   &   & -1 & -5 & -12 & -16 \\
\hline
   & 1 & 5 & 12 & 16 & 4
\end{array}
\qquad
\begin{array}{r|rrrrr}
-2 & 1 & 6 & 17 & 28 & 20 \\
   &   & -2 & -8 & -18 & -20 \\
\hline
   & 1 & 4 & 9 & 10 & 0
\end{array} \quad \Rightarrow x = -2 \text{ is a zero.}
$$

$P(x) = x^4 + 6x^3 + 17x^2 + 28x + 20 = (x + 2)\left(x^3 + 4x^2 + 9x + 10\right)$. Continuing with the quotient, we have

$$
\begin{array}{r|rrrr}
-2 & 1 & 4 & 9 & 10 \\
   &   & -2 & -4 & -10 \\
\hline
   & 1 & 2 & 5 & 0
\end{array} \quad \Rightarrow x = -2 \text{ is a zero.}
$$

Thus $P(x) = x^4 + 6x^3 + 17x^2 + 28x + 20 = (x + 2)^2\left(x^2 + 2x + 5\right)$. Now $x^2 + 2x + 5 = 0$ when $x = \frac{-2 \pm \sqrt{4 - 4(5)(1)}}{2} = \frac{-2 \pm 4i}{2} = -1 \pm 2i$. Thus, the zeros are -2 (multiplicity 2) and $-1 \pm 2i$.

45. $P(x) = x^5 - 3x^4 - x^3 + 11x^2 - 12x + 4$ has possible rational zeros $\pm 1, \pm 2, \pm 4$.

$$
\begin{array}{r|rrrrrr}
1 & 1 & -3 & -1 & 11 & -12 & 4 \\
  &   & 1 & -2 & -3 & 8   & -4 \\
\hline
  & 1 & -2 & -3 & 8 & -4 & 0
\end{array}
\quad \Rightarrow x = 1 \text{ is a zero.}
$$

$P(x) = x^5 - 3x^4 - x^3 + 11x^2 - 12x + 4 = (x-1)(x^4 - 2x^3 - 3x^2 + 8x - 4)$. Continuing with the quotient, we have

$$
\begin{array}{r|rrrrr}
1 & 1 & -2 & -3 & 8 & -4 \\
  &   & 1 & -1 & -4 & 4 \\
\hline
  & 1 & -1 & -4 & 4 & 0
\end{array}
\quad \Rightarrow x = 1 \text{ is a zero.}
$$

$$
\begin{aligned}
x^5 - 3x^4 - x^3 + 11x^2 - 12x + 4 &= (x-1)^2 (x^3 - x^2 - 4x + 4) = (x-1)^3 (x^2 - 4) \\
&= (x-1)^3 (x-2)(x+2)
\end{aligned}
$$

Therefore, the zeros are 1 (multiplicity 3), -2, and 2.

47. $P(x) = x^6 - 64 = (x^3 - 8)(x^3 + 8) = (x-2)(x^2 + 2x + 4)(x+2)(x^2 - 2x + 4)$. Now using the quadratic formula to find the zeros of $x^2 + 2x + 4$, we have

$x = \dfrac{-2 \pm \sqrt{4 - 4(4)(1)}}{2} = \dfrac{-2 \pm 2\sqrt{3}i}{2} = -1 \pm \sqrt{3}i$, and using the quadratic formula to find the zeros of $x^2 - 2x + 4$, we have

$x = \dfrac{2 \pm \sqrt{4 - 4(4)(1)}}{2} = \dfrac{2 \pm 2\sqrt{3}i}{2} = 1 \pm \sqrt{3}i$. Therefore, the zeros are $2, -2, 1 \pm \sqrt{3}i$, and $-1 \pm \sqrt{3}i$.

49. $P(x) = 6x^4 - 18x^3 + 6x^2 - 30x + 36 = 6(x^4 - 3x^3 + x^2 - 5x + 6)$ has possible rational zeros $\pm 1, \pm 2, \pm 3, \pm 6$.

$$
\begin{array}{r|rrrrr}
1 & 6 & -18 & 6 & -30 & 36 \\
  &   & 6   & -12 & -6 & -36 \\
\hline
  & 6 & -12 & -6 & -36 & 0
\end{array}
\quad \Rightarrow x = 1 \text{ is a zero.}
$$

So $P(x) = 6x^4 - 18x^3 + 6x^2 - 30x + 36 = (x-1)(6x^3 - 12x^2 - 6x - 36) = 6(x-1)(x^3 - 2x^2 - x - 6)$. Continuing with the quotient we have

$$
\begin{array}{r|rrrr}
1 & 1 & -2 & -1 & -6 \\
  &   & 1 & -1 & -2 \\
\hline
  & 1 & -1 & -2 & -8
\end{array}
\qquad
\begin{array}{r|rrrr}
2 & 1 & -2 & -1 & -6 \\
  &   & 2 & 0 & -2 \\
\hline
  & 1 & 0 & -1 & -8
\end{array}
\qquad
\begin{array}{r|rrrr}
3 & 1 & -2 & -1 & -6 \\
  &   & 3 & 3 & 6 \\
\hline
  & 1 & 1 & 2 & 0
\end{array}
\quad \Rightarrow x = 3 \text{ is a zero.}
$$

So $P(x) = 6x^4 - 18x^3 + 6x^2 - 30x + 36 = 6(x-1)(x-3)(x^2 + x + 2)$. Now $x^2 + x + 2 = 0$ when

$x = \dfrac{-1 \pm \sqrt{1 - 4(1)(2)}}{2} = \dfrac{-1 \pm \sqrt{7}i}{2}$, and so the zeros are $1, 3$, and $\dfrac{-1 \pm \sqrt{7}i}{2}$.

51. $2x^2 = 5x + 3 \quad \Leftrightarrow \quad 2x^2 - 5x - 3 = 0$. The solutions are $x = -0.5, 3$.

53. $x^4 - 3x^3 - 3x^2 - 9x - 2 = 0$ has solutions $x \approx -0.24$, 4.24.

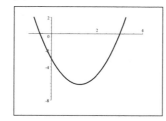

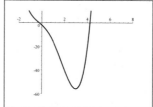

55. $r(x) = \dfrac{3x - 12}{x + 1}$. When $x = 0$, we have $r(0) = \dfrac{-12}{1} = -12$, so the y-intercept

is -12. Since $y = 0$, when $3x - 12 = 0 \quad\Leftrightarrow\quad x = 4$, the x-intercept is 4. The

vertical asymptote is $x = -1$. Because the degree of the denominator and

numerator are the same, the horizontal asymptote is $y = \frac{3}{1} = 3$.

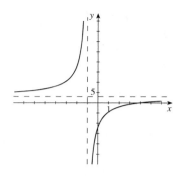

57. $r(x) = \dfrac{x - 2}{x^2 - 2x - 8} = \dfrac{x - 2}{(x + 2)(x - 4)}$. When $x = 0$, we have

$r(0) = \frac{-2}{-8} = \frac{1}{4}$, so the y-intercept is $\frac{1}{4}$. When $y = 0$, we have $x - 2 = 0 \quad\Leftrightarrow\quad$

$x = 2$, so the x-intercept is 2. There are vertical asymptotes at $x = -2$ and $x = 4$.

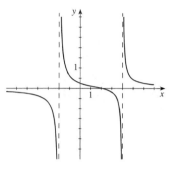

59. $r(x) = \dfrac{x^2 - 9}{2x^2 + 1} = \dfrac{(x + 3)(x - 3)}{2x^2 + 1}$. When $x = 0$, we have $r(0) = \frac{-9}{1}$, so the

y-intercept is -9. When $y = 0$, we have $x^2 - 9 = 0 \quad\Leftrightarrow\quad x = \pm 3$ so the

x-intercepts are -3 and 3. Since $2x^2 + 1 > 0$, the denominator is never zero so

there are no vertical asymptotes. The horizontal asymptote is at $y = \frac{1}{2}$ because the

degree of the denominator and numerator are the same.

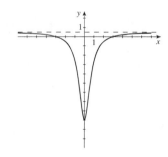

61. $r(x) = \dfrac{x - 3}{2x + 6}$. From the graph we see that the x-intercept is 3, the y-intercept is

-0.5, there is a vertical asymptote at $x = -3$ and a horizontal asymptote at

$y = 0.5$, and there is no local extremum.

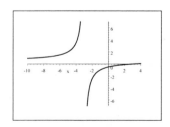

63. $r(x) = \dfrac{x^3 + 8}{x^2 - x - 2}$. From the graph we see that the x-intercept is -2, the

y-intercept is -4, there are vertical asymptotes at $x = -1$ and $x = 2$ and a

horizontal asymptote at $y = 0.5$. r has a local maximum of $(0.425, -3.599)$ and a

local minimum of $(4.216, 7.175)$. By using long division, we see that

$$f(x) = x + 1 + \dfrac{10 - x}{x^2 - x - 2}, \text{ so } f \text{ has a slant asymptote of } y = x + 1.$$

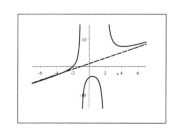

65. The graphs of $y = x^4 + x^2 + 24x$ and $y = 6x^3 + 20$ intersect when $x^4 + x^2 + 24x = 6x^3 + 20$ $\Leftrightarrow$ $x^4 - 6x^3 + x^2 + 24x - 20 = 0$. The possible rational zeros are $\pm 1, \pm 2, \pm 4, \pm 5, \pm 10, \pm 20$.

$$
\begin{array}{r|rrrrr}
1 & 1 & -6 & 1 & 24 & -20 \\
  &   & 1 & -5 & -4 & 20 \\
\hline
  & 1 & -5 & -4 & 20 & 0 \quad \Rightarrow x = 1 \text{ is a zero.}
\end{array}
$$

So $x^4 - 6x^3 + x^2 + 24x - 20 = (x-1)\left(x^3 - 5x^2 - 4x + 20\right) = 0$. Continuing with the quotient:

$$
\begin{array}{r|rrrr}
1 & 1 & -5 & -4 & 20 \\
  &   & 1 & -4 & -8 \\
\hline
  & 1 & -4 & -8 & 12
\end{array}
\qquad
\begin{array}{r|rrrr}
2 & 1 & -5 & -4 & 20 \\
  &   & 2 & -6 & -20 \\
\hline
  & 1 & -3 & -10 & 0 \quad \Rightarrow x = 2 \text{ is a zero.}
\end{array}
$$

So

$$
\begin{aligned}
x^4 - 6x^3 + x^2 + 24x - 20 &= (x-1)(x-2)\left(x^2 - 3x - 10\right) \\
&= (x-1)(x-2)(x-5)(x+2) = 0
\end{aligned}
$$

Hence, the points of intersection are $(1, 26)$, $(2, 68)$, $(5, 770)$, and $(-2, -28)$.

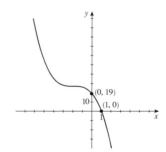

Chapter 4 Test

1. $f(x) = -(x+2)^3 + 27$ has y-intercept $y = -2^3 + 27 = 19$ and x-intercept where $-(x+2)^3 = -27$ $\Leftrightarrow$ $x = 1$.

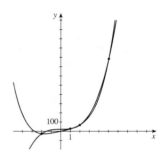

2. (a)

$$
\begin{array}{r|rrrrr}
2 & 1 & 0 & -4 & 2 & 5 \\
  &   & 2 & 4 & 0 & 4 \\
\hline
  & 1 & 2 & 0 & 2 & 9
\end{array}
$$

Therefore, the quotient is
$Q(x) = x^3 + 2x^2 + 2$, and the remainder is
$R(x) = 9$.

(b)

$$
\begin{array}{r}
x^3 + 2x^2 \qquad\quad + \tfrac{1}{2} \\
2x^2 - 1 \overline{\big)\ 2x^5 + 4x^4 - x^3 - x^2 + 0x + 7} \\
\underline{2x^5 \qquad\quad - x^3} \\
4x^4 \qquad\quad - x^2 \\
\underline{4x^4 \qquad\quad - 2x^2} \\
x^2 \qquad + 7 \\
\underline{x^2 \qquad - \tfrac{1}{2}} \\
\tfrac{15}{2}
\end{array}
$$

Therefore, the quotient is $Q(x) = x^3 + 2x^2 + \tfrac{1}{2}$ and the remainder is $R(x) = \tfrac{15}{2}$.

3. (a) Possible rational zeros are: $\pm 1, \pm 3, \pm\frac{1}{2}, \pm\frac{3}{2}$.

(b)

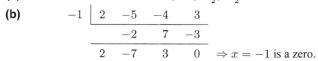

$$-1 \;\big|\; \begin{array}{cccc} 2 & -5 & -4 & 3 \\ & -2 & 7 & -3 \\ \hline 2 & -7 & 3 & 0 \end{array} \quad \Rightarrow x = -1 \text{ is a zero.}$$

$$\begin{aligned} P(x) &= (x+1)\left(2x^2 - 7x + 3\right) = (x+1)(2x-1)(x-3) \\ &= 2(x+1)\left(x - \tfrac{1}{2}\right)(x-3) \end{aligned}$$

(c) The zeros of P are $x = -1, 3, \frac{1}{2}$.

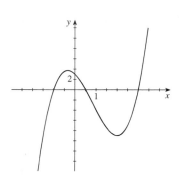

4. $P(x) = x^3 - x^2 - 4x - 6$. Possible rational zeros are: $\pm 1, \pm 2, \pm 3, \pm 6$.

$$1 \;\big|\; \begin{array}{cccc} 1 & -1 & -4 & -6 \\ & 1 & 0 & -4 \\ \hline 1 & 0 & -4 & -10 \end{array} \qquad 2 \;\big|\; \begin{array}{cccc} 1 & -1 & -4 & -6 \\ & 2 & 2 & -4 \\ \hline 1 & 1 & -2 & -10 \end{array} \qquad 3 \;\big|\; \begin{array}{cccc} 1 & -1 & -4 & -6 \\ & 3 & 6 & 6 \\ \hline 1 & 2 & 2 & 0 \end{array} \Rightarrow x = 3 \text{ is a zero.}$$

So $P(x) = (x-3)\left(x^2 + 2x + 2\right)$. Using the quadratic formula on the second factor, we have

$$x = \frac{-2 \pm \sqrt{2^2 - 4(1)(2)}}{2(1)} = \frac{-2 \pm \sqrt{-4}}{2} = \frac{-2 \pm 2\sqrt{-1}}{2} = -1 \pm i.$$ So zeros of $P(x)$ are 3, $-1 - i$, and $-1 + i$.

5. $P(x) = x^4 - 2x^3 + 5x^2 - 8x + 4$. The possible rational zeros of P are: $\pm 1, \pm 2$, and ± 4. Since there are four changes in sign, P has 4, 2, or 0 positive real zeros.

$$1 \;\big|\; \begin{array}{ccccc} 1 & -2 & 5 & -8 & 4 \\ & 1 & -1 & 4 & -4 \\ \hline 1 & -1 & 4 & -4 & 0 \end{array}$$

So $P(x) = (x-1)\left(x^3 - x^2 + 4x - 4\right)$. Factoring the second factor by grouping, we have

$$P(x) = (x-1)\left[x^2(x-1) + 4(x-1)\right] = (x-1)\left(x^2 + 4\right)(x-1) = (x-1)^2 (x-2i)(x+2i).$$

6. Since $3i$ is a zero of $P(x)$, $-3i$ is also a zero of $P(x)$. And since -1 is a zero of multiplicity 2,

$$P(x) = (x+1)^2 (x-3i)(x+3i) = \left(x^2 + 2x + 1\right)\left(x^2 + 9\right) = x^4 + 2x^3 + 10x^2 + 18x + 9.$$

7. $P(x) = 2x^4 - 7x^3 + x^2 - 18x + 3$.

(a) Since $P(x)$ has 4 variations in sign, $P(x)$ can have 4, 2, or 0 positive real zeros. Since
$P(-x) = 2x^4 + 7x^3 + x^2 + 18x + 3$ has no variations in sign, there are no negative real zeros.

(b)

$$4 \;\big|\; \begin{array}{ccccc} 2 & -7 & 1 & -18 & 3 \\ & 8 & 4 & 20 & 8 \\ \hline 2 & 1 & 5 & 2 & 11 \end{array}$$

Since the last row contains no negative entry, 4 is an upper bound for the real zeros of $P(x)$.

$$-1 \;\big|\; \begin{array}{ccccc} 2 & -7 & 1 & -18 & 3 \\ & -1 & 9 & -10 & 28 \\ \hline 2 & -9 & 10 & -28 & 31 \end{array}$$

Since the last row alternates in sign, -1 is a lower bound for the real zeros of $P(x)$.

(c) Using the upper and lower limit from part (b), we graph $P(x)$ in the viewing rectangle $[-1, 4]$ by $[-1, 1]$. The two real zeros are 0.17 and 3.93.

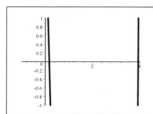

(d) Local minimum $(2.8, -70.3)$.

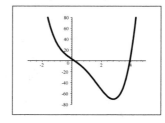

8. $r(x) = \dfrac{2x-1}{x^2 - x - 2}$, $s(x) = \dfrac{x^3 + 27}{x^2 + 4}$, $t(x) = \dfrac{x^3 - 9x}{x+2}$ and $u(x) = \dfrac{x^2 + x - 6}{x^2 - 25}$.

(a) $r(x)$ has the horizontal asymptote $y = 0$ because the degree of the denominator is greater than the degree of the numerator. $u(x)$ has the horizontal asymptote $y = \frac{1}{1} = 1$ because the degree of the numerator and the denominator are the same.

(b) The degree of the numerator of $s(x)$ is one more than the degree of the denominator, so s has a slant asymptote.

(c) The denominator of $s(x)$ is never 0, so s has no vertical asymptote.

(d) $u(x) = \dfrac{x^2 + x - 6}{x^2 - 25} = \dfrac{(x+3)(x-2)}{(x-5)(x+5)}$. When $x = 0$, we have $u(x) = \frac{-6}{-25} = \frac{6}{25}$, so the y-intercept is $y = \frac{6}{25}$. When $y = 0$, we have $x = -3$ or $x = 2$, so the x-intercepts are -3 and 2. The vertical asymptotes are $x = -5$ and $x = 5$. The horizontal asymptote occurs at $y = \frac{1}{1} = 1$ because the degree of the denominator and numerator are the same.

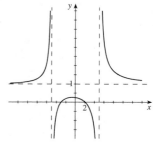

(e)

$$
\begin{array}{r}
x^2 - 2x - 5 \\
x + 2 \enclose{longdiv}{x^3 + 0x^2 - 9x + 0} \\
\underline{x^3 + 2x^2} \\
-2x^2 - 9x \\
\underline{-2x^2 - 4x} \\
-5x + 0 \\
\underline{-5x - 10} \\
-10
\end{array}
$$

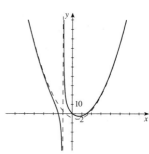

Thus $P(x) = x^2 - 2x - 5$ and $t(x) = \dfrac{x^3 - 9x}{x+2}$ have the same end behavior.

Focus on Modeling: Fitting Polynomial Curves to Data

1. (a) Using a graphing calculator, we obtain the quadratic polynomial $y = -0.275428x^2 + 19.7485x - 273.5523$ (where miles are measured in thousands).

(c) Moving the cursor along the path of the polynomial, we find that 35.85 lb/in^2 gives the longest tire life.

(b)

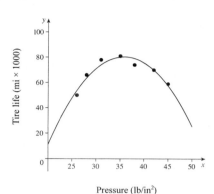

3. (a) Using a graphing calculator, we obtain the cubic polynomial $y = 0.00203709x^3 - 0.104522x^2 + 1.966206x + 1.45576$.

(c) Moving the cursor along the path of the polynomial, we find that the subjects could name about 43 vegetables in 40 seconds.

(d) Moving the cursor along the path of the polynomial, we find that the subjects could name 5 vegetables in about 2.0 seconds.

(b)

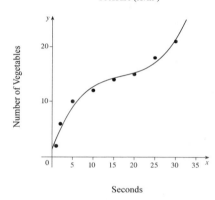

5. (a)

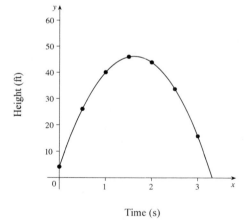

Time (s)

A quadratic model seems appropriate.

(b) Using a graphing calculator, we obtain the quadratic polynomial $y = -16.0x^2 + 51.8429x + 4.20714$.

(c) Moving the cursor along the path of the polynomial, we find that the ball is 20 ft. above the ground 0.3 seconds and 2.9 seconds after it is thrown upward.

(d) Again, moving the cursor along the path of the polynomial, we find that the maximum height is 46.2 ft.

5 Exponential and Logarithmic Functions

5.1 Exponential Functions

1. $f(x) = 4^x$; $f(0.5) = 2$, $f(\sqrt{2}) = 7.103$, $f(\pi) = 77.880$, $f\left(\frac{1}{3}\right) = 1.587$

3. $g(x) = \left(\frac{2}{3}\right)^{x-1}$; $g(1.3) = 0.885$, $g(\sqrt{5}) = 0.606$, $g(2\pi) = 0.117$, $g\left(-\frac{1}{2}\right) = 1.837$

5. $f(x) = 2^x$

x	y
-4	$\frac{1}{16}$
-2	$\frac{1}{4}$
0	1
2	4
4	16

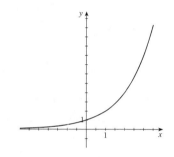

7. $f(x) = \left(\frac{1}{3}\right)^x$

x	y
-2	9
-1	3
0	1
1	$\frac{1}{3}$
2	$\frac{1}{9}$

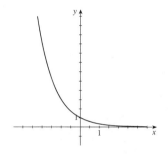

9. $f(x) = 3e^x$

x	y
-2	0.406
-1	1.104
0	3
0.5	4.946
1	8.155
1.5	13.445
2	22.167

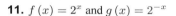

11. $f(x) = 2^x$ and $g(x) = 2^{-x}$

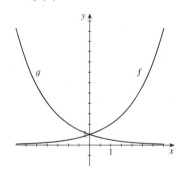

13. $f(x) = 4^x$ and $g(x) = 7^x$.

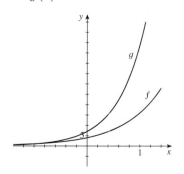

15. From the graph, $f(2) = a^2 = 9$, so $a = 3$. Thus $f(x) = 3^x$.

17. From the graph, $f(2) = a^2 = \frac{1}{16}$, so $a = \frac{1}{4}$. Thus $f(x) = \left(\frac{1}{4}\right)^x$.

19. III   **21.** I   **23.** II

25. The graph of $f(x) = -3^x$ is obtained by reflecting the graph of $y = 3^x$ about the x-axis. Domain: $(-\infty, \infty)$. Range: $(-\infty, 0)$. Asymptote: $y = 0$.

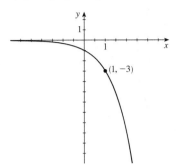

27. $g(x) = 2^x - 3$. The graph of g is obtained by shifting the graph of $y = 2^x$ downward 3 units. Domain: $(-\infty, \infty)$. Range: $(-3, \infty)$. Asymptote: $y = -3$.

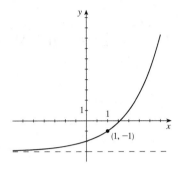

29. $h(x) = 4 + \left(\frac{1}{2}\right)^x$. The graph of h is obtained by shifting the graph of $y = \left(\frac{1}{2}\right)^x$ upward 4 units. Domain: $(-\infty, \infty)$. Range: $(4, \infty)$. Asymptote: $y = 4$.

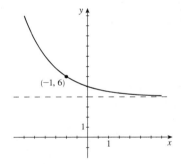

31. $f(x) = 10^{x+3}$. The graph of f is obtained by shifting the graph of $y = 10^x$ to the left 3 units. Domain: $(-\infty, \infty)$. Range: $(0, \infty)$. Asymptote: $y = 0$.

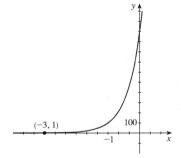

33. $y = -e^x$. The graph of $y = -e^x$ is obtained from the graph of $y = e^x$ by reflecting it about the x-axis. Domain: $(-\infty, \infty)$. Range: $(-\infty, 0)$. Asymptote: $y = 0$.

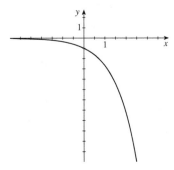

35. $y = e^{-x} - 1$. The graph of $y = e^{-x} - 1$ is obtained from the graph of $y = e^x$ by reflecting it about the y-axis then shifting downward 1 unit. Domain: $(-\infty, \infty)$. Range: $(-1, \infty)$. Asymptote: $y = -1$.

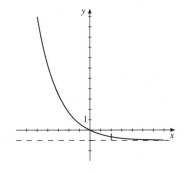

37. $y = e^{x-2}$. The graph of $y = e^{x-2}$ is obtained from the graph of $y = e^x$ by shifting it to the right 2 units. Domain: $(-\infty, \infty)$. Range: $(0, \infty)$. Asymptote: $y = 0$.

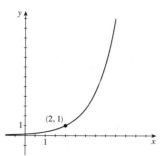

39. Using the points $(0, 3)$ and $(2, 12)$, we have
$$f(0) = Ca^0 = 3 \quad \Leftrightarrow \quad C = 3. \text{ We also have}$$
$$f(2) = 3a^2 = 12 \quad \Leftrightarrow \quad a^2 = 4 \quad \Leftrightarrow \quad a = 2 \text{ (recall}$$
that for an exponential function $f(x) = a^x$ we require $a > 0$). Thus $f(x) = 3 \cdot 2^x$.

41. (a)

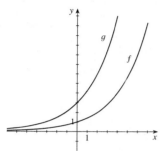

(b) Since $g(x) = 3(2^x) = 3f(x)$ and $f(x) > 0$, the height of the graph of $g(x)$ is always three times the height of the graph of $f(x) = 2^x$, so the graph of g is steeper than the graph of f.

43. $f(x) = 10^x$, so $\dfrac{f(x+h) - f(x)}{h} = \dfrac{10^{x+h} - 10^x}{h} = \dfrac{10^x \cdot 10^h - 10^x}{h} = 10^x \dfrac{10^h - 1}{h}$.

45.

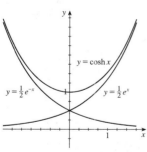

47. $\cosh(-x) = \dfrac{e^{-x} + e^{-(-x)}}{2} = \dfrac{e^{-x} + e^x}{2} = \dfrac{e^x + e^{-x}}{2}$
$$= \cosh x$$

49. $(\cosh x)^2 - (\sinh x)^2 = \left(\dfrac{e^x + e^{-x}}{2}\right)^2 - \left(\dfrac{e^x - e^{-x}}{2}\right)^2 = \frac{1}{4}\left(e^{2x} + 2 + e^{-2x}\right) - \frac{1}{4}\left(e^{2x} - 2 + e^{-2x}\right) = \frac{2}{4} + \frac{2}{4} = 1$

51. (a) From the graphs below, we see that the graph of f ultimately increases much more quickly than the graph of g.

(i) $[0, 5]$ by $[0, 20]$

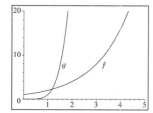

(ii) $[0, 25]$ by $[0, 10^7]$

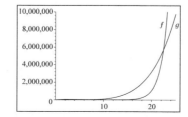

(iii) $[0, 50]$ by $[0, 10^8]$

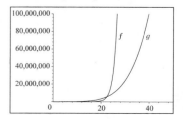

(b) From the graphs in parts (a)(i) and (a)(ii), we see that the approximate solutions are $x \approx 1.2$ and $x \approx 22.4$.

53.

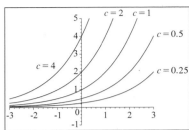

The larger the value of c, the more rapidly the graph of $f(x) = c2^x$ increases. Also notice that the graphs are just shifted horizontally 1 unit. This is because of our choice of c; each c in this exercise is of the form 2^k. So $f(x) = 2^k \cdot 2^x = 2^{x+k}$.

55.

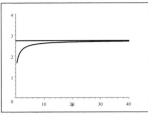

Note from the graph that $y = [1 + (1/x)]^x$ approaches e as x get large.

57. (a)

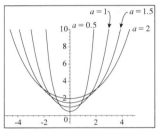

(b) As a increases the curve $y = \dfrac{a}{2}\left(e^{x/a} + e^{-x/a}\right)$ flattens out and the y intercept increases.

59. $y = \dfrac{e^x}{x}$ has vertical asymptote $x = 0$ and horizontal asymptote $y = 0$. As $x \to -\infty$, $y \to 0$, and as $x \to \infty$, $y \to \infty$.

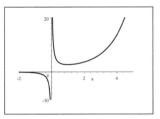

61. $g(x) = e^x + e^{-3x}$. The graph of $g(x)$ is shown in the viewing rectangle $[-4, 4]$ by $[0, 20]$. From the graph, we see that there is a local minimum of approximately 1.75 when $x \approx 0.27$.

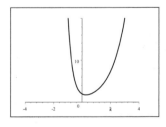

63. $y = xe^{-x}$

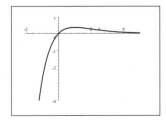

(a) From the graph, we see that the function $f(x) = xe^x$ is increasing on $(-\infty, 1]$ and decreasing on $[1, \infty)$.

(b) From the graph, we see that the range is approximately $(-\infty, 0.37)$.

65. $m(t) = 13e^{-0.015t}$

(a) $m(0) = 13$ kg.

(b) $m(45) = 13e^{-0.015(45)} = 13e^{-0.675} = 6.619$ kg. Thus the mass of the radioactive substance after 45 days is about 6.6 kg.

67. $v(t) = 80\left(1 - e^{-0.2t}\right)$

(a) $v(0) = 80\left(1 - e^{0}\right) = 80\left(1 - 1\right) = 0.$

(b) $v(5) = 80\left(1 - e^{-0.2(5)}\right) \approx 80(0.632) = 50.57$ ft/s. So the velocity after 5 s is about 50.6 ft/s.

$v(10) = 80\left(1 - e^{-0.2(10)}\right) \approx 80(0.865) = 69.2$ ft/s. So the velocity after 10 s is about 69.2 ft/s.

(d) The terminal velocity is 80 ft/s.

(c)

69. $P(t) = \dfrac{1200}{1 + 11e^{-0.2t}}$

(a) $P(0) = \dfrac{1200}{1 + 11e^{-0.2(0)}} = \dfrac{1200}{1 + 11} = 100.$

(b) $P(10) = \dfrac{1200}{1 + 11e^{-0.2(10)}} \approx 482.$ $P(20) = \dfrac{1200}{1 + 11e^{-0.2(20)}} \approx 999.$ $P(30) = \dfrac{1200}{1 + 11e^{-0.2(30)}} \approx 1168.$

(c) As $t \to \infty$ we have $e^{-0.2t} \to 0$, so $P(t) \to \dfrac{1200}{1+0} = 1200.$ The graph shown confirms this.

71. $D(t) = \dfrac{5.4}{1 + 2.9e^{-0.01t}}.$ So $D(20) = \dfrac{5.4}{1 + 2.9e^{-0.01(20)}} \approx 1.600$ ft.

73. Using the formula $A(t) = P(1+i)^k$ with

$P = 5000$, $i = 4\%$ per year $= \dfrac{0.04}{12}$ per month,

and $k = 12 \cdot$ number of years, we fill in the table.

Time (years)	Amount
1	$5203.71
2	$5415.71
3	$5636.36
4	$5865.99
5	$6104.98
6	$6353.71

75. $P = 10{,}000$, $r = 0.10$, and $n = 2$. So $A(t) = 10{,}000\left(1 + \frac{0.10}{2}\right)^{2t} = 10{,}000 \cdot 1.05^{2t}.$

(a) $A(5) = 10{,}000 \cdot 1.05^{10} \approx 16{,}288.95$, and so the value of the investment is $16,288.95.

(b) $A(10) = 10{,}000 \cdot 1.05^{20} \approx 26{,}532.98$, and so the value of the investment is $26,532.98.

(c) $A(15) = 10{,}000 \cdot 1.05^{30} \approx 43{,}219.42$, and so the value of the investment is $43,219.42.

77. $P = 3000$ and $r = 0.09$. Then we have $A(t) = 3000\left(1 + \dfrac{0.09}{n}\right)^{nt}$, and so $A(5) = 3000\left(1 + \dfrac{0.09}{n}\right)^{5n}.$

(a) If $n = 1$, $A(5) = 3000\left(1 + \frac{0.09}{1}\right)^{5} = 3000 \cdot 1.09^{5} \approx \$4{,}615.87.$

(b) If $n = 2$, $A(5) = 3000\left(1 + \frac{0.09}{2}\right)^{10} = 3000 \cdot 1.045^{10} \approx \$4{,}658.91.$

(c) If $n = 12$, $A(5) = 3000\left(1 + \frac{0.09}{12}\right)^{60} = 3000 \cdot 1.0075^{60} \approx \$4{,}697.04.$

(d) If $n = 52$, $A(5) = 3000\left(1 + \frac{0.09}{52}\right)^{260} \approx \$4{,}703.11.$

(e) If $n = 365$, $A(5) = 3000\left(1 + \frac{0.09}{365}\right)^{1825} \approx \$4{,}704.68.$

(f) If $n = 24 \cdot 365 = 8760$, $A(5) = 3000\left(1 + \frac{0.09}{8760}\right)^{43800} \approx \$4{,}704.93.$

(g) If interest is compounded continuously, $A(5) = 3000 \cdot e^{0.45} \approx \$4{,}704.94.$

79. We find the effective rate with $P = 1$ and $t = 1$. So $A = \left(1 + \dfrac{r}{n}\right)^n$

   (i) $n = 2$, $r = 0.085$; $A(2) = \left(1 + \dfrac{0.085}{2}\right)^2 = (1.0425)^2 \approx 1.0868$.

   (ii) $n = 4$, $r = 0.0825$; $A(4) = \left(1 + \dfrac{0.0825}{4}\right)^4 = (1.020625)^4 \approx 1.0851$.

   (iii) Continuous compounding: $r = 0.08$; $A(1) = e^{0.08} \approx 1.0833$.

   Since (i) is larger than the others, the best investment is the one at 8.5% compounded semiannually.

81. (a) We must solve for P in the equation $10000 = P\left(1 + \dfrac{0.09}{2}\right)^{2(3)} = P(1.045)^6 \quad \Leftrightarrow \quad 10000 = 1.3023P \quad \Leftrightarrow$
   $P = 7678.96$. Thus the present value is $\$7{,}678.96$.

   (b) We must solve for P in the equation $100000 = P\left(1 + \dfrac{0.08}{12}\right)^{12(5)} = P(1.00667)^{60} \quad \Leftrightarrow \quad 100000 = 1.4898P$
   $\Leftrightarrow \quad P = \$67{,}121.04$.

83. (a) In this case the payment is $\$1$ million.

   (b) In this case the total pay is $2 + 2^2 + 2^3 + \cdots + 2^{30} > 2^{30}$ cents $= \$10{,}737{,}418.24$. Since this is much more than method (a), method (b) is more profitable.

5.2   Logarithmic Functions

1.

Logarithmic form	Exponential form
$\log_8 8 = 1$	$8^1 = 8$
$\log_8 64 = 2$	$8^2 = 64$
$\log_8 4 = \frac{2}{3}$	$8^{2/3} = 4$
$\log_8 512 = 3$	$8^3 = 512$
$\log_8 \frac{1}{8} = -1$	$8^{-1} = \frac{1}{8}$
$\log_8 \frac{1}{64} = -2$	$8^{-2} = \frac{1}{64}$

3. (a) $5^2 = 25$
   (b) $5^0 = 1$

5. (a) $8^{1/3} = 2$
   (b) $2^{-3} = \frac{1}{8}$

7. (a) $e^x = 5$
   (b) $e^5 = y$

9. (a) $\log_5 125 = 3$
   (b) $\log_{10} 0.0001 = -4$

11. (a) $\log_8 \frac{1}{8} = -1$
   (b) $\log_2 \left(\frac{1}{8}\right) = -3$

13. (a) $\ln 2 = x$
   (b) $\ln y = 3$

15. (a) $\log_3 3 = 1$
   (b) $\log_3 1 = \log_3 3^0 = 0$
   (c) $\log_3 3^2 = 2$

17. (a) $\log_6 36 = \log_6 6^2 = 2$
   (b) $\log_9 81 = \log_9 9^2 = 2$
   (c) $\log_7 7^{10} = 10$

19. (a) $\log_3 \left(\frac{1}{27}\right) = \log_3 3^{-3} = -3$
   (b) $\log_{10} \sqrt{10} = \log_{10} 10^{1/2} = \frac{1}{2}$
   (c) $\log_5 0.2 = \log_5 \left(\frac{1}{5}\right) = \log_5 5^{-1} = -1$

21. (a) $2^{\log_2 37} = 37$
   (b) $3^{\log_3 8} = 8$
   (c) $e^{\ln \sqrt{5}} = \sqrt{5}$

23. (a) $\log_8 0.25 = \log_8 8^{-2/3} = -\frac{2}{3}$
   (b) $\ln e^4 = 4$
   (c) $\ln \left(\dfrac{1}{e}\right) = \ln e^{-1} = -1$

25. (a) $\log_2 x = 5 \quad \Leftrightarrow \quad x = 2^5 = 32$
   (b) $x = \log_2 16 = \log_2 2^4 = 4$

27. (a) $x = \log_3 243 = \log_3 3^5 = 5$
   (b) $\log_3 x = 3 \quad \Leftrightarrow \quad x = 3^3 = 27$

29. (a) $\log_{10} x = 2 \quad \Leftrightarrow \quad x = 10^2 = 100$
   (b) $\log_5 x = 2 \quad \Leftrightarrow \quad x = 5^2 = 25$

31. (a) $\log_x 16 = 4 \quad \Leftrightarrow \quad x^4 = 16 \quad \Leftrightarrow \quad x = 2$

(b) $\log_x 8 = \dfrac{3}{2} \quad \Leftrightarrow \quad x^{3/2} = 8 \quad \Leftrightarrow \quad x = 8^{2/3} = 4$

33. (a) $\log 2 \approx 0.3010$

(b) $\log 35.2 \approx 1.5465$

(c) $\log \left(\frac{2}{3}\right) \approx -0.1761$

35. (a) $\ln 5 \approx 1.6094$

(b) $\ln 25.3 \approx 3.2308$

(c) $\ln \left(1 + \sqrt{3}\right) \approx 1.0051$

37. Since the point $(5, 1)$ is on the graph, we have $1 = \log_a 5$ $\Leftrightarrow \quad a^1 = 5$. Thus the function is $y = \log_5 x$.

39. Since the point $\left(3, \frac{1}{2}\right)$ is on the graph, we have $\frac{1}{2} = \log_a 3 \quad \Leftrightarrow \quad a^{1/2} = 3 \quad \Leftrightarrow \quad a = 9$. Thus the function is $y = \log_9 x$.

41. II

43. III

45. VI

47. The graph of $y = \log_4 x$ is obtained from the graph of $y = 4^x$ by reflecting it about the line $y = x$.

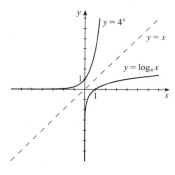

49. $f(x) = \log_2 (x - 4)$. The graph of f is obtained from the graph of $y = \log_2 x$ by shifting it to the right 4 units. Domain: $(4, \infty)$. Range: $(-\infty, \infty)$. Vertical asymptote: $x = 4$.

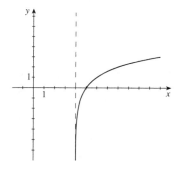

51. $g(x) = \log_5 (-x)$. The graph of g is obtained from the graph of $y = \log_5 x$ by reflecting it about the y-axis. Domain: $(-\infty, 0)$. Range: $(-\infty, \infty)$. Vertical asymptote: $x = 0$.

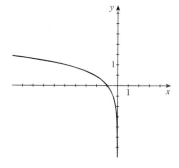

53. $y = 2 + \log_3 x$. The graph of $y = 2 + \log_3 x$ is obtained from the graph of $y = \log_3 x$ by shifting it upward 2 units. Domain: $(0, \infty)$. Range: $(-\infty, \infty)$. Vertical asymptote: $x = 0$.

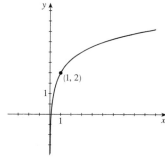

55. $y = 1 - \log_{10} x$. The graph of $y = 1 - \log_{10} x$ is obtained from the graph of $y = \log_{10} x$ by reflecting it about the x-axis, and then shifting it upward 1 unit. Domain: $(0, \infty)$. Range: $(-\infty, \infty)$. Vertical asymptote: $x = 0$.

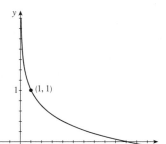

57. $y = |\ln x|$. The graph of $y = |\ln x|$ is obtained from the graph of $y = \ln x$ by reflecting the part of the graph for $0 < x < 1$ about the x-axis. Domain: $(0, \infty)$. Range: $[0, \infty)$. Vertical asymptote: $x = 0$.

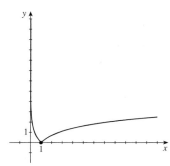

59. $f(x) = \log_{10}(x + 3)$. We require that $x + 3 > 0 \quad \Leftrightarrow \quad x > -3$, so the domain is $(-3, \infty)$.

61. $g(x) = \log_3(x^2 - 1)$. We require that $x^2 - 1 > 0 \quad \Leftrightarrow \quad x^2 > 1 \Rightarrow x < -1$ or $x > 1$, so the domain is $(-\infty, -1) \cup (1, \infty)$.

63. $h(x) = \ln x + \ln(2 - x)$. We require that $x > 0$ and $2 - x > 0 \quad \Leftrightarrow \quad x > 0$ and $x < 2 \quad \Leftrightarrow \quad 0 < x < 2$, so the domain is $(0, 2)$.

65. $y = \log_{10}(1 - x^2)$ has domain $(-1, 1)$, vertical asymptotes $x = -1$ and $x = 1$, and local maximum $y = 0$ at $x = 0$.

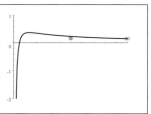

67. $y = x + \ln x$ has domain $(0, \infty)$, vertical asymptote $x = 0$, and no local maximum or minimum.

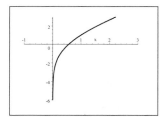

69. $y = \dfrac{\ln x}{x}$ has domain $(0, \infty)$, vertical asymptote $x = 0$, horizontal asymptote $y = 0$, and local maximum $y \approx 0.37$ at $x \approx 2.72$.

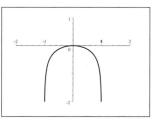

71. The graph of $g(x) = \sqrt{x}$ grows faster than the graph of $f(x) = \ln x$.

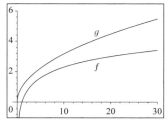

73. (a)

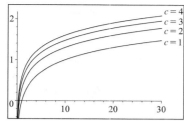

(b) Notice that $f(x) = \log(cx) = \log c + \log x$, so as c increases, the graph of $f(x) = \log(cx)$ is shifted upward $\log c$ units.

77. (a) $f(x) = \dfrac{2^x}{1+2^x}$. $y = \dfrac{2^x}{1+2^x}$ $\Leftrightarrow$

$y + y2^x = 2^x$ $\Leftrightarrow$

$y = 2^x - y2^x = 2^x(1-y)$ $\Leftrightarrow$

$2^x = \dfrac{y}{1-y}$ $\Leftrightarrow$ $x = \log_2\left(\dfrac{y}{1-y}\right)$. Thus

$f^{-1}(x) = \log_2\left(\dfrac{x}{1-x}\right)$.

75. (a) $f(x) = \log_2(\log_{10} x)$. Since the domain of $\log_2 x$ is the positive real numbers, we have: $\log_{10} x > 0$ $\Leftrightarrow$ $x > 10^0 = 1$. Thus the domain of $f(x)$ is $(1, \infty)$.

(b) $y = \log_2(\log_{10} x)$ $\Leftrightarrow$ $2^y = \log_{10} x$ $\Leftrightarrow$ $10^{2^y} = x$. Thus $f^{-1}(x) = 10^{2^x}$.

(b) $\dfrac{x}{1-x} > 0$. Solving this using the methods from Chapter 1, we start with the endpoints, 0 and 1.

Interval	$(-\infty, 0)$	$(0, 1)$	$(1, \infty)$
Sign of x	$-$	$+$	$+$
Sign of $1 - x$	$+$	$+$	$-$
Sign of $\dfrac{x}{1-x}$	$-$	$+$	$-$

Thus the domain of $f^{-1}(x)$ is $(0, 1)$.

79. Using $D = 0.73 D_0$ we have $A = -8267 \ln\left(\dfrac{D}{D_0}\right) = -8267 \ln 0.73 \approx 2601$ years.

81. When $r = 6\%$ we have $t = \dfrac{\ln 2}{0.06} \approx 11.6$ years. When $r = 7\%$ we have $t = \dfrac{\ln 2}{0.07} \approx 9.9$ years. And when $r = 8\%$ we have $t = \dfrac{\ln 2}{0.08} \approx 8.7$ years.

83. Using $A = 100$ and $W = 5$ we find the ID to be $\dfrac{\log(2A/W)}{\log 2} = \dfrac{\log(2 \cdot 100/5)}{\log 2} = \dfrac{\log 40}{\log 2} \approx 5.32$. Using $A = 100$ and $W = 10$ we find the ID to be $\dfrac{\log(2A/W)}{\log 2} = \dfrac{\log(2 \cdot 100/10)}{\log 2} = \dfrac{\log 20}{\log 2} \approx 4.32$. So the smaller icon is $\dfrac{5.23}{4.32} \approx 1.23$ times harder.

85. $\log(\log 10^{100}) = \log 100 = 2$

$\log(\log(\log 10^{\text{googol}})) = \log(\log(\text{googol})) = \log(\log 10^{100}) = \log(100) = 2$

87. The numbers between 1000 and 9999 (inclusive) each have 4 digits, while $\log 1000 = 3$ and $\log 10{,}000 = 4$. Since $[\![\log x]\!] = 3$ for all integers x where $1000 \le x < 10{,}000$, the number of digits is $[\![\log x]\!] + 1$. Likewise, if x is an integer where $10^{n-1} \le x < 10^n$, then x has n digits and $[\![\log x]\!] = n - 1$. Since $[\![\log x]\!] = n - 1$ $\Leftrightarrow$ $n = [\![\log x]\!] + 1$, the number of digits in x is $[\![\log x]\!] + 1$.

5.3 Laws of Logarithms

1. $\log_3 \sqrt{27} = \log_3 3^{3/2} = \dfrac{3}{2}$

3. $\log 4 + \log 25 = \log(4 \cdot 25) = \log 100 = 2$

5. $\log_4 192 - \log_4 3 = \log_4 \dfrac{192}{3} = \log_4 64 = \log_4 4^3 = 3$

7. $\log_2 6 - \log_2 15 + \log_2 20 = \log_2 \dfrac{6}{15} + \log_2 20 = \log_2\left(\dfrac{2}{5} \cdot 20\right) = \log_2 8 = \log_2 2^3 = 3$

9. $\log_4 16^{100} = \log_4 (4^2)^{100} = \log_4 4^{200} = 200$

11. $\log\left(\log 10^{10,000}\right) = \log\left(10,000 \log 10\right) = \log\left(10,000 \cdot 1\right) = \log\left(10,000\right) = \log 10^4 = 4 \log 10 = 4$

13. $\log_2 2x = \log_2 2 + \log_2 x = 1 + \log_2 x$

15. $\log_2 \left[x\left(x-1\right)\right] = \log_2 x + \log_2 \left(x-1\right)$

17. $\log 6^{10} = 10 \log 6$

19. $\log_2 \left(AB^2\right) = \log_2 A + \log_2 B^2 = \log_2 A + 2 \log_2 B$

21. $\log_3 \left(x\sqrt{y}\right) = \log_3 x + \log_3 \sqrt{y} = \log_3 x + \frac{1}{2} \log_3 y$

23. $\log_5 \sqrt[3]{x^2+1} = \frac{1}{3} \log_5 \left(x^2+1\right)$

25. $\ln \sqrt{ab} = \frac{1}{2} \ln ab = \frac{1}{2} \left(\ln a + \ln b\right)$

27. $\log \left(\dfrac{x^3 y^4}{z^6}\right) = \log \left(x^3 y^4\right) - \log z^6 = 3 \log x + 4 \log y - 6 \log z$

29. $\log_2 \left(\dfrac{x\left(x^2+1\right)}{\sqrt{x^2-1}}\right) = \log_2 x + \log_2 \left(x^2+1\right) - \frac{1}{2} \log_2 \left(x^2-1\right)$

31. $\ln \left(x\sqrt{\dfrac{y}{z}}\right) = \ln x + \frac{1}{2} \ln \left(\dfrac{y}{z}\right) = \ln x + \frac{1}{2} \left(\ln y - \ln z\right)$

33. $\log \sqrt[4]{x^2+y^2} = \frac{1}{4} \log \left(x^2+y^2\right)$

35. $\log \sqrt{\dfrac{x^2+4}{\left(x^2+1\right)\left(x^3-7\right)^2}} = \frac{1}{2} \log \dfrac{x^2+4}{\left(x^2+1\right)\left(x^3-7\right)^2} = \frac{1}{2} \left[\log \left(x^2+4\right) - \log \left(x^2+1\right)\left(x^3-7\right)^2\right]$

$$= \frac{1}{2} \left[\log \left(x^2+4\right) - \log \left(x^2+1\right) - 2 \log \left(x^3-7\right)\right]$$

37. $\ln \dfrac{x^3 \sqrt{x-1}}{3x+4} = \ln \left(x^3 \sqrt{x-1}\right) - \ln \left(3x+4\right) = 3 \ln x + \frac{1}{2} \ln \left(x-1\right) - \ln \left(3x+4\right)$

39. $\log_3 5 + 5 \log_3 2 = \log_3 5 + \log_3 2^5 = \log_3 \left(5 \cdot 2^5\right) = \log_3 160$

41. $\log_2 A + \log_2 B - 2 \log_2 C = \log_2 \left(AB\right) - \log_2 \left(C^2\right) = \log_2 \left(\dfrac{AB}{C^2}\right)$

43. $4 \log x - \frac{1}{3} \log \left(x^2+1\right) + 2 \log \left(x-1\right) = \log x^4 - \log \sqrt[3]{x^2+1} + \log \left(x-1\right)^2$

$$= \log \left(\dfrac{x^4}{\sqrt[3]{x^2+1}}\right) + \log \left(x-1\right)^2 = \log \left(\dfrac{x^4 \left(x-1\right)^2}{\sqrt[3]{x^2+1}}\right)$$

45. $\ln 5 + 2 \ln x + 3 \ln \left(x^2+5\right) = \ln \left(5x^2\right) + \ln \left(x^2+5\right)^3 = \ln \left[5x^2 \left(x^2+5\right)^3\right]$

47. $\frac{1}{3} \log \left(2x+1\right) + \frac{1}{2} \left[\log \left(x-4\right) - \log \left(x^4-x^2-1\right)\right] = \log \sqrt[3]{2x+1} + \frac{1}{2} \log \dfrac{x-4}{x^4-x^2-1}$

$$= \log \left(\sqrt[3]{2x+1} \cdot \sqrt{\dfrac{x-4}{x^4-x^2-1}}\right)$$

49. $\log_2 5 = \dfrac{\log 5}{\log 2} \approx 2.321928$

51. $\log_3 16 = \dfrac{\log 16}{\log 3} \approx 2.523719$

53. $\log_7 2.61 = \dfrac{\log 2.61}{\log 7} \approx 0.493008$

55. $\log_4 125 = \dfrac{\log 125}{\log 4} \approx 3.482892$

57. $\log_3 x = \dfrac{\log_e x}{\log_e 3} = \dfrac{\ln x}{\ln 3} = \dfrac{1}{\ln 3} \ln x$. The graph of $y = \dfrac{1}{\ln 3} \ln x$ is

shown in the viewing rectangle $[-1, 4]$ by $[-3, 2]$.

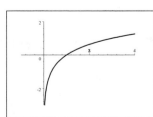

59. $\log e = \dfrac{\ln e}{\ln 10} = \dfrac{1}{\ln 10}$

61. $-\ln\left(x - \sqrt{x^2 - 1}\right) = \ln\left(\dfrac{1}{x - \sqrt{x^2 - 1}}\right) = \ln\left(\dfrac{1}{x - \sqrt{x^2 - 1}} \cdot \dfrac{x + \sqrt{x^2 - 1}}{x + \sqrt{x^2 - 1}}\right) = \ln\left(\dfrac{x + \sqrt{x^2 - 1}}{x^2 - (x^2 - 1)}\right)$

$\qquad = \ln\left(x + \sqrt{x^2 - 1}\right)$

63. (a) $\log P = \log c - k \log W \quad \Leftrightarrow \quad \log P = \log c - \log W^k \quad \Leftrightarrow \quad \log P = \log\left(\dfrac{c}{W^k}\right) \quad \Leftrightarrow \quad P = \dfrac{c}{W^k}$.

(b) Using $k = 2.1$ and $c = 8000$, when $W = 2$ we have $P = \dfrac{8000}{2^{2.1}} \approx 1866$ and when $W = 10$ we have $P = \dfrac{8000}{10^{2.1}} \approx 64$.

65. (a) $M = -2.5 \log(B/B_0) = -2.5 \log B + 2.5 \log B_0$.

(b) Suppose B_1 and B_2 are the brightness of two stars such that $B_1 < B_2$ and let M_1 and M_2 be their respective magnitudes. Since $\log$ is an increasing function, we have $\log B_1 < \log B_2$. Then
$\log B_1 < \log B_2 \quad \Leftrightarrow \quad \log B_1 - \log B_0 < \log B_2 - \log B_0 \quad \Leftrightarrow \quad \log(B_1/B_0) < \log(B_2/B_0) \quad \Leftrightarrow$
$-2.5 \log(B_1/B_0) > -2.5 \log(B_2/B_0) \quad \Leftrightarrow \quad M_1 > M_2$. Thus the brighter star has less magnitudes.

(c) Let B_1 be the brightness of the star Albiero. Then $100 B_1$ is the brightness of Betelgeuse, and its magnitude is
$M = -2.5 \log(100 B_1/B_0) = -2.5\left[\log 100 + \log(B_1/B_0)\right] = -2.5\left[2 + \log(B_1/B_0)\right] = -5 - 2.5 \log(B_1/B_0)$
$\qquad = -5 + \text{magnitude of Albiero}$

67. The error is on the first line: $\log 0.1 < 0$, so $2 \log 0.1 < \log 0.1$.

5.4   Exponential and Logarithmic Equations

1. $10^x = 25 \quad \Leftrightarrow \quad \log 10^x = \log 25 \quad \Leftrightarrow \quad x \log 10 = \log 25 \quad \Leftrightarrow \quad x \approx 1.398$

3. $e^{-2x} = 7 \quad \Leftrightarrow \quad \ln e^{-2x} = \ln 7 \quad \Leftrightarrow \quad -2x \ln e = \ln 7 \quad \Leftrightarrow \quad -2x = \ln 7 \quad \Leftrightarrow \quad x = -\frac{1}{2}\ln 7 \approx -0.9730$

5. $2^{1-x} = 3 \quad \Leftrightarrow \quad \log 2^{1-x} = \log 3 \quad \Leftrightarrow \quad (1-x)\log 2 = \log 3 \quad \Leftrightarrow \quad 1 - x = \dfrac{\log 3}{\log 2} \quad \Leftrightarrow$

$x = 1 - \dfrac{\log 3}{\log 2} \approx -0.5850$

7. $3e^x = 10 \quad \Leftrightarrow \quad e^x = \frac{10}{3} \quad \Leftrightarrow \quad x = \ln\left(\frac{10}{3}\right) \approx 1.2040$

9. $e^{1-4x} = 2 \quad \Leftrightarrow \quad 1 - 4x = \ln 2 \quad \Leftrightarrow \quad -4x = -1 + \ln 2 \quad \Leftrightarrow \quad x = \dfrac{1 - \ln 2}{4} = 0.0767$

11. $4 + 3^{5x} = 8 \quad \Leftrightarrow \quad 3^{5x} = 4 \quad \Leftrightarrow \quad \log 3^{5x} = \log 4 \quad \Leftrightarrow \quad 5x \log 3 = \log 4 \quad \Leftrightarrow \quad 5x = \dfrac{\log 4}{\log 3} \quad \Leftrightarrow$

$x = \dfrac{\log 4}{5 \log 3} \approx 0.2524$

13. $8^{0.4x} = 5 \quad \Leftrightarrow \quad \log 8^{0.4x} = \log 5 \quad \Leftrightarrow \quad 0.4x \log 8 = \log 5 \quad \Leftrightarrow \quad 0.4x = \dfrac{\log 5}{\log 8} \quad \Leftrightarrow \quad x = \dfrac{\log 5}{0.4 \log 8} \approx 1.9349$

15. $5^{-x/100} = 2 \quad \Leftrightarrow \quad \log 5^{-x/100} = \log 2 \quad \Leftrightarrow \quad -\dfrac{x}{100}\log 5 = \log 2 \quad \Leftrightarrow \quad x = -\dfrac{100 \log 2}{\log 5} \approx -43.0677$

17. $e^{2x+1} = 200 \quad \Leftrightarrow \quad 2x + 1 = \ln 200 \quad \Leftrightarrow \quad 2x = -1 + \ln 200 \quad \Leftrightarrow \quad x = \dfrac{-1 + \ln 200}{2} \approx 2.1492$

19. $5^x = 4^{x+1} \quad \Leftrightarrow \quad \log 5^x = \log 4^{x+1} \quad \Leftrightarrow \quad x \log 5 = (x+1)\log 4 = x \log 4 + \log 4 \quad \Leftrightarrow \quad x \log 5 - x \log 4 = \log 4$
$\Leftrightarrow \quad x(\log 5 - \log 4) = \log 4 \quad \Leftrightarrow \quad x = \dfrac{\log 4}{\log 5 - \log 4} \approx 6.2126$

21. $2^{3x+1} = 3^{x-2} \quad \Leftrightarrow \quad \log 2^{3x+1} = \log 3^{x-2} \quad \Leftrightarrow \quad (3x+1)\log 2 = (x-2)\log 3 \quad \Leftrightarrow \quad 3x \log 2 +$
$\log 2 = x \log 3 - 2 \log 3 \quad \Leftrightarrow \quad 3x \log 2 - x \log 3 = -\log 2 - 2 \log 3 \quad \Leftrightarrow \quad x(3 \log 2 - \log 3) = -(\log 2 + 2 \log 3)$
$\Leftrightarrow \quad s = -\dfrac{\log 2 + 2 \log 3}{3 \log 2 - \log 3} \approx -2.9469$

23. $\dfrac{50}{1 + e^{-x}} = 4 \quad \Leftrightarrow \quad 50 = 4 + 4e^{-x} \quad \Leftrightarrow \quad 46 = 4e^{-x} \quad \Leftrightarrow \quad 11.5 = e^{-x} \quad \Leftrightarrow \quad \ln 11.5 = -x \quad \Leftrightarrow$
$x = -\ln 11.5 \approx -2.4423$

25. $100 \, (1.04)^{2t} = 300 \quad \Leftrightarrow \quad 1.04^{2t} = 3 \quad \Leftrightarrow \quad \log 1.04^{2t} = \log 3 \quad \Leftrightarrow \quad 2t \log 1.04 = \log 3 \quad \Leftrightarrow$
$t = \dfrac{\log 3}{2 \log 1.04} \approx 14.0055$

27. $x^2 2^x - 2^x = 0 \quad \Leftrightarrow \quad 2^x \, (x^2 - 1) = 0 \Rightarrow 2^x = 0$ (never) or $x^2 - 1 = 0$. If $x^2 - 1 = 0$, then $x^2 = 1 \Rightarrow x = \pm 1$. So the only solutions are $x = \pm 1$.

29. $4x^3 e^{-3x} - 3x^4 e^{-3x} = 0 \quad \Leftrightarrow \quad x^3 e^{-3x} \, (4 - 3x) = 0 \Rightarrow x = 0$ or $e^{-3x} = 0$ (never) or $4 - 3x = 0$. If $4 - 3x = 0$, then $3x = 4 \quad \Leftrightarrow \quad x = \frac{4}{3}$. So the solutions are $x = 0$ and $x = \frac{4}{3}$.

31. $e^{2x} - 3e^x + 2 = 0 \quad \Leftrightarrow \quad (e^x - 1)(e^x - 2) = 0 \Rightarrow e^x - 1 = 0$ or $e^x - 2 = 0$. If $e^x - 1 = 0$, then $e^x = 1 \quad \Leftrightarrow$ $x = \ln 1 = 0$. If $e^x - 2 = 0$, then $e^x = 2 \quad \Leftrightarrow \quad x = \ln 2 \approx 0.6931$. So the solutions are $x = 0$ and $x \approx 0.6931$.

33. $e^{4x} + 4e^{2x} - 21 = 0 \quad \Leftrightarrow \quad (e^{2x} + 7)(e^{2x} - 3) = 0 \Rightarrow e^{2x} = -7$ or $e^{2x} = 3$. Now $e^{2x} = -7$ has no solution, since $e^{2x} > 0$ for all x. But we can solve $e^{2x} = 3 \quad \Leftrightarrow \quad 2x = \ln 3 \quad \Leftrightarrow \quad x = \frac{1}{2} \ln 3 \approx 0.5493$. So the only solution is $x \approx 0.5493$.

35. $\ln x = 10 \quad \Leftrightarrow \quad x = e^{10} \approx 22026$

37. $\log x = -2 \quad \Leftrightarrow \quad x = 10^{-2} = 0.01$

39. $\log (3x + 5) = 2 \quad \Leftrightarrow \quad 3x + 5 = 10^2 = 100 \quad \Leftrightarrow \quad 3x = 95 \quad \Leftrightarrow \quad x = \frac{95}{3} \approx 31.6667$

41. $2 - \ln (3 - x) = 0 \quad \Leftrightarrow \quad 2 = \ln (3 - x) \quad \Leftrightarrow \quad e^2 = 3 - x \quad \Leftrightarrow \quad x = 3 - e^2 \approx -4.3891$

43. $\log_2 3 + \log_2 x = \log_2 5 + \log_2 (x - 2) \quad \Leftrightarrow \quad \log_2 (3x) = \log_2 (5x - 10) \quad \Leftrightarrow \quad 3x = 5x - 10 \quad \Leftrightarrow \quad 2x = 10$
$\Leftrightarrow \quad x = 5$

45. $\log x + \log (x - 1) = \log (4x) \quad \Leftrightarrow \quad \log [x \, (x - 1)] = \log (4x) \quad \Leftrightarrow \quad x^2 - x = 4x \quad \Leftrightarrow \quad x^2 - 5x = 0 \quad \Leftrightarrow$
$x \, (x - 5) = 0 \Rightarrow x = 0$ or $x = 5$. So the possible solutions are $x = 0$ and $x = 5$. However, when $x = 0$, $\log x$ is undefined. Thus the only solution is $x = 5$.

47. $\log_5 (x + 1) - \log_5 (x - 1) = 2 \quad \Leftrightarrow \quad \log_5 \left(\dfrac{x + 1}{x - 1} \right) = 2 \quad \Leftrightarrow \quad \dfrac{x + 1}{x - 1} = 5^2 \quad \Leftrightarrow \quad x + 1 = 25x - 25 \quad \Leftrightarrow$
$24x = 26 \quad \Leftrightarrow \quad x = \frac{13}{12}$

49. $\log_9 (x - 5) + \log_9 (x + 3) = 1 \quad \Leftrightarrow \quad \log_9 [(x - 5)(x + 3)] = 1 \quad \Leftrightarrow \quad (x - 5)(x + 3) = 9^1 \quad \Leftrightarrow$
$x^2 - 2x - 24 = 0 \quad \Leftrightarrow \quad (x - 6)(x + 4) = 0 \Rightarrow x = 6$ or -4. However, $x = -4$ is inadmissible, so $x = 6$ is the only solution.

51. $\log (x + 3) = \log x + \log 3 \quad \Leftrightarrow \quad \log (x + 3) = \log (3x) \quad \Leftrightarrow \quad x + 3 = 3x \quad \Leftrightarrow \quad 2x = 3 \quad \Leftrightarrow \quad x = \frac{3}{2}$

53. $2^{2/\log_5 x} = \frac{1}{16} \quad \Leftrightarrow \quad \log_2 2^{2/\log_5 x} = \log_2 \left(\frac{1}{16} \right) \quad \Leftrightarrow \quad \dfrac{2}{\log_5 x} = -4 \quad \Leftrightarrow \quad \log_5 x = -\frac{1}{2} \quad \Leftrightarrow$
$x = 5^{-1/2} = \frac{1}{\sqrt{5}} \approx 0.4472$

55. $\ln x = 3 - x$ $\Leftrightarrow$ $\ln x + x - 3 = 0$. Let $f(x) = \ln x + x - 3$. We need to solve the equation $f(x) = 0$. From the graph of f, we get $x \approx 2.21$.

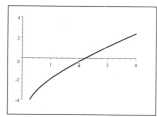

57. $x^3 - x = \log_{10}(x+1)$ $\Leftrightarrow$ $x^3 - x - \log_{10}(x+1) = 0$. Let $f(x) = x^3 - x - \log_{10}(x+1)$. We need to solve the equation $f(x) = 0$. From the graph of f, we get $x = 0$ or $x \approx 1.14$.

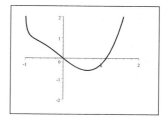

59. $e^x = -x$ $\Leftrightarrow$ $e^x + x = 0$. Let $f(x) = e^x + x$. We need to solve the equation $f(x) = 0$. From the graph of f, we get $x \approx -0.57$.

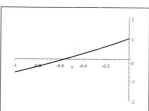

61. $4^{-x} = \sqrt{x}$ $\Leftrightarrow$ $4^{-x} - \sqrt{x} = 0$. Let $f(x) = 4^{-x} - \sqrt{x}$. We need to solve the equation $f(x) = 0$. From the graph of f, we get $x \approx 0.36$.

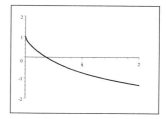

63. $\log(x-2) + \log(9-x) < 1$ $\Leftrightarrow$ $\log[(x-2)(9-x)] < 1$ $\Leftrightarrow$ $\log(-x^2 + 11x - 18) < 1 \Rightarrow$ $-x^2 + 11x - 18 < 10^1$ $\Leftrightarrow$ $0 < x^2 - 11x + 28$ $\Leftrightarrow$ $0 < (x-7)(x-4)$. Also, since the domain of a logarithm is positive we must have $0 < -x^2 + 11x - 18$ $\Leftrightarrow$ $0 < (x-2)(9-x)$. Using the methods from Chapter 1 with the endpoints 2, 4, 7, 9 for the intervals, we make the following table:

Interval	$(-\infty, 2)$	$(2, 4)$	$(4, 7)$	$(7, 9)$	$(9, \infty)$
Sign of $x - 7$	$-$	$-$	$-$	$+$	$+$
Sign of $x - 4$	$-$	$-$	$+$	$+$	$+$
Sign of $x - 2$	$-$	$+$	$+$	$+$	$+$
Sign of $9 - x$	$+$	$+$	$+$	$+$	$-$
Sign of $(x-7)(x-4)$	$+$	$+$	$-$	$+$	$+$
Sign of $(x-2)(9-x)$	$-$	$+$	$+$	$+$	$-$

Thus the solution is $(2, 4) \cup (7, 9)$.

65. $2 < 10^x < 5$ $\Leftrightarrow$ $\log 2 < x < \log 5$ $\Leftrightarrow$ $0.3010 < x < 0.6990$. Hence the solution to the inequality is approximately the interval $(0.3010, 0.6990)$.

67. (a) $A(3) = 5000\left(1 + \dfrac{0.085}{4}\right)^{4(3)} = 5000\left(1.02125^{12}\right) = 6435.09$. Thus the amount after 3 years is \$6,435.09.

(b) $10000 = 5000\left(1 + \dfrac{0.085}{4}\right)^{4t} = 5000\left(1.02125^{4t}\right)$ $\Leftrightarrow$ $2 = 1.02125^{4t}$ $\Leftrightarrow$ $\log 2 = 4t \log 1.02125$ $\Leftrightarrow$

$t = \dfrac{\log 2}{4 \log 1.02125} \approx 8.24$ years. Thus the investment will double in about 8.24 years.

69. $8000 = 5000 \left(1 + \dfrac{0.075}{4}\right)^{4t} = 5000\left(1.01875^{4t}\right) \quad\Leftrightarrow\quad 1.6 = 1.01875^{4t} \quad\Leftrightarrow\quad \log 1.6 = 4t \log 1.01875 \quad\Leftrightarrow$

$t = \dfrac{\log 1.6}{4 \log 1.01875} \approx 6.33$ years. The investment will increase to \$8000 in approximately 6 years and 4 months.

71. $2 = e^{0.085t} \quad\Leftrightarrow\quad \ln 2 = 0.085t \quad\Leftrightarrow\quad t = \dfrac{\ln 2}{0.085} \approx 8.15$ years. Thus the investment will double in about 8.15 years.

73. $r_{\text{eff}} = \left(1 + \dfrac{r}{n}\right)^n - 1$. Here $r = 0.08$ and $n = 12$, so $r_{\text{eff}} = \left(1 + \dfrac{0.08}{12}\right)^{12} - 1 = (1.0066667)^{12} - 1 = 8.30\%$.

75. $15 e^{-0.087t} = 5 \quad\Leftrightarrow\quad e^{-0.087t} = \frac{1}{3} \quad\Leftrightarrow\quad -0.087t = \ln\left(\frac{1}{3}\right) = -\ln 3 \quad\Leftrightarrow\quad t = \dfrac{\ln 3}{0.087} \approx 12.6277$. So only
5 grams remain after approximately 13 days.

77. (a) $P(3) = \dfrac{10}{1 + 4e^{-0.8(3)}} = 7.337$ So there are approximately 7337 fish after 3 years.

(b) We solve for t. $\dfrac{10}{1 + 4e^{-0.8t}} = 5 \quad\Leftrightarrow\quad 1 + 4e^{-0.8t} = \frac{10}{5} = 2 \quad\Leftrightarrow\quad 4e^{-0.8t} = 1 \quad\Leftrightarrow\quad e^{-0.8t} = 0.25 \quad\Leftrightarrow$

$-0.8t = \ln 0.25 \quad\Leftrightarrow\quad t = \dfrac{\ln 0.25}{-0.8} = 1.73$. So the population will reach 5000 fish in about 1 year and 9 months.

79. (a) $\ln\left(\dfrac{P}{P_0}\right) = -\dfrac{h}{k} \quad\Leftrightarrow\quad \dfrac{P}{P_0} = e^{-h/k} \quad\Leftrightarrow\quad P = P_0 e^{-h/k}$. Substituting $k = 7$ and $P_0 = 100$ we get
$P = 100 e^{-h/7}$.

(b) When $h = 4$ we have $P = 100 e^{-4/7} \approx 56.47$ kPa.

81. (a) $I = \frac{60}{13}\left(1 - e^{-13t/5}\right) \quad\Leftrightarrow\quad \frac{13}{60}I = 1 - e^{-13t/5} \quad\Leftrightarrow\quad e^{-13t/5} = 1 - \frac{13}{60}I \quad\Leftrightarrow\quad -\frac{13}{5}t = \ln\left(1 - \frac{13}{60}I\right) \quad\Leftrightarrow$
$t = -\frac{5}{13}\ln\left(1 - \frac{13}{60}I\right)$.

(b) Substituting $I = 2$, we have $t = -\frac{5}{13}\ln\left[1 - \frac{13}{60}(2)\right] \approx 0.218$ seconds.

83. Since $9^1 = 9$, $9^2 = 81$, and $9^3 = 729$, the solution of $9^x = 20$ must be between 1 and 2 (because 20 is between 9 and 81),
whereas the solution to $9^x = 100$ must be between 2 and 3 (because 100 is between 81 and 729).

85. (a) $(x-1)^{\log(x-1)} = 100(x-1) \quad\Leftrightarrow\quad \log\left((x-1)^{\log(x-1)}\right) = \log(100(x-1)) \quad\Leftrightarrow$
$[\log(x-1)]\log(x-1) = \log 100 + \log(x-1) \quad\Leftrightarrow\quad [\log(x-1)]^2 - \log(x-1) - 2 = 0 \quad\Leftrightarrow$
$[\log(x-1) - 2][\log(x-1) + 1] = 0$. Thus either $\log(x-1) = 2 \quad\Leftrightarrow\quad x = 101$ or $\log(x-1) = -1 \quad\Leftrightarrow$
$x = \frac{11}{10}$.

(b) $\log_2 x + \log_4 x + \log_8 x = 11 \quad\Leftrightarrow\quad \log_2 x + \log_2 \sqrt{x} + \log_2 \sqrt[3]{x} = 11 \quad\Leftrightarrow\quad \log_2\left(x\sqrt{x}\sqrt[3]{x}\right) = 11 \quad\Leftrightarrow$
$\log_2\left(x^{11/6}\right) = 11 \quad\Leftrightarrow\quad \frac{11}{6}\log_2 x = 11 \quad\Leftrightarrow\quad \log_2 x = 6 \quad\Leftrightarrow\quad x = 2^6 = 64$

(c) $4^x - 2^{x+1} = 3 \quad\Leftrightarrow\quad (2^x)^2 - 2(2^x) - 3 = 0 \quad\Leftrightarrow\quad (2^x - 3)(2^x + 1) = 0 \quad\Leftrightarrow\quad$ either $2^x = 3 \quad\Leftrightarrow$
$x = \dfrac{\ln 3}{\ln 2}$ or $2^x = -1$, which has no real solution. So $x = \dfrac{\ln 3}{\ln 2}$ is the only real solution.

5.5   Modeling with Exponential and Logarithmic Functions

1. (a) $n(0) = 500$.

(b) The relative growth rate is $0.45 = 45\%$.

(c) $n(3) = 500 e^{0.45(3)} \approx 1929$.

(d) $10{,}000 = 500 e^{0.45t} \quad\Leftrightarrow\quad 20 = e^{0.45t} \quad\Leftrightarrow\quad 0.45t = \ln 20$
$\Leftrightarrow\quad t = \dfrac{\ln 20}{0.45} \approx 6.66$ hours, or 6 hours 40 minutes.

3. (a) $r = 0.08$ and $n(0) = 18000$. Thus the population is given by the
formula $n(t) = 18{,}000e^{0.08t}$.

(b) $t = 2008 - 2000 = 8$. Then we have
$n(8) = 18000e^{0.08(8)} = 18000e^{0.64} \approx 34{,}137$. Thus there should
be 34,137 foxes in the region by the year 2008.

(c)

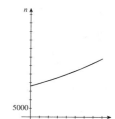

5. (a) $n(t) = 112{,}000e^{0.04t}$.

(b) $t = 2000 - 1994 = 6$ and $n(6) = 112{,}000e^{0.04(6)} \approx 142380$. The projected population is about142,000.

(c) $200{,}000 = 112{,}000e^{0.04t}$ $\Leftrightarrow$ $\frac{25}{14} = e^{0.04t}$ $\Leftrightarrow$ $0.04t = \ln\left(\frac{25}{14}\right)$ $\Leftrightarrow$ $t = 25\ln\left(\frac{25}{14}\right) \approx 14.5$. Since
$1994 + 14.5 = 2008.5$, the population will reach 200,000 during the year 2008.

7. (a) The deer population in 1996 was 20,000.

(b) Using the model $n(t) = 20{,}000e^{rt}$ and the point $(4, 31000)$, we have $31{,}000 = 20{,}000e^{4r}$ $\Leftrightarrow$ $1.55 = e^{4r}$ $\Leftrightarrow$
$4r = \ln 1.55$ $\Leftrightarrow$ $r = \frac{1}{4}\ln 1.55 \approx 0.1096$. Thus $n(t) = 20{,}000e^{0.1096t}$

(c) $n(8) = 20{,}000e^{0.1096(8)} \approx 48{,}218$, so the projected deer population in 2004 is about 48,000.

(d) $100{,}000 = 20{,}000e^{0.1096t}$ $\Leftrightarrow$ $5 = e^{0.1096t}$ $\Leftrightarrow$ $0.1096t = \ln 5$ $\Leftrightarrow$ $t = \dfrac{\ln 5}{0.1096} \approx 14.63$. Since
$1996 + 14.63 = 2010.63$, the deer population will reach 100,000 during the year 2010.

9. (a) Using the formula $n(t) = n_0e^{rt}$ with $n_0 = 8600$ and $n(1) = 10000$, we solve for r, giving $10000 = n(1) = 8600e^r$
$\Leftrightarrow$ $\frac{50}{43} = e^r$ $\Leftrightarrow$ $r = \ln\left(\frac{50}{43}\right) \approx 0.1508$. Thus $n(t) = 8600e^{0.1508t}$.

(b) $n(2) = 8600e^{0.1508(2)} \approx 11627$. Thus the number of bacteria after two hours is about 11,600.

(c) $17200 = 8600e^{0.1508t}$ $\Leftrightarrow$ $2 = e^{0.1508t}$ $\Leftrightarrow$ $0.1508t = \ln 2$ $\Leftrightarrow$ $t = \dfrac{\ln 2}{0.1508} \approx 4.596$. Thus the number
of bacteria will double in about 4.6 hours.

11. (a) $2n_0 = n_0e^{0.02t}$ $\Leftrightarrow$ $2 = e^{0.02t}$ $\Leftrightarrow$ $0.02t = \ln 2$ $\Leftrightarrow$ $t = 50\ln 2 \approx 34.65$. So we have
$t = 1995 + 34.65 = 2029.65$, and hence at the current growth rate the population will double by the year 2029.

(b) $3n_0 = n_0e^{0.02t}$ $\Leftrightarrow$ $3 = e^{0.02t}$ $\Leftrightarrow$ $0.02t = \ln 3$ $\Leftrightarrow$ $t = 50\ln 3 \approx 54.93$. So we have
$t = 1995 + 54.93 = 2049.93$, and hence at the current growth rate the population will triple by the year 2050.

13. $n(t) = n_0e^{2t}$. When $n_0 = 1$, the critical level is $n(24) = e^{2(24)} = e^{48}$.We solve the equation
$e^{48} = n_0e^{2t}$, where $n_0 = 10$. This gives$e^{48} = 10e^{2t}$ $\Leftrightarrow$ $48 = \ln 10 + 2t$ $\Leftrightarrow$ $2t = 48 - \ln 10$ $\Leftrightarrow$
$t = \frac{1}{2}(48 - \ln 10) \approx 22.85$ hours.

15. (a) Using $m(t) = m_0e^{-rt}$ with $m_0 = 10$ and $h = 30$, we have $r = \dfrac{\ln 2}{h} = \dfrac{\ln 2}{30} \approx 0.0231$. Thus $m(t) = 10e^{-0.0231t}$.

(b) $m(80) = 10e^{-0.0231(80)} \approx 1.6$ grams.

(c) $2 = 10e^{-0.0231t}$ $\Leftrightarrow$ $\frac{1}{5} = e^{-0.0231t}$ $\Leftrightarrow$ $\ln\left(\frac{1}{5}\right) = -0.0231t$ $\Leftrightarrow$ $t = \dfrac{-\ln 5}{-0.0231} \approx 70$ years.

17. By the formula in the text, $m(t) = m_0e^{-rt}$ where $r = \dfrac{\ln 2}{h}$, so $m(t) = 50e^{-[(\ln 2)/28]t}$. We need to solve

for t in the equation $32 = 50e^{-[(\ln 2)/28]t}$. This gives $e^{-[(\ln 2)/28]t} = \frac{32}{50}$ $\Leftrightarrow$ $-\dfrac{\ln 2}{28}t = \ln\left(\frac{32}{50}\right)$ $\Leftrightarrow$

$t = -\dfrac{28}{\ln 2} \cdot \ln\left(\frac{32}{50}\right) \approx 18.03$, so it takes about 18 years.

19. By the formula for radioactive decay, we have $m(t) = m_0 e^{-rt}$, where $r = \dfrac{\ln 2}{h}$, in other words $m(t) = m_0 e^{-[(\ln 2)/h]t}$.

In this exercise we have to solve for h in the equation $200 = 250 e^{-[(\ln 2)/h]\cdot 48}$ $\Leftrightarrow$ $0.8 = e^{-[(\ln 2)/h]\cdot 48}$ $\Leftrightarrow$

$\ln(0.8) = -\dfrac{\ln 2}{h} \cdot 48$ $\Leftrightarrow$ $h = -\dfrac{\ln 2}{\ln 0.8} \cdot 48 \approx 149.1$ hours. So the half-life is approximately 149 hours.

21. By the formula in the text, $m(t) = m_0 e^{-[(\ln 2)/h]\cdot t}$, so we have $0.65 = 1 \cdot e^{-[(\ln 2)/5730]\cdot t}$ $\Leftrightarrow$ $\ln(0.65) = -\dfrac{\ln 2}{5730} t$

$\Leftrightarrow$ $t = -\dfrac{5730 \ln 0.65}{\ln 2} \approx 3561$. Thus the artifact is about 3560 years old.

23. (a) $T(0) = 65 + 145 e^{-0.05(0)} = 65 + 145 = 210°$ F.

(b) $T(10) = 65 + 145 e^{-0.05(10)} \approx 152.9$. Thus the temperature after 10 minutes is about 153° F.

(c) $100 = 65 + 145 e^{-0.05t}$ $\Leftrightarrow$ $35 = 145 e^{-0.05t}$ $\Leftrightarrow$ $0.2414 = e^{-0.05t}$ $\Leftrightarrow$ $\ln 0.2414 = -0.05t$ $\Leftrightarrow$

$t = -\dfrac{\ln 0.2414}{0.05} \approx 28.4$. Thus the temperature will be 100° F in about 28 minutes.

25. Using Newton's Law of Cooling, $T(t) = T_s + D_0 e^{-kt}$ with $T_s = 75$ and $D_0 = 185 - 75 = 110$. So $T(t) = 75 + 110 e^{-kt}$.

(a) Since $T(30) = 150$, we have $T(30) = 75 + 110 e^{-30k} = 150$ $\Leftrightarrow$ $110 e^{-30k} = 75$ $\Leftrightarrow$ $e^{-30k} = \frac{15}{22}$ $\Leftrightarrow$

$-30k = \ln\left(\frac{15}{22}\right)$ $\Leftrightarrow$ $k = -\frac{1}{30}\ln\left(\frac{15}{22}\right)$. Thus we have $T(45) = 75 + 110 e^{(45/30)\ln(15/22)} \approx 136.9$, and so the

temperature of the turkey after 45 minutes is about 137° F.

(b) The temperature will be 100°F when $75 + 110 e^{(t/30)\ln(15/22)} = 100$ $\Leftrightarrow$ $e^{(t/30)\ln(15/22)} = \dfrac{25}{110} = \frac{5}{22}$ $\Leftrightarrow$

$\left(\dfrac{t}{30}\right)\ln\left(\frac{15}{22}\right) = \ln\left(\frac{5}{22}\right)$ $\Leftrightarrow$ $t = 30\dfrac{\ln\left(\frac{5}{22}\right)}{\ln\left(\frac{15}{22}\right)} \approx 116.1$. So the temperature will be 100° F after 116 minutes.

27. (a) $\text{pH} = -\log\left[\text{H}^+\right] = -\log\left(5.0 \times 10^{-3}\right) \approx 2.3$

(b) $\text{pH} = -\log\left[\text{H}^+\right] = -\log\left(3.2 \times 10^{-4}\right) \approx 3.5$

(c) $\text{pH} = -\log\left[\text{H}^+\right] = -\log\left(5.0 \times 10^{-9}\right) \approx 8.3$

29. (a) $\text{pH} = -\log\left[\text{H}^+\right] = 3.0$ $\Leftrightarrow$ $\left[\text{H}^+\right] = 10^{-3}$ M

(b) $\text{pH} = -\log\left[\text{H}^+\right] = 6.5$ $\Leftrightarrow$ $\left[\text{H}^+\right] = 10^{-6.5} \approx 3.2 \times 10^{-7}$ M

31. $4.0 \times 10^{-7} \le \left[\text{H}^+\right] \le 1.6 \times 10^{-5}$ $\Leftrightarrow$ $\log\left(4.0 \times 10^{-7}\right) \le \log\left[\text{H}^+\right] \le \log\left(1.6 \times 10^{-5}\right)$ $\Leftrightarrow$

$-\log\left(4.0 \times 10^{-7}\right) \ge \text{pH} \ge -\log\left(1.6 \times 10^{-5}\right)$ $\Leftrightarrow$ $6.4 \ge \text{pH} \ge 4.8$. Therefore the range of pH readings for cheese

is approximately 4.8 to 6.4.

33. Let I_0 be the intensity of the smaller earthquake and I_1 the intensity of the larger earthquake. Then $I_1 = 20 I_0$. Notice

that $M_0 = \log\left(\dfrac{I_0}{S}\right) = \log I_0 - \log S$ and $M_1 = \log\left(\dfrac{I_1}{S}\right) = \log\left(\dfrac{20 I_0}{S}\right) = \log 20 + \log I_0 - \log S$. Then

$M_1 - M_0 = \log 20 + \log I_0 - \log S - \log I_0 + \log S = \log 20 \approx 1.3$. Therefore the magnitude is 1.3 times larger.

35. Let the subscript A represent the Alaska earthquake and S represent the San Francisco earthquake. Then

$M_A = \log\left(\dfrac{I_A}{S}\right) = 8.6$ $\Leftrightarrow$ $I_A = S \cdot 10^{8.6}$; also, $M_S = \log\left(\dfrac{I_S}{S}\right) = 8.3$ $\Leftrightarrow$ $I_S = S \cdot 10^{8.6}$. So

$\dfrac{I_A}{I_S} = \dfrac{S \cdot 10^{8.6}}{S \cdot 10^{8.3}} = 10^{0.3} \approx 1.995$, and hence the Alaskan earthquake was roughly twice as intense as the San Francisco

earthquake.

37. Let the subscript M represent the Mexico City earthquake, and T represent the Tangshan earthquake. We have

$\dfrac{I_T}{I_M} = 1.26$ $\Leftrightarrow$ $\log 1.26 = \log\dfrac{I_T}{I_M} = \log\dfrac{I_T/S}{I_M/S} = \log\dfrac{I_T}{S} - \log\dfrac{I_M}{S} = M_T - M_M$. Therefore

$M_T = M_M + \log 1.26 \approx 8.1 + 0.1 = 8.2$. Thus the magnitude of the Tangshan earthquake was roughly 8.2.

39. $98 = 10 \log \left(\dfrac{I}{10^{-12}} \right) \quad \Leftrightarrow \quad \log \left(I \cdot 10^{12} \right) = 9.8 \quad \Leftrightarrow \quad \log I = 9.8 - \log 10^{12} = -2.2 \quad \Leftrightarrow$

$I = 10^{-2.2} \approx 6.3 \times 10^{-3}$. So the intensity was 6.3×10^{-3} watts/m^2.

41. (a) $\beta_1 = 10 \log \left(\dfrac{I_1}{I_0} \right)$ and $I_1 = \dfrac{k}{d_1^2} \Leftrightarrow \beta_1 = 10 \log \left(\dfrac{k}{d_1^2 I_0} \right) = 10 \left[\log \left(\dfrac{k}{I_0} \right) - 2 \log d_1 \right] = 10 \log \left(\dfrac{k}{I_0} \right) - 20 \log d_1$.

Similarly, $\beta_2 = 10 \log \left(\dfrac{k}{I_0} \right) - 20 \log d_2$. Substituting the expression for β_1 gives

$\beta_2 = 10 \log \left(\dfrac{k}{I_0} \right) - 20 \log d_1 + 20 \log d_1 - 20 \log d_2 = \beta_1 + 20 \log d_1 - 20 \log d_2 = \beta_1 + 20 \log \left(\dfrac{d_1}{d_2} \right)$.

(b) $\beta_1 = 120$, $d_1 = 2$, and $d_2 = 10$. Then $\beta_2 = \beta_1 + 20 \log \left(\dfrac{d_1}{d_2} \right) = 120 + 20 \log \left(\frac{2}{10} \right) = 120 + 20 \log 0.2 \approx 106$, and

so the intensity level at 10 m is approximately 106 dB.

Chapter 5 Review

1. $f(x) = 2^{-x+1}$. Domain $(-\infty, \infty)$, range $(0, \infty)$, asymptote $y = 0$.

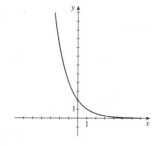

3. $g(x) = 3 + 2^x$. Domain $(-\infty, \infty)$, range $(3, \infty)$, asymptote $y = 3$.

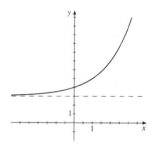

5. $f(x) = \log_3 (x - 1)$. Domain $(1, \infty)$, range $(-\infty, \infty)$, asymptote $x = 1$.

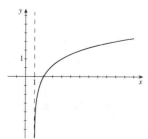

7. $f(x) = 2 - \log_2 x$. Domain $(0, \infty)$, range $(-\infty, \infty)$, asymptote $x = 0$.

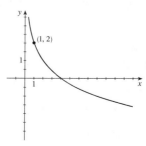

9. $F(x) = e^x - 1$. Domain $(-\infty, \infty)$, range $(-1, \infty)$, asymptote $y = -1$.

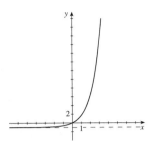

11. $g(x) = 2 \ln x$. Domain $(0, \infty)$, range $(-\infty, \infty)$, asymptote $x = 0$.

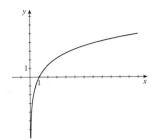

13. $f(x) = 10^{x^2} + \log(1 - 2x)$. Since $\log u$ is defined only for $u > 0$, we require $1 - 2x > 0 \quad \Leftrightarrow \quad -2x > -1 \quad \Leftrightarrow$ $x < \frac{1}{2}$, and so the domain is $\left(-\infty, \frac{1}{2}\right)$.

15. $h(x) = \ln(x^2 - 4)$. We must have $x^2 - 4 > 0$ (since $\ln y$ is defined only for $y > 0$) $\quad \Leftrightarrow \quad x^2 - 4 > 0 \quad \Leftrightarrow$ $(x - 2)(x + 2) > 0$. The endpoints of the intervals are -2 and 2.

Interval	$(-\infty, -2)$	$(-2, 2)$	$(2, \infty)$
Sign of $x - 2$	$-$	$-$	$+$
Sign of $x + 2$	$-$	$+$	$+$
Sign of $(x - 2)(x + 2)$	$+$	$-$	$+$

Thus the domain is $(-\infty, -2) \cup (2, \infty)$.

17. $\log_2 1024 = 10 \quad \Leftrightarrow \quad 2^{10} = 1024$

19. $\log x = y \quad \Leftrightarrow \quad 10^y = x$

21. $2^6 = 64 \quad \Leftrightarrow \quad \log_2 64 = 6$

23. $10^x = 74 \quad \Leftrightarrow \quad \log_{10} 74 = x \quad \Leftrightarrow \quad \log 74 = x$

25. $\log_2 128 = \log_2 (2^7) = 7$

27. $10^{\log 45} = 45$

29. $\ln(e^6) = 6$

31. $\log_3 \frac{1}{27} = \log_3 3^{-3} = -3$

33. $\log_5 \sqrt{5} = \log_5 5^{1/2} = \frac{1}{2}$

35. $\log 25 + \log 4 = \log(25 \cdot 4) = \log 10^2 = 2$

37. $\log_2 (16^{23}) = \log_2 (2^4)^{23} = \log_2 2^{92} = 92$

39. $\log_8 6 - \log_8 3 + \log_8 2 = \log_8 \left(\frac{6}{3} \cdot 2\right) = \log_8 4 = \log_8 8^{2/3} = \frac{2}{3}$

41. $\log(AB^2C^3) = \log A + 2\log B + 3\log C$

43. $\ln \sqrt{\dfrac{x^2 - 1}{x^2 + 1}} = \frac{1}{2} \ln\left(\dfrac{x^2 - 1}{x^2 + 1}\right) = \frac{1}{2}\left[\ln(x^2 - 1) - \ln(x^2 + 1)\right]$

45. $\log_5 \left(\dfrac{x^2 (1 - 5x)^{3/2}}{\sqrt{x^3 - x}}\right) = \log_5 x^2 (1 - 5x)^{3/2} - \log_5 \sqrt{x(x^2 - 1)} = 2\log_5 x + \frac{3}{2}\log_5 (1 - 5x) - \frac{1}{2}\log_5 (x^3 - x)$

47. $\log 6 + 4\log 2 = \log 6 + \log 2^4 = \log(6 \cdot 2^4) = \log 96$

49. $\frac{3}{2}\log_2 (x - y) - 2\log_2 (x^2 + y^2) = \log_2 (x - y)^{3/2} - \log_2 (x^2 + y^2)^2 = \log_2 \left(\dfrac{(x - y)^{3/2}}{(x^2 + y^2)^2}\right)$

51. $\log(x - 2) + \log(x + 2) - \frac{1}{2}\log(x^2 + 4) = \log[(x - 2)(x + 2)] - \log\sqrt{x^2 + 4} = \log\left(\dfrac{x^2 - 4}{\sqrt{x^2 + 4}}\right)$

53. $\log_2 (1 - x) = 4 \quad \Leftrightarrow \quad 1 - x = 2^4 \quad \Leftrightarrow \quad x = 1 - 2^4 = -15$

55. $5^{5 - 3x} = 26 \quad \Leftrightarrow \quad \log_5 26 = 5 - 3x \quad \Leftrightarrow \quad 3x = 5 - \log_5 26 \quad \Leftrightarrow \quad x = \frac{1}{3}(5 - \log_5 26) \approx 0.99$

57. $e^{3x/4} = 10 \quad \Leftrightarrow \quad \ln e^{3x/4} = \ln 10 \quad \Leftrightarrow \quad \dfrac{3x}{4} = \ln 10 \quad \Leftrightarrow \quad x = \frac{4}{3}\ln 10 \approx 3.07$

59. $\log x + \log (x+1) = \log 12 \quad \Leftrightarrow \quad \log [x(x+1)] = \log 12 \quad \Leftrightarrow \quad x(x+1) = 12 \quad \Leftrightarrow \quad x^2 + x - 12 = 0 \quad \Leftrightarrow$
$(x+4)(x-3) = 0 \Rightarrow x = 3$ or -4. Because $\log x$ and $\log(x+1)$ are undefined at $x = -4$, it follows that $x = 3$ is the only solution.

61. $x^2 e^{2x} + 2x e^{2x} = 8 e^{2x} \quad \Leftrightarrow \quad e^{2x}(x^2 + 2x - 8) = 0 \quad \Leftrightarrow \quad x^2 + 2x - 8 = 0$ (since $e^{2x} > 0$ for all x) $\quad \Leftrightarrow$
$(x+4)(x-2) = 0 \quad \Leftrightarrow \quad x = 2, -4$

63. $5^{-2x/3} = 0.63 \quad \Leftrightarrow \quad \dfrac{-2x}{3} \log 5 = \log 0.63 \quad \Leftrightarrow \quad x = -\dfrac{3 \log 0.63}{2 \log 5} \approx 0.430618$

65. $5^{2x+1} = 3^{4x-1} \quad \Leftrightarrow \quad (2x+1) \log 5 = (4x-1) \log 3 \quad \Leftrightarrow \quad 2x \log 5 + \log 5 = 4x \log 3 - \log 3 \quad \Leftrightarrow$
$x(2 \log 5 - 4 \log 3) = -\log 3 - \log 5 \quad \Leftrightarrow \quad x = \dfrac{\log 3 + \log 5}{4 \log 3 - 2 \log 5} \approx 2.303600$

67. $y = e^{x/(x+2)}$. Vertical asymptote $x = -2$, horizontal asymptote $y = 2.72$, no maximum or minimum.

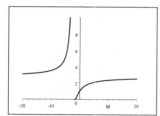

69. $y = \log(x^3 - x)$. Vertical asymptotes $x = -1$, $x = 0$, $x = 1$, no horizontal asymptote, local maximum of about -0.41 when $x \approx -0.58$.

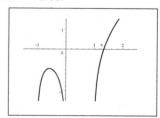

71. $3 \log x = 6 - 2x$. We graph $y = 3 \log x$ and $y = 6 - 2x$ in the same viewing rectangle. The solution occurs where the two graphs intersect. From the graphs, we see that the solution is $x \approx 2.42$.

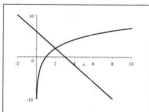

73. $\ln x > x - 2$. We graph the function $f(x) = \ln x - x + 2$, and we see that the graph lies above the x-axis for $0.16 < x < 3.15$. So the approximate solution of the given inequality is $0.16 < x < 3.15$.

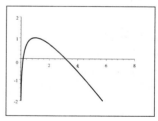

75. $f(x) = e^x - 3e^{-x} - 4x$. We graph the function $f(x)$, and we see that the function is increasing on $(-\infty, 0]$ and $[1.10, \infty)$ and decreasing on $[0, 1.10]$.

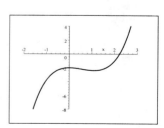

77. $\log_4 15 = \dfrac{\log 15}{\log 4} = 1.953445$

79. Notice that $\log_4 258 > \log_4 256 = \log_4 4^4 = 4$ and so $\log_4 258 > 4$. Also $\log_5 620 < \log_5 625 = \log_5 5^4 = 4$ and so $\log_5 620 < 4$. Then $\log_4 258 > 4 > \log_5 620$ and so $\log_4 258$ is larger.

81. $P = 12,000$, $r = 0.10$, and $t = 3$. Then $A = P\left(1 + \dfrac{r}{n}\right)^{nt}$.

   (a) For $n = 2$, $A = 12,000\left(1 + \dfrac{0.10}{2}\right)^{2(3)} = 12,000\left(1.05^6\right) \approx \$16,081.15$.

   (b) For $n = 12$, $A = 12,000\left(1 + \dfrac{0.10}{12}\right)^{12(3)} \approx \$16,178.18$.

   (c) For $n = 365$, $A = 12,000\left(1 + \dfrac{0.10}{365}\right)^{365(3)} \approx \$16,197.64$.

   (d) For $n = \infty$, $A = Pe^{rt} = 12,000e^{0.10(3)} \approx \$16,198.31$.

83. (a) Using the model $n(t) = n_0 e^{rt}$, with $n_0 = 30$ and $r = 0.15$, we have the formula $n(t) = 30e^{0.15t}$.

   (b) $n(4) = 30e^{0.15(4)} \approx 55$.

   (c) $500 = 30e^{0.15t}$ $\Leftrightarrow$ $\dfrac{50}{3} = e^{0.15t}$ $\Leftrightarrow$ $0.15t = \ln\left(\dfrac{50}{3}\right)$ $\Leftrightarrow$ $t = \dfrac{1}{0.15}\ln\left(\dfrac{50}{3}\right) \approx 18.76$. So the stray cat population will reach 500 in about 19 years.

85. (a) From the formula for radioactive decay, we have $m(t) = 10e^{-rt}$, where $r = -\dfrac{\ln 2}{2.7 \times 10^5}$. So after 1000 years the amount remaining is $m(1000) = 10 \cdot e^{\left[-\ln 2/\left(2.7 \times 10^5\right)\right]\cdot 1000} = 10e^{-(\ln 2)/\left(2.7 \times 10^2\right)} = 10e^{-(\ln 2)/270} \approx 9.97$. Therefore the amount remaining is about 9.97 mg.

   (b) We solve for t in the equation $7 = 10e^{-\left[\ln 2/\left(2.7 \times 10^5\right)\right]\cdot t}$. We have $7 = 10e^{-\left[\ln 2/\left(2.7 \times 10^5\right)\right]\cdot t}$ $\Leftrightarrow$ $0.7 = e^{-\left[\ln 2/\left(2.7 \times 10^5\right)\right]\cdot t}$ $\Leftrightarrow$ $\ln 0.7 = -\dfrac{\ln 2}{2.7 \times 10^5}\cdot t$ $\Leftrightarrow$ $t = -\dfrac{\ln 0.7}{\ln 2}\cdot 2.7 \times 10^5 \approx 138,934.75$. Thus it takes about 139,000 years.

87. (a) From the formula for radioactive decay, $r = \dfrac{\ln 2}{1590} \approx 0.0004359$ and $n(t) = 150 \cdot e^{-0.0004359t}$.

   (b) $n(1000) = 150 \cdot e^{-0.0004359 \cdot 1000} \approx 97.00$, and so the amount remaining is about 97.00 mg.

   (c) Find t so that $50 = 150 \cdot e^{-0.0004359t}$. We have $50 = 150 \cdot e^{-0.0004359t}$ $\Leftrightarrow$ $\dfrac{1}{3} = e^{-0.0004359t}$ $\Leftrightarrow$ $t = -\dfrac{1}{0.0004359}\ln\left(\dfrac{1}{3}\right) \approx 2520$. Thus only 50 mg remain after about 2520 years.

89. (a) Using $n_0 = 1500$ and $n(5) = 3200$ in the formula $n(t) = n_0 e^{rt}$, we have $3200 = n(5) = 1500e^{5r}$ $\Leftrightarrow$ $e^{5r} = \dfrac{32}{15}$ $\Leftrightarrow$ $5r = \ln\left(\dfrac{32}{15}\right)$ $\Leftrightarrow$ $r = \dfrac{1}{5}\ln\left(\dfrac{32}{15}\right) \approx 0.1515$. Thus $n(t) = 1500 \cdot e^{0.1515t}$.

   (b) We have $t = 1999 - 1988 = 11$ so $n(11) = 1500e^{0.1515 \cdot 11} \approx 7940$. Thus in 1999 the bird population should be about 7940.

91. $\left[H^+\right] = 1.3 \times 10^{-8}$ M. Then $\text{pH} = -\log\left[H^+\right] = -\log\left(1.3 \times 10^{-8}\right) \approx 7.9$, and so fresh egg whites are basic.

93. Let I_0 be the intensity of the smaller earthquake and I_1 be the intensity of the larger earthquake. Then $I_1 = 35I_0$. Since $M = \log\left(\dfrac{I}{S}\right)$, we have $M_0 = \log\left(\dfrac{I_0}{S}\right) = 6.5$ and $M_1 = \log\left(\dfrac{I_1}{S}\right) = \log\left(\dfrac{35I_0}{S}\right) = \log 35 + \log\left(\dfrac{I_0}{S}\right) = \log 35 + M_0 = \log 35 + 6.5 \approx 8.04$. So the magnitude on the Richter scale of the larger earthquake is approximately 8.0.

Chapter 5 Test

1. $y = 2^x$ and $y = \log_2 x$.

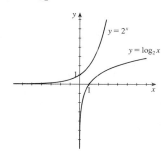

2. $f(x) = \log(x+1)$ has domain $(-1, \infty)$, range $(-\infty, \infty)$, and vertical asymptote $x = -1$.

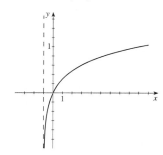

3. (a) $\log_3 \sqrt{27} = \log_3 \left(3^3\right)^{1/2} = \log_3 3^{3/2} = \frac{3}{2}$

(b) $\log_2 80 - \log_2 10 = \log_2 \left(\frac{80}{10}\right) = \log_2 8 = \log_2 2^3 = 3$

(c) $\log_8 4 = \log_8 8^{2/3} = \frac{2}{3}$

(d) $\log_6 4 + \log_6 9 = \log_6 (4 \cdot 9) = \log_6 6^2 = 2$

4. $\log \sqrt[3]{\dfrac{x+2}{x^4 \left(x^2 + 4\right)}} = \dfrac{1}{3} \log \left(\dfrac{x+2}{x^4 \left(x^2 + 4\right)}\right) = \frac{1}{3} \left[\log (x+2) - \left(4 \log x + \log \left(x^2 + 4\right)\right)\right]$

$\qquad = \frac{1}{3} \log (x+2) - \frac{4}{3} \log x - \frac{1}{3} \log \left(x^2 + 4\right)$

5. $\ln x - 2 \ln \left(x^2 + 1\right) + \frac{1}{2} \ln \left(3 - x^4\right) = \ln \left(x\sqrt{3 - x^4}\right) - \ln \left(x^2 + 1\right)^2 = \ln \left(\dfrac{x\sqrt{3 - x^4}}{\left(x^2 + 1\right)^2}\right)$

6. (a) $2^{x-1} = 10 \quad \Leftrightarrow \quad \log 2^{x-1} = \log 10 = 1 \quad \Leftrightarrow \quad (x-1) \log 2 = 1 \quad \Leftrightarrow \quad x - 1 = \dfrac{1}{\log 2} \quad \Leftrightarrow$

$\qquad x = 1 + \dfrac{1}{\log 2} \approx 4.32$

(b) $5 \ln (3 - x) = 4 \quad \Leftrightarrow \quad \ln (3 - x) = \frac{4}{5} \quad \Leftrightarrow \quad e^{\ln(3-x)} = e^{4/5} \quad \Leftrightarrow \quad 3 - x = e^{4/5} \quad \Leftrightarrow$

$\qquad x = 3 - e^{4/5} \approx 0.77$

(c) $10^{x+3} = 6^{2x} \quad \Leftrightarrow \quad \log 10^{x+3} = \log 6^{2x} \quad \Leftrightarrow \quad x + 3 = 2x \log 6 \quad \Leftrightarrow \quad 2x \log 6 - x = 3 \quad \Leftrightarrow$

$\qquad x (2 \log 6 - 1) = 3 \quad \Leftrightarrow \quad x = \dfrac{3}{2 \log 6 - 1} \approx 5.39$

(d) $\log_2 (x+2) + \log_2 (x-1) = 2 \quad \Leftrightarrow \quad \log_2 ((x+2)(x-1)) = 2 \quad \Leftrightarrow \quad x^2 + x - 2 = 2^2 \quad \Leftrightarrow$

$\qquad x^2 + x - 6 = 0 \quad \Leftrightarrow \quad (x+3)(x-2) = 0 \Rightarrow x = -3 \text{ or } x = 2.$ However, both logarithms are undefined at $x = -3$, so the only solution is $x = 2$.

7. (a) From the formula for population growth, we have $8000 = 1000e^{r \cdot 1}$

$\qquad \Leftrightarrow \quad 8 = e^r \quad \Leftrightarrow \quad r = \ln 8 \approx 2.07944.$ Thus
$\qquad n(t) = 1000e^{2.07944t}$.

(b) $n(1.5) = 1000e^{2.07944(1.5)} \approx 22{,}627$

(c) $15000 = 1000e^{2.07944t} \quad \Leftrightarrow \quad 15 = e^{2.07944t} \quad \Leftrightarrow$

$\qquad \ln 15 = 2.07944t \quad \Leftrightarrow \quad t = \dfrac{\ln 15}{2.07944} \approx 1.3.$ Thus the population

$\qquad$ will reach $15{,}000$ after approximately 1.3 hours.

(d)

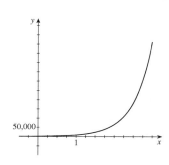

8. (a) $A(t) = 12{,}000 \left(1 + \frac{0.056}{12}\right)^{12t}$, where t is in years.

(b) $A(t) = 12{,}000 \left(1 + \frac{0.056}{365}\right)^{365t}$. So $A(3) = 12{,}000 \left(1 + \frac{0.056}{365}\right)^{365(3)} = \$14{,}195.06$.

(c) $A(t) = 12{,}000 \left(1 + \frac{0.056}{2}\right)^{2t} = 12{,}000 (1.028)^{2t}$. So $20{,}000 = 12{,}000 (1.028)^{2t} \quad \Leftrightarrow \quad 1.6667 = 1.028^{2t} \quad \Leftrightarrow$
$\ln 1.6667 = \ln 1.028^{2t} \quad \Leftrightarrow \quad \ln 1.6667 = 2t \ln 1.028 \quad \Leftrightarrow \quad t = \frac{\ln 1.6667}{2 \ln 1.028} \approx 9.25$ years.

9. $f(x) = \dfrac{e^x}{x^3}$

(a)

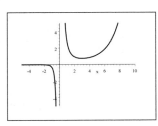

(b) f has vertical asymptote $x = 0$ and horizontal asymptote $y = 0$.

(c) f has a local minimum of about 0.74 when $x \approx 3.00$.

(d) The range of f is approximately $(-\infty, 0) \cup [0.74, \infty)$.

(e) $\dfrac{e^x}{x^3} = 2x + 1$.

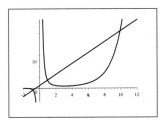

We see that the graphs intersect at $x \approx -0.85$, 0.96, and 9.92.

Focus on Modeling: Fitting Exponential and Power Curves to Data

1. (a)

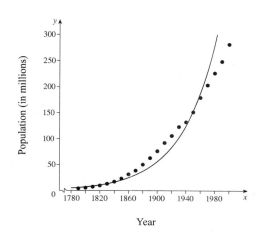

Year

(b) Using a graphing calculator, we obtain the model $y = ab^t$, where $a = 1.1806094 \times 10^{-15}$ and $b = 1.0204139$, and y is the population (in millions) in the year t.

(c) Substituting $t = 2010$ into the model of part (b), we get $y = ab^{2010} \approx 515.9$ million.

(d) According to the model, the population in 1965 should have been about $y = ab^{1965} \approx 207.8$ million.

(e) The values given by the model are clearly much too large. This means that an exponential model is *not* appropriate for these data.

3. (a) Yes.

(b)

Year t	Health Expenditures E ($bn)	$\ln E$
1970	74.3	4.30811
1980	251.1	5.52585
1985	434.5	6.07420
1987	506.2	6.22693
1990	696.6	6.54621
1992	820.3	6.70967
1994	937.2	6.84290
1996	1039.4	6.94640
1998	1150.0	7.04752
2000	1310.0	7.17778
2001	1424.5	7.26158

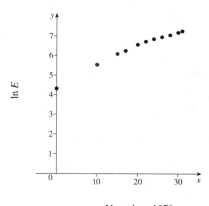

Year since 1970

Yes, the scatter plot appears to be roughly linear.

(c) Let t be the number of years elapsed since 1970 . Then $\ln E = 4.551437 + 0.09238268t$, where E is expenditure in billions of dollars.

(d) $E = e^{4.551437+0.09238268t} = 94.76849e^{0.09238268t}$

(e) In 2009 we have $t = 2009 - 1970 = 39$, so the estimated 2009 health-care expenditures are $94.76849e^{0.09238268(39)} \approx 3478.5$ billion dollars.

5. (a) Using a graphing calculator, we find that
$I_0 = 22.7586444$ and $k = 0.1062398$.

(c) We solve $0.15 = 22.7586444e^{-0.1062398x}$ for x:
$0.15 = 22.7586444e^{-0.1062398x} \quad \Leftrightarrow$
$0.006590902 = e^{-0.1062398x} \quad \Leftrightarrow$
$-5.022065 = -0.1062398x \quad \Leftrightarrow \quad x \approx 47.27$. So
light intensity drops below 0.15 lumens below around
47.27 feet.

(b)

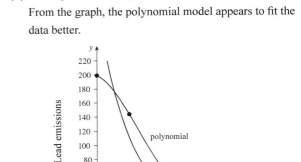

Depth (ft)

7. (a) Let t be the number of years elapsed since 1970, and let
y be the number of millions of tons of lead emissions in
year t. Using a graphing calculator, we obtain the
exponential model $y = ab^t$, where $a = 301.813054$
and $b = 0.819745$.

(b) Using a graphing calculator, we obtain the
fourth-degree polynomial model
$y = at^4 + bt^3 + ct^2 + dt + e$, where $a = -0.002430$,
$b = 0.135159$, $c = -2.014322$, $d = -4.055294$, and
$e = 199.092227$.

(d) Using the exponential model, we estimate emissions in
1972 to be $y(2) \approx 202.8$ million metric tons and
emissions in 1982 to be $y(12) \approx 27.8$ million metric
tons.
Using the polynomial model, we estimate emissions in
1972 to be $y(2) \approx 184.0$ million metric tons and
emissions in 1982 to be $y(12) \approx 43.5$ million metric
tons.

(c) The exponential and polynomial models are shown.
From the graph, the polynomial model appears to fit the
data better.

Year since 1970

9. (a)

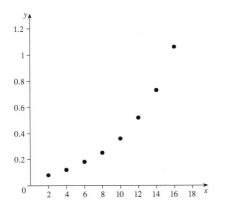

(b)

x	y	$\ln x$	$\ln y$
2	0.08	0.69315	-2.52573
4	0.12	1.38629	-2.12026
6	0.18	1.79176	-1.71480
8	0.25	2.07944	-1.38629
10	0.36	2.30259	-1.02165
12	0.52	2.48491	-0.65393
14	0.73	2.63906	-0.31471
16	1.06	2.77259	0.05827

(b)

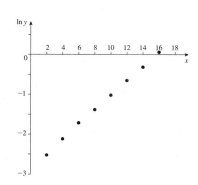

(c) The exponential function.

(d) $y = a \cdot b^x$ where $a = 0.057697$ and $b = 1.200236$.

11. (a) Using the `Logistic` command on a TI-83 we get $y = \dfrac{c}{1 + ae^{-bx}}$ where $a = 49.10976596$, $b = 0.4981144989$, and $c = 500.855793$.

(b) Using the model $N = \dfrac{c}{1 + ae^{-bt}}$ we solve for t. So $N = \dfrac{c}{1 + ae^{-bt}}$ $\Leftrightarrow$ $1 + ae^{-bt} = \dfrac{c}{N}$

$\Leftrightarrow$ $ae^{-bt} = \left(\dfrac{c}{N}\right) - 1 = \dfrac{c - N}{N}$ $\Leftrightarrow$ $e^{-bt} = \dfrac{c - N}{aN}$ $\Leftrightarrow$ $-bt = \ln(c - N) - \ln aN$

$\Leftrightarrow$ $t = \dfrac{1}{b}\left[\ln aN - \ln(c - N)\right]$. Substituting the values for a, b, and c, with $N = 400$ we have

$t = \frac{1}{0.4981144989}\left(\ln 19643.90638 - \ln 100.855793\right) \approx 10.58$ days.

6 Trigonometric Functions of Angles

6.1 Angle Measure

1. $72° = 72° \cdot \frac{\pi}{180°}$ rad $= \frac{2\pi}{5}$ rad ≈ 1.257 rad

3. $-45° = -45° \cdot \frac{\pi}{180°}$ rad $= -\frac{\pi}{4}$ rad ≈ -0.785 rad

5. $-75° = -75° \cdot \frac{\pi}{180°}$ rad $= -\frac{5\pi}{12}$ rad ≈ -1.309 rad

7. $1080° = 1080° \cdot \frac{\pi}{180°}$ rad $= 6\pi$ rad ≈ 18.850 rad

9. $96° = 96° \cdot \frac{\pi}{180°}$ rad $= \frac{8\pi}{15}$ rad ≈ 1.676 rad

11. $7.5° = 7.5° \cdot \frac{\pi}{180°}$ rad $= \frac{\pi}{24}$ rad ≈ 0.131 rad

13. $\frac{7\pi}{6} = \frac{7\pi}{6} \cdot \frac{180°}{\pi} = 210°$

15. $-\frac{5\pi}{4} = -\frac{5\pi}{4} \cdot \frac{180°}{\pi} = -225°$

17. $3 = 3 \cdot \frac{180°}{\pi} = \frac{540°}{\pi} \approx 171.9°$

19. $-1.2 = -1.2 \cdot \frac{180°}{\pi} = -\frac{216°}{\pi} = -68.8°$

21. $\frac{\pi}{10} = \frac{\pi}{10} \cdot \frac{180°}{\pi} = 18°$

23. $-\frac{2\pi}{15} = -\frac{2\pi}{15} \cdot \frac{180°}{\pi} = -24°$

25. $50°$ is coterminal with $50° + 360° = 410°$, $50° + 720° = 770°$, $50° - 360° = -310°$, and $50° - 720° = -670°$. (Other answers are possible.)

27. $\frac{3\pi}{4}$ is coterminal with $\frac{3\pi}{4} + 2\pi = \frac{11\pi}{4}$, $\frac{3\pi}{4} + 4\pi = \frac{19\pi}{4}$, $\frac{3\pi}{4} - 2\pi = -\frac{5\pi}{4}$, and $\frac{3\pi}{4} - 4\pi = -\frac{13\pi}{4}$. (Other answers are possible.)

29. $-\frac{\pi}{4}$ is coterminal with $-\frac{\pi}{4} + 2\pi = \frac{7\pi}{4}$, $-\frac{\pi}{4} + 4\pi = \frac{15\pi}{4}$, $-\frac{\pi}{4} - 2\pi = -\frac{9\pi}{4}$, and $-\frac{\pi}{4} - 4\pi = -\frac{17\pi}{4}$. (Other answers are possible.)

31. Since $430° - 70° = 360°$, the angles are coterminal.

33. Since $\frac{17\pi}{6} - \frac{5\pi}{6} = \frac{12\pi}{6} = 2\pi$; the angles are coterminal.

35. Since $875° - 155° = 720° = 2 \cdot 360°$, the angles are coterminal.

37. Since $733° - 2 \cdot 360° = 13°$, the angles $733°$ and $13°$ are coterminal.

39. Since $1110° - 3 \cdot 360° = 30°$, the angles $1110°$ and $30°$ are coterminal.

41. Since $-800° + 3 \cdot 360° = 280°$, the angles $-800°$ and $280°$ are coterminal.

43. Since $\frac{17\pi}{6} - 2\pi = \frac{5\pi}{6}$, the angles $\frac{17\pi}{6}$ and $\frac{5\pi}{6}$ are coterminal.

45. Since $87\pi - 43 \cdot 2\pi = \pi$, the angles 87π and π are coterminal.

47. Since $\frac{17\pi}{4} - 2 \cdot 2\pi = \frac{\pi}{4}$, the angles $\frac{17\pi}{4}$ and $\frac{\pi}{4}$ are coterminal.

49. Using the formula $s = \theta r$, the length of the arc is $s = \left(220° \cdot \frac{\pi}{180°}\right) \cdot 5 = \frac{55\pi}{9} \approx 19.2$.

51. Solving for r we have $r = \frac{s}{\theta}$, so the radius of the circle is $r = \frac{8}{2} = 4$.

53. Using the formula $s = \theta r$, the length of the arc is $s = 2 \cdot 2 = 4$ mi.

55. Solving for θ, we have $\theta = \frac{s}{r}$, so the measure of the central angle is $\theta = \frac{100}{50} = 2$ rad. Converting to degrees we have $\theta = 2 \cdot \frac{180°}{\pi} \approx 114.6°$

57. Solving for r, we have $r = \frac{s}{\theta}$, so the radius of the circle is $r = \frac{6}{\pi/6} = \frac{36}{\pi} \approx 11.46$ m.

59. (a) $A = \frac{1}{2}r^2\theta = \frac{1}{2} \cdot 8^2 \cdot 80° \cdot \frac{\pi}{180°} = 32 \cdot \frac{4\pi}{9} = \frac{128\pi}{9} \approx 44.68$

(b) $A = \frac{1}{2}r^2\theta = \frac{1}{2} \cdot 10^2 \cdot 0.5 = 25$

61. $A = \frac{1}{2}r^2\theta = \frac{1}{2} \cdot 10^2 \cdot 1 = 50$ m^2

63. $\theta = 2$ rad, $A = 16$ m^2. Since $A = \frac{1}{2}r^2\theta$, we have $r = \sqrt{2A/\theta} = \sqrt{2 \cdot 16/2} = \sqrt{16} = 4$ m.

65. Since the area of the circle is 72 cm^2, the radius of the circle is $r = \sqrt{A/\pi} = \sqrt{72/\pi}$. Then the area of the sector is $A = \frac{1}{2}r^2\theta = \frac{1}{2} \cdot \frac{72}{\pi} \cdot \frac{\pi}{6} = 6$ cm^2.

67. The circumference of each wheel is $\pi d = 28\pi$ in. If the wheels revolve 10,000 times, the distance traveled is $10{,}000 \cdot 28\pi$ in. $\cdot \dfrac{1 \text{ ft}}{12 \text{ in.}} \cdot \dfrac{1 \text{ mi}}{5280 \text{ ft}} \approx 13.88$ mi.

69. We find the measure of the angle in degrees and then convert to radians. $\theta = 40.5° - 25.5° = 15°$ and $15 \cdot \frac{\pi}{180°}$ rad $= \frac{\pi}{12}$ rad. Then using the formula $s = \theta r$, we have $s = \frac{\pi}{12} \cdot 3960 = 330\pi \approx 1036.725$ and so the distance between the two cities is roughly 1037 mi.

71. In one day, the earth travels $\frac{1}{365}$ of its orbit which is $\frac{2\pi}{365}$ rad. Then $s = \theta r = \frac{2\pi}{365} \cdot 93{,}000{,}000 \approx 1{,}600{,}911.3$, so the distance traveled is approximately 1.6 million miles.

73. The central angle is 1 minute $= \left(\frac{1}{60}\right)° = \frac{1}{60} \cdot \frac{\pi}{180°}$ rad $= \frac{\pi}{10{,}800}$ rad. Then $s = \theta r = \frac{\pi}{10{,}800} \cdot 3960 \approx 1.152$, and so a nautical mile is approximately 1.152 mi.

75. The area is equal to the area of the large sector (with radius 34 in.) minus the area of the small sector (with radius 14 in.) Thus, $A = \frac{1}{2}r_1^2\theta - \frac{1}{2}r_2^2\theta = \frac{1}{2}\left(34^2 - 14^2\right)\left(135° \cdot \frac{\pi}{180°}\right) \approx 1131$ in.^{2}.

77. $v = \dfrac{8 \cdot 2\pi \cdot 2}{15} = \dfrac{32\pi}{15} \approx 6.702$ ft/s.

79. (a) The angular speed is $\omega = \dfrac{1000 \cdot 2\pi \text{ rad}}{1 \text{ min}} = 2000\pi$ rad/min.

  (b) The linear speed is $v = \dfrac{1000 \cdot 2\pi \cdot \frac{6}{12}}{60} = \dfrac{50\pi}{3}$ ft/s ≈ 52.4 ft/s.

81. $v = \dfrac{600 \cdot 2\pi \cdot \frac{11}{12} \text{ ft}}{1 \text{ min}} \cdot \dfrac{1 \text{ mi}}{5280 \text{ ft}} \cdot \dfrac{60 \text{ min}}{1 \text{ hr}} = 12.5\pi$ mi/h ≈ 39.3 mi/h.

83. $v = \dfrac{100 \cdot 2\pi \cdot 0.20 \text{ m}}{60 \text{ s}} = \dfrac{2\pi}{3} \approx 2.09$ m/s.

85. (a) The circumference of the opening is the length of the arc subtended by the angle θ on the flat piece of paper, that is, $C = s = r\theta = 6 \cdot \frac{5\pi}{3} = 10\pi \approx 31.4$ cm.

  (b) Solving for r, we find $r = \dfrac{C}{2\pi} = \dfrac{10\pi}{2\pi} = 5$ cm.

  (c) By the Pythagorean Theorem, $h^2 = 6^2 - 5^2 = 11$, so $h = \sqrt{11} \approx 3.3$ cm.

  (d) The volume of a cone is $V = \frac{1}{3}\pi r^2 h$. In this case $V = \frac{1}{3}\pi \cdot 5^2 \cdot \sqrt{11} \approx 86.8$ cm^3.

87. Answers will vary, although of course everyone prefers radians.

6.2   Trigonometry of Right Angles

1. $\sin\theta = \frac{4}{5}$, $\cos\theta = \frac{3}{5}$, $\tan\theta = \frac{4}{3}$, $\csc\theta = \frac{5}{4}$, $\sec\theta = \frac{5}{3}$, $\cot\theta = \frac{3}{4}$

3. The remaining side is obtained by the Pythagorean Theorem: $\sqrt{41^2 - 40^2} = \sqrt{81} = 9$. Then $\sin\theta = \frac{40}{41}$, $\cos\theta = \frac{9}{41}$, $\tan\theta = \frac{40}{9}$, $\csc\theta = \frac{41}{40}$, $\sec\theta = \frac{41}{9}$, $\cot\theta = \frac{9}{40}$

5. The remaining side is obtained by the Pythagorean Theorem: $\sqrt{3^2 + 2^2} = \sqrt{13}$. Then $\sin\theta = \frac{2}{\sqrt{13}} = \frac{2\sqrt{13}}{13}$, $\cos\theta = \frac{3}{\sqrt{13}} = \frac{3\sqrt{13}}{13}$, $\tan\theta = \frac{2}{3}$, $\csc\theta = \frac{\sqrt{13}}{2}$, $\sec\theta = \frac{\sqrt{13}}{3}$, $\cot\theta = \frac{3}{2}$

7. $c = \sqrt{5^2 + 3^2} = \sqrt{34}$

  (a) $\sin\alpha = \cos\beta = \frac{3}{\sqrt{34}} = \frac{3\sqrt{34}}{34}$ **(b)** $\tan\alpha = \cot\beta = \frac{3}{5}$ **(c)** $\sec\alpha = \csc\beta = \frac{\sqrt{34}}{5}$

9. Since $\sin 30° = \dfrac{x}{25}$, we have $x = 25\sin 30° = 25 \cdot \frac{1}{2} = \frac{25}{2}$.

11. Since $\sin 60° = \dfrac{x}{13}$, we have $x = 13\sin 60° = 13 \cdot \frac{\sqrt{3}}{2} = \frac{13\sqrt{3}}{2}$.

13. Since $\tan 36° = \dfrac{12}{x}$, we have $x = \dfrac{12}{\tan 36°} \approx 16.51658$.

15. $\dfrac{x}{28} = \cos\theta \quad\Leftrightarrow\quad x = 28\cos\theta$, and $\dfrac{y}{28} = \sin\theta \quad\Leftrightarrow\quad y = 28\sin\theta$.

17. $\sin\theta = \frac{3}{5}$. Then the third side is $x = \sqrt{5^2 - 3^2} = 4$. The other five ratios are $\cos\theta = \frac{4}{5}$, $\tan\theta = \frac{3}{4}$, $\csc\theta = \frac{5}{3}$, $\sec\theta = \frac{5}{4}$, and $\cot\theta = \frac{4}{3}$.

19. $\cot\theta = 1$. Then the third side is $r = \sqrt{1^2 + 1^2} = \sqrt{2}$. The other five ratios are $\sin\theta = \frac{1}{\sqrt{2}} = \frac{\sqrt{2}}{2}$, $\cos\theta = \frac{1}{\sqrt{2}} = \frac{\sqrt{2}}{2}$, $\tan\theta = 1$, $\csc\theta = \sqrt{2}$, and $\sec\theta = \sqrt{2}$.

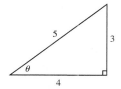

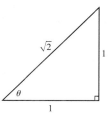

21. $\sec\theta = 7$. The third side is $y = \sqrt{7^2 - 2^2} = \sqrt{45} = 3\sqrt{5}$ The other five ratios are $\sin\theta = \frac{3\sqrt{5}}{7}$, $\cos\theta = \frac{2}{7}$, $\tan\theta = \frac{3\sqrt{5}}{2}$, $\csc\theta = \frac{7}{3\sqrt{5}} = \frac{7\sqrt{5}}{15}$, and $\cot\theta = \frac{2}{3\sqrt{5}} = \frac{2\sqrt{5}}{15}$.

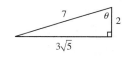

23. $\sin\frac{\pi}{6} + \cos\frac{\pi}{6} = \frac{1}{2} + \frac{\sqrt{3}}{2} = \frac{1+\sqrt{3}}{2}$

25. $\sin 30° \cos 60° + \sin 60° \cos 30° = \frac{1}{2} \cdot \frac{1}{2} + \frac{\sqrt{3}}{2} \cdot \frac{\sqrt{3}}{2} = \frac{1}{4} + \frac{3}{4} = 1$

27. $(\cos 30°)^2 - (\sin 30°)^2 = \left(\frac{\sqrt{3}}{2}\right)^2 - \left(\frac{1}{2}\right)^2 = \frac{3}{4} - \frac{1}{4} = \frac{1}{2}$

29. This is an isosceles right triangle, so the other leg has length $16\tan 45° = 16$, the hypotenuse has length $\dfrac{16}{\sin 45°} = 16\sqrt{2} \approx 22.63$, and the other angle is $90° - 45° = 45°$.

31. The other leg has length $35\tan 52° \approx 44.79$, the hypotenuse has length $\dfrac{35}{\cos 52°} \approx 56.85$, and the other angle is $90° - 52° = 38°$.

33. The adjacent leg has length $33.5\cos\frac{\pi}{8} \approx 30.95$, the opposite leg has length $33.5\sin\frac{\pi}{8} \approx 12.82$, and the other angle is $\frac{\pi}{2} - \frac{\pi}{8} = \frac{3\pi}{8}$.

35. The adjacent leg has length $\dfrac{106}{\tan\frac{\pi}{5}} \approx 145.90$, the hypotenuse has length $\dfrac{106}{\sin\frac{\pi}{5}} \approx 180.34$, and the other angle is $\frac{\pi}{2} - \frac{\pi}{5} = \frac{3\pi}{10}$.

37. $\sin\theta \approx \frac{1}{2.24} \approx 0.45$. $\cos\theta \approx \frac{2}{2.24} \approx 0.89$, $\tan\theta = \frac{1}{2}$, $\csc\theta \approx 2.24$, $\sec\theta \approx \frac{2.24}{2} \approx 1.12$, $\cot\theta \approx 2.00$.

39. $x = \dfrac{100}{\tan 60°} + \dfrac{100}{\tan 30°} \approx 230.9$

41. Let h be the length of the shared side. Then $\sin 60° = \dfrac{50}{h} \quad\Leftrightarrow\quad h = \dfrac{50}{\sin 60°} \approx 57.735 \quad\Leftrightarrow\quad \sin 65° = \dfrac{h}{x} \Leftrightarrow$ $x = \dfrac{h}{\sin 65°} \approx 63.7$

43.

From the diagram, $\sin\theta = \dfrac{x}{y}$ and $\tan\theta = \dfrac{y}{10}$, so $x = y\sin\theta = 10\sin\theta\tan\theta$.

45. Let h be the height, in feet, of the Empire State Building. Then $\tan 11° = \dfrac{h}{5280}$ $\Leftrightarrow$ $h = 5280 \cdot \tan 11° \approx 1026$ ft.

47. **(a)** Let h be the distance, in miles, that the beam has diverged. Then $\tan 0.5° = \dfrac{h}{240{,}000}$ $\Leftrightarrow$

$h = 240{,}000 \cdot \tan 0.5° \approx 2100$ mi.

(b) Since the deflection is about 2100 mi whereas the radius of the moon is about 1000 mi, the beam will not strike the moon.

49. Let h represent the height, in feet, that the ladder reaches on the building. Then $\sin 72° = \dfrac{h}{20}$ $\Leftrightarrow$

$h = 20\sin 72° \approx 19$ ft.

51. Let θ be the angle of elevation of the sun. Then $\tan\theta = \dfrac{96}{120} = 0.8$ $\Leftrightarrow$ $\theta = \tan^{-1} 0.8 \approx 0.675 \approx 38.7°$.

53. Let h be the height, in feet, of the kite above the ground. Then $\sin 50° = \dfrac{h}{450}$ $\Leftrightarrow$ $h = 450\sin 50° \approx 345$ ft.

55. Let h_1 be the height of the window in feet and h_2 be the height from the window to the top of the tower. Then

$\tan 25° = \dfrac{h_1}{325}$ $\Leftrightarrow$ $h_1 = 325 \cdot \tan 25° \approx 152$ ft. Also, $\tan 39° = \dfrac{h_2}{325} \Leftrightarrow h_2 = 325 \cdot \tan 39° \approx 263$ ft. Therefore,

the height of the window is approximately 152 ft and the height of the tower is approximately $152 + 263 = 415$ ft.

57. Let d_1 be the distance, in feet, between a point directly below the plane and one car, and d_2 be the distance, in feet, between

the same point and the other car. Then $\tan 52° = \dfrac{d_1}{5150} \Leftrightarrow d_1 = 5150 \cdot \tan 52° \approx 6591.7$ ft. Also, $\tan 38° = \dfrac{d_2}{5150}$ $\Leftrightarrow$

$d_2 = 5150 \cdot \tan 38° \approx 4023.6$ ft. So in this case, the distance between the two cars is about 2570 ft.

59. Let x be the horizontal distance, in feet, between a point on the ground directly below the top of the mountain and

the point on the plain closest to the mountain. Let h be the height, in feet, of the mountain. Then $\tan 35° = \dfrac{h}{x}$ and

$\tan 32° = \dfrac{h}{x + 1000}$. So $h = x\tan 35° = (x + 1000)\tan 32°$ $\Leftrightarrow$ $x = \dfrac{1000 \cdot \tan 32°}{\tan 35° - \tan 32°} \approx 8294.2$. Thus

$h \approx 8294.2 \cdot \tan 35° \approx 5808$ ft.

61. Let d be the distance, in miles, from the earth to the sun. Then $\tan 89.95° = \dfrac{d}{240{,}000}$ $\Leftrightarrow$

$d = 240{,}000 \cdot \tan 89.95° \approx 91.7$ million miles.

63. Let r represent the radius, in miles, of the earth. Then $\sin 60.276° = \dfrac{r}{r+600}$ $\Leftrightarrow$ $(r + 600)\sin 60.276° = r$ $\Leftrightarrow$

$600\sin 60.276° = r\left(1 - \sin 60.276°\right)$ $\Leftrightarrow$ $r = \dfrac{600\sin 60.276°}{1 - \sin 60.276°} \approx 3960.099$. So the radius of the earth is approximately

3960 mi.

65. Let d be the distance, in AU, between Venus and the sun. Then $\sin 46.3° = \dfrac{d}{1} = d$, so $d = \sin 46.3° \approx 0.723$ AU.

6.3 Trigonometric Functions of Angles

1. (a) The reference angle for $150°$ is $180° - 150° = 30°$.

(b) The reference angle for $330°$ is $360° - 330° = 30°$.

(c) The reference angle for $-30°$ is $-(-30°) = 30°$.

3. (a) The reference angle for $225°$ is $225° - 180° = 45°$.

(b) The reference angle for $810°$ is $810° - 720° = 90°$.

(c) The reference angle for $-105°$ is $180° - 105° = 75°$.

5. (a) The reference angle for $\frac{11\pi}{4}$ is $3\pi - \frac{11\pi}{4} = \frac{\pi}{4}$.

(b) The reference angle for $-\frac{11\pi}{6}$ is $2\pi - \frac{11\pi}{6} = \frac{\pi}{6}$.

(c) The reference angle for $\frac{11\pi}{3}$ is $4\pi - \frac{11\pi}{3} = \frac{\pi}{3}$.

7. (a) The reference angle for $\frac{5\pi}{7}$ is $\pi - \frac{5\pi}{7} = \frac{2\pi}{7}$.

(b) The reference angle for -1.4π is $1.4\pi - \pi = 0.4\pi$.

(c) The reference angle for 1.4 is 1.4 because $1.4 < \frac{\pi}{2}$.

9. $\sin 150° = \sin 30° = \frac{1}{2}$

11. $\cos 135° = -\cos 45° = -\frac{1}{\sqrt{2}} = -\frac{\sqrt{2}}{2}$

13. $\tan(-60°) = -\tan 60° = -\sqrt{3}$

15. $\csc(-630°) = \csc 90° = \dfrac{1}{\sin 90°} = 1$

17. $\cos 570° = -\cos 30° = -\frac{\sqrt{3}}{2}$

19. $\tan 750° = \tan 30° = \frac{1}{\sqrt{3}} = \frac{\sqrt{3}}{3}$

21. $\sin \frac{2\pi}{3} = \sin \frac{\pi}{3} = \frac{\sqrt{3}}{2}$

23. $\sin \frac{3\pi}{2} = -\sin \frac{\pi}{2} = -1$

25. $\cos\left(-\frac{7\pi}{3}\right) = \cos \frac{\pi}{3} = \frac{1}{2}$

27. $\sec \frac{17\pi}{3} = \sec \frac{\pi}{3} = \dfrac{1}{\cos \frac{\pi}{3}} = 2$

29. $\cot\left(-\frac{\pi}{4}\right) = -\cot \frac{\pi}{4} = \dfrac{-1}{\tan \frac{\pi}{4}} = -1$

31. $\tan \frac{5\pi}{2} = \tan \frac{\pi}{2}$ which is undefined.

33. Since $\sin\theta < 0$ and $\cos\theta < 0$, θ is in quadrant III.

35. $\sec\theta > 0 \implies \cos\theta > 0$. Also $\tan\theta < 0 \implies \dfrac{\sin\theta}{\cos\theta} < 0 \iff \sin\theta < 0$ (since $\cos\theta > 0$). Since $\sin\theta < 0$ and $\cos\theta > 0$, θ is in quadrant IV.

37. $\sec^2\theta = 1 + \tan^2\theta \iff \tan^2\theta = \dfrac{1}{\cos^2\theta} - 1 \iff$

$\tan\theta = \sqrt{\dfrac{1}{\cos^2\theta} - 1} = \sqrt{\dfrac{1 - \cos^2\theta}{\cos^2\theta}} = \dfrac{\sqrt{1 - \cos^2\theta}}{|\cos\theta|} = \dfrac{\sqrt{1 - \cos^2\theta}}{-\cos\theta}$ (since $\cos\theta < 0$ in quadrant III,

$|\cos\theta| = -\cos\theta$). Thus $\tan\theta = -\dfrac{\sqrt{1 - \cos^2\theta}}{\cos\theta}$.

39. $\cos^2\theta + \sin^2\theta = 1 \iff \cos\theta = \sqrt{1 - \sin^2\theta}$ because $\cos\theta > 0$ in quadrant IV.

41. $\sec^2\theta = 1 + \tan^2\theta \iff \sec\theta = -\sqrt{1 + \tan^2\theta}$ because $\sec\theta < 0$ in quadrant II.

43. $\sin\theta = \frac{3}{5}$. Then $x = -\sqrt{5^2 - 3^2} = -\sqrt{16} = -4$, since θ is in quadrant II. Thus, $\cos\theta = -\frac{4}{5}$, $\tan\theta = -\frac{3}{4}$, $\csc\theta = \frac{5}{3}$, $\sec\theta = -\frac{5}{4}$, and $\cot\theta = -\frac{4}{3}$.

45. $\tan\theta = -\frac{3}{4}$. Then $r = \sqrt{3^2 + 4^2} = 5$, and so $\sin\theta = -\frac{3}{5}$, $\cos\theta = \frac{4}{5}$, $\csc\theta = -\frac{5}{3}$, $\sec\theta = \frac{5}{4}$, and $\cot\theta = -\frac{4}{3}$.

47. $\csc\theta = 2$. Then $\sin\theta = \frac{1}{2}$ and $x = \sqrt{2^2 - 1^2} = \sqrt{3}$. So $\sin\theta = \frac{1}{2}$, $\cos\theta = \frac{\sqrt{3}}{2}$, $\tan\theta = \frac{1}{\sqrt{3}} = \frac{\sqrt{3}}{3}$, $\sec\theta = \frac{2}{\sqrt{3}} = \frac{2\sqrt{3}}{3}$, and $\cot\theta = \sqrt{3}$.

49. $\cos\theta = -\frac{2}{7}$. Then $y = \sqrt{7^2 - 2^2} = \sqrt{45} = 3\sqrt{5}$, and so $\sin\theta = \frac{3\sqrt{5}}{7}$, $\tan\theta = -\frac{3\sqrt{5}}{2}$, $\csc\theta = \frac{7}{3\sqrt{5}} = \frac{7\sqrt{5}}{15}$, $\sec\theta = -\frac{7}{2}$, and $\cot\theta = -\frac{2}{3\sqrt{5}} = -\frac{2\sqrt{5}}{15}$.

51. (a) $\sin 2\theta = \sin\left(2 \cdot \frac{\pi}{3}\right) = \sin \frac{2\pi}{3} = \sin \frac{\pi}{3} = \frac{\sqrt{3}}{2}$, while $2\sin\theta = 2\sin \frac{\pi}{3} = 2 \cdot \frac{\sqrt{3}}{2} = \sqrt{3}$.

(b) $\sin \frac{1}{2}\theta = \sin\left(\frac{1}{2} \cdot \frac{\pi}{3}\right) = \sin \frac{\pi}{6} = \frac{1}{2}$, while $\frac{1}{2}\sin\theta = \frac{1}{2}\sin \frac{\pi}{3} = \frac{1}{2} \cdot \frac{\sqrt{3}}{2} = \frac{\sqrt{3}}{4}$.

(c) $\sin^2\theta = \left(\sin \frac{\pi}{3}\right)^2 = \left(\frac{\sqrt{3}}{2}\right)^2 = \frac{3}{4}$, while $\sin\left(\theta^2\right) = \sin\left(\frac{\pi}{3}\right)^2 = \sin \frac{\pi^2}{9} \approx 0.88967$.

53. $a = 10$, $b = 22$, and $\theta = 10°$. Thus, the area of the triangle is $A = \frac{1}{2}(10)(22)\sin 10° = 110\sin 10° \approx 19.1$.

55. $A = 16$, $a = 5$, and $b = 7$. So $\sin \theta = \dfrac{2A}{ab} = \dfrac{2 \cdot 16}{5 \cdot 7} = \dfrac{32}{35}$ $\Leftrightarrow$ $\theta = \sin^{-1} \frac{32}{35} \approx 66.1°$.

57. For the sector defined by the two sides, $A_1 = \frac{1}{2} r^2 \theta = \frac{1}{2} \cdot 2^2 \cdot 120° \cdot \frac{\pi}{180°} = \frac{4\pi}{3}$. For the triangle defined by the two sides,

$A_2 = \frac{1}{2} ab \sin \theta = \frac{1}{2} \cdot 2 \cdot 2 \cdot \sin 120° = 2 \sin 60° = \sqrt{3}$. Thus the area of the region is $A_1 - A_2 = \frac{4\pi}{3} - \sqrt{3} \approx 2.46$.

59. $\sin^2 \theta + \cos^2 \theta = 1$ $\Leftrightarrow$ $\left(\sin^2 \theta + \cos^2 \theta \right) \cdot \dfrac{1}{\cos^2 \theta} = 1 \cdot \dfrac{1}{\cos^2 \theta}$ $\Leftrightarrow$ $\tan^2 \theta + 1 = \sec^2 \theta$

61. (a) $\tan \theta = \dfrac{h}{1 \text{ mile}}$, so $h = \tan \theta \cdot 1 \text{ mile} \cdot \dfrac{5280 \text{ ft}}{1 \text{ mile}} = 5280 \tan \theta$ ft.

(b)

θ	$20°$	$60°$	$80°$	$85°$
h	1922	9145	29,944	60,351

63. (a) From the figure in the text, we express depth and width in terms of θ.

Since $\sin \theta = \dfrac{\text{depth}}{20}$ and $\cos \theta = \dfrac{\text{width}}{20}$, we have depth $= 20 \sin \theta$

and width $= 20 \cos \theta$. Thus, the cross-section area of the beam is

$A(\theta) = (\text{depth}) (\text{width}) = (20 \cos \theta)(20 \sin \theta) = 400 \cos \theta \sin \theta$.

(b)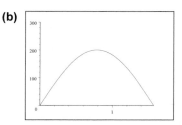

(c) The beam with the largest cross-sectional area is the square beam,

$10\sqrt{2}$ by $10\sqrt{2}$ (about 14.14 by 14.14).

65. (a) On Earth, the range is $R = \dfrac{v_0^2 \sin(2\theta)}{g} = \dfrac{12^2 \sin \frac{\pi}{3}}{32} = \dfrac{9\sqrt{3}}{4} \approx 3.897$ ft and the height is

$H = \dfrac{v_0^2 \sin^2 \theta}{2g} = \dfrac{12^2 \sin^2 \frac{\pi}{6}}{2 \cdot 32} = \dfrac{9}{16} = 0.5625$ ft.

(b) On the moon, $R = \dfrac{12^2 \sin \frac{\pi}{3}}{5.2} \approx 23.982$ ft and $H = \dfrac{12^2 \sin^2 \frac{\pi}{6}}{2 \cdot 5.2} \approx 3.462$ ft

67. (a) $W = 3.02 - 0.38 \cot \theta + 0.65 \csc \theta$

(b) From the graph, it appears that W has its minimum value at about

$\theta = 0.946 \approx 54.2°$.

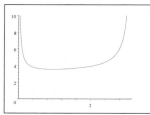

69. We have $\sin \alpha = k \sin \beta$, where $\alpha = 59.4°$ and $k = 1.33$. Substituting, $\sin 59.4° = 1.33 \sin \beta$

$\Rightarrow \sin \beta = \dfrac{\sin 59.4°}{1.33} \approx 0.6472$. Using a calculator, we find that $\beta \approx \sin^{-1} 0.6472 \approx 40.3°$, so

$\theta = 4\beta - 2\alpha \approx 4(40.3°) - 2(59.4°) = 42.4°$.

71. $\cos \theta = \dfrac{\text{adj}}{\text{hyp}} = \dfrac{|OP|}{|OR|} = \dfrac{|OP|}{1} = |OP|$. Since QS is tangent to the circle at R, $\triangle ORQ$ is a right triangle. Then

$\tan \theta = \dfrac{\text{opp}}{\text{adj}} = \dfrac{|RQ|}{|OR|} = |RQ|$ and $\sec \theta = \dfrac{\text{hyp}}{\text{adj}} = \dfrac{|OQ|}{|OR|} = |OQ|$. Since $\angle SOQ$ is a right angle $\triangle SOQ$ is a right

triangle and $\angle OSR = \theta$. Then $\csc \theta = \dfrac{\text{hyp}}{\text{opp}} = \dfrac{|OS|}{|OR|} = |OS|$ and $\cot \theta = \dfrac{\text{adj}}{\text{opp}} = \dfrac{|SR|}{|OR|} = |SR|$. Summarizing, we have

$\sin \theta = |PR|$, $\cos \theta = |OP|$, $\tan \theta = |RQ|$, $\sec \theta = |OQ|$, $\csc \theta = |OS|$, and $\cot \theta = |SR|$.

6.4   The Law of Sines

1. $\angle C = 180° - 98.4° - 24.6° = 57°$. $x = \dfrac{376 \sin 57°}{\sin 98.4°} \approx 318.75$.

3. $\angle C = 180° - 52° - 70° = 58°$. $x = \dfrac{26.7 \sin 52°}{\sin 58°} \approx 24.8$.

5. $\sin C = \dfrac{36 \sin 120°}{45} \approx 0.693 \Leftrightarrow \angle C \approx \sin^{-1} 0.693 \approx 44°$.

7. $\angle C = 180° - 46° - 20° = 114°$. Then $a = \dfrac{65 \sin 46°}{\sin 114°} \approx 51$ and $b = \dfrac{65 \sin 20°}{\sin 114°} \approx 24$.

9. $\angle B = 68°$, so $\angle A = 180° - 68° - 68° = 44°$ and $a = \dfrac{12 \sin 44°}{\sin 68°} \approx 8.99$.

11. $\angle C = 180° - 50° - 68° = 62°$. Then

$a = \dfrac{230 \sin 50°}{\sin 62°} \approx 200$ and $b = \dfrac{230 \sin 68°}{\sin 62°} \approx 242$.

13. $\angle B = 180° - 30° - 65° = 85°$. Then

$a = \dfrac{10 \sin 30°}{\sin 85°} \approx 5.0$ and $c = \dfrac{10 \sin 65°}{\sin 85°} \approx 9$.

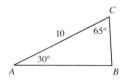

15. $\angle A = 180° - 51° - 29° = 100°$. Then

$a = \dfrac{44 \sin 100°}{\sin 29°} \approx 89$ and $c = \dfrac{44 \sin 51°}{\sin 29°} \approx 71$.

17. Since $\angle A > 90°$ there is only one triangle.

$\sin B = \dfrac{15 \sin 110°}{28} \approx 0.503 \quad \Leftrightarrow$

$\angle B \approx \sin^{-1} 0.503 \approx 30°$. Then

$\angle C \approx 180° - 110° - 30° = 40°$, and so

$c = \dfrac{28 \sin 40°}{\sin 110°} \approx 19$. Thus $\angle B \approx 30°$, $\angle C \approx 40°$, and

$c \approx 19$.

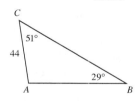

19. $\angle A = 125°$ is the largest angle, but since side a is not the longest side, there can be no such triangle.

21. $\sin C = \dfrac{30 \sin 25°}{25} \approx 0.507 \Leftrightarrow \angle C_1 \approx \sin^{-1} 0.507 \approx 30.47°$ or $\angle C_2 \approx 180° - 39.47° = 149.53°$.

If $\angle C_1 = 30.47°$, then $\angle A_1 \approx 180° - 25° - 30.47° = 124.53°$ and $a_1 = \dfrac{25 \sin 124.53°}{\sin 25°} \approx 48.73$.

If $\angle C_2 = 149.53°$, then $\angle A_2 \approx 180° - 25° - 149.53° = 5.47°$ and $a_2 = \dfrac{25 \sin 5.47°}{\sin 25°} \approx 5.64$.

Thus, one triangle has $\angle A_1 \approx 125°$, $\angle C_1 \approx 30°$, and $a_1 \approx 49$; the other has $\angle A_2 \approx 5°$, $\angle C_2 \approx 150°$, and $a_2 \approx 5.6$.

23. $\sin B = \dfrac{100 \sin 50°}{50} \approx 1.532$. Since $|\sin \theta| \leq 1$ for all θ, there can be no such angle B, and thus no such triangle.

25. $\sin A = \dfrac{26 \sin 29°}{15} \approx 0.840 \Leftrightarrow \angle A_1 \approx \sin^{-1} 0.840 \approx 57.2°$ or $\angle A_2 \approx 180° - 57.2° = 122.8°$.

If $\angle A_1 \approx 57.2°$, then $\angle B_1 = 180° - 29° - 57.2° = 93.8°$ and $b_1 \approx \dfrac{15 \sin 93.8°}{\sin 29°} \approx 30.9$.

If $\angle A_2 \approx 122.8°$, then $\angle B_2 = 180 - 29° - 122.8° = 28.2°$ and $b_2 \approx \dfrac{15 \sin 28.1°}{\sin 29°} \approx 14.6$.

Thus, one triangle has $\angle A_1 \approx 57.2°$, $\angle B_1 \approx 93.8°$, and $b_1 \approx 30.9$; the other has $\angle A_2 \approx 122.8°$, $\angle B_2 \approx 28.2°$, and $b_2 \approx 14.6$.

27. (a) From $\triangle ABC$ and the Law of Sines we get $\dfrac{\sin 30^\circ}{20} = \dfrac{\sin B}{28} \Leftrightarrow \sin B = \dfrac{28 \sin 30^\circ}{20} = 0.7$,

so $\angle B \approx \sin^{-1} 0.7 \approx 44.427^\circ$. Since $\triangle BCD$ is isosceles, $\angle B = \angle BDC \approx 44.427^\circ$. Thus,

$\angle BCD = 180^\circ - 2\angle B \approx 91.146^\circ \approx 91.1^\circ$.

(b) From $\triangle ABC$ we get $\angle BCA = 180^\circ - \angle A - \angle B \approx 180^\circ - 30^\circ - 44.427^\circ = 105.573^\circ$. Hence

$\angle DCA = \angle BCA - \angle BCD \approx 105.573^\circ - 91.146^\circ = 14.4^\circ$.

29. (a) $\sin B = \dfrac{20 \sin 40^\circ}{15} \approx 0.857 \Leftrightarrow \angle B_1 \approx \sin^{-1} 0.857 \approx 58.99^\circ$ or $\angle B_2 \approx 180^\circ - 58.99^\circ \approx 121.01^\circ$.

If $\angle B_1 = 30.47^\circ$, then $\angle C_1 \approx 180^\circ - 15^\circ - 58.99^\circ = 106.01^\circ$ and $c_1 = \dfrac{15 \sin 106.01^\circ}{\sin 40^\circ} \approx 22.43$.

If $\angle B_2 = 121.01^\circ$, then $\angle C_2 \approx 180^\circ - 15^\circ - 121.01^\circ = 43.99^\circ$ and $c_2 = \dfrac{15 \sin 43.99^\circ}{\sin 40^\circ} \approx 16.21$. Thus there are

two triangles.

(b) By the area formula given in Section 6.3, $\dfrac{\text{Area of } \triangle ABC}{\text{Area of } \triangle A'B'C'} = \dfrac{\frac{1}{2}ab \sin C}{\frac{1}{2}ab \sin C'} = \dfrac{\sin C}{\sin C'}$, since a and b are the same in

both triangles.

31. (a) Let a be the distance from satellite to the tracking station A in miles. Then the subtended angle at the satellite is

$\angle C = 180^\circ - 93^\circ - 84.2^\circ = 2.8^\circ$, and so $a = \dfrac{50 \sin 84.2^\circ}{\sin 2.8^\circ} \approx 1018$ mi.

(b) Let d be the distance above the ground in miles. Then $d = 1018.3 \sin 87^\circ \approx 1017$ mi.

33. $\angle C = 180^\circ - 82^\circ - 52^\circ = 46^\circ$, so by the Law of Sines, $\dfrac{|AC|}{\sin 52^\circ} = \dfrac{|AB|}{\sin 46^\circ} \Leftrightarrow |AC| = \dfrac{|AB| \sin 52^\circ}{\sin 46^\circ}$, so

substituting we have $|AC| = \dfrac{200 \sin 52^\circ}{\sin 46^\circ} \approx 219$ ft.

35.

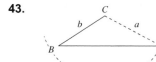

We draw a diagram. A is the position of the tourist and C is the top of the tower.

$\angle B = 90^\circ - 5.6^\circ = 84.4^\circ$ and so $\angle C = 180^\circ - 29.2^\circ - 84.4^\circ = 66.4^\circ$. Thus,

by the Law of Sines, the length of the tower is $|BC| = \dfrac{105 \sin 29.2^\circ}{\sin 66.4^\circ} \approx 55.9$ m.

37. The angle subtended by the top of the tree and the sun's rays is $\angle A = 180^\circ - 90^\circ - 52^\circ = 38^\circ$. Thus the height of the tree

is $h = \dfrac{215 \sin 30^\circ}{\sin 38^\circ} \approx 175$ ft.

39. Call the balloon's position R. Then in $\triangle PQR$, we see that $\angle P = 62^\circ - 32^\circ = 30^\circ$, and $\angle Q = 180^\circ - 71^\circ + 32^\circ = 141^\circ$.

Therefore, $\angle R = 180^\circ - 30^\circ - 141^\circ = 9^\circ$. So by the Law of Sines, $\dfrac{|QR|}{\sin 30^\circ} = \dfrac{|PQ|}{\sin 9^\circ} \Leftrightarrow |QR| = 60 \cdot \dfrac{\sin 30^\circ}{\sin 9^\circ} \approx 192$ m.

41. Let d be the distance from the earth to Venus, and let β be the angle formed by sun, Venus, and earth. By the Law

of Sines, $\dfrac{\sin \beta}{1} = \dfrac{\sin 39.4^\circ}{0.723} \approx 0.878$, so either $\beta \approx \sin^{-1} 0.878 \approx 61.4^\circ$ or $\beta \approx 180^\circ - \sin^{-1} 0.878 \approx 118.6^\circ$.

In the first case, $\dfrac{d}{\sin(180^\circ - 39.4^\circ - 61.4^\circ)} = \dfrac{0.723}{\sin 39.4^\circ} \Leftrightarrow d \approx 1.119$ AU; in the second case,

$\dfrac{d}{\sin(180^\circ - 39.4^\circ - 118.6^\circ)} = \dfrac{0.723}{\sin 39.4^\circ} \Leftrightarrow d \approx 0.427$ AU.

43.

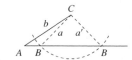 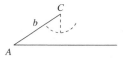

$b > a > b \sin A$: Two solutions $a = b \sin A$: One solution $a < b \sin A$: No solution

$a \geq b$: One solution

$\angle A = 30^\circ$, $b = 100$, $\sin A = \frac{1}{2}$. If $a \geq b = 100$ then there is one triangle. If $100 > a > 100 \sin 30^\circ = 50$, then there are

two possible triangles. If $a = 50$, then there is one (right) triangle. And if $a < 50$, then no triangle is possible.

6.5   The Law of Cosines

1. $x^2 = 21^2 + 42^2 - 2 \cdot 21 \cdot 42 \cdot \cos 39° = 441 + 1764 - 1764 \cos 39° \approx 834.115$ and so $x \approx \sqrt{834.115} \approx 28.9$.

3. $x^2 = 25^2 + 25^2 - 2 \cdot 25 \cdot 25 \cdot \cos 140° = 625 + 625 - 1250 \cos 140° \approx 2207.556$ and so $x \approx \sqrt{2207.556} \approx 47$.

5. $37.83^2 = 68.01^2 + 42.15^2 - 2 \cdot 68.01 \cdot 42.15 \cdot \cos \theta$. Then $\cos \theta = \dfrac{37.83^2 - 68.01^2 - 42.15^2}{-2 \cdot 68.01 \cdot 42.15} \approx 0.867 \quad \Leftrightarrow$

$\theta \approx \cos^{-1} 0.867 \approx 29.89°$.

7. $x^2 = 24^2 + 30^2 - 2 \cdot 24 \cdot 30 \cdot \cos 30° = 576 + 900 - 1440 \cos 30° \approx 228.923$ and so $x \approx \sqrt{228.923} \approx 15$.

9. $c^2 = 10^2 + 18^2 - 2 \cdot 10 \cdot 18 \cdot \cos 120° = 100 + 324 - 360 \cos 120° = 604$ and so $c \approx \sqrt{604} \approx 24.576$. Then

$\sin A \approx \dfrac{18 \sin 120°}{24.576} \approx 0.634295 \quad \Leftrightarrow \quad \angle A \approx \sin^{-1} 0.634295 \approx 39.4°$, and $\angle B \approx 180° - 120° - 39.4° = 20.6°$.

11. $c^2 = 3^2 + 4^2 - 2 \cdot 3 \cdot 4 \cdot \cos 53° = 9 + 16 - 24 \cos 53° \approx 10.556 \quad \Leftrightarrow \quad c \approx \sqrt{10.556} \approx 3.2$. Then

$\sin B = \dfrac{4 \sin 53°}{3.25} \approx 0.983 \quad \Leftrightarrow \quad \angle B \approx \sin^{-1} 0.983 \approx 79°$ and $\angle A \approx 180° - 53° - 79° = 48°$.

13. $20^2 = 25^2 + 22^2 - 2 \cdot 25 \cdot 22 \cdot \cos A \Leftrightarrow \cos A = \dfrac{20^2 - 25^2 - 22^2}{-2 \cdot 25 \cdot 22} \approx 0.644 \quad \Leftrightarrow \quad \angle A \approx \cos^{-1} 0.644 \approx 50°$. Then

$\sin B \approx \dfrac{25 \sin 49.9°}{20} \approx 0.956 \quad \Leftrightarrow \quad \angle B \approx \sin^{-1} 0.956 \approx 73°$, and so $\angle C \approx 180° - 50° - 73° = 57°$.

15. $\sin C = \dfrac{162 \sin 40°}{125} \approx 0.833 \Leftrightarrow \angle C_1 \approx \sin^{-1} 0.833 \approx 56.4°$ or $\angle C_2 \approx 180° - 56.4° \approx 123.6°$.

If $\angle C_1 \approx 56.4°$, then $\angle A_1 \approx 180° - 40° - 56.4° = 83.6°$ and $a_1 = \dfrac{125 \sin 83.6°}{\sin 40°} \approx 193$.

If $\angle C_2 \approx 123.6°$, then $\angle A_2 \approx 180° - 40° - 123.6° = 16.4°$ and $a_2 = \dfrac{125 \sin 16.4°}{\sin 40°} \approx 54.9$.

Thus, one possible triangle has $\angle A \approx 83.6°$, $\angle C \approx 56.4°$, and $a \approx 193$; the other has $\angle A \approx 16.4°$, $\angle C \approx 123.6°$, and $a \approx 54.9$.

17. $\sin B = \dfrac{65 \sin 55°}{50} \approx 1.065$. Since $|\sin \theta| \leq 1$ for all θ, there is no such $\angle B$, and hence there is no such triangle.

19. $\angle B = 180° - 35° - 85° = 60°$. Then $x = \dfrac{3 \sin 35°}{\sin 60°} \approx 2$.

21. $x = \dfrac{50 \sin 30°}{\sin 100°} \approx 25.4$

23. $b^2 = 110^2 + 138^2 - 2 (110)(138) \cdot \cos 38° = 12{,}100 + 19{,}044 - 30{,}360 \cos 38° \approx 7220.0$ and so $b \approx 85.0$. Therefore,

using the Law of Cosines again, we have $\cos \theta = \frac{110^2 + 85^2 - 128^2}{2(110)(138)} \quad \Leftrightarrow \quad \theta \approx 89.15°$.

25. $x^2 = 38^2 + 48^2 - 2 \cdot 38 \cdot 48 \cdot \cos 30° = 1444 + 2304 - 3648 \cos 30° \approx 588.739$ and so $x \approx 24.3$.

27. The semiperimeter is $s = \dfrac{9 + 12 + 15}{2} = 18$, so by Heron's Formula the area is

$A = \sqrt{18 (18 - 9)(18 - 12)(18 - 15)} = \sqrt{2916} = 54$.

29. The semiperimeter is $s = \dfrac{7 + 8 + 9}{2} = 12$, so by Heron's Formula the area is

$A = \sqrt{12 (12 - 7)(12 - 8)(12 - 9)} = \sqrt{720} = 12\sqrt{5} \approx 26.8$.

31. The semiperimeter is $s = \dfrac{3 + 4 + 6}{2} = \dfrac{13}{2}$, so by Heron's Formula the area is

$A = \sqrt{\dfrac{13}{2} \left(\dfrac{13}{2} - 3\right) \left(\dfrac{13}{2} - 4\right) \left(\dfrac{13}{2} - 6\right)} = \sqrt{\dfrac{455}{16}} = \dfrac{\sqrt{455}}{4} \approx 5.33$.

33. We draw a diagonal connecting the vertices adjacent to the $100°$ angle. This forms two triangles. Consider the triangle with sides of length 5 and 6 containing the $100°$ angle. The area of this triangle is $A_1 = \frac{1}{2}(5)(6)\sin 100° \approx 14.77$. To use Heron's Formula to find the area of the second triangle, we need to find the length of the diagonal using the Law of Cosines: $c^2 = a^2 + b^2 - 2ab\cos C = 5^2 + 6^2 - 2 \cdot 5 \cdot 6\cos 100° \approx 71.419 \Rightarrow c \approx 8.45$. Thus the second triangle has semiperimeter $s = \dfrac{8 + 7 + 8.45}{2} \approx 11.7255$ and area $A_2 = \sqrt{11.7255(11.7255 - 8)(11.7255 - 7)(11.7255 - 8.45)} \approx 26.00$. The area of the quadrilateral is the sum of the areas of the two triangles: $A = A_1 + A_2 \approx 14.77 + 26.00 = 40.77$.

35.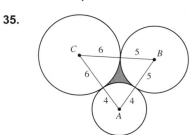

Label the centers of the circles A, B, and C, as in the figure. By the Law of Cosines, $\cos A = \dfrac{AB^2 + AC^2 - BC^2}{2(AB)(AC)} = \dfrac{9^2 + 10^2 - 11^2}{2(9)(10)} = \dfrac{1}{3} \Rightarrow$ $\angle A \approx 70.53°$. Now, by the Law of Sines, $\dfrac{\sin 70.53°}{11} = \dfrac{\sin B}{AC} = \dfrac{\sin C}{AB}$. So $\sin B = \frac{10}{11}\sin 70.53° \approx 0.85710 \Rightarrow B \approx \sin^{-1} 0.85710 \approx 58.99°$ and $\sin C = \frac{9}{11}\sin 70.53° \approx 0.77139 \Rightarrow C \approx \sin^{-1} 0.77139 \approx 50.48°$. The area of $\triangle ABC$ is $\frac{1}{2}(AB)(AC)\sin A = \frac{1}{2}(9)(10)(\sin 70.53°) \approx 42.426$.

The area of sector A is given by $S_A = \pi R^2 \cdot \dfrac{\theta}{360°} = \pi(4)^2 \cdot \dfrac{70.53°}{360°} \approx 9.848$. Similarly, the areas of sectors B and C are $S_B \approx 12.870$ and $S_C \approx 15.859$. Thus, the area enclosed between the circles is $A = \triangle ABC - S_A - S_B - S_C \Rightarrow A \approx 42.426 - 9.848 - 12.870 - 15.859 \approx 3.85 \text{ cm}^2$.

37. Let c be the distance across the lake, in miles. Then $c^2 = 2.82^2 + 3.56^2 - 2(2.82)(3.56) \cdot \cos 40.3° \approx 5.313 \quad \Leftrightarrow$ $c \approx 2.30$ mi.

39. In half an hour, the faster car travels 25 miles while the slower car travels 15 miles. The distance between them is given by the Law of Cosines: $d^2 = 25^2 + 15^2 - 2(25)(15) \cdot \cos 65° \Rightarrow$ $d = \sqrt{25^2 + 15^2 - 2(25)(15) \cdot \cos 65°} = 5\sqrt{25 + 9 - 30 \cdot \cos 65°} \approx 23.1$ mi.

41. The pilot travels a distance of $625 \cdot 1.5 = 937.5$ miles in her original direction and $625 \cdot 2 = 1250$ miles in the new direction. Since she makes a course correction of $10°$ to the right, the included angle is $180° - 10° = 170°$. From the figure, we use the Law of Cosines to get

the expression $d^2 = 937.5^2 + 1250^2 - 2(937.5)(1250) \cdot \cos 170° \approx 4{,}749{,}549.42$, so $d \approx 2179$ miles. Thus, the pilot's distance from her original position is approximately 2179 miles.

43. (a) The angle subtended at Egg Island is $100°$. Thus using the Law of Cosines, the distance from Forrest Island to the fisherman's home port is

$$x^2 = 30^2 + 50^2 - 2 \cdot 30 \cdot 50 \cdot \cos 100°$$
$$= 900 + 2500 - 3000\cos 100° \approx 3920.945$$

and so $x \approx \sqrt{3920.945} \approx 62.62$ miles.

(b) Let θ be the angle shown in the figure. Using the Law of Sines, $\sin\theta = \dfrac{50\sin 100°}{62.62} \approx 0.7863 \Leftrightarrow \theta \approx \sin^{-1} 0.7863 \approx 51.8°$. Then $\gamma = 90° - 20° - 51.8° = 18.2°$. Thus the bearing to his home port is S $18.2°$ E.

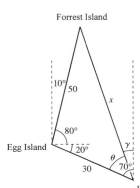

45. The largest angle is the one opposite the longest side; call this angle θ. Then by the Law of Cosines,

$$44^2 = 36^2 + 22^2 - 2\,(36)\,(22)\cdot\cos\theta \quad\Leftrightarrow\quad \cos\theta = \frac{36^2 + 22^2 - 44^2}{2\,(36)\,(22)} = -0.09848 \;\Rightarrow\; \theta \approx \cos^{-1}(-0.09848) \approx 96°.$$

47. Let d be the distance between the kites; then $d^2 \approx 380^2 + 420^2 - 2\,(380)\,(420)\cdot\cos 30° \;\Rightarrow$
$d \approx \sqrt{380^2 + 420^2 - 2\,(380)\,(420)\cdot\cos 30°} \approx 211$ ft.

49. *Solution 1:* From the figure, we see that $\gamma = 106°$ and $\sin 74° = \dfrac{3400}{b}$ $\quad\Leftrightarrow$

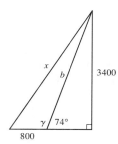

$b = \dfrac{3400}{\sin 74°} \approx 3537$. Thus, $x^2 = 800^2 + 3537^2 - 2\,(800)\,(3537)\cos 106°$ $\quad\Rightarrow$

$x = \sqrt{800^2 + 3537^2 - 2\,(800)\,(3537)\cos 106°}$ $\quad\Rightarrow\quad x \approx 3835$ ft.

Solution 2: Notice that $\tan 74° = \dfrac{3400}{a}$ $\quad\Leftrightarrow\quad a = \dfrac{3400}{\tan 74°} \approx 974.9$. By the

Pythagorean theorem, $x^2 = (a+800)^2 + 3400^2$. So

$x = \sqrt{(974.9 + 800)^2 + 3400^2} \approx 3835$ ft.

51. By Heron's formula, $A = \sqrt{s\,(s-a)\,(s-b)\,(s-c)}$, where $s = \dfrac{a+b+c}{2} = \dfrac{112 + 148 + 190}{2} = 225$. Thus,
$A = \sqrt{225\,(225 - 112)\,(225 - 148)\,(225 - 190)} \approx 8277.7$ ft^2. Since the land value is \$20 per square foot, the value of the lot is approximately $8277.7 \cdot 20 = \$165,554$.

Chapter 6 Review

1. (a) $60° = 60 \cdot \frac{\pi}{180} = \frac{\pi}{3} \approx 1.05$ rad

(b) $330° = 330 \cdot \frac{\pi}{180} = \frac{11\pi}{6} \approx 5.76$ rad

(c) $-135° = -135 \cdot \frac{\pi}{180} = -\frac{3\pi}{4} \approx -3.36$ rad

(d) $-90° = -90 \cdot \frac{\pi}{180} = -\frac{\pi}{2} \approx -1.57$ rad

3. (a) $\frac{5\pi}{2}$ rad $= \frac{5\pi}{2} \cdot \frac{180}{\pi} = 450°$

(b) $-\frac{\pi}{6}$ rad $= -\frac{\pi}{6} \cdot \frac{180}{\pi} = -30°$

(c) $\frac{9\pi}{4}$ rad $= \frac{9\pi}{4} \cdot \frac{180}{\pi} = 405°$

(d) 3.1 rad $= 3.1 \cdot \frac{180}{\pi} = \frac{558}{\pi} \approx 177.6°$

5. $r = 8$ m, $\theta = 1$ rad. Then $s = r\theta = 8 \cdot 1 = 8$ m.

7. $s = 100$ ft, $\theta = 70° = 70 \cdot \frac{\pi}{180} = \frac{7\pi}{18}$ rad. Then $r = \frac{s}{\theta} = 100 \cdot \frac{18}{7\pi} = \frac{1800}{7\pi} \approx 82$ ft.

9. $r = 3960$ miles, $s = 2450$ miles. Then $\theta = \frac{s}{r} = \frac{2450}{3960} \approx 0.619$ rad $= 0.619 \cdot \frac{\pi}{180} \approx 35.448°$ and so the angle is approximately $35.4°$.

11. $A = \frac{1}{2}r^2\theta = \frac{1}{2}\,(200)^2\left(52° \cdot \frac{\pi}{180°}\right) \approx 18{,}151$ ft^2

13. The angular speed is $\omega = \dfrac{150 \cdot 2\pi \text{ rad}}{1 \text{ min}} = 300\pi$ rad/min ≈ 942.5 rad/min. The linear speed is
$v = \dfrac{150 \cdot 2\pi \cdot 8}{1} = 2400\pi$ in./min ≈ 7539.8 in./min.

15. $r = \sqrt{5^2 + 7^2} = \sqrt{74}$. Then $\sin\theta = \frac{5}{\sqrt{74}}$, $\cos\theta = \frac{7}{\sqrt{74}}$, $\tan\theta = \frac{5}{7}$, $\csc\theta = \frac{\sqrt{74}}{5}$, $\sec\theta = \frac{\sqrt{74}}{7}$, and $\cot\theta = \frac{7}{5}$.

17. $\frac{x}{5} = \cos 40° \quad\Leftrightarrow\quad x = 5\cos 40° \approx 3.83$, and $\frac{y}{5} = \sin 40° \quad\Leftrightarrow\quad y = 5\sin 40° \approx 3.21$.

19. $\frac{1}{x} = \sin 20° \quad\Leftrightarrow\quad x = \frac{1}{\sin 20°} \approx 2.92$, and $\frac{x}{y} = \cos 20° \quad\Leftrightarrow\quad y = \frac{x}{\cos 20°} \approx \frac{2.924}{0.9397} \approx 3.11$.

21. $w = 3\sin 20° \approx 1.026$,
$v = 3\cos 20° \approx 2.819$.

23. $\tan\theta = \dfrac{1}{a} \quad\Leftrightarrow\quad a = \dfrac{1}{\tan\theta} = \cot\theta$, $\sin\theta = \dfrac{1}{b} \quad\Leftrightarrow$

$b = \dfrac{1}{\sin\theta} = \csc\theta$

25. 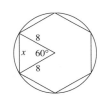 One side of the hexagon together with radial line segments through its endpoints forms a triangle with two sides of length 8 m and subtended angle $60°$. Let x be the length of one such side (in meters). By the Law of Cosines,
$$x^2 = 8^2 + 8^2 - 2 \cdot 8 \cdot 8 \cdot \cos 60° = 64 \quad \Leftrightarrow \quad x = 8.$$ Thus the perimeter of the hexagon is $6x = 6 \cdot 8 = 48$ m.

27. Let r represent the radius, in miles, of the moon. Then $\tan \dfrac{\theta}{2} = \dfrac{r}{r + |AB|}$, $\theta = 0.518°$

$\Leftrightarrow \quad r = (r + 236{,}900) \cdot \tan 0.259° \quad \Leftrightarrow \quad r (1 - \tan 0.259°) = 236{,}900 \cdot \tan 0.259° \Leftrightarrow$

$r = \dfrac{236{,}900 \cdot \tan 0.259°}{1 - \tan 0.259°} \approx 1076$ and so the radius of the moon is roughly 1076 miles.

29. $\sin 315° = -\sin 45° = -\dfrac{1}{\sqrt{2}} = -\dfrac{\sqrt{2}}{2}$

31. $\tan (-135°) = \tan 45° = 1$

33. $\cot \left(-\dfrac{22\pi}{3}\right) = \cot \dfrac{2\pi}{3} = \cot \dfrac{\pi}{3} = -\dfrac{1}{\sqrt{3}} = -\dfrac{\sqrt{3}}{3}$

35. $\cos 585° = \cos 225° = -\cos 45° = -\dfrac{1}{\sqrt{2}} = -\dfrac{\sqrt{2}}{2}$

37. $\csc \dfrac{8\pi}{3} = \csc \dfrac{2\pi}{3} = \csc \dfrac{\pi}{3} = \dfrac{2}{\sqrt{3}} = \dfrac{2\sqrt{3}}{3}$

39. $\cot (-390°) = \cot (-30°) = -\cot 30° = -\sqrt{3}$

41. $r = \sqrt{(-5)^2 + 12^2} = \sqrt{169} = 13$. Then $\sin \theta = \dfrac{12}{13}$, $\cos \theta = -\dfrac{5}{13}$, $\tan \theta = -\dfrac{12}{5}$, $\csc \theta = \dfrac{13}{12}$, $\sec \theta = -\dfrac{13}{5}$, and $\cot \theta = -\dfrac{5}{12}$.

43. $y - \sqrt{3}x + 1 = 0 \Leftrightarrow y = \sqrt{3}x - 1$, so the slope of the line is $m = \sqrt{3}$. Then $\tan \theta = m = \sqrt{3} \Leftrightarrow \theta = 60°$.

45. $\sec^2 \theta = 1 + \tan^2 \theta \quad \Leftrightarrow \quad \tan^2 \theta = \dfrac{1}{\cos^2 \theta} - 1 \quad \Leftrightarrow$

$\tan \theta = \sqrt{\dfrac{1}{\cos^2 \theta} - 1} = \sqrt{\dfrac{1 - \cos^2 \theta}{\cos^2 \theta}} = \dfrac{\sqrt{1 - \cos^2 \theta}}{|\cos \theta|} = \dfrac{\sqrt{1 - \cos^2 \theta}}{-\cos \theta}$ (since $\cos \theta < 0$ in quadrant III,

$|\cos \theta| = -\cos \theta$). Thus $\tan \theta = -\dfrac{\sqrt{1 - \cos^2 \theta}}{\cos \theta}$.

47. $\tan^2 \theta = \dfrac{\sin^2 \theta}{\cos^2 \theta} = \dfrac{\sin^2 \theta}{1 - \sin^2 \theta}$

49. $\tan \theta = \dfrac{\sqrt{7}}{3}$, $\sec \theta = \dfrac{4}{3}$. Then $\cos \theta = \dfrac{3}{4}$ and $\sin \theta = \tan \theta \cdot \cos \theta = \dfrac{\sqrt{7}}{4}$, $\csc \theta = \dfrac{4}{\sqrt{7}} = \dfrac{4\sqrt{7}}{7}$, and $\cot \theta = \dfrac{3}{\sqrt{7}} = \dfrac{3\sqrt{7}}{7}$.

51. $\sin \theta = \dfrac{3}{5}$. Since $\cos \theta < 0$, θ is in quadrant II. Thus, $x = -\sqrt{5^2 - 3^2} = -\sqrt{16} = -4$ and so $\cos \theta = -\dfrac{4}{5}$, $\tan \theta = -\dfrac{3}{4}$, $\csc \theta = \dfrac{5}{3}$, $\sec \theta = -\dfrac{5}{4}$, $\cot \theta = -\dfrac{4}{3}$.

53. $\tan \theta = -\dfrac{1}{2}$. $\sec^2 \theta = 1 + \tan^2 \theta = 1 + \dfrac{1}{4} = \dfrac{5}{4} \quad \Leftrightarrow \quad \cos^2 \theta = \dfrac{4}{5} \quad \Rightarrow \quad \cos \theta = -\sqrt{\dfrac{4}{5}} = -\dfrac{2}{\sqrt{5}}$ since

$\cos \theta < 0$ in quadrant II. But $\tan \theta = \dfrac{\sin \theta}{\cos \theta} = -\dfrac{1}{2} \quad \Leftrightarrow \quad \sin \theta = -\dfrac{1}{2} \cos \theta = -\dfrac{1}{2}\left(-\dfrac{2}{\sqrt{5}}\right) = \dfrac{1}{\sqrt{5}}$. Therefore,

$\sin \theta + \cos \theta = \dfrac{1}{\sqrt{5}} + \left(-\dfrac{2}{\sqrt{5}}\right) = -\dfrac{1}{\sqrt{5}} = -\dfrac{\sqrt{5}}{5}$.

55. By the Pythagorean Theorem, $\sin^2 \theta + \cos^2 \theta = 1$ for any angle θ.

57. $\angle B = 180° - 30° - 80° = 70°$, and so by the Law of Sines, $x = \dfrac{10 \sin 30°}{\sin 70°} \approx 5.32$.

59. $x^2 = 100^2 + 210^2 - 2 \cdot 100 \cdot 210 \cdot \cos 40° \approx 21{,}926.133 \quad \Leftrightarrow \quad x \approx 148.07$

61. $\sin B = \dfrac{20 \sin 60°}{70} \approx 0.247 \Leftrightarrow \angle B \approx \sin^{-1} 0.247 \approx 14.33°$. Then $\angle C \approx 180° - 60° - 14.33° = 105.67°$, and so

$x \approx \dfrac{70 \sin 105.67°}{\sin 60°} \approx 77.82$.

63. After 2 hours the ships have traveled distances $d_1 = 40$ mi and $d_2 = 56$ mi. The subtended angle is $180° - 32° - 42° = 106°$. Let d be the distance between the two ships in miles. Then by the Law of Cosines, $d^2 = 40^2 + 56^2 - 2(40)(56) \cos 106° \approx 5970.855 \quad \Leftrightarrow \quad d \approx 77.3$ miles.

65. Let d be the distance, in miles, between the points A and B . Then by the Law of Cosines,

$d^2 = 3.2^2 + 5.6^2 - 2\,(3.2)\,(5.6)\cos 42° \approx 14.966 \quad \Leftrightarrow \quad d \approx 3.9$ mi.

67. $A = \frac{1}{2}ab\sin\theta = \frac{1}{2}\,(8)\,(14)\sin 35° \approx 32.12$

Chapter 6 Test

1. $330° = 330 \cdot \frac{\pi}{180} = \frac{11\pi}{6}$ rad. $-135° = -135 \cdot \frac{\pi}{180} = -\frac{3\pi}{4}$ rad.

2. $\frac{4\pi}{3}$ rad $= \frac{4\pi}{3} \cdot \frac{180}{\pi} = 240°$. -1.3 rad $= -1.3 \cdot \frac{180}{\pi} = -\frac{234}{\pi} \approx -74.5°$

3. (a) The angular speed is $\omega = \frac{120 \cdot 2\pi \text{ rad}}{1 \text{ min}} = 240\pi$ rad/min.

    (b) The linear speed is

$$v = \frac{120 \cdot 2\pi \cdot 16}{1} = 3840\pi \text{ ft/min}$$

$$\approx 12{,}063.7 \text{ ft/min}$$

$$\approx 137 \text{ mi/h}.$$

4. (a) $\sin 405° = \sin 45° = \frac{1}{\sqrt{2}} = \frac{\sqrt{2}}{2}$

    (b) $\tan(-150°) = \tan 30° = \frac{1}{\sqrt{3}} = \frac{\sqrt{3}}{3}$

    (c) $\sec\frac{5\pi}{3} = \sec\frac{\pi}{3} = 2$

    (d) $\csc\frac{5\pi}{2} = \csc\frac{\pi}{2} = 1$

5. $r = \sqrt{3^2 + 2^2} = \sqrt{13}$. Then $\tan\theta + \sin\theta = \frac{2}{3} + \frac{2}{\sqrt{13}} = \frac{2\left(\sqrt{13}+3\right)}{3\sqrt{13}} = \frac{26 + 6\sqrt{13}}{39}$.

6. $\sin\theta = \dfrac{a}{24} \quad \Leftrightarrow \quad a = 24\sin\theta$. Also, $\cos\theta = \dfrac{b}{24} \quad \Leftrightarrow \quad b = 24\cos\theta$.

7. $\cos\theta = -\frac{1}{3}$ and θ is in quadrant III, so $r = 3$, $x = -1$, and $y = -\sqrt{3^2 - 1^2} = -2\sqrt{2}$. Then

$\tan\theta\cot\theta + \csc\theta = \tan\theta \cdot \dfrac{1}{\tan\theta} + \csc\theta = 1 - \dfrac{3}{2\sqrt{2}} = \dfrac{2\sqrt{2}-3}{2\sqrt{2}} = \dfrac{4 - 3\sqrt{2}}{4}$.

8. $\sin\theta = \frac{5}{13}$, $\tan\theta = -\frac{5}{12}$. Then $\sec\theta = \dfrac{1}{\cos\theta} = \dfrac{1}{\cos\theta} \cdot \dfrac{\sin\theta}{\sin\theta} = \tan\theta \cdot \dfrac{1}{\sin\theta} = -\dfrac{5}{12} \cdot \dfrac{13}{5} = -\dfrac{13}{12}$.

9. $\sec^2\theta = 1 + \tan^2\theta \quad \Leftrightarrow \quad \tan\theta = \pm\sqrt{\sec^2\theta - 1}$. Thus, $\tan\theta = -\sqrt{\sec^2\theta - 1}$ since $\tan\theta < 0$ in quadrant II.

10. $\tan 73° = \dfrac{h}{6} \quad \Rightarrow \quad h = 6\tan 73° \approx 19.6$ ft.

11. By the Law of Cosines, $x^2 = 10^2 + 12^2 - 2\,(10)\,(12) \cdot \cos 48° \approx 8.409 \quad \Leftrightarrow \quad x \approx 9.1$.

12. $\angle C = 180° - 52° - 69° = 59°$. Then by the Law of Sines, $x = \frac{230\sin 69°}{\sin 59°} \approx 250.5$.

13. Let h be the height of the shorter altitude. Then $\tan 20° = \dfrac{h}{50} \quad \Leftrightarrow \quad h = 50\tan 20°$ and $\tan 28° = \dfrac{x+h}{50} \quad \Leftrightarrow$

$x + h = 50\tan 28° \quad \Leftrightarrow \quad x = 50\tan 28° - h = 50\tan 28° - 50\tan 20° \approx 8.4$.

14. Let $\angle A$ and $\angle X$ be the other angles in the triangle. Then $\sin A = \frac{15\sin 108°}{28} \approx 0.509 \Leftrightarrow \angle A \approx 30.63°$. Then

$\angle X \approx 180° - 108° - 30.63° \approx 41.37°$, and so $x \approx \frac{28\sin 41.37°}{\sin 108°} \approx 19.5$.

15. (a) A (sector) $= \frac{1}{2}r^2\theta = \frac{1}{2} \cdot 10^2 \cdot 72 \cdot \frac{\pi}{180} = 50 \cdot \frac{72\pi}{180}$. A (triangle) $= \frac{1}{2}r \cdot r\sin\theta = \frac{1}{2} \cdot 10^2\sin 72°$. Thus, the area of the

    shaded region is A (shaded) $= A$ (sector) $- A$ (triangle) $= 50\left(\frac{72\pi}{180} - \sin 72°\right) \approx 15.3$ m^2.

    (b) The shaded region is bounded by two pieces: one piece is part of the triangle, the other is part of the circle. The

    first part has length $l = \sqrt{10^2 + 10^2 - 2\,(10)\,(10) \cdot \cos 72°} = 10\sqrt{2 - 2 \cdot \cos 72°}$. The second has length

    $s = 10 \cdot 72 \cdot \frac{\pi}{180} = 4\pi$. Thus, the perimeter of the shaded region is $p = l + s = 10\sqrt{2 - 2\cos 72°} + 4\pi \approx 24.3$ m.

16. (a) If θ is the angle opposite the longest side, then by the Law of Cosines $\cos\theta = \frac{9^2+13^2-20^2}{2(9)(20)} = -0.6410$. Therefore, $\theta = \cos^{-1}(-0.6410) \approx 129.9°$.

(b) From part (a), $\theta \approx 129.9°$, so the area of the triangle is $A = \frac{1}{2}(9)(13)\sin 129.9° \approx 44.9$ units^2. Another way to find the area is to use Heron's Formula: $A = \sqrt{s(s-a)(s-b)(s-c)}$, where $s = \frac{a+b+c}{2} = \frac{9+13+20}{2} = 21$. Thus, $A = \sqrt{21(21-20)(21-13)(21-9)} = \sqrt{2016} \approx 44.9$ units^2.

17. Label the figure as shown. Now $\angle\beta = 85° - 75° = 10°$, so by the Law of Sines,

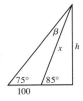

$$\frac{x}{\sin 75°} = \frac{100}{\sin 10°} \quad \Leftrightarrow \quad x = 100 \cdot \frac{\sin 75°}{\sin 10°}. \text{ Now } \sin 85° = \frac{h}{x} \quad \Leftrightarrow$$

$$h = x\sin 85° = 100 \cdot \frac{\sin 75°}{\sin 10°}\sin 85° \approx 554.$$

Focus on Modeling: Surveying

1. Let x be the distance between the church and City Hall. To apply the Law of Sines to the triangle with vertices at City Hall, the church, and the first bridge, we first need the measure of the angle at the first bridge, which is $180° - 25° - 30° = 125°$. Then $\dfrac{x}{\sin 125°} = \dfrac{0.86}{\sin 30°} \Leftrightarrow x = \dfrac{0.86 \sin 125°}{\sin 30°} \approx 1.4089$. So the distance between the church and City Hall is about 1.41 miles.

3. First notice that $\angle DBC = 180° - 20° - 95° - 65°$ and $\angle DAC = 180° - 60° - 45° = 75°$. From $\triangle ACD$ we get $\dfrac{|AC|}{\sin 45°} = \dfrac{20}{\sin 75°} \Leftrightarrow |AC| = \dfrac{20 \sin 45°}{\sin 75°} \approx 14.6°$. From $\triangle BCD$ we get $\dfrac{|BC|}{\sin 95°} = \dfrac{20}{\sin 65°} \Leftrightarrow |BC| = \dfrac{20 \sin 95°}{\sin 65°} \approx 22.0$. By applying the Law of Cosines to $\triangle ABC$ we get $|AB|^2 = |AC|^2 + |BC|^2 - 2|AC||BC|\cos 40° \approx 14.6^2 + 22.0^2 - 2 \cdot 14.6 \cdot 22.0 \cdot \cos 40° \approx 205$, so $|AB| \approx \sqrt{205} \approx 14.3$ m. Therefore, the distance between A and B is approximately 14.3 m.

5. (a) In $\triangle ABC$, $\angle B = 180° - \beta$, so $\angle C = 180° - \alpha - (180° - \beta) = \beta - \alpha$. By the Law of Sines, $\dfrac{|BC|}{\sin \alpha} = \dfrac{|AB|}{\sin (\beta - \alpha)}$

$$\Rightarrow |BC| = |AB| \frac{\sin \alpha}{\sin (\beta - \alpha)} = \frac{d \sin \alpha}{\sin (\beta - \alpha)}.$$

(b) From part (a) we know that $|BC| = \dfrac{d \sin \alpha}{\sin (\beta - \alpha)}$. But $\sin \beta = \dfrac{h}{|BC|} \Leftrightarrow |BC| = \dfrac{h}{\sin \beta}$. Therefore,

$$|BC| = \frac{d \sin \alpha}{\sin (\beta - \alpha)} = \frac{h}{\sin \beta} \Rightarrow h = \frac{d \sin \alpha \sin \beta}{\sin (\beta - \alpha)}.$$

(c) $h = \dfrac{d \sin \alpha \sin \beta}{\sin (\beta - \alpha)} = \dfrac{800 \sin 25° \sin 29°}{\sin 4°} \approx 2350$ ft

7. We start by labeling the edges and calculating the remaining angles, as shown in the first figure. Using the Law of Sines, we find the following: $\dfrac{a}{\sin 29°} = \dfrac{150}{\sin 60°} \Leftrightarrow a = \dfrac{150 \sin 29°}{\sin 60°} \approx 83.97$, $\dfrac{b}{\sin 91°} = \dfrac{150}{\sin 60°} \Leftrightarrow b = \dfrac{150 \sin 91°}{\sin 60°} \approx 173.18$,

$\dfrac{c}{\sin 32°} = \dfrac{173.18}{\sin 87°} \Leftrightarrow c = \dfrac{173.18 \sin 32°}{\sin 87°} \approx 91.90$, $\dfrac{d}{\sin 61°} = \dfrac{173.18}{\sin 87°} \Leftrightarrow e = \dfrac{173.18 \sin 61°}{\sin 87°} \approx 151.67$,

$\dfrac{e}{\sin 41°} = \dfrac{151.67}{\sin 51°} \Leftrightarrow e = \dfrac{151.67 \sin 41°}{\sin 51°} \approx 128.04$, $\dfrac{f}{\sin 88°} = \dfrac{151.67}{\sin 51°} \Leftrightarrow f = \dfrac{151.67 \sin 88°}{\sin 51°} \approx 195.04$,

$\dfrac{g}{\sin 50°} = \dfrac{195.04}{\sin 92°} \Leftrightarrow g = \dfrac{195.04 \sin 50°}{\sin 92°} \approx 149.50$, and $\dfrac{h}{\sin 38°} = \dfrac{195.04}{\sin 92°} \Leftrightarrow h = \dfrac{195.04 \sin 38°}{\sin 92°} \approx 120.15$. Note that we used two decimal places throughout our calculations. Our results are shown (to one decimal place) in the second figure.

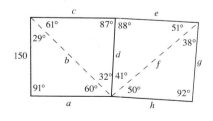
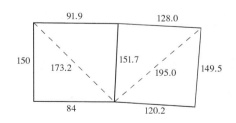

7 Trigonometric Functions of Real Numbers

7.1 The Unit Circle

1. Since $\left(\frac{4}{5}\right)^2 + \left(-\frac{3}{5}\right)^2 = \frac{16}{25} + \frac{9}{25} = 1$, $P\left(\frac{4}{5}, -\frac{3}{5}\right)$ lies on the unit circle.

3. Since $\left(\frac{7}{25}\right)^2 + \left(\frac{24}{25}\right)^2 = \frac{49}{625} + \frac{576}{625} = 1$, $P\left(\frac{7}{25}, \frac{24}{25}\right)$ lies on the unit circle.

5. Since $\left(-\frac{\sqrt{5}}{3}\right)^2 + \left(\frac{2}{3}\right)^2 = \frac{5}{9} + \frac{4}{9} = 1$, $P\left(-\frac{\sqrt{5}}{3}, \frac{2}{3}\right)$ lies on the unit circle.

7. $\left(-\frac{3}{5}\right)^2 + y^2 = 1 \Leftrightarrow y^2 = 1 - \frac{9}{25} \Leftrightarrow y^2 = \frac{16}{25} \Leftrightarrow y = \pm\frac{4}{5}$. Since $P(x, y)$ is in quadrant III, y is negative, so the point is $P\left(-\frac{3}{5}, -\frac{4}{5}\right)$.

9. $x^2 + \left(\frac{1}{3}\right)^2 = 1 \Leftrightarrow x^2 = 1 - \frac{1}{9} \Leftrightarrow x^2 = \frac{8}{9} \Leftrightarrow x = \pm\frac{2\sqrt{2}}{3}$. Since P is in quadrant II, x is negative, so the point is $P\left(-\frac{2\sqrt{2}}{3}, \frac{1}{3}\right)$.

11. $x^2 + \left(-\frac{2}{7}\right)^2 = 1 \Leftrightarrow x^2 = 1 - \frac{4}{49} \Leftrightarrow x^2 = \frac{45}{49} \Leftrightarrow x = \pm\frac{3\sqrt{5}}{7}$. Since $P(x, y)$ is in quadrant IV, x is positive, so the point is $P\left(\frac{3\sqrt{5}}{7}, -\frac{2}{7}\right)$.

13. $\left(\frac{4}{5}\right)^2 + y^2 = 1 \Leftrightarrow y^2 = 1 - \frac{16}{25} \Leftrightarrow y^2 = \frac{9}{25} \Leftrightarrow y = \pm\frac{3}{5}$. Since its y-coordinate is positive, the point is $P\left(\frac{4}{5}, \frac{3}{5}\right)$.

15. $x^2 + \left(\frac{2}{3}\right)^2 = 1 \Leftrightarrow x^2 = 1 - \frac{4}{9} \Leftrightarrow x^2 = \frac{5}{9} \Leftrightarrow x = \pm\frac{\sqrt{5}}{3}$. Since its x-coordinate is negative, the point is $P\left(-\frac{\sqrt{5}}{3}, \frac{2}{3}\right)$.

17. $\left(-\frac{\sqrt{2}}{3}\right)^2 + y^2 = 1 \Leftrightarrow y^2 = 1 - \frac{2}{9} \Leftrightarrow y^2 = \frac{7}{9} \Leftrightarrow y = \pm\frac{\sqrt{7}}{3}$. Since P lies below the x-axis, its y-coordinate is negative, so the point is $P\left(-\frac{\sqrt{2}}{3}, -\frac{\sqrt{7}}{3}\right)$.

19.

t	Terminal Point
0	$(1, 0)$
$\frac{\pi}{4}$	$\left(\frac{\sqrt{2}}{2}, \frac{\sqrt{2}}{2}\right)$
$\frac{\pi}{2}$	$(0, 1)$
$\frac{3\pi}{4}$	$\left(-\frac{\sqrt{2}}{2}, \frac{\sqrt{2}}{2}\right)$
π	$(-1, 0)$

t	Terminal Point
π	$(-1, 0)$
$\frac{5\pi}{4}$	$\left(-\frac{\sqrt{2}}{2}, -\frac{\sqrt{2}}{2}\right)$
$\frac{3\pi}{2}$	$(0, -1)$
$\frac{7\pi}{4}$	$\left(\frac{\sqrt{2}}{2}, -\frac{\sqrt{2}}{2}\right)$
2π	$(1, 0)$

21. $P(x, y) = (0, 1)$

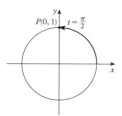

23. $P(x, y) = \left(-\frac{\sqrt{3}}{2}, \frac{1}{2}\right)$

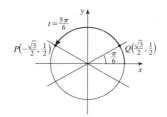

25. $P(x, y) = \left(\frac{1}{2}, -\frac{\sqrt{3}}{2}\right)$

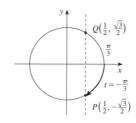

27. $P(x, y) = \left(-\frac{1}{2}, \frac{\sqrt{3}}{2}\right)$

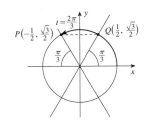

29. $P(x, y) = \left(-\frac{\sqrt{2}}{2}, -\frac{\sqrt{2}}{2}\right)$

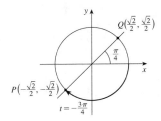

31. Let $Q(x, y) = \left(\frac{3}{5}, \frac{4}{5}\right)$ be the terminal point determined by t.

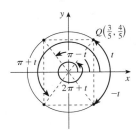

(a) $\pi - t$ determines the point $P(-x, y) = \left(-\frac{3}{5}, \frac{4}{5}\right)$.

(b) $-t$ determines the point $P(x, -y) = \left(\frac{3}{5}, -\frac{4}{5}\right)$.

(c) $\pi + t$ determines the point $P(-x, -y) = \left(-\frac{3}{5}, -\frac{4}{5}\right)$.

(d) $2\pi + t$ determines the point $P(x, y) = \left(\frac{3}{5}, \frac{4}{5}\right)$.

33. (a) $\bar{t} = \frac{5\pi}{4} - \pi = \frac{\pi}{4}$

(b) $\bar{t} = \frac{7\pi}{3} - 2\pi = \frac{\pi}{3}$

(c) $\bar{t} = \frac{4\pi}{3} - \pi = \frac{\pi}{3}$

(d) $\bar{t} = \frac{\pi}{6}$

35. (a) $\bar{t} = \pi - \frac{5\pi}{7} = \frac{2\pi}{7}$

(b) $\bar{t} = \pi - \frac{7\pi}{9} = \frac{2\pi}{9}$

(c) $\bar{t} = \pi - 3 \approx 0.142$

(d) $\bar{t} = 2\pi - 5 \approx 1.283$

37. (a) $\bar{t} = \pi - \frac{2\pi}{3} = \frac{\pi}{3}$

(b) $P\left(-\frac{1}{2}, \frac{\sqrt{3}}{2}\right)$

39. (a) $\bar{t} = \pi - \frac{3\pi}{4} = \frac{\pi}{4}$

(b) $P\left(-\frac{\sqrt{2}}{2}, \frac{\sqrt{2}}{2}\right)$

41. (a) $\bar{t} = \pi - \frac{2\pi}{3} = \frac{\pi}{3}$

(b) $P\left(-\frac{1}{2}, -\frac{\sqrt{3}}{2}\right)$

43. (a) $\bar{t} = \frac{13\pi}{4} - 3\pi = \frac{\pi}{4}$

(b) $P\left(-\frac{\sqrt{2}}{2}, -\frac{\sqrt{2}}{2}\right)$

45. (a) $\bar{t} = \frac{7\pi}{6} - \pi = \frac{\pi}{6}$

(b) $P\left(-\frac{\sqrt{3}}{2}, -\frac{1}{2}\right)$

47. (a) $\bar{t} = 4\pi - \frac{11\pi}{3} = \frac{\pi}{3}$

(b) $P\left(\frac{1}{2}, \frac{\sqrt{3}}{2}\right)$

49. (a) $\bar{t} = \frac{16\pi}{3} - 5\pi = \frac{\pi}{3}$

(b) $P\left(-\frac{1}{2}, -\frac{\sqrt{3}}{2}\right)$

51. $t = 1 \quad \Rightarrow \quad (0.5, 0.8)$

53. $t = -1.1 \quad \Rightarrow \quad (0.5, -0.9)$

55. The distances PQ and PR are equal because they both subtend arcs of length $\frac{\pi}{3}$. Since $P(x, y)$ is a point on the unit circle, $x^2 + y^2 = 1$. Now $d(P, Q) = \sqrt{(x - x)^2 + (y - (-y))^2} = 2y$ and

$d(R, S) = \sqrt{(x - 0)^2 + (y - 1)^2} = \sqrt{x^2 + y^2 - 2y + 1} = \sqrt{2 - 2y}$ (using the fact that $x^2 + y^2 = 1$). Setting these equal gives $2y = \sqrt{2 - 2y} \quad \Rightarrow \quad 4y^2 = 2 - 2y \Leftrightarrow 4y^2 + 2y - 2 = 0 \quad \Leftrightarrow 2(2y - 1)(y + 1) = 0$. So $y = -1$ or $y = \frac{1}{2}$.

Since P is in quadrant I, $y = \frac{1}{2}$ is the only viable solution. Again using $x^2 + y^2 = 1$ we have $x^2 + \left(\frac{1}{2}\right)^2 = 1 \Leftrightarrow x^2 = \frac{3}{4}$

$\Rightarrow \quad x = \pm\frac{\sqrt{3}}{2}$. Again, since P is in quadrant I the coordinates must be $\left(\frac{\sqrt{3}}{2}, \frac{1}{2}\right)$.

7.2   Trigonometric Functions of Real Numbers

1.

t	$\sin t$	$\cos t$
0	0	1
$\frac{\pi}{4}$	$\frac{\sqrt{2}}{2}$	$\frac{\sqrt{2}}{2}$
$\frac{\pi}{2}$	1	0
$\frac{3\pi}{4}$	$\frac{\sqrt{2}}{2}$	$-\frac{\sqrt{2}}{2}$
π	0	-1
$\frac{5\pi}{4}$	$-\frac{\sqrt{2}}{2}$	$-\frac{\sqrt{2}}{2}$
$\frac{3\pi}{2}$	-1	0
$\frac{7\pi}{4}$	$-\frac{\sqrt{2}}{2}$	$\frac{\sqrt{2}}{2}$
2π	0	1

3. (a) $\sin \frac{2\pi}{3} = \frac{\sqrt{3}}{2}$

    (b) $\cos \frac{2\pi}{3} = -\frac{1}{2}$

    (c) $\tan \frac{2\pi}{3} = -\sqrt{3}$

5. (a) $\sin \frac{7\pi}{6} = -\frac{1}{2}$

    (b) $\sin \left(-\frac{\pi}{6}\right) = -\frac{1}{2}$

    (c) $\sin \frac{11\pi}{6} = -\frac{1}{2}$

7. (a) $\cos \frac{3\pi}{4} = -\frac{\sqrt{2}}{2}$

    (b) $\cos \frac{5\pi}{4} = -\frac{\sqrt{2}}{2}$

    (c) $\cos \frac{7\pi}{4} = \frac{\sqrt{2}}{2}$

9. (a) $\sin \frac{7\pi}{3} = \frac{\sqrt{3}}{2}$

    (b) $\csc \frac{7\pi}{3} = \frac{2\sqrt{3}}{3}$

    (c) $\cot \frac{7\pi}{3} = \frac{\sqrt{3}}{3}$

11. (a) $\sin \left(-\frac{\pi}{2}\right) = -1$

    (b) $\cos \left(-\frac{\pi}{2}\right) = 0$

    (c) $\cot \left(-\frac{\pi}{2}\right) = 0$

13. (a) $\sec \frac{11\pi}{3} = 2$

    (b) $\csc \frac{11\pi}{3} = -\frac{2\sqrt{3}}{3}$

    (c) $\sec \left(-\frac{\pi}{3}\right) = 2$

15. (a) $\tan \frac{5\pi}{6} = -\frac{\sqrt{3}}{3}$

    (b) $\tan \frac{7\pi}{6} = \frac{\sqrt{3}}{3}$

    (c) $\tan \frac{11\pi}{6} = -\frac{\sqrt{3}}{3}$

17. (a) $\cos \left(-\frac{\pi}{4}\right) = \frac{\sqrt{2}}{2}$

    (b) $\csc \left(-\frac{\pi}{4}\right) = -\sqrt{2}$

    (c) $\cot \left(-\frac{\pi}{4}\right) = -1$

19. (a) $\csc \left(-\frac{\pi}{2}\right) = -1$

    (b) $\csc \frac{\pi}{2} = 1$

    (c) $\csc \frac{3\pi}{2} = -1$

21. (a) $\sin 13\pi = 0$

    (b) $\cos 14\pi = 1$

    (c) $\tan 15\pi = 0$

23. $t = 0 \ \Rightarrow \ \sin t = 0, \cos t = 1, \tan t = 0, \sec t = 1,$ $\csc t$ and $\cot t$ are undefined.

25. $t = \pi \ \Rightarrow \ \sin t = 0, \cos t = -1, \tan t = 0, \sec t = -1, \csc t$ and $\cot t$ are undefined.

27. $\left(\frac{3}{5}\right)^2 + \left(\frac{4}{5}\right)^2 = \frac{9}{25} + \frac{1}{2} = 1$. So $\sin t = \frac{4}{5}$, $\cos t = \frac{3}{5}$, and $\tan t = \frac{\frac{4}{5}}{\frac{3}{5}} = \frac{4}{3}$.

29. $\left(\frac{\sqrt{5}}{4}\right)^2 + \left(-\frac{\sqrt{11}}{4}\right)^2 = \frac{5}{16} + \frac{11}{16} = 1$. So $\sin t = -\frac{\sqrt{11}}{4}$, $\cos t = \frac{\sqrt{5}}{4}$, and $\tan t = \frac{-\frac{\sqrt{11}}{4}}{\frac{\sqrt{5}}{4}} = -\frac{\sqrt{11}}{\sqrt{5}} = -\frac{\sqrt{55}}{5}$.

31. $\left(-\frac{6}{7}\right)^2 + \left(\frac{\sqrt{13}}{7}\right)^2 = \frac{36}{49} + \frac{13}{49} = 1$. So $\sin t = \frac{\sqrt{13}}{7}$, $\cos t = -\frac{6}{7}$, and $\tan t = \frac{\frac{\sqrt{13}}{7}}{-\frac{6}{7}} = -\frac{\sqrt{13}}{6}$.

33. $\left(-\frac{5}{13}\right)^2 + \left(-\frac{12}{13}\right)^2 = \frac{25}{169} + \frac{144}{169} = 1$. So $\sin t = -\frac{12}{13}$, $\cos t = -\frac{5}{13}$, and $\tan t \frac{-\frac{12}{13}}{-\frac{5}{13}} = \frac{12}{5}$.

35. $\left(-\frac{20}{29}\right)^2 + \left(\frac{21}{29}\right)^2 = \frac{400}{841} + \frac{441}{841} = 1$. So $\sin t = \frac{21}{29}$, $\cos t = -\frac{20}{29}$, and $\tan t = \frac{\frac{21}{29}}{-\frac{20}{29}} = -\frac{21}{20}$.

37. (a) 0.8 **39. (a)** 0.9 **41. (a)** 1.0 **43. (a)** -0.6

(b) 0.84147 **(b)** 0.93204 **(b)** 1.02964 **(b)** -0.57482

45. $\sin t \cdot \cos t$. Since $\sin t$ is positive in quadrant II and $\cos t$ is negative in quadrant II, their product is negative.

47. $\dfrac{\tan t \cdot \sin t}{\cot t} = \tan t \cdot \dfrac{1}{\cot t} \cdot \sin t = \tan t \cdot \tan t \cdot \sin t = \tan^2 t \cdot \sin t$. Since $\tan^2 t$ is always positive and $\sin t$ is negative in quadrant III, the expression is negative in quadrant III.

49. Quadrant II

51. Quadrant II

53. $\sin t = \sqrt{1 - \cos^2 t}$

55. $\tan t = \dfrac{\sin t}{\cos t} = \dfrac{\sin t}{\sqrt{1 - \sin^2 t}}$

57. $\sec t = -\sqrt{1 + \tan^2 t}$

59. $\tan t = \sqrt{\sec^2 t - 1}$

61. $\tan^2 t = \dfrac{\sin^2 t}{\cos^2 t} = \dfrac{\sin^2 t}{1 - \sin^2 t}$

63. $\sin t = \frac{3}{5}$ and t is in quadrant II, so the terminal point determined by t is $P\left(x, \frac{3}{5}\right)$. Since P is on the unit circle $x^2 + \left(\frac{3}{5}\right)^2 = 1$. Solving for x gives $x = \pm\sqrt{1 - \frac{9}{25}} = \pm\sqrt{\frac{1}{2}} = \pm\frac{4}{5}$. Since t is in quadrant III, $x = -\frac{4}{5}$. Thus the terminal point is $P\left(-\frac{4}{5}, \frac{3}{5}\right)$. Thus, $\cos t = -\frac{4}{5}$, $\tan t = -\frac{3}{4}$, $\csc t = \frac{5}{3}$, $\sec t = -\frac{5}{4}$, $\cot t = -\frac{4}{3}$.

65. $\sec t = 3$ and t lies in quadrant IV. Thus, $\cos t = \frac{1}{3}$ and the terminal point determined by t is $P\left(\frac{1}{3}, y\right)$. Since P is on the unit circle $\left(\frac{1}{3}\right)^2 + y^2 = 1$. Solving for y gives $y = \pm\sqrt{1 - \frac{1}{9}} = \pm\sqrt{\frac{8}{9}} = \pm\frac{2\sqrt{2}}{3}$. Since t is in quadrant IV, $y = -\frac{2\sqrt{2}}{3}$. Thus the terminal point is $P\left(\frac{1}{3}, -\frac{2\sqrt{2}}{3}\right)$. Therefore, $\sin t = -\frac{2\sqrt{2}}{3}$, $\cos t = \frac{1}{3}$, $\tan t = -2\sqrt{2}$, $\csc t = -\dfrac{3}{2\sqrt{2}} = -\dfrac{3\sqrt{2}}{4}$, $\cot t = -\dfrac{1}{2\sqrt{2}} = -\dfrac{\sqrt{2}}{4}$.

67. $\tan t = -\frac{3}{4}$ and $\cos t > 0$, so t is in quadrant IV. Since $\sec^2 t = \tan^2 t + 1$ we have $\sec^2 t = \left(-\frac{3}{4}\right)^2 + 1 = \frac{9}{16} + 1 = \frac{25}{16}$. Thus $\sec t = \pm\sqrt{\frac{25}{16}} = \pm\frac{5}{4}$. Since $\cos t > 0$, we have $\cos t = \dfrac{1}{\sec t} = \dfrac{1}{\frac{5}{4}} = \frac{4}{5}$. Let $P\left(\frac{4}{5}, y\right)$. Since $\tan t \cdot \cos t = \sin t$ we have $\sin t = \left(-\frac{3}{4}\right)\left(\frac{4}{5}\right) = -\frac{3}{5}$. Thus, the terminal point determined by t is $P\left(\frac{4}{5}, -\frac{3}{5}\right)$, and so $\sin t = -\frac{3}{5}$, $\cos t = \frac{4}{5}$, $\csc t = -\frac{5}{3}$, $\sec t = \frac{5}{4}$, $\cot t = -\frac{4}{3}$.

69. $\sin t = -\frac{1}{4}$, $\sec t < 0$, so t is in quadrant III. So the terminal point determined by t is $P\left(x, -\frac{1}{4}\right)$. Since P is on the unit circle $x^2 + \left(-\frac{1}{4}\right)^2 = 1$. Solving for x gives $x = \pm\sqrt{1 - \frac{1}{16}} = \pm\sqrt{\frac{15}{16}} = \pm\frac{\sqrt{15}}{4}$. Since t is in quadrant III, $x = -\dfrac{\sqrt{15}}{4}$. Thus, the terminal point determined by t is $P\left(-\frac{\sqrt{15}}{4}, -\frac{1}{4}\right)$, and so $\cos t = -\frac{\sqrt{15}}{4}$, $\tan t = \frac{1}{\sqrt{15}} = \frac{\sqrt{15}}{15}$, $\csc t = -4$, $\sec t = -\frac{4}{\sqrt{15}} = -\frac{4\sqrt{15}}{15}$, $\cot t = \sqrt{15}$.

71. $f(-x) = (-x)^2 \sin(-x) = -x^2 \sin x = -f(x)$, so f is odd.

73. $f(-x) = \sin(-x)\cos(-x) = -\sin x \cos x = -f(x)$, so f is odd.

75. $f(-x) = |-x|\cos(-x) = |x|\cos x = f(x)$, so f is even.

77. $f(-x) = (-x)^3 + \cos(-x) = -x^3 + \cos x$ which is neither $f(x)$ nor $-f(x)$, so f is neither even nor odd.

79.

t	$y(t)$
0	4
0.25	-2.83
0.50	0
0.75	2.83
1.00	-4
1.25	2.83

81. (a) $I(0.1) = 0.8e^{-0.3}\sin 1 \approx 0.499$ A

(b) $I(0.5) = 0.8e^{-1.5}\sin 5 \approx -0.171$ A

83. Notice that if $P(t) = (x, y)$, then $P(t+\pi) = (-x, -y)$. Thus,

(a) $\sin(t+\pi) = -y$ and $\sin t = y$. Therefore, $\sin(t+\pi) = -\sin t$.

(b) $\cos(t+\pi) = -x$ and $\cos t = x$. Therefore, $\cos(t+\pi) = -\cos t$.

(c) $\tan(t+\pi) = \dfrac{\sin(t+\pi)}{\cos(t+\pi)} = \dfrac{-y}{-x} = \dfrac{y}{x} = \dfrac{\sin t}{\cos t} = \tan t$.

7.3 Trigonometric Graphs

1. $f(x) = 1 + \cos x$

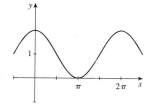

3. $f(x) = -\sin x$

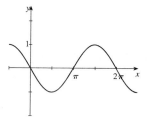

5. $f(x) = -2 + \sin x$

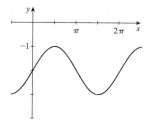

7. $g(x) = 3\cos x$

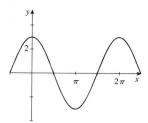

9. $g(x) = -\frac{1}{2}\sin x$

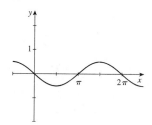

11. $g(x) = 3 + 3\cos x$

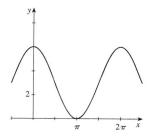

13. $h(x) = |\cos x|$

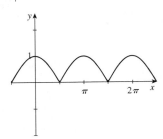

15. $y = \cos 2x$ has amplitude 1 and period π.

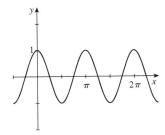

17. $y = -3 \sin 3x$ has amplitude 3 and period $\frac{2\pi}{3}$.

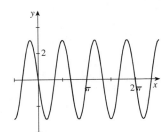

19. $y = 10 \sin \frac{1}{2}x$ has amplitude 10 and period 4π.

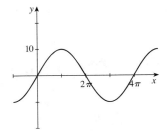

21. $y = -\frac{1}{3} \cos \frac{1}{3}x$ has amplitude $\frac{1}{3}$ and period 6π.

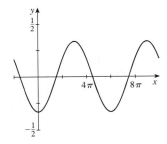

23. $y = -2 \sin 2\pi x$ has amplitude 2 and period 1.

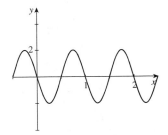

25. $y = 1 + \frac{1}{2} \cos \pi x$ has amplitude $\frac{1}{2}$ and period 2.

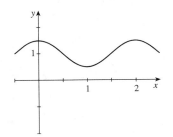

27. $y = \cos\left(x - \frac{\pi}{2}\right)$ has amplitude 1, period 2π, and phase shift $\frac{\pi}{2}$.

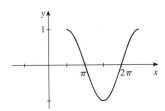

29. $y = -2\sin\left(x - \frac{\pi}{6}\right)$ has amplitude 2, period 2π, and phase shift $\frac{\pi}{6}$.

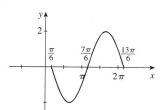

31. $y = -4\sin 2\left(x + \frac{\pi}{2}\right)$ has amplitude 4, period π, and phase shift $-\frac{\pi}{2}$.

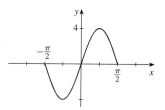

33. $y = 5\cos\left(3x - \frac{\pi}{4}\right) = 5\cos 3\left(x - \frac{\pi}{12}\right)$ has amplitude 5, period $\frac{2\pi}{3}$, and phase shift $\frac{\pi}{12}$.

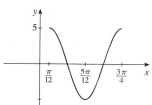

35. $y = \frac{1}{2} - \frac{1}{2}\cos\left(2x - \frac{\pi}{3}\right) = \frac{1}{2} - \frac{1}{2}\cos 2\left(x - \frac{\pi}{6}\right)$ has amplitude $\frac{1}{2}$, period π, and phase shift $\frac{\pi}{6}$.

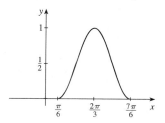

37. $y = 3\cos\pi\left(x + \frac{1}{2}\right)$ has amplitude 3, period 2, and phase shift $-\frac{1}{2}$.

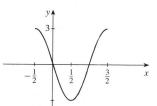

39. $y = \sin(3x + \pi) = \sin 3\left(x + \frac{\pi}{3}\right)$ has amplitude 1, period $\frac{2\pi}{3}$, and phase shift $-\frac{\pi}{3}$

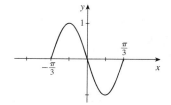

41. (a) This function has amplitude $a = 4$, period $\frac{2\pi}{k} = 2\pi$, and phase shift $b = 0$ as a sine curve.

  (b) $y = a\sin k(x - b) = 4\sin x$

43. (a) This curve has amplitude $a = \frac{3}{2}$, period $\frac{2\pi}{k} = \frac{2\pi}{3}$, and phase shift $b = 0$ as a cosine curve.

  (b) $y = a\cos k(x - b) = \frac{3}{2}\cos 3x$

45. (a) This curve has amplitude $a = \frac{1}{2}$, period $\frac{2\pi}{k} = \pi$, and phase shift $b = -\frac{\pi}{3}$ as a cosine curve.

  (b) $y = -\frac{1}{2}\cos 2\left(x + \frac{\pi}{3}\right)$

47. (a) This curve has amplitude $a = 4$, period $\frac{2\pi}{k} = \frac{3}{2}$, and phase shift $b = -\frac{1}{2}$ as a sine curve.

  (b) $y = 4\sin\frac{4\pi}{3}\left(x + \frac{1}{2}\right)$

49. $f(x) = \cos 100x$, $[-0.1, 0.1]$ by $[-1.5, 1.5]$

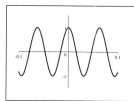

51. $f(x) = \sin\dfrac{x}{40}$, $[-250, 250]$ by $[-1.5, 1.5]$

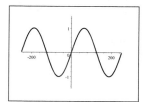

53. $y = \tan 25x$, $[-0.2, 0.2]$ by $[-3, 3]$

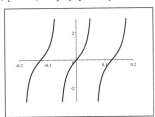

55. $y = \sin^2 20x$, $[-0.5, 0.5]$ by $[-0.2, 1.2]$

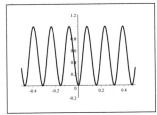

57. $f(x) = x$, $g(x) = \sin x$

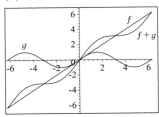

59. $y = x^2 \sin x$ is a sine curve that lies between the graphs of $y = x^2$ and $y = -x^2$.

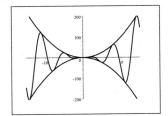

61. $y = \sqrt{x} \sin 5\pi x$ is a sine curve that lies between the graphs of $y = \sqrt{x}$ and $y = -\sqrt{x}$.

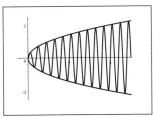

63. $y = \cos 3\pi x \cos 21\pi x$ is a cosine curve that lies between the graphs of $y = \cos 3\pi x$ and $y = -\cos 3\pi x$.

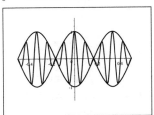

65. $y = \sin x + \sin 2x$. The period is 2π, so we graph the function over one period, $(-\pi, \pi)$. Maximum value 1.76 when $x \approx 0.94 + 2n\pi$, minimum value -1.76 when $x \approx -0.94 + 2n\pi$, n any integer.

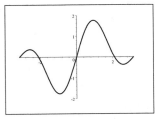

67. $y = 2\sin x + \sin^2 x$. The period is 2π, so we graph the function over one period, $(-\pi, \pi)$. Maximum value 3.00 when $x \approx 1.57 + 2n\pi$, minimum value -1.00 when $x \approx -1.57 + 2n\pi$, n any integer.

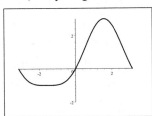

69. $\cos x = 0.4$, $x \in [0, \pi]$. The solution is $x \approx 1.16$.

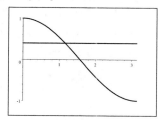

71. $\csc x = 3$, $x \in [0, \pi]$. The solutions are $x \approx 0.34, 2.80$.

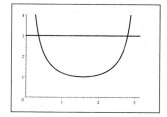

73. $f(x) = \dfrac{1 - \cos x}{x}$

(c)

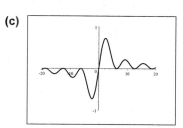

    (a) Since $f(-x) = \dfrac{1 - \cos(-x)}{-x} = \dfrac{1 - \cos x}{-x} = -f(x)$, the function

    is odd.

    (b) The x-intercepts occur when $1 - \cos x = 0$ $\Leftrightarrow$ $\cos x = 1$ $\Leftrightarrow$

    $x = 0, \pm 2\pi, \pm 4\pi, \pm 6\pi, \ldots$

    (d) As $x \to \pm\infty$, $f(x) \to 0$.

    (e) As $x \to 0$, $f(x) \to 0$.

75. (a) The period of the wave is $\dfrac{2\pi}{\pi/10} = 20$ seconds.

    (b) Since $h(0) = 3$ and $h(10) = -3$, the wave height is $3 - (-3) = 6$ feet.

77. (a) The period of p is $\dfrac{2\pi}{160\pi} = \dfrac{1}{80}$ minute.

    (b) Since each period represents a heart beat, there are 80 heart beats per

    minute.

    (d) The maximum (or systolic) is $115 + 25 = 140$ and the minimum (or

    diastolic) is $115 - 25 = 90$. The read would be $140/90$ which higher

    than normal.

(c)

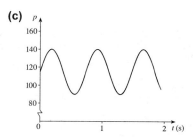

79. (a) $y = \sin\left(\sqrt{x}\right)$. This graph looks like a sine function
which has been stretched horizontally (stretched
more for larger values of x). It is defined only for
$x \geq 0$, so it is neither even nor odd.

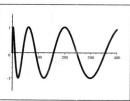

(b) $y = \sin\left(x^2\right)$. This graph looks like a graph of $\sin|x|$ which
has been shrunk for $|x| > 1$ (shrunk more for larger values of
x) and stretched for $|x| < 1$. It is an even function, whereas
$\sin x$ is odd.

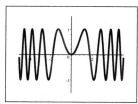

81. (a) The graph of $y = |\sin x|$ is shown in the viewing rectangle $[-6.28, 6.28]$ by $[-0.5, 1.5]$. This function is periodic with period π.

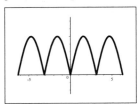

(b) The graph of $y = \sin |x|$ is shown in the viewing rectangle $[-10, 10]$ by $[-1.5, 1.5]$. The function is not periodic. Note that while $\sin |x + 2\pi| = \sin |x|$ for many values of x, it is false for $x \in (-2\pi, 0)$. For example $\sin \left|-\frac{\pi}{2}\right| = \sin \frac{\pi}{2} = 1$ while $\sin \left|-\frac{\pi}{2} + 2\pi\right| = \sin \frac{3\pi}{2} = -1$.

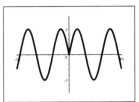

(c) The graph of $y = 2^{\cos x}$ is shown in the viewing rectangle $[-10, 10]$ by $[-1, 3]$. This function is periodic with period $= 2\pi$.

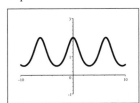

(d) The graph of $y = x - [\![x]\!]$ is shown in the viewing rectangle $[-7.5, 7.5]$ by $[-0.5, 1.5]$. This function is periodic with period 1. Be sure to turn off "connected" mode when graphing functions with gaps in their graph.

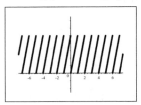

7.4 More Trigonometric Graphs

1. $f(x) = \tan\left(x + \frac{\pi}{4}\right)$ corresponds to Graph II. f is undefined at $x = \frac{\pi}{4}$ and $x = \frac{3\pi}{4}$, and Graph II has the shape of a graph of a tangent function.

3. $f(x) = \cot 2x$ corresponds to Graph VI.

5. $f(x) = 2\sec x$ corresponds to Graph IV.

7. $y = 4\tan x$ has period π.

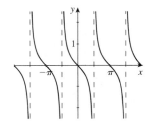

9. $y = -\frac{1}{2}\tan x$ has period π.

11. $y = -\cot x$ has period π.

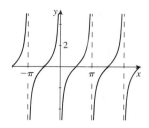

13. $y = 2\csc x$ has period 2π.

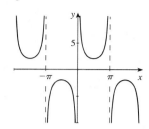

15. $y = 3\sec x$ has period 2π.

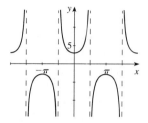

17. $y = \tan\left(x + \frac{\pi}{2}\right)$ has period π.

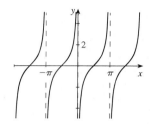

19. $y = \csc\left(x - \frac{\pi}{2}\right)$ has period 2π.

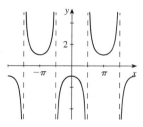

21. $y = \cot\left(x + \frac{\pi}{4}\right)$ has period π.

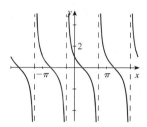

23. $y = \frac{1}{2}\sec\left(x - \frac{\pi}{6}\right)$ has period 2π.

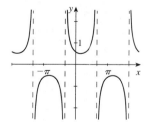

25. $y = \tan 2x$ has period $\frac{\pi}{2}$.

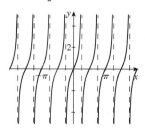

27. $y = \tan\left(\frac{\pi}{4}x\right)$ has period $\frac{\pi}{\frac{\pi}{4}} = 4$.

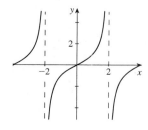

29. $y = \sec 2x$ has period $\frac{2\pi}{2} = \pi$.

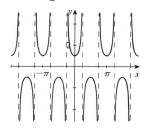

31. $y = \csc 2x$ has period $\frac{2\pi}{2} = \pi$.

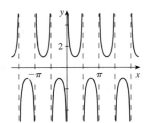

33. $y = 2\tan 3\pi x$ has period $\frac{1}{3}$.

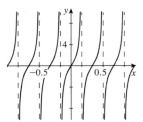

35. $y = 5\csc \frac{3\pi}{2}x$ has period $\frac{2\pi}{\frac{3\pi}{2}} = \frac{4}{3}$.

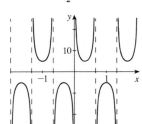

37. $y = \tan 2\left(x + \frac{\pi}{2}\right)$ has period $\frac{\pi}{2}$.

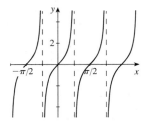

39. $y = \tan 2\left(x - \pi\right) = \tan 2x$ has period $\frac{\pi}{2}$.

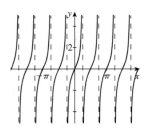

41. $y = \cot\left(2x - \frac{\pi}{2}\right) = \cot 2\left(x - \frac{\pi}{4}\right)$ has period $\frac{\pi}{2}$.

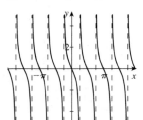

43. $y = 2\csc\left(\pi x - \frac{\pi}{3}\right) = 2\csc \pi\left(x - \frac{1}{3}\right)$ has period $\frac{2\pi}{\pi} = 2$.

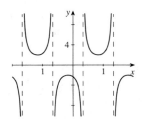

45. $y = 5\sec\left(3x - \frac{\pi}{2}\right) = 5\sec 3\left(x - \frac{\pi}{6}\right)$ has period $\frac{2\pi}{3}$.

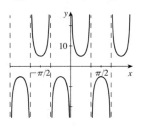

47. $y = \tan\left(\frac{2}{3}x - \frac{\pi}{6}\right) = \tan\frac{2}{3}\left(x - \frac{\pi}{4}\right)$ has period $\pi / \left(\frac{2}{3}\right) = \frac{3\pi}{2}$.

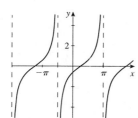

49. $y = 3\sec\pi\left(x + \frac{1}{2}\right)$ has period $\frac{2\pi}{\pi} = 2$.

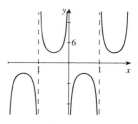

51. $y = -2\tan\left(2x - \frac{\pi}{3}\right) = -2\tan 2\left(x - \frac{\pi}{6}\right)$ has period $\frac{\pi}{2}$.

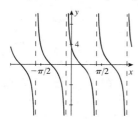

53. (a) If f is periodic with period p, then by the definition of a period, $f(x+p) = f(x)$ for all x in the domain of f. Therefore, $\dfrac{1}{f(x+p)} = \dfrac{1}{f(x)}$ for all $f(x) \neq 0$.

Thus, $\dfrac{1}{f}$ is also periodic with period p.

(b) Since $\sin x$ has period 2π, it follows from part (a) that $\csc x = \dfrac{1}{\sin x}$ also has period 2π. Similarly, since $\cos x$ has period 2π, we conclude $\sec x = \dfrac{1}{\cos x}$ also has period 2π.

55. (a) $d(t) = 3\tan\pi t$, so $d(0.15) \approx 1.53$, $d(0.25) \approx 3.00$, and $d(0.45) \approx 18.94$.

(b)

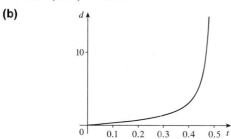

(c) $d \to \infty$ as $t \to \frac{1}{2}$.

57. The graph of $y = -\cot x$ is the same as the graph of $y = \tan x$ shifted $\frac{\pi}{2}$ units to the right, and the graph of $y = \csc x$ is the same as the graph of $y = \sec x$ shifted $\frac{\pi}{2}$ units to the right.

7.5 Modeling Harmonic Motion

1. $y = 2 \sin 3t$

  (a) Amplitude 2, period $\frac{2\pi}{3}$, frequency $\frac{1}{\text{period}} = \frac{3}{2\pi}$.

  (b)

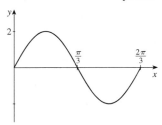

3. $y = -\cos 0.3t$

  (a) Amplitude 1, period $\frac{2\pi}{0.3} = \frac{20\pi}{3}$, frequency $\frac{3}{20\pi}$.

  (b)

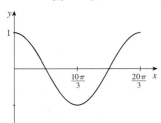

5. $y = -0.25 \cos \left(1.5t - \frac{\pi}{3}\right) = -0.25 \cos \left(\frac{3}{2}t - \frac{\pi}{3}\right)$
$= -0.25 \cos \frac{3}{2} \left(t - \frac{2\pi}{9}\right)$

  (a) Amplitude 0.25, period $\frac{2\pi}{\frac{3}{2}} = \frac{4\pi}{3}$, frequency $\frac{3}{4\pi}$.

  (b)

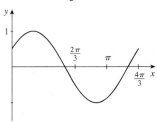

7. $y = 5 \cos \left(\frac{2}{3}t + \frac{3}{4}\right) = 5 \cos \frac{2}{3} \left(t + \frac{9}{8}\right)$

  (a) Amplitude 5, period $\frac{2\pi}{\frac{2}{3}} = 3\pi$, frequency $\frac{1}{3\pi}$.

  (b)

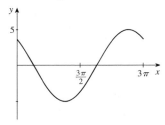

9. The amplitude is $a = 10$ cm, the period is $\frac{2\pi}{k} = 3$ s, and $f(0) = 0$, so $f(x) = 10 \sin \frac{2\pi}{3}t$.

11. The amplitude is 6 in., the frequency is $\frac{k}{2\pi} = \frac{5}{\pi}$ Hz, and $f(0) = 0$, so $f(x) = 6 \sin 10t$.

13. The amplitude is 60 ft, the period is $\frac{2\pi}{k} = 0.5$ min, and $f(0) = 60$, so $f(x) = 60 \cos 4\pi t$.

15. The amplitude is 2.4 m, the frequency is $\frac{k}{2\pi} = 750$ Hz, and $f(0) = 2.4$, so $f(x) = 2.4 \cos 1500\pi t$.

17. (a) $k = 2$, $c = 1.5$, and $f = 3 \Rightarrow \omega = 6\pi$, so we have
$y = 2e^{-1.5t} \cos 6\pi t$.

  (b)

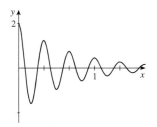

19. (a) $k = 100$, $c = 0.05$, and $p = 4 \Rightarrow \omega = \frac{\pi}{2}$, so we have
$y = 100e^{-0.05t} \cos \frac{\pi}{2}t$.

  (b)

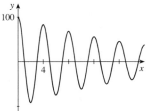
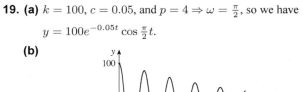

21. (a) $k = 7$, $c = 10$, and $p = \frac{\pi}{6} \Rightarrow \omega = 12$, so we have

$y = 7e^{-10t} \sin 12t$.

(b)

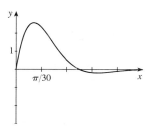

23. (a) $k = 0.3$, $c = 0.2$, and $f = 20 \Rightarrow \omega = 40\pi$, so we

have $y = 0.3e^{-0.2t} \sin 40\pi t$.

(b)

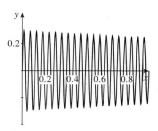

25. $y = 0.2 \cos 20\pi t + 8$

(a) The frequency is $\frac{20\pi}{2\pi} = 10$ cycles/min.

(b)

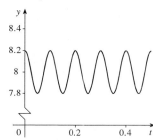

(c) Since $y = 0.2 \cos 20\pi t + 8 \le 0.2\,(1) + 8 = 8.2$ and

when $t = 0$, $y = 8.2$, the maximum displacement is

8.2 m.

27. (a) When $t = 0$, π, 2π, ... we have $\cos 2t = 1$ so y is

maximized at $y = 8900$.

(b) The length of time between successive periods of

maximum population is the length of a period which is

$\frac{2\pi}{2} = \pi \approx 3.14$ years.

29. The graph resembles a sine wave with an amplitude of 5, a period of $\frac{2}{5}$, and no phase shift. Therefore, $a = 5$, $\dfrac{2\pi}{\omega} = \frac{2}{5}$ $\Leftrightarrow$

$\omega = 5\pi$, and a formula is $d\,(t) = 5 \sin 5\pi t$.

31. $a = 21$, $f = \frac{1}{12}$ cycle/hour $\Rightarrow$ $\dfrac{\omega}{2\pi} = \frac{1}{12}$ $\Leftrightarrow$ $\omega = \frac{\pi}{6}$. So,

$y = 21 \sin\left(\frac{\pi}{6} t\right)$ (assuming the tide is at mean level and rising when $t = 0$).

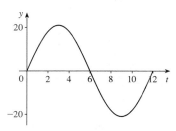

33. Since the mass travels from its highest point (compressed spring) to its lowest point in $\frac{1}{2}$ s, it completes half a period in $\frac{1}{2}$ s.

So, $\frac{1}{2}$ (one period) $= \frac{1}{2}$ s $\Rightarrow$ $\dfrac{1}{2} \cdot \dfrac{2\pi}{\omega} = \frac{1}{2}$ $\Leftrightarrow$ $\omega = 2\pi$. Also, $a = 5$. So $y = 5 \cos 2\pi t$.

35. Since the Ferris wheel has a radius of 10 m and the bottom of the wheel is 1 m above the ground, the minimum height is

1 m and the maximum height is 21 m. Then $a = 10$ and $\dfrac{2\pi}{\omega} = 20$ s $\Leftrightarrow$ $\omega = \frac{\pi}{10}$, and so $y = 11 + 10 \sin\left(\frac{\pi}{10} t\right)$, where t

is in seconds.

37. $a = 0.2$, $\dfrac{2\pi}{\omega} = 10$ $\Leftrightarrow$ $\omega = \frac{\pi}{5}$. Then $y = 3.8 + 0.2 \sin\left(\frac{\pi}{5} t\right)$.

39. $E_0 = 310$, frequency is 100 $\Rightarrow$ $\dfrac{\omega}{2\pi} = 100$ $\Leftrightarrow$ $\omega = 200\pi$. Then, $E\,(t) = 310 \cos 200\pi t$. The maximum voltage

produced occurs when $\cos 2\pi t = 1$, and hence is $E_{\max} = 310$ V. The rms voltage is $\dfrac{310}{\sqrt{2}} \approx 219$ V.

41. (a) The maximum voltage is the amplitude, that is, $V_{\max} = a = 45$ V.

(b) From the graph we see that 4 cycles are completed every 0.1 seconds, or equivalently, 40 cycles are completed every second, so $f = 40$.

(c) The number of revolutions per second of the armature is the frequency, that is, $\dfrac{\omega}{2\pi} = f = 40$.

(d) $a = 45$, $f = \dfrac{\omega}{2\pi} = 40 \;\Leftrightarrow\; \omega = 80\pi$. Then $V(t) = 45\cos 80\pi t$.

43. $k = 1$, $c = 0.9$, and $\dfrac{\omega}{2\pi} = \dfrac{1}{2} \Leftrightarrow \omega = \pi$. Since $f(0) = 0$, $f(t) = e^{-0.9t}\sin \pi t$.

45. $\dfrac{ke^{-ct}}{ke^{-c(t+3)}} = 4 \Leftrightarrow e^{-ct + c(t+3)} = 4 \quad\Leftrightarrow\quad e^{3c} = 4 \quad\Leftrightarrow\quad 3c = \ln 4 \quad\Leftrightarrow\quad c = \tfrac{1}{3}\ln 4 \approx 0.46$.

Chapter 7 Review

1. (a) Since $\left(-\dfrac{\sqrt{3}}{2}\right)^2 + \left(\dfrac{1}{2}\right)^2 = \dfrac{3}{4} + \dfrac{1}{4} = 1$, the point $P\left(-\dfrac{\sqrt{3}}{2}, \dfrac{1}{2}\right)$ lies on the unit circle.

(b) $\sin t = \dfrac{1}{2}$, $\cos t = -\dfrac{\sqrt{3}}{2}$, $\tan t = \dfrac{\frac{1}{2}}{-\frac{\sqrt{3}}{2}} = -\dfrac{\sqrt{3}}{3}$.

3. $t = \dfrac{2\pi}{3}$

(a) $\bar{t} = \pi - \dfrac{2\pi}{3} = \dfrac{\pi}{3}$

(b) $P\left(-\dfrac{1}{2}, \dfrac{\sqrt{3}}{2}\right)$

(c) $\sin t = \dfrac{\sqrt{3}}{2}$, $\cos t = -\dfrac{1}{2}$, $\tan t = -\sqrt{3}$, $\csc t = \dfrac{2\sqrt{3}}{3}$,

$\sec t = -2$, and $\cot t = -\dfrac{\sqrt{3}}{3}$.

5. $t = -\dfrac{11\pi}{4}$

(a) $\bar{t} = 3\pi + \left(-\dfrac{11\pi}{4}\right) = \dfrac{\pi}{4}$

(b) $P\left(-\dfrac{\sqrt{2}}{2}, -\dfrac{\sqrt{2}}{2}\right)$

(c) $\sin t = -\dfrac{\sqrt{2}}{2}$, $\cos t = -\dfrac{\sqrt{2}}{2}$, $\tan t = 1$, $\csc t = -\sqrt{2}$,

$\sec t = -\sqrt{2}$, and $\cot t = 1$.

7. (a) $\sin \dfrac{3\pi}{4} = \sin \dfrac{\pi}{4} = \dfrac{\sqrt{2}}{2}$

(b) $\cos \dfrac{3\pi}{4} = -\cos \dfrac{\pi}{4} = -\dfrac{\sqrt{2}}{2}$

9. (a) $\sin 1.1 \approx 0.89121$

(b) $\cos 1.1 \approx 0.45360$

11. (a) $\cos \dfrac{9\pi}{2} = \cos \dfrac{\pi}{2} = 0$

(b) $\sec \dfrac{9\pi}{2}$ is undefined

13. (a) $\tan \dfrac{5\pi}{2}$ is undefined

(b) $\cot \dfrac{5\pi}{2} = \cot \dfrac{\pi}{2} = 0$

15. (a) $\tan \dfrac{5\pi}{6} = -\dfrac{\sqrt{3}}{3}$

(b) $\cot \dfrac{5\pi}{6} = -\sqrt{3}$

17. $\dfrac{\tan t}{\cos t} = \dfrac{\frac{\sin t}{\cos t}}{\cos t} = \dfrac{\sin t}{\cos^2 t} = \dfrac{\sin t}{1 - \sin^2 t}$

19. $\tan t = \dfrac{\sin t}{\cos t} = \dfrac{\sin t}{\pm\sqrt{1 - \sin^2 t}} = \dfrac{\sin t}{\sqrt{1 - \sin^2 t}}$ (since t is in quadrant IV, $\cos t$ is positive)

21. $\sin t = \dfrac{5}{13}$, $\cos t = -\dfrac{12}{13}$. Then $\tan t = \dfrac{\frac{5}{13}}{-\frac{12}{13}} = -\dfrac{5}{12}$, $\csc t = \dfrac{13}{5}$, $\sec t = -\dfrac{13}{12}$, and $\cot t = -\dfrac{12}{5}$.

23. $\cot t = -\dfrac{1}{2}$, $\csc t = \dfrac{\sqrt{5}}{2}$. Since $\csc t = \dfrac{1}{\sin t}$, we know $\sin t = \dfrac{2}{\sqrt{5}} = \dfrac{2\sqrt{5}}{5}$. Now $\cot t = \dfrac{\cos t}{\sin t}$, so

$\cos t = \sin t \cdot \cot t = \dfrac{2\sqrt{5}}{5} \cdot \left(-\dfrac{1}{2}\right) = -\dfrac{\sqrt{5}}{5}$, and $\tan t = \dfrac{1}{\left(-\frac{1}{2}\right)} = -2$ while $\sec t = \dfrac{1}{\cos t} = \dfrac{1}{\left(-\frac{\sqrt{5}}{5}\right)} = -\dfrac{5}{\sqrt{5}} = -\sqrt{5}$.

25. $\tan t = \dfrac{1}{4}$, t is in quadrant III $\Rightarrow$ $\sec t + \cot t = -\sqrt{\tan^2 t + 1} + \dfrac{1}{\tan t} = -\sqrt{\left(\dfrac{1}{4}\right)^2 + 1} + 4 = -\sqrt{\dfrac{17}{16}} + 4 = 4 -$

$\dfrac{\sqrt{17}}{4} = \dfrac{16 - \sqrt{17}}{4}$

27. $\cos t = \frac{3}{5}$, t is in quadrant I $\Rightarrow$ $\tan t + \sec t = \dfrac{\sin t}{\cos t} + \dfrac{1}{\cos t} = \dfrac{\sqrt{1-\cos^2 t}}{\cos t} + \dfrac{1}{\cos t} = \dfrac{\sqrt{1-\left(\frac{3}{5}\right)^2}}{\frac{3}{5}} + \dfrac{5}{3} = \dfrac{\sqrt{\frac{1}{2}}}{\frac{3}{5}} +$

$\dfrac{5}{3} = \dfrac{4}{5} \cdot \dfrac{5}{3} + \dfrac{5}{3} = \dfrac{9}{3} = 3$

29. $y = 10\cos\frac{1}{2}x$

    (a) This function has amplitude 10, period $\dfrac{2\pi}{\frac{1}{2}} = 4\pi$, and

    phase shift 0.

    (b)

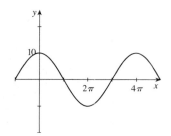

31. $y = -\sin\frac{1}{2}x$

    (a) This function has amplitude 1, period $\dfrac{2\pi}{\left(\frac{1}{2}\right)} = 4\pi$, and

    phase shift 0.

    (b)

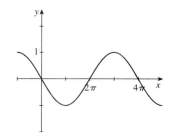

33. $y = 3\sin(2x - 2) = 3\sin 2(x - 1)$

    (a) This function has amplitude 3, period $\frac{2\pi}{2} = \pi$, and
    phase shift 1.

    (b)

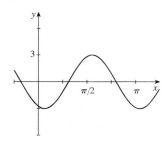

35. $y = -\cos\left(\frac{\pi}{2}x + \frac{\pi}{6}\right) = -\cos\frac{\pi}{2}\left(x + \frac{1}{3}\right)$

    (a) This function has amplitude 1, period $\dfrac{2\pi}{\left(\frac{\pi}{2}\right)} = 4$, and

    phase shift $-\frac{1}{3}$.

    (b)

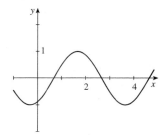

37. From the graph we see that the amplitude is 5, the period is $\frac{\pi}{2}$, and there is no phase shift. Therefore, the function is
$y = 5\sin 4x$.

39. From the graph we see that the amplitude is $\frac{1}{2}$, the period is 1, and there is a phase shift of $-\frac{1}{3}$. Therefore, the function is
$y = \frac{1}{2}\sin 2\pi\left(x + \frac{1}{3}\right)$.

41. $y = 3\tan x$ has period π.

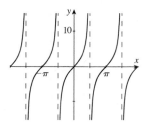

43. $y = 2\cot\left(x - \frac{\pi}{2}\right)$ has period π.

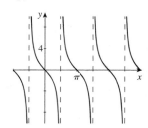

45. $y = 4\csc(2x + \pi) = 4\csc 2\left(x + \frac{\pi}{2}\right)$ has period $\frac{2\pi}{2} = \pi$.

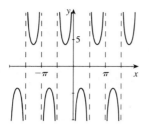

47. $y = \tan\left(\frac{1}{2}x - \frac{\pi}{8}\right) = \tan\frac{1}{2}\left(x - \frac{\pi}{4}\right)$ has period $\frac{\pi}{\frac{1}{2}} = 2\pi$.

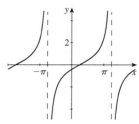

49. (a) $y = |\cos x|$

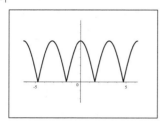

(b) This function has period π.

(c) This function is even.

51. (a) $y = \cos\left(2^{0.1x}\right)$

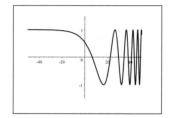

(b) This function is not periodic.

(c) This function is neither even nor odd.

53. (a) $y = |x|\cos 3x$

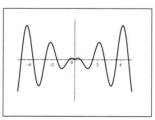

(b) This function is not periodic.

(c) This function is even.

55. $y = x\sin x$ is a sine function whose graph lies between those of $y = x$ and $y = -x$.

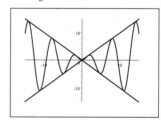

57. $y = x + \sin 4x$ is the sum of the two functions $y = x$ and $y = \sin 4x$.

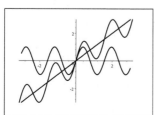

59. $y = \cos x + \sin 2x$. Since the period is 2π, we graph over the interval $[-\pi, \pi]$. The maximum value is 1.76 when $x \approx 0.63 \pm 2n\pi$, the minimum value is -1.76 when $x \approx 2.51 \pm 2n\pi$, n an integer.

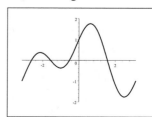

61. We want to find solutions to $\sin x = 0.3$ in the interval $[0, 2\pi]$, so we plot the functions $y = \sin x$ and $y = 0.3$ and look for their intersection. We see that $x \approx 0.305$ or $x \approx 2.837$.

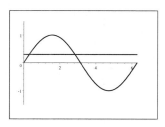

63. $f(x) = \dfrac{\sin^2 x}{x}$

(a) The function is odd.

(b) The graph intersects the x-axis at $x = 0, \pm\pi, \pm 2\pi, \pm 3\pi, \ldots$

(d) As $x \to \pm\infty$, $f(x) \to 0$.

(e) As $x \to 0$, $f(x) \to 0$.

(c)

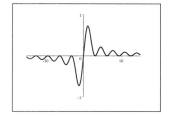

65. The amplitude is $a = 50$ cm. The frequency is 8 Hz, so $\omega = 8\,(2\pi) = 16\pi$. Since the mass is at its maximum displacement when $t = 0$, the motion follows a cosine curve. So a function describing the motion of P is $f(t) = 50\cos 16\pi t$.

67. From the graph, we see that the amplitude is 4 ft, the period is 12 hours, and there is no phase shift. Thus, the variation in water level is described by $y = 4\cos\frac{\pi}{6}t$.

Chapter 7 Test

1. Since $P(x, y)$ lies on the unit circle, $x^2 + y^2 = 1 \Rightarrow y = \pm\sqrt{1 - \left(\frac{\sqrt{11}}{6}\right)^2} = \pm\sqrt{\frac{25}{36}} = \pm\frac{5}{6}$. But $P(x, y)$ lies in the fourth quadrant. Therefore y is negative $\Rightarrow y = -\frac{5}{6}$.

2. Since P is on the unit circle, $x^2 + y^2 = 1 \Leftrightarrow x^2 = 1 - y^2$. Thus, $x^2 = 1 - \left(\frac{4}{5}\right)^2 = \frac{9}{25}$, and so $x = \pm\frac{3}{5}$. From the diagram, x is clearly negative, so $x = -\frac{3}{5}$. Therefore, P is the point $\left(-\frac{3}{5}, \frac{4}{5}\right)$.

(a) $\sin t = \frac{4}{5}$

(b) $\cos t = -\frac{3}{5}$

(c) $\tan t = \dfrac{\frac{4}{5}}{-\frac{3}{5}} = -\frac{4}{3}$

(d) $\sec(t) = -\frac{5}{3}$

3. (a) $\sin \frac{7\pi}{6} = -0.5$

(b) $\cos \dfrac{13\pi}{4} = -\dfrac{\sqrt{2}}{2}$

(c) $\tan\left(-\frac{5\pi}{3}\right) = \sqrt{3}$

(d) $\csc\left(\frac{3\pi}{2}\right) = -1$

4. $\tan t = \dfrac{\sin t}{\cos t} = \dfrac{\sin t}{\pm\sqrt{1 - \sin^2 t}}$. But t is in quadrant II $\Rightarrow \cos t$ is negative, so we choose the negative square root.

Thus, $\tan t = \dfrac{\sin t}{-\sqrt{1 - \sin^2 t}}$.

5. $\cos t = -\frac{8}{17}$, t in quadrant III $\Rightarrow \tan t \cdot \cot t + \csc t = 1 + \dfrac{1}{-\sqrt{1 - \cos^2 t}}$ (since t is in

quadrant III)$= 1 - \dfrac{1}{-\sqrt{1 - \frac{64}{289}}} = 1 - \dfrac{1}{\frac{15}{17}} = -\frac{2}{15}$.

6. $y = -5 \cos 4x$

   (a) This function has amplitude 5, period $\frac{2\pi}{4} = \frac{\pi}{2}$, and phase shift 0.

   (b)

   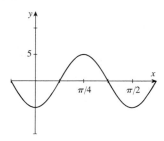

7. $y = 2 \sin\left(\frac{1}{2}x - \frac{\pi}{6}\right) = \sin \frac{1}{2}\left(x - \frac{\pi}{3}\right)$

   (a) This function has amplitude 2, period $\dfrac{2\pi}{\frac{1}{2}} = 4\pi$, and phase shift $\frac{\pi}{3}$.

   (b)

   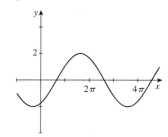

8. $y = -\csc 2x$ has period $\frac{2\pi}{2} = \pi$.

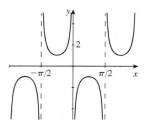

9. $y = \tan 2\left(x - \frac{\pi}{4}\right)$ has period $\frac{\pi}{2}$.

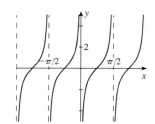

10. From the graph, we see that the amplitude is 2 and the phase shift is $-\frac{\pi}{3}$. Also, the period is π, so $\frac{2\pi}{k} = \pi$ $\Rightarrow$ $k = \frac{2\pi}{\pi} = 2$. Thus, the function is $y = 2 \sin 2\left(x + \frac{\pi}{3}\right)$.

11. $y = \dfrac{\cos x}{1 + x^2}$

   (b) The function is even.

   (c) The function has a minimum value of approximately -0.11 when $x \approx \pm 2.54$ and a maximum value of 1 when $x = 0$.

   (a)

   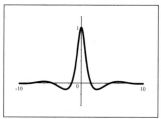

12. The amplitude is $\frac{1}{2}(10) = 5$ cm and the frequency is 2 Hz. Assuming that the mass is at its rest position and moving upward when $t = 0$, a function describing the distance of the mass from its rest position is $f(t) = 5 \sin 4\pi t$.

13. (a) The initial amplitude is 16 in and the frequency is 12 Hz, so a function describing the motion is

   $y = 16e^{-0.1t} \cos 24\pi t$.

   (b)

   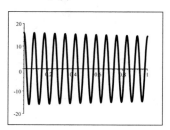

Focus on Modeling: Fitting Sinusoidal Curves to Data

1. (a) See the graph in part (c).

(b) Using the method of Example 1, we find the vertical shift
$b = \frac{1}{2}$ (maximum value + minimum value) $= \frac{1}{2}(2.1 - 2.1) = 0$, the amplitude
$a = \frac{1}{2}$ (maximum value − minimum value) $= \frac{1}{2}(2.1 - (-2.1)) = 2.1$, the period $\dfrac{2\pi}{\omega} = 2(6-0) = 12$ (so
$\omega \approx 0.5236$), and the phase shift $c = 0$. Thus, our model is $y = 2.1\cos\frac{\pi}{6}t$.

(c)

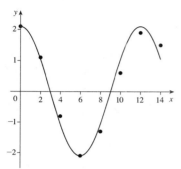

The curve fits the data quite well.

(d) Using the `SinReg` command on the TI-83, we find
$$y = 2.048714222\sin(0.5030795477t + 1.551856108)$$
$$- 0.0089616507.$$

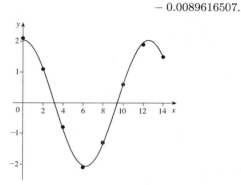

(e) Our model from part (b) is equivalent to $y = 2.1\sin(0.5236t + 1.5708)$, which is close to the model in part (d).

3. (a) See the graph in part (c).

(b) Using the method of Example 1, we find the vertical shift
$b = \frac{1}{2}$ (maximum value + minimum value) $= \frac{1}{2}(25.1 + 1.0) = 13.05$, the amplitude
$a = \frac{1}{2}$ (maximum value − minimum value) $= \frac{1}{2}(25.1 - 1.0) = 12.05$, the period $\dfrac{2\pi}{\omega} = 2(1.5 - 0.9) = 1.2$ (so
$\omega \approx 5.236$), and the phase shift $c = 0.3$. Thus, our model is $y = 12.05\cos(5.236(t - 0.3)) + 13.05$.

(c)

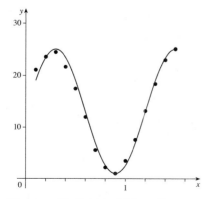

The curve fits the data fairly well.

(d) Using the `SinReg` command on the TI-83, we find
$$y = 11.71905062\sin(5.048853286t + 0.2388957877)$$
$$+ 12.96070536.$$

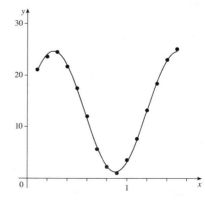

(e) Our model from part (b) is equivalent to $y = 12.05\sin 5.236t + 13.05$, which is close to the model in part (d).

5. (a) See the graph in part (c).

(b) Let t be the time (in months) from January. We find a function of the form $y = a \cos \omega (t - c) + b$, where y is temperature in °F.

$a = \frac{1}{2} (85.8 - 40) = 22.9$. The period is $2 (\text{Jul} - \text{Jan}) = 2 (6 - 0) = 12$ and so $\omega = \frac{2\pi}{12} \approx 0.52$.

$b = \frac{1}{2} (85.8 + 40) = 62.9$. Since the maximum value occurs in July, $c = 6$. Thus the function is $y = 22.9 \cos 0.52 (t - 6) + 62.9$.

(c)

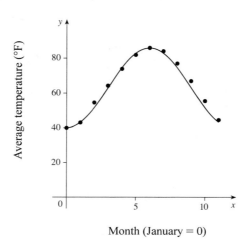

Month (January = 0)

(d) Using the `SinReg` command on the TI-83 we find that for the function $y = a \sin (bt + c) + d$, where $a = 23.4$, $b = 0.48$, $c = -1.36$, and $d = 62.2$. Thus we get the model $y = 23.4 \sin (0.48t - 1.36) + 62.2$.

7. (a) See the graph in part (c).

(b) Let t be the time years. We find a function of the form $y = a \sin \omega (t - c) + b$, where y is the owl population.

$a = \frac{1}{2} (80 - 20) = 30$. The period is $2 (9 - 3) = 12$ and so $\omega = \frac{2\pi}{12} \approx 0.52$. $b = \frac{1}{2} (80 + 20) = 50$. Since the values start at the middle we have $c = 0$. Thus the function is $y = 30 \sin 0.52t + 50$.

(c)

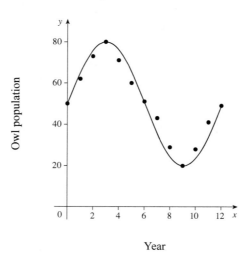

Year

(d) Using the `SinReg` command on the TI-83 we find that for the function $y = a \sin (bt + c) + d$, where $a = 25.8$, $b = 0.52$, $c = -0.02$, and $d = 50.6$. Thus we get the model $y = 25.8 \sin (0.52t - 0.02) + 50.6$.

9. (a) See the graph in part (c).

(b) Let t be the time since 1975. We find a function of the form $y = a \cos \omega (t - c) + b$. $a = \frac{1}{2} (158 - 9) = 74.5$. The first period seems to stretch from 1980 to 1989 (maximum to maximum) and the second from 1989 to 2000, so the period is about 10 years and so $\omega = \frac{2\pi}{10} \approx 0.628$. $b = \frac{1}{2} (158 + 9) = 83.5$. The first maximum value occurs in the fifth year, so $c = 5$. Thus, our model is $y = 74.5 \cos 0.628 (t - 5) + 83.5$.

(c)

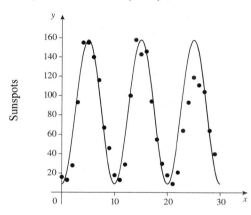

Year since 1975

(d) Using the `SinReg` command on the TI-83, we find that for the function $y = a \sin (bt + c) + d$, where $a = 67.65094323$, $b = 0.6205550572$, $c = -1.654463632$, and $d = 74.50460325$. Thus we get the model $y = 67.65 \sin (0.62t - 1.65) + 74.5$. This model is more accurate for more recent years than the one we found in part (b).

8 Analytic Trigonometry

8.1 Trigonometric Identities

1. $\cos t \tan t = \cos t \cdot \dfrac{\sin t}{\cos t} = \sin t$

3. $\sin \theta \sec \theta = \sin \theta \cdot \dfrac{1}{\cos \theta} = \tan \theta$

5. $\tan^2 x - \sec^2 x = \dfrac{\sin^2 x}{\cos^2 x} - \dfrac{1}{\cos^2 x} = \dfrac{\sin^2 x - 1}{\cos^2 x} = \dfrac{-\cos^2 x}{\cos^2 x} = -1$

7. $\sin u + \cot u \cos u = \sin u + \dfrac{\cos u}{\sin u} \cdot \cos u = \dfrac{\sin^2 u + \cos^2 u}{\sin u} = \dfrac{1}{\sin u} = \csc u$

9. $\dfrac{\sec \theta - \cos \theta}{\sin \theta} = \dfrac{\dfrac{1}{\cos \theta} - \cos \theta}{\sin \theta} = \dfrac{1 - \cos^2 \theta}{\sin \theta \cos \theta} = \dfrac{\sin^2 \theta}{\sin \theta \cos \theta} = \dfrac{\sin \theta}{\cos \theta} = \tan \theta$

11. $\dfrac{\sin x \sec x}{\tan x} = \dfrac{\sin x \cdot \dfrac{1}{\cos x}}{\dfrac{\cos x}{\sin x}} = 1$

13. $\dfrac{1 + \cos y}{1 + \sec y} = \dfrac{1 + \cos y}{1 + \dfrac{1}{\cos y}} = \dfrac{1 + \cos y}{\dfrac{\cos y + 1}{\cos y}} = \dfrac{1 + \cos y}{1} \cdot \dfrac{\cos y}{\cos y + 1} = \cos y$

15. $\dfrac{\sec^2 x - 1}{\sec^2 x} = \dfrac{\tan^2 x}{\sec^2 x} = \dfrac{\sin^2 x}{\cos^2 x} \cdot \cos^2 x = \sin^2 x.$ *Another method:* $\dfrac{\sec^2 x - 1}{\sec^2 x} = 1 - \dfrac{1}{\sec^2 x} = 1 - \cos^2 x = \sin^2 x$

17. $\dfrac{1 + \csc x}{\cos x + \cot x} = \dfrac{1 + \dfrac{1}{\sin x}}{\cos x + \dfrac{\cos x}{\sin x}} = \dfrac{1 + \dfrac{1}{\sin x}}{\cos x + \dfrac{\cos x}{\sin x}} \cdot \dfrac{\sin x}{\sin x} = \dfrac{\sin x + 1}{\cos x (\sin x + 1)} = \dfrac{1}{\cos x} = \sec x$

19. $\dfrac{1 + \sin u}{\cos u} + \dfrac{\cos u}{1 + \sin u} = \dfrac{(1 + \sin u)^2 + \cos^2 u}{\cos u (1 + \sin u)} = \dfrac{1 + 2 \sin u + \sin^2 u + \cos^2 u}{\cos u (1 + \sin u)} = \dfrac{1 + 2 \sin u + 1}{\cos u (1 + \sin u)}$

$= \dfrac{2 + 2 \sin u}{\cos u (1 + \sin u)} = \dfrac{2 (1 + \sin u)}{\cos u (1 + \sin u)} = \dfrac{2}{\cos u} = 2 \sec u$

21. $\dfrac{2 + \tan^2 x}{\sec^2 x} - 1 = \dfrac{1 + 1 + \tan^2 x}{\sec^2 x} - 1 = \dfrac{1}{\sec^2 x} + \dfrac{1 + \tan^2 x}{\sec^2 x} - 1 = \dfrac{1}{\sec^2 x} + \dfrac{\sec^2 x}{\sec^2 x} - 1$

$= \dfrac{1}{\sec^2 x} + 1 - 1 = \dfrac{1}{\sec^2 x} = \cos^2 x$

23. $\tan \theta + \cos (-\theta) + \tan (-\theta) = \tan \theta + \cos \theta - \tan \theta = \cos \theta$

25. $\dfrac{\sin \theta}{\tan \theta} = \dfrac{\sin \theta}{\dfrac{\sin \theta}{\cos \theta}} = \sin \theta \cdot \dfrac{\cos \theta}{\sin \theta} = \cos \theta$

27. $\dfrac{\cos u \sec u}{\tan u} = \cos u \dfrac{1}{\cos u} \cot u = \cot u$

29. $\dfrac{\tan y}{\csc y} = \dfrac{\sin y}{\cos y} \sin y = \dfrac{\sin^2 y}{\cos y} = \dfrac{1 - \cos^2 y}{\cos y} = \sec y - \cos y$

31. $\sin B + \cos B \cot B = \sin B + \cos B \dfrac{\cos B}{\sin B} = \dfrac{\sin^2 B + \cos^2 B}{\sin B} = \dfrac{1}{\sin B} = \csc B$

33. $\cot (-\alpha) \cos (-\alpha) + \sin (-\alpha) = -\dfrac{\cos \alpha}{\sin \alpha} \cos \alpha - \sin \alpha = \dfrac{-\cos^2 \alpha - \sin^2 \alpha}{\sin \alpha} = \dfrac{-1}{\sin \alpha} = -\csc \alpha$

35. $\tan\theta + \cot\theta = \dfrac{\sin\theta}{\cos\theta} + \dfrac{\cos\theta}{\sin\theta} = \dfrac{\sin^2\theta + \cos^2\theta}{\cos\theta\sin\theta} = \dfrac{1}{\cos\theta\sin\theta} = \sec\theta\csc\theta$

37. $(1-\cos\beta)(1+\cos\beta) = 1 - \cos^2\beta = \sin^2\beta = \dfrac{1}{\csc^2\beta}$

39. $\dfrac{(\sin x + \cos x)^2}{\sin^2 x - \cos^2 x} = \dfrac{(\sin x + \cos x)^2}{(\sin x + \cos x)(\sin x - \cos x)} = \dfrac{\sin x + \cos x}{\sin x - \cos x} = \dfrac{(\sin x + \cos x)(\sin x - \cos x)}{(\sin x - \cos x)(\sin x - \cos x)}$

$$= \dfrac{\sin^2 x - \cos^2 x}{(\sin x - \cos x)^2}$$

41. $\dfrac{\sec t - \cos t}{\sec t} = \dfrac{\dfrac{1}{\cos t} - \cos t}{\dfrac{1}{\cos t}} = \dfrac{\dfrac{1}{\cos t} - \cos t}{\dfrac{1}{\cos t}} \cdot \dfrac{\cos t}{\cos t} = \dfrac{1 - \cos^2 t}{1} = \sin^2 t$

43. $\dfrac{1}{1 - \sin^2 y} = \dfrac{1}{\cos^2 y} = \sec^2 y = 1 + \tan^2 y$

45. $(\cot x - \csc x)(\cos x + 1) = \cot x\cos x + \cot x - \csc x\cos x - \csc x = \dfrac{\cos^2 x}{\sin x} + \dfrac{\cos x}{\sin x} - \dfrac{\cos x}{\sin x} - \dfrac{1}{\sin x}$

$$= \dfrac{\cos^2 x - 1}{\sin x} = \dfrac{-\sin^2 x}{\sin x} = -\sin x$$

47. $(1 - \cos^2 x)(1 + \cot^2 x) = \sin^2 x\left(1 + \dfrac{\cos^2 x}{\sin^2 x}\right) = \sin^2 x + \cos^2 x = 1$

49. $2\cos^2 x - 1 = 2(1 - \sin^2 x) - 1 = 2 - 2\sin^2 x - 1 = 1 - 2\sin^2 x$

51. $\dfrac{1 - \cos\alpha}{\sin\alpha} = \dfrac{1 - \cos\alpha}{\sin\alpha} \cdot \dfrac{1 + \cos\alpha}{1 + \cos\alpha} = \dfrac{1 - \cos^2\alpha}{\sin\alpha(1 + \cos\alpha)} = \dfrac{\sin^2\alpha}{\sin\alpha(1 + \cos\alpha)} = \dfrac{\sin\alpha}{1 + \cos\alpha}$

53. $\tan^2\theta\sin^2\theta = \tan^2\theta(1 - \cos^2\theta) = \tan^2\theta - \dfrac{\sin^2\theta}{\cos^2\theta}\cos^2\theta = \tan^2\theta - \sin^2\theta$

55. $\dfrac{\sin x - 1}{\sin x + 1} = \dfrac{\sin x - 1}{\sin x + 1} \cdot \dfrac{\sin x + 1}{\sin x + 1} = \dfrac{\sin^2 x - 1}{(\sin x + 1)^2} = \dfrac{-\cos^2 x}{(\sin x + 1)^2}$

57. $\dfrac{(\sin t + \cos t)^2}{\sin t\cos t} = \dfrac{\sin^2 t + 2\sin t\cos t + \cos^2 t}{\sin t\cos t} = \dfrac{\sin^2 t + \cos^2 t}{\sin t\cos t} + \dfrac{2\sin t\cos t}{\sin t\cos t} = \dfrac{1}{\sin t\cos t} + 2 = 2 + \sec t\cos t$

59. $\dfrac{1 + \tan^2 u}{1 - \tan^2 u} = \dfrac{1 + \dfrac{\sin^2 u}{\cos^2 u}}{1 - \dfrac{\sin^2 u}{\cos^2 u}} = \dfrac{1 + \dfrac{\sin^2 u}{\cos^2 u}}{1 - \dfrac{\sin^2 u}{\cos^2 u}} \cdot \dfrac{\cos^2 u}{\cos^2 u} = \dfrac{\cos^2 u + \sin^2 u}{\cos^2 u - \sin^2 u} = \dfrac{1}{\cos^2 u - \sin^2 u}$

61. $\dfrac{\sec x}{\sec x - \tan x} = \dfrac{\sec x}{\sec x - \tan x} \cdot \dfrac{\sec x + \tan x}{\sec x + \tan x} = \dfrac{\sec x(\sec x + \tan x)}{\sec^2 x - \tan^2 x} = \dfrac{\sec x(\sec x + \tan x)}{1}$

$$= \sec x(\sec x + \tan x)$$

63. $\sec v - \tan v = (\sec v - \tan v) \cdot \dfrac{\sec v + \tan v}{\sec v + \tan v} = \dfrac{\sec^2 v - \tan^2 v}{\sec v + \tan v} = \dfrac{1}{\sec v + \tan v}$

65. $\dfrac{\sin x + \cos x}{\sec x + \csc x} = \dfrac{\sin x + \cos x}{\dfrac{1}{\cos x} + \dfrac{1}{\sin x}} = \dfrac{\sin x + \cos x}{\dfrac{\sin x + \cos x}{\cos x\sin x}} = (\sin x + \cos x)\dfrac{\cos x\sin x}{\sin x + \cos x} = \cos x\sin x$

67. $\dfrac{\csc x - \cot x}{\sec x - 1} = \dfrac{\dfrac{1}{\sin x} - \dfrac{\cos x}{\sin x}}{\dfrac{1}{\cos x} - 1} = \dfrac{\dfrac{1}{\sin x} - \dfrac{\cos x}{\sin x}}{\dfrac{1}{\cos x} - 1} \cdot \dfrac{\sin x\cos x}{\sin x\cos x} = \dfrac{\cos x(1 - \cos x)}{\sin x(1 - \cos x)} = \dfrac{\cos x}{\sin x} = \cot x$

69. $\tan^2 u - \sin^2 u = \dfrac{\sin^2 u}{\cos^2 u} - \dfrac{\sin^2 u\cos^2 u}{\cos^2 u} = \dfrac{\sin^2 u}{\cos^2 u}(1 - \cos^2 u) = \tan^2 u\sin^2 u$

71. $\sec^4 x - \tan^4 x = (\sec^2 x - \tan^2 x)(\sec^2 x + \tan^2 x) = 1(\sec^2 x + \tan^2 x) = \sec^2 x + \tan^2 x$

73. $\dfrac{\sin\theta-\csc\theta}{\cos\theta-\cot\theta}=\dfrac{\sin\theta-\dfrac{1}{\sin\theta}}{\cos\theta-\dfrac{\cos\theta}{\sin\theta}}=\dfrac{\dfrac{\sin^2\theta-1}{\sin\theta}}{\dfrac{\cos\theta\sin\theta-\cos\theta}{\sin\theta}}=\dfrac{\cos^2\theta}{\cos\theta\,(\sin\theta-1)}=\dfrac{\cos\theta}{\sin\theta-1}$

75. $\dfrac{\cos^2 t+\tan^2 t-1}{\sin^2 t}=\dfrac{-\sin^2 t+\tan^2 t}{\sin^2 t}=-1+\dfrac{\sin^2 t}{\cos^2 t}\cdot\dfrac{1}{\sin^2 t}=-1+\sec^2 t=\tan^2 t$

77. $\dfrac{1}{\sec x+\tan x}+\dfrac{1}{\sec x-\tan x}=\dfrac{\sec x-\tan x+\sec x+\tan x}{(\sec x+\tan x)\,(\sec x-\tan x)}=\dfrac{2\sec x}{\sec^2 x-\tan^2 x}=\dfrac{2\sec x}{1}=2\sec x$

79. $(\tan x+\cot x)^2=\tan^2 x+2\tan x\cot x+\cot^2 x=\tan^2 x+2+\cot^2 x=\left(\tan^2 x+1\right)+\left(\cot^2 x+1\right)=\sec^2 x+\csc^2 x$

81. $\dfrac{\sec u-1}{\sec u+1}=\dfrac{\dfrac{1}{\cos u}-1}{\dfrac{1}{\cos u}+1}\cdot\dfrac{\cos u}{\cos u}=\dfrac{1-\cos u}{1+\cos u}$

83. $\dfrac{\sin^3 x+\cos^3 x}{\sin x+\cos x}=\dfrac{(\sin x+\cos x)\left(\sin^2 x-\sin x\cos x+\cos^2 x\right)}{\sin x+\cos x}=\sin^2-\sin x\cos x+\cos^2 x=1-\sin x\cos x$

85. $\dfrac{1+\sin x}{1-\sin x}=\dfrac{1+\sin x}{1-\sin x}\cdot\dfrac{1+\sin x}{1+\sin x}=\dfrac{(1+\sin x)^2}{1-\sin^2 x}=\dfrac{(1+\sin x)^2}{\cos^2 x}=\left(\dfrac{1+\sin x}{\cos x}\right)^2=(\tan x+\sec x)^2$

87. $(\tan x+\cot x)^4=\left(\dfrac{\sin x}{\cos x}+\dfrac{\cos x}{\sin x}\right)^4=\left(\dfrac{\sin^2 x+\cos^2 x}{\sin x\cos x}\right)^4=\left(\dfrac{1}{\sin x\cos x}\right)^4=\sec^4 x\csc^4 x$

89. $x=\sin\theta$; then $\dfrac{x}{\sqrt{1-x^2}}=\dfrac{\sin\theta}{\sqrt{1-\sin^2\theta}}=\dfrac{\sin\theta}{\sqrt{\cos^2\theta}}=\dfrac{\sin\theta}{\cos\theta}=\tan\theta$ (since $\cos\theta\ge 0$ for $0\le\theta\le\frac{\pi}{2}$).

91. $x=\sec\theta$; then $\sqrt{x^2-1}=\sqrt{\sec^2\theta-1}=\sqrt{(\tan^2\theta+1)-1}=\sqrt{\tan^2\theta}=\tan\theta$ (since $\tan\theta\ge 0$ for $0\le\theta<\frac{\pi}{2}$)

93. $x=3\sin\theta$; then $\sqrt{9-x^2}=\sqrt{9-(3\sin\theta)^2}=\sqrt{9-9\sin^2\theta}=\sqrt{9\left(1-\sin^2\theta\right)}=3\sqrt{\cos^2\theta}=3\cos\theta$ (since $\cos\theta\ge 0$ for $0\le\theta<\frac{\pi}{2}$).

95. $f\left(x\right)=\cos^2 x-\sin^2 x$, $g\left(x\right)=1-2\sin^2 x$. From the graph, $f\left(x\right)=g\left(x\right)$ this appears to be an identity. *Proof:*

$f\left(x\right)=\cos^2 x-\sin^2 x=\cos^2 x+\sin^2 x-2\sin^2 x=1-2\sin^2 x=g\left(x\right)$.

Since $f\left(x\right)=g\left(x\right)$ for all x, this is an identity.

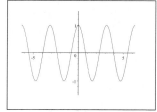

97. $f\left(x\right)=(\sin x+\cos x)^2$, $g\left(x\right)=1$. From the graph, $f\left(x\right)=g\left(x\right)$ does not appear to be an identity. In order to show this, we can set $x=\frac{\pi}{4}$. Then we have

$f\left(\frac{\pi}{4}\right)=\left(\frac{1}{\sqrt{2}}+\frac{1}{\sqrt{2}}\right)^2=\left(\frac{2}{\sqrt{2}}\right)^2=\left(\sqrt{2}\right)^2=2\ne 1=g\left(\frac{\pi}{4}\right)$. Since

$f\left(\frac{\pi}{4}\right)\ne g\left(\frac{\pi}{4}\right)$, this is not an identity.

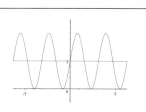

99. (a) Choose $x=\frac{\pi}{2}$. Then $\sin 2x=\sin\pi=0$ whereas $2\sin x=2\sin\frac{\pi}{2}=2$.

(b) Choose $x=\frac{\pi}{4}$ and $y=\frac{\pi}{4}$. Then $\sin\left(x+y\right)=\sin\frac{\pi}{2}=1$ whereas $\sin x+\sin y=\sin\frac{\pi}{4}+\sin\frac{\pi}{4}=\frac{1}{\sqrt{2}}+\frac{1}{\sqrt{2}}=\frac{2}{\sqrt{2}}$. Since these are not equal, the equation is not an identity.

(c) Choose $\theta=\frac{\pi}{4}$. Then $\sec^2\theta+\csc^2\theta=\left(\sqrt{2}\right)^2+\left(\sqrt{2}\right)^2=4\ne 1$.

(d) Choose $x=\frac{\pi}{4}$. Then $\dfrac{1}{\sin x+\cos x}=\dfrac{1}{\sin\frac{\pi}{4}+\cos\frac{\pi}{4}}=\dfrac{1}{\frac{1}{\sqrt{2}}+\frac{1}{\sqrt{2}}}=\frac{1}{\sqrt{2}}$ whereas

$\csc x+\sec x=\csc\frac{\pi}{4}+\sec\frac{\pi}{4}=\sqrt{2}+\sqrt{2}$. Since these are not equal, the equation is not an identity.

101. No. All this proves is that $f(x) = g(x)$ for x in the range of the viewing rectangle. It does not prove that these functions are equal for all values of x. For example, let $f(x) = 1 - \dfrac{x^2}{2} + \dfrac{x^4}{24} - \dfrac{x^6}{720}$ and $g(x) = \cos x$. In the first viewing rectangle the graphs of these two functions appear identical. However, when the domain is expanded in the second viewing rectangle, you can see that these two functions are not identical.

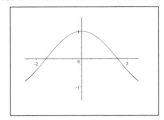

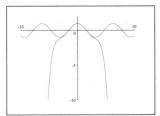

8.2 Addition and Subtraction Formulas

1. $\sin 75^\circ = \sin(45^\circ + 30^\circ) = \sin 45^\circ \cos 30^\circ + \cos 45^\circ \sin 30^\circ = \frac{\sqrt{2}}{2} \cdot \frac{\sqrt{3}}{2} + \frac{\sqrt{2}}{2} \cdot \frac{1}{2} = \frac{\sqrt{6} + \sqrt{2}}{4}$

3. $\cos 105^\circ = \cos(60^\circ + 45^\circ) = \cos 60^\circ \cos 45^\circ - \sin 60^\circ \sin 45^\circ = \frac{1}{2} \cdot \frac{\sqrt{2}}{2} - \frac{\sqrt{3}}{2} \cdot \frac{\sqrt{2}}{2} = \frac{\sqrt{2} - \sqrt{6}}{4}$

5. $\tan 15^\circ = \tan(45^\circ - 30^\circ) = \dfrac{\tan 45^\circ - \tan 30^\circ}{1 + \tan 45^\circ \tan 30^\circ} = \dfrac{1 - \frac{\sqrt{3}}{3}}{1 + 1 \cdot \frac{\sqrt{3}}{3}} = \dfrac{3 - \sqrt{3}}{3 + \sqrt{3}} = 2 - \sqrt{3}$

7. $\sin \frac{19\pi}{12} = -\sin \frac{7\pi}{12} = -\sin\left(\frac{\pi}{4} + \frac{\pi}{3}\right) = -\sin \frac{\pi}{4} \cos \frac{\pi}{3} - \cos \frac{\pi}{4} \sin \frac{\pi}{3} = -\frac{\sqrt{2}}{2} \cdot \frac{1}{2} - \frac{\sqrt{2}}{2} \cdot \frac{\sqrt{3}}{2} = -\frac{\sqrt{6} + \sqrt{2}}{4}$

9. $\tan\left(-\frac{\pi}{12}\right) = -\tan \frac{\pi}{12} = -\tan\left(\frac{\pi}{3} - \frac{\pi}{4}\right) = -\dfrac{\tan \frac{\pi}{3} - \tan \frac{\pi}{4}}{1 + \tan \frac{\pi}{3} \tan \frac{\pi}{4}} = \dfrac{1 - \sqrt{3}}{1 + \sqrt{3}} = \sqrt{3} - 2$

11. $\cos \frac{11\pi}{12} = -\cos \frac{\pi}{12} = -\cos\left(\frac{\pi}{3} - \frac{\pi}{4}\right) = -\cos \frac{\pi}{3} \cos \frac{\pi}{4} - \sin \frac{\pi}{3} \sin \frac{\pi}{4} = -\frac{\sqrt{3}}{2} \cdot \frac{\sqrt{2}}{2} - \frac{1}{2} \cdot \frac{\sqrt{2}}{2} = -\frac{\sqrt{6} + \sqrt{2}}{4}$

13. $\sin 18^\circ \cos 27^\circ + \cos 18^\circ \sin 27^\circ = \sin(18^\circ + 27^\circ) = \sin 45^\circ = \frac{1}{\sqrt{2}} = \frac{\sqrt{2}}{2}$

15. $\cos \frac{3\pi}{7} \cos \frac{2\pi}{21} + \sin \frac{3\pi}{7} \sin \frac{2\pi}{21} = \cos\left(\frac{3\pi}{7} - \frac{2\pi}{21}\right) = \cos \frac{7\pi}{21} = \cos \frac{\pi}{3} = \frac{1}{2}$

17. $\dfrac{\tan 73^\circ - \tan 13^\circ}{1 + \tan 73^\circ \tan 13^\circ} = \tan(73^\circ - 13^\circ) = \tan 60^\circ = \sqrt{3}$

19. $\tan\left(\frac{\pi}{2} - u\right) = \dfrac{\sin\left(\frac{\pi}{2} - u\right)}{\cos\left(\frac{\pi}{2} - u\right)} = \dfrac{\sin \frac{\pi}{2} \cos u - \cos \frac{\pi}{2} \sin u}{\cos \frac{\pi}{2} \cos u + \sin \frac{\pi}{2} \sin u} = \dfrac{1 \cdot \cos u - 0 \cdot \sin u}{0 \cdot \cos u + 1 \cdot \sin u} = \dfrac{\cos u}{\sin u} = \cot u$

21. $\sec\left(\frac{\pi}{2} - u\right) = \dfrac{1}{\cos\left(\frac{\pi}{2} - u\right)} = \dfrac{1}{\cos \frac{\pi}{2} \cos u + \sin \frac{\pi}{2} \sin u} = \dfrac{1}{0 \cdot \cos u + 1 \cdot \sin u} = \dfrac{1}{\sin u} = \csc u$

23. $\sin\left(x - \frac{\pi}{2}\right) = \sin x \cos \frac{\pi}{2} - \cos x \sin \frac{\pi}{2} = 0 \cdot \sin x - 1 \cdot \cos x = -\cos x$

25. $\sin(x - \pi) = \sin x \cos \pi - \cos x \sin \pi = -1 \cdot \sin x - 0 \cdot \cos x = -\sin x$

27. $\tan(x - \pi) = \dfrac{\tan x - \tan \pi}{1 + \tan x \tan \pi} = \dfrac{\tan x - 0}{1 + \tan x \cdot 0} = \tan x$

29. $\cos\left(x + \frac{\pi}{6}\right) + \sin\left(x - \frac{\pi}{3}\right) = \cos x \cos \frac{\pi}{6} - \sin x \sin \frac{\pi}{6} + \sin x \cos \frac{\pi}{3} - \cos x \sin \frac{\pi}{3}$

31. $\sin(x + y) - \sin(x - y) = \sin x \cos y + \cos x \sin y - (\sin x \cos y - \cos x \sin y) = 2 \cos x \sin y$

33. $\cot(x - y) = \dfrac{1}{\tan(x - y)} = \dfrac{1 + \tan x \tan y}{\tan x - \tan y} = \dfrac{1 + \frac{1}{\cot x} \frac{1}{\cot y}}{\frac{1}{\cot x} - \frac{1}{\cot y}} \cdot \dfrac{\cot x \cot y}{\cot x \cot y} = \dfrac{\cot x \cot y + 1}{\cot y - \cot x}$

35. $\tan x - \tan y = \dfrac{\sin x}{\cos x} - \dfrac{\sin y}{\cos y} = \dfrac{\sin x \cos y - \cos x \sin y}{\cos x \cos y} = \dfrac{\sin(x - y)}{\cos x \cos y}$

37. $\dfrac{\sin(x + y) - \sin(x - y)}{\cos(x + y) + \cos(x - y)} = \dfrac{\sin x \cos y + \cos x \sin y - (\sin x \cos y - \cos x \sin y)}{\cos x \cos y - \sin x \sin y + \cos x \cos y + \sin x \sin y} = \dfrac{2 \cos x \sin y}{2 \cos x \cos y} = \tan y$

39. $\sin(x + y + z) = \sin((x + y) + z) = \sin(x + y)\cos z + \cos(x + y)\sin z$

$$= \cos z\,(\sin x \cos y + \cos x \sin y) + \sin z\,(\cos x \cos y - \sin x \sin y)$$

$$= \sin x \cos y \cos z + \cos x \sin y \cos z + \cos x \cos y \sin z - \sin x \sin y \sin z$$

41. $k = \sqrt{A^2 + B^2} = \sqrt{\left(-\sqrt{3}\right)^2 + 1^2} = \sqrt{4} = 2$. $\sin\phi = \frac{1}{2}$ and $\cos\phi = \dfrac{-\sqrt{3}}{2} \;\Rightarrow\; \phi = \frac{5\pi}{6}$. Therefore, $-\sqrt{3}$

$\sin x + \cos x = k\sin(x + \phi) = 2\sin\left(x + \frac{5\pi}{6}\right)$.

43. $k = \sqrt{A^2 + B^2} = \sqrt{5^2 + (-5)^2} = \sqrt{50} = 5\sqrt{2}$. $\sin\phi = -\frac{5}{5\sqrt{2}} = -\frac{1}{\sqrt{2}}$ and $\cos\phi = \frac{5}{5\sqrt{2}} = \frac{1}{\sqrt{2}} \;\Rightarrow\; \phi = \frac{7\pi}{4}$.

Therefore, $5\,(\sin 2x - \cos 2x) = k\sin(2x + \phi) = 5\sqrt{2}\sin\left(2x + \frac{7\pi}{4}\right)$.

45. (a) $f(x) = \sin x + \cos x \Rightarrow k = \sqrt{1^2 + 1^2} = \sqrt{2}$,

and ϕ satisfies $\sin\phi = \cos\phi = \frac{1}{\sqrt{2}} \Rightarrow \phi = \frac{\pi}{4}$.

Thus, we can write

$f(x) = k\sin(x + \phi) = \sqrt{2}\sin\left(x + \frac{\pi}{4}\right)$.

(b) This is a sine curve with amplitude $\sqrt{2}$, period 2π, and phase shift $-\frac{\pi}{4}$.

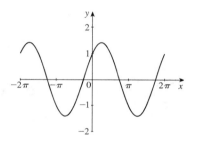

47. If $\beta - \alpha = \frac{\pi}{2}$, then $\beta = \alpha + \frac{\pi}{2}$. Now if we let $y = x + \alpha$, then $\sin(x + \alpha) + \cos\left(x + \alpha + \frac{\pi}{2}\right) = \sin y + \cos\left(y + \frac{\pi}{2}\right) = \sin y + (-\sin y) = 0$. Therefore, $\sin(x + \alpha) + \cos(x + \beta) = 0$.

49. Let $\angle A$ and $\angle B$ be the two angles shown in the diagram. Then

$180° = \gamma + A + B$, $90° = \alpha + A$, and $90° = \beta + B$.

Subtracting the second and third equation from the first, we get

$180° - 90° - 90° = \gamma + A + B - (\alpha + A) - (\beta + B) \quad\Leftrightarrow$

$\alpha + \beta = \gamma$. Then

$\tan\gamma = \tan(\alpha + \beta) = \dfrac{\tan\alpha + \tan\beta}{1 - \tan\alpha\tan\beta} = \dfrac{\frac{4}{6} + \frac{3}{4}}{1 - \frac{4}{6}\cdot\frac{3}{4}} = \dfrac{\frac{8}{12} + \frac{9}{12}}{1 - \frac{1}{2}} = 2\cdot\frac{17}{12} = \frac{17}{6}$.

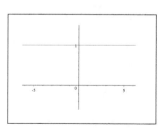

51. (a) $y = \sin^2\left(x + \frac{\pi}{4}\right) + \sin^2\left(x - \frac{\pi}{4}\right)$. From the graph we see that the value of y seems to always be equal to 1.

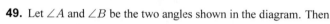

(b) $y = \sin^2\left(x + \frac{\pi}{4}\right) + \sin^2\left(x - \frac{\pi}{4}\right) = \left(\sin x \cos\frac{\pi}{4} + \cos x \sin\frac{\pi}{4}\right)^2 + \left(\sin x \cos\frac{\pi}{4} - \cos x \sin\frac{\pi}{4}\right)^2$

$= \left[\frac{1}{\sqrt{2}}(\sin x + \cos x)\right]^2 + \left[\frac{1}{\sqrt{2}}(\sin x - \cos x)\right]^2 = \frac{1}{2}\left[(\sin x + \cos x)^2 + (\sin x - \cos x)^2\right]$

$= \frac{1}{2}\left[(\sin^2 x + 2\sin x \cos x + \cos^2 x) + (\sin^2 x - 2\sin x \cos x + \cos^2 x)\right]$

$= \frac{1}{2}\left[(1 + 2\sin x \cos x) + (1 - 2\sin x \cos x)\right] = \frac{1}{2}\cdot 2 = 1$

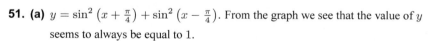

53. Clearly $C = \frac{\pi}{4}$. Now $\tan(A+B) = \dfrac{\tan A + \tan B}{1 - \tan A \tan B} = \dfrac{\frac{1}{3} + \frac{1}{2}}{1 - \frac{1}{3} \cdot \frac{1}{2}} = 1$. Thus $A+B = \frac{\pi}{4}$, so $A+B+C = \frac{\pi}{4} + \frac{\pi}{4} = \frac{\pi}{2}$.

55. (a) $f(t) = C \sin \omega t + C \sin(\omega t + \alpha) = C \sin \omega t + C(\sin \omega t \cos \alpha + \cos \omega t \sin \alpha)$

$\qquad = C(1 + \cos \alpha) \sin \omega t + C \sin \alpha \cos \omega t = A \sin \omega t + B \cos \omega t$

where $A = C(1 + \cos \alpha)$ and $B = C \sin \alpha$.

(b) In this case, $f(t) = 10\left(1 + \cos \frac{\pi}{3}\right) \sin \omega t + 10 \sin \frac{\pi}{3} \cos \omega t = 15 \sin \omega t + 5\sqrt{3} \cos \omega t$. Thus

$k = \sqrt{15^2 + \left(5\sqrt{3}\right)^2} = 10\sqrt{3}$ and ϕ has $\cos \phi = \frac{15}{10\sqrt{3}} = \frac{\sqrt{3}}{2}$ and $\sin \phi = \frac{5\sqrt{3}}{10\sqrt{3}} = \frac{1}{2}$, so $\phi = \frac{\pi}{6}$. Therefore,

$f(t) = 10\sqrt{3} \sin\left(\omega t + \frac{\pi}{6}\right)$.

57. $\tan(s+t) = \dfrac{\sin(s+t)}{\cos(s+t)} = \dfrac{\sin s \cos t + \cos s \sin t}{\cos s \cos t + \sin s \sin t}$

$\qquad = \dfrac{\sin s \cos t + \cos s \sin t}{\cos s \cos t - \sin s \sin t} \cdot \dfrac{\frac{1}{\cos s \cos t}}{\frac{1}{\cos s \cos t}} = \dfrac{\frac{\sin s}{\cos s} + \frac{\sin t}{\cos t}}{1 - \frac{\sin s}{\cos s} \cdot \frac{\sin t}{\cos t}} = \dfrac{\tan s + \tan t}{1 - \tan s \tan t}$

8.3 Double-Angle, Half-Angle, and Product-Sum Formulas

1. $\sin x = \frac{5}{13}$, x in quadrant I $\Rightarrow$ $\cos x = \frac{12}{13}$ and $\tan x = \frac{5}{12}$. Thus, $\sin 2x = 2 \sin x \cos x = 2\left(\frac{5}{13}\right)\left(\frac{12}{13}\right) = \frac{120}{169}$,

$\cos 2x = \cos^2 x - \sin^2 x = \left(\frac{12}{13}\right)^2 - \left(\frac{5}{13}\right)^2 = \frac{144 - 25}{169} = \frac{119}{169}$, and $\tan 2x = \dfrac{\sin 2x}{\cos 2x} = \dfrac{\frac{120}{169}}{\frac{119}{169}} = \frac{120}{169} \cdot \frac{169}{119} = \frac{120}{119}$.

3. $\cos x = \frac{4}{5}$. Then $\sin x = -\frac{3}{5}$ ($\csc x < 0$) and $\tan x = -\frac{3}{4}$. Thus, $\sin 2x = 2 \sin x \cos x = 2\left(-\frac{3}{5}\right) \cdot \frac{4}{5} = -\frac{24}{25}$,

$\cos 2x = \cos^2 x - \sin^2 x = \left(\frac{4}{5}\right)^2 - \left(-\frac{3}{5}\right)^2 = \frac{16 - 9}{25} = \frac{7}{25}$, and $\tan 2x = \dfrac{\sin 2x}{\cos 2x} = \dfrac{-\frac{24}{25}}{\frac{7}{25}} = -\frac{24}{25} \cdot \frac{25}{7} = -\frac{24}{7}$.

5. $\sin x = -\frac{3}{5}$. Then, $\cos x = -\frac{4}{5}$ and $\tan x = \frac{3}{4}$ (x is in quadrant III). Thus, $\sin 2x = 2 \sin x \cos x = 2\left(-\frac{3}{5}\right)\left(-\frac{4}{5}\right) = \frac{24}{25}$,

$\cos 2x = \cos^2 x - \sin^2 x = \left(-\frac{4}{5}\right)^2 - \left(-\frac{3}{5}\right)^2 = \frac{16 - 9}{25} = \frac{7}{25}$, and $\tan 2x = \dfrac{\sin 2x}{\cos 2x} = \dfrac{\frac{24}{25}}{\frac{7}{25}} = \frac{24}{25} \cdot \frac{25}{7} = \frac{24}{7}$.

7. $\tan x = -\frac{1}{3}$ and $\cos x > 0$, so $\sin x < 0$. Thus, $\sin x = -\frac{1}{\sqrt{10}}$ and $\cos x = \frac{3}{\sqrt{10}}$. Thus,

$\sin 2x = 2 \sin x \cos x = 2\left(-\frac{1}{\sqrt{10}}\right)\left(\frac{3}{\sqrt{10}}\right) = -\frac{6}{10} = -\frac{3}{5}$, $\cos 2x = \cos^2 x - \sin^2 x = \left(\frac{3}{\sqrt{10}}\right)^2 - \left(-\frac{1}{\sqrt{10}}\right)^2 = \frac{8}{10} = \frac{4}{5}$,

and $\tan 2x = \dfrac{\sin 2x}{\cos 2x} = \dfrac{-\frac{3}{5}}{\frac{4}{5}} = -\frac{3}{5} \cdot \frac{5}{4} = -\frac{3}{4}$.

9. $\sin^4 x = \left(\sin^2 x\right)^2 = \left(\dfrac{1 - \cos 2x}{2}\right)^2 = \frac{1}{4} - \frac{1}{2} \cos 2x + \frac{1}{4} \cos^2 2x$

$\qquad = \frac{1}{4} - \frac{1}{2} \cos 2x + \frac{1}{4} \cdot \dfrac{1 + \cos 4x}{2} = \frac{1}{4} - \frac{1}{2} \cos 2x + \frac{1}{8} + \frac{1}{8} \cos 4x = \frac{3}{8} - \frac{1}{2} \cos 2x + \frac{1}{8} \cos 4x$

$\qquad = \frac{1}{2}\left(\frac{3}{4} - \cos 2x + \frac{1}{4} \cos 4x\right)$

11. We use the result of Example 4 to get

$\cos^2 x \sin^4 x = \left(\sin^2 x \cos^2 x\right) \sin^2 x = \left(\frac{1}{8} - \frac{1}{8} \cos 4x\right) \cdot \left(\frac{1}{2} - \frac{1}{2} \cos 2x\right)$

$\qquad\qquad\quad = \frac{1}{16}\left(1 - \cos 2x - \cos 4x + \cos 2x \cos 4x\right)$

13. Since $\sin^4 x \cos^4 x = \left(\sin^2 x \cos^2 x\right)^2$ we can use the result of Example 4 to get

$$\sin^4 x \cos^4 x = \left(\tfrac{1}{8} - \tfrac{1}{8}\cos 4x\right)^2 = \tfrac{1}{64} - \tfrac{1}{32}\cos 4x + \tfrac{1}{64}\cos^2 4x$$

$$= \tfrac{1}{64} - \tfrac{1}{32}\cos 4x + \tfrac{1}{64}\cdot\tfrac{1}{2}\left(1 + \cos 8x\right) = \tfrac{1}{64} - \tfrac{1}{32}\cos 4x + \tfrac{1}{128} + \tfrac{1}{128}\cos 8x$$

$$= \tfrac{3}{128} - \tfrac{1}{32}\cos 4x + \tfrac{1}{128}\cos 8x = \tfrac{1}{32}\left(\tfrac{3}{4} - \cos 4x + \tfrac{1}{4}\cos 8x\right)$$

15. $\sin 15° = \sqrt{\tfrac{1}{2}\left(1 - \cos 30°\right)} = \sqrt{\tfrac{1}{2}\left(1 - \tfrac{\sqrt{3}}{2}\right)} = \sqrt{\tfrac{1}{4}\left(2 - \sqrt{3}\right)} = \tfrac{1}{2}\sqrt{2 - \sqrt{3}}$

17. $\tan 22.5° = \dfrac{1 - \cos 45°}{\sin 45°} = \dfrac{1 - \tfrac{\sqrt{2}}{2}}{\tfrac{\sqrt{2}}{2}} = \sqrt{2} - 1$

19. $\cos 165° = -\sqrt{\tfrac{1}{2}\left(1 + \cos 330°\right)} = -\sqrt{\tfrac{1}{2}\left(1 + \cos 30°\right)} = -\sqrt{\tfrac{1}{2}\left(1 + \tfrac{\sqrt{3}}{2}\right)} = -\tfrac{1}{2}\sqrt{2 + \sqrt{3}}$

21. $\tan\dfrac{\pi}{8} = \dfrac{1 - \cos\tfrac{\pi}{4}}{\sin\tfrac{\pi}{4}} = \dfrac{1 - \tfrac{\sqrt{2}}{2}}{\tfrac{\sqrt{2}}{2}} = \sqrt{2} - 1$

23. $\cos\dfrac{\pi}{12} = \sqrt{\tfrac{1}{2}\left(1 + \cos\tfrac{\pi}{6}\right)} = \sqrt{\tfrac{1}{2}\left(1 + \tfrac{\sqrt{3}}{2}\right)} = \tfrac{1}{2}\sqrt{2 + \sqrt{3}}$

25. $\sin\dfrac{9\pi}{8} = -\sqrt{\tfrac{1}{2}\left(1 - \cos\tfrac{9\pi}{4}\right)} = -\sqrt{\tfrac{1}{2}\left(1 - \tfrac{\sqrt{2}}{2}\right)} = -\tfrac{1}{2}\sqrt{2 - \sqrt{2}}$. We have chosen the negative root because $\dfrac{9\pi}{8}$ is in

quadrant III, so $\sin\dfrac{9\pi}{8} < 0$.

27. (a) $2\sin 18° \cos 18° = \sin 36°$ **29. (a)** $\cos^2 34° - \sin^2 34° = \cos 68°$

    (b) $2\sin 3\theta \cos 3\theta = \sin 6\theta$ **(b)** $\cos^2 5\theta - \sin^2 5\theta = \cos 10\theta$

31. (a) $\dfrac{\sin 8°}{1 + \cos 8°} = \tan\dfrac{8°}{2} = \tan 4°$ **33.** $\sin\left(x + x\right) = \sin x \cos x + \cos x \sin x = 2\sin x \cos x$

    (b) $\dfrac{1 - \cos 4\theta}{\sin 4\theta} = \tan\dfrac{4\theta}{2} = \tan 2\theta$

35. $\sin x = \tfrac{3}{5}$. Since x is in quadrant I, $\cos x = \tfrac{4}{5}$ and $\dfrac{x}{2}$ is also in quadrant I. Thus,

$\sin\dfrac{x}{2} = \sqrt{\tfrac{1}{2}\left(1 - \cos x\right)} = \sqrt{\tfrac{1}{2}\left(1 - \tfrac{4}{5}\right)} = \tfrac{1}{\sqrt{10}} = \dfrac{\sqrt{10}}{10}$, $\cos\dfrac{x}{2} = \sqrt{\tfrac{1}{2}\left(1 + \cos x\right)} = \sqrt{\tfrac{1}{2}\left(1 + \tfrac{4}{5}\right)} = \tfrac{3}{\sqrt{10}} = \dfrac{3\sqrt{10}}{10}$, and

$\tan\dfrac{x}{2} = \dfrac{\sin\tfrac{x}{2}}{\cos\tfrac{x}{2}} = \dfrac{1}{\sqrt{10}}\cdot\dfrac{\sqrt{10}}{3} = \tfrac{1}{3}$.

37. $\csc x = 3$. Then, $\sin x = \tfrac{1}{3}$ and since x is in quadrant II, $\cos x = -\dfrac{2\sqrt{2}}{3}$. Since $90° \le x \le 180°$, we have

$45° \le \dfrac{x}{2} \le 90°$ and so $\dfrac{x}{2}$ is in quadrant I. Thus, $\sin\dfrac{x}{2} = \sqrt{\tfrac{1}{2}\left(1 - \cos x\right)} = \sqrt{\tfrac{1}{2}\left(1 + \tfrac{2\sqrt{2}}{3}\right)} = \sqrt{\tfrac{1}{6}\left(3 + 2\sqrt{2}\right)}$,

$\cos\dfrac{x}{2} = \sqrt{\tfrac{1}{2}\left(1 + \cos x\right)} = \sqrt{\tfrac{1}{2}\left(1 - \tfrac{2\sqrt{2}}{3}\right)} = \sqrt{\tfrac{1}{6}\left(3 - 2\sqrt{2}\right)}$, and $\tan\dfrac{x}{2} = \dfrac{\sin\tfrac{x}{2}}{\cos\tfrac{x}{2}} = \sqrt{\dfrac{3 + 2\sqrt{2}}{3 - 2\sqrt{2}}} = 3 + 2\sqrt{2}$.

39. $\sec x = \tfrac{3}{2}$. Then $\cos x = \tfrac{2}{3}$ and since x is in quadrant IV, $\sin x = -\dfrac{\sqrt{5}}{3}$. Since $270° \le x \le 360°$, we have

$135° \le \dfrac{x}{2} \le 180°$ and so $\dfrac{x}{2}$ is in quadrant II. Thus, $\sin\dfrac{x}{2} = \sqrt{\tfrac{1}{2}\left(1 - \cos x\right)} = \sqrt{\tfrac{1}{2}\left(1 - \tfrac{2}{3}\right)} = \tfrac{1}{\sqrt{6}} = \dfrac{\sqrt{6}}{6}$,

$\cos\dfrac{x}{2} = -\sqrt{\tfrac{1}{2}\left(1 + \cos x\right)} = -\sqrt{\tfrac{1}{2}\left(1 + \tfrac{2}{3}\right)} = -\dfrac{\sqrt{5}}{\sqrt{6}} = -\dfrac{\sqrt{30}}{6}$, and $\tan\dfrac{x}{2} = \dfrac{\sin\tfrac{x}{2}}{\cos\tfrac{x}{2}} = \dfrac{1}{\sqrt{6}}\cdot\dfrac{\sqrt{6}}{-\sqrt{5}} = -\dfrac{1}{\sqrt{5}} = -\dfrac{\sqrt{5}}{5}$.

41. $\sin 2x \cos 3x = \tfrac{1}{2}\left[\sin\left(2x + 3x\right) + \sin\left(2x - 3x\right)\right] = \tfrac{1}{2}\left(\sin 5x - \sin x\right)$

43. $\cos x \sin 4x = \tfrac{1}{2}\left[\sin\left(4x + x\right) + \sin\left(4x - x\right)\right] = \tfrac{1}{2}\left(\sin 5x + \sin 3x\right)$

45. $3\cos 4x \cos 7x = 3\cdot\tfrac{1}{2}\left[\cos\left(4x + 7x\right) + \cos\left(4x - 7x\right)\right] = \tfrac{3}{2}\left(\cos 11x + \cos 3x\right)$

47. $\sin 5x + \sin 3x = 2 \sin \left(\dfrac{5x + 3x}{2} \right) \cos \left(\dfrac{5x - 3x}{2} \right) = 2 \sin 4x \cos x$

49. $\cos 4x - \cos 6x = -2 \sin \left(\dfrac{4x + 6x}{2} \right) \sin \left(\dfrac{4x - 6x}{2} \right) = -2 \sin 5x \sin (-x) = 2 \sin 5x \sin x$

51. $\sin 2x - \sin 7x = 2 \cos \left(\dfrac{2x + 7x}{2} \right) \sin \left(\dfrac{2x - 7x}{2} \right) = 2 \cos \dfrac{9x}{2} \sin \left(-\dfrac{5x}{2} \right) = -2 \cos \dfrac{9x}{2} \sin \dfrac{5x}{2}$

53. $2 \sin 52.5° \sin 97.5° = 2 \cdot \frac{1}{2} \left[\cos (52.5° - 97.5°) - \cos (52.5° + 97.5°) \right] = \cos (-45°) - \cos 150°$

$\qquad = \cos 45° - \cos 150° = \dfrac{\sqrt{2}}{2} + \dfrac{\sqrt{3}}{2} = \frac{1}{2} \left(\sqrt{2} + \sqrt{3} \right)$

55. $\cos 37.5° \sin 7.5° = \frac{1}{2} \left(\sin 45° - \sin 30° \right) = \frac{1}{2} \left(\dfrac{\sqrt{2}}{2} - \dfrac{1}{2} \right) = \frac{1}{4} \left(\sqrt{2} - 1 \right)$

57. $\cos 255° - \cos 195° = -2 \sin \left(\dfrac{255° + 195°}{2} \right) \sin \left(\dfrac{255° - 195°}{2} \right) = -2 \sin 225° \sin 30° = -2 \left(-\dfrac{\sqrt{2}}{2} \right) \frac{1}{2} = \dfrac{\sqrt{2}}{2}$

59. $\cos^2 5x - \sin^2 5x = \cos (2 \cdot 5x) = \cos 10x$

61. $(\sin x + \cos x)^2 = \sin^2 x + 2 \sin x \cos x + \cos^2 x = 1 + 2 \sin x \cos x = 1 + \sin 2x$

63. $\dfrac{\sin 4x}{\sin x} = \dfrac{2 \sin 2x \cos 2x}{\sin x} = \dfrac{2 (2 \sin x \cos x) (\cos 2x)}{\sin x} = 4 \cos x \cos 2x$

65. $\dfrac{2 (\tan x - \cot x)}{\tan^2 x - \cot^2 x} = \dfrac{2 (\tan x - \cot x)}{(\tan x + \cot x) (\tan x - \cot x)} = \dfrac{2}{\tan x + \cot x} = \dfrac{2}{\dfrac{\sin x}{\cos x} + \dfrac{\cos x}{\sin x}}$

$\qquad = \dfrac{2}{\dfrac{\sin x}{\cos x} + \dfrac{\cos x}{\sin x}} \cdot \dfrac{\sin x \cos x}{\sin x \cos x} = \dfrac{2 \sin x \cos x}{\sin^2 x + \cos^2 x} = 2 \sin x \cos x = \sin 2x$

67. $\tan 3x = \tan (2x + x) = \dfrac{\tan 2x + \tan x}{1 - \tan 2x \tan x} = \dfrac{\dfrac{2 \tan x}{1 - \tan^2 x} + \tan x}{1 - \dfrac{2 \tan x}{1 - \tan^2 x} \tan x} = \dfrac{2 \tan x + \tan x \left(1 - \tan^2 x \right)}{1 - \tan^2 x - 2 \tan x \tan x}$

$\qquad = \dfrac{3 \tan x - \tan^3 x}{1 - 3 \tan^2 x}$

69. $\cos^4 x - \sin^4 x = \left(\cos^2 x + \sin^2 x \right) \left(\cos^2 x - \sin^2 x \right) = \cos^2 x - \sin^2 x = \cos 2x$

71. $\dfrac{\sin x + \sin 5x}{\cos x + \cos 5x} = \dfrac{2 \sin 3x \cos 2x}{2 \cos 3x \cos 2x} = \dfrac{\sin 3x}{\cos 3x} = \tan 3x$

73. $\dfrac{\sin 10x}{\sin 9x + \sin x} = \dfrac{2 \sin 5x \cos 5x}{2 \sin 5x \cos 4x} = \dfrac{\cos 5x}{\cos 4x}$

75. $\dfrac{\sin x + \sin y}{\cos x + \cos y} = \dfrac{2 \sin \left(\dfrac{x + y}{2} \right) \cos \left(\dfrac{x - y}{2} \right)}{2 \cos \left(\dfrac{x + y}{2} \right) \cos \left(\dfrac{x - y}{2} \right)} = \dfrac{\sin \left(\dfrac{x + y}{2} \right)}{\cos \left(\dfrac{x + y}{2} \right)} = \tan \left(\dfrac{x + y}{2} \right)$

77. $\sin 130° - \sin 110° = 2 \cos \dfrac{130° + 110°}{2} \sin \dfrac{130° - 110°}{2} = 2 \cos 120° \sin 10° = 2 \left(-\frac{1}{2} \right) \sin 10° = -\sin 10°$

79. $\sin 45° + \sin 15° = 2 \sin \left(\dfrac{45° + 15°}{2} \right) \cos \left(\dfrac{45° - 15°}{2} \right) = 2 \sin 30° \cos 15° = 2 \cdot \frac{1}{2} \cdot \cos 15°$

$\qquad = \cos 15° = \sin (90° - 15°) = \sin 75° \text{ (applying the cofunction identity)}$

81. $\dfrac{\sin x + \sin 2x + \sin 3x + \sin 4x + \sin 5x}{\cos x + \cos 2x + \cos 3x + \cos 4x + \cos 5x} = \dfrac{(\sin x + \sin 5x) + (\sin 2x + \sin 4x) + \sin 3x}{(\cos x + \cos 5x) + (\cos 2x + \cos 4x) + \cos 3x}$

$\qquad = \dfrac{2 \sin 3x \cos 2x + 2 \sin 3x \cos x + \sin 3x}{2 \cos 3x \cos 2x + 2 \cos 3x \cos x + \cos 3x} = \dfrac{\sin 3x (2 \cos 2x + 2 \cos x + 1)}{\cos 3x (2 \cos 2x + 2 \cos x + 1)} = \tan 3x$

83. (a) $f(x) = \dfrac{\sin 3x}{\sin x} - \dfrac{\cos 3x}{\cos x}$

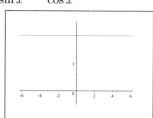

The function appears to have a constant value of 2
wherever it is defined.

(b) $f(x) = \dfrac{\sin 3x}{\sin x} - \dfrac{\cos 3x}{\cos x}$

$= \dfrac{\sin 3x \cos x - \cos 3x \sin x}{\sin x \cos x} = \dfrac{\sin(3x - x)}{\sin x \cos x}$

$= \dfrac{\sin 2x}{\sin x \cos x} = \dfrac{2 \sin x \cos x}{\sin x \cos x} = 2$

for all x for which the function is defined.

85. (a) $y = \sin 6x + \sin 7x$

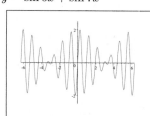

(b) By a sum-to-product formula,

$y = \sin 6x + \sin 7x$

$= 2 \sin\left(\dfrac{6x + 7x}{2}\right) \cos\left(\dfrac{6x - 7x}{2}\right)$

$= 2 \sin\left(\tfrac{13}{2}x\right) \cos\left(-\tfrac{1}{2}x\right)$

$= 2 \sin \tfrac{13}{2}x \cos \tfrac{1}{2}x$

(c) We graph $y = \sin 6x + \sin 7x$,

$y = 2\cos\left(\tfrac{1}{2}x\right)$, and

$y = -2\cos\left(\tfrac{1}{2}x\right)$.

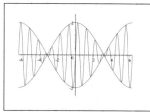

The graph of $y = f(x)$ lies
between the other two graphs.

87. (a) $\cos 4x = \cos(2x + 2x) = 2\cos^2 2x - 1 = 2\left(2\cos^2 x - 1\right)^2 - 1 = 8\cos^4 x - 8\cos^2 x + 1$. Thus the desired
polynomial is $P(t) = 8t^4 - 8t^2 + 1$.

(b) $\cos 5x = \cos(4x + x) = \cos 4x \cos x - \sin 4x \sin x = \cos x \left(8\cos^4 x - 8\cos^2 x + 1\right) - 2\sin 2x \cos 2x \sin x$

$= 8\cos^5 x - 8\cos^3 x + \cos x - 4\sin x \cos x \left(2\cos^2 x - 1\right) \sin x$ [from part (a)]

$= 8\cos^5 x - 8\cos^3 x + \cos x - 4\cos x \left(2\cos^2 x - 1\right) \sin^2 x$

$= 8\cos^5 x - 8\cos^3 x + \cos x - 4\cos x \left(2\cos^2 x - 1\right) \left(1 - \cos^2 x\right)$

$= 8\cos^5 x - 8\cos^3 x + \cos x + 8\cos^5 x - 12\cos^3 x + 4\cos x = 16\cos^5 x - 20\cos^3 x + 5\cos x$

Thus, the desired polynomial is $P(t) = 16t^5 - 20t^3 + 5t$.

89. Using a product-to-sum formula,
RHS $= 4\sin A \sin B \sin C = 4\sin A \left\{\tfrac{1}{2}\left[\cos(B - C) - \cos(B + C)\right]\right\} = 2\sin A \cos(B - C) - 2\sin A \cos(B + C)$.
Using another product-to-sum formula, this is equal to
$2\left\{\tfrac{1}{2}\left[\sin(A + B - C) + \sin(A - B + C)\right]\right\} - 2\left\{\tfrac{1}{2}\left[\sin(A + B + C) + \sin(A - B - C)\right]\right\}$

$= \sin(A + B - C) + \sin(A - B + C) - \sin(A + B + C) - \sin(A - B - C)$

Now $A + B + C = \pi$, so $A + B - C = \pi - 2C$, $A - B + C = \pi - 2B$, and $A - B - C = 2A - \pi$. Thus our expression
simplifies to
$\sin(A + B - C) + \sin(A - B + C) - \sin(A + B + C) - \sin(A - B - C)$

$= \sin(\pi - 2C) + \sin(\pi - 2B) + 0 - \sin(2A - \pi) = \sin 2C + \sin 2B + \sin 2A = $ LHS

91. (a) In both logs the length of the adjacent side is $20\cos\theta$ and the length of the opposite side is $20\sin\theta$.
Thus, the cross-sectional area of the beam is modeled by
$$A(\theta) = (20\cos\theta)(20\sin\theta) = 400\sin\theta\cos\theta = 200(2\sin\theta\cos\theta) = 200\sin 2\theta.$$
(b) The function $y = \sin u$ is maximized when $u = \frac{\pi}{2}$. So $2\theta = \frac{\pi}{2} \iff \theta = \frac{\pi}{4}$. Thus the maximum cross-sectional area is $A\left(\frac{\pi}{4}\right) = 200\sin 2\left(\frac{\pi}{4}\right) = 200$.

93. (a) $y = f_1(t) + f_2(t) = \cos 11t + \cos 13t$

(b) Using the identity
$$\cos\alpha + \cos y = 2\cdot\cos\left(\frac{\alpha+y}{2}\right)\cos\left(\frac{\alpha-y}{2}\right),\text{ we have}$$
$$f(t) = \cos 11t + \cos 13t = 2\cdot\cos\left(\frac{11t+13t}{2}\right)\cos\left(\frac{11t-13t}{2}\right)$$
$$= 2\cdot\cos 12t\cdot\cos(-t) = 2\cos 12t\cos t$$

(c) We graph $y = \cos 11t + \cos 13t$,
$y = 2\cos t$, and $y = -2\cos t$.

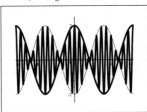

The graph of f lies between the graphs of $y = 2\cos t$ and $y = -2\cos t$. Thus, the loudness of the sound varies between $y = \pm 2\cos t$.

95. We find the area of $\triangle ABC$ in two different ways. First, let AB be the base and CD be the height. Since $\angle BOC = 2\theta$ we see that $CD = \sin 2\theta$. So the area is $\frac{1}{2}(\text{base})(\text{height}) = \frac{1}{2}\cdot 2\cdot\sin 2\theta = \sin 2\theta$. On the other hand, in $\triangle ABC$ we see that $\angle C$ is a right angle. So $BC = 2\sin\theta$ and $AC = 2\cos\theta$, and the area is $\frac{1}{2}(\text{base})(\text{height}) = \frac{1}{2}\cdot(2\sin\theta)(2\cos\theta) = 2\sin\theta\cos\theta$. Equating the two expressions for the area of $\triangle ABC$, we get $\sin 2\theta = 2\sin\theta\cos\theta$.

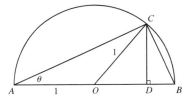

8.4   Inverse Trigonometric Functions

1. (a) $\sin^{-1}\frac{1}{2} = \frac{\pi}{6}$
(b) $\cos^{-1}\frac{1}{2} = \frac{\pi}{3}$
(c) $\cos^{-1} 2$ is not defined.

3. (a) $\sin^{-1}\frac{\sqrt{2}}{2} = \frac{\pi}{4}$
(b) $\cos^{-1}\frac{\sqrt{2}}{2} = \frac{\pi}{4}$
(c) $\sin^{-1}\left(-\frac{\sqrt{2}}{2}\right) = -\frac{\pi}{4}$

5. (a) $\sin^{-1} 1 = \frac{\pi}{2}$
(b) $\cos^{-1} 1 = 0$
(c) $\cos^{-1}(-1) = \pi$

7. (a) $\tan^{-1}\frac{\sqrt{3}}{3} = \frac{\pi}{6}$
(b) $\tan^{-1}\left(-\frac{\sqrt{3}}{3}\right) = -\frac{\pi}{6}$
(c) $\sin^{-1}(-2)$ is not defined.

9. (a) $\sin^{-1}(0.13844) \approx 0.13889$
(b) $\cos^{-1}(-0.92761) \approx 2.75876$

11. (a) $\tan^{-1}(1.23456) \approx 0.88998$
(b) $\sin^{-1}(1.23456)$ is not defined.

13. $\sin\left(\sin^{-1}\frac{1}{4}\right) = \frac{1}{4}$

15. $\tan\left(\tan^{-1} 5\right) = 5$

17. $\cos^{-1}\left(\cos\frac{\pi}{3}\right) = \frac{\pi}{3}$

19. $\sin^{-1}\left(\sin\left(-\frac{\pi}{6}\right)\right) = -\frac{\pi}{6}$

21. $\tan^{-1}\left(\tan\frac{2\pi}{3}\right) = \tan^{-1}\left(\tan\left(\frac{2\pi}{3} - \pi\right)\right)$
$= \tan^{-1}\left(\tan\left(-\frac{\pi}{3}\right)\right) = -\frac{\pi}{3}$ since $\frac{2\pi}{3} > \frac{\pi}{2}$.

23. $\tan\left(\sin^{-1}\frac{1}{2}\right) = \tan\frac{\pi}{6} = \frac{\sqrt{3}}{3}$

25. $\cos\left(\sin^{-1}\frac{\sqrt{3}}{2}\right) = \cos\frac{\pi}{3} = \frac{1}{2}$

27. $\tan^{-1}\left(2\sin\frac{\pi}{3}\right) = \tan^{-1}\left(2\cdot\frac{\sqrt{3}}{2}\right) = \tan^{-1}\sqrt{3} = \frac{\pi}{3}$

29. Let $u = \cos^{-1} \frac{3}{5}$, so $\cos u = \frac{3}{5}$. Then from the triangle,
$\sin\left(\cos^{-1} \frac{3}{5}\right) = \sin u = \frac{4}{5}$.

31. Let $u = \tan^{-1} \frac{12}{5}$, so $\tan u = \frac{12}{5}$. Then from the triangle,
$\sin\left(\tan^{-1} \frac{12}{5}\right) = \sin u = \frac{12}{13}$.

33. Let $\theta = \sin^{-1} \frac{12}{13}$, so $\sin \theta = \frac{12}{13}$. Then from the triangle,
$\sec\left(\sin^{-1} \frac{12}{13}\right) = \sec \theta = \frac{13}{5}$.

35. Let $u = \tan^{-1} 2$, so $\tan u = 2$. Then from the triangle,
$\cos\left(\tan^{-1} 2\right) = \cos u = \frac{1}{\sqrt{5}} = \frac{\sqrt{5}}{5}$.

37. Let $u = \cos^{-1} \frac{3}{5}$, so $\cos u = \frac{3}{5}$. From the triangle,
$\sin u = \frac{4}{5}$, so
$$\sin\left(2\cos^{-1} \frac{3}{5}\right) = \sin(2u) = 2 \sin u \cos u$$
$$= 2 \cdot \frac{4}{5} \cdot \frac{3}{5} = \frac{24}{25}$$

39. Let $u = \cos^{-1} \frac{1}{2}$ and $v = \sin^{-1} \frac{1}{2}$, so $\cos u = \frac{1}{2}$ and
$\sin v = \frac{1}{2}$. From the triangle,
$$\sin\left(\sin^{-1} \frac{1}{2} + \cos^{-1} \frac{1}{2}\right) = \sin(v + u)$$
$$= \sin v \cos u + \cos v \sin u = \frac{1}{2} \cdot \frac{1}{2} + \frac{\sqrt{3}}{2} \cdot \frac{\sqrt{3}}{2} = 1$$
Another method:
$$\sin\left(\sin^{-1} \frac{1}{2} + \cos^{-1} \frac{1}{2}\right) = \sin\left(\frac{\pi}{6} + \frac{\pi}{3}\right) = \sin \frac{\pi}{2} = 1$$

41. Let $u = \sin^{-1} x$, so $\sin u = x$. From the triangle,
$\cos\left(\sin^{-1} x\right) = \cos u = \sqrt{1 - x^2}$.

43. Let $u = \sin^{-1} x$, so $\sin u = x$. From the triangle,
$\tan\left(\sin^{-1} x\right) = \tan u = \dfrac{x}{\sqrt{1 - x^2}}$.

45. Let $u = \tan^{-1} x$, so $\tan u = x$. From the triangle,
$$\cos\left(2\tan^{-1} x\right) = \cos 2u = 2\cos^2 u - 1 = \frac{2}{1 + x^2} - 1 = \frac{1 - x^2}{1 + x^2}$$

47. Let $u = \cos^{-1} x$ and $v = \sin^{-1} x$, so $\cos u = x$ and $\sin v = x$. From the triangle,

$$\cos\left(\cos^{-1} x + \sin^{-1} x\right) = \cos\left(\cos^{-1} x\right)\cos\left(\sin^{-1} x\right) - \sin\left(\cos^{-1} x\right)\sin\left(\sin^{-1} x\right)$$

$$= \cos u \cos v - \sin u \sin v = x\sqrt{1-x^2} - x\sqrt{1-x^2} = 0$$

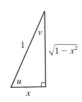

49. (a) $y = \sin^{-1} x + \cos^{-1} x$. Note the domain of both $\sin^{-1} x$ and $\cos^{-1} x$ is $[-1, 1]$.

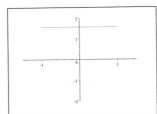

Conjecture: $y = \frac{\pi}{2}$ for $-1 \le x \le 1$.

(b) To prove this conjecture, let $\sin^{-1} x = u \quad \Leftrightarrow \quad \sin u = x$, $-\frac{\pi}{2} \le u \le \frac{\pi}{2}$. Using the cofunction identity

$\sin u = \cos\left(\frac{\pi}{2} - u\right) = x$, $\cos^{-1} x = \frac{\pi}{2} - u$. Therefore,

$\sin^{-1} x + \cos^{-1} x = u + \left(\frac{\pi}{2} - u\right) = \frac{\pi}{2}$.

51. (a) $\tan^{-1} x + \tan^{-1} 2x = \frac{\pi}{4}$

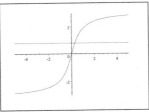

From the graph, the solution is $x \approx 0.28$.

(b) To solve the equation exactly, we set $\tan^{-1} x + \tan^{-1} 2x = \frac{\pi}{4} \quad \Leftrightarrow$

$\tan^{-1} x = \frac{\pi}{4} - \tan^{-1} 2x \Rightarrow \tan\left(\tan^{-1} x\right) = \tan\left(\frac{\pi}{4} - \tan^{-1} 2x\right)$

$$\Rightarrow x = \frac{\tan\frac{\pi}{4} - \tan\left(\tan^{-1} 2x\right)}{1 + \tan\frac{\pi}{4} \cdot \tan\left(\tan^{-1} 2x\right)} = \frac{1 - 2x}{1 + (1)2x} \Rightarrow$$

$x(1 + 2x) = 1 - 2x \quad \Leftrightarrow \quad x + 2x^2 = 1 - 2x$. Thus,

$2x^2 + 3x - 1 = 0$. We use the quadratic formula to solve for x:

$x = \frac{-3 \pm \sqrt{3^2 - 4(2)(-1)}}{2(2)} = \frac{-3 \pm \sqrt{17}}{4}$. Substituting into the original

equation, we see that $x = \frac{-3 - \sqrt{17}}{4}$ is not a solution, and so

$x = \frac{-3 + \sqrt{17}}{4} \approx 0.28$ is the only root.

53. (a) Solving $\tan\theta = h/2$ for h we have $h = 2\tan\theta$.

(b) Solving $\tan\theta = h/2$ for θ we have $\theta = \tan^{-1}(h/2)$.

55. (a) Solving $\sin\theta = h/680$ for θ we have $\theta = \sin^{-1}(h/680)$.

(b) Set $h = 500$ to get $\theta = \sin^{-1}\left(\frac{500}{680}\right) \approx 0.826$ rad.

57. (a) $\theta = \sin^{-1}\left(\frac{1}{(2 \cdot 3 + 1)\tan 10°}\right) = \sin^{-1}\left(\frac{1}{7\tan 10°}\right) \approx \sin^{-1} 0.8102 \approx 54.1°$

(b) For $n = 2$, $\theta = \sin^{-1}\left(\frac{1}{5\tan 15°}\right) \approx 48.3°$. For $n = 3$, $\theta = \sin^{-1}\left(\frac{1}{7\tan 15°}\right) \approx 32.2°$. For $n = 4$,

$\theta = \sin^{-1}\left(\frac{1}{9\tan 15°}\right) \approx 24.5°$. $n = 0$ and $n = 1$ are outside of the domain for $\beta = 15°$, because $\frac{1}{\tan 15°} \approx 3.732$

and $\frac{1}{3\tan 15°} \approx 1.244$, neither of which is in the domain of $\sin^{-1}$.

59. (a) Let $\theta = \sec^{-1} x$. Then $\sec\theta = x$, as shown in the figure. Then $\cos\theta = \dfrac{1}{x}$, so

$$\theta = \cos^{-1}\left(\frac{1}{x}\right). \text{ Thus } \sec^{-1} x = \cos^{-1}\left(\frac{1}{x}\right), \ x \geq 1.$$

(b) Let $\gamma = \csc^{-1} x$. Then $\csc\gamma = x$, as shown in the figure. Then $\sin\gamma = \dfrac{1}{x}$, so

$$\gamma = \sin^{-1}\left(\frac{1}{x}\right). \text{ Thus } \csc^{-1} x = \sin^{-1}\left(\frac{1}{x}\right), \ x \geq 1.$$

(c) Let $\beta = \cot^{-1} x$. Then $\cot\beta = x$, as shown in the figure. Then $\tan\beta = \dfrac{1}{x}$, so

$$\beta = \tan^{-1}\left(\frac{1}{x}\right). \text{ Thus } \cot^{-1} x = \tan^{-1}\left(\frac{1}{x}\right), \ x \geq 1.$$

8.5   Trigonometric Equations

1. $\cos x + 1 = 0 \iff \cos x = -1$. In the interval $[0, 2\pi)$ the only solution is $x = \pi$. Thus the solutions are $x = (2k+1)\pi$ for any integer k.

3. $2\sin x - 1 = 0 \iff 2\sin x = 1 \iff \sin x = \frac{1}{2}$. In the interval $[0, 2\pi)$ the solutions are $x = \frac{\pi}{6}, \frac{5\pi}{6}$. Therefore, the solutions are $x = \frac{\pi}{6} + 2k\pi, \frac{5\pi}{6} + 2k\pi, k = 0, \pm 1, \pm 2, \ldots$.

5. $\sqrt{3}\tan x + 1 = 0 \iff \sqrt{3}\tan x = -1 \iff \tan x = -\frac{\sqrt{3}}{3}$. In the interval $[0, \pi)$ the only solution is $x = \frac{5\pi}{6}$. Therefore, the solutions are $x = \frac{5\pi}{6} + k\pi, k = 0, \pm 1, \pm 2, \ldots$.

7. $4\cos^2 x - 1 = 0 \iff 4\cos^2 x = 1 \iff 4\cos^2 x = 1 \iff \cos^2 x = \frac{1}{4} \iff \cos x = \pm\frac{1}{2}$. In the interval $[0, 2\pi)$ the solutions are $x = \frac{\pi}{3}, \frac{2\pi}{3}, \frac{4\pi}{3}, \frac{5\pi}{3}$. So the solutions are $x = \frac{\pi}{3} + 2k\pi, \frac{2\pi}{3} + 2k\pi, \frac{4\pi}{3} + 2k\pi, \frac{5\pi}{3} + 2k\pi$, which can be expressed more simply as $x = \frac{\pi}{3} + k\pi, \frac{2\pi}{3} + k\pi$ for any integer k.

9. $\sec^2 x - 2 = 0 \iff \sec^2 x = 2 \iff \sec x = \pm\sqrt{2}$. In the interval $[0, 2\pi)$ the solutions are $x = \frac{\pi}{4}, \frac{3\pi}{4}, \frac{5\pi}{4}, \frac{7\pi}{4}$. Thus, the solutions are $x = (2k+1)\frac{\pi}{4}$ for any integer k.

11. $3\csc^2 x - 4 = 0 \iff \csc^2 x = \frac{4}{3} \iff \csc x = \pm\frac{2}{\sqrt{3}} \iff \sin x = \pm\frac{\sqrt{3}}{2} \iff x = \frac{\pi}{3}, \frac{2\pi}{3}, \frac{4\pi}{3}, \frac{5\pi}{3}$ for x in $[0, 2\pi)$. Thus, the solutions are $x = \frac{\pi}{3} + k\pi, \frac{2\pi}{3} + k\pi$ for any integer k.

13. $\cos x\,(2\sin x + 1) = 0 \iff \cos x = 0$ or $2\sin x + 1 = 0 \iff \sin x = -\frac{1}{2}$. On $[0, 2\pi)$, $\cos x = 0 \iff x = \frac{\pi}{2}, \frac{3\pi}{2}$ and $\sin x = -\frac{1}{2} \iff x = \frac{7\pi}{6}, \frac{11\pi}{6}$. Thus the solutions are $x = \frac{\pi}{2} + k\pi, x = \frac{7\pi}{6} + 2k\pi, x = \frac{11\pi}{6} + 2k\pi, k = 0, \pm 1, \pm 2, \ldots$.

15. $\left(\tan x + \sqrt{3}\right)\left(\cos x + 2\right) = 0 \iff \tan x + \sqrt{3} = 0$ or $\cos x + 2 = 0$. Since $|\cos x| \leq 1$ for all x, there is no solution for $\cos x + 2 = 0$. Hence, $\tan x + \sqrt{3} = 0 \iff \tan x = -\sqrt{3} \iff x = -\frac{\pi}{3}$ on $\left(-\frac{\pi}{2}, \frac{\pi}{2}\right)$. Thus the solutions are $x = -\frac{\pi}{3} + k\pi, k = 0, \pm 1, \pm 2, \ldots$.

17. $\cos x \sin x - 2\cos x = 0 \iff \cos x\,(\sin x - 2) = 0 \iff \cos x = 0$ or $\sin x - 2 = 0$. Since $|\sin x| \leq 1$ for all x, there is no solution for $\sin x - 2 = 0$. Hence, $\cos x = 0 \iff x = \frac{\pi}{2} + 2k\pi, \frac{3\pi}{2} + 2k\pi \iff x = \frac{\pi}{2} + k\pi, k = 0, \pm 1, \pm 2, \ldots$.

19. $4\cos^2 x - 4\cos x + 1 = 0 \iff (2\cos x - 1)^2 = 0 \iff 2\cos x - 1 = 0 \iff \cos x = \frac{1}{2} \iff x = \frac{\pi}{3} + 2k\pi, \frac{5\pi}{3} + 2k\pi, k = 0, \pm 1, \pm 2, \ldots$.

21. $\sin^2 x = 2\sin x + 3$ $\Leftrightarrow$ $\sin^2 x - 2\sin x - 3 = 0$ $\Leftrightarrow$ $(\sin x - 3)(\sin x + 1) = 0$ $\Leftrightarrow$ $\sin x - 3 = 0$ or $\sin x + 1 = 0$. Since $|\sin x| \leq 1$ for all x, there is no solution for $\sin x - 3 = 0$. Hence $\sin x + 1 = 0$ $\Leftrightarrow$ $\sin x = -1$ $\Leftrightarrow$ $x = \frac{3\pi}{2} + 2k\pi$, $k = 0, \pm1, \pm2, \ldots$.

23. $\sin^2 x = 4 - 2\cos^2 x$ $\Leftrightarrow$ $\sin^2 x + \cos^2 x + \cos^2 x = 4$ $\Leftrightarrow$ $1 + \cos^2 x = 4$ $\Leftrightarrow$ $\cos^2 x = 3$ Since $|\cos x| \leq 1$ for all x, it follows that $\cos^2 x \leq 1$ and so there is no solution for $\cos^2 x = 3$.

25. $2\sin 3x + 1 = 0$ $\Leftrightarrow$ $2\sin 3x = -1$ $\Leftrightarrow$ $\sin 3x = -\frac{1}{2}$. In the interval $[0, 6\pi)$ the solution are $3x = \frac{7\pi}{6}, \frac{11\pi}{6}, \frac{19\pi}{6}$, $\frac{23\pi}{6}, \frac{31\pi}{6}, \frac{35\pi}{6}$ $\Leftrightarrow$ $x = \frac{7\pi}{18}, \frac{11\pi}{18}, \frac{19\pi}{18}, \frac{23\pi}{18}, \frac{25\pi}{18}, \frac{29\pi}{18}$. So $x = \frac{7\pi}{18} + 2k\frac{\pi}{3}, \frac{11\pi}{18} + 2k\frac{\pi}{3}$ for any integer k.

27. $\sec 4x - 2 = 0$ $\Leftrightarrow$ $\sec 4x = 2$ $\Leftrightarrow$ $4x = \frac{\pi}{3} + 2k\pi, \frac{5\pi}{3} + 2k\pi$ $\Leftrightarrow$ $x = \frac{\pi}{12} + \frac{1}{2}k\pi, -\frac{\pi}{12} + \frac{1}{2}k\pi$ for any integer k.

29. $\sqrt{3}\sin 2x = \cos 2x$ $\Leftrightarrow$ $\tan 2x = \frac{1}{\sqrt{3}}$ (if $\cos 2x \neq 0$) $\Leftrightarrow$ $2x = \frac{\pi}{6} + k\pi$ $\Leftrightarrow$ $x = \frac{\pi}{12} + \frac{1}{2}k\pi$ for any integer k.

31. $\cos\frac{x}{2} - 1 = 0$ $\Leftrightarrow$ $\cos\frac{x}{2} = 1$ $\Leftrightarrow$ $\frac{x}{2} = 2k\pi$ $\Leftrightarrow$ $x = 4k\pi$ for any integer k.

33. $\tan\frac{x}{4} + \sqrt{3} = 0$ $\Leftrightarrow$ $\tan\frac{x}{4} = -\sqrt{3}$ $\Leftrightarrow$ $\frac{x}{4} = \frac{2\pi}{3} + k\pi$ $\Leftrightarrow$ $x = \frac{8\pi}{3} + 4k\pi$ for any integer k.

35. $\tan^5 x - 9\tan x = 0$ $\Leftrightarrow$ $\tan x (\tan^4 x - 9) = 0$ $\Leftrightarrow$ $\tan x = 0$ or $\tan^4 x = 9$ $\Leftrightarrow$ $\tan x = 0$ or $\tan x = \pm\sqrt{3}$ $\Leftrightarrow$ $x = 0, \pi, \frac{\pi}{3}, \frac{2\pi}{3}, \frac{4\pi}{3}, \frac{5\pi}{3}$ in $[0, 2\pi)$. Thus, $x = \frac{k\pi}{3}$ for any integer k.

37. $4\sin x \cos x + 2\sin x - 2\cos x - 1 = 0$ $\Leftrightarrow$ $(2\sin x - 1)(2\cos x + 1) = 0$ $\Leftrightarrow$ $2\sin x - 1 = 0$ or $2\cos x + 1 = 0$ $\Leftrightarrow$ $\sin x = \frac{1}{2}$ or $\cos x = -\frac{1}{2}$ $\Leftrightarrow$ $x = \frac{\pi}{6}, \frac{5\pi}{6}, \frac{2\pi}{3}, \frac{4\pi}{3}$ in $[0, 2\pi)$. Thus $x = \frac{\pi}{6} + 2k\pi, \frac{5\pi}{6} + 2k\pi, \frac{2\pi}{3} + 2k\pi, \frac{4\pi}{3} + 2k\pi$ for any integer k.

39. $\cos^2 2x - \sin^2 2x = 0$ $\Leftrightarrow$ $(\cos 2x - \sin 2x)(\cos 2x + \sin 2x) = 0$ $\Leftrightarrow$ $\cos 2x = \pm\sin 2x$ $\Leftrightarrow$ $\tan 2x = \pm1$ $\Leftrightarrow$ $2x = \frac{\pi}{4}, \frac{3\pi}{4}, \frac{5\pi}{4}, \frac{7\pi}{4}, \frac{9\pi}{4}, \frac{11\pi}{4}, \frac{13\pi}{4}, \frac{15\pi}{4}$ in $[0, 4\pi)$ $\Leftrightarrow$ $x = \frac{\pi}{8}, \frac{3\pi}{8}, \frac{5\pi}{8}, \frac{7\pi}{8}, \frac{9\pi}{8}, \frac{11\pi}{8}, \frac{13\pi}{8}, \frac{15\pi}{8}$ in $[0, 2\pi)$. So the solution can be expressed as the odd multiples of $\frac{\pi}{8}$ which are $x = \frac{\pi}{8} + \frac{k\pi}{4}$ for any integer k.

41. $2\cos 3x = 1$ $\Leftrightarrow$ $\cos 3x = \frac{1}{2}$ $\Rightarrow$ $3x = \frac{\pi}{3}, \frac{5\pi}{3}, \frac{7\pi}{3}, \frac{11\pi}{3}, \frac{13\pi}{3}, \frac{17\pi}{3}$ on $[0, 6\pi)$ $\Leftrightarrow$ $x = \frac{\pi}{9}, \frac{5\pi}{9}, \frac{7\pi}{9}, \frac{11\pi}{9}, \frac{13\pi}{9}, \frac{17\pi}{9}$ on $[0, 2\pi)$.

43. $2\sin x \tan x - \tan x = 1 - 2\sin x$ $\Leftrightarrow$ $2\sin x \tan x - \tan x + 2\sin x - 1 = 0$ $\Leftrightarrow$ $(2\sin x - 1)(\tan x + 1) = 0$ $\Leftrightarrow$ $2\sin x - 1 = 0$ or $\tan x + 1 = 0$ $\Leftrightarrow$ $\sin x = \frac{1}{2}$ or $\tan x = -1$ $\Leftrightarrow$ $x = \frac{\pi}{6}, \frac{5\pi}{6}$ or $x = \frac{3\pi}{4}, \frac{7\pi}{4}$. Thus, the solutions in $[0, 2\pi)$ are $\frac{\pi}{6}, \frac{3\pi}{4}, \frac{5\pi}{6}, \frac{7\pi}{4}$.

45. $\tan x - 3\cot x = 0$ $\Leftrightarrow$ $\frac{\sin x}{\cos x} - \frac{3\cos x}{\sin x} = 0$ $\Leftrightarrow$ $\frac{\sin^2 x - 3\cos^2 x}{\cos x \sin x} = 0$ $\Leftrightarrow$ $\frac{\sin^2 x + \cos^2 x - 4\cos^2 x}{\cos x \sin x} = 0$ $\Leftrightarrow$ $\frac{1 - 4\cos^2 x}{\cos x \sin x} = 0$ $\Leftrightarrow$ $1 - 4\cos^2 x = 0$ $\Leftrightarrow$ $4\cos^2 x = 1$ $\Leftrightarrow$ $\cos x = \pm\frac{1}{2}$ $\Leftrightarrow$ $x = \frac{\pi}{3}, \frac{2\pi}{3}, \frac{4\pi}{3}, \frac{5\pi}{3}$ in $[0, 2\pi)$.

47. $\tan 3x + 1 = \sec 3x$ $\Rightarrow$ $(\tan 3x + 1)^2 = \sec^2 3x$ $\Leftrightarrow$ $\tan^2 3x + 2\tan 3x + 1 = \sec^2 3x$ $\Leftrightarrow$ $\sec^2 3x + 2\tan 3x = \sec^2 3x$ $\Leftrightarrow$ $2\tan 3x = 0$ $\Leftrightarrow$ $3x = 0, \pi, 2\pi, 3\pi, 4\pi, 5\pi$ in $[0, 6\pi)$ $\Leftrightarrow$ $x = 0, \frac{\pi}{3}, \frac{2\pi}{3}, \pi, \frac{4\pi}{3}, \frac{5\pi}{3}$ in $[0, 2\pi)$. Since squaring both sides is an operation that can introduce extraneous solutions, we must check each of the possible solution in the original equation. We see that $x = \frac{\pi}{3}, \pi, \frac{5\pi}{3}$ are not solutions. Hence the only solutions are $x = 0, \frac{2\pi}{3}, \frac{4\pi}{3}$.

49. (a) Since the period of cosine is 2π the solutions are $x \approx 1.15928 + 2k\pi$ and $x \approx 5.12391 + 2k\pi$ for any integer k.

   (b) $\cos x = 0.4$ $\Rightarrow$ $x = \cos^{-1} 0.4 \approx 1.15928$. The other solution is $2\pi - \cos^{-1} 0.4 \approx 5.12391$.

51. (a) Since the period of secant is 2π the solutions are $x \approx 1.36943 \pm 2k\pi$ and $\approx 4.91375 \pm 2k\pi$ for any integer k.

   (b) $\sec x = 5$ $\Leftrightarrow$ $\cos x = \frac{1}{5}$ $\Rightarrow$ $x = \cos^{-1}\frac{1}{5} \approx 1.36944$. The other solution is $2\pi - \cos^{-1}\frac{1}{5} \approx 4.91375$.

53. (a) Since $\sin(\pi + x) = -\sin x$, we can express the general solutions as $x \approx 0.46365 + k\pi$ and $x \approx 2.67795 + k\pi$ for any integer k.

   (b) $5\sin^2 x - 1 = 0$ $\Rightarrow$ $\sin^2 x = \frac{1}{5}$ $\Rightarrow$ $\sin x = \pm\frac{1}{\sqrt{5}}$. Now $\sin x = \frac{1}{\sqrt{5}}$ $\Rightarrow$ $x = \sin^{-1}\frac{1}{\sqrt{5}} \approx 0.46365$, and $x \approx \pi - 0.46365 \approx 2.67795$. Also, $\sin x = -\frac{1}{\sqrt{5}}$ $\Rightarrow$ $x = \sin^{-1}-\frac{1}{\sqrt{5}} \approx -0.46365$, so in the interval $[0, 2\pi)$, $x \approx \pi - (-0.46365) \approx 3.60524$ and $x \approx 2\pi - 0.46365 \approx 5.81954$.

55. (a) Since the period for sine is 2π, the general solution is of the form $x \approx 0.33984 + 2k\pi$ and $x \approx 2.80176 + 2k\pi$ for any integer k.

(b) $3\sin^2 x - 7\sin x + 2 = 0 \Rightarrow (3\sin x - 1)(\sin x - 2) = 0 \Rightarrow 3\sin x - 1 = 0$ or $\sin x - 2 = 0$. Since $|\sin x| \le 1$, $\sin x - 2 = 0$ has no solution. Thus $3\sin x - 1 = 0 \Rightarrow \sin x = \frac{1}{3} \Rightarrow x \approx 0.33984$ and $x \approx \pi - 0.33984 \approx 2.80176$.

57. $f(x) = 3\cos x + 1$; $g(x) = \cos x - 1$. $f(x) = g(x)$ when
$3\cos x + 1 = \cos x - 1 \Leftrightarrow 2\cos x = -2 \Leftrightarrow \cos x = -1 \Leftrightarrow$
$x = \pi + 2k\pi = (2k + 1)\pi$. The points of intersection are $((2k + 1)\pi, -2)$ for any integer k.

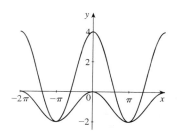

59. $f(x) = \tan x$; $g(x) = \sqrt{3}$. $f(x) = g(x)$ when $\tan x = \sqrt{3} \Leftrightarrow$
$x = \frac{\pi}{3} + k\pi$. The intersection points are $\left(\frac{\pi}{3} + k\pi, \sqrt{3}\right)$ for any integer k.

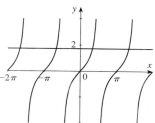

61. $\cos x \cos 3x - \sin x \sin 3x = 0 \Leftrightarrow \cos(x + 3x) = 0 \Leftrightarrow \cos 4x = 0 \Leftrightarrow 4x = \frac{\pi}{2}, \frac{3\pi}{2}, \frac{5\pi}{2}, \frac{7\pi}{2}, \frac{9\pi}{2}, \frac{11\pi}{2},$
$\frac{13\pi}{2}, \frac{15\pi}{2}$ in $[0, 8\pi) \Leftrightarrow x = \frac{\pi}{8}, \frac{3\pi}{8}, \frac{5\pi}{8}, \frac{7\pi}{8}, \frac{9\pi}{8}, \frac{11\pi}{8}, \frac{13\pi}{8}, \frac{15\pi}{8}$ in $[0, 2\pi)$.

63. $\sin 2x \cos x + \cos 2x \sin x = \frac{\sqrt{3}}{2} \Leftrightarrow \sin(2x + x) = \frac{\sqrt{3}}{2} \Leftrightarrow \sin 3x = \frac{\sqrt{3}}{2} \Leftrightarrow 3x = \frac{\pi}{3}, \frac{2\pi}{3}, \frac{7\pi}{3}, \frac{8\pi}{3}, \frac{13\pi}{3}, \frac{14\pi}{3}$
in $[0, 6\pi) \Leftrightarrow x = \frac{\pi}{9}, \frac{2\pi}{9}, \frac{7\pi}{9}, \frac{8\pi}{9}, \frac{13\pi}{9}, \frac{14\pi}{9}$ in $[0, 2\pi)$.

65. $\sin 2x + \cos x = 0 \Leftrightarrow 2\sin x \cos x + \cos x = 0 \Leftrightarrow \cos x(2\sin x + 1) = 0 \Leftrightarrow \cos x = 0$ or $\sin x = -\frac{1}{2}$
$\Leftrightarrow x = \frac{\pi}{2}, \frac{7\pi}{6}, \frac{3\pi}{2}, \frac{11\pi}{6}$.

67. $\cos 2x + \cos x = 2 \Leftrightarrow 2\cos^2 x - 1 + \cos x - 2 = 0 \Leftrightarrow 2\cos^2 x + \cos x - 3 = 0 \Leftrightarrow$
$(2\cos x + 3)(\cos x - 1) = 0 \Leftrightarrow 2\cos x + 3 = 0$ or $\cos x - 1 = 0 \Leftrightarrow \cos x = -\frac{3}{2}$ (which is impossible) or
$\cos x = 1 \Leftrightarrow x = 0$ in $[0, 2\pi)$.

69. $\sin x + \sin 3x = 0 \Leftrightarrow 2\sin 2x \cos(-x) = 0 \Leftrightarrow 2\sin 2x \cos x = 0 \Leftrightarrow \sin 2x = 0$ or $\cos x = 0 \Leftrightarrow$
$2x = k\pi$ or $x = k\frac{\pi}{2} \Leftrightarrow x = k\frac{\pi}{2}$ for any integer k.

71. $\cos 4x + \cos 2x = \cos x \Leftrightarrow 2\cos 3x \cos x = \cos x \Leftrightarrow \cos x(2\cos 3x - 1) = 0 \Leftrightarrow \cos x = 0$ or $\cos 3x = \frac{1}{2}$
$\Leftrightarrow x = \frac{\pi}{2}$ or $3x = \frac{\pi}{3} + 2k\pi, \frac{5\pi}{3} + 2k\pi, \frac{7\pi}{3} + 2k\pi, \frac{11\pi}{3} + 2k\pi, \frac{13\pi}{3} + 2k\pi, \frac{17\pi}{3} + 2k\pi \Leftrightarrow x = \frac{\pi}{2} + k\pi, \frac{\pi}{9} + \frac{2}{3}k\pi,$
$\frac{5\pi}{9} + \frac{2}{3}k\pi$.

73. $\sin 2x = x$

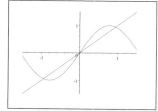

75. $2^{\sin x} = x$

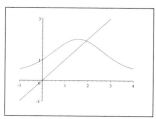

The three solutions are $x = 0$ and $x \approx \pm 0.95$.

The only solution is $x \approx 1.92$.

77. $\dfrac{\cos x}{1 + x^2} = x^2$

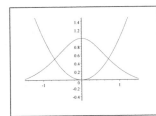

The two solutions are $x \approx \pm 0.71$.

79. We substitute $v_0 = 2200$ and $R(\theta) = 5000$ and solve for θ. So $5000 = \dfrac{(2200)^2 \sin 2\theta}{32}$ $\Leftrightarrow$

$5000 = 151250 \sin 2\theta$ $\Leftrightarrow$ $\sin 2\theta = 0.03308$ $\Rightarrow$
$2\theta = 1.89442°$ or $2\theta = 180° - 1.89442° = 178.10558°$.
If $2\theta = 1.89442°$, then $\theta = 0.94721°$, and if
$2\theta = 178.10558°$, then $\theta = 89.05279°$.

81. We substitute $\theta_1 = 70°$ and $\dfrac{v_1}{v_2} = 1.33$ into Snell's Law to get $\dfrac{\sin 70°}{\sin \theta_2} = 1.33$ $\Leftrightarrow$ $\sin \theta_2 = \dfrac{\sin 70°}{1.33} = 0.7065$ $\Rightarrow$
$\theta_2 \approx 44.95°$.

83. (a) $10 = 12 + 2.83 \sin\left(\frac{2\pi}{3}(t - 80)\right)$ $\Leftrightarrow$ $2.83 \sin\left(\frac{2\pi}{3}(t - 80)\right) = -2$ $\Leftrightarrow$ $\sin\left(\frac{2\pi}{3}(t - 80)\right) = -0.70671$.
Now $\sin \theta = -0.70671$ and $\theta = -0.78484$. If $\frac{2\pi}{3}(t - 80) = -0.78484$ $\Leftrightarrow$ $t - 80 = 45.6$ $\Leftrightarrow$ $t = 34.4$.
Now in the interval $[0, 2\pi)$, we have $\theta = \pi + 0.78484 \approx 3.92644$ and $\theta = 2\pi - 0.78484 \approx 5.49834$. If
$\frac{2\pi}{3}(t - 80) = 3.92644$ $\Leftrightarrow$ $t - 80 = 228.1$ $\Leftrightarrow$ $t = 308.1$. And if $\frac{2\pi}{3}(t - 80) = 5.49834$ $\Leftrightarrow$
$t - 80 = 319.4$ $\Leftrightarrow$ $t = 399.4$ $(399.4 - 365 = 34.4)$. So according to this model, there should be 10 hours of
sunshine on the 34th day (February 3) and on the 308th day (November 4).

(b) Since $L(t) = 12 + 2.83 \sin\left(\frac{2\pi}{3}(t - 80)\right) \geq 10$ for $t \in [34, 308]$, the number of days with more than 10 hours of
daylight is $308 - 34 + 1 = 275$ days.

85. (a) First note that $\alpha = \dfrac{\pi}{2} - \dfrac{\theta}{2}$. The part of the belt touching the larger
pulley has length $2(\pi - \alpha)R = (\theta + \pi)R$ and similarly the part
touching the smaller belt has length $(\theta + \pi)r$. To calculate a and

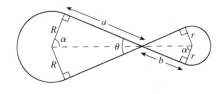

b, we write $\cot \dfrac{\theta}{2} = \dfrac{a}{R} = \dfrac{b}{r} \Rightarrow a = R \cot \dfrac{\theta}{2}$ and $b = r \cot \dfrac{\theta}{2}$, so
the length of the straight parts of the belt is $2a + 2b = 2(R + r)\cot \dfrac{\theta}{2}$. Thus, the total length of the belt is

$$L = (\theta + \pi)R + (\theta + \pi)r + 2(R + r)\cot \dfrac{\theta}{2} = (R + r)\left(\theta + \pi + 2\cot \dfrac{\theta}{2}\right) \text{ and so } \theta + \pi + 2\cot \dfrac{\theta}{2} = \dfrac{L}{R + r} \quad \Leftrightarrow$$

$$\theta + 2\cot \dfrac{\theta}{2} = \dfrac{L}{R + r} - \pi.$$

(b) We plot $\theta + 2\cot \dfrac{\theta}{2}$ and $\dfrac{L}{R + r} - \pi = \dfrac{27.78}{2.42 + 1.21} - \pi \approx 4.5113$ in the same
viewing rectangle. The solution is $\theta \approx 1.047$ rad $\approx 60°$.

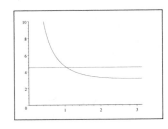

87. $\sin(\cos x)$ is a function of a function, that is, a composition of trigonometric functions (see Section 2.7). Most of the other equations involve sums, products, differences, or quotients of trigonometric functions.

$\sin(\cos x) = 0 \quad \Leftrightarrow \quad \cos x = 0$ or $\cos x = \pi$. However, since $|\cos x| \le 1$, the only solution is $\cos x = 0 \Rightarrow x = \frac{\pi}{2} + k\pi$. The graph of $f(x) = \sin(\cos x)$ is shown.

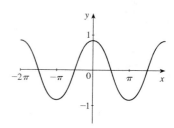

Chapter 8 Review

1. $\sin\theta(\cot\theta + \tan\theta) = \sin\theta\left(\dfrac{\cos\theta}{\sin\theta} + \dfrac{\sin\theta}{\cos\theta}\right) = \cos\theta + \dfrac{\sin^2\theta}{\cos\theta} = \dfrac{\cos^2\theta + \sin^2\theta}{\cos\theta} = \dfrac{1}{\cos\theta} = \sec\theta$

3. $\cos^2 x\csc x - \csc x = (1 - \sin^2 x)\csc x - \csc x = \csc x - \sin^2 x\csc x - \csc x = -\sin^2 x \cdot \dfrac{1}{\sin x} = -\sin x$

5. $\dfrac{\cos^2 x - \tan^2 x}{\sin^2 x} = \dfrac{\cos^2 x}{\sin^2 x} - \dfrac{\tan^2 x}{\sin^2 x} = \cot^2 x - \dfrac{1}{\cos^2 x} = \cot^2 x - \sec^2 x$

7. $\dfrac{\cos^2 x}{1 - \sin x} = \dfrac{\cos x}{\dfrac{1}{\cos x}(1 - \sin x)} = \dfrac{\cos x}{\dfrac{1}{\cos x} - \dfrac{\sin x}{\cos x}} = \dfrac{\cos x}{\sec x - \tan x}$

9. $\sin^2 x\cot^2 x + \cos^2 x\tan^2 x = \sin^2 x \cdot \dfrac{\cos^2 x}{\sin^2 x} + \cos^2 x \cdot \dfrac{\sin^2 x}{\cos^2 x} = \cos^2 x + \sin^2 x = 1$

11. $\dfrac{\sin 2x}{1 + \cos 2x} = \dfrac{2\sin x\cos x}{1 + 2\cos^2 x - 1} = \dfrac{2\sin x\cos x}{2\cos^2 x} = \dfrac{2\sin x}{2\cos x} = \tan x$

13. $\tan\dfrac{x}{2} = \dfrac{1 - \cos x}{\sin x} = \dfrac{1}{\sin x} - \dfrac{\cos x}{\sin x} = \csc x - \cot x$

15. $\sin(x + y)\sin(x - y) = \frac{1}{2}\left[\cos\left((x+y) - (x-y)\right) - \cos\left((x+y) + (x-y)\right)\right] = \frac{1}{2}\left(\cos 2y - \cos 2x\right)$
$\qquad = \frac{1}{2}\left[1 - 2\sin^2 y - (1 - 2\sin^2 x)\right] = \frac{1}{2}\left(2\sin^2 x - 2\sin^2 y\right) = \sin^2 x - \sin^2 y$

17. $1 + \tan x\tan\dfrac{x}{2} = 1 + \dfrac{\sin x}{\cos x} \cdot \dfrac{1 - \cos x}{\sin x} = 1 + \dfrac{1 - \cos x}{\cos x} = 1 + \dfrac{1}{\cos x} - 1 = \dfrac{1}{\cos x} = \sec x$

19. $\left(\cos\dfrac{x}{2} - \sin\dfrac{x}{2}\right)^2 = \cos^2\dfrac{x}{2} - 2\sin\dfrac{x}{2}\cos\dfrac{x}{2} + \sin^2\dfrac{x}{2} = \sin^2\dfrac{x}{2} + \cos^2\dfrac{x}{2} - 2\sin\dfrac{x}{2}\cos\dfrac{x}{2} = 1 - \sin\left(2 \cdot \dfrac{x}{2}\right) = 1 - \sin x$

21. $\dfrac{\sin 2x}{\sin x} - \dfrac{\cos 2x}{\cos x} = \dfrac{2\sin x\cos x}{\sin x} - \dfrac{2\cos^2 x - 1}{\cos x} = 2\cos x - 2\cos x + \dfrac{1}{\cos x} = \sec x$

23. $\tan\left(x + \dfrac{\pi}{4}\right) = \dfrac{\tan x + \tan\dfrac{\pi}{4}}{1 - \tan x\tan\dfrac{\pi}{4}} = \dfrac{1 + \tan x}{1 - \tan x}$

25. (a) $f(x) = 1 - \left(\cos\dfrac{x}{2} - \sin\dfrac{x}{2}\right)^2$, $g(x) = \sin x$

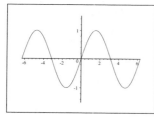

(b) The graphs suggest that $f(x) = g(x)$ is an identity. To prove this, expand $f(x)$ and simplify, using the double-angle formula for sine:

$$\begin{aligned}
f(x) &= 1 - \left(\cos\tfrac{x}{2} - \sin\tfrac{x}{2}\right)^2 \\
&= 1 - \left(\cos^2\tfrac{x}{2} - 2\cos\tfrac{x}{2}\sin\tfrac{x}{2} + \sin^2\tfrac{x}{2}\right) \\
&= 1 + 2\cos\tfrac{x}{2}\sin\tfrac{x}{2} - \left(\cos^2\tfrac{x}{2} + \sin^2\tfrac{x}{2}\right) \\
&= 1 + \sin x - (1) = \sin x = g(x)
\end{aligned}$$

27. (a) $f(x) = \tan x \tan \frac{x}{2}$, $g(x) = \dfrac{1}{\cos x}$

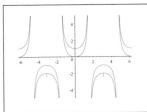

(b) The graphs suggest that $f(x) \neq g(x)$ in general. For example, choose $x = \frac{\pi}{3}$ and evaluate: $f\left(\frac{\pi}{3}\right) = \tan \frac{\pi}{3}$

$\tan \frac{\pi}{6} = \sqrt{3} \cdot \frac{1}{\sqrt{3}} = 1$, whereas $g\left(\frac{\pi}{3}\right) = \frac{1}{\frac{1}{2}} = 2$, so

$f(x) \neq g(x)$.

29. (a) $f(x) = 2\sin^2 3x + \cos 6x$

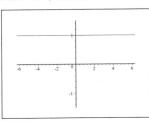

(b) The graph suggests that $f(x) = 1$ for all x. To prove this, we use the double angle formula to note that

$\cos 6x = \cos(2(3x)) = 1 - 2\sin^2 3x$, so

$f(x) = 2\sin^2 3x + (1 - 2\sin^2 3x) = 1$.

31. $\cos x \sin x - \sin x = 0 \quad \Leftrightarrow \quad \sin x (\cos x - 1) = 0 \quad \Leftrightarrow \quad \sin x = 0$ or $\cos x = 1 \quad \Leftrightarrow \quad x = 0, \pi$ or $x = 0$.
Therefore, the solutions are $x = 0$ and π.

33. $2\sin^2 x - 5\sin x + 2 = 0 \quad \Leftrightarrow \quad (2\sin x - 1)(\sin x - 2) = 0 \quad \Leftrightarrow \quad \sin x = \frac{1}{2}$ or $\sin x = 2$ (which is inadmissible)
$\Leftrightarrow \quad x = \frac{\pi}{6}, \frac{5\pi}{6}$. Thus, the solutions in $[0, 2\pi)$ are $x = \frac{\pi}{6}$ and $\frac{5\pi}{6}$.

35. $2\cos^2 x - 7\cos x + 3 = 0 \quad \Leftrightarrow \quad (2\cos x - 1)(\cos x - 3) = 0 \quad \Leftrightarrow \quad \cos x = \frac{1}{2}$ or $\cos x = 3$ (which is inadmissible)
$\Leftrightarrow \quad x = \frac{\pi}{3}, \frac{5\pi}{3}$. Therefore, the solutions in $[0, 2\pi)$ are $x = \frac{\pi}{3}, \frac{5\pi}{3}$.

37. Note that $x = \pi$ is not a solution because the denominator is zero. $\dfrac{1 - \cos x}{1 + \cos x} = 3 \quad \Leftrightarrow \quad 1 - \cos x = 3 + 3\cos x \quad \Leftrightarrow$
$-4\cos x = 2 \quad \Leftrightarrow \quad \cos x = -\frac{1}{2} \quad \Leftrightarrow \quad x = \frac{2\pi}{3}, \frac{4\pi}{3}$ in $[0, 2\pi)$.

39. Factor by grouping: $\tan^3 x + \tan^2 x - 3\tan x - 3 = 0 \quad \Leftrightarrow \quad (\tan x + 1)(\tan^2 x - 3) = 0 \quad \Leftrightarrow \quad \tan x = -1$ or
$\tan x = \pm\sqrt{3} \quad \Leftrightarrow \quad x = \frac{3\pi}{4}, \frac{7\pi}{4}$ or $x = \frac{\pi}{3}, \frac{2\pi}{3}, \frac{4\pi}{3}, \frac{5\pi}{3}$. Therefore, the solutions in $[0, 2\pi)$ are $x = \frac{\pi}{3}, \frac{2\pi}{3}, \frac{3\pi}{4}, \frac{4\pi}{3}, \frac{5\pi}{3}$,
$\frac{7\pi}{4}$.

41. $\tan \frac{1}{2}x + 2\sin 2x = \csc x \quad \Leftrightarrow \quad \dfrac{1 - \cos x}{\sin x} + 4\sin x \cos x = \dfrac{1}{\sin x} \quad \Leftrightarrow \quad 1 - \cos x + 4\sin^2 x \cos x = 1 \quad \Leftrightarrow$
$4\sin^2 x \cos x - \cos x = 0 \quad \Leftrightarrow \quad \cos x (4\sin^2 x - 1) = 0 \quad \Leftrightarrow \quad \cos x = 0$ or $\sin x = \pm\frac{1}{2} \quad \Leftrightarrow \quad x = \frac{\pi}{2}, \frac{3\pi}{2}$ or
$x = \frac{\pi}{6}, \frac{5\pi}{6}, \frac{7\pi}{6}, \frac{11\pi}{6}$. Thus, the solutions in $[0, 2\pi)$ are $x = \frac{\pi}{6}, \frac{\pi}{2}, \frac{5\pi}{6}, \frac{7\pi}{6}, \frac{3\pi}{2}, \frac{11\pi}{6}$.

43. $\tan x + \sec x = \sqrt{3} \quad \Leftrightarrow \quad \dfrac{\sin x}{\cos x} + \dfrac{1}{\cos x} = \sqrt{3} \quad \Leftrightarrow \quad \sin x + 1 = \sqrt{3}\cos x \quad \Leftrightarrow \quad \sqrt{3}\cos x - \sin x = 1$
$\Leftrightarrow \quad \frac{\sqrt{3}}{2}\cos x - \frac{1}{2}\sin x = \frac{1}{2} \quad \Leftrightarrow \quad \cos \frac{\pi}{6}\cos x - \sin \frac{\pi}{6}\sin x = \frac{1}{2} \quad \Leftrightarrow \quad \cos\left(x + \frac{\pi}{6}\right) = \frac{1}{2} \quad \Leftrightarrow \quad x + \frac{\pi}{6} = \frac{\pi}{3}, \frac{5\pi}{3} \quad \Leftrightarrow$
$x = \frac{\pi}{6}, \frac{3\pi}{2}$. However, $x = \frac{3\pi}{2}$ is inadmissible because $\sec \frac{3\pi}{2}$ is undefined. Thus, the only solution in $[0, 2\pi)$ is $x = \frac{\pi}{6}$.

45. We graph $f(x) = \cos x$ and $g(x) = x^2 - 1$ in the viewing rectangle $[0, 6.5]$ by $[-2, 2]$. The two functions intersect at only one point, $x \approx 1.18$.

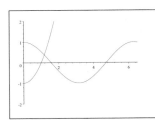

47. (a) $2000 = \dfrac{(400)^2 \sin^2 \theta}{64} \quad \Leftrightarrow \quad \sin^2 \theta = 0.8 \quad \Leftrightarrow \quad \sin \theta \approx 0.8944 \quad \Leftrightarrow \quad \theta \approx 63.4°$

(b) $\dfrac{(400)^2 \sin^2 \theta}{64} = 2500 \sin^2 \theta \leq 2500$. Therefore it is impossible for the projectile to reach a height of 3000 ft.

(c) The function $M(\theta) = 2500 \sin^2 \theta$ is maximized when $\sin^2 \theta = 1$, so $\theta = 90°$. The projectile will travel the highest when it is shot straight up.

49. Since $15°$ is in quadrant I, $\cos 15° = \sqrt{\dfrac{1 + \cos 30°}{2}} = \sqrt{\dfrac{2 + \sqrt{3}}{4}} = \frac{1}{2}\sqrt{2 + \sqrt{3}}.$

51. $\tan \dfrac{\pi}{8} = \dfrac{1 - \cos \frac{\pi}{4}}{\sin \frac{\pi}{4}} = \dfrac{1 - \frac{1}{\sqrt{2}}}{\frac{1}{\sqrt{2}}} = \left(1 - \frac{1}{\sqrt{2}}\right)\sqrt{2} = \sqrt{2} - 1$

53. $\sin 5° \cos 40° + \cos 5° \sin 40° = \sin(5° + 40°) = \sin 45° = \frac{1}{\sqrt{2}} = \frac{\sqrt{2}}{2}$

55. $\cos^2 \dfrac{\pi}{8} - \sin^2 \dfrac{\pi}{8} = \cos\left(2\left(\frac{\pi}{8}\right)\right) = \cos \frac{\pi}{4} = \frac{1}{\sqrt{2}} = \frac{\sqrt{2}}{2}$

57. We use a product-to-sum formula: $\cos 37.5° \cos 7.5° = \frac{1}{2}(\cos 45° + \cos 30°) = \frac{1}{2}\left(\frac{\sqrt{2}}{2} + \frac{\sqrt{3}}{2}\right) = \frac{1}{4}\left(\sqrt{2} + \sqrt{3}\right).$

In Solutions 59–63, x and y are in quadrant I, so we know that $\sec x = \frac{3}{2} \;\Rightarrow\; \cos x = \frac{2}{3}$, **so** $\sin x = \frac{\sqrt{5}}{3}$ **and** $\tan x = \frac{\sqrt{5}}{2}$. **Also,** $\csc y = 3 \;\Rightarrow\; \sin y = \frac{1}{3}$, **and so** $\cos y = \frac{2\sqrt{2}}{3}$, **and** $\tan y = \frac{1}{2\sqrt{2}} = \frac{\sqrt{2}}{4}$.

59. $\sin(x + y) = \sin x \cos y + \cos x \sin y = \frac{\sqrt{5}}{3} \cdot \frac{2\sqrt{2}}{3} + \frac{2}{3} \cdot \frac{1}{3} = \frac{2}{9}\left(1 + \sqrt{10}\right).$

61. $\tan(x + y) = \dfrac{\tan x + \tan y}{1 - \tan x \tan y} = \dfrac{\frac{\sqrt{5}}{2} + \frac{\sqrt{2}}{4}}{1 - \left(\frac{\sqrt{5}}{2}\right)\left(\frac{\sqrt{2}}{4}\right)} = \dfrac{\frac{\sqrt{5}}{2} + \frac{\sqrt{2}}{4}}{1 - \left(\frac{\sqrt{5}}{2}\right)\left(\frac{\sqrt{2}}{4}\right)} \cdot \dfrac{8}{8} = \dfrac{2\left(2\sqrt{5} + \sqrt{2}\right)}{8 - \sqrt{10}} \cdot \dfrac{8 + \sqrt{10}}{8 + \sqrt{10}}$

$= \frac{2}{3}\left(\sqrt{2} + \sqrt{5}\right)$

63. $\cos \dfrac{y}{2} = \sqrt{\dfrac{1 + \cos y}{2}} = \sqrt{\dfrac{1 + \left(\frac{2\sqrt{2}}{3}\right)}{2}} = \sqrt{\dfrac{3 + 2\sqrt{2}}{6}}$ (since cosine is positive in quadrant I)

65. $\sin^{-1} \dfrac{\sqrt{3}}{2} = \dfrac{\pi}{3}$

67. $\cos\left(\tan^{-1}\sqrt{3}\right) = \cos \dfrac{\pi}{3} = \dfrac{1}{2}$

69. Let $u = \sin^{-1}\frac{2}{5}$ and so $\sin u = \frac{2}{5}$. Then from the triangle, $\tan\left(\sin^{-1}\frac{2}{5}\right) = \tan u = \frac{2}{\sqrt{21}}.$

73. Let $\theta = \tan^{-1} x \;\Leftrightarrow\; \tan \theta = x$. Then from the triangle, we have $\sin\left(\tan^{-1} x\right) = \sin \theta = \dfrac{x}{\sqrt{1 + x^2}}.$

71. $\cos\left(2\sin^{-1}\frac{1}{3}\right) = 1 - 2\sin^2\left(\sin^{-1}\frac{1}{3}\right) = 1 - 2\left(\frac{1}{3}\right)^2 = \frac{7}{9}$

75. $\cos \theta = \dfrac{x}{3} \;\Rightarrow\; \theta = \cos^{-1}\left(\dfrac{x}{3}\right)$

77. (a) $\tan \theta = \dfrac{10}{x} \;\Leftrightarrow\; \theta = \tan^{-1}\left(\dfrac{10}{x}\right)$

(b) $\theta = \tan^{-1}\left(\dfrac{10}{x}\right)$, for $x > 0$. Since the road sign can first be seen when $\theta = 2°$, we have $2° = \tan^{-1}\left(\dfrac{10}{x}\right) \;\Leftrightarrow\; x = \dfrac{10}{\tan 2°} \approx 286.4$ ft. Thus, the sign can first be seen at a height of 286.4 ft.

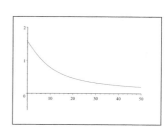

Chapter 8 Test

1. (a) $\tan\theta\sin\theta + \cos\theta = \dfrac{\sin\theta}{\cos\theta}\sin\theta + \cos\theta = \dfrac{\sin^2\theta}{\cos\theta} + \dfrac{\cos^2\theta}{\cos\theta} = \dfrac{1}{\cos\theta} = \sec\theta$

(b) $\dfrac{\tan x}{1-\cos x} = \dfrac{\tan x}{1-\cos x}\cdot\dfrac{1+\cos x}{1+\cos x} = \dfrac{\tan x\,(1+\cos x)}{1-\cos^2 x} = \dfrac{\dfrac{\sin x}{\cos x}\,(1+\cos x)}{\sin^2 x} = \dfrac{1}{\sin x}\cdot$

$\dfrac{1+\cos x}{\cos x} = \csc x\,(1+\sec x)$

(c) $\dfrac{2\tan x}{1+\tan^2 x} = \dfrac{2\tan x}{\sec^2 x} = \dfrac{2\sin x}{\cos x}\cdot\cos^2 x = 2\sin x\cos x = \sin 2x$

2. $\dfrac{x}{\sqrt{4-x^2}} = \dfrac{2\sin\theta}{\sqrt{4-(2\sin\theta)^2}} = \dfrac{2\sin\theta}{\sqrt{4-4\sin^2\theta}} = \dfrac{2\sin\theta}{2\sqrt{1-\sin^2\theta}} = \dfrac{\sin\theta}{\sqrt{\cos^2\theta}} = \dfrac{\sin\theta}{|\cos\theta|} = \dfrac{\sin\theta}{\cos\theta} = \tan\theta$ (because

$\cos\theta > 0$ for $-\dfrac{\pi}{2} < \theta < \dfrac{\pi}{2}$)

3. (a) $\sin 8° \cos 22° + \cos 8° \sin 22° = \sin(8° + 22°) = \sin 30° = \dfrac{1}{2}$

(b) $\sin 75° = \sin(45° + 30°) = \sin 45° \cos 30° + \cos 45° \sin 30° = \dfrac{\sqrt{2}}{2}\cdot\dfrac{\sqrt{3}}{2} + \dfrac{\sqrt{2}}{2}\cdot\dfrac{1}{2} = \dfrac{1}{4}\left(\sqrt{6}+\sqrt{2}\right)$

(c) $\sin\dfrac{\pi}{12} = \sin\left(\dfrac{\frac{\pi}{6}}{2}\right) = \sqrt{\dfrac{1-\cos\frac{\pi}{6}}{2}} = \sqrt{\dfrac{1-\frac{\sqrt{3}}{2}}{2}} = \sqrt{\dfrac{2-\sqrt{3}}{4}} = \dfrac{1}{2}\sqrt{2-\sqrt{3}}$

4. From the figures, we have $\cos(\alpha+\beta) = \cos\alpha\cos\beta - \sin\alpha\sin\beta = \dfrac{2}{\sqrt{5}}\cdot\dfrac{\sqrt{5}}{3} - \dfrac{1}{\sqrt{5}}\cdot\dfrac{2}{3} = \dfrac{2\sqrt{5}-2}{3\sqrt{5}} = \dfrac{10-2\sqrt{5}}{15}$.

5. (a) $\sin 3x\cos 5x = \dfrac{1}{2}[\sin(3x+5x) + \sin(3x-5x)] = \dfrac{1}{2}(\sin 8x - \sin 2x)$

(b) $\sin 2x - \sin 5x = 2\cos\left(\dfrac{2x+5x}{2}\right)\sin\left(\dfrac{2x-5x}{2}\right) = -2\cos\dfrac{7x}{2}\sin\dfrac{3x}{2}$

6. $\sin\theta = -\dfrac{4}{5}$. Since θ is in quadrant III, $\cos\theta = -\dfrac{3}{5}$. Then $\tan\dfrac{\theta}{2} = \dfrac{1-\cos\theta}{\sin\theta} = \dfrac{1-\left(-\frac{3}{5}\right)}{-\frac{4}{5}} = -\dfrac{1+\frac{3}{5}}{\frac{4}{5}} = -\dfrac{5+3}{4} = -2$.

7. $y = \sin x$ has domain $(-\infty, \infty)$ and $y = \sin^{-1} x$ has domain $[-1, 1]$.

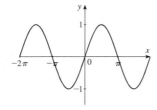
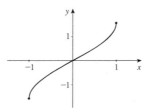

8. (a) $\tan\theta = \dfrac{x}{4} \Rightarrow \theta = \tan^{-1}\left(\dfrac{x}{4}\right)$ **(b)** $\cos\theta = \dfrac{3}{x} \Rightarrow \theta = \cos^{-1}\left(\dfrac{3}{x}\right)$

9. (a) $2\cos^2 x + 5\cos x + 2 = 0 \Leftrightarrow (2\cos x + 1)(\cos x + 2) = 0 \Leftrightarrow \cos x = -\dfrac{1}{2}$ or $\cos x = -2$ (which is impossible). So in the interval $[0, 2\pi)$, the solutions are $x = \dfrac{2\pi}{3}, \dfrac{4\pi}{3}$.

(b) $\sin 2x - \cos x = 0 \Leftrightarrow 2\sin x\cos x - \cos x = 0 \Leftrightarrow \cos x(2\sin x - 1) = 0 \Leftrightarrow \cos x = 0$ or $\sin x = \dfrac{1}{2}$

$\Leftrightarrow x = \dfrac{\pi}{2}, \dfrac{3\pi}{2}$ or $x = \dfrac{\pi}{6}, \dfrac{5\pi}{6}$. Therefore, the solutions in $[0, 2\pi)$ are $x = \dfrac{\pi}{6}, \dfrac{\pi}{2}, \dfrac{5\pi}{6}, \dfrac{3\pi}{2}$.

10. $5\cos 2x = 2 \Leftrightarrow \cos 2x = \dfrac{2}{5} \Leftrightarrow 2x = \cos^{-1} 0.4 \approx 1.159279$. The solutions in $[0, 4\pi)$ are $2x \approx 1.159279$,

$2\pi - 1.159279,\ 2\pi + 1.159279,\ 4\pi - 1.159279 \Leftrightarrow 2x \approx 1.159279,\ 5.123906,\ 7.442465,\ 11.407091 \Leftrightarrow$

$x \approx 0.57964,\ 2.56195,\ 3.72123,\ 5.70355$ in $[0, 2\pi)$.

11. Let $u = \tan^{-1}\dfrac{9}{40}$ so $\tan u = \dfrac{9}{40}$. From the triangle, $r = \sqrt{9^2 + 40^2} = 41$. So

$\cos\left(\tan^{-1}\dfrac{9}{40}\right) = \cos u = \dfrac{40}{41}$.

Focus on Modeling: Traveling and Standing Waves

1. (a) Substituting $x = 0$, we get $y(0, t) = 5 \sin \left(2 \cdot 0 - \frac{\pi}{2}t\right) = 5 \sin \left(-\frac{\pi}{2}t\right) = -5 \sin \frac{\pi}{2}t$.

(b)

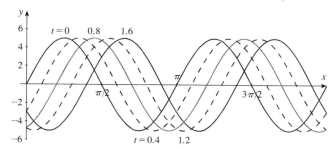

(c) We express the function in the standard form $y(x, t) = A \sin k(x - vt)$: $y(x, t) = 5 \sin \left(2x - \frac{\pi}{2}t\right) = 5 \sin 2\left(x - \frac{\pi}{4}t\right)$. Comparing this to the standard form, we see that the velocity of the wave is $v = \frac{\pi}{4}$.

3. From the graph, we see that the amplitude is $A = 2.7$ and the period is 9.2, so $k = \frac{2\pi}{9.2} \approx 0.68$. Since $v = 6$, we have $kv = \frac{2\pi}{9.2} \cdot 6 \approx 4.10$, so the equation we seek is $y(x, t) = 2.7 \sin (0.68x - 4.10t)$.

5. From the graphs, we see that the amplitude is $A = 0.6$. The nodes occur at $x = 0, 1, 2, 3$. Since $\sin \alpha x = 0$ when $\alpha x = k\pi$ (k any integer), we have $\alpha = \pi$. Then since the frequency is $\beta/2\pi$, we get $20 = \beta/2\pi \quad \Leftrightarrow \quad \beta = 40\pi$. Thus, an equation for this model is $f(x, t) = 0.6 \sin \pi x \cos 40\pi t$.

7. (a) The first standing wave has $\alpha = 1$, the second has $\alpha = 2$, the third has $\alpha = 3$, and the fourth has $\alpha = 4$.

(b) α is equal to the number of nodes minus 1. The first string has two nodes and $\alpha = 1$; the second string has three nodes and $\alpha = 2$, and so forth.

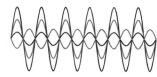

(c) Since the frequency is $\beta/2\pi$, we have $440 = \beta/2\pi \quad \Leftrightarrow \quad \beta = 880\pi$.

(d) The first standing wave has equation $y = \sin x \cos 880\pi t$, the second has equation $y = \sin 2x \cos 880\pi t$, the third has equation $y = \sin 3x \cos 880\pi t$, and the fourth has equation $y = \sin 4x \cos 880\pi t$.

9 Polar Coordinates and Vectors

9.1 Polar Coordinates

1.

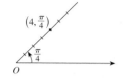

3.

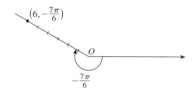

5.

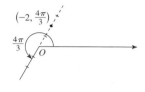

Answers to Exercises 7–11 will vary.

7. $\left(3, \frac{\pi}{2}\right)$ also has polar coordinates $\left(3, \frac{5\pi}{2}\right)$ or $\left(-3, \frac{3\pi}{2}\right)$

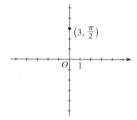

9. $\left(-1, \frac{7\pi}{6}\right)$ also has polar coordinates $\left(1, \frac{\pi}{6}\right)$ or $\left(-1, -\frac{5\pi}{6}\right)$.

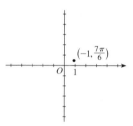

11. $(-5, 0)$ also has polar coordinates $(5, \pi)$ or $(-5, 2\pi)$.

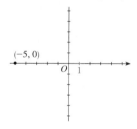

13. Q has coordinates $\left(4, \frac{3\pi}{4}\right)$.

15. Q has coordinates $\left(-4, -\frac{\pi}{4}\right) = \left(4, \frac{3\pi}{4}\right)$.

17. P has coordinates $\left(4, -\frac{23\pi}{4}\right) = \left(4, \frac{\pi}{4}\right)$.

19. P has coordinates $\left(-4, \frac{101\pi}{4}\right) = \left(-4, \frac{5\pi}{4}\right) = \left(4, \frac{\pi}{4}\right)$.

21. $P = (-3, 3)$ in rectangular coordinates, so $r^2 = x^2 + y^2 = (-3)^2 + 3^2 = 18$ and we can take $r = 3\sqrt{2}$. $\tan\theta = \frac{y}{x} = \frac{3}{-3} = -1$, so since P is in quadrant 2 we take $\theta = \frac{3\pi}{4}$. Thus, polar coordinates for P are $\left(3\sqrt{2}, \frac{3\pi}{4}\right)$.

23. Here $r = 5$ and $\theta = -\frac{2\pi}{3}$, so $x = r\cos\theta = 5\cos\left(-\frac{2\pi}{3}\right) = -\frac{5}{2}$ and $y = r\sin\theta = 5\sin\left(-\frac{2\pi}{3}\right) = -\frac{5\sqrt{3}}{2}$. R has rectangular coordinates $\left(-\frac{5}{2}, -\frac{5\sqrt{3}}{2}\right)$.

25. $(r, \theta) = \left(4, \frac{\pi}{6}\right)$. So $x = r\cos\theta = 4\cos\frac{\pi}{6} = 4 \cdot \frac{\sqrt{3}}{2} = 2\sqrt{3}$ and $y = r\sin\theta = 4\sin\frac{\pi}{6} = 4 \cdot \frac{1}{2} = 2$. Thus, the rectangular coordinates are $\left(2\sqrt{3}, 2\right)$.

27. $(r, \theta) = \left(\sqrt{2}, -\frac{\pi}{4}\right)$. So $x = r\cos\theta = \sqrt{2}\cos\left(-\frac{\pi}{4}\right) = \sqrt{2} \cdot \frac{1}{\sqrt{2}} = 1$, and $y = r\sin\theta = \sqrt{2}\sin\left(-\frac{\pi}{4}\right) = \sqrt{2}\left(-\frac{1}{\sqrt{2}}\right) = -1$. Thus, the rectangular coordinates are $(1, -1)$.

29. $(r, \theta) = (5, 5\pi)$. So $x = r\cos\theta = 5\cos 5\pi = -5$, and $y = r\sin\theta = 5\sin 5\pi = 0$. Thus, the rectangular coordinates are $(-5, 0)$.

31. $(r, \theta) = \left(6\sqrt{2}, \frac{11\pi}{6}\right)$. So $x = r\cos\theta = 6\sqrt{2}\cos\frac{11\pi}{6} = 3\sqrt{6}$ and $y = r\sin\theta = 6\sqrt{2}\sin\frac{11\pi}{6} = -3\sqrt{2}$. Thus, the rectangular coordinates are $\left(3\sqrt{6}, -3\sqrt{2}\right)$.

33. $(x, y) = (-1, 1)$. Since $r^2 = x^2 + y^2$, we have $r^2 = (-1)^2 + 1^2 = 2$, so $r = \sqrt{2}$. Now $\tan\theta = \frac{y}{x} = \frac{1}{-1} = -1$, so, since the point is in the second quadrant, $\theta = \frac{3\pi}{4}$. Thus, polar coordinates are $\left(\sqrt{2}, \frac{3\pi}{4}\right)$.

35. $(x, y) = \left(\sqrt{8}, \sqrt{8}\right)$. Since $r^2 = x^2 + y^2$, we have $r^2 = \left(\sqrt{8}\right)^2 + \left(\sqrt{8}\right)^2 = 16$, so $r = 4$. Now $\tan\theta = \frac{y}{x} = \frac{\sqrt{8}}{\sqrt{8}} = 1$, so, since the point is in the first quadrant, $\theta = \frac{\pi}{4}$. Thus, polar coordinates are $\left(4, \frac{\pi}{4}\right)$.

37. $(x, y) = (3, 4)$. Since $r^2 = x^2 + y^2$, we have $r^2 = 3^2 + 4^2 = 25$, so $r = 5$. Now $\tan\theta = \frac{y}{x} = \frac{4}{3}$, so, since the point is in the first quadrant, $\theta = \tan^{-1}\frac{4}{3}$. Thus, polar coordinates are $\left(5, \tan^{-1}\frac{4}{3}\right)$.

39. $(x, y) = (-6, 0)$. $r^2 = \left(-6^2\right) = 36$, so $r = 6$. Now $\tan\theta = \frac{y}{x} = 0$, so since the point is on the negative x-axis, $\theta = \pi$. Thus, polar coordinates are $(6, \pi)$.

41. $x = y \quad \Leftrightarrow \quad r\cos\theta = r\sin\theta \quad \Leftrightarrow \quad \tan\theta = 1$, and so $\theta = \frac{\pi}{4}$.

43. $y = x^2$. We substitute and then solve for r: $r\sin\theta = (r\cos\theta)^2 = r^2\cos^2\theta \quad \Leftrightarrow \quad \sin\theta = r\cos^2\theta \quad \Leftrightarrow$

$r = \dfrac{\sin\theta}{\cos^2\theta} = \tan\theta\sec\theta$.

45. $x = 4$. We substitute and then solve for r: $r\cos\theta = 4 \quad \Leftrightarrow \quad r = \dfrac{4}{\cos\theta} = 4\sec\theta$.

47. $r = 7$. But $r^2 = x^2 + y^2$, so $x^2 + y^2 = r^2 = 49$. Hence, the equivalent equation in rectangular coordinates is $x^2 + y^2 = 49$.

49. $r\cos\theta = 6$. But $x = r\cos\theta$, and so $x = 6$ is the equation.

51. $r^2 = \tan\theta$. Substituting $r^2 = x^2 + y^2$ and $\tan\theta = \frac{y}{x}$, we get $x^2 + y^2 = \frac{y}{x}$.

53. $r = \dfrac{1}{\sin\theta - \cos\theta} \quad \Rightarrow \quad r(\sin\theta - \cos\theta) = 1 \quad \Leftrightarrow \quad r\sin\theta - r\cos\theta = 1$, and since $r\cos\theta = x$ and $r\sin\theta = y$, we get $y - x = 1$.

55. $r = 1 + \cos\theta$. If we multiply both sides of this equation by r we get $r^2 = r + r\cos\theta$. Thus $r^2 - r\cos\theta = r$, and squaring both sides gives $\left(r^2 - r\cos\theta\right)^2 = r^2 \quad \Leftrightarrow \quad \left(x^2 + y^2 - x\right)^2 = x^2 + y^2$

57. $r = 2\sec\theta \quad \Leftrightarrow \quad r = 2\cdot\dfrac{1}{\cos\theta} \quad \Leftrightarrow \quad r\cos\theta = 2 \quad \Leftrightarrow \quad x = 2$.

59. $\sec\theta = 2 \quad \Leftrightarrow \quad \cos\theta = \frac{1}{2} \quad \Leftrightarrow \quad \theta = \pm\frac{\pi}{3} \quad \Leftrightarrow \quad \tan\theta = \pm\sqrt{3} \quad \Leftrightarrow \quad \frac{y}{x} = \pm\sqrt{3} \quad \Leftrightarrow \quad y = \pm\sqrt{3}x$.

61. (a) In rectangular coordinates, the points (r_1, θ_1) and (r_2, θ_2) are $(x_1, y_1) = (r_1\cos\theta_1, r_1\sin\theta_1)$ and $(x_2, y_2) = (r_2\cos\theta_2, r_2\sin\theta_2)$. Then, the distance between the points is

$$\begin{aligned} D &= \sqrt{(x_1 - x_2)^2 + (y_1 - y_2)^2} = \sqrt{(r_1\cos\theta_1 - r_2\cos\theta_2)^2 + (r_1\sin\theta_1 - r_2\sin\theta_2)^2} \\ &= \sqrt{r_1^2\left(\cos^2\theta_1 + \sin^2\theta_1\right) + r_2^2\left(\cos^2\theta_2 + \sin^2\theta_2\right) - 2r_1 r_2\left(\cos\theta_1\cos\theta_2 + \sin\theta_1\sin\theta_2\right)} \\ &= \sqrt{r_1^2 + r_2^2 - 2r_1 r_2\cos(\theta_2 - \theta_1)} \end{aligned}$$

(b) The distance between the points $\left(3, \frac{3\pi}{4}\right)$ and $\left(-1, \frac{7\pi}{6}\right)$ is

$$D = \sqrt{3^2 + (-1)^2 - 2(3)(-1)\cos\left(\frac{7\pi}{6} - \frac{3\pi}{4}\right)} = \sqrt{9 + 1 + 6\cos\frac{5\pi}{12}} \approx 3.40$$

Graphs of Polar Equations

1. VI

3. II

5. I

7. Polar axis: $2 - \sin(-\theta) = 2 + \sin\theta \neq r$, so the graph is not symmetric about the polar axis.

Pole: $2 - \sin(\theta + \pi) = 2 - (\sin\pi\cos\theta + \cos\pi\sin\theta) = 2 - (-\sin\theta) = 2 + \sin\theta \neq r$, so the graph is not symmetric about the pole.

Line $\theta = \frac{\pi}{2}$: $2 - \sin(\pi - \theta) = 2 - (\sin\pi\cos\theta - \cos\pi\sin\theta) = 2 - \sin\theta = r$, so the graph is symmetric about $\theta = \frac{\pi}{2}$.

9. Polar axis: $3\sec(-\theta) = 3\sec\theta = r$, so the graph is symmetric about the polar axis.

Pole: $3\sec(\theta + \pi) = \dfrac{3}{\cos(\theta + \pi)} = \dfrac{1}{\cos\pi\cos\theta - \sin\pi\sin\theta} = \dfrac{3}{-\cos\theta} = -3\sec\theta \neq r$, so the graph is not symmetric about the pole.

Line $\theta = \frac{\pi}{2}$: $3\sec(\pi - \theta) = \dfrac{3}{\cos(\pi - \theta)} = \dfrac{1}{\cos\pi\cos\theta + \sin\pi\sin\theta} = \dfrac{3}{-\cos\theta} = -3\sec\theta \neq r$, so the graph is not symmetric about $\theta = \frac{\pi}{2}$.

11. Polar axis: $\dfrac{4}{3 - 2\sin(-\theta)} = \dfrac{4}{3 + 2\sin\theta} \neq r$, so the graph is not symmetric about the polar axis.

Pole: $\dfrac{4}{3 - 2\sin(\theta + \pi)} = \dfrac{4}{3 - 2(\sin\pi\cos\theta + \cos\pi\sin\theta)} = \dfrac{4}{3 - 2(-\sin\theta)} = \dfrac{4}{3 + 2\sin\theta} \neq r$, so the graph is not symmetric about the pole.

Line $\theta = \frac{\pi}{2}$: $\dfrac{4}{3 - 2\sin(\pi - \theta)} = \dfrac{4}{3 - 2(\sin\pi\cos\theta - \cos\pi\sin\theta)} = \dfrac{4}{3 - 2\sin\theta} = r$, so the graph is symmetric about $\theta = \frac{\pi}{2}$.

13. Polar axis: $4\cos 2(-\theta) = 4\cos 2\theta = r^2$, so the graph is symmetric about the polar axis.

Pole: $(-r)^2 = r^2$, so the graph is symmetric about the pole.

Line $\theta = \frac{\pi}{2}$: $4\cos 2(\pi - \theta) = 4\cos(2\pi - 2\theta) = 4\cos(-2\theta) = 4\cos 2\theta = r^2$, so the graph is symmetric about $\theta = \frac{\pi}{2}$.

15. $r = 2$. Circle.

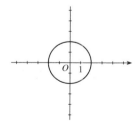

17. $\theta = -\frac{\pi}{2}$. Line.

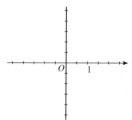

19. $r = 6\sin\theta$. Circle.

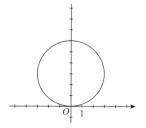

21. $r = -2\cos\theta$. Circle.

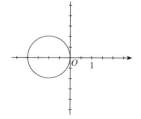

23. $r = 2 - 2\cos\theta$. Cardioid.

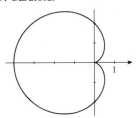

25. $r = -3\left(1 + \sin\theta\right)$. Cardioid.

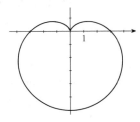

27. $r = \theta, \theta \geq 0$

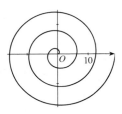

29. $r = \sin 2\theta$

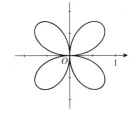

31. $r^2 = \cos 2\theta$

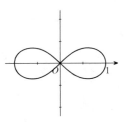

33. $r = 2 + \sin\theta$

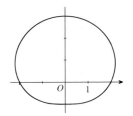

35. $r = 2 + \sec\theta$

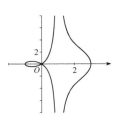

37. $r = \cos\left(\dfrac{\theta}{2}\right), \theta \in [0, 4\pi]$

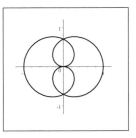

39. $r = 1 + 2\sin\left(\dfrac{\theta}{2}\right), \theta \in [0, 4\pi]$

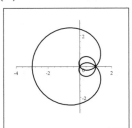

41. $r = 1 + \sin n\theta$. The number of loops is n.

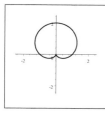 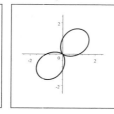 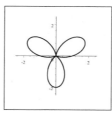 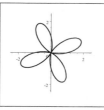 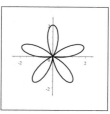

$n = 1$ $n = 2$ $n = 3$ $n = 4$ $n = 5$

43. The graph of $r = \sin\left(\dfrac{\theta}{2}\right)$ is IV, since the graph must contain the points $(0,0)$, $\left(\dfrac{1}{\sqrt{2}}, \dfrac{\pi}{2}\right)$, $(1, \pi)$, and so on.

45. The graph of $r = \theta \sin \theta$ is III, since for $\theta = \dfrac{\pi}{2}, \dfrac{5\pi}{2}, \dfrac{7\pi}{2}, \ldots$ the values of r are also $\dfrac{\pi}{2}, \dfrac{5\pi}{2}, \dfrac{7\pi}{2}, \ldots$. Thus the graph must cross the vertical axis at an infinite number of points.

47. $\left(x^2 + y^2\right)^3 = 4x^2y^2 \quad \Leftrightarrow \quad \left(r^2\right)^3 = 4\left(r\cos\theta\right)^2\left(r\sin\theta\right)^2 \quad \Leftrightarrow$
$r^6 = 4r^4 \cos^2\theta \sin^2\theta \quad \Leftrightarrow \quad r^2 = 4\cos^2\theta \sin^2\theta \quad \Leftrightarrow$
$r = 2\cos\theta\sin\theta = \sin 2\theta$. The equation is $r = \sin 2\theta$, a rose.

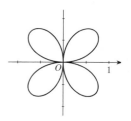

49. $\left(x^2 + y^2\right)^2 = x^2 - y^2 \quad \Leftrightarrow \quad \left(r^2\right)^2 = \left(r\cos\theta\right)^2 - \left(r\sin\theta\right)^2 \quad \Leftrightarrow$
$r^4 = r^2 \cos^2\theta - r^2 \sin^2\theta \quad \Leftrightarrow \quad r^4 = r^2\left(\cos^2\theta - \sin^2\theta\right) \quad \Leftrightarrow$
$r^2 = \cos^2\theta - \sin^2\theta = \cos 2\theta$. The graph is $r^2 = \cos 2\theta$, a leminiscate.

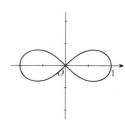

51. $r = a\cos\theta + b\sin\theta \quad \Leftrightarrow \quad r^2 = ar\cos\theta + br\sin\theta \quad \Leftrightarrow \quad x^2 + y^2 = ax + by \quad \Leftrightarrow \quad x^2 - ax + y^2 - by = 0 \quad \Leftrightarrow$
$x^2 - ax + \frac{1}{4}a^2 + y^2 - by + \frac{1}{4}b^2 = \frac{1}{4}a^2 + \frac{1}{4}b^2 \quad \Leftrightarrow$
$\left(x - \frac{1}{2}a\right)^2 + \left(y - \frac{1}{2}b\right)^2 = \frac{1}{4}\left(a^2 + b^2\right)$. Thus, in rectangular coordinates the center is $\left(\frac{1}{2}a, \frac{1}{2}b\right)$ and the radius is $\frac{1}{2}\sqrt{a^2 + b^2}$.

53. (a)

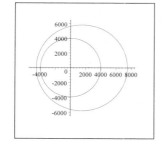

At $\theta = 0$, the satellite is at the "rightmost" point in its orbit, $(5625, 0)$. As θ increases, it travels counterclockwise. Note that it is moving fastest when $\theta = \pi$.

(b) The satellite is closest to earth when $\theta = \pi$. Its height above the earth's surface at this point is
$22500/\left(4 - \cos\pi\right) - 3960 = 4500 - 3960 = 540$ mi.

55. The graphs of $r = 1 + \sin\left(\theta - \frac{\pi}{6}\right)$ and $r = 1 + \sin\left(\theta - \frac{\pi}{3}\right)$ have the same shape as $r = 1 + \sin\theta$, rotated through angles of $\frac{\pi}{6}$ and $\frac{\pi}{3}$, respectively. Similarly, the graph of $r = f(\theta - \alpha)$ is the graph of $r = f(\theta)$ rotated by the angle α.

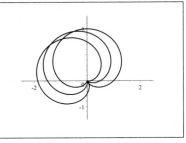

57. $y = 2 \quad \Leftrightarrow \quad r\sin\theta = 2 \quad \Leftrightarrow \quad r = 2\csc\theta$. The rectangular coordinate system gives the simpler equation here. It is easier to study lines in rectangular coordinates.

9.3 Polar Form of Complex Numbers; DeMoivre's Theorem

1. $|4i| = \sqrt{0^2 + 4^2} = 4$

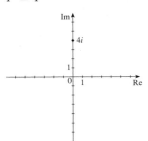

3. $|-2| = \sqrt{4 + 0} = 2$

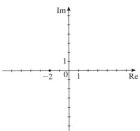

5. $|5 + 2i| = \sqrt{5^2 + 2^2} = \sqrt{29}$

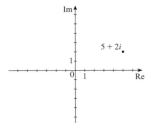

7. $\left|\sqrt{3} + i\right| = \sqrt{3 + 1} = 2$

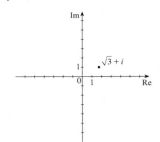

9. $\left|\frac{3 + 4i}{5}\right| = \sqrt{\frac{9}{25} + \frac{16}{25}} = 1$

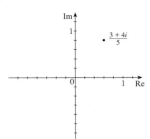

11. $z = 1 + i,\, 2z = 2 + 2i,\, -z = -1 - i,\, \frac{1}{2}z = \frac{1}{2} + \frac{1}{2}i$

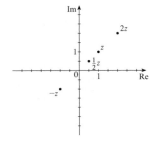

13. $z = 8 + 2i, \overline{z} = 8 - 2i$

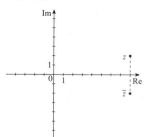

15. $z_1 = 2 - i$, $z_2 = 2 + i$, $z_1 + z_2 = 2 - i + 2 + i = 4$,
$z_1 z_2 = (2 - i)(2 + i) = 4 - i^2 = 5$

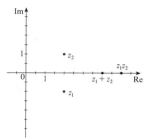

17. $\{z = a + bi \mid a \leq 0, b \geq 0\}$

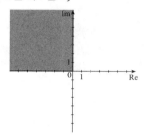

19. $\{z \mid |z| = 3\}$

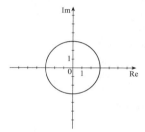

21. $\{z \mid |z| < 2\}$

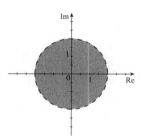

23. $\{z = a + bi \mid a + b < 2\}$

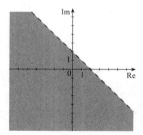

25. $1 + i$. Then $\tan\theta = \frac{1}{1} = 1$ with θ in quadrant I $\Rightarrow$ $\theta = \frac{\pi}{4}$, and $r = \sqrt{1^2 + 1^2} = \sqrt{2}$. Hence, $1 + i = \sqrt{2}\left(\cos\frac{\pi}{4} + i\sin\frac{\pi}{4}\right)$.

27. $\sqrt{2} - \sqrt{2}i$. Then $\tan\theta = \frac{\sqrt{2}}{\sqrt{2}} = -1$ with θ in quadrant IV $\Rightarrow$ $\theta = \frac{7\pi}{4}$, and $r = \sqrt{2 + 2} = 2$. Hence, $\sqrt{2} - \sqrt{2}i = 2\left(\cos\frac{7\pi}{4} + i\sin\frac{7\pi}{4}\right)$.

29. $2\sqrt{3} - 2i$. Then $\tan\theta = \frac{-2}{2\sqrt{3}} = -\frac{1}{\sqrt{3}}$ with θ in quadrant IV $\Rightarrow$ $\theta = \frac{11\pi}{6}$, and $r = \sqrt{12 + 4} = 4$. Hence, $2\sqrt{3} - 2i = 4\left(\cos\frac{11\pi}{6} + i\sin\frac{11\pi}{6}\right)$.

31. $-3i$. Then $\theta = \frac{3\pi}{2}$, and $r = \sqrt{0 + 9} = 3$. Hence, $3i = 3\left(\cos\frac{3\pi}{2} + i\sin\frac{3\pi}{2}\right)$.

33. $5 + 5i$. Then $\tan\theta = \frac{5}{5} = 1$ with θ in quadrant I $\Rightarrow$ $\theta = \frac{\pi}{4}$, and $r = \sqrt{25 + 25} = 5\sqrt{2}$. Hence, $5 + 5i = 5\sqrt{2}\left(\cos\frac{\pi}{4} + i\sin\frac{\pi}{4}\right)$.

35. $4\sqrt{3} - 4i$. Then $\tan\theta = \frac{-4}{4\sqrt{3}} = -\frac{1}{\sqrt{3}}$ with θ in quadrant IV $\Rightarrow$ $\theta = \frac{11\pi}{6}$, and $r = \sqrt{48 + 16} = 8$. Hence, $4\sqrt{3} - 4i = 8\left(\cos\frac{11\pi}{6} + i\sin\frac{11\pi}{6}\right)$.

37. -20. Then $\theta = \pi$, and $r = 20$. Hence, $-20 = 20\left(\cos\pi + i\sin\pi\right)$.

39. $3 + 4i$. Then $\tan\theta = \frac{4}{3}$ with θ in quadrant I $\Rightarrow$ $\theta = \tan^{-1}\frac{4}{3}$, and $r = \sqrt{9 + 16} = 5$. Hence, $3 + 4i = 5\left[\cos\left(\tan^{-1}\frac{4}{3}\right) + i\sin\left(\tan^{-1}\frac{4}{3}\right)\right]$.

41. $3i(1+i) = -3 + 3i$. Then $\tan\theta = \frac{3}{-3} = -1$ with θ in quadrant II $\Rightarrow$ $\theta = \frac{3\pi}{4}$, and $r = \sqrt{9+9} = 3\sqrt{2}$. Hence, $3i(1+i) = 3\sqrt{2}\left(\cos\frac{3\pi}{4} + i\sin\frac{3\pi}{4}\right)$.

43. $4\left(\sqrt{3} + i\right) = 4\sqrt{3} + 4i$. Then $\tan\theta = \frac{4}{4\sqrt{3}} = \frac{1}{\sqrt{3}}$ with θ in quadrant I $\Rightarrow$ $\theta = \frac{\pi}{6}$, and $r = \sqrt{48+16} = 8$. Hence, $4\left(\sqrt{3}+i\right) = 8\left(\cos\frac{\pi}{6} + i\sin\frac{\pi}{6}\right)$.

45. $2 + i$. Then $\tan\theta = \frac{1}{2}$ with θ in quadrant I $\Rightarrow$ $\theta = \tan^{-1}\frac{1}{2}$, and $r = \sqrt{4+1} = \sqrt{5}$. Hence, $2 + i = \sqrt{5}\left[\cos\left(\tan^{-1}\frac{1}{2}\right) + i\sin\left(\tan^{-1}\frac{1}{2}\right)\right]$.

47. $\sqrt{2} + \sqrt{2}i$. Then $\tan\theta = \frac{\sqrt{2}}{\sqrt{2}} = 1$ with θ in quadrant I $\Rightarrow$ $\theta = \frac{\pi}{4}$, and $r = \sqrt{2+2} = 2$. Hence, $2 + \sqrt{2}i = 2\left(\cos\frac{\pi}{4} + i\sin\frac{\pi}{4}\right)$.

49. $z_1 = \cos\pi + i\sin\pi$, $z_2 = \cos\frac{\pi}{3} + i\sin\frac{\pi}{3}$, $z_1z_2 = \cos\left(\pi + \frac{\pi}{3}\right) + i\sin\left(\pi + \frac{\pi}{3}\right) = \cos\frac{4\pi}{3} + i\sin\frac{4\pi}{3}$, $z_1/z_2 = \cos\left(\pi - \frac{\pi}{3}\right) + i\sin\left(\pi - \frac{\pi}{3}\right) = \cos\frac{2\pi}{3} + i\sin\frac{2\pi}{3}$

51. $z_1 = 3\left(\cos\frac{\pi}{6} + i\sin\frac{\pi}{6}\right)$, $z_2 = 5\left(\cos\frac{4\pi}{3} + i\sin\frac{4\pi}{3}\right)$,
$z_1z_2 = 3\cdot 5\left[\cos\left(\frac{\pi}{6} + \frac{4\pi}{3}\right) + i\sin\left(\frac{\pi}{6} + \frac{4\pi}{3}\right)\right] = 15\left(\cos\frac{9\pi}{6} + i\sin\frac{9\pi}{6}\right) = 15\left(\cos\frac{3\pi}{2} + i\sin\frac{3\pi}{2}\right)$,
$z_1/z_2 = \frac{3}{5}\left[\cos\left(\frac{\pi}{6} - \frac{4\pi}{3}\right) + i\sin\left(\frac{\pi}{6} - \frac{4\pi}{3}\right)\right] = \frac{3}{5}\left[\cos\left(-\frac{7\pi}{6}\right) + i\sin\left(-\frac{7\pi}{6}\right)\right] = \frac{3}{5}\left[\cos\left(\frac{7\pi}{6}\right) - i\sin\left(\frac{7\pi}{6}\right)\right]$

53. $z_1 = 4\left(\cos 120° + i\sin 120°\right)$, $z_2 = 2\left(\cos 30° + i\sin 30°\right)$,
$z_1z_2 = 4\cdot 2\left[\cos\left(120° + 30°\right) + i\sin\left(120° + 30°\right)\right] = 8\left(\cos 150° + i\sin 150°\right)$,
$z_1/z_2 = \frac{4}{2}\left[\cos\left(120° - 30°\right) + i\sin\left(120° - 30°\right)\right] = 2\left(\cos 90° + i\sin 90°\right)$

55. $z_1 = 4\left(\cos 200° + i\sin 200°\right)$, $z_2 = 25\left(\cos 150° + i\sin 150°\right)$,
$z_1z_2 = 4\cdot 25\left[\cos\left(200° + 150°\right) + i\sin\left(200° + 150°\right)\right] = 100\left(\cos 350° + i\sin 350°\right)$,
$z_1/z_2 = \frac{4}{25}\left[\cos\left(200° - 150°\right) + i\sin\left(200° - 150°\right)\right] = \frac{4}{25}\left(\cos 50° + i\sin 50°\right)$

57. $z_1 = \sqrt{3} + i$, so $\tan\theta_1 = \frac{1}{\sqrt{3}}$ with θ_1 in quadrant I $\Rightarrow$ $\theta_1 = \frac{\pi}{6}$, and $r_1 = \sqrt{3+1} = 2$.
$z_2 = 1 + \sqrt{3}i$, so $\tan\theta_2 = \sqrt{3}$ with θ_2 in quadrant I $\Rightarrow$ $\theta_2 = \frac{\pi}{3}$, and $r_1 = \sqrt{1+3} = 2$.
Hence, $z_1 = 2\left(\cos\frac{\pi}{6} + i\sin\frac{\pi}{6}\right)$ and $z_2 = 2\left(\cos\frac{\pi}{3} + i\sin\frac{\pi}{3}\right)$.
Thus, $z_1z_2 = 2\cdot 2\left[\cos\left(\frac{\pi}{6} + \frac{\pi}{3}\right) + i\sin\left(\frac{\pi}{6} + \frac{\pi}{3}\right)\right] = 4\left(\cos\frac{\pi}{2} + i\sin\frac{\pi}{2}\right)$,
$z_1/z_2 = \frac{2}{2}\left[\cos\left(\frac{\pi}{6} - \frac{\pi}{3}\right) + i\sin\left(\frac{\pi}{6} - \frac{\pi}{3}\right)\right] = \cos\left(-\frac{\pi}{6}\right) + i\sin\left(-\frac{\pi}{6}\right) = \cos\frac{\pi}{6} - i\sin\frac{\pi}{6}$, and
$1/z_1 = \frac{1}{2}\left[\cos\left(-\frac{\pi}{6}\right) + i\sin\left(-\frac{\pi}{6}\right)\right] = \frac{1}{2}\left(\cos\frac{\pi}{6} - i\sin\frac{\pi}{6}\right)$.

59. $z_1 = 2\sqrt{3} - 2i$, so $\tan\theta_1 = \frac{-2}{2\sqrt{3}} = -\frac{1}{\sqrt{3}}$ with θ_1 in quadrant IV $\Rightarrow$ $\theta_1 = \frac{11\pi}{6}$, and $r_1 = \sqrt{12+4} = 4$.
$z_2 = -1 + i$, so $\tan\theta_2 = -1$ with θ_2 in quadrant II $\Rightarrow$ $\theta_2 = \frac{3\pi}{4}$, and $r_2 = \sqrt{1+1} = \sqrt{2}$.
Hence, $z_1 = 4\left(\cos\frac{11\pi}{6} + i\sin\frac{11\pi}{6}\right)$ and $z_2 = \sqrt{2}\left(\cos\frac{3\pi}{4} + i\sin\frac{3\pi}{4}\right)$.
Thus, $z_1z_2 = 4\cdot\sqrt{2}\left[\cos\left(\frac{11\pi}{6} + \frac{3\pi}{4}\right) + i\sin\left(\frac{11\pi}{6} + \frac{3\pi}{4}\right)\right] = 4\sqrt{2}\left(\cos\frac{7\pi}{12} + i\sin\frac{7\pi}{12}\right)$,
$z_1/z_2 = \frac{4}{\sqrt{2}}\left[\cos\left(\frac{11\pi}{6} - \frac{3\pi}{4}\right) + i\sin\left(\frac{11\pi}{6} - \frac{3\pi}{4}\right)\right] = 2\sqrt{2}\left(\cos\frac{13\pi}{12} + i\sin\frac{13\pi}{12}\right)$, and
$1/z_1 = \frac{1}{4}\left(\cos\left(-\frac{11\pi}{6}\right) + i\sin\left(-\frac{11\pi}{6}\right)\right) = \frac{1}{4}\left(\cos\frac{11\pi}{6} - i\sin\frac{11\pi}{6}\right)$.

61. $z_1 = 5 + 5i$, so $\tan\theta_1 = \frac{5}{5} = 1$ with θ_1 in quadrant I $\Rightarrow$ $\theta_1 = \frac{\pi}{4}$, and $r_1 = \sqrt{25+25} = 5\sqrt{2}$.
$z_2 = 4$, so $\theta_2 = 0$, and $r_2 = 4$.
Hence, $z_1 = 5\sqrt{2}\left(\cos\frac{\pi}{4} + i\sin\frac{\pi}{4}\right)$ and $z_2 = 4\left(\cos 0 + i\sin 0\right)$.
Thus, $z_1z_2 = 5\sqrt{2}\cdot 4\left[\cos\left(\frac{\pi}{4} + 0\right) + i\sin\left(\frac{\pi}{4} + 0\right)\right] = 20\sqrt{2}\left(\cos\frac{\pi}{4} + i\sin\frac{\pi}{4}\right)$, $z_1/z_2 = \frac{5\sqrt{2}}{4}\left(\cos\frac{\pi}{4} + i\sin\frac{\pi}{4}\right)$, and
$1/z_1 = \frac{1}{5\sqrt{2}}\left(\cos\left(-\frac{\pi}{4}\right) + i\sin\left(-\frac{\pi}{4}\right)\right) = \frac{\sqrt{2}}{10}\left(\cos\frac{\pi}{4} - i\sin\frac{\pi}{4}\right)$.

63. $z_1 = -20$, so $\theta_1 = \pi$, and $r_1 = 20$.

$z_2 = \sqrt{3} + i$, so $\tan \theta_2 = \frac{1}{\sqrt{3}}$ with θ_2 in quadrant I $\Rightarrow$ $\theta_2 = \frac{\pi}{6}$, and $r_2 = \sqrt{3+1} = 2$.

Hence, $z_1 = 20 \left(\cos \pi + i \sin \pi \right)$ and $z_2 = 2 \left(\cos \frac{\pi}{6} + i \sin \frac{\pi}{6} \right)$.

Thus, $z_1 z_2 = 20 \cdot 2 \left[\cos \left(\pi + \frac{\pi}{6} \right) + i \sin \left(\pi + \frac{\pi}{6} \right) \right] = 40 \left(\cos \frac{7\pi}{6} + i \sin \frac{7\pi}{6} \right)$,

$z_1 / z_2 = \frac{20}{2} \left[\cos \left(\pi - \frac{\pi}{6} \right) + i \sin \left(\pi - \frac{\pi}{6} \right) \right] = 10 \left(\cos \frac{5\pi}{6} + i \sin \frac{5\pi}{6} \right)$, and

$1 / z_1 = \frac{1}{20} \left[\cos \left(-\pi \right) + i \sin \left(-\pi \right) \right] = \frac{1}{20} \left(\cos \pi - i \sin \pi \right)$.

65. From Exercise 25, $1 + i = \sqrt{2} \left(\cos \frac{\pi}{4} + i \sin \frac{\pi}{4} \right)$.   Thus,

$(1+i)^{20} = \left(\sqrt{2} \right)^{20} \left[\cos 20 \left(\frac{\pi}{4} \right) + i \sin 20 \left(\frac{\pi}{4} \right) \right] = \left(2^{1/2} \right)^{20} \left(\cos 5\pi + i \sin 5\pi \right) = 2^{10} \left(-1 + 0i \right) = -1024$.

67. $r = \sqrt{12 + 4} = 4$ and $\tan \theta = \frac{2}{2\sqrt{3}} = \frac{1}{\sqrt{3}}$ $\Rightarrow$ $\theta = \frac{\pi}{6}$. Thus, $2\sqrt{3} + 2i = 4 \left(\cos \frac{\pi}{6} + i \sin \frac{\pi}{6} \right)$.   So

$\left(2\sqrt{3} + 2i \right)^5 = 4^5 \left(\cos \frac{5\pi}{6} + i \sin \frac{5\pi}{6} \right) = 1024 \left(-\frac{\sqrt{3}}{2} + \frac{1}{2} i \right) = 512 \left(-\sqrt{3} + i \right)$.

69. $r = \sqrt{\frac{1}{2} + \frac{1}{2}} = 1$ and $\tan \theta = 1$ $\Rightarrow$ $\theta = \frac{\pi}{4}$. Thus $\frac{\sqrt{2}}{2} + \frac{\sqrt{2}}{2} i = \cos \frac{\pi}{4} + i \sin \frac{\pi}{4}$. Therefore,

$\left(\frac{\sqrt{2}}{2} + \frac{\sqrt{2}}{2} i \right)^{12} = \cos 12 \left(\frac{\pi}{4} \right) + i \sin 12 \left(\frac{\pi}{4} \right) = \cos 3\pi + i \sin 3\pi = -1$.

71. $r = \sqrt{4+4} = 4\sqrt{2}$ and $\tan \theta = -1$ with θ in quadrant IV $\Rightarrow$ $\theta = \frac{7\pi}{4}$. Thus $2 - 2i = 2\sqrt{2} \left(\cos \frac{7\pi}{4} + i \sin \frac{7\pi}{4} \right)$, so

$(2 - 2i)^8 = \left(2\sqrt{2} \right)^8 \left(\cos 14\pi + i \sin 14\pi \right) = 4096 \left(1 - 0i \right) = 4096$.

73. $r = \sqrt{1+1} = \sqrt{2}$ and $\tan \theta = 1$ with θ in quadrant III $\Rightarrow$ $\theta = \frac{5\pi}{4}$. Thus $-1 - i = \sqrt{2} \left(\cos \frac{5\pi}{4} + i \sin \frac{5\pi}{4} \right)$, so

$(-1 - i)^7 = \left(\sqrt{2} \right)^7 \left(\cos \frac{35\pi}{4} + i \sin \frac{35\pi}{4} \right) = 8\sqrt{2} \left(\cos \frac{3\pi}{4} + i \sin \frac{3\pi}{4} \right) = 8\sqrt{2} \left(\frac{1}{\sqrt{2}} - i \frac{1}{\sqrt{2}} \right) = 8 \left(-1 + i \right)$.

75. $r = \sqrt{12 + 4} = 4$ and $\tan \theta = \frac{2}{2\sqrt{3}} = \frac{1}{\sqrt{3}}$ $\Rightarrow$ $\theta = \frac{\pi}{6}$. Thus $2\sqrt{3} + 2i = 4 \left(\cos \frac{\pi}{6} + i \sin \frac{\pi}{6} \right)$, so

$\left(2\sqrt{3} + 2i \right)^{-5} = \left(\frac{1}{4} \right)^5 \left(\cos \frac{-5\pi}{6} + i \sin \frac{-5\pi}{6} \right) = \frac{1}{1024} \left(-\frac{\sqrt{3}}{2} - \frac{1}{2} i \right) = \frac{1}{2048} \left(-\sqrt{3} - i \right)$

77. $r = \sqrt{48 + 16} = 8$ and $\tan \theta = \frac{4}{4\sqrt{3}} = \frac{1}{\sqrt{3}} \Rightarrow \theta = \frac{\pi}{6}$. Thus

$4\sqrt{3} + 4i = 8 \left(\cos \frac{\pi}{6} + i \sin \frac{\pi}{6} \right)$. So,

$\left(4\sqrt{3} + 4i \right)^{1/2} = \sqrt{8} \left[\cos \left(\frac{\pi/6 + 2k\pi}{2} \right) + i \sin \left(\frac{\pi/6 + 2k\pi}{2} \right) \right]$ for $k = 0, 1$.

Thus the two roots are $w_0 = 2\sqrt{2} \left(\cos \frac{\pi}{12} + i \sin \frac{\pi}{12} \right)$ and

$w_1 = 2\sqrt{2} \left(\cos \frac{13\pi}{12} + i \sin \frac{13\pi}{12} \right)$.

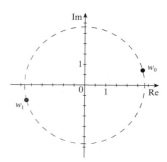

79. $-81i = 81 \left(\cos \frac{3\pi}{2} + i \sin \frac{3\pi}{2} \right)$. Thus,

$(-81i)^{1/4} = 81^{1/4} \left[\cos \left(\frac{3\pi/2 + 2k\pi}{4} \right) + i \sin \left(\frac{3\pi/2 + 2k\pi}{4} \right) \right]$ for $k = 0, 1$,

2, 3. The four roots are $w_0 = 3 \left(\cos \frac{3\pi}{8} + i \sin \frac{3\pi}{8} \right)$, $w_1 = 3 \left(\cos \frac{7\pi}{8} + i \sin \frac{7\pi}{8} \right)$,

$w_2 = 3 \left(\cos \frac{11\pi}{8} + i \sin \frac{11\pi}{8} \right)$, and $w_3 = 3 \left(\cos \frac{15\pi}{8} + i \sin \frac{15\pi}{8} \right)$.

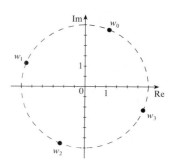

81. $1 = \cos 0 + i \sin 0$. Thus, $1^{1/8} = \cos \dfrac{2k\pi}{8} + i \sin \dfrac{2k\pi}{8}$, for $k = 0, 1, 2, 3, 4, 5, 6,$

7. So the eight roots are $w_0 = \cos 0 + i \sin 0 = 1$,

$w_1 = \cos \frac{\pi}{4} + i \sin \frac{\pi}{4} = \frac{\sqrt{2}}{2} + i \frac{\sqrt{2}}{2}$, $w_2 = \cos \frac{\pi}{2} + i \sin \frac{\pi}{2} = i$,

$w_3 = \cos \frac{3\pi}{4} + i \sin \frac{3\pi}{4} = -\frac{\sqrt{2}}{2} + i \frac{\sqrt{2}}{2}$, $w_4 = \cos \pi + i \sin \pi = -1$,

$w_5 = \cos \frac{5\pi}{4} + i \sin \frac{5\pi}{4} = -\frac{\sqrt{2}}{2} - i \frac{\sqrt{2}}{2}$, $w_6 = \cos \frac{3\pi}{2} + i \sin \frac{3\pi}{2} = -i$, and

$w_7 = \cos \frac{7\pi}{4} + i \sin \frac{7\pi}{4} = \frac{\sqrt{2}}{2} - i \frac{\sqrt{2}}{2}$.

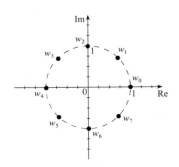

83. $i = \cos \frac{\pi}{2} + i \sin \frac{\pi}{2}$, so $i^{1/3} = \cos \left(\dfrac{\pi/2 + 2k\pi}{3} \right) + i \sin \left(\dfrac{\pi/2 + 2k\pi}{3} \right)$ for

$k = 0, 1, 2$. Thus the three roots are $w_0 = \cos \frac{\pi}{6} + i \sin \frac{\pi}{6} = \frac{\sqrt{3}}{2} + \frac{1}{2}i$,

$w_1 = \cos \frac{5\pi}{6} + i \sin \frac{5\pi}{6} = -\frac{\sqrt{3}}{2} + \frac{1}{2}i$, and $w_2 = \cos \frac{3\pi}{2} + i \sin \frac{3\pi}{2} = -i$.

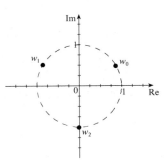

85. $-1 = \cos \pi + i \sin \pi$. Then $(-1)^{1/4} = \cos \left(\dfrac{\pi + 2k\pi}{4} \right) + i \sin \left(\dfrac{\pi + 2k\pi}{4} \right)$ for

$k = 0, 1, 2, 3$. So the four roots are $w_0 = \cos \frac{\pi}{4} + i \sin \frac{\pi}{4} = \frac{\sqrt{2}}{2} + i \frac{\sqrt{2}}{2}$,

$w_1 = \cos \frac{3\pi}{4} + i \sin \frac{3\pi}{4} = -\frac{\sqrt{2}}{2} + i \frac{\sqrt{2}}{2}$, $w_2 = \cos \frac{5\pi}{4} + i \sin \frac{5\pi}{4} = -\frac{\sqrt{2}}{2} - i \frac{\sqrt{2}}{2}$,

and $w_3 = \cos \frac{7\pi}{4} + i \sin \frac{7\pi}{4} = \frac{\sqrt{2}}{2} - i \frac{\sqrt{2}}{2}$.

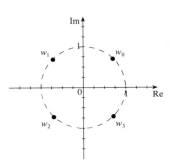

87. $z^4 + 1 = 0 \quad \Leftrightarrow \quad z = (-1)^{1/4} = \frac{\sqrt{2}}{2}(\pm 1 \pm i)$ (from Exercise 85)

89. $z^3 - 4\sqrt{3} - 4i = 0 \quad \Leftrightarrow \quad z = \left(4\sqrt{3} + 4i\right)^{1/3}$. Since $4\sqrt{3} + 4i = 8\left(\cos \frac{\pi}{6} + i \sin \frac{\pi}{6}\right)$,

$\left(4\sqrt{3} + 4i\right)^{1/3} = 8^{1/3} \left[\cos \left(\dfrac{\pi/6 + 2k\pi}{3} \right) + i \sin \left(\dfrac{\pi/6 + 2k\pi}{3} \right) \right]$, for $k = 0, 1, 2$. Thus the three roots are

$z = 2\left(\cos \frac{\pi}{18} + i \sin \frac{\pi}{18}\right)$, $z = 2\left(\cos \frac{13\pi}{18} + i \sin \frac{13\pi}{8}\right)$, and $z = 2\left(\cos \frac{25\pi}{18} + i \sin \frac{25\pi}{18}\right)$.

91. $z^3 + 1 = -i \quad \Rightarrow \quad z = (-1 - i)^{1/3}$. Since $-1 - i = \sqrt{2}\left(\cos \frac{5\pi}{4} + i \sin \frac{5\pi}{4}\right)$,

$z = (-1 - i)^{1/3} = 2^{1/6} \left[\cos \left(\dfrac{5\pi/4 + 2k\pi}{3} \right) + i \sin \left(\dfrac{5\pi/4 + 2k\pi}{3} \right) \right]$ for $k = 0, 1, 2$. Thus the three solutions to this

equation are $z = 2^{1/6}\left(\cos \frac{5\pi}{12} + i \sin \frac{5\pi}{12}\right)$, $2^{1/6}\left(\cos \frac{13\pi}{12} + i \sin \frac{13\pi}{12}\right)$, and $2^{1/6}\left(\cos \frac{21\pi}{12} + i \sin \frac{21\pi}{12}\right)$.

93. (a) $w = \cos \dfrac{2\pi}{n} + i \sin \dfrac{2\pi}{n}$ for a positive integer n. Then, $w^k = \cos \dfrac{2k\pi}{n} + i \sin \dfrac{2k\pi}{n}$. Now $w^0 = \cos 0 + i \sin 0 = 1$

and for $k \neq 0$, $\left(w^k\right)^n = \cos 2k\pi + i \sin 2k\pi = 1$. So the nth roots of 1 are $\cos \dfrac{2k\pi}{n} + i \sin \dfrac{2k\pi}{n} = w^k$ for

$k = 0, 1, 2, \dots, n - 1$. In other words, the nth roots of 1 are $w^0, w^1, w^2, w^3, \dots, w^{n-1}$ or $1, w, w^2, w^3, \dots, w^{n-1}$.

(b) For $k = 0, 1, \dots, n - 1$, we have $\left(sw^k\right)^n = s^n \left(w^k\right)^n = z \cdot 1 = z$, so sw^k are nth roots of z for $k = 0, 1, \dots, n - 1$.

95. The cube roots of 1 are $w^0 = 1$, $w^1 = \cos \frac{2\pi}{3} + i \sin \frac{2\pi}{3}$, and $w^2 = \cos \frac{4\pi}{3} + i \sin \frac{4\pi}{3}$, so their product is

$w^0 \cdot w^1 \cdot w^2 = (1)\left(\cos \frac{2\pi}{3} + i \sin \frac{2\pi}{3}\right)\left(\cos \frac{4\pi}{3} + i \sin \frac{4\pi}{3}\right) = \cos 2\pi + i \sin 2\pi = 1$.

The fourth roots of 1 are $w^0 = 1$, $w^1 = i$, $w^2 = -1$, and $w^3 = -i$, so their product is

$w^0 \cdot w^1 \cdot w^2 \cdot w^3 = (1) \cdot (i) \cdot (-1) \cdot (-i) = i^2 = -1$.

The fifth roots of 1 are $w^0 = 1$, $w^1 = \cos \frac{2\pi}{5} + i \sin \frac{2\pi}{5}$, $w^2 = \cos \frac{4\pi}{5} + i \sin \frac{4\pi}{5}$, $w^3 = \cos \frac{6\pi}{5} + i \sin \frac{6\pi}{5}$, and $w^4 = \cos \frac{8\pi}{5} + i \sin \frac{8\pi}{5}$, so their product is $1\left(\cos \frac{2\pi}{5} + i \sin \frac{2\pi}{5}\right)\left(\cos \frac{4\pi}{5} + i \sin \frac{4\pi}{5}\right)\left(\cos \frac{6\pi}{5} + i \sin \frac{6\pi}{5}\right)\left(\cos \frac{8\pi}{5} + i \sin \frac{8\pi}{5}\right) = \cos 4\pi + i \sin 4\pi = 1$.

The sixth roots of 1 are $w^0 = 1$, $w^1 = \cos \frac{\pi}{3} + i \sin \frac{\pi}{3}$, $w^2 = \cos \frac{2\pi}{3} + i \sin \frac{2\pi}{3} = -\frac{1}{2} + \frac{\sqrt{3}}{2}i$, $w^3 = -1$, $w^4 = \cos \frac{4\pi}{3} + i \sin \frac{4\pi}{3} = -\frac{1}{2} - \frac{\sqrt{3}}{2}i$, and $w^5 = \cos \frac{5\pi}{3} + i \sin \frac{5\pi}{3} = \frac{1}{2} - \frac{\sqrt{3}}{2}i$, so their product is $1\left(\cos \frac{\pi}{3} + i \sin \frac{\pi}{3}\right)\left(\cos \frac{2\pi}{3} + i \sin \frac{2\pi}{3}\right)(-1)\left(\cos \frac{4\pi}{3} + i \sin \frac{4\pi}{3}\right)\left(\cos \frac{5\pi}{3} + i \sin \frac{5\pi}{3}\right) = \cos 5\pi + i \sin 5\pi = -1$.

The eight roots of 1 are $w^0 = 1$, $w^1 = \cos \frac{\pi}{4} + i \sin \frac{\pi}{4}$, $w^2 = i$, $w^3 = \cos \frac{3\pi}{4} + i \sin \frac{3\pi}{4}$, $w^4 = -1$, $w^5 = \cos \frac{5\pi}{4} + i \sin \frac{5\pi}{4}$, $w^6 = -i$, $w^7 = \cos \frac{7\pi}{4} + i \sin \frac{7\pi}{4}$, so their product is $1\left(\cos \frac{\pi}{4} + i \sin \frac{\pi}{4}\right) i \left(\cos \frac{3\pi}{4} + i \sin \frac{3\pi}{4}\right)(-1)\left(\cos \frac{5\pi}{4} + i \sin \frac{5\pi}{4}\right)(-i)\left(\cos \frac{7\pi}{4} + i \sin \frac{7\pi}{4}\right) = i^2 \cdot (\cos 2\pi + i \sin 2\pi) = -1$.

The product of the nth roots of 1 is -1 if n is even and 1 if n is odd.

The proof requires the fact that the sum of the first m integers is $\dfrac{m(m+1)}{2}$.

Let $w = \cos \dfrac{2\pi}{n} + i \sin \dfrac{2\pi}{n}$. Then $w^k = \cos \dfrac{2k\pi}{n} + i \sin \dfrac{2k\pi}{n}$ for $k = 0, 1, 2, \ldots, n-1$. The argument of the product of the n roots of unity can be found by adding the arguments of each w^k. So the argument of the product is

$\theta = 0 + \dfrac{2(1)\pi}{n} + \dfrac{2(2)\pi}{n} + \dfrac{2(3)\pi}{n} + \cdots + \dfrac{2(n-2)\pi}{n} + \dfrac{2(n-1)\pi}{n} = \dfrac{2\pi}{n}[0 + 1 + 2 + 3 + \cdots + (n-2) + (n-1)]$.

Since this is the sum of the first $n-1$ integers, this sum is $\dfrac{2\pi}{n} \cdot \dfrac{(n-1)n}{2} = (n-1)\pi$. Thus the product of the n roots of unity is $\cos((n-1)\pi) + i \sin((n-1)\pi) = -1$ if n is even and 1 if n is odd.

9.4 Vectors

1. $2\mathbf{u} = 2\langle -2, 3\rangle = \langle -4, 6\rangle$

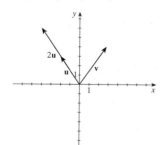

3. $\mathbf{u} + \mathbf{v} = \langle -2, 3\rangle + \langle 3, 4\rangle$
$= \langle -2 + 3, 3 + 4\rangle = \langle 1, 7\rangle$

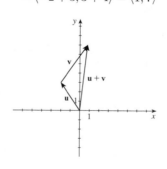

5. $\mathbf{v} - 2\mathbf{u} = \langle 3, 4\rangle - 2\langle -2, 3\rangle$
$= \langle 3 - 2(-2), 4 - 2(3)\rangle$
$= \langle 7, -2\rangle$

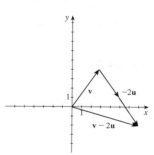

In Solutions 7–15, v represents the vector with initial point P and terminal point Q.

7. $P(2, 1)$, $Q(5, 4)$. $\mathbf{v} = \langle 5 - 2, 4 - 1\rangle = \langle 3, 3\rangle$

9. $P(1, 2)$, $Q(4, 1)$. $\mathbf{v} = \langle 4 - 1, 1 - 2\rangle = \langle 3, -1\rangle$

11. $P(3, 2)$, $Q(8, 9)$. $\mathbf{v} = \langle 8 - 3, 9 - 2\rangle = \langle 5, 7\rangle$

13. $P(5, 3)$, $Q(1, 0)$. $\mathbf{v} = \langle 1 - 5, 0 - 3\rangle = \langle -4, -3\rangle$

15. $P(-1, -1)$, $Q(-1, 1)$. $\mathbf{v} = \langle -1 - (-1), 1 - (-1)\rangle = \langle 0, 2\rangle$

17. $\mathbf{u} = \langle 2, 7 \rangle$, $\mathbf{v} = \langle 3, 1 \rangle$. $2\mathbf{u} = 2 \cdot \langle 2, 7 \rangle = \langle 4, 14 \rangle$; $-3\mathbf{v} = -3 \cdot \langle 3, 1 \rangle = \langle -9, -3 \rangle$; $\mathbf{u} + \mathbf{v} = \langle 2, 7 \rangle + \langle 3, 1 \rangle = \langle 5, 8 \rangle$;
$3\mathbf{u} - 4\mathbf{v} = \langle 6, 21 \rangle - \langle 12, 4 \rangle = \langle -6, 17 \rangle$

19. $\mathbf{u} = \langle 0, -1 \rangle$, $\mathbf{v} = \langle -2, 0 \rangle$. $2\mathbf{u} = 2 \cdot \langle 0, -1 \rangle = \langle 0, -2 \rangle$; $-3\mathbf{v} = -3 \cdot \langle -2, 0 \rangle = \langle 6, 0 \rangle$;
$\mathbf{u} + \mathbf{v} = \langle 0, -1 \rangle + \langle -2, 0 \rangle = \langle -2, -1 \rangle$; $3\mathbf{u} - 4\mathbf{v} = \langle 0, -3 \rangle - \langle -8, 0 \rangle = \langle 8, -3 \rangle$

21. $\mathbf{u} = 2\mathbf{i}$, $\mathbf{v} = 3\mathbf{i} - 2\mathbf{j}$. $2\mathbf{u} = 2 \cdot 2\mathbf{i} = 4\mathbf{i}$; $-3\mathbf{v} = -3(3\mathbf{i} - 2\mathbf{j}) = -9\mathbf{i} + 6\mathbf{j}$; $\mathbf{u} + \mathbf{v} = 2\mathbf{i} + 3\mathbf{i} - 2\mathbf{j} = 5\mathbf{i} - 2\mathbf{j}$;
$3\mathbf{u} - 4\mathbf{v} = 3 \cdot 2\mathbf{i} - 4(3\mathbf{i} - 2\mathbf{j}) = -6\mathbf{i} + 8\mathbf{j}$

23. $\mathbf{u} = 2\mathbf{i} + \mathbf{j}$, $\mathbf{v} = 3\mathbf{i} - 2\mathbf{j}$. Then $|\mathbf{u}| = \sqrt{2^2 + 1^2} = \sqrt{5}$; $|\mathbf{v}| = \sqrt{3^2 + 2^2} = \sqrt{13}$; $2\mathbf{u} = 4\mathbf{i} + 2\mathbf{j}$;
$|2\mathbf{u}| = \sqrt{4^2 + 2^2} = 2\sqrt{5}$; $\frac{1}{2}\mathbf{v} = \frac{3}{2}\mathbf{i} - \mathbf{j}$; $\left|\frac{1}{2}\mathbf{v}\right| = \sqrt{\left(\frac{3}{2}\right)^2 + 1^2} = \frac{1}{2}\sqrt{13}$; $\mathbf{u} + \mathbf{v} = 5\mathbf{i} - \mathbf{j}$; $|\mathbf{u} + \mathbf{v}| = \sqrt{5^2 + 1^2} = \sqrt{26}$;
$\mathbf{u} - \mathbf{v} = 2\mathbf{i} + \mathbf{j} - 3\mathbf{i} + 2\mathbf{j} = -\mathbf{i} + 3\mathbf{j}$; $\quad |\mathbf{u} - \mathbf{v}| = \sqrt{1^2 + 3^2} = \sqrt{10}$; $|\mathbf{u}| - |\mathbf{v}| = \sqrt{5} - \sqrt{13}$

25. $\mathbf{u} = \langle 10, -1 \rangle$, $\mathbf{v} = \langle -2, -2 \rangle$. Then $|\mathbf{u}| = \sqrt{10^2 + 1^2} = \sqrt{101}$; $|\mathbf{v}| = \sqrt{(-2)^2 + (-2)^2} = 2\sqrt{2}$; $2\mathbf{u} = \langle 20, -2 \rangle$;
$|2\mathbf{u}| = \sqrt{20^2 + 2^2} = \sqrt{404} = 2\sqrt{101}$; $\frac{1}{2}\mathbf{v} = \langle -1, -1 \rangle$; $\left|\frac{1}{2}\mathbf{v}\right| = \sqrt{(-1)^2 + (-1)^2} = \sqrt{2}$; $\mathbf{u} + \mathbf{v} = \langle 8, -3 \rangle$;
$|\mathbf{u} + \mathbf{v}| = \sqrt{8^2 + 3^2} = \sqrt{73}$; $\mathbf{u} - \mathbf{v} = \langle 12, 1 \rangle$; $|\mathbf{u} - \mathbf{v}| = \sqrt{12^2 + 1^2} = \sqrt{145}$; $|\mathbf{u}| - |\mathbf{v}| = \sqrt{101} - 2\sqrt{2}$

In Solutions 27–31, x represents the horizontal component and y the vertical component.

27. $|\mathbf{v}| = 40$, direction $\theta = 30°$. $x = 40\cos 30° = 20\sqrt{3}$ and $y = 40\sin 30° = 20$. Thus, $\mathbf{v} = x\mathbf{i} + y\mathbf{j} = 20\sqrt{3}\mathbf{i} + 20\mathbf{j}$.

29. $|\mathbf{v}| = 1$, direction $\theta = 225°$. $x = \cos 225° = -\frac{1}{\sqrt{2}}$ and $y = \sin 225° = -\frac{1}{\sqrt{2}}$. Thus,
$\mathbf{v} = x\mathbf{i} + y\mathbf{j} = -\frac{1}{\sqrt{2}}\mathbf{i} - \frac{1}{\sqrt{2}}\mathbf{j} = -\frac{\sqrt{2}}{2}\mathbf{i} - \frac{\sqrt{2}}{2}\mathbf{j}$.

31. $|\mathbf{v}| = 4$, direction $\theta = 10°$. $x = 4\cos 10° \approx 3.94$ and $y = 4\sin 10° \approx 0.69$. Thus,
$\mathbf{v} = x\mathbf{i} + y\mathbf{j} = (4\cos 10°)\mathbf{i} + (4\sin 10°)\mathbf{j} \approx 3.94\mathbf{i} + 0.69\mathbf{j}$.

33. $\mathbf{v} = \langle 3, 4 \rangle$. The magnitude is $|\mathbf{v}| = \sqrt{3^2 + 4^2} = 5$. The direction is θ where $\tan \theta = \frac{4}{3} \quad \Leftrightarrow \quad \theta = \tan^{-1}\left(\frac{4}{3}\right) \approx 53.13°$.

35. $\mathbf{v} = \langle -12, 5 \rangle$. The magnitude is $|\mathbf{v}| = \sqrt{(-12)^2 + 5^2} = \sqrt{169} = 13$. The direction is θ where $\tan \theta = -\frac{5}{12}$ with θ in
quadrant II $\quad \Leftrightarrow \quad \theta = \tan^{-1}\left(-\frac{5}{12}\right) \approx 157.38°$.

37. $\mathbf{v} = \mathbf{i} + \sqrt{3}\mathbf{j}$. The magnitude is $|\mathbf{v}| = \sqrt{1^2 + \left(\sqrt{3}\right)^2} = 2$. The direction is θ where $\tan \theta = \sqrt{3}$ with θ in quadrant I $\quad \Leftrightarrow$
$\theta = \tan^{-1} \sqrt{3} = 60°$.

39. $|\mathbf{v}| = 30$, direction $\theta = 30°$. $x = 30\cos 30° = 30 \cdot \frac{\sqrt{3}}{2} \approx 25.98$, $y = 30\sin 30° = 15$. So the horizontal component of
force is $15\sqrt{3}$ lb and the vertical component is -15 lb.

41. The flow of the river can be represented by the vector $\mathbf{v} = -3\mathbf{j}$ and the swimmer can be represented by the vector $\mathbf{u} = 2\mathbf{i}$.
Therefore the true velocity is $\mathbf{u} + \mathbf{v} = 2\mathbf{i} - 3\mathbf{j}$.

43. (a) The velocity of the wind is $40\mathbf{j}$.

(b) The velocity of the jet relative to the air is $425\mathbf{i}$.

(c) The true velocity of the jet is $\mathbf{v} = 425\mathbf{i} + 40\mathbf{j} = \langle 425, 40 \rangle$.

(d) The true speed of the jet is $|\mathbf{v}| = \sqrt{425^2 + 40^2} \approx 427$ mi/h, and the true direction is $\theta = \tan^{-1}\left(\frac{40}{425}\right) \approx 5.4° \quad \Rightarrow$
θ is N 84.6° E.

45. If the direction of the plane is N 30° W, the airplane's velocity is $\mathbf{u} = \langle u_x, u_y \rangle$ where $u_x = -765\cos 60° = -382.5$,
and $u_y = 765\sin 60° \approx 662.51$. If the direction of the wind is N 30° E, the wind velocity is
$\mathbf{w} = \langle w_x, w_y \rangle$ where $w_x = 55\cos 60° = 27.5$, and $w_y = 55\sin 60° \approx 47.63$. Thus, the actual flight
path is $\mathbf{v} = \mathbf{u} + \mathbf{w} = \langle -382.5 + 27.5, 662.51 + 47.63 \rangle = \langle -355, 710.14 \rangle$, and so the true speed is
$|\mathbf{v}| = \sqrt{355^2 + 710.14^2} \approx 794$ mi/h, and the true direction is $\theta = \tan^{-1}\left(-\frac{710.14}{355}\right) \approx 116.6°$ so θ is N 26.6° W.

47. (a) The velocity of the river is represented by the vector $\mathbf{r} = \langle 10, 0 \rangle$.

(b) Since the boater direction is $60°$ from the shore at 20 mi/h, the velocity of the boat is represented by the vector $\mathbf{b} = \langle 20 \cos 60°, 20 \sin 60° \rangle \approx \langle 10, 17.32 \rangle$.

(c) $\mathbf{w} = \mathbf{r} + \mathbf{b} = \langle 10 + 10, 0 + 17.32 \rangle = \langle 20, 17.32 \rangle$

(d) The true speed of the boat is $|\mathbf{w}| = \sqrt{20^2 + 17.32^2} \approx 26.5$ mi/h, and the true direction is $\theta = \tan^{-1}\left(\frac{17.32}{20}\right) \approx 40.9° \approx \text{N } 49.1° \text{ E}$.

49. (a) Let $\mathbf{b} = \langle b_x, b_y \rangle$ represent the velocity of the boat relative to the water. Then $\mathbf{b} = \langle 24 \cos 18°, 24 \sin 18° \rangle$.

(b) Let $\mathbf{w} = \langle w_x, w_y \rangle$ represent the velocity of the water. Then $\mathbf{w} = \langle 0, w \rangle$ where w is the speed of the water. So the true velocity of the boat is $\mathbf{b} + \mathbf{w} = \langle 24 \cos 18°, 24 \sin 18° - w \rangle$. For the direction to be due east, we must have $24 \sin 18° - w = 0 \iff w = 7.42$ mi/h. Therefore, the true speed of the water is 7.4 mi/h. Since $\mathbf{b} + \mathbf{w} = \langle 24 \cos 18°, 0 \rangle$, the true speed of the boat is $|\mathbf{b} + \mathbf{w}| = 24 \cos 18° \approx 22.8$ mi/h.

51. $\mathbf{F}_1 = \langle 2, 5 \rangle$ and $\mathbf{F}_2 = \langle 3, -8 \rangle$.

(a) $\mathbf{F}_1 + \mathbf{F}_2 = \langle 2 + 3, 5 - 8 \rangle = \langle 5, -3 \rangle$

(b) The additional force required is $\mathbf{F}_3 = \langle 0, 0 \rangle - \langle 5, -3 \rangle = \langle -5, 3 \rangle$.

53. $\mathbf{F}_1 = 4\mathbf{i} - \mathbf{j}$, $\mathbf{F}_2 = 3\mathbf{i} - 7\mathbf{j}$, $\mathbf{F}_3 = -8\mathbf{i} + 3\mathbf{j}$, and $\mathbf{F}_4 = \mathbf{i} + \mathbf{j}$.

(a) $\mathbf{F}_1 + \mathbf{F}_2 + \mathbf{F}_3 + \mathbf{F}_4 = (4 + 3 - 8 + 1)\mathbf{i} + (-1 - 7 + 3 + 1)\mathbf{j} = 0\mathbf{i} - 4\mathbf{j}$

(b) The additional force required is $\mathbf{F}_5 = 0\mathbf{i} + 0\mathbf{j} - (0\mathbf{i} - 4\mathbf{j}) = 4\mathbf{j}$.

55. $\mathbf{F}_1 = \langle 10 \cos 60°, 10 \sin 60° \rangle = \langle 5, 5\sqrt{3} \rangle$, $\mathbf{F}_2 = \langle -8 \cos 30°, 8 \sin 30° \rangle = \langle -4\sqrt{3}, 4 \rangle$, and $\mathbf{F}_3 = \langle -6 \cos 20°, -6 \sin 20° \rangle \approx \langle -5.638, -2.052 \rangle$.

(a) $\mathbf{F}_1 + \mathbf{F}_2 + \mathbf{F}_3 = \langle 5 - 4\sqrt{3} - 5.638, 5\sqrt{3} + 4 - 2.052 \rangle \approx \langle -7.57, 10.61 \rangle$.

(b) The additional force required is $\mathbf{F}_4 = \langle 0, 0 \rangle - \langle -7.57, 10.61 \rangle = \langle 7.57, -10.61 \rangle$.

57. From the figure we see that $\mathbf{T}_1 = -|\mathbf{T}_1| \cos 50° \mathbf{i} + |\mathbf{T}_1| \sin 50° \mathbf{j}$ and $\mathbf{T}_2 = |\mathbf{T}_2| \cos 30° \mathbf{i} + |\mathbf{T}_2| \sin 30° \mathbf{j}$. Since $\mathbf{T}_1 + \mathbf{T}_2 = 100\mathbf{j}$ we get $-|\mathbf{T}_1| \cos 50° + |\mathbf{T}_2| \cos 30° = 0$ and $|\mathbf{T}_1| \sin 50° + |\mathbf{T}_2| \sin 30° = 100$. From the first equation, $|\mathbf{T}_2| = |\mathbf{T}_1| \dfrac{\cos 50°}{\cos 30°}$, and substituting into the second equation gives $|\mathbf{T}_1| \sin 50° + |\mathbf{T}_1| \dfrac{\cos 50° \sin 30°}{\cos 30°} = 100$

$\iff |\mathbf{T}_1| (\sin 50° \cos 30° + \cos 50° \sin 30°) = 100 \cos 30° \iff |\mathbf{T}_1| \sin (50° + 30°) = 100 \cos 30° \iff$

$|\mathbf{T}_1| = 100 \dfrac{\cos 30°}{\sin 80°} \approx 87.9385$.

Similarly, solving for $|\mathbf{T}_1|$ in the first equation gives $|\mathbf{T}_1| = |\mathbf{T}_2| \dfrac{\cos 30°}{\cos 50°}$ and substituting gives

$|\mathbf{T}_2| \dfrac{\cos 30° \sin 50°}{\cos 50°} + |\mathbf{T}_2| \sin 30° = 100 \iff |\mathbf{T}_2| (\cos 30° \sin 50° + \cos 50° \sin 30°) = 100 \cos 50° \iff$

$|\mathbf{T}_2| = \dfrac{100 \cos 50°}{\sin 80°} \approx 65.2704$. Thus, $\mathbf{T}_1 \approx (-87.9385 \cos 50°) \mathbf{i} + (87.9385 \sin 50°) \mathbf{j} \approx -56.5\mathbf{i} + 67.4\mathbf{j}$ and $\mathbf{T}_2 \approx (65.2704 \cos 30°) \mathbf{i} + (65.2704 \sin 30°) \mathbf{j} \approx 56.5\mathbf{i} + 32.6\mathbf{j}$.

59. When we add two (or more vectors), the resultant vector can be found by first placing the initial point of the second vector at the terminal point of the first vector. The resultant vector can then found by using the new terminal point of the second vector and the initial point of the first vector. When the n vectors are placed head to tail in the plane so that they form a polygon, the initial point and the terminal point are the same. Thus the sum of these n vectors is the zero vector.

9.5 The Dot Product

1. (a) $\mathbf{u} \cdot \mathbf{v} = \langle 2, 0 \rangle \cdot \langle 1, 1 \rangle = 2 + 0 = 2$
   (b) $\cos \theta = \dfrac{\mathbf{u} \cdot \mathbf{v}}{|\mathbf{u}| \, |\mathbf{v}|} = \dfrac{2}{2 \cdot \sqrt{2}} = \dfrac{1}{\sqrt{2}} \Rightarrow \theta = 45°$

3. (a) $\mathbf{u} \cdot \mathbf{v} = \langle 2, 7 \rangle \cdot \langle 3, 1 \rangle = 6 + 7 = 13$
   (b) $\cos \theta = \dfrac{\mathbf{u} \cdot \mathbf{v}}{|\mathbf{u}| \, |\mathbf{v}|} = \dfrac{13}{\sqrt{53} \cdot \sqrt{10}} \Rightarrow \theta \approx 56°$

5. (a) $\mathbf{u} \cdot \mathbf{v} = \langle 3, -2 \rangle \cdot \langle 1, 2 \rangle = 3 + (-4) = -1$
   (b) $\cos \theta = \dfrac{\mathbf{u} \cdot \mathbf{v}}{|\mathbf{u}| \, |\mathbf{v}|} = \dfrac{-1}{\sqrt{13} \cdot \sqrt{5}} \Rightarrow \theta \approx 97°$

7. (a) $\mathbf{u} \cdot \mathbf{v} = \langle 0, -5 \rangle \cdot \langle -1, -\sqrt{3} \rangle = 0 + 5\sqrt{3} = 5\sqrt{3}$
   (b) $\cos \theta = \dfrac{\mathbf{u} \cdot \mathbf{v}}{|\mathbf{u}| \, |\mathbf{v}|} = \dfrac{5\sqrt{3}}{5 \cdot 2} = \dfrac{\sqrt{3}}{2} \Rightarrow \theta = 30°$

9. $\mathbf{u} \cdot \mathbf{v} = -12 + 12 = 0 \Rightarrow$ vectors are orthogonal

11. $\mathbf{u} \cdot \mathbf{v} = -8 + 12 = 4 \neq 0 \Rightarrow$ vectors are not orthogonal

13. $\mathbf{u} \cdot \mathbf{v} = -24 + 24 = 0 \Rightarrow$ vectors are orthogonal

15. $\mathbf{u} \cdot \mathbf{v} + \mathbf{u} \cdot \mathbf{w} = \langle 2, 1 \rangle \cdot \langle 1, -3 \rangle + \langle 2, 1 \rangle \cdot \langle 3, 4 \rangle$
$= 2 - 3 + 6 + 4 = 9$

17. $(\mathbf{u} + \mathbf{v}) \cdot (\mathbf{u} - \mathbf{v}) = [\langle 2, 1 \rangle + \langle 1, -3 \rangle] \cdot [\langle 2, 1 \rangle - \langle 1, -3 \rangle]$
$= \langle 3, -2 \rangle \cdot \langle 1, 4 \rangle = 3 - 8 = -5$

19. $x = \dfrac{\mathbf{u} \cdot \mathbf{v}}{|\mathbf{v}|} = \dfrac{12 - 24}{5} = -\dfrac{12}{5}$

21. $x = \dfrac{\mathbf{u} \cdot \mathbf{v}}{|\mathbf{v}|} = \dfrac{0 - 24}{1} = -24$

23. (a) $\mathbf{u}_1 = \text{proj}_{\mathbf{v}} \, \mathbf{u} = \left(\dfrac{\mathbf{u} \cdot \mathbf{v}}{|\mathbf{v}|^2} \right) \mathbf{v} = \left(\dfrac{\langle -2, 4 \rangle \cdot \langle 1, 1 \rangle}{1^2 + 1^2} \right) \langle 1, 1 \rangle = \langle 1, 1 \rangle.$

   (b) $\mathbf{u}_2 = \mathbf{u} - \mathbf{u}_1 = \langle -2, 4 \rangle - \langle 1, 1 \rangle = \langle -3, 3 \rangle$

25. (a) $\mathbf{u}_1 = \text{proj}_{\mathbf{v}} \, \mathbf{u} = \left(\dfrac{\mathbf{u} \cdot \mathbf{v}}{|\mathbf{v}|^2} \right) \mathbf{v} = \left(\dfrac{\langle 1, 2 \rangle \cdot \langle 1, -3 \rangle}{1^2 + (-3)^2} \right) \langle 1, -3 \rangle = -\dfrac{1}{2} \langle 1, -3 \rangle = \left\langle -\dfrac{1}{2}, \dfrac{3}{2} \right\rangle$

   (b) $\mathbf{u}_2 = \mathbf{u} - \mathbf{u}_1 = \langle 1, 2 \rangle - \left\langle -\dfrac{1}{2}, \dfrac{3}{2} \right\rangle = \left\langle \dfrac{3}{2}, \dfrac{1}{2} \right\rangle$

27. (a) $\mathbf{u}_1 = \text{proj}_{\mathbf{v}} \, \mathbf{u} = \left(\dfrac{\mathbf{u} \cdot \mathbf{v}}{|\mathbf{v}|^2} \right) \mathbf{v} = \left(\dfrac{\langle 2, 9 \rangle \cdot \langle -3, 4 \rangle}{(-3)^3 + 4^2} \right) \langle -3, 4 \rangle = \dfrac{6}{5} \langle -3, 4 \rangle = \left\langle -\dfrac{18}{5}, \dfrac{24}{5} \right\rangle$

   (b) $\mathbf{u}_2 = \mathbf{u} - \mathbf{u}_1 = \langle 2, 9 \rangle - \left\langle -\dfrac{18}{5}, \dfrac{24}{5} \right\rangle = \left\langle \dfrac{28}{5}, \dfrac{21}{5} \right\rangle$

29. $W = \mathbf{F} \cdot \mathbf{d} = \langle 4, -5 \rangle \cdot \langle 3, 8 \rangle = -28$

31. $W = \mathbf{F} \cdot \mathbf{d} = \langle 10, 3 \rangle \cdot \langle 4, -5 \rangle = 25$

33. Let $\mathbf{u} = \langle u_1, u_2 \rangle$ and $\mathbf{v} = \langle v_1, v_2 \rangle$. Then
$\mathbf{u} \cdot \mathbf{v} = \langle u_1, u_2 \rangle \cdot \langle v_1, v_2 \rangle = u_1 v_1 + u_2 v_2 = v_1 u_1 + v_2 u_2 = \langle v_1, v_2 \rangle \cdot \langle u_1, u_2 \rangle = \mathbf{v} \cdot \mathbf{u}$

35. Let $\mathbf{u} = \langle u_1, u_2 \rangle$, $\mathbf{v} = \langle v_1, v_2 \rangle$, and $\mathbf{w} = \langle w_1, w_2 \rangle$. Then
$(\mathbf{u} + \mathbf{v}) \cdot \mathbf{w} = (\langle u_1, u_2 \rangle + \langle v_1, v_2 \rangle) \cdot \langle w_1, w_2 \rangle = \langle u_1 + v_1, u_2 + v_2 \rangle \cdot \langle w_1, w_2 \rangle$
$= u_1 w_1 + v_1 w_1 + u_2 w_2 + v_2 w_2 = u_1 w_1 + u_2 w_2 + v_1 w_1 + v_2 w_2$
$= \langle u_1, u_2 \rangle \cdot \langle w_1, w_2 \rangle + \langle v_1, v_2 \rangle \cdot \langle w_1, w_2 \rangle = \mathbf{u} \cdot \mathbf{w} + \mathbf{v} \cdot \mathbf{w}$

37. We use the definition that $\text{proj}_{\mathbf{v}} \, \mathbf{u} = \left(\dfrac{\mathbf{u} \cdot \mathbf{v}}{|\mathbf{v}|^2} \right) \mathbf{v}$. Then

$\text{proj}_{\mathbf{v}} \, \mathbf{u} \cdot (\mathbf{u} - \text{proj}_{\mathbf{v}} \, \mathbf{u}) = \left(\dfrac{\mathbf{u} \cdot \mathbf{v}}{|\mathbf{v}|^2} \right) \mathbf{v} \cdot \left[\mathbf{u} - \left(\dfrac{\mathbf{u} \cdot \mathbf{v}}{|\mathbf{v}|^2} \right) \mathbf{v} \right] = \left(\dfrac{\mathbf{u} \cdot \mathbf{v}}{|\mathbf{v}|^2} \right) (\mathbf{v} \cdot \mathbf{u}) - \left(\dfrac{\mathbf{u} \cdot \mathbf{v}}{|\mathbf{v}|^2} \right) \mathbf{v} \cdot \left(\dfrac{\mathbf{u} \cdot \mathbf{v}}{|\mathbf{v}|^2} \right) \mathbf{v}$

$= \dfrac{(\mathbf{u} \cdot \mathbf{v})^2}{|\mathbf{v}|^2} - \dfrac{(\mathbf{u} \cdot \mathbf{v})^2}{|\mathbf{v}|^4} |\mathbf{v}|^2 = \dfrac{(\mathbf{u} \cdot \mathbf{v})^2}{|\mathbf{v}|^2} - \dfrac{(\mathbf{u} \cdot \mathbf{v})^2}{|\mathbf{v}|^2} = 0$

Thus $\mathbf{u}$ and $\mathbf{u} - \text{proj}_{\mathbf{v}} \, \mathbf{u}$ are orthogonal.

39. $W = \mathbf{F} \cdot \mathbf{d} = \langle 4, -7 \rangle \cdot \langle 4, 0 \rangle = 16$ ft-lb

41. The distance vector is $\mathbf{D} = \langle 200, 0 \rangle$ and the force vector is $\mathbf{F} = \langle 50 \cos 30°, 50 \sin 30° \rangle$. Hence, the work done is
$W = \mathbf{F} \cdot \mathbf{D} = \langle 200, 0 \rangle \cdot \langle 50 \cos 30°, 50 \sin 30° \rangle = 200 \cdot 50 \cos 30° \approx 8660$ ft-lb.

43. Since the weight of the car is 2755 lb, the force exerted perpendicular to the earth is 2755 lb. Resolving this into a force **u** perpendicular to the driveway gives $|\mathbf{u}| = 2766 \cos 65° \approx 1164$ lb. Thus, a force of about 1164 lb is required.

45. Since the force required parallel to the plane is 80 lb and the weight of the package is 200 lb, it follows that $80 = 200 \sin \theta$, where θ is the angle of inclination of the plane. Then $\theta = \sin^{-1}\left(\frac{80}{200}\right) \approx 23.58°$, and so the angle of inclination is approximately $23.6°$.

47. (a) $2(0) + 4(2) = 8$, so $Q(0, 2)$ lies on L. $2(2) + 4(1) = 4 + 4 = 8$, so $R(2, 1)$ lies on L.

(b) $\mathbf{u} = \overrightarrow{QP} = \langle 0, 2 \rangle - \langle 3, 4 \rangle = \langle -3, -2 \rangle$.

$\mathbf{v} = \overrightarrow{QR} = \langle 0, 2 \rangle - \langle 2, 1 \rangle = \langle -2, 1 \rangle$.

$\mathbf{w} = \text{proj}_{\mathbf{v}}\, \mathbf{u} = \left(\dfrac{\mathbf{u} \cdot \mathbf{v}}{|\mathbf{v}|^2}\right) \mathbf{v} = \dfrac{\langle -3, -2 \rangle \cdot \langle 2, 1 \rangle}{(-2)^2 + 1^2} \langle -2, 1 \rangle$

$= -\frac{8}{5} \langle -2, 1 \rangle = \langle \frac{16}{5}, -\frac{8}{5} \rangle$

(c) From the graph, we can see that $\mathbf{u} - \mathbf{w}$ is orthogonal to $\mathbf{v}$ (and thus to L). Thus, the distance from P to L is $|\mathbf{u} - \mathbf{w}|$.

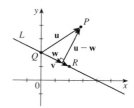

Chapter 9 Review

1. (a)

(b) $x = 12 \cos \frac{\pi}{6} = 12 \cdot \frac{\sqrt{3}}{2} = 6\sqrt{3}$,

$y = 12 \sin \frac{\pi}{6} = 12 \cdot \frac{1}{2} = 6$. Thus, the rectangular coordinates of P are $\left(6\sqrt{3}, 6\right)$.

3. (a)

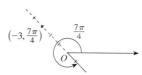

(b) $x = -3 \cos \frac{7\pi}{4} = -3\left(\frac{\sqrt{2}}{2}\right) = -\frac{3\sqrt{2}}{2}$,

$y = -3 \sin \frac{7\pi}{4} = -3\left(-\frac{\sqrt{2}}{2}\right) = \frac{3\sqrt{2}}{2}$. Thus, the rectangular coordinates of P are $\left(-\frac{3\sqrt{2}}{2}, \frac{3\sqrt{2}}{2}\right)$.

5. (a)

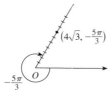

(b) $x = 4\sqrt{3} \cos\left(-\frac{5\pi}{3}\right) = 4\sqrt{3}\left(\frac{1}{2}\right) = 2\sqrt{3}$,

$y = 4\sqrt{3} \sin\left(-\frac{5\pi}{3}\right) = 4\sqrt{3}\left(\frac{\sqrt{3}}{2}\right) = 6$. Thus, the rectangular coordinates of P are $\left(2\sqrt{3}, 6\right)$.

7. (a)

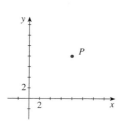

(b) $r = \sqrt{8^2 + 8^2} = \sqrt{128} = 8\sqrt{2}$ and $\bar{\theta} = \tan^{-1}\frac{8}{8}$. Since P is in quadrant I, $\theta = \frac{\pi}{4}$. Polar coordinates for P are $\left(8\sqrt{2}, \frac{\pi}{4}\right)$.

(c) $\left(-8\sqrt{2}, \frac{5\pi}{4}\right)$

9. (a)

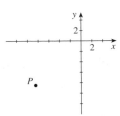

(b) $r = \sqrt{\left(-6\sqrt{2}\right)^2 + \left(-6\sqrt{2}\right)^2} = \sqrt{144} = 12$ and
$\overline{\theta} = \tan^{-1}\frac{-6\sqrt{2}}{-6\sqrt{2}} = \frac{\pi}{4}$. Since P is in quadrant III,
$\theta = \frac{5\pi}{4}$. Polar coordinates for P are $\left(12, \frac{5\pi}{4}\right)$.

(c) $\left(-12, \frac{\pi}{4}\right)$

13. (a) $x + y = 4 \quad \Leftrightarrow \quad r\cos\theta + r\sin\theta = 4 \quad \Leftrightarrow$
$r\left(\cos\theta + \sin\theta\right) = 4 \quad \Leftrightarrow \quad r = \dfrac{4}{\cos\theta + \sin\theta}$

(b) The rectangular equation is easier to graph.

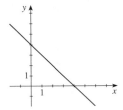

17. (a)

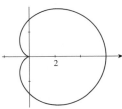

(b) $r = 3 + 3\cos\theta \quad \Leftrightarrow \quad r^2 = 3r + 3r\cos\theta$, which
gives $x^2 + y^2 = 3\sqrt{x^2 + y^2} + 3x \quad \Leftrightarrow$
$x^2 - 3x + y^2 = 3\sqrt{x^2 + y^2}$. Squaring both sides
gives $\left(x^2 - 3x + y^2\right)^2 = 9\left(x^2 + y^2\right)$.

21. (a)

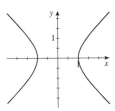

(b) $r^2 = \sec 2\theta = \dfrac{1}{\cos 2\theta} = \dfrac{1}{\cos^2\theta - \sin^2\theta} \quad \Leftrightarrow$
$r^2\left(\cos^2\theta - \sin^2\theta\right) = 1 \quad \Leftrightarrow$
$r^2\cos^2\theta - r^2\sin^2\theta = 1 \quad \Leftrightarrow$
$\left(r\cos\theta\right)^2 - \left(r\sin\theta\right)^2 = 1 \quad \Leftrightarrow \quad x^2 - y^2 = 1.$

11. (a)

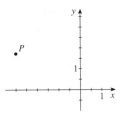

(b) $r = \sqrt{(-3)^2 + \left(\sqrt{3}\right)^2} = \sqrt{12} = 2\sqrt{3}$ and
$\overline{\theta} = \tan^{-1}\frac{\sqrt{3}}{-3}$. Since P is in quadrant II, $\theta = \frac{5\pi}{6}$.
Polar coordinates for P are $\left(2\sqrt{3}, \frac{5\pi}{6}\right)$.

(c) $\left(-2\sqrt{3}, -\frac{\pi}{6}\right)$

15. (a) $x^2 + y^2 = 4x + 4y \quad \Leftrightarrow \quad r^2 = 4r\cos\theta + 4r\sin\theta$
$\Leftrightarrow \quad r^2 = r\left(4\cos\theta + 4\sin\theta\right) \quad \Leftrightarrow$
$r = 4\cos\theta + 4\sin\theta$

(b) The polar equation is easier to graph.

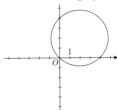

19. (a)

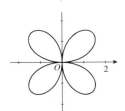

(b) $r = 2\sin 2\theta \quad \Leftrightarrow \quad r = 2 \cdot 2\sin\theta\cos\theta \quad \Leftrightarrow$
$r^3 = 4r^2\sin\theta\cos\theta \quad \Leftrightarrow$
$\left(r^2\right)^{3/2} = 4\left(r\sin\theta\right)\left(r\cos\theta\right)$ and so, since
$x = r\cos\theta$ and $y = r\sin\theta$, we get
$\left(x^2 + y^2\right)^3 = 16x^2y^2.$

23. (a)

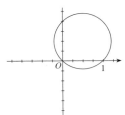

(b) $r = \sin\theta + \cos\theta \quad \Leftrightarrow \quad r^2 = r\sin\theta + r\cos\theta$, so
$x^2 + y^2 = y + x \quad \Leftrightarrow$
$\left(x^2 - x + \frac{1}{4}\right) + \left(y^2 - y + \frac{1}{4}\right) = \frac{1}{2} \quad \Leftrightarrow$
$\left(x - \frac{1}{2}\right)^2 + \left(y - \frac{1}{2}\right)^2 = \frac{1}{2}.$

25. $r = \cos(\theta/3)$, $\theta \in [0, 3\pi]$.

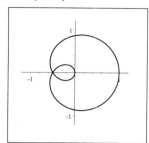

27. $r = 1 + 4\cos(\theta/3)$, $\theta \in [0, 6\pi]$.

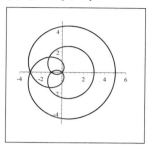

29. (a)

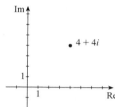

(b) $4 + 4i$ has $r = \sqrt{16 + 16} = 4\sqrt{2}$, and $\theta = \tan^{-1}\frac{4}{4} = \frac{\pi}{4}$ (in quadrant I).

(c) $4 + 4i = 4\sqrt{2}\left(\cos\frac{\pi}{4} + i\sin\frac{\pi}{4}\right)$

31. (a)

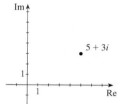

(b) $5 + 3i$. Then $r = \sqrt{25 + 9} = \sqrt{34}$, and $\theta = \tan^{-1}\frac{3}{5}$.

(c) $5 + 3i = \sqrt{34}\left[\cos\left(\tan^{-1}\frac{3}{5}\right) + i\sin\left(\tan^{-1}\frac{3}{5}\right)\right]$

33. (a)

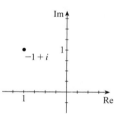

(b) $-1 + i$ has $r = \sqrt{1 + 1} = \sqrt{2}$ and $\tan\theta = \frac{1}{-1}$ with

θ in quadrant II $\quad\Leftrightarrow\quad \theta = \frac{3\pi}{4}$.

(c) $-1 + i = \sqrt{2}\left(\cos\frac{3\pi}{4} + i\sin\frac{3\pi}{4}\right)$

35. $1 - \sqrt{3}i$ has $r = \sqrt{1 + 3} = 2$ and $\tan\theta = \frac{-\sqrt{3}}{1} = -\sqrt{3}$ with θ in quadrant III $\quad\Leftrightarrow\quad \theta = \frac{5\pi}{3}$. Therefore,

$1 - \sqrt{3}i = 2\left(\cos\frac{5\pi}{3} + i\sin\frac{5\pi}{3}\right)$, and so

$\left(1 - \sqrt{3}i\right)^4 = 2^4\left(\cos\frac{20\pi}{3} + i\sin\frac{20\pi}{3}\right)$

$= 16\left(\cos\frac{2\pi}{3} + i\sin\frac{2\pi}{3}\right)$

$= 16\left(-\frac{1}{2} + i\frac{\sqrt{3}}{2}\right) = 8\left(-1 + i\sqrt{3}\right)$

37. $\sqrt{3} + i$ has $r = \sqrt{3 + 1} = 2$ and $\tan\theta = \frac{1}{\sqrt{3}}$ with θ in quadrant I $\quad\Leftrightarrow\quad \theta = \frac{\pi}{6}$. Therefore, $\sqrt{3} + i = 2\left(\cos\frac{\pi}{6} + i\sin\frac{\pi}{6}\right)$, and so

$\left(\sqrt{3} + i\right)^{-4} = 2^{-4}\left(\cos\frac{-4\pi}{6} + i\sin\frac{-4\pi}{6}\right) = \frac{1}{16}\left(\cos\frac{2\pi}{3} - i\sin\frac{2\pi}{3}\right) = \frac{1}{16}\left(-\frac{1}{2} - i\frac{\sqrt{3}}{2}\right)$

$= \frac{1}{32}\left(-1 - i\sqrt{3}\right) = -\frac{1}{32}\left(1 + i\sqrt{3}\right)$

39. $-16i$ has $r = 16$ and $\theta = \frac{3\pi}{2}$. Thus, $-16i = 16\left(\cos\frac{3\pi}{2} + i\sin\frac{3\pi}{2}\right)$ and so

$(-16i)^{1/2} = 16^{1/2}\left[\cos\left(\frac{3\pi + 4k\pi}{2}\right) + i\sin\left(\frac{3\pi + 4k\pi}{2}\right)\right]$ for $k = 0, 1$. Thus

the roots are $w_0 = 4\left(\cos\frac{3\pi}{4} + i\sin\frac{3\pi}{4}\right) = 4\left(-\frac{1}{\sqrt{2}} + i\frac{1}{\sqrt{2}}\right) = 2\sqrt{2}\left(-1 + i\right)$ and

$w_1 = 4\left(\cos\frac{7\pi}{4} + i\sin\frac{7\pi}{4}\right) = 4\left(\frac{1}{\sqrt{2}} - i\frac{1}{\sqrt{2}}\right) = 2\sqrt{2}\left(1 - i\right)$.

41. $1 = \cos 0 + i\sin 0$. Then $1^{1/6} = 1\left(\cos\frac{2k\pi}{6} + i\sin\frac{2k\pi}{6}\right)$ for $k = 0, 1, 2, 3, 4, 5$. Thus the six roots are

$w_0 = 1\left(\cos 0 + i\sin 0\right) = 1$, $w_1 = 1\left(\cos\frac{\pi}{3} + i\sin\frac{\pi}{3}\right) = \frac{1}{2} + i\frac{\sqrt{3}}{2}$, $w_2 = 1\left(\cos\frac{2\pi}{3} + i\sin\frac{2\pi}{3}\right) = -\frac{1}{2} + i\frac{\sqrt{3}}{2}$,

$w_3 = 1\left(\cos\pi + i\sin\pi\right) = -1$, $w_4 = 1\left(\cos\frac{4\pi}{3} + i\sin\frac{4\pi}{3}\right) = -\frac{1}{2} - i\frac{\sqrt{3}}{2}$, and $w_5 = 1\left(\cos\frac{5\pi}{3} + i\sin\frac{5\pi}{3}\right) = \frac{1}{2} - i\frac{\sqrt{3}}{2}$.

43. $\mathbf{u} = \langle -2, 3 \rangle$ and $\mathbf{v} = \langle 8, 1 \rangle$. Then $|\mathbf{u}| = \sqrt{(-2)^2 + 3^2} = \sqrt{13}$, $\mathbf{u} + \mathbf{v} = \langle -2 + 8, 3 + 1 \rangle = \langle 6, 4 \rangle$,

$\mathbf{u} - \mathbf{v} = \langle -2 - 8, 3 - 1 \rangle = \langle -10, 2 \rangle$, $2\mathbf{u} = \langle -4, 6 \rangle$, and $3\mathbf{u} - 2\mathbf{v} = \langle -6, 9 \rangle - \langle 16, 2 \rangle = \langle -6 - 16, 9 - 2 \rangle = \langle -22, 7 \rangle$.

45. $P(0, 3)$ and $Q(3, -1)$. $\mathbf{u} = \langle 3 - 0, -1 - 3 \rangle = \langle 3, -4 \rangle = 3\mathbf{i} - 4\mathbf{j}$

47. Let $Q(x, y)$ be the terminal point. Then $\langle x - 5, y - 6 \rangle = \langle 5, -8 \rangle$ ⟺ $x - 5 = 5$ and $y - 6 = -8$ ⟺ $x = 10$ and $y = -2$. Therefore, the terminal point is $Q(10, -2)$.

49. (a) The resultant force $\mathbf{r}$ is the sum of the forces; to find this we first resolve the two vectors:

$\mathbf{v}_1 = 2.0 \times 10^4$ lb N $50°$ E $= \langle 2.0 \times 10^4 \cdot \cos 40°, 2.0 \times 10^4 \cdot \sin 40° \rangle \approx \langle 15321, 12856 \rangle$ and

$\mathbf{v}_2 = 3.4 \times 10^4$ lb S $75°$ E $= \langle 3.4 \times 10^4 \cdot \cos(-15°), 3.4 \times 10^4 \cdot \sin(-15°) \rangle \approx \langle 32841, -8800 \rangle$. Thus

$\mathbf{r} = \mathbf{v}_1 + \mathbf{v}_2 \approx \langle 15321, 12856 \rangle + \langle 32841, -8800 \rangle = \langle 48163, 4056 \rangle \approx \langle 4.82, 0.41 \rangle \times 10^4 = (4.82\mathbf{i} + 0.41\mathbf{j}) \times 10^4$.

(b) The magnitude is $|\mathbf{r}| = \sqrt{48163^2 + 4056^2} \approx 4.8 \times 10^4$ lb. We have $\theta \approx \tan^{-1}\left(\frac{4056}{48163}\right) \approx 4.8°$. Thus the direction of $\mathbf{r}$ is approximately N $85.2°$ E.

51. $\mathbf{u} = \langle 4, -3 \rangle$ and $\mathbf{v} = \langle 9, -8 \rangle$. Then $|\mathbf{u}| = \sqrt{(4)^2 + (-3)^2} = \sqrt{25} = 5$,

$\mathbf{u} \cdot \mathbf{u} = \langle 4, -3 \rangle \cdot \langle 4, -3 \rangle = (4)(4) + (-3)(-3) = 16 + 9 = 25$, and

$\mathbf{u} \cdot \mathbf{v} = \langle 4, -3 \rangle \cdot \langle 9, -8 \rangle = (4)(9) + (-3)(-8) = 36 + 24 = 60$.

53. $\mathbf{u} = -2\mathbf{i} + 2\mathbf{j}$ and $\mathbf{v} = \mathbf{i} + \mathbf{j}$. Then $|\mathbf{u}| = \sqrt{(-2)^2 + (2)^2} = \sqrt{8} = 2\sqrt{2}$,

$\mathbf{u} \cdot \mathbf{u} = -2\mathbf{i} + 2\mathbf{j} \cdot -2\mathbf{i} + 2\mathbf{j} = (-2)(-2) + (2)(2) = 4 + 4 = 8$, and

$\mathbf{u} \cdot \mathbf{v} = -2\mathbf{i} + 2\mathbf{j} \cdot \mathbf{i} + \mathbf{j} = (-2)(1) + (2)(1) = -2 + 2 = 0$.

55. $\mathbf{u} = \langle -4, 2 \rangle$ and $\mathbf{v} = \langle 3, 6 \rangle$. Since $\mathbf{u} \cdot \mathbf{v} = -12 + 12 = 0$, the vectors are orthogonal.

57. $\mathbf{u} = 2\mathbf{i} + \mathbf{j}$ and $\mathbf{v} = \mathbf{i} + 3\mathbf{j}$. Then $\cos\theta = \dfrac{\mathbf{u} \cdot \mathbf{v}}{|\mathbf{u}||\mathbf{v}|} = \dfrac{2 + 3}{\sqrt{5}\sqrt{10}} = \dfrac{\sqrt{2}}{2}$. Thus, $\theta = \cos^{-1}\dfrac{\sqrt{2}}{2} = 45°$, so the vectors are not orthogonal.

59. (a) $\mathbf{u} = \langle 3, 1 \rangle$ and $\mathbf{v} = \langle 6, -1 \rangle$. Then the component of $\mathbf{u}$ along $\mathbf{v}$ is $\dfrac{\mathbf{u} \cdot \mathbf{v}}{|\mathbf{v}|} = \dfrac{\langle 3, 1 \rangle \cdot \langle 6, -1 \rangle}{\sqrt{\langle 6, -1 \rangle \cdot \langle 6, -1 \rangle}} = \dfrac{18 - 1}{\sqrt{36 + 1}} = \dfrac{17\sqrt{37}}{37}$.

(b) $\mathbf{u}_1 = \text{proj}_{\mathbf{v}}\,\mathbf{u} = \left(\dfrac{\mathbf{u} \cdot \mathbf{v}}{|\mathbf{v}|^2}\right)\mathbf{v} = \left(\dfrac{\langle 3, 1 \rangle \cdot \langle 6, -1 \rangle}{6^2 + (-1)^2}\right)\langle 6, -1 \rangle = \dfrac{17}{37}\langle 6, -1 \rangle = \left\langle \dfrac{102}{37}, -\dfrac{17}{37} \right\rangle$.

(c) $\mathbf{u}_2 = \mathbf{u} - \mathbf{u}_1 = \langle 3, 1 \rangle - \left\langle \dfrac{102}{37}, -\dfrac{17}{37} \right\rangle = \left\langle \dfrac{9}{37}, \dfrac{54}{37} \right\rangle$.

61. The displacement is $\mathbf{D} = \langle 7 - 1, -1 - 1 \rangle = \langle 6, -2 \rangle$. Thus, $W = \mathbf{F} \cdot \mathbf{D} = \langle 2, 9 \rangle \cdot \langle 6, -2 \rangle = -6$.

Chapter 9 Test

1. (a) $x = 8\cos\dfrac{5\pi}{4} = 8\left(-\dfrac{\sqrt{2}}{2}\right) = -4\sqrt{2}$, $y = 8\sin\dfrac{5\pi}{4} = 8\left(-\dfrac{\sqrt{2}}{2}\right) = -4\sqrt{2}$. So the point has rectangular coordinates $\left(-4\sqrt{2}, -4\sqrt{2}\right)$.

(b) $P = \left(-6, 2\sqrt{3}\right)$ in rectangular coordinates. So $\tan\theta = \dfrac{2\sqrt{3}}{-6}$ and the reference angle is $\bar{\theta} = \dfrac{\pi}{6}$. Since P is in quadrant II, we have $\theta = \dfrac{5\pi}{6}$. Next, $r^2 = (-6)^2 + \left(2\sqrt{3}\right)^2 = 36 + 12 = 48$, so $r = 4\sqrt{3}$. Thus, polar coordinates for the point are $\left(4\sqrt{3}, \dfrac{5\pi}{6}\right)$ or $\left(-4\sqrt{3}, \dfrac{11\pi}{6}\right)$.

2. (a)

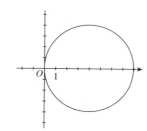

(b) $r = 8\cos\theta \quad\Leftrightarrow\quad r^2 = 8r\cos\theta \quad\Leftrightarrow\quad x^2 + y^2 = 8x \quad\Leftrightarrow$

$x^2 - 8x + y^2 = 0 \quad\Leftrightarrow\quad x^2 - 8x + 16 + y^2 = 16 \quad\Leftrightarrow$

$(x-4)^2 + y^2 = 16$

3. (a)

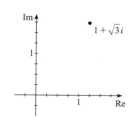

(b) $1 + \sqrt{3}i$ has $r = \sqrt{1+3} = 2$ and $\theta = \tan^{-1}\left(\sqrt{3}\right) = \frac{\pi}{3}$. So, in trigonometric form, $1 + \sqrt{3}i = 2\left(\cos\frac{\pi}{3} + i\sin\frac{\pi}{3}\right)$.

(c) $z = 1 + \sqrt{3}i = 2\left(\cos\frac{\pi}{3} + i\sin\frac{\pi}{3}\right) \quad\Rightarrow$

$z^9 = 2^9\left(\cos\frac{9\pi}{3} + i\sin\frac{9\pi}{3}\right) = 512\left(\cos 3\pi + i\sin 3\pi\right)$

$= 512\left(-1 + i\left(0\right)\right) = -512$

4. $z_1 = 4\left(\cos\frac{7\pi}{12} + i\sin\frac{7\pi}{12}\right)$ and $z_2 = 2\left(\cos\frac{5\pi}{12} + i\sin\frac{5\pi}{12}\right)$.

Then $z_1 z_2 = 4\cdot 2\left[\cos\left(\dfrac{7\pi + 5\pi}{12}\right) + i\sin\left(\dfrac{7\pi + 5\pi}{12}\right)\right] = 8\left(\cos\pi + i\sin\pi\right) = -8$ and

$z_1/z_2 = \frac{4}{2}\left[\cos\left(\dfrac{7\pi - 5\pi}{12}\right) + i\sin\left(\dfrac{7\pi - 5\pi}{12}\right)\right] = 2\left(\cos\frac{\pi}{6} + i\sin\frac{\pi}{6}\right) = 2\left(\frac{\sqrt{3}}{2} + \frac{1}{2}i\right) = \sqrt{3} + i$.

5. $27i$ has $r = 27$ and $\theta = \frac{\pi}{2}$, so $27i = 27\left(\cos\frac{\pi}{2} + i\sin\frac{\pi}{2}\right)$. Thus,

$(27i)^{1/3} = \sqrt[3]{27}\left[\cos\left(\dfrac{\frac{\pi}{2} + 2k\pi}{3}\right) + i\sin\left(\dfrac{\frac{\pi}{2} + 2k\pi}{3}\right)\right]$

$= 3\left[\cos\left(\dfrac{\pi + 4k\pi}{6}\right) + i\sin\left(\dfrac{\pi + 4k\pi}{6}\right)\right]$

for $k = 0, 1, 2$. Thus, the three roots are

$w_0 = 3\left(\cos\frac{\pi}{6} + i\sin\frac{\pi}{6}\right) = 3\left(\frac{\sqrt{3}}{2} + \frac{1}{2}i\right) = \frac{3}{2}\left(\sqrt{3} + i\right)$,

$w_1 = 3\left(\cos\frac{5\pi}{6} + i\sin\frac{5\pi}{6}\right) = 3\left(-\frac{\sqrt{3}}{2} + \frac{1}{2}i\right) = \frac{3}{2}\left(-\sqrt{3} + i\right)$, and

$w_2 = 3\left(\cos\frac{9\pi}{6} + i\sin\frac{9\pi}{6}\right) = -3i$.

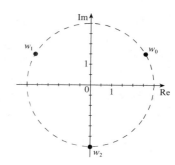

6. (a) $\mathbf{u}$ has initial point $P\left(3, -1\right)$ and terminal point $Q\left(-3, 9\right)$. Therefore,

$\mathbf{u} = \langle -3 - 3, 9 - \left(-1\right)\rangle = \langle -6, 10\rangle = -6\mathbf{i} + 10\mathbf{j}$.

(b) $\mathbf{u} = \langle -6, 10\rangle \quad\Rightarrow\quad |\mathbf{u}| = \sqrt{\left(-6\right)^2 + 10^2} = \sqrt{4\cdot\left(9 + 25\right)} = 2\sqrt{34}$

7. $\mathbf{u} = \langle 1, 3\rangle$ and $\mathbf{v} = \langle -6, 2\rangle$.

(a) $\mathbf{u} - 3\mathbf{v} = \langle 1, 3\rangle - \langle -18, 6\rangle = \langle 19, -3\rangle$

(b) $|\mathbf{u} + \mathbf{v}| = |\langle 1 - 6, 3 + 2\rangle| = |\langle -5, 5\rangle| = \sqrt{\left(-5\right)^2 + 5^2} = 5\sqrt{2}$

(c) $\mathbf{u}\cdot\mathbf{v} = \left(1\right)\left(-6\right) + \left(3\right)\left(2\right) = -6 + 6 = 0$

(d) Since $\mathbf{u}\cdot\mathbf{v} = 0$, it is true that $\mathbf{u}$ and $\mathbf{v}$ are perpendicular.

8. (a)

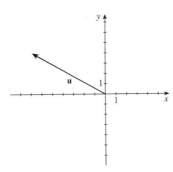

(b) $|\mathbf{u}| = \sqrt{\left(-4\sqrt{3}\right)^2 + 4^2} = \sqrt{48 + 16} = 8$ and $\tan\theta = \frac{4}{-4\sqrt{3}} = -\frac{\sqrt{3}}{3}$, so $\theta = \frac{5\pi}{6}$ (since the terminal point of $\mathbf{u}$ is in
quadrant III).

9. Let $\mathbf{r}$ represent the velocity of the river and $\mathbf{b}$ the velocity of the boat. Then $\mathbf{r} = \langle 8, 0 \rangle$
and $\mathbf{b} = \langle 12\sin 30°, 12\cos 30° \rangle = \langle 6, 6\sqrt{3} \rangle$.

(a) The resultant (or true) velocity is $\mathbf{v} = \mathbf{r} + \mathbf{b} = \langle 14, 6\sqrt{3} \rangle = 14\mathbf{i} + 6\sqrt{3}\mathbf{j}$.

(b) $|\mathbf{v}| = \sqrt{14^2 + \left(6\sqrt{3}\right)^2} = \sqrt{196 + 108} = \sqrt{304} \approx 17.44$ and $\theta = \tan^{-1}\left(\frac{6\sqrt{3}}{14}\right) = \tan^{-1}\left(\frac{3\sqrt{3}}{7}\right) \approx 36.6°$.

Therefore, the true speed of the boat is approximately 17.4 mi/h and the true direction is approximately N 53.4° E.

10. (a) $\cos\theta = \dfrac{\langle 3, 2 \rangle \cdot \langle 5, -1 \rangle}{|\langle 3, 2 \rangle|\,|\langle 5, -1 \rangle|} = \dfrac{15 - 2}{\sqrt{13}\sqrt{26}} = \dfrac{13}{13\sqrt{2}} = \dfrac{1}{\sqrt{2}} \quad \Rightarrow \quad \theta = \frac{\pi}{4}$ rad $= 45°$.

(b) The component of $\mathbf{u}$ along $\mathbf{v}$ is $\dfrac{\mathbf{u} \cdot \mathbf{v}}{|\mathbf{v}|} = \dfrac{\langle 3, 2 \rangle \cdot \langle 5, -1 \rangle}{\sqrt{26}} = \dfrac{15 - 2}{\sqrt{26}} = \dfrac{13}{\sqrt{26}} = \dfrac{\sqrt{26}}{2}.$

(c) $\text{proj}_{\mathbf{v}}\,\mathbf{u} = \left(\dfrac{\mathbf{u} \cdot \mathbf{v}}{|\mathbf{v}|^2}\right)\mathbf{v} = \left(\dfrac{\langle 3, 2 \rangle \cdot \langle 5, -1 \rangle}{5^2 + (-1)^2}\right)\langle 5, -1 \rangle = \frac{13}{26}\langle 5, -1 \rangle = \frac{1}{2}\langle 5, -1 \rangle = \langle \frac{5}{2}, -\frac{1}{2} \rangle.$

11. The displacement vector is $\mathbf{D} = \langle 7, -13 \rangle - \langle 2, 2 \rangle = \langle 5, -15 \rangle$. Therefore, the work done is
$W = \mathbf{F} \cdot \mathbf{D} = \langle 3, -5 \rangle \cdot \langle 5, -15 \rangle = 15 + 75 = 90.$

Focus on Modeling: Mapping the World

1. (a) Since we want $x = 36$ when $\alpha = 360°$, we substitute these values and solve for R: $36 = \frac{\pi}{180} \cdot 360R$ $\Leftrightarrow$ $R = \frac{36}{2\pi} \approx 5.73$ inches.

(b) The circumference of the earth at the equator is $2\pi (3960)$. Since the length of the equator on the map is 36 inches, each inch represents $\frac{1}{36}$ of that length, or $\frac{1}{36} \cdot 7920\pi \approx 691.2$ miles.

3. (a) $x = \frac{\pi}{180} (-122.3) (5.73) \approx -12.33$; $y = 5.73 \tan 47.6° \approx 6.28$.

(b) $x = \frac{\pi}{180} (37.6) (5.73) \approx 3.76$; $y = 5.73 \tan 55.8° \approx 8.43$

(c) $x = \frac{\pi}{180} (151.2) (5.73) \approx 15.12$; $y = 5.73 \tan (-33.9°) \approx -3.85$

(d) $x = \frac{\pi}{180} (-43.1) (5.73) \approx -4.31$; $y = 5.73 \tan (-22.9°) \approx -2.42$

5. The formula for the length of a circular arc is $s = r\theta$, where θ is measured in radians. Using this formula, we find that the distance along a meridian between the latitudes $\beta°$ and $(\beta + 1)°$ is $\frac{\pi}{180} R \approx 0.01745R$.

(a) The projected distance on the cylinder is $R \tan 21° - R \tan 10° \approx 0.01989R$. Thus, the ratio is $\dfrac{0.01989R}{0.01745R} \approx 1.14$.

(b) The projected distance on the cylinder is $R \tan 41° - R \tan 40° \approx 0.03019R$. Thus, the ratio is $\dfrac{0.03019R}{0.01745R} \approx 1.73$.

(c) The projected distance on the cylinder is $R \tan 81° - R \tan 80° \approx 0.64247R$. Thus, the ratio is $\dfrac{0.64247R}{0.01745R} \approx 36.82$.

7. The projected distance on the plane between latitude $\beta_1°$ and $\beta_2°$ is

$$2R \tan \left(\tfrac{1}{2}\beta_1 + 45° \right) - 2R \tan \left(\tfrac{1}{2}\beta_2 + 45° \right) = 2R \left[\tan \left(\tfrac{1}{2}\beta_1 + 45° \right) - \tan \left(\tfrac{1}{2}\beta_2 + 45° \right) \right]$$

In Problem 5 we found that on the sphere, the distance along a meridian between the latitudes $\beta°$ and $(\beta + 1)°$ is $\frac{\pi}{180} R \approx 0.01745R$.

(a) The projected distance on the plane is

$$2R \left[\tan \left(\tfrac{1}{2} (-20°) + 45° \right) - \tan \left(\tfrac{1}{2} (-21°) + 45° \right) \right] = 2R (\tan 35° - R \tan 34.5°)$$
$$\approx 0.02583R$$

Thus, the ratio is $\dfrac{0.02583R}{0.01745R} \approx 1.48$.

(b) The projected distance on the plane is

$$2R \left[\tan \left(\tfrac{1}{2} (-40°) + 45° \right) - \tan \left(\tfrac{1}{2} (-41°) + 45° \right) \right] = 2R (\tan 25° - R \tan 24.5°)$$
$$\approx 0.02116R$$

Thus, the ratio is $\dfrac{0.02116R}{0.01745R} \approx 1.21$.

(c) The projected distance on the plane is

$$2R \left[\tan \left(\tfrac{1}{2} (-80°) + 45° \right) - \tan \left(\tfrac{1}{2} (-81°) + 45° \right) \right] = 2R (\tan 5° - R \tan 4.5°)$$
$$\approx 0.01757R$$

Thus, the ratio is $\dfrac{0.01757R}{0.01745R} \approx 1.01$.

10 Systems of Equations and Inequalities

10.1 Systems of Equations

1. $\begin{cases} x - y = 2 \\ 2x + 3y = 9 \end{cases}$ Solving the first equation for x, we get $x = y + 2$, and substituting this into the second equation gives

$2(y+2) + 3y = 9 \quad \Leftrightarrow \quad 5y + 4 = 9 \quad \Leftrightarrow \quad 5y = 5 \quad \Leftrightarrow \quad y = 1$. Substituting for y we get $x = y + 2 = (1) + 2 = 3$.
Thus, the solution is $(3, 1)$.

3. $\begin{cases} y = x^2 \\ y = x + 12 \end{cases}$ Substituting $y = x^2$ into the second equation gives $x^2 = x + 12 \quad \Leftrightarrow$

$0 = x^2 - x - 12 = (x-4)(x+3) \Rightarrow x = 4$ or $x = -3$. So since $y = x^2$, the solutions are $(-3, 9)$ and $(4, 16)$.

5. $\begin{cases} x^2 + y^2 = 8 \\ x + y = 0 \end{cases}$ Solving the second equation for y gives $y = -x$, and substituting this into the first equation gives

$x^2 + (-x)^2 = 8 \quad \Leftrightarrow \quad 2x^2 = 8 \quad \Leftrightarrow \quad x = \pm 2$. So since $y = -x$, the solutions are $(2, -2)$ and $(-2, 2)$.

7. $\begin{cases} x + y^2 = 0 \\ 2x + 5y^2 = 75 \end{cases}$ Solving the first equation for x gives $x = -y^2$, and substituting this into the second equation

gives $2(-y^2) + 5y^2 = 75 \quad \Leftrightarrow \quad 3y^2 = 75 \quad \Leftrightarrow \quad y^2 = 25 \quad \Leftrightarrow \quad y = \pm 5$. So since $x = -y^2$, the solutions are
$(-25, -5)$ and $(-25, 5)$.

9. $\begin{cases} x + 2y = 5 \\ 2x + 3y = 8 \end{cases}$ Multiplying the first equation by 2 and the second by -1 gives the system

$\begin{cases} 2x + 4y = 10 \\ -2x - 3y = -8 \end{cases}$ Adding, we get $y = 2$, and substituting into the first equation in the original system gives

$x + 2(2) = 5 \Leftrightarrow x + 4 = 5 \quad \Leftrightarrow \quad x = 1$. The solution is $(1, 2)$.

11. $\begin{cases} x^2 - 2y = 1 \\ x^2 + 5y = 29 \end{cases}$ Subtracting the first equation from the second equation gives $7y = 28 \Rightarrow y = 4$. Substituting $y = 4$ into

the first equation of the original system gives $x^2 - 2(4) = 1 \quad \Leftrightarrow \quad x^2 = 9 \quad \Leftrightarrow \quad x = \pm 3$. The solutions are $(3, 4)$ and
$(-3, 4)$.

13. $\begin{cases} 3x^2 - y^2 = 11 \\ x^2 + 4y^2 = 8 \end{cases}$ Multiplying the first equation by 4 gives the system $\begin{cases} 12x^2 - 4y^2 = 44 \\ x^2 + 4y^2 = 8 \end{cases}$ Adding the equations

gives $13x^2 = 52 \Leftrightarrow x = \pm 2$. Substituting into the first equation we get $3(4) - y^2 = 11 \quad \Leftrightarrow \quad y = \pm 1$. Thus, the
solutions are $(2, 1)$, $(2, -1)$, $(-2, 1)$, and $(-2, -1)$.

15. $\begin{cases} x - y^2 + 3 = 0 \\ 2x^2 + y^2 - 4 = 0 \end{cases}$ Adding the two equations gives $2x^2 + x - 1 = 0$. Using the quadratic formula we have

$x = \dfrac{-1 \pm \sqrt{1 - 4(2)(-1)}}{2(2)} = \dfrac{-1 \pm \sqrt{9}}{4} = \dfrac{-1 \pm 3}{4}$. So $x = \dfrac{-1 - 3}{4} = -1$ or $x = \dfrac{-1 + 3}{4} = \frac{1}{2}$. Substituting $x = -1$

into the first equation gives $-1 - y^2 + 3 = 0 \Leftrightarrow y^2 = 2 \quad \Leftrightarrow \quad y = \pm\sqrt{2}$. Substituting $x = \frac{1}{2}$ into the first equation

gives $\frac{1}{2} - y^2 + 3 = 0 \quad \Leftrightarrow \quad y^2 = \frac{7}{2} \quad \Leftrightarrow \quad y = \pm\sqrt{\frac{7}{2}}$. Thus the solutions are $\left(-1, \pm\sqrt{2}\right)$ and $\left(\frac{1}{2}, \pm\sqrt{\frac{7}{2}}\right)$.

17. $\begin{cases} 2x + y = -1 \\ x - 2y = -8 \end{cases}$ By inspection of the graph, it appears that $(-2, 3)$ is the solution to the system. We check this in both

equations to verify that it is a solution. $2(-2) + 3 = -4 + 3 = -1$ and $-2 - 2(3) = -2 - 6 = -8$. Since both equations are satisfied, the solution is $(-2, 3)$.

19. $\begin{cases} x^2 + y = 8 \\ x - 2y = -6 \end{cases}$ By inspection of the graph, it appears that $(2, 4)$ is a solution, but is difficult to get accurate values

for the other point. Multiplying the first equation by 2 gives the system $\begin{cases} 2x^2 + 2y = 16 \\ x - 2y = -6 \end{cases}$ Adding the equations

gives $2x^2 + x = 10 \Leftrightarrow 2x^2 + x - 10 = 0 \quad \Leftrightarrow \quad (2x + 5)(x - 2) = 0$. So $x = -\frac{5}{2}$ or $x = 2$. If $x = -\frac{5}{2}$, then

$-\frac{5}{2} - 2y = -6 \quad \Leftrightarrow \quad -2y = -\frac{7}{2} \Leftrightarrow y = \frac{7}{4}$, and if $x = 2$, then $2 - 2y = -6 \Leftrightarrow -2y = -8 \quad \Leftrightarrow \quad y = 4$. Hence, the

solutions are $\left(-\frac{5}{2}, \frac{7}{4}\right)$ and $(2, 4)$.

21. $\begin{cases} x^2 + y = 0 \\ x^3 - 2x - y = 0 \end{cases}$ By inspection of the graph, it appears that $(-2, -4)$, $(0, 0)$, and $(1, -1)$ are solutions to the

system. We check each point in both equations to verify that it is a solution.
For $(-2, -4)$: $(-2)^2 + (-4) = 4 - 4 = 0$ and $(-2)^3 - 2(-2) - (-4) = -8 + 4 + 4 = 0$.
For $(0, 0)$: $(0)^2 + (0) = 0$ and $(0)^3 - 2(0) - (0) = 0$.
For $(1, -1)$: $(1)^2 + (-1) = 1 - 1 = 0$ and $(1)^3 - 2(1) - (-1) = 1 - 2 + 1 = 0$.
Thus, the solutions are $(-2, -4)$, $(0, 0)$, and $(1, -1)$.

23. $\begin{cases} y + x^2 = 4x \\ y + 4x = 16 \end{cases}$ Subtracting the second equation from the first equation gives $x^2 - 4x = 4x - 16 \quad \Leftrightarrow$

$x^2 - 8x + 16 = 0 \Leftrightarrow (x - 4)^2 = 0 \quad \Leftrightarrow \quad x = 4$. Substituting this value for x into either of the original equations gives

$y = 0$. Therefore, the solution is $(4, 0)$.

25. $\begin{cases} x - 2y = 2 \\ y^2 - x^2 = 2x + 4 \end{cases}$ Now $x - 2y = 2 \quad \Leftrightarrow \quad x = 2y + 2$. Substituting for x gives $y^2 - x^2 = 2x + 4 \quad \Leftrightarrow$

$y^2 - (2y + 2)^2 = 2(2y + 2) + 4 \quad \Leftrightarrow \quad y^2 - 4y^2 - 8y - 4 = 4y + 4 + 4 \quad \Leftrightarrow \quad y^2 + 4y + 4 = 0 \Leftrightarrow (y + 2)^2 = 0$
$\Leftrightarrow \quad y = -2$. Since $x = 2y + 2$, we have $x = 2(-2) + 2 = -2$. Thus, the solution is $(-2, -2)$.

27. $\begin{cases} x - y = 4 \\ xy = 12 \end{cases}$ Now $x - y = 4 \quad \Leftrightarrow \quad x = 4 + y$. Substituting for x gives $xy = 12 \quad \Leftrightarrow$

$(4 + y)y = 12 \Leftrightarrow y^2 + 4y - 12 = 0 \quad \Leftrightarrow \quad (y + 6)(y - 2) = 0 \quad \Leftrightarrow \quad y = -6, y = 2$. Since $x = 4 + y$, the solutions

are $(-2, -6)$ and $(6, 2)$.

29. $\begin{cases} x^2y = 16 \\ x^2 + 4y + 16 = 0 \end{cases}$ Now $x^2y = 16 \Leftrightarrow x^2 = \dfrac{16}{y}$.   Substituting for x^2 gives

$\dfrac{16}{y} + 4y + 16 = 0 \Rightarrow 4y^2 + 16y + 16 = 0 \Leftrightarrow y^2 + 4y + 4 = 0 \Leftrightarrow (y+2)^2 = 0 \Leftrightarrow y = -2$. Therefore,

$x^2 = \dfrac{16}{-2} = -8$, which has no real solution, and so the system has no solution.

31. $\begin{cases} x^2 + y^2 = 9 \\ x^2 - y^2 = 1 \end{cases}$ Adding the equations gives $2x^2 = 10 \Leftrightarrow x^2 = 5 \Leftrightarrow x = \pm\sqrt{5}$.   Now

$x = \pm\sqrt{5} \Rightarrow y^2 = 9 - 5 = 4 \Leftrightarrow y = \pm 2$, and so the solutions are $(\sqrt{5}, 2)$, $(\sqrt{5}, -2)$, $(-\sqrt{5}, 2)$, and $(-\sqrt{5}, -2)$.

33. $\begin{cases} 2x^2 - 8y^3 = 19 \\ 4x^2 + 16y^3 = 34 \end{cases}$ Multiplying the first equation by 2 gives the system $\begin{cases} 4x^2 - 16y^3 = 38 \\ 4x^2 + 16y^3 = 34 \end{cases}$ Adding the two

equations gives $8x^2 = 72 \Leftrightarrow x = \pm 3$, and then substituting into the first equation we have $2(9) - 8y^3 = 19 \Leftrightarrow$

$y^3 = -\frac{1}{8} \Leftrightarrow y = -\frac{1}{2}$. Therefore, the solutions are $\left(3, -\frac{1}{2}\right)$ and $\left(-3, -\frac{1}{2}\right)$.

35. $\begin{cases} \dfrac{2}{x} - \dfrac{3}{y} = 1 \\ -\dfrac{4}{x} + \dfrac{7}{y} = 1 \end{cases}$ If we let $u = \dfrac{1}{x}$ and $v = \dfrac{1}{y}$, the system is equivalent to $\begin{cases} 2u - 3v = 1 \\ -4u + 7v = 1 \end{cases}$ Multiplying the first

equation by 4 gives the system $\begin{cases} 4u - 6v = 2 \\ -4u + 7v = 1 \end{cases}$ Adding the equations gives $v = 3$, and then substituting into the first

equation gives $2u - 9 = 1 \Leftrightarrow u = 5$. Thus, the solution is $\left(\frac{1}{5}, \frac{1}{3}\right)$.

37. $\begin{cases} y = 2x + 6 \\ y = -x + 5 \end{cases}$

The solution is approximately $(-0.33, 5.33)$.

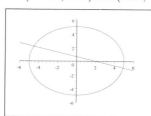

39. $\begin{cases} y = x^2 + 8x \\ y = 2x + 16 \end{cases}$

The solutions are $(-8, 0)$ and $(2, 20)$.

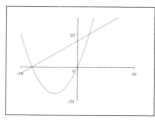

41. $\begin{cases} x^2 + y^2 = 25 \\ x + 3y = 2 \end{cases} \Leftrightarrow \begin{cases} y = \pm\sqrt{25 - x^2} \\ y = -\frac{1}{3}x + \frac{2}{3} \end{cases}$

The solutions are $(-4.51, 2.17)$ and $(4.91, -0.97)$.

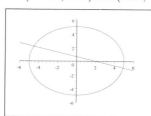

43. $\begin{cases} \dfrac{x^2}{9} + \dfrac{y^2}{18} = 1 \\ y = -x^2 + 6x - 2 \end{cases} \Leftrightarrow \begin{cases} y = \pm\sqrt{18 - 2x^2} \\ y = -x^2 + 6x - 2 \end{cases}$

The solutions are $(1.23, 3.87)$ and $(-0.35, -4.21)$.

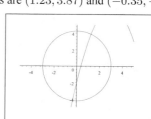

45. $\begin{cases} x^4 + 16y^4 = 32 \\ x^2 + 2x + y = 0 \end{cases} \Leftrightarrow \begin{cases} y = \pm\dfrac{\sqrt[4]{32 - x^4}}{2} \\ y = -x^2 - 2x \end{cases}$

The solutions are $(-2.30, -0.70)$ and $(0.48, -1.19)$.

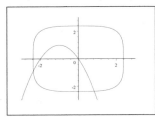

47. Let w and l be the lengths of the sides, in cm. Then we

have the system $\begin{cases} lw = 180 \\ 2l + 2w = 54 \end{cases}$ We solve the

second equation for w giving, $w = 27 - l$, and substitute

into the first equation to get

$l\,(27 - l) = 180 \Leftrightarrow l^2 - 27l + 180 = 0 \quad \Leftrightarrow$

$(l - 15)\,(l - 12) = 0 \Rightarrow l = 15$ or $l = 12$. If $l = 15$,

then $w = 27 - 15 = 12$, and if $l = 12$, then

$w = 27 - 12 = 15$. Therefore, the dimensions of the

rectangle are 12 cm by 15 cm.

49. Let l and w be the length and width, respectively, of the rectangle. Then, the system of equations

is $\begin{cases} 2l + 2w = 70 \\ \sqrt{l^2 + w^2} = 25 \end{cases}$ Solving the first equation for l, we have $l = 35 - w$, and substituting into

the second gives $\sqrt{l^2 + w^2} = 25 \Leftrightarrow l^2 + w^2 = 625 \quad \Leftrightarrow \quad (35 - w)^2 + w^2 = 625 \quad \Leftrightarrow$

$1225 - 70w + w^2 + w^2 = 625 \Leftrightarrow 2w^2 - 70w + 600 = 0 \quad \Leftrightarrow \quad (w - 15)\,(w - 20) = 0 \Rightarrow w = 15$ or $w = 20$. So the

dimensions of the rectangle are 15 and 20.

51. At the points where the rocket path and the hillside meet, we have $\begin{cases} y = \frac{1}{2}x \\ y = -x^2 + 401x \end{cases}$ Substituting for y in the second

equation gives $\frac{1}{2}x = -x^2 + 401x \quad \Leftrightarrow \quad x^2 - \frac{801}{2}x = 0 \Leftrightarrow x\left(x - \frac{801}{2}\right) = 0 \Rightarrow x = 0, x = \frac{801}{2}$. When $x = 0$, the

rocket has not left the pad. When $x = \frac{801}{2}$, then $y = \frac{1}{2}\left(\frac{801}{2}\right) = \frac{801}{4}$. So the rocket lands at the point $\left(\frac{801}{2}, \frac{801}{4}\right)$. The

distance from the base of the hill is $\sqrt{\left(\frac{801}{2}\right)^2 + \left(\frac{801}{4}\right)^2} \approx 447.77$ meters.

53. The point P is at an intersection of the circle of radius 26 centered at $A\,(22, 32)$

and the circle of radius 20 centered at $B\,(28, 20)$. We have the system

$\begin{cases} (x - 22)^2 + (y - 32)^2 = 26^2 \\ (x - 28)^2 + (y - 20)^2 = 20^2 \end{cases} \quad \Leftrightarrow$

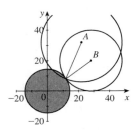

$\begin{cases} x^2 - 44x + 484 + y^2 - 64y + 1024 = 676 \\ x^2 - 56x + 784 + y^2 - 40y + 400 = 400 \end{cases} \quad \Leftrightarrow$

$\begin{cases} x^2 - 44x + y^2 - 64y = -832 \\ x^2 - 56x + y^2 - 40y = -784 \end{cases}$ Subtracting the two equations, we get $12x - 24y = -48 \Leftrightarrow x - 2y = -4$,

which is the equation of a line. Solving for x, we have $x = 2y - 4$. Substituting into the first equation gives

$(2y - 4)^2 - 44\,(2y - 4) + y^2 - 64y = -832 \quad \Leftrightarrow \quad 4y^2 - 16y + 16 - 88y + 176 + y^2 - 64y = -832$

$\Leftrightarrow \quad 5y^2 - 168y + 192 = -832 \quad \Leftrightarrow \quad 5y^2 - 168y + 1024 = 0$. Using the quadratic formula, we have

$y = \frac{168 \pm \sqrt{168^2 - 4(5)(1024)}}{2(5)} = \frac{168 \pm \sqrt{7744}}{10} = \frac{168 \pm 88}{10} \quad \Leftrightarrow \quad y = 8$ or $y = 25.60$. Since the y-coordinate of the point P

must be less than that of point A, we have $y = 8$. Then $x = 2\,(8) - 4 = 12$. So the coordinates of P are $(12, 8)$.

To solve graphically, we must solve each equation for y. This gives $(x - 22)^2 + (y - 32)^2 = 26^2$ $\Leftrightarrow (y - 32)^2 = 26^2 - (x - 22)^2 \Rightarrow y - 32 = \pm\sqrt{676 - (x - 22)^2} \Leftrightarrow y = 32 \pm \sqrt{676 - (x - 22)^2}$. We use the function $y = 32 - \sqrt{676 - (x - 22)^2}$ because the intersection we at interested in is below the point A. Likewise, solving the second equation for y, we would get the function $y = 20 - \sqrt{400 - (x - 28)^2}$. In a three-dimensional situation, you would need a minimum of three satellites, since a point on the earth can be uniquely specified as the intersection of three spheres centered at the satellites.

55. (a) $\begin{cases} \log x + \log y = \frac{3}{2} \\ 2\log x - \log y = 0 \end{cases}$ Adding the two equations gives $3\log x = \frac{3}{2} \Leftrightarrow \log x = \frac{1}{2} \Leftrightarrow x = \sqrt{10}$. Substituting

into the second equation we get $2\log 10^{1/2} - \log y = 0 \Leftrightarrow \log 10 - \log y = 0 \Leftrightarrow \log y = 1 \Leftrightarrow y = 10$. Thus, the solution is $(\sqrt{10}, 10)$.

(b) $\begin{cases} 2^x + 2^y = 10 \\ 4^x + 4^y = 68 \end{cases} \Leftrightarrow \begin{cases} 2^x + 2^y = 10 \\ 2^{2x} + 2^{2y} = 68 \end{cases}$ If we let $u = 2^x$ and $v = 2^y$, the system becomes $\begin{cases} u + v = 10 \\ u^2 + v^2 = 68 \end{cases}$

Solving the first equation for u, and substituting this into the second equation gives $u + v = 10 \Leftrightarrow u = 10 - v$, so $(10 - v)^2 + v^2 = 68 \Leftrightarrow 100 - 20v + v^2 + v^2 = 68 \Leftrightarrow v^2 - 10v + 16 = 0 \Leftrightarrow (v - 8)(v - 2) = 0 \Rightarrow v = 2$ or $v = 8$. If $v = 2$, then $u = 8$, and so $y = 1$ and $x = 3$. If $v = 8$, then $u = 2$, and so $y = 3$ and $x = 1$. Thus, the solutions are $(1, 3)$ and $(3, 1)$.

(c) $\begin{cases} x - y = 3 \\ x^3 - y^3 = 387 \end{cases}$ Solving the first equation for x gives $x = 3 + y$ and using the hint,

$x^3 - y^3 = 387 \Leftrightarrow (x - y)(x^2 + xy + y^2) = 387$. Next, substituting for x, we get

$3\left[(3 + y)^2 + y(3 + y) + y^2\right] = 387 \Leftrightarrow 9 + 6y + y^2 + 3y + y^2 + y^2 = 129 \Leftrightarrow 3y^2 + 9y + 9 = 129 \Leftrightarrow$

$(y + 8)(y - 5) = 0 \Rightarrow y = -8$ or $y = 5$. If $y = -8$, then $x = 3 + (-8) = -5$, and if $y = 5$, then $x = 3 + 5 = 8$. Thus the solutions are $(-5, -8)$ and $(8, 5)$.

(d) $\begin{cases} x^2 + xy = 1 \\ xy + y^2 = 3 \end{cases}$ Adding the equations gives $x^2 + xy + xy + y^2 = 4 \Leftrightarrow x^2 + 2xy + y^2 = 4 \Leftrightarrow$

$(x + y)^2 = 4 \Rightarrow x + y = \pm 2$. If $x + y = 2$, then from the first equation we get $x(x + y) = 1 \Rightarrow x \cdot 2 = 1 \Rightarrow x = \frac{1}{2}$, and so $y = 2 - \frac{1}{2} = \frac{3}{2}$. If $x + y = -2$, then from the first equation we get $x(x + y) = 1 \Rightarrow$ $x \cdot (-2) = 1 \Rightarrow x = -\frac{1}{2}$, and so $y = -2 - \left(-\frac{1}{2}\right) = -\frac{3}{2}$. Thus the solutions are $\left(\frac{1}{2}, \frac{3}{2}\right)$ and $\left(-\frac{1}{2}, -\frac{3}{2}\right)$.

10.2   Systems of Linear Equations in Two Variables

1. $\begin{cases} x + y = 4 \\ 2x - y = 2 \end{cases}$

The solution is $x = 2$, $y = 2$.

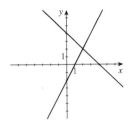

3. $\begin{cases} 2x - 3y = 12 \\ -x + \frac{3}{2}y = 4 \end{cases}$

The lines are parallel, so there is no intersection and hence no solution.

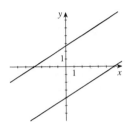

5. $\begin{cases} -x + \frac{1}{2}y = -5 \\ 2x - y = 10 \end{cases}$

There are infinitely many solutions.

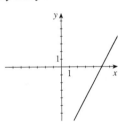

7. $\begin{cases} x + y = 4 \\ -x + y = 0 \end{cases}$ Adding the two equations gives

$2y = 4 \Leftrightarrow y = 2$. Substituting for y in the first equation gives $x + 2 = 4 \Leftrightarrow x = 2$. Hence, the solution is $(2, 2)$.

9. $\begin{cases} 2x - 3y = 9 \\ 4x + 3y = 9 \end{cases}$ Adding the two equations gives $6x = 18 \Leftrightarrow x = 3$. Substituting for x in the second equation gives

$4(3) + 3y = 9 \quad \Leftrightarrow \quad 12 + 3y = 9 \quad \Leftrightarrow \quad 3y = -3 \quad \Leftrightarrow \quad x = -1$. Hence, the solution is $(3, -1)$.

11. $\begin{cases} x + 3y = 5 \\ 2x - y = 3 \end{cases}$ Solving the first equation for x gives $x = -3y + 5$. Substituting for x in the second equation gives

$2(-3y + 5) - y = 3 \Leftrightarrow -6y + 10 - y = 3 \quad \Leftrightarrow \quad -7y = -7 \Leftrightarrow y = 1$. Then $x = -3(1) + 5 = 2$. Hence, the solution is $(2, 1)$.

13. $-x + y = 2 \quad \Leftrightarrow \quad y = x + 2$. Substituting for y into $4x - 3y = -3$ gives $4x - 3(x + 2) = -3 \quad \Leftrightarrow$ $4x - 3x - 6 = -3 \Leftrightarrow x = 3$, and so $y = (3) + 2 = 5$. Hence, the solution is $(3, 5)$.

15. $x + 2y = 7 \quad \Leftrightarrow \quad x = 7 - 2y$. Substituting for x into $5x - y = 2$ gives $5(7 - 2y) - y = 2$ $\Leftrightarrow \quad 35 - 10y - y = 2 \Leftrightarrow -11y = -33 \quad \Leftrightarrow \quad y = 3$, and so $x = 7 - 2(3) = 1$. Hence, the solution is $(1, 3)$.

17. $\frac{1}{2}x + \frac{1}{3}y = 2 \quad \Leftrightarrow \quad x + \frac{2}{3}y = 4 \quad \Leftrightarrow \quad x = 4 - \frac{2}{3}y$. Substituting for x into $\frac{1}{5}x - \frac{2}{3}y = 8$ gives $\frac{1}{5}\left(4 - \frac{2}{3}y\right) - \frac{2}{3}y = 8$ $\Leftrightarrow \quad \frac{4}{5} - \frac{2}{15}y - \frac{10}{15}y = 8 \Leftrightarrow 12 - 2y - 10y = 120 \quad \Leftrightarrow \quad y = -9$, and so $x = 4 - \frac{2}{3}(-9) = 10$. Hence, the solution is $(10, -9)$.

19. Adding twice the first equation to the second gives $0 = 32$, which is false. Thus, the system has no solution.

21. $\begin{cases} x + 4y = 8 \\ 3x + 12y = 2 \end{cases}$ Adding -3 times the first equation to the second equation gives $0 = -22$, which is never true. Thus,

the system has no solution.

23. $\begin{cases} 2x - 6y = 10 \\ -3x + 9y = -15 \end{cases}$ Adding 3 times the first equation to 2 times the second equation gives $0 = 0$. Writing the equation

in slope-intercept form, we have $2x - 6y = 10 \quad \Leftrightarrow \quad -6y = -2x + 10 \quad \Leftrightarrow \quad y = \frac{1}{3}x - \frac{5}{3}$, so the solutions are all

pairs of the form $\left(x, \frac{1}{3}x - \frac{5}{3}\right)$ where x is a real number.

25. $\begin{cases} 6x + 4y = 12 \\ 9x + 6y = 18 \end{cases}$ Adding 3 times the first equation to -2 times the second equation gives $0 = 0$. Writing the equation in

slope-intercept form, we have $6x + 4y = 12 \quad \Leftrightarrow \quad 4y = -6x + 12 \quad \Leftrightarrow \quad y = -\frac{3}{2}x + 3$, so the solutions are all pairs

of the form $\left(x, -\frac{3}{2}x + 3\right)$ where x is a real number.

27. $\begin{cases} 8s - 3t = -3 \\ 5s - 2t = -1 \end{cases}$ Adding 2 times the first equation to 3 times the second equation gives $s = -3$, so

$8(-3) - 3t = -3 \Leftrightarrow -24 - 3t = -3 \quad \Leftrightarrow \quad t = -7$. Thus, the solution is $(-3, -7)$.

29. $\begin{cases} \frac{1}{2}x + \frac{3}{5}y = 3 \\ \frac{5}{3}x + 2y = 10 \end{cases}$ Adding 10 times the first equation to -3 times the second equation gives $0 = 0$. Writing the equation

in slope-intercept form, we have $\frac{1}{2}x + \frac{3}{5}y = 3 \quad \Leftrightarrow \quad \frac{3}{5}y = -\frac{1}{2}x + 3 \quad \Leftrightarrow \quad y = -\frac{5}{6}x + 5$, so the solutions are all pairs

of the form $\left(x, -\frac{5}{6}x + 5\right)$ where x is a real number.

31. $\begin{cases} 0.4x + 1.2y = 14 \\ 12x - 5y = 10 \end{cases}$ Adding 30 times the first equation to -1 times the second equation gives $41y = 410 \quad \Leftrightarrow$

$y = 10$, so $12x - 5(10) = 10 \quad \Leftrightarrow \quad 12x = 60 \quad \Leftrightarrow \quad x = 5$. Thus, the solution is $(5, 10)$.

33. $\begin{cases} \frac{1}{3}x - \frac{1}{4}y = 2 \\ -8x + 6y = 10 \end{cases}$ Adding 24 times the first equation to the second equation gives $0 = 58$, which is never true. Thus,

the system has no solution.

35. $\begin{cases} 0.21x + 3.17y = 9.51 \\ 2.35x - 1.17y = 5.89 \end{cases}$

The solution is approximately $(3.87, 2.74)$.

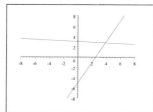

37. $\begin{cases} 2371x - 6552y = 13{,}591 \\ 9815x + 992y = 618{,}555 \end{cases}$

The solution is approximately $(61.00, 20.00)$.

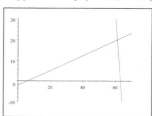

39. Subtracting the first equation from the second, we get $ay - y = 1 \Leftrightarrow y(a - 1) = 1 \quad \Leftrightarrow \quad y = \dfrac{1}{a - 1}$, $a \neq 1$. So

$x + \left(\dfrac{1}{a - 1}\right) = 0 \quad \Leftrightarrow \quad x = \dfrac{1}{1 - a} = -\dfrac{1}{a - 1}$. Thus, the solution is $\left(-\dfrac{1}{a - 1}, \dfrac{1}{a - 1}\right)$.

41. Subtracting b times the first equation from a times the second, we get $(a^2 - b^2) y = a - b$ $\Leftrightarrow$ $y = \dfrac{a - b}{a^2 - b^2} = \dfrac{1}{a + b}$,

$a^2 - b^2 \neq 0$. So $ax + \dfrac{b}{a + b} = 1$ $\Leftrightarrow$ $ax = \dfrac{a}{a + b}$ $\Leftrightarrow$ $x = \dfrac{1}{a + b}$. Thus, the solution is $\left(\dfrac{1}{a + b}, \dfrac{1}{a + b} \right)$.

43. Let the two numbers be x and y. Then $\begin{cases} x + y = 34 \\ x - y = 10 \end{cases}$ Adding these two equations gives $2x = 44 \Leftrightarrow x = 22$. So

$22 + y = 34$ $\Leftrightarrow$ $y = 12$. Therefore, the two numbers are 22 and 12.

45. Let d be the number of dimes and q be the number of quarters. This gives $\begin{cases} d + q = 14 \\ 0.10d + 0.25q = 2.75 \end{cases}$ Subtracting the

first equation from 10 times the second gives $1.5q = 13.5$ $\Leftrightarrow$ $q = 9$. So $d + 9 = 14 \Leftrightarrow d = 5$. Thus, the number of dimes is 5 and the number of quarters is 9.

47. Let x be the speed of the plane in still air and y be the speed of the wind. This gives $\begin{cases} 2x - 2y = 180 \\ 1.2x + 1.2y = 180 \end{cases}$ Subtracting

6 times the first equation from 10 times the second gives $24x = 2880$ $\Leftrightarrow$ $x = 120$, so $2(120) - 2y = 180$ $\Leftrightarrow$ $-2y = -60$ $\Leftrightarrow$ $y = 30$. Therefore, the speed of the plane is 120 mi/h and the wind speed is 30 mi/h.

49. Let x be the cycling speed and y be the running speed. (Remember to divide by 60 to convert minutes to decimal hours.)

We have $\begin{cases} 0.5x + 0.5y = 12.5 \\ 0.75x + 0.2y = 16 \end{cases}$ Subtracting 2 times the first equation from 5 times the second, we get $2.75x = 55$

$\Leftrightarrow$ $x = 20$, so $20 + y = 25$ $\Leftrightarrow$ $y = 5$. Thus, the cycling speed is 20 mi/h and the running speed is 5 mi/h.

51. Let a and b be the number of grams of food A and food B. Then $\begin{cases} 0.12a + 0.20b = 32 \\ 100a + 50b = 22{,}000 \end{cases}$ Subtracting 250 times the

first equation from the second, we get $70a = 14{,}000$ $\Leftrightarrow$ $a = 200$, so $0.12(200) + 0.20b = 32$ $\Leftrightarrow$ $0.20b = 8$ $\Leftrightarrow$ $b = 40$. Thus, she should use 200 grams of food A and 40 grams of food B.

53. Let x and y be the sulfuric acid concentrations in the first and second containers.

$\begin{cases} 300x + 600y = 900\,(0.15) \\ 100x + 500y = 600\,(0.125) \end{cases}$ Subtracting the first equation from 3 times the second gives $900y = 90$ $\Leftrightarrow$

$y = 0.10$, so $100x + 500(0.10) = 75$ $\Leftrightarrow$ $x = 0.25$. Thus, the concentrations of sulfuric acid are 25% in the first container and 10% in the second.

55. Let x be the amount invested at 6% and y the amount invested at 10%. The ratio of the amounts invested gives $x = 2y$. Then the interest earned is $0.06x + 0.10y = 3520$ $\Leftrightarrow$ $6x + 10y = 352{,}000$. Substituting gives $6(2y) + 10y = 352{,}000 \Leftrightarrow 22y = 352{,}000$ $\Leftrightarrow$ $y = 16{,}000$. Then $x = 2(16{,}000) = 32{,}000$. Thus, he invests $32,000 at 6% and $16,000 at 10%.

57. Let x be the tens digit and y be the ones digit of the number. $\begin{cases} x + y = 7 \\ 10y + x = 27 + 10x + y \end{cases}$ Adding 9 times the first

equation to the second gives $18x = 36$ $\Leftrightarrow$ $x = 2$, so $2 + y = 7$ $\Leftrightarrow$ $y = 5$. Thus, the number is 25.

59. $n = 5$, so $\sum_{k=1}^{n} x_k = 1 + 2 + 3 + 5 + 7 = 18$,

$\sum_{k=1}^{n} y_k = 3 + 5 + 6 + 6 + 9 = 29$,

$\sum_{k=1}^{n} x_k y_k = 1(3) + 2(5) + 3(6) + 5(6) + 7(9) = 124$, and

$\sum_{k=1}^{n} x_k^2 = 1^2 + 2^2 + 3^2 + 5^2 + 7^2 = 88$. Thus we get the system

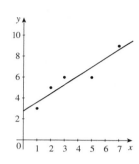

$$\begin{cases} 18a + 5b = 29 \\ 88a + 18b = 124 \end{cases}$$ Subtracting 18 times the first equation from 5 times the

second, we get $116a = 98$ $\Leftrightarrow$ $a \approx 0.845$. Then

$b = \frac{1}{5}[-18(0.845) + 29] \approx 2.758$. So the regression line is $y = 0.845x + 2.758$.

10.3 Systems of Linear Equations in Several Variables

1. The equation $6x - \sqrt{3}y + \frac{1}{2}z = 0$ is linear.

3. The system $\begin{cases} xy - 3y + z = 5 \\ x - y^2 + 5z = 0 \\ 2x + yz = 3 \end{cases}$ is not a linear system, since the first equation contains a product of variables. In fact both the second and the third equation are not linear.

5. $\begin{cases} x - 2y + 4z = 3 \\ y + 2z = 7 \\ z = 2 \end{cases}$ Substituting $z = 2$ into the second equation gives $y + 2(2) = 7$ $\Leftrightarrow$ $y = 3$. Substituting $z = 2$

and $y = 3$ into the first equation gives $x - 2(3) + 4(2) = 3 \Leftrightarrow x = 1$. Thus, the solution is $(1, 3, 2)$.

7. $\begin{cases} x + 2y + z = 7 \\ -y + 3z = 9 \\ 2z = 6 \end{cases}$ Solving we get $2z = 6$ $\Leftrightarrow$ $z = 3$. Substituting $z = 3$ into the second equation gives

$-y + 3(3) = 9 \Leftrightarrow y = 0$. Substituting $z = 3$ and $y = 0$ into the first equation gives $x + 2(0) + 3 = 7$ $\Leftrightarrow$ $x = 4$.
Thus, the solution is $(4, 0, 3)$.

9. $\begin{cases} 2x - y + 6z = 5 \\ y + 4z = 0 \\ -2z = 1 \end{cases}$ Solving we get $-2z = 1$ $\Leftrightarrow$ $z = -\frac{1}{2}$. Substituting $z = -\frac{1}{2}$ into the second equation gives

$y + 4\left(-\frac{1}{2}\right) = 0$ $\Leftrightarrow$ $y = 2$. Substituting $z = -\frac{1}{2}$ and $y = 2$ into the first equation gives $2x - (2) + 6\left(-\frac{1}{2}\right) = 5$ $\Leftrightarrow$
$x = 5$. Thus, the solution is $\left(5, 2, -\frac{1}{2}\right)$.

11. $\begin{cases} x - 2y - z = 4 \\ x - y + 3z = 0 \\ 2x + y + z = 0 \end{cases}$ Subtract the first equation from the second equation: $\begin{cases} x - 2y - z = 4 \\ y + 4z = -4 \\ 2x + y + z = 0 \end{cases}$

Or, subtract $\frac{1}{2}$ times the third equation from the second equation: $\begin{cases} x - 2y - z = 4 \\ -\frac{3}{2}y + \frac{5}{2}z = 0 \\ 2x + y + z = 0 \end{cases}$

13. $\begin{cases} 2x - y + 3z = 2 \\ x + 2y - z = 4 \\ -4x + 5y + z = 10 \end{cases}$ Add 2 times the first equation to the third equation: $\begin{cases} 2x - y + 3z = 2 \\ x + 2y - z = 4 \\ 3y + 7z = 14 \end{cases}$

Or, add 4 times the second equation to the third equation: $\begin{cases} 2x - y + 3z = 2 \\ x + 2y - z = 4 \\ 13y - 3z = 26 \end{cases}$

15. $\begin{cases} x + y + z = 4 \\ x + 3y + 3z = 10 \\ 2x + y - z = 3 \end{cases}$ $\Leftrightarrow$ $\begin{cases} x + y + z = 4 \\ 2y + 2z = 6 \\ y + 3z = 5 \end{cases}$ $\begin{array}{l} (-1) \times \text{Eq. 1} + \text{Eq. 2} \\ 2 \times \text{Eq. 1} + (-1) \times \text{Eq. 3} \end{array}$ $\Leftrightarrow$ $\begin{cases} x + y + z = 4 \\ y + 3z = 5 \quad \text{Eq. 3} \\ 2y + 2z = 6 \quad \text{Eq. 2} \end{cases}$

$\Leftrightarrow$ $\begin{cases} x + y + z = 4 \\ y + 3z = 5 \\ -4z = -4 \quad (-2) \times \text{Eq. 2} + \text{Eq. 3} \end{cases}$ $\Leftrightarrow$ $z = 1$ and $y + 3(1) = 8 \Leftrightarrow y = 2$. Then $x + 2 + 1 = 4$ $\Leftrightarrow$

$x = 1$. So the solution is $(1, 2, 1)$.

17. $\begin{cases} x - 4z = 1 \\ 2x - y - 6z = 4 \\ 2x + 3y - 2z = 8 \end{cases}$ $\Leftrightarrow$ $\begin{cases} x - 4z = 1 \\ -y + 2z = 2 \quad (-2) \times \text{Eq. 1} + \text{Eq. 2} \\ 3y + 6z = 6 \quad (-2) \times \text{Eq. 1} + \text{Eq. 3} \end{cases}$ $\Leftrightarrow$ $\begin{cases} x - 4z = 1 \\ -y + 2z = 2 \\ 12z = 12 \quad 3 \times \text{Eq. 2} + \text{Eq. 3} \end{cases}$

So $z = 1$ and $-y + 2(1) = 2 \Leftrightarrow y = 0$. Then $x - 4(1) = 1$ $\Leftrightarrow$ $x = 5$. So the solution is $(5, 0, 1)$.

19. $\begin{cases} 2x + 4y - z = 2 \\ x + 2y - 3z = -4 \\ 3x - y + z = 1 \end{cases}$ $\Leftrightarrow$ $\begin{cases} x + 2y - 3z = -4 \\ 2x + 4y - z = 2 \\ 3x - y + z = 1 \end{cases}$ $\Leftrightarrow$ $\begin{cases} x + 2y - 3z = -4 \\ -7y + 10z = 13 \quad \text{So } z = 2 \text{ and} \\ 5z = 10 \end{cases}$

$-7y + 10(2) = 13$ $\Leftrightarrow$ $y = 1$. Then $x + 2(1) - 3(2) = -4$ $\Leftrightarrow$ $x = 0$. So the solution is $(0, 1, 2)$.

21. $\begin{cases} y - 2z = 0 \\ 2x + 3y = 2 \\ -x - 2y + z = -1 \end{cases}$ $\Leftrightarrow$ $\begin{cases} -x - 2y + z = -1 \\ y - 2z = 0 \\ 2x + 3y = 2 \end{cases}$ $\Leftrightarrow$ $\begin{cases} -x - 2y + z = -1 \\ y - 2z = 0 \\ -y + 2z = 0 \end{cases}$ $\Leftrightarrow$ $\begin{cases} -x - 2y + z = -1 \\ y - 2z = 0 \\ 0 = 0 \end{cases}$

The system is dependent, so when $z = t$ we solve for y to get $y - 2t = 0$ $\Leftrightarrow$ $y = 2t$. Then
$-x - 2(2t) + t = -1 \Leftrightarrow -x - 3t = -1$ $\Leftrightarrow$ $x = -3t + 1$. So the solutions are $(-3t + 1, 2t, t)$, where t is any real number.

23. $\begin{cases} x + 2y - z = 1 \\ 2x + 3y - 4z = -3 \\ 3x + 6y - 3z = 4 \end{cases}$ $\Leftrightarrow$ $\begin{cases} x + 2y - z = 1 \\ -y - 2z = -5 \\ 0 = 1 \end{cases}$ Since $0 = 1$ is false, this system is inconsistent.

25. $\begin{cases} 2x + 3y - z = 1 \\ x + 2y = 3 \\ x + 3y + z = 4 \end{cases}$ $\Leftrightarrow$ $\begin{cases} x + 2y = 3 \\ 2x + 3y - z = 1 \\ x + 3y + z = 4 \end{cases}$ $\Leftrightarrow$ $\begin{cases} x + 2y = 3 \\ -y - z = -5 \\ y + z = 1 \end{cases}$ $\Leftrightarrow$ $\begin{cases} x + 2y = 3 \\ -y - z = -5 \\ 0 = -4 \end{cases}$

Since $0 = -4$ is false, this system is inconsistent.

27. $\begin{cases} x + y - z = 0 \\ x + 2y - 3z = -3 \\ 2x + 3y - 4z = -3 \end{cases}$ $\Leftrightarrow$ $\begin{cases} x + y - z = 0 \\ y - 2z = -3 \\ y - 2z = -3 \end{cases}$ $\Leftrightarrow$ $\begin{cases} x + y - z = 0 \\ y - 2z = -3 \\ 0 = 0 \end{cases}$

So $z = t$ and $y - 2t = -3$ $\Leftrightarrow$ $y = 2t - 3$. Then $x + (2t - 3) - t = 0$ $\Leftrightarrow$ $x = -t + 3$. So the solutions are
$(-t + 3, 2t - 3, t)$, where t is any real number.

29. $\begin{cases} x + 3y - 2z = 0 \\ 2x \quad\quad + 4z = 4 \\ 4x + 6y \quad\quad = 4 \end{cases}$ $\Leftrightarrow$ $\begin{cases} x + 3y - 2z = 0 \\ -6y + 8z = 4 \\ -6y + 8z = 4 \end{cases}$ $\Leftrightarrow$ $\begin{cases} x + 3y - 2z = 0 \\ -6y + 8z = 4 \\ 0 = 0 \end{cases}$

So $z = t$ and $-6y + 8t = 4$ $\Leftrightarrow$ $-6y = -8t + 4 \Leftrightarrow y = \frac{4}{3}t - \frac{2}{3}$. Then $x + 3\left(\frac{4}{3}t - \frac{2}{3}\right) - 2t = 0$ $\Leftrightarrow$ $x = -2t + 2$.

So the solutions are $\left(-2t + 2, \frac{4}{3}t - \frac{2}{3}, t\right)$, where t is any real number.

31. $\begin{cases} x \quad\quad + z + 2w = 6 \\ y - 2z \quad\quad = -3 \\ x + 2y - z \quad\quad = -2 \\ 2x + y + 3z - 2w = 0 \end{cases}$ $\Leftrightarrow$ $\begin{cases} x \quad\quad + z + 2w = 6 \\ y - 2z \quad\quad = -3 \\ 2y - 2z - 2w = -8 \left(R_2 - 2\right) + R_3 \\ y + z - 6w = -12 \end{cases}$ $\Leftrightarrow$ $\begin{cases} x \quad\quad + z + 2w = 6 \\ y - 2z \quad\quad = -3 \\ z - w = -1 \\ 3z - 6w = -9 \end{cases}$ $\Leftrightarrow$

$\begin{cases} x \quad + z + 2w = 6 \\ y - 2z \quad\quad = -3 \\ z - w = -1 \\ -3w = -6 \end{cases}$

$2y + 4z = 6$
$+2y - 2z - 2w = -8$
$2z - 2w = -2$

So $w = 2$ and $z - 2 = -1$ $\Leftrightarrow$ $z = 1$. Then $y - 2(1) = -3$ $\Leftrightarrow$ $y = -1$ and

$x + 1 + 2(2) = 6$ $\Leftrightarrow$ $x = 1$. Thus, the solution is $(1, -1, 1, 2)$.

33. Let x be the amount invested at 4%, y the amount invested at 5%, and z the amount invested at 6%. We set up

a model and get the following equations: $\begin{cases} \text{Total money:} & x + y + z = 100{,}000 \\ \text{Annual income:} & 0.04x + 0.05y + 0.06z = 0.051\,(100{,}000) \\ \text{Equal amounts:} & x = y \end{cases}$ $\Leftrightarrow$

$\begin{cases} x + y + z = 100{,}000 \\ 4x + 5y + 6z = 510{,}000 \\ x - y = 0 \end{cases}$ $\Leftrightarrow$ $\begin{cases} x + y + z = 100{,}000 \\ y + 2z = 110{,}000 \\ -2y - z = -100{,}000 \end{cases}$ $\Leftrightarrow$ $\begin{cases} x + y + z = 100{,}000 \\ y + 2z = 110{,}000 \\ 3z = 120{,}000 \end{cases}$ So

$z = 40{,}000$ and $y + 2(40{,}000) = 110{,}000$ $\Leftrightarrow$ $y = 30{,}000$. Since $x = y$, $x = 30{,}000$. She must invest $30,000 in short-term bonds, $30,000 in intermediate-term bonds, and $40,000 in long-term bonds.

35. Let a, b, and c be the number of ounces of Type A, Type B, and Type C pellets used. The requirements for the different

vitamins gives the following system: $\begin{cases} 2a + 3b + c = 9 \\ 3a + b + 3c = 14 \\ 8a + 5b + 7c = 32 \end{cases}$ $\Leftrightarrow$ $\begin{cases} 2a + 3b + c = 9 \\ -7b + 3c = 1 \\ -7b + 3c = -4 \end{cases}$

Equations 2 and 3 are inconsistent, so there is no solution.

37. Let x, y, and z be the number of acres of land planted with corn, wheat, and soybeans. We set up a model and

get the following equations: $\begin{cases} \text{Total acres:} & x + y + z = 1200 \\ \text{Market demand:} & 2x = y \\ \text{Total cost:} & 45x + 60y + 50z = 63{,}750 \end{cases}$ Substituting $2x$ for y, we get

$\begin{cases} x + 2x + z = 1200 \\ 2x = y \\ 45x + 60\,(2x) + 50z = 63{,}750 \end{cases}$ $\Leftrightarrow$ $\begin{cases} 3x + z = 1200 \\ 2x - y = 0 \\ 165x + 50z = 63{,}750 \end{cases}$ $\Leftrightarrow$ $\begin{cases} 3x + z = 1200 \\ 2x - y = 0 \\ 15x = 3750 \end{cases}$ So

$15x = 3{,}750$ $\Leftrightarrow$ $x = 250$ and $y = 2\,(250) = 500$. Substituting into the original equation, we have

$250 + 500 + z = 1200$ $\Leftrightarrow$ $z = 450$. Thus the farmer should plant 250 acres of corn, 500 acres of wheat, and 450 acres of soybeans.

39. (a) We begin by substituting $\dfrac{x_0 + x_1}{2}$, $\dfrac{y_0 + y_1}{2}$, and $\dfrac{z_0 + z_1}{2}$ into the left-hand side of the first equation:

$$a_1 \left(\frac{x_0 + x_1}{2}\right) + b_1 \left(\frac{y_0 + y_1}{2}\right) + c_1 \left(\frac{z_0 + z_1}{2}\right) = \tfrac{1}{2}\left[(a_1 x_0 + b_1 y_0 + c_1 z_0) + (a_1 x_1 + b_1 y_1 + c_1 z_1)\right]$$

$$= \tfrac{1}{2}\left[d_1 + d_1\right] = d_1$$

Thus the given ordered triple satisfies the first equation. We can show that it satisfies the second and the third in exactly the same way. Thus it is a solution of the system.

(b) We have shown in part (a) that if the system has two different solutions, we can find a third one by averaging the two solutions. But then we can find a fourth and a fifth solution by averaging the new one with each of the previous two. Then we can find four more by repeating this process with these new solutions, and so on. Clearly this process can continue indefinitely, so there are infinitely many solutions.

10.4   Systems of Linear Equations: Matrices

1. 3×2

3. 2×1

5. 1×3

7. (a) Yes, this matrix is in row-echelon form.

(b) Yes, this matrix is in reduced row-echelon form.

(c) $\begin{cases} x = -3 \\ y = 5 \end{cases}$

9. (a) Yes, this matrix is in row-echelon form.

(b) No, this matrix is not in reduced row-echelon form, since the leading 1 in the second row does not have a zero above it.

(c) $\begin{cases} x + 2y + 8z = 0 \\ y + 3z = 2 \\ 0 = 0 \end{cases}$

11. (a) No, this matrix is not in row-echelon form, since the row of zeros is not at the bottom.

(b) No, this matrix is not in reduced row-echelon form.

(c) $\begin{cases} x = 0 \\ 0 = 0 \\ y + 5z = 1 \end{cases}$

13. (a) Yes, this matrix is in row-echelon form.

(b) Yes, this matrix is in reduced row-echelon form.

(c) $\begin{cases} x + 3y - w = 0 \\ z + 2w = 2 \\ 0 = 1 \\ 0 = 0 \end{cases}$

Notice that this system has no solution.

15. $\begin{bmatrix} 1 & -2 & 1 & 1 \\ 0 & 1 & 2 & 5 \\ 1 & 1 & 3 & 8 \end{bmatrix} \xrightarrow{R_3 - R_1 \to R_3} \begin{bmatrix} 1 & -2 & 1 & 1 \\ 0 & 1 & 2 & 5 \\ 0 & 3 & 2 & 7 \end{bmatrix} \xrightarrow{R_3 - 3R_2 \to R_3} \begin{bmatrix} 1 & -2 & 1 & 1 \\ 0 & 1 & 2 & 5 \\ 0 & 0 & -4 & -8 \end{bmatrix}$. Thus, $-4z = -8 \iff$

$z = 2$; $y + 2(2) = 5 \iff y = 1$; and $x - 2(1) + (2) = 1 \iff x = 1$. Therefore, the solution is $(1, 1, 2)$.

17. $\begin{bmatrix} 1 & 1 & 1 & 2 \\ 2 & -3 & 2 & 4 \\ 4 & 1 & -3 & 1 \end{bmatrix} \begin{array}{c} \xrightarrow{R_2 - 2R_1 \to R_2} \\ \xrightarrow{R_3 - 4R_1 \to R_3} \end{array} \begin{bmatrix} 1 & 1 & 1 & 2 \\ 0 & -5 & 0 & 0 \\ 0 & -3 & -7 & -7 \end{bmatrix} \xrightarrow{R_3 - \frac{3}{5}R_2 \to R_3} \begin{bmatrix} 1 & 1 & 1 & 2 \\ 0 & -5 & 0 & 0 \\ 0 & 0 & -7 & -7 \end{bmatrix}$. Thus, $-7z = -7$

$\iff z = 1$; $-5y = 0 \iff y = 0$; and $x + 0 + 1 = 2 \iff x = 1$. Therefore, the solution is $(1, 0, 1)$.

19. $\begin{bmatrix} 1 & 2 & -1 & -2 \\ 1 & 0 & 1 & 0 \\ 2 & -1 & -1 & -3 \end{bmatrix}$ $\xrightarrow[R_3 - 2R_1 \to R_3]{R_2 - R_1 \to R_2}$ $\begin{bmatrix} 1 & 2 & -1 & -2 \\ 0 & -2 & 2 & 2 \\ 0 & -5 & 1 & 1 \end{bmatrix}$ $\xrightarrow{-\frac{1}{2}R_2}$ $\begin{bmatrix} 1 & 2 & -1 & -2 \\ 0 & 1 & 1 & 1 \\ 0 & -5 & 1 & 1 \end{bmatrix}$ $\xrightarrow{R_3 + 5R_2 \to R_3}$

$\begin{bmatrix} 1 & 2 & -1 & -2 \\ 0 & 1 & 1 & 1 \\ 0 & 0 & 6 & 6 \end{bmatrix}$. Thus, $6z = 6 \iff z = 1$; $y + (1) = 1 \iff y = 0$; and $x + 2(0) - (1) = -2 \iff x = -1$.

Therefore, the solution is $(-1, 0, 1)$.

21. $\begin{bmatrix} 1 & 2 & -1 & 9 \\ 2 & 0 & -1 & -2 \\ 3 & 5 & 2 & 22 \end{bmatrix}$ $\xrightarrow[R_3 - 3R_1 \to R_3]{R_2 - 2R_1 \to R_2}$ $\begin{bmatrix} 1 & 2 & -1 & 9 \\ 0 & -4 & 1 & -20 \\ 0 & -1 & 5 & -5 \end{bmatrix}$ $\xrightarrow{4R_3 - R_2 \to R_3}$ $\begin{bmatrix} 1 & 2 & -1 & 9 \\ 0 & -4 & 1 & -20 \\ 0 & 0 & 19 & 0 \end{bmatrix}$ Thus, $19x_3 = 0$

$\iff x_3 = 0$; $-4x_2 = -20 \iff x_2 = 5$; and $x_1 + 2(5) = 9 \iff x_1 = -1$. Therefore, the solution is $(-1, 5, 0)$.

23. $\begin{bmatrix} 2 & -3 & -1 & 13 \\ -1 & 2 & -5 & 6 \\ 5 & -1 & -1 & 49 \end{bmatrix}$ $\xrightarrow[2R_3 - 5R_1 \to R_3]{2R_2 + R_1 \to R_2}$ $\begin{bmatrix} 2 & -3 & -1 & 13 \\ 0 & 1 & -11 & 25 \\ 0 & 13 & 3 & 33 \end{bmatrix}$ $\xrightarrow{R_3 - 13R_2 \to R_3}$ $\begin{bmatrix} 2 & -3 & -1 & 13 \\ 0 & 1 & -11 & 25 \\ 0 & 0 & 146 & -292 \end{bmatrix}$ Thus,

$146z = -292 \iff z = -2$; $y - 11(-2) = 25 \iff y = 3$; and $2x - 3 \cdot 3 + 2 = 13 \iff x = 10$. Therefore, the solution is $(10, 3, -2)$.

25. $\begin{bmatrix} 1 & 1 & 1 & 2 \\ 0 & 1 & -3 & 1 \\ 2 & 1 & 5 & 0 \end{bmatrix}$ $\xrightarrow{R_3 - 2R_1 \to R_3}$ $\begin{bmatrix} 1 & 1 & 1 & 2 \\ 0 & 1 & -3 & 1 \\ 0 & -1 & 3 & -4 \end{bmatrix}$ $\xrightarrow{R_3 + R_2 \to R_3}$ $\begin{bmatrix} 1 & 1 & 1 & 3 \\ 0 & 1 & -3 & 1 \\ 0 & 0 & 0 & -3 \end{bmatrix}$. The third row of the

matrix states $0 = -3$, which is impossible. Hence, the system is inconsistent, and there is no solution.

27. $\begin{bmatrix} 2 & -3 & -9 & -5 \\ 1 & 0 & 3 & 2 \\ -3 & 1 & -4 & -3 \end{bmatrix}$ $\xrightarrow{R_1 \leftrightarrow R_2}$ $\begin{bmatrix} 1 & 0 & 3 & 2 \\ 2 & -3 & -9 & -5 \\ -3 & 1 & -4 & -3 \end{bmatrix}$ $\xrightarrow[R_3 + 3R_1 \to R_3]{R_2 - 2R_1 \to R_2}$ $\begin{bmatrix} 1 & 0 & 3 & 2 \\ 0 & -3 & -15 & -9 \\ 0 & 1 & 5 & 3 \end{bmatrix}$ $\xrightarrow{-\frac{1}{3}R_2}$

$\begin{bmatrix} 1 & 0 & 3 & 2 \\ 0 & 1 & 5 & 3 \\ 0 & 1 & 5 & 3 \end{bmatrix}$ $\xrightarrow{R_3 - R_2 \to R_3}$ $\begin{bmatrix} 1 & 0 & 3 & 2 \\ 0 & 1 & 5 & 3 \\ 0 & 0 & 0 & 0 \end{bmatrix}$. Therefore, this system has infinitely many solutions, given by $x + 3t = 2$

$\iff x = 2 - 3t$, and $y + 5t = 3 \iff y = 3 - 5t$. Hence, the solutions are $(2 - 3t, 3 - 5t, t)$, where t is any real number.

29. $\begin{bmatrix} 1 & -1 & 3 & 3 \\ 4 & -8 & 32 & 24 \\ 2 & -3 & 11 & 4 \end{bmatrix}$ $\xrightarrow[R_3 - 2R_1 \to R_3]{R_2 - 4R_1 \to R_2}$ $\begin{bmatrix} 1 & -1 & 3 & 3 \\ 0 & -4 & 20 & 12 \\ 0 & -1 & 5 & -2 \end{bmatrix}$ $\xrightarrow[R_3 + R_2 \to R_3]{-\frac{1}{4}R_2}$ $\begin{bmatrix} 1 & -1 & 3 & 3 \\ 0 & 1 & -5 & -3 \\ 0 & 0 & 0 & -5 \end{bmatrix}$ The third row of the

matrix states $0 = -5$, which is impossible. Hence, the system is inconsistent, and there is no solution.

31. $\begin{bmatrix} 1 & 4 & -2 & -3 \\ 2 & -1 & 5 & 12 \\ 8 & 5 & 11 & 30 \end{bmatrix}$ $\xrightarrow[R_3 - 8R_1 \to R_3]{R_2 - 2R_1 \to R_2}$ $\begin{bmatrix} 1 & 4 & -2 & -3 \\ 0 & -9 & 9 & 18 \\ 0 & -27 & 27 & 54 \end{bmatrix}$ $\xrightarrow{R_3 - 3R_2 \to R_3}$ $\begin{bmatrix} 1 & 4 & -2 & -3 \\ 0 & -9 & 9 & 18 \\ 0 & 0 & 0 & 0 \end{bmatrix}$.

Therefore, this system has infinitely many solutions, given by $-9y + 9t = 18 \iff y = -2 + t$, and $x + 4(-2 + t) - 2t = -3 \iff x = 5 - 2t$. Hence, the solutions are $(5 - 2t, -2 + t, t)$, where t is any real number.

33. $\begin{bmatrix} 2 & 1 & -2 & 12 \\ -1 & -\frac{1}{2} & 1 & -6 \\ 3 & \frac{3}{2} & -3 & 18 \end{bmatrix} \xrightarrow[-R_1]{R_1 \leftrightarrow R_2} \begin{bmatrix} 1 & \frac{1}{2} & -1 & 6 \\ 2 & 1 & -2 & 12 \\ 3 & \frac{3}{2} & -3 & 18 \end{bmatrix} \xrightarrow[R_3 - 3R_1 \rightarrow R_3]{R_2 - 2R_1 \rightarrow R_2} \begin{bmatrix} 1 & \frac{1}{2} & -1 & 6 \\ 0 & 0 & 0 & 0 \\ 0 & 0 & 0 & 0 \end{bmatrix}$ Therefore, this system

has infinitely many solutions, given by $x + \frac{1}{2}s - t = 6 \quad \Leftrightarrow \quad x = 6 - \frac{1}{2}s + t$. Hence, the solutions are $\left(6 - \frac{1}{2}s + t, s, t\right)$, where s and t are any real numbers.

35. $\begin{bmatrix} 4 & -3 & 1 & -8 \\ -2 & 1 & -3 & -4 \\ 1 & -1 & 2 & 3 \end{bmatrix} \xrightarrow{R_1 \leftrightarrow R_3} \begin{bmatrix} 1 & -1 & 2 & 3 \\ -2 & 1 & -3 & -4 \\ 4 & -3 & 1 & -8 \end{bmatrix} \xrightarrow[R_3 - 4R_1 \rightarrow R_3]{R_2 + 2R_1 \rightarrow R_2} \begin{bmatrix} 1 & -1 & 2 & 3 \\ 0 & -1 & 1 & 2 \\ 0 & 1 & -7 & -20 \end{bmatrix} \xrightarrow{-R_2}$

$\begin{bmatrix} 1 & -1 & 2 & 3 \\ 0 & 1 & -1 & -2 \\ 0 & 1 & -7 & -20 \end{bmatrix} \xrightarrow{R_3 - R_2 \rightarrow R_3} \begin{bmatrix} 1 & -1 & 2 & 3 \\ 0 & 1 & -1 & -2 \\ 0 & 0 & -6 & -18 \end{bmatrix}$. Therefore, $-6z = -18 \quad \Leftrightarrow \quad z = 3; y - (3) = -2$

$\Leftrightarrow \quad y = 1$; and $x - (1) + 2(3) = 3 \quad \Leftrightarrow \quad x = -2$. Hence, the solution is $(-2, 1, 3)$.

37. $\begin{bmatrix} 1 & 2 & -3 & -5 \\ -2 & -4 & -6 & 10 \\ 3 & 7 & -2 & -13 \end{bmatrix} \xrightarrow[R_3 - 3R_1 \rightarrow R_3]{R_2 + 2R_1 \rightarrow R_2} \begin{bmatrix} 1 & 2 & -3 & -5 \\ 0 & 0 & -12 & 0 \\ 0 & 1 & 7 & 2 \end{bmatrix} \xrightarrow{R_2 \leftrightarrow R_3} \begin{bmatrix} 1 & 2 & -3 & -5 \\ 0 & 1 & 7 & 2 \\ 0 & 0 & -12 & 0 \end{bmatrix}$. Therefore, $-12z = 0$

$\Leftrightarrow \quad z = 0; y + 7(0) = 2 \quad \Leftrightarrow \quad y = 2$; and $x + 2(2) - 3(0) = -5 \Leftrightarrow x = -9$. Hence, the solution is $(-9, 2, 0)$.

39. $\begin{bmatrix} -1 & 2 & 1 & -3 & 3 \\ 3 & -4 & 1 & 1 & 9 \\ -1 & -1 & 1 & 1 & 0 \\ 2 & 1 & 4 & -2 & 3 \end{bmatrix} \xrightarrow{-R_1} \begin{bmatrix} 1 & -2 & -1 & 3 & -3 \\ 3 & -4 & 1 & 1 & 9 \\ -1 & -1 & 1 & 1 & 0 \\ 2 & 1 & 4 & -2 & 3 \end{bmatrix} \xrightarrow[\substack{R_3 + R_1 \rightarrow R_3 \\ R_4 - 2R_1 \rightarrow R_4}]{R_2 - 3R_1 \rightarrow R_2} \begin{bmatrix} 1 & -2 & -1 & 3 & -3 \\ 0 & 2 & 4 & -8 & 18 \\ 0 & -3 & 0 & 4 & -3 \\ 0 & 5 & 6 & -8 & 9 \end{bmatrix} \xrightarrow{\frac{1}{2}R_2}$

$\begin{bmatrix} 1 & -2 & -1 & 3 & -3 \\ 0 & 1 & 2 & -4 & 9 \\ 0 & -3 & 0 & 4 & -3 \\ 0 & 5 & 6 & -8 & 9 \end{bmatrix} \xrightarrow[R_4 - 5R_2 \rightarrow R_4]{R_3 + 3R_2 \rightarrow R_3} \begin{bmatrix} 1 & -2 & -1 & 3 & -3 \\ 0 & 1 & 2 & -4 & 9 \\ 0 & 0 & 6 & -8 & 24 \\ 0 & 0 & -4 & 12 & -36 \end{bmatrix} \xrightarrow{3R_4 + 2R_3 \rightarrow R_4} \begin{bmatrix} 1 & -2 & -1 & 3 & -3 \\ 0 & 1 & 2 & -4 & 9 \\ 0 & 0 & 6 & -8 & 24 \\ 0 & 0 & 0 & 20 & -60 \end{bmatrix}$.

Therefore, $20w = -60 \quad \Leftrightarrow \quad w = -3; 6z + 24 = 24 \Leftrightarrow z = 0$. Then $y + 12 = 9 \quad \Leftrightarrow \quad y = -3$ and $x + 6 - 9 = -3$

$\Leftrightarrow \quad x = 0$. Hence, the solution is $(0, -3, 0, -3)$.

41. $\begin{bmatrix} 1 & 1 & 2 & -1 & -2 \\ 0 & 3 & 1 & 2 & 2 \\ 1 & 1 & 0 & 3 & 2 \\ -3 & 0 & 1 & 2 & 5 \end{bmatrix} \xrightarrow[R_4 + 3R_1 \rightarrow R_4]{R_3 - R_1 \rightarrow R_3} \begin{bmatrix} 1 & 1 & 2 & -1 & -2 \\ 0 & 3 & 1 & 2 & 2 \\ 0 & 0 & -2 & 4 & 4 \\ 0 & 3 & 7 & -1 & -1 \end{bmatrix} \xrightarrow{R_4 - R_2 \rightarrow R_4} \begin{bmatrix} 1 & 1 & 2 & -1 & -2 \\ 0 & 3 & 1 & 2 & 2 \\ 0 & 0 & -2 & 4 & 4 \\ 0 & 0 & 6 & -3 & -3 \end{bmatrix}$

$\xrightarrow{R_4 + 3R_3 \rightarrow R_4} \begin{bmatrix} 1 & 1 & 2 & -1 & -2 \\ 0 & 3 & 1 & 2 & 2 \\ 0 & 0 & -2 & 4 & 4 \\ 0 & 0 & 0 & 9 & 9 \end{bmatrix}$. Therefore, $9w = 9 \quad \Leftrightarrow \quad w = 1; -2z + 4(1) = 4 \quad \Leftrightarrow \quad z = 0$. Then

$3y + (0) + 2(1) = 2 \quad \Leftrightarrow \quad y = 0$ and $x + (0) + 2(0) - (1) = -2 \quad \Leftrightarrow \quad x = -1$. Hence, the solution is $(-1, 0, 0, 1)$.

43.
$$\begin{bmatrix} 1 & 0 & 1 & 1 & 4 \\ 0 & 1 & -1 & 0 & -4 \\ 1 & -2 & 3 & 1 & 12 \\ 2 & 0 & -2 & 5 & -1 \end{bmatrix} \xrightarrow[\substack{R_3 - R_1 \to R_3 \\ R_4 - 2R_1 \to R_4}]{} \begin{bmatrix} 1 & 0 & 1 & 1 & 4 \\ 0 & 1 & -1 & 0 & -4 \\ 0 & -2 & 2 & 0 & 8 \\ 0 & 0 & -4 & 3 & -9 \end{bmatrix} \xrightarrow[\substack{R_3 + 2R_2 \to R_3}]{} \begin{bmatrix} 1 & 0 & 1 & 1 & 4 \\ 0 & 1 & -1 & 0 & -4 \\ 0 & 0 & 0 & 0 & 0 \\ 0 & 0 & -4 & 3 & -9 \end{bmatrix} \xrightarrow[\substack{R_3 \leftrightarrow -R_4}]{}$$

$$\begin{bmatrix} 1 & 0 & 1 & 1 & 4 \\ 0 & 1 & -1 & 0 & -4 \\ 0 & 0 & 4 & -3 & 9 \\ 0 & 0 & 0 & 0 & 0 \end{bmatrix}.$$ Therefore, $4z - 3t = 9$ $\Leftrightarrow$ $4z = 9 + 3t \Leftrightarrow z = \frac{9}{4} + \frac{3}{4}t$. Then we have $y - \left(\frac{9}{4} + \frac{3}{4}t\right) = -4$

$\Leftrightarrow$ $y = \frac{-7}{4} + \frac{3}{4}t$ and $x + \left(\frac{9}{4} + \frac{3}{4}t\right) + t = 4$ $\Leftrightarrow$ $x = \frac{7}{4} - \frac{7}{4}t$. Hence, the solutions are $\left(\frac{7}{4} - \frac{7}{4}t, -\frac{7}{4} + \frac{3}{4}t, \frac{9}{4} + \frac{3}{4}t, t\right)$, where t is any real number.

45.
$$\begin{bmatrix} 1 & -1 & 0 & 1 & 0 \\ 3 & 0 & -1 & 2 & 0 \\ 1 & -4 & 1 & 2 & 0 \end{bmatrix} \xrightarrow[\substack{R_2 - 3R_1 \to R_2 \\ R_3 - R_1 \to R_3}]{} \begin{bmatrix} 1 & -1 & 0 & 1 & 0 \\ 0 & 3 & -1 & -1 & 0 \\ 0 & -3 & 1 & 1 & 0 \end{bmatrix} \xrightarrow[\substack{R_3 + R_2 \to R_3}]{} \begin{bmatrix} 1 & -1 & 0 & 1 & 0 \\ 0 & 3 & -1 & -1 & 0 \\ 0 & 0 & 0 & 0 & 0 \end{bmatrix}.$$

Therefore, the system has infinitely many solutions, given by $3y - s - t = 0$ $\Leftrightarrow$ $y = \frac{1}{3}(s + t)$ and $x - \frac{1}{3}(s + t) + t = 0$
$\Leftrightarrow$ $x = \frac{1}{3}(s - 2t)$. So the solutions are $\left(\frac{1}{3}(s - 2t), \frac{1}{3}(s + t), s, t\right)$, where s and t are any real numbers.

47. Let x, y, z represent the number of VitaMax, Vitron, and VitaPlus pills taken daily. The matrix representation for the system of equations is

$$\begin{bmatrix} 5 & 10 & 15 & 50 \\ 15 & 20 & 0 & 50 \\ 10 & 10 & 10 & 50 \end{bmatrix} \xrightarrow[\substack{\frac{1}{5}R_1 \\ \frac{1}{5}R_2 \\ \frac{1}{5}R_3}]{} \begin{bmatrix} 1 & 2 & 3 & 10 \\ 3 & 4 & 0 & 10 \\ 2 & 2 & 2 & 10 \end{bmatrix} \xrightarrow[\substack{R_2 - 3R_1 \to R_2 \\ R_3 - 2R_1 \to R_3}]{} \begin{bmatrix} 1 & 2 & 3 & 10 \\ 0 & -2 & -9 & -20 \\ 0 & -2 & -4 & -10 \end{bmatrix} \xrightarrow[\substack{R_3 - R_2 \to R_3}]{} \begin{bmatrix} 1 & 2 & 3 & 10 \\ 0 & -2 & -9 & -20 \\ 0 & 0 & 5 & 10 \end{bmatrix}.$$

Thus, $5z = 10$ $\Leftrightarrow$ $z = 2$; $-2y - 18 = -20 \Leftrightarrow y = 1$; and $x + 2 + 6 = 10$ $\Leftrightarrow$ $x = 2$. Hence, he should take
2 VitaMax, 1 Vitron, and 2 VitaPlus pills daily.

49. Let x, y, and z represent the distance, in miles, of the run, swim, and cycle parts of the race respectively. Then, since
time $= \dfrac{\text{distance}}{\text{speed}}$, we get the following equations from the three contestants' race times:

$$\begin{cases} \left(\frac{x}{10}\right) + \left(\frac{y}{4}\right) + \left(\frac{z}{20}\right) = 2.5 \\ \left(\frac{x}{7.5}\right) + \left(\frac{y}{6}\right) + \left(\frac{z}{15}\right) = 3 \\ \left(\frac{x}{15}\right) + \left(\frac{y}{3}\right) + \left(\frac{z}{40}\right) = 1.75 \end{cases} \Leftrightarrow \begin{cases} 2x + 5y + z = 50 \\ 4x + 5y + 2z = 90 \\ 8x + 40y + 3z = 210 \end{cases}$$ which has the following matrix representation:

$$\begin{bmatrix} 2 & 5 & 1 & 50 \\ 4 & 5 & 2 & 90 \\ 8 & 40 & 3 & 210 \end{bmatrix} \xrightarrow[\substack{R_2 - 2R_1 \to R_2 \\ R_3 - 4R_1 \to R_3}]{} \begin{bmatrix} 2 & 5 & 1 & 50 \\ 0 & -5 & 0 & -10 \\ 0 & 20 & -1 & 10 \end{bmatrix} \xrightarrow[\substack{R_3 + 4R_2 \to R_3}]{} \begin{bmatrix} 2 & 5 & 1 & 50 \\ 0 & -5 & 0 & -10 \\ 0 & 0 & -1 & -30 \end{bmatrix}.$$

Thus, $-z = -30$ $\Leftrightarrow$ $z = 30$; $-5y = -10$ $\Leftrightarrow$ $y = 2$; and $2x + 10 + 30 = 50$ $\Leftrightarrow$ $x = 5$. So the race has a
5 mile run, 2 mile swim, and 30 mile cycle.

51. Let t be the number of tables produced, c the number of chairs, and a the number of armoires. Then, the system of equations

is $\begin{cases} \frac{1}{2}t + c + a = 300 \\ \frac{1}{2}t + \frac{3}{2}c + a = 400 \\ t + \frac{3}{2}c + 2a = 590 \end{cases} \Leftrightarrow \begin{cases} t + 2c + 2a = 600 \\ t + 3c + 2a = 800 \\ 2t + 3c + 4a = 1180 \end{cases}$ and a matrix representation is

$\begin{bmatrix} 1 & 2 & 2 & 600 \\ 1 & 3 & 2 & 800 \\ 2 & 3 & 4 & 1180 \end{bmatrix} \xrightarrow[R_3 - 2R_1 \to R_3]{R_2 - R_1 \to R_2} \begin{bmatrix} 1 & 2 & 2 & 600 \\ 0 & 1 & 0 & 200 \\ 0 & -1 & 0 & -20 \end{bmatrix} \xrightarrow{R_3 + R_2 \to R_3} \begin{bmatrix} 1 & 2 & 2 & 600 \\ 0 & 1 & 0 & 200 \\ 0 & 0 & 0 & 180 \end{bmatrix}$. The third row states

$0 = 180$, which is impossible, and so the system is inconsistent. Therefore, it is impossible to use all of the available labor-hours.

53. *Line containing the points* $(0, 0)$ *and* $(1, 12)$: Using the general form of a line, $y = ax + b$, we substitute for x and y and solve for a and b. The point $(0, 0)$ gives $0 = a(0) + b \Rightarrow b = 0$; the point $(1, 12)$ gives $12 = a(1) + b \Rightarrow a = 12$. Since $a = 12$ and $b = 0$, the equation of the line is $y = 12x$.

Quadratic containing the points $(0,0)$, $(1, 12)$, *and* $(3, 6)$: Using the general form of a quadratic, $y = ax^2 + bx + c$, we substitute for x and y and solve for a, b, and c. The point $(0, 0)$ gives $0 = a(0)^2 + b(0) + c \Rightarrow c = 0$; the point $(1, 12)$ gives $12 = a(1)^2 + b(1) + c \Rightarrow a + b = 12$; the point $(3, 6)$ gives $6 = a(3)^2 + b(3) + c \Rightarrow 9a + 3b = 6$. Subtracting the third equation from -3 times the third gives $6a = -30 \Leftrightarrow a = -5$. So $a + b = 12 \Leftrightarrow b = 12 - a \Rightarrow b = 17$. Since $a = -5$, $b = 17$, and $c = 0$, the equation of the quadratic is $y = -5x^2 + 17x$.

Cubic containing the points $(0, 0)$, $(1, 12)$, $(2, 40)$, *and* $(3, 6)$: Using the general form of a cubic, $y = ax^3 + bx^2 + cx + d$, we substitute for x and y and solve for a, b, c, and d. The point $(0, 0)$ gives $0 = a(0)^3 + b(0)^2 + c(0) + d \Rightarrow d = 0$; the point the point $(1, 12)$ gives $12 = a(1)^3 + b(1)^2 + c(1) + d \Rightarrow a + b + c + d = 12$; the point $(2, 40)$ gives $40 = a(2)^3 + b(2)^2 + c(2) + d \Rightarrow 8a + 4b + 2c + d = 40$; the point $(3, 6)$ gives $6 = a(3)^3 + b(3)^2 + c(3) + d$

$\Rightarrow 27a + 9b + 3c + d = 6$. Since $d = 0$, the system reduces to $\begin{cases} a + b + c = 12 \\ 8a + 4b + 2c = 40 \\ 27a + 9b + 3c = 6 \end{cases}$ which has representation

$\begin{bmatrix} 1 & 1 & 1 & 12 \\ 8 & 4 & 2 & 40 \\ 27 & 9 & 3 & 6 \end{bmatrix} \xrightarrow[R_3 - 27R_1 \to R_3]{R_2 - 8R_1 \to R_2} \begin{bmatrix} 1 & 1 & 1 & 12 \\ 0 & -4 & -6 & -56 \\ 0 & -18 & -24 & -318 \end{bmatrix} \xrightarrow[-\frac{1}{6}R_3]{-\frac{1}{2}R_2} \begin{bmatrix} 1 & 1 & 1 & 12 \\ 0 & 2 & 3 & 28 \\ 0 & 3 & 4 & 53 \end{bmatrix} \xrightarrow{2R_3 - 3R_2 \to R_3} \begin{bmatrix} 1 & 1 & 1 & 12 \\ 0 & 2 & 3 & 28 \\ 0 & 0 & -1 & 22 \end{bmatrix}$.

So $c = -22$ and back-substituting we have $2b + 3(-22) = 28 \Leftrightarrow b = 47$ and $a + 47 + (-22) = 0 \Leftrightarrow a = -13$. So the cubic is $y = -13x^3 + 47x^2 - 22x$.

Fourth-degree polynomial containing the points $(0,0)$, $(1, 12)$, $(2, 40)$, $(3, 6)$, *and* $(-1, -14)$: Using the general form of a fourth-degree polynomial, $y = ax^4 + bx^3 + cx^2 + dx + e$, we substitute for x and y and solve for a, b, c, d, and e. The point $(0, 0)$ gives $0 = a(0)^4 + b(0)^3 + c(0)^2 + d(0) + e \Rightarrow e = 0$; the point $(1, 12)$ gives $12 = a(1)^4 + b(1)^3 + c(1)^2 + d(1) + e$; the point $(2, 40)$ gives $40 = a(2)^4 + b(2)^3 + c(2)^2 + d(2) + e$; the point $(3, 6)$ gives $6 = a(3)^4 + b(3)^3 + c(3)^2 + d(3) + e$; the point $(-1, -14)$ gives $-14 = a(-1)^4 + b(-1)^3 + c(-1)^2 + d(-1) + e$. Since the first equation is $e = 0$, we eliminate e from the other equations to get the system $\begin{cases} a + b + c + d = 12 \\ 16a + 8b + 4c + 2d = 40 \\ 81a + 27b + 9c + 3d = 6 \\ a - b + c - d = -14 \end{cases}$

which has the matrix representation

$$\begin{bmatrix} 1 & 1 & 1 & 1 & 12 \\ 16 & 8 & 4 & 2 & 40 \\ 81 & 27 & 9 & 3 & 6 \\ 1 & -1 & 1 & -1 & -14 \end{bmatrix} \xrightarrow[\substack{R_2 - 16R_1 \to R_2 \\ R_3 - 81R_1 \to R_3 \\ R_4 - R_1 \to R_4}]{} \begin{bmatrix} 1 & 1 & 1 & 1 & 12 \\ 0 & -8 & -12 & -14 & -152 \\ 0 & -54 & -72 & -78 & -966 \\ 0 & -2 & 0 & -2 & -26 \end{bmatrix} \xrightarrow[\substack{-\frac{1}{8}R_4 \to R_2 \\ R_2 \to R_3 \\ R_3 \to R_4}]{}$$

$$\begin{bmatrix} 1 & 1 & 1 & 1 & 12 \\ 0 & 1 & 0 & 1 & 13 \\ 0 & -8 & -12 & -14 & -152 \\ 0 & -54 & -72 & -78 & -966 \end{bmatrix} \xrightarrow[\substack{R_3 + 8R_2 \to R_3 \\ R_4 + 54R_2 \to R_4}]{} \begin{bmatrix} 1 & 1 & 1 & 1 & 12 \\ 0 & 1 & 0 & 1 & 13 \\ 0 & 0 & -12 & -6 & -48 \\ 0 & 0 & -72 & -24 & -264 \end{bmatrix} \xrightarrow[R_4 - 6R_3 \to R_4]{} \begin{bmatrix} 1 & 1 & 1 & 1 & 12 \\ 0 & 1 & 0 & 1 & 13 \\ 0 & 0 & -12 & -6 & -48 \\ 0 & 0 & 0 & 12 & 24 \end{bmatrix}.$$

So $d = 2$. Then $-12c - 6(2) = -48 \quad \Leftrightarrow \quad c = 3$ and $b + 2 = 13 \quad \Leftrightarrow$ $b = 11$. Finally, $a + 11 + 3 + 2 = 12 \quad \Leftrightarrow \quad a = -4$. So the fourth-degree polynomial containing these points is $y = -4x^4 + 11x^3 + 3x^2 + 2x$.

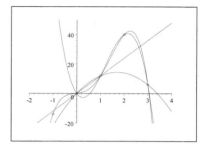

10.5   The Algebra of Matrices

1. The matrices have different dimensions, so they cannot be equal.

3. $\begin{bmatrix} 2 & 6 \\ -5 & 3 \end{bmatrix} + \begin{bmatrix} -1 & -3 \\ 6 & 2 \end{bmatrix} = \begin{bmatrix} 1 & 3 \\ 1 & 5 \end{bmatrix}$

5. $3\begin{bmatrix} 1 & 2 \\ 4 & -1 \\ 1 & 0 \end{bmatrix} = \begin{bmatrix} 3 & 6 \\ 12 & -3 \\ 3 & 0 \end{bmatrix}$

7. $\begin{bmatrix} 2 & 6 \\ 1 & 3 \\ 2 & 4 \end{bmatrix} \begin{bmatrix} 1 & -2 \\ 3 & 6 \\ -2 & 0 \end{bmatrix}$ is undefined because these matrices have incompatible dimensions.

9. $\begin{bmatrix} 1 & 2 \\ -1 & 4 \end{bmatrix} \begin{bmatrix} 1 & -2 & 3 \\ 2 & 2 & -1 \end{bmatrix} = \begin{bmatrix} 5 & 2 & 1 \\ 7 & 10 & -7 \end{bmatrix}$

11. $2X + A = B \quad \Leftrightarrow \quad X = \frac{1}{2}(B - A) = \frac{1}{2}\left(\begin{bmatrix} 2 & 5 \\ 3 & 7 \end{bmatrix} - \begin{bmatrix} 4 & 6 \\ 1 & 3 \end{bmatrix}\right) = \frac{1}{2}\begin{bmatrix} -2 & -1 \\ 2 & 4 \end{bmatrix} = \begin{bmatrix} -1 & -\frac{1}{2} \\ 1 & 2 \end{bmatrix}.$

13. $2(B - X) = D$. Since B is a 2×2 matrix, $B - X$ is defined only when X is a 2×2 matrix, so $2(B - X)$ is a 2×2 matrix. But D is a 3×2 matrix. Thus, there is no solution.

15. $\frac{1}{5}(X + D) = C \quad \Leftrightarrow \quad X + D = 5C \quad \Leftrightarrow$

$$X = 5C - D = 5\begin{bmatrix} 2 & 3 \\ 1 & 0 \\ 0 & 2 \end{bmatrix} - \begin{bmatrix} 10 & 20 \\ 30 & 20 \\ 10 & 0 \end{bmatrix} = \begin{bmatrix} 10 & 15 \\ 5 & 0 \\ 0 & 10 \end{bmatrix} - \begin{bmatrix} 10 & 20 \\ 30 & 20 \\ 10 & 0 \end{bmatrix} = \begin{bmatrix} 0 & -5 \\ -25 & -20 \\ -10 & 10 \end{bmatrix}.$$

In Solutions 17–37, the matrices A, B, C, D, E, F, and G are defined as follows:

$$A = \begin{bmatrix} 2 & -5 \\ 0 & 7 \end{bmatrix} \qquad B = \begin{bmatrix} 3 & \frac{1}{2} & 5 \\ 1 & -1 & 3 \end{bmatrix} \qquad C = \begin{bmatrix} 2 & -\frac{5}{2} & 0 \\ 0 & 2 & -3 \end{bmatrix}$$

$$D = \begin{bmatrix} 7 & 3 \end{bmatrix} \qquad E = \begin{bmatrix} 1 \\ 2 \\ 0 \end{bmatrix} \qquad F = \begin{bmatrix} 1 & 0 & 0 \\ 0 & 1 & 0 \\ 0 & 0 & 1 \end{bmatrix} \qquad G = \begin{bmatrix} 5 & -3 & 10 \\ 6 & 1 & 0 \\ -5 & 2 & 2 \end{bmatrix}$$

17. $B + C = \begin{bmatrix} 3 & \frac{1}{2} & 5 \\ 1 & -1 & 3 \end{bmatrix} + \begin{bmatrix} 2 & -\frac{5}{2} & 0 \\ 0 & 2 & -3 \end{bmatrix} = \begin{bmatrix} 5 & -2 & 5 \\ 1 & 1 & 0 \end{bmatrix}$

19. $C - B = \begin{bmatrix} 2 & -\frac{5}{2} & 0 \\ 0 & 2 & -3 \end{bmatrix} - \begin{bmatrix} 3 & \frac{1}{2} & 5 \\ 1 & -1 & 3 \end{bmatrix} = \begin{bmatrix} -1 & -3 & -5 \\ -1 & 3 & -6 \end{bmatrix}$

21. $3B + 2C = 3 \begin{bmatrix} 3 & \frac{1}{2} & 5 \\ 1 & -1 & 3 \end{bmatrix} + 2 \begin{bmatrix} 2 & -\frac{5}{2} & 0 \\ 0 & 2 & -3 \end{bmatrix} = \begin{bmatrix} 13 & -\frac{7}{2} & 15 \\ 3 & 1 & 3 \end{bmatrix}$

23. $2C - 6B = 2 \begin{bmatrix} 2 & -\frac{5}{2} & 0 \\ 0 & 2 & -3 \end{bmatrix} - 6 \begin{bmatrix} 3 & \frac{1}{2} & 5 \\ 1 & -1 & 3 \end{bmatrix} = \begin{bmatrix} -14 & -8 & -30 \\ -6 & 10 & -24 \end{bmatrix}$

25. AD is undefined because A (2×2) and D (1×2) have incompatible dimensions.

27. $BF = \begin{bmatrix} 3 & \frac{1}{2} & 5 \\ 1 & -1 & 3 \end{bmatrix} \begin{bmatrix} 1 & 0 & 0 \\ 0 & 1 & 0 \\ 0 & 0 & 1 \end{bmatrix} = \begin{bmatrix} 3 & \frac{1}{2} & 5 \\ 1 & -1 & 3 \end{bmatrix}$

29. $(DA)B = \begin{bmatrix} 7 & 3 \end{bmatrix} \begin{bmatrix} 2 & -5 \\ 0 & 7 \end{bmatrix} \begin{bmatrix} 3 & \frac{1}{2} & 5 \\ 1 & -1 & 3 \end{bmatrix} = \begin{bmatrix} 14 & -14 \end{bmatrix} \begin{bmatrix} 3 & \frac{1}{2} & 5 \\ 1 & -1 & 3 \end{bmatrix} = \begin{bmatrix} 28 & 21 & 28 \end{bmatrix}$

31. $GE = \begin{bmatrix} 5 & -3 & 10 \\ 6 & 1 & 0 \\ -5 & 2 & 2 \end{bmatrix} \begin{bmatrix} 1 \\ 2 \\ 0 \end{bmatrix} = \begin{bmatrix} -1 \\ 8 \\ -1 \end{bmatrix}$

33. $A^3 = \begin{bmatrix} 2 & -5 \\ 0 & 7 \end{bmatrix} \begin{bmatrix} 2 & -5 \\ 0 & 7 \end{bmatrix} \begin{bmatrix} 2 & -5 \\ 0 & 7 \end{bmatrix} = \begin{bmatrix} 4 & -45 \\ 0 & 49 \end{bmatrix} \begin{bmatrix} 2 & -5 \\ 0 & 7 \end{bmatrix} = \begin{bmatrix} 8 & -335 \\ 0 & 343 \end{bmatrix}$

35. B^2 is undefined because the dimensions 2×3 and 2×3 are incompatible.

37. $BF + FE$ is undefined; the dimensions $(2 \times 3) \cdot (3 \times 3) = (2 \times 3)$ and $(3 \times 3) \cdot (3 \times 1) = (3 \times 1)$ are incompatible.

39. $\begin{bmatrix} x & 2y \\ 4 & 6 \end{bmatrix} = \begin{bmatrix} 2 & -2 \\ 2x & -6y \end{bmatrix}$. Thus we must solve the system $\begin{cases} x = 2,\, 2y = -2 \\ 4 = 2x,\, 6 = -6y \end{cases}$ So $x = 2$ and $2y = -2$ $\Leftrightarrow$

$y = -1$. Since these values for x and y also satisfy the last two equations, the solution is $x = 2$, $y = -1$.

41. $2 \begin{bmatrix} x & y \\ x+y & x-y \end{bmatrix} = \begin{bmatrix} 2 & -4 \\ -2 & 6 \end{bmatrix}$. Since $2 \begin{bmatrix} x & y \\ x+y & x-y \end{bmatrix} = \begin{bmatrix} 2x & 2y \\ 2(x+y) & 2(x-y) \end{bmatrix}$, Thus we must solve the

system $\begin{cases} 2x = 2,\, 2y = -4 \\ 2(x+y) = -2 \\ 2(x-y) = 6 \end{cases}$ So $x = 1$ and $y = -2$. Since these values for x and y also satisfy the last two equations,

the solution is $x = 1$, $y = -2$.

43. $\begin{cases} 2x - 5y = 7 \\ 3x + 2y = 4 \end{cases}$ written as a matrix equation is $\begin{bmatrix} 2 & -5 \\ 3 & 2 \end{bmatrix} \begin{bmatrix} x \\ y \end{bmatrix} = \begin{bmatrix} 7 \\ 4 \end{bmatrix}$.

45. $\begin{cases} 3x_1 + 2x_2 - x_3 + x_4 = 0 \\ x_1 \quad\quad - x_3 \quad\quad = 5 \\ \quad\quad 3x_2 + x_3 - x_4 = 4 \end{cases}$ written as a matrix equation is $\begin{bmatrix} 3 & 2 & -1 & 1 \\ 1 & 0 & -1 & 0 \\ 0 & 3 & 1 & -1 \end{bmatrix} \begin{bmatrix} x_1 \\ x_2 \\ x_3 \\ x_4 \end{bmatrix} = \begin{bmatrix} 0 \\ 5 \\ 4 \end{bmatrix}$.

47. $A = \begin{bmatrix} 1 & 0 & 6 & -1 \\ 2 & \frac{1}{2} & 4 & 0 \end{bmatrix}$, $B = \begin{bmatrix} 1 & 7 & -9 & 2 \end{bmatrix}$, and $C = \begin{bmatrix} 1 \\ 0 \\ -1 \\ -2 \end{bmatrix}$. ABC is undefined because the dimensions of A (2×4)

and B (1×4) are not compatible. $ACB = \begin{bmatrix} -3 \\ -2 \end{bmatrix} \begin{bmatrix} 1 & 7 & -9 & 2 \end{bmatrix} = \begin{bmatrix} -3 & -21 & 27 & -6 \\ -2 & -14 & 18 & -4 \end{bmatrix}$. BAC is undefined

because the dimensions of B (1×4) and A (2×4) are not compatible. BCA is undefined because the dimensions of C (4×1) and A (2×4) are not compatible. CAB is undefined because the dimensions of C (4×1) and A (2×4) are not compatible. CBA is undefined because the dimensions of B (1×4) and A (2×4) are not compatible.

49. (a) $BA = \begin{bmatrix} \$0.90 & \$0.80 & \$1.10 \end{bmatrix} \begin{bmatrix} 4000 & 1000 & 3500 \\ 400 & 300 & 200 \\ 700 & 500 & 9000 \end{bmatrix} = \begin{bmatrix} \$4690 & \$1690 & \$13{,}210 \end{bmatrix}$

(b) The entries in the product matrix represent the total food sales in Santa Monica, Long Beach, and Anaheim, respectively.

51. (a) $AB = \begin{bmatrix} 6 & 10 & 14 & 28 \end{bmatrix} \begin{bmatrix} 2000 & 2500 \\ 3000 & 1500 \\ 2500 & 1000 \\ 1000 & 500 \end{bmatrix} = \begin{bmatrix} 105{,}000 & 58{,}000 \end{bmatrix}$

(b) That day they canned 105,000 ounces of tomato sauce and 58,000 ounces of tomato paste.

53. (a) $\begin{bmatrix} 1 & 0 & 1 & 0 & 1 & 1 \\ 0 & 3 & 0 & 1 & 2 & 1 \\ 1 & 2 & 0 & 0 & 3 & 0 \\ 1 & 3 & 2 & 3 & 2 & 0 \\ 0 & 3 & 0 & 0 & 2 & 1 \\ 1 & 2 & 0 & 1 & 3 & 1 \end{bmatrix}$
(b) $\begin{bmatrix} 2 & 1 & 2 & 1 & 2 & 2 \\ 1 & 3 & 1 & 2 & 3 & 2 \\ 2 & 3 & 1 & 1 & 3 & 1 \\ 2 & 3 & 3 & 3 & 3 & 1 \\ 1 & 3 & 1 & 1 & 3 & 2 \\ 2 & 3 & 1 & 2 & 3 & 2 \end{bmatrix}$
(c) $\begin{bmatrix} 2 & 3 & 2 & 3 & 2 & 2 \\ 3 & 0 & 3 & 2 & 1 & 2 \\ 2 & 1 & 3 & 3 & 0 & 3 \\ 2 & 0 & 1 & 0 & 1 & 3 \\ 3 & 0 & 3 & 3 & 1 & 2 \\ 2 & 1 & 3 & 2 & 0 & 2 \end{bmatrix}$

(d) $\begin{bmatrix} 3 & 3 & 3 & 3 & 3 & 3 \\ 3 & 0 & 3 & 3 & 0 & 3 \\ 3 & 0 & 3 & 3 & 0 & 3 \\ 3 & 0 & 0 & 0 & 0 & 3 \\ 3 & 0 & 3 & 3 & 0 & 3 \\ 3 & 0 & 3 & 3 & 0 & 3 \end{bmatrix}$

(e)

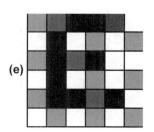

55. $A = \begin{bmatrix} 1 & 1 \\ 0 & 1 \end{bmatrix}$; $A^2 = \begin{bmatrix} 1 & 1 \\ 0 & 1 \end{bmatrix}\begin{bmatrix} 1 & 1 \\ 0 & 1 \end{bmatrix} = \begin{bmatrix} 1 & 2 \\ 0 & 1 \end{bmatrix}$; $A^3 = A \cdot A^2 = \begin{bmatrix} 1 & 1 \\ 0 & 1 \end{bmatrix}\begin{bmatrix} 1 & 2 \\ 0 & 1 \end{bmatrix} = \begin{bmatrix} 1 & 3 \\ 0 & 1 \end{bmatrix}$;

$A^4 = A \cdot A^3 = \begin{bmatrix} 1 & 1 \\ 0 & 1 \end{bmatrix}\begin{bmatrix} 1 & 3 \\ 0 & 1 \end{bmatrix} = \begin{bmatrix} 1 & 4 \\ 0 & 1 \end{bmatrix}$. Therefore, it seems that $A^n = \begin{bmatrix} 1 & n \\ 0 & 1 \end{bmatrix}$.

57. Let $A = \begin{bmatrix} a & b \\ c & d \end{bmatrix}$. For the first matrix, we have $A^2 = \begin{bmatrix} a & b \\ c & d \end{bmatrix}$.

$\begin{bmatrix} a & b \\ c & d \end{bmatrix} = \begin{bmatrix} a^2 + bc & ab + bd \\ ac + cd & bc + d^2 \end{bmatrix} = \begin{bmatrix} a^2 + bc & b(a+d) \\ c(a+d) & bc + d^2 \end{bmatrix}$. So $A^2 = \begin{bmatrix} 4 & 0 \\ 0 & 9 \end{bmatrix} \Leftrightarrow \begin{cases} a^2 + bc = 4 \\ b(a+d) = 0 \\ c(a+d) = 0 \\ bc + d^2 = 9 \end{cases}$

If $a + d = 0$, then $a = -d$, so $4 = a^2 + bc = (-d)^2 + bc = d^2 + bc = 9$, which is a contradiction. Thus $a + d \neq 0$. Since $b(a+d) = 0$ and $c(a+d) = 0$, we must have $b = 0$ and $c = 0$. So the first equation becomes $a^2 = 4 \Rightarrow a = \pm 2$, and the fourth equation becomes $d^2 = 9 \Rightarrow d = \pm 3$.

Thus the square roots of $\begin{bmatrix} 4 & 0 \\ 0 & 9 \end{bmatrix}$ are $A_1 = \begin{bmatrix} 2 & 0 \\ 0 & 3 \end{bmatrix}$, $A_2 = \begin{bmatrix} 2 & 0 \\ 0 & -3 \end{bmatrix}$, $A_3 = \begin{bmatrix} -2 & 0 \\ 0 & 3 \end{bmatrix}$, and $A_4 = \begin{bmatrix} -2 & 0 \\ 0 & -3 \end{bmatrix}$.

For the second matrix, we have $A^2 = \begin{bmatrix} 1 & 5 \\ 0 & 9 \end{bmatrix} \Leftrightarrow$

$\begin{cases} a^2 + bc = 1 \\ b(a+d) = 5 \\ c(a+d) = 0 \\ bc + d^2 = 9 \end{cases}$ Since $a + d \neq 0$ and $c(a+d) = 0$, $c = 0$. Thus, $\begin{cases} a^2 = 1 \\ b(a+d) = 5 \\ d^2 = 9 \end{cases} \Rightarrow \begin{cases} a = \pm 1 \\ b(a+d) = 5 \\ d = \pm 3 \end{cases}$

We consider the four possible values of a and d. If $a = 1$ and $d = 3$, then $b(a+d) = 5 \Rightarrow b(4) = 5 \Rightarrow b = \frac{5}{4}$. If $a = 1$ and $d = -3$, then $b(a+d) = 5 \Rightarrow b(-2) = 5 \Rightarrow b = -\frac{5}{2}$. If $a = -1$ and $d = 3$, then $b(a+d) = 5 \Rightarrow b(2) = 5 \Rightarrow b = \frac{5}{2}$. If $a = -1$ and $d = -3$, then $b(a+d) = 5 \Rightarrow b(-4) = 5 \Rightarrow b = -\frac{5}{4}$. Thus,

the square roots of $\begin{bmatrix} 1 & 5 \\ 0 & 9 \end{bmatrix}$ are $A_1 = \begin{bmatrix} 1 & \frac{5}{4} \\ 0 & 3 \end{bmatrix}$, $A_2 = \begin{bmatrix} 1 & -\frac{5}{2} \\ 0 & -3 \end{bmatrix}$, $A_3 = \begin{bmatrix} -1 & \frac{5}{2} \\ 0 & 3 \end{bmatrix}$, and $A_4 = \begin{bmatrix} -1 & -\frac{5}{4} \\ 0 & -3 \end{bmatrix}$.

10.6   Inverses of Matrices and Matrix Equations

1. $A = \begin{bmatrix} 4 & 1 \\ 7 & 2 \end{bmatrix}$, $B = \begin{bmatrix} 2 & -1 \\ -7 & 4 \end{bmatrix}$; $AB = \begin{bmatrix} 4 & 1 \\ 7 & 2 \end{bmatrix}\begin{bmatrix} 2 & -1 \\ -7 & 4 \end{bmatrix} = \begin{bmatrix} 1 & 0 \\ 0 & 1 \end{bmatrix}$; $BA = \begin{bmatrix} 2 & -1 \\ -7 & 4 \end{bmatrix}\begin{bmatrix} 4 & 1 \\ 7 & 2 \end{bmatrix} = \begin{bmatrix} 1 & 0 \\ 0 & 1 \end{bmatrix}$

3. $A = \begin{bmatrix} 1 & 3 & -1 \\ 1 & 4 & 0 \\ -1 & -3 & 2 \end{bmatrix}$; $B = \begin{bmatrix} 8 & -3 & 4 \\ -2 & 1 & -1 \\ 1 & 0 & 1 \end{bmatrix}$. $AB = \begin{bmatrix} 1 & 3 & -1 \\ 1 & 4 & 0 \\ -1 & -3 & 2 \end{bmatrix}\begin{bmatrix} 8 & -3 & 4 \\ -2 & 1 & -1 \\ 1 & 0 & 1 \end{bmatrix} = \begin{bmatrix} 1 & 0 & 0 \\ 0 & 1 & 0 \\ 0 & 0 & 1 \end{bmatrix}$ and

$BA = \begin{bmatrix} 8 & -3 & 4 \\ -2 & 1 & -1 \\ 1 & 0 & 1 \end{bmatrix}\begin{bmatrix} 1 & 3 & -1 \\ 1 & 4 & 0 \\ -1 & -3 & 2 \end{bmatrix} = \begin{bmatrix} 1 & 0 & 0 \\ 0 & 1 & 0 \\ 0 & 0 & 1 \end{bmatrix}$

5. $A = \begin{bmatrix} 7 & 4 \\ 3 & 2 \end{bmatrix} \Leftrightarrow A^{-1} = \dfrac{1}{14-12}\begin{bmatrix} 2 & -4 \\ -3 & 7 \end{bmatrix} = \begin{bmatrix} 1 & -2 \\ -\frac{3}{2} & \frac{7}{2} \end{bmatrix}$.   Then,

$AA^{-1} = \begin{bmatrix} 7 & 4 \\ 3 & 2 \end{bmatrix}\begin{bmatrix} 1 & -2 \\ -\frac{3}{2} & \frac{7}{2} \end{bmatrix} = \begin{bmatrix} 1 & 0 \\ 0 & 1 \end{bmatrix}$ and $A^{-1}A = \begin{bmatrix} 1 & -2 \\ -\frac{3}{2} & \frac{7}{2} \end{bmatrix}\begin{bmatrix} 7 & 4 \\ 3 & 2 \end{bmatrix} = \begin{bmatrix} 1 & 0 \\ 0 & 1 \end{bmatrix}$.

7. $\begin{bmatrix} 5 & 3 \\ 3 & 2 \end{bmatrix}^{-1} = \dfrac{1}{10-9}\begin{bmatrix} 2 & -3 \\ -3 & 5 \end{bmatrix} = \begin{bmatrix} 2 & -3 \\ -3 & 5 \end{bmatrix}$

9. $\begin{bmatrix} 2 & 5 \\ -5 & -13 \end{bmatrix}^{-1} = \dfrac{1}{-26+25}\begin{bmatrix} -13 & -5 \\ 5 & 2 \end{bmatrix} = \begin{bmatrix} 13 & 5 \\ -5 & -2 \end{bmatrix}$

11. $\begin{bmatrix} 6 & -3 \\ -8 & 4 \end{bmatrix}^{-1} = \dfrac{1}{24-24}\begin{bmatrix} 4 & 3 \\ 8 & 6 \end{bmatrix}$, which is not defined, and so there is no inverse.

13. $\begin{bmatrix} 0.4 & -1.2 \\ 0.3 & 0.6 \end{bmatrix}^{-1} = \dfrac{1}{0.24+0.36}\begin{bmatrix} 0.6 & 1.2 \\ -0.3 & 0.4 \end{bmatrix} = \begin{bmatrix} 1 & 2 \\ -\frac{1}{2} & \frac{2}{3} \end{bmatrix}$

15. $\begin{bmatrix} 2 & 4 & 1 & 1 & 0 & 0 \\ -1 & 1 & -1 & 0 & 1 & 0 \\ 1 & 4 & 0 & 0 & 0 & 1 \end{bmatrix} \xrightarrow[2R_3-R_1\to R_3]{2R_2+R_1\to R_2} \begin{bmatrix} 2 & 4 & 1 & 1 & 0 & 0 \\ 0 & 6 & -1 & 1 & 2 & 0 \\ 0 & 4 & -1 & -1 & 0 & 2 \end{bmatrix} \xrightarrow[3R_1-2R_2\to R_1]{3R_3-2R_2\to R_3} \begin{bmatrix} 6 & 0 & 5 & 1 & -4 & 0 \\ 0 & 6 & -1 & 1 & 2 & 0 \\ 0 & 0 & -1 & -5 & -4 & 6 \end{bmatrix}$

$\xrightarrow[R_2-R_3\to R_2]{R_1+5R_3\to R_1} \begin{bmatrix} 6 & 0 & 0 & -24 & -24 & 30 \\ 0 & 6 & 0 & 6 & 6 & -6 \\ 0 & 0 & -1 & -5 & -4 & 6 \end{bmatrix} \xrightarrow[\frac{1}{6}R_2,\,-R_3]{\frac{1}{6}R_1} \begin{bmatrix} 1 & 0 & 0 & -4 & -4 & 5 \\ 0 & 1 & 0 & 1 & 1 & -1 \\ 0 & 0 & 1 & 5 & 4 & -6 \end{bmatrix}$.

Therefore, the inverse matrix is $\begin{bmatrix} -4 & -4 & 5 \\ 1 & 1 & -1 \\ 5 & 4 & -6 \end{bmatrix}$.

17. $\begin{bmatrix} 1 & 2 & 3 & 1 & 0 & 0 \\ 4 & 5 & -1 & 0 & 1 & 0 \\ 1 & -1 & -10 & 0 & 0 & 1 \end{bmatrix} \xrightarrow[R_3-R_1\to R_3]{R_2-4R_1\to R_2} \begin{bmatrix} 1 & 2 & 3 & 1 & 0 & 0 \\ 0 & -3 & -13 & -4 & 1 & 0 \\ 0 & -3 & -13 & -1 & 0 & 1 \end{bmatrix} \xrightarrow{R_3-R_2\to R_3} \begin{bmatrix} 1 & 2 & 3 & 1 & 0 & 0 \\ 0 & -3 & -13 & -4 & 1 & 0 \\ 0 & 0 & 0 & 3 & -1 & 1 \end{bmatrix}$.

Since the left half of the last row consists entirely of zeros, there is no inverse matrix.

19. $\begin{bmatrix} 0 & -2 & 2 & 1 & 0 & 0 \\ 3 & 1 & 3 & 0 & 1 & 0 \\ 1 & -2 & 3 & 0 & 0 & 1 \end{bmatrix} \xrightarrow{R_1\leftrightarrow R_3} \begin{bmatrix} 1 & -2 & 3 & 0 & 0 & 1 \\ 3 & 1 & 3 & 0 & 1 & 0 \\ 0 & -2 & 2 & 1 & 0 & 0 \end{bmatrix} \xrightarrow{R_2-3R_1\to R_2} \begin{bmatrix} 1 & -2 & 3 & 0 & 0 & 1 \\ 0 & 7 & -6 & 0 & 1 & -3 \\ 0 & -2 & 2 & 1 & 0 & 0 \end{bmatrix}$

$\xrightarrow[R_2+3R_3\to R_2]{R_1-R_3\to R_1} \begin{bmatrix} 1 & 0 & 1 & -1 & 0 & 1 \\ 0 & 1 & 0 & 3 & 1 & -3 \\ 0 & -2 & 2 & 1 & 0 & 0 \end{bmatrix} \xrightarrow{R_3+2R_2\to R_3} \begin{bmatrix} 1 & 0 & 1 & -1 & 0 & 1 \\ 0 & 1 & 0 & 3 & 1 & -3 \\ 0 & 0 & 2 & 7 & 2 & -6 \end{bmatrix} \xrightarrow{\frac{1}{2}R_3}$

$\begin{bmatrix} 1 & 0 & 1 & -1 & 0 & 1 \\ 0 & 1 & 0 & 3 & 1 & -3 \\ 0 & 0 & 1 & \frac{7}{2} & 1 & -3 \end{bmatrix} \xrightarrow{R_1-R_3\to R_1} \begin{bmatrix} 1 & 0 & 0 & -\frac{9}{2} & -1 & 4 \\ 0 & 1 & 0 & 3 & 1 & -3 \\ 0 & 0 & 1 & \frac{7}{2} & 1 & -3 \end{bmatrix}$. Therefore, the inverse matrix is $\begin{bmatrix} -\frac{9}{2} & -1 & 4 \\ 3 & 1 & -3 \\ \frac{7}{2} & 1 & -3 \end{bmatrix}$.

21.
$$\begin{bmatrix} 1 & 2 & 0 & 3 & 1 & 0 & 0 & 0 \\ 0 & 1 & 1 & 1 & 0 & 1 & 0 & 0 \\ 0 & 1 & 0 & 1 & 0 & 0 & 1 & 0 \\ 1 & 2 & 0 & 2 & 0 & 0 & 0 & 1 \end{bmatrix} \xrightarrow[R_4 - R_1 \to R_4]{R_3 - R_2 \to R_3} \begin{bmatrix} 1 & 2 & 0 & 3 & 1 & 0 & 0 & 0 \\ 0 & 1 & 1 & 1 & 0 & 1 & 0 & 0 \\ 0 & 0 & -1 & 0 & 0 & -1 & 1 & 0 \\ 0 & 0 & 0 & -1 & -1 & 0 & 0 & 1 \end{bmatrix} \xrightarrow[-R_4]{-R_3} \begin{bmatrix} 1 & 2 & 0 & 3 & 1 & 0 & 0 & 0 \\ 0 & 1 & 1 & 1 & 0 & 1 & 0 & 0 \\ 0 & 0 & 1 & 0 & 0 & 1 & -1 & 0 \\ 0 & 0 & 0 & 1 & 1 & 0 & 0 & -1 \end{bmatrix}$$

$$\xrightarrow[R_2 - R_3 \to R_2]{R_1 - 2R_2 \to R_1} \begin{bmatrix} 1 & 0 & -2 & 1 & 1 & -2 & 0 & 0 \\ 0 & 1 & 0 & 1 & 0 & 0 & 1 & 0 \\ 0 & 0 & 1 & 0 & 0 & 1 & -1 & 0 \\ 0 & 0 & 0 & 1 & 1 & 0 & 0 & -1 \end{bmatrix} \xrightarrow[R_2 - R_4 \to R_2]{R_1 + 2R_3 \to R_1} \begin{bmatrix} 1 & 0 & 0 & 1 & 1 & 0 & -2 & 0 \\ 0 & 1 & 0 & 0 & -1 & 0 & 1 & 1 \\ 0 & 0 & 1 & 0 & 0 & 1 & -1 & 0 \\ 0 & 0 & 0 & 1 & 1 & 0 & 0 & -1 \end{bmatrix} \xrightarrow{R_1 \to R_1 - R_4}$$

$$\begin{bmatrix} 1 & 0 & 0 & 0 & 0 & 0 & -2 & 1 \\ 0 & 1 & 0 & 0 & -1 & 0 & 1 & 1 \\ 0 & 0 & 1 & 0 & 0 & 1 & -1 & 0 \\ 0 & 0 & 0 & 1 & 1 & 0 & 0 & -1 \end{bmatrix}. \text{ Therefore, the inverse matrix is } \begin{bmatrix} 0 & 0 & -2 & 1 \\ -1 & 0 & 1 & 1 \\ 0 & 1 & -1 & 0 \\ 1 & 0 & 0 & -1 \end{bmatrix}.$$

23. $\begin{cases} 5x + 3y = 4 \\ 3x + 2y = 0 \end{cases}$ is equivalent to the matrix equation $\begin{bmatrix} 5 & 3 \\ 3 & 2 \end{bmatrix} \begin{bmatrix} x \\ y \end{bmatrix} = \begin{bmatrix} 4 \\ 0 \end{bmatrix}$. Using the inverse from Exercise 3,

$\begin{bmatrix} x \\ y \end{bmatrix} = \begin{bmatrix} 2 & -3 \\ -3 & 5 \end{bmatrix} \begin{bmatrix} 4 \\ 0 \end{bmatrix} = \begin{bmatrix} 8 \\ -12 \end{bmatrix}$. Therefore, $x = 8$ and $y = -12$.

25. $\begin{cases} 2x + 5y = 2 \\ -5x - 13y = 20 \end{cases}$ is equivalent to the matrix equation $\begin{bmatrix} 2 & 5 \\ -5 & -13 \end{bmatrix} \begin{bmatrix} x \\ y \end{bmatrix} = \begin{bmatrix} 2 \\ 20 \end{bmatrix}$. Using the inverse from

Exercise 5, $\begin{bmatrix} x \\ y \end{bmatrix} = \begin{bmatrix} 13 & 5 \\ -5 & -2 \end{bmatrix} \begin{bmatrix} 2 \\ 20 \end{bmatrix} = \begin{bmatrix} 126 \\ -50 \end{bmatrix}$. Therefore, $x = 126$ and $y = -50$.

27. $\begin{cases} 2x + 4y + z = 7 \\ -x + y - z = 0 \\ x + 4y = -2 \end{cases}$ is equivalent to the matrix equation $\begin{bmatrix} 2 & 4 & 1 \\ -1 & 1 & -1 \\ 1 & 4 & 0 \end{bmatrix} \begin{bmatrix} x \\ y \\ z \end{bmatrix} = \begin{bmatrix} 7 \\ 0 \\ -2 \end{bmatrix}$. Using the inverse from

Exercise 11, $\begin{bmatrix} x \\ y \\ z \end{bmatrix} = \begin{bmatrix} -4 & -4 & 5 \\ 1 & 1 & -1 \\ 5 & 4 & -6 \end{bmatrix} \begin{bmatrix} 7 \\ 0 \\ -2 \end{bmatrix} = \begin{bmatrix} -38 \\ 9 \\ 47 \end{bmatrix}$. Therefore, $x = -38$, $y = 9$, and $z = 47$.

29. $\begin{cases} -2y + 2z = 12 \\ 3x + y + 3z = -2 \\ x - 2y + 3z = 8 \end{cases}$ is equivalent to the matrix equation $\begin{bmatrix} 0 & -2 & 2 \\ 3 & 1 & 3 \\ 1 & -2 & 3 \end{bmatrix} \begin{bmatrix} x \\ y \\ z \end{bmatrix} = \begin{bmatrix} 12 \\ -2 \\ 8 \end{bmatrix}$. Using the inverse from

Exercise 15, $\begin{bmatrix} x \\ y \\ z \end{bmatrix} = \begin{bmatrix} -\frac{9}{2} & -1 & 4 \\ 3 & 1 & -3 \\ \frac{7}{2} & 1 & -3 \end{bmatrix} \begin{bmatrix} 12 \\ -2 \\ 8 \end{bmatrix} = \begin{bmatrix} -20 \\ 10 \\ 16 \end{bmatrix}$. Therefore, $x = -20$, $y = 10$, and $z = 16$.

31. Using a calculator, we get the result $(3, 2, 1)$.

33. Using a calculator, we get the result $(3, -2, 2)$.

35. Using a calculator, we get the result $(8, 1, 0, 3)$.

37. This has the form $MX = C$, so $M^{-1}(MX) = M^{-1}C$ and $M^{-1}(MX) = (M^{-1}M)X = X$.

Now $M^{-1} = \begin{bmatrix} 3 & -2 \\ -4 & 3 \end{bmatrix}^{-1} = \frac{1}{9-8}\begin{bmatrix} 3 & 2 \\ 4 & 3 \end{bmatrix} = \begin{bmatrix} 3 & 2 \\ 4 & 3 \end{bmatrix}$. Since $X = M^{-1}C$, we get

$$\begin{bmatrix} x & y & z \\ u & v & w \end{bmatrix} = \begin{bmatrix} 3 & 2 \\ 4 & 3 \end{bmatrix}\begin{bmatrix} 1 & 0 & -1 \\ 2 & 1 & 3 \end{bmatrix} = \begin{bmatrix} 7 & 2 & 3 \\ 10 & 3 & 5 \end{bmatrix}.$$

39. $\begin{bmatrix} a & -a \\ a & a \end{bmatrix}^{-1} = \frac{1}{a^2-(-a^2)}\begin{bmatrix} a & a \\ -a & a \end{bmatrix} = \frac{1}{2a^2}\begin{bmatrix} a & a \\ -a & a \end{bmatrix} = \frac{1}{2a}\begin{bmatrix} 1 & 1 \\ -1 & 1 \end{bmatrix}.$

41. $\begin{bmatrix} 2 & x \\ x & x^2 \end{bmatrix}^{-1} = \frac{1}{2x^2-x^2}\begin{bmatrix} x^2 & -x \\ -x & 2 \end{bmatrix} = \frac{1}{x^2}\begin{bmatrix} x^2 & -x \\ -x & 2 \end{bmatrix} = \begin{bmatrix} 1 & -1/x \\ -1/x & 2/x^2 \end{bmatrix}.$ The inverse does not exist when $x = 0$.

43. $\begin{bmatrix} 1 & e^x & 0 & 1 & 0 & 0 \\ e^x & -e^{2x} & 0 & 0 & 1 & 0 \\ 0 & 0 & 2 & 0 & 0 & 1 \end{bmatrix} \xrightarrow{R_2 - e^x R_1 \to R_2} \begin{bmatrix} 1 & e^x & 0 & 1 & 0 & 0 \\ 0 & -2e^{2x} & 0 & -e^x & 1 & 0 \\ 0 & 0 & 2 & 0 & 0 & 1 \end{bmatrix} \xrightarrow[\frac{1}{2}R_3]{-\frac{1}{2}e^{-2x}R_2}$

$\begin{bmatrix} 1 & e^x & 0 & 1 & 0 & 0 \\ 0 & 1 & 0 & \frac{1}{2}e^{-x} & -\frac{1}{2}e^{-2x} & 0 \\ 0 & 0 & 1 & 0 & 0 & \frac{1}{2} \end{bmatrix} \xrightarrow{R_1 - e^x R_2 \to R_1} \begin{bmatrix} 1 & 0 & 0 & \frac{1}{2} & \frac{1}{2}e^{-x} & 0 \\ 0 & 1 & 0 & \frac{1}{2}e^{-x} & -\frac{1}{2}e^{-2x} & 0 \\ 0 & 0 & 1 & 0 & 0 & \frac{1}{2} \end{bmatrix}.$

Therefore, the inverse matrix is $\begin{bmatrix} \frac{1}{2} & \frac{1}{2}e^{-x} & 0 \\ \frac{1}{2}e^{-x} & -\frac{1}{2}e^{-2x} & 0 \\ 0 & 0 & \frac{1}{2} \end{bmatrix}.$ The inverse exists for all x.

45. $\begin{bmatrix} \cos x & \sin x \\ -\sin x & \cos x \end{bmatrix}^{-1} = \frac{1}{\cos^2 x + \sin^2 x}\begin{bmatrix} \cos x & -\sin x \\ \sin x & \cos x \end{bmatrix} = \begin{bmatrix} \cos x & -\sin x \\ \sin x & \cos x \end{bmatrix}.$ The inverse exists for all x.

47. (a) $\begin{bmatrix} 3 & 1 & 3 & 1 & 0 & 0 \\ 4 & 2 & 4 & 0 & 1 & 0 \\ 3 & 2 & 4 & 0 & 0 & 1 \end{bmatrix} \xrightarrow[R_1 \leftrightarrow R_2]{R_3 - R_1 \to R_3} \begin{bmatrix} 4 & 2 & 4 & 0 & 1 & 0 \\ 3 & 1 & 3 & 1 & 0 & 0 \\ 0 & 1 & 1 & -1 & 0 & 1 \end{bmatrix} \xrightarrow{R_1 - R_2 \to R_1} \begin{bmatrix} 1 & 1 & 1 & -1 & 1 & 0 \\ 3 & 1 & 3 & 1 & 0 & 0 \\ 0 & 1 & 1 & -1 & 0 & 1 \end{bmatrix}$

$\xrightarrow{R_2 - 3R_1 \to R_2} \begin{bmatrix} 1 & 1 & 1 & -1 & 1 & 0 \\ 0 & -2 & 0 & 4 & -3 & 0 \\ 0 & 1 & 1 & -1 & 0 & 1 \end{bmatrix} \xrightarrow[\substack{-\frac{1}{2}R_2 \\ R_3 + \frac{1}{2}R_2 \to R_3}]{R_1 + \frac{1}{2}R_2 \to R_1} \begin{bmatrix} 1 & 0 & 1 & 1 & -\frac{1}{2} & 0 \\ 0 & 1 & 0 & -2 & \frac{3}{2} & 0 \\ 0 & 0 & 1 & 1 & -\frac{3}{2} & 1 \end{bmatrix} \xrightarrow{R_1 - R_3 \to R_1}$

$\begin{bmatrix} 1 & 0 & 0 & 0 & 1 & -1 \\ 0 & 1 & 0 & -2 & \frac{3}{2} & 0 \\ 0 & 0 & 1 & 1 & -\frac{3}{2} & 1 \end{bmatrix}.$ Therefore, the inverse of the matrix is $\begin{bmatrix} 0 & 1 & -1 \\ -2 & \frac{3}{2} & 0 \\ 1 & -\frac{3}{2} & 1 \end{bmatrix}.$

(b) $\begin{bmatrix} A \\ B \\ C \end{bmatrix} = \begin{bmatrix} 0 & 1 & -1 \\ -2 & \frac{3}{2} & 0 \\ 1 & -\frac{3}{2} & 1 \end{bmatrix}\begin{bmatrix} 10 \\ 14 \\ 13 \end{bmatrix} = \begin{bmatrix} 1 \\ 1 \\ 2 \end{bmatrix}.$

Therefore, he should feed the rats 1 oz of food A, 1 oz of food B, and 2 oz of food C.

(c) $\begin{bmatrix} A \\ B \\ C \end{bmatrix} = \begin{bmatrix} 0 & 1 & -1 \\ -2 & \frac{3}{2} & 0 \\ 1 & -\frac{3}{2} & 1 \end{bmatrix} \begin{bmatrix} 9 \\ 12 \\ 10 \end{bmatrix} = \begin{bmatrix} 2 \\ 0 \\ 1 \end{bmatrix}$.

Therefore, he should feed the rats 2 oz of food A, no food B, and 1 oz of food C.

(d) $\begin{bmatrix} A \\ B \\ C \end{bmatrix} = \begin{bmatrix} 0 & 1 & -1 \\ -2 & \frac{3}{2} & 0 \\ 1 & -\frac{3}{2} & 1 \end{bmatrix} \begin{bmatrix} 2 \\ 4 \\ 11 \end{bmatrix} = \begin{bmatrix} -7 \\ 2 \\ 7 \end{bmatrix}$.

Since $A < 0$, there is no combination of foods giving the required supply.

49. (a) $\begin{cases} x + y + 2z = 675 \\ 2x + y + z = 600 \\ x + 2y + z = 625 \end{cases}$
(b) $\begin{bmatrix} 1 & 1 & 2 \\ 2 & 1 & 1 \\ 1 & 2 & 1 \end{bmatrix} \begin{bmatrix} x \\ y \\ z \end{bmatrix} = \begin{bmatrix} 675 \\ 600 \\ 625 \end{bmatrix}$

(c) $\begin{bmatrix} 1 & 1 & 2 & 1 & 0 & 0 \\ 2 & 1 & 1 & 0 & 1 & 0 \\ 1 & 2 & 1 & 0 & 0 & 1 \end{bmatrix} \xrightarrow[R_3 - R_1 \to R_3]{R_2 - 2R_1 \to R_2} \begin{bmatrix} 1 & 1 & 2 & 1 & 0 & 0 \\ 0 & -1 & -3 & -2 & 1 & 0 \\ 0 & 1 & -1 & -1 & 0 & 1 \end{bmatrix} \xrightarrow[R_3 + R_2 \to R_3]{R_1 + R_2 \to R_1} \begin{bmatrix} 1 & 0 & -1 & -1 & 1 & 0 \\ 0 & -1 & -3 & -2 & 1 & 0 \\ 0 & 0 & -4 & -3 & 1 & 1 \end{bmatrix}$

$\xrightarrow[-\frac{1}{4}R_3]{-R_2} \begin{bmatrix} 1 & 0 & -1 & -1 & 1 & 0 \\ 0 & 1 & 3 & 2 & -1 & 0 \\ 0 & 0 & 1 & \frac{3}{4} & -\frac{1}{4} & -\frac{1}{4} \end{bmatrix} \xrightarrow[R_2 - 3R_3 \to R_2]{R_1 + R_3 \to R_1} \begin{bmatrix} 1 & 0 & 0 & -\frac{1}{4} & \frac{3}{4} & -\frac{1}{4} \\ 0 & 1 & 0 & -\frac{1}{4} & -\frac{1}{4} & \frac{3}{4} \\ 0 & 0 & 1 & \frac{3}{4} & -\frac{1}{4} & -\frac{1}{4} \end{bmatrix}$. Therefore, the inverse of the

matrix is $\begin{bmatrix} -\frac{1}{4} & \frac{3}{4} & -\frac{1}{4} \\ -\frac{1}{4} & -\frac{1}{4} & \frac{3}{4} \\ \frac{3}{4} & -\frac{1}{4} & -\frac{1}{4} \end{bmatrix}$ and $\begin{bmatrix} x \\ y \\ z \end{bmatrix} = \begin{bmatrix} -\frac{1}{4} & \frac{3}{4} & -\frac{1}{4} \\ -\frac{1}{4} & -\frac{1}{4} & \frac{3}{4} \\ \frac{3}{4} & -\frac{1}{4} & -\frac{1}{4} \end{bmatrix} \begin{bmatrix} 675 \\ 600 \\ 625 \end{bmatrix} = \begin{bmatrix} 125 \\ 150 \\ 200 \end{bmatrix}$. Thus, he earns \$125 on a

standard set, \$150 on a deluxe set, and \$200 on a leather-bound set.

10.7 Determinants and Cramer's Rule

1. The matrix $\begin{bmatrix} 2 & 0 \\ 0 & 3 \end{bmatrix}$ has determinant

$|D| = (2)(3) - (0)(0) = 6$.

3. The matrix $\begin{bmatrix} 4 & 5 \\ 0 & -1 \end{bmatrix}$ has determinant

$|D| = (4)(-1) - (5)(0) = -4$.

5. The matrix $\begin{bmatrix} 2 & 5 \end{bmatrix}$ does not have a determinant because the matrix is not square.

7. The matrix $\begin{bmatrix} \frac{1}{2} & \frac{1}{8} \\ 1 & \frac{1}{2} \end{bmatrix}$ has determinant

$|D| = \frac{1}{2} \cdot \frac{1}{2} - 1 \cdot \frac{1}{8} = \frac{1}{4} - \frac{1}{8} = \frac{1}{8}$.

In Solutions 9–13, $A = \begin{bmatrix} 1 & 0 & \frac{1}{2} \\ -3 & 5 & 2 \\ 0 & 0 & 4 \end{bmatrix}$.

9. $M_{11} = 5 \cdot 4 - 0 \cdot 2 = 20$, $A_{11} = (-1)^2 M_{11} = 20$

11. $M_{12} = -3 \cdot 4 - 0 \cdot 2 = -12$, $A_{12} = (-1)^3 M_{12} = 12$

13. $M_{23} = 1 \cdot 0 - 0 \cdot 0 = 0$, $A_{23} = (-1)^5 M_{23} = 0$

15. $M = \begin{bmatrix} 2 & 1 & 0 \\ 0 & -2 & 4 \\ 0 & 1 & -3 \end{bmatrix}$. Therefore, expanding by the first column, $|M| = 2 \begin{vmatrix} -2 & 4 \\ 1 & -3 \end{vmatrix} = 2\,(6 - 4) = 4$. Since $|M| \neq 0$,

the matrix has an inverse.

17. $M = \begin{bmatrix} 1 & 3 & 7 \\ 2 & 0 & -1 \\ 0 & 2 & 6 \end{bmatrix}$. Therefore, expanding by the third row,

$|M| = -2 \begin{vmatrix} 1 & 7 \\ 2 & -1 \end{vmatrix} + 6 \begin{vmatrix} 1 & 3 \\ 2 & 0 \end{vmatrix} = -2\,(-1 - 14) + 6\,(0 - 6) = 30 - 36 = -6$. Since $|M| \neq 0$, the matrix has an

inverse.

19. $M = \begin{bmatrix} 30 & 0 & 20 \\ 0 & -10 & -20 \\ 40 & 0 & 10 \end{bmatrix}$. Therefore, expanding by the first row,

$|M| = 30 \begin{vmatrix} -10 & -20 \\ 0 & 10 \end{vmatrix} + 20 \begin{vmatrix} 0 & -10 \\ 40 & 0 \end{vmatrix} = 30\,(-100 + 0) + 20\,(0 + 400) = -3000 + 8000 = 5000$, and so M^{-1}

exists.

21. $M = \begin{bmatrix} 1 & 3 & 3 & 0 \\ 0 & 2 & 0 & 1 \\ -1 & 0 & 0 & 2 \\ 1 & 6 & 4 & 1 \end{bmatrix}$. Therefore, expanding by the third row,

$|M| = -1 \begin{vmatrix} 3 & 3 & 0 \\ 2 & 0 & 1 \\ 6 & 4 & 1 \end{vmatrix} - 2 \begin{vmatrix} 1 & 3 & 3 \\ 0 & 2 & 0 \\ 1 & 6 & 4 \end{vmatrix} = 1 \begin{vmatrix} 3 & 3 \\ 6 & 4 \end{vmatrix} - 1 \begin{vmatrix} 3 & 3 \\ 2 & 0 \end{vmatrix} - 4 \begin{vmatrix} 1 & 3 \\ 1 & 4 \end{vmatrix} = -6 + 6 - 4 = -4$, and so M^{-1} exists.

23. $|M| = \begin{vmatrix} 0 & 0 & 4 & 6 \\ 2 & 1 & 1 & 3 \\ 2 & 1 & 2 & 3 \\ 3 & 0 & 1 & 7 \end{vmatrix} = \begin{vmatrix} 0 & 0 & 4 & 6 \\ 2 & 1 & 1 & 3 \\ 0 & 0 & 1 & 0 \\ 3 & 0 & 1 & 7 \end{vmatrix}$, by replacing R_3 with $R_3 - R_2$. Then, expanding by the third row,

$|M| = 1 \begin{vmatrix} 0 & 0 & 6 \\ 2 & 1 & 3 \\ 3 & 0 & 7 \end{vmatrix} = 6 \begin{vmatrix} 2 & 1 \\ 3 & 0 \end{vmatrix} = 6\,(2 \cdot 0 - 3 \cdot 1) = -18$.

25. $M = \begin{bmatrix} 1 & 2 & 3 & 4 & 5 \\ 0 & 2 & 4 & 6 & 8 \\ 0 & 0 & 3 & 6 & 9 \\ 0 & 0 & 0 & 4 & 8 \\ 0 & 0 & 0 & 0 & 5 \end{bmatrix}$, so $|M| = 5 \begin{vmatrix} 1 & 2 & 3 & 4 \\ 0 & 2 & 4 & 6 \\ 0 & 0 & 3 & 6 \\ 0 & 0 & 0 & 4 \end{vmatrix} = 5 \cdot 4 \begin{vmatrix} 1 & 2 & 3 \\ 0 & 2 & 4 \\ 0 & 0 & 3 \end{vmatrix} = 20 \cdot 3 \begin{vmatrix} 1 & 2 \\ 0 & 2 \end{vmatrix} = 60 \cdot 2 = 120$.

27. $B = \begin{bmatrix} 4 & 1 & 0 \\ -2 & -1 & 1 \\ 4 & 0 & 3 \end{bmatrix}$

(a) $|B| = 2 \begin{vmatrix} 1 & 0 \\ 0 & 3 \end{vmatrix} - 1 \begin{vmatrix} 4 & 0 \\ 4 & 3 \end{vmatrix} - 1 \begin{vmatrix} 4 & 1 \\ 4 & 0 \end{vmatrix} = 6 - 12 + 4 = -2$

(b) $|B| = -1 \begin{vmatrix} 4 & 1 \\ 4 & 0 \end{vmatrix} + 3 \begin{vmatrix} 4 & 1 \\ -2 & -1 \end{vmatrix} = 4 - 6 = -2$

(c) Yes, as expected, the results agree.

29. $\begin{cases} 2x - y = -9 \\ x + 2y = 8 \end{cases}$ Then $|D| = \begin{vmatrix} 2 & -1 \\ 1 & 2 \end{vmatrix} = 5, |D_x| = \begin{vmatrix} -9 & -1 \\ 8 & 2 \end{vmatrix} = -10$, and $|D_y| = \begin{vmatrix} 2 & -9 \\ 1 & 8 \end{vmatrix} = 25$.

Hence, $x = \dfrac{|D_x|}{|D|} = \dfrac{-10}{5} = -2$, $y = \dfrac{|D_y|}{|D|} = \dfrac{25}{5} = 5$, and so the solution is $(-2, 5)$.

31. $\begin{cases} x - 6y = 3 \\ 3x + 2y = 1 \end{cases}$ Then, $|D| = \begin{vmatrix} 1 & -6 \\ 3 & 2 \end{vmatrix} = 20, |D_x| = \begin{vmatrix} 3 & -6 \\ 1 & 2 \end{vmatrix} = 12$, and $|D_y| = \begin{vmatrix} 1 & 3 \\ 3 & 1 \end{vmatrix} = -8$.

Hence, $x = \dfrac{|D_x|}{|D|} = \dfrac{12}{20} = 0.6$, $y = \dfrac{|D_y|}{|D|} = \dfrac{-8}{20} = -0.4$, and so the solution is $(0.6, -0.4)$.

33. $\begin{cases} 0.4x + 1.2y = 0.4 \\ 1.2x + 1.6y = 3.2 \end{cases}$ Then, $|D| = \begin{vmatrix} 0.4 & 1.2 \\ 1.2 & 1.6 \end{vmatrix} = -0.8, |D_x| = \begin{vmatrix} 0.4 & 1.2 \\ 3.2 & 1.6 \end{vmatrix} = -3.2$, and $|D_y| = \begin{vmatrix} 0.4 & 0.4 \\ 1.2 & 3.2 \end{vmatrix} = 0.8$.

Hence, $x = \dfrac{|D_x|}{|D|} = \dfrac{-3.2}{-0.8} = 4$, $y = \dfrac{|D_y|}{|D|} = \dfrac{0.8}{-0.8} = -1$, and so the solution is $(4, -1)$.

35. $\begin{cases} x - y + 2z = 0 \\ 3x + z = 11 \\ -x + 2y = 0 \end{cases}$ Then expanding by the second row,

$|D| = \begin{vmatrix} 1 & -1 & 2 \\ 3 & 0 & 1 \\ -1 & 2 & 0 \end{vmatrix} = -3 \begin{vmatrix} -1 & 2 \\ 2 & 0 \end{vmatrix} - 1 \begin{vmatrix} 1 & -1 \\ -1 & 2 \end{vmatrix} = 12 - 1 = 11, |D_x| = \begin{vmatrix} 0 & -1 & 2 \\ 11 & 0 & 1 \\ 0 & 2 & 0 \end{vmatrix} = -11 \begin{vmatrix} -1 & 2 \\ 2 & 0 \end{vmatrix} = 44,$

$|D_y| = \begin{vmatrix} 1 & 0 & 2 \\ 3 & 11 & 1 \\ -1 & 0 & 0 \end{vmatrix} = 11 \begin{vmatrix} 1 & 2 \\ -1 & 0 \end{vmatrix} = 22$, and $|D_z| = \begin{vmatrix} 1 & -1 & 0 \\ 3 & 0 & 11 \\ -1 & 2 & 0 \end{vmatrix} = -11 \begin{vmatrix} 1 & -1 \\ -1 & 2 \end{vmatrix} = -11.$

Therefore, $x = \dfrac{44}{11} = 4$, $y = \dfrac{22}{11} = 2$, $z = \dfrac{-11}{11} = -1$, and so the solution is $(4, 2, -1)$.

37. $\begin{cases} 2x_1 + 3x_2 - 5x_3 = 1 \\ x_1 + x_2 - x_3 = 2 \\ 2x_2 + x_3 = 8 \end{cases}$

Then, expanding by the third row,

$$|D| = \begin{vmatrix} 2 & 3 & -5 \\ 1 & 1 & -1 \\ 0 & 2 & 1 \end{vmatrix} = -2\begin{vmatrix} 2 & -5 \\ 1 & -1 \end{vmatrix} + \begin{vmatrix} 2 & 3 \\ 1 & 1 \end{vmatrix} = -6 - 1 = -7,$$

$$|D_{x_1}| = \begin{vmatrix} 1 & 3 & -5 \\ 2 & 1 & -1 \\ 8 & 2 & 1 \end{vmatrix} = \begin{vmatrix} 1 & -1 \\ 2 & 1 \end{vmatrix} - 3\begin{vmatrix} 2 & -1 \\ 8 & 1 \end{vmatrix} - 5\begin{vmatrix} 2 & 1 \\ 8 & 2 \end{vmatrix} = 3 - 30 + 20 = -7,$$

$$|D_{x_2}| = \begin{vmatrix} 2 & 1 & -5 \\ 1 & 2 & -1 \\ 0 & 8 & 1 \end{vmatrix} = -8\begin{vmatrix} 2 & -5 \\ 1 & -1 \end{vmatrix} + \begin{vmatrix} 2 & 1 \\ 1 & 2 \end{vmatrix} = -24 + 3 = -21, \text{ and}$$

$$|D_{x_3}| = \begin{vmatrix} 2 & 3 & 1 \\ 1 & 1 & 2 \\ 0 & 2 & 8 \end{vmatrix} = -2\begin{vmatrix} 2 & 1 \\ 1 & 2 \end{vmatrix} + 8\begin{vmatrix} 2 & 3 \\ 1 & 1 \end{vmatrix} = -6 - 8 = -14.$$

Thus, $x_1 = \frac{-7}{-7} = 1$, $x_2 = \frac{-21}{-7} = 3$, $x_3 = \frac{-14}{-7} = 2$, and so the solution is $(1, 3, 2)$.

39. $\begin{cases} \frac{1}{3}x - \frac{1}{5}y + \frac{1}{2}z = \frac{7}{10} \\ -\frac{2}{3}x + \frac{2}{5}y + \frac{3}{2}z = \frac{11}{10} \\ x - \frac{4}{5}y + z = \frac{9}{5} \end{cases}$ $\Leftrightarrow$ $\begin{cases} 10x - 6y + 15z = 21 \\ -20x + 12y + 45z = 33 \\ 5x - 4y + 5z = 9 \end{cases}$ Then

$$|D| = \begin{vmatrix} 10 & -6 & 15 \\ -20 & 12 & 45 \\ 5 & -4 & 5 \end{vmatrix} = 10\begin{vmatrix} 12 & 45 \\ -4 & 5 \end{vmatrix} + 6\begin{vmatrix} -20 & 45 \\ 5 & 5 \end{vmatrix} + 15\begin{vmatrix} -20 & 12 \\ 5 & -4 \end{vmatrix} = 2400 - 1950 + 300 = 750,$$

$$|D_x| = \begin{vmatrix} 21 & -6 & 15 \\ 33 & 12 & 45 \\ 9 & -4 & 5 \end{vmatrix} = 21\begin{vmatrix} 12 & 45 \\ -4 & 5 \end{vmatrix} + 6\begin{vmatrix} 33 & 45 \\ 9 & 5 \end{vmatrix} + 15\begin{vmatrix} 33 & 12 \\ 9 & -4 \end{vmatrix} = 5040 - 1440 - 3600 = 0,$$

$$|D_y| = \begin{vmatrix} 10 & 21 & 15 \\ -20 & 33 & 45 \\ 5 & 9 & 5 \end{vmatrix} = 10\begin{vmatrix} 33 & 45 \\ 9 & 5 \end{vmatrix} - 21\begin{vmatrix} -20 & 45 \\ 5 & 5 \end{vmatrix} + 15\begin{vmatrix} -20 & 33 \\ 5 & 9 \end{vmatrix} = -2400 + 6825 - 5175 = -750, \text{ and}$$

$$|D_z| = \begin{vmatrix} 10 & -6 & 21 \\ -20 & 12 & 33 \\ 5 & -4 & 9 \end{vmatrix} = 10\begin{vmatrix} 12 & 33 \\ -4 & 9 \end{vmatrix} + 6\begin{vmatrix} -20 & 33 \\ 5 & 9 \end{vmatrix} + 21\begin{vmatrix} -20 & 12 \\ 5 & -4 \end{vmatrix} = 2400 - 2070 + 420 = 750.$$

Therefore, $x = 0$, $y = -1$, $z = 1$, and so the solution is $(0, -1, 1)$.

41. $\begin{cases} 3y + 5z = 4 \\ 2x - z = 10 \\ 4x + 7y = 0 \end{cases}$ Then $|D| = \begin{vmatrix} 0 & 3 & 5 \\ 2 & 0 & -1 \\ 4 & 7 & 0 \end{vmatrix} = -3 \begin{vmatrix} 2 & -1 \\ 4 & 0 \end{vmatrix} + 5 \begin{vmatrix} 2 & 0 \\ 4 & 7 \end{vmatrix} = -12 + 70 = 58,$

$|D_x| = \begin{vmatrix} 4 & 3 & 5 \\ 10 & 0 & -1 \\ 0 & 7 & 0 \end{vmatrix} = -7 \begin{vmatrix} 4 & 5 \\ 10 & -1 \end{vmatrix} = 378, \quad |D_y| = \begin{vmatrix} 0 & 4 & 5 \\ 2 & 10 & -1 \\ 4 & 0 & 0 \end{vmatrix} = 4 \begin{vmatrix} 4 & 5 \\ 10 & -1 \end{vmatrix} = -216, \text{ and}$

$|D_z| = \begin{vmatrix} 0 & 3 & 4 \\ 2 & 0 & 10 \\ 4 & 7 & 0 \end{vmatrix} = 4 \begin{vmatrix} 3 & 4 \\ 0 & 10 \end{vmatrix} - 7 \begin{vmatrix} 0 & 4 \\ 2 & 10 \end{vmatrix} = 120 + 56 = 176.$

Thus, $x = \frac{189}{29}$, $y = -\frac{108}{29}$, and $z = \frac{88}{29}$, and so the solution is $\left(\frac{189}{29}, -\frac{108}{29}, \frac{88}{29} \right)$.

43. $\begin{cases} x + y + z + w = 0 \\ 2z + w = 0 \\ y - z = 0 \\ x + 2z = 1 \end{cases}$ Then

$|D| = \begin{vmatrix} 1 & 1 & 1 & 1 \\ 2 & 0 & 0 & 1 \\ 0 & 1 & -1 & 0 \\ 1 & 0 & 2 & 0 \end{vmatrix} = -1 \begin{vmatrix} 2 & 0 & 1 \\ 0 & -1 & 0 \\ 1 & 2 & 0 \end{vmatrix} - 1 \begin{vmatrix} 1 & 1 & 1 \\ 2 & 0 & 1 \\ 1 & 2 & 0 \end{vmatrix} = - \left(2 \begin{vmatrix} -1 & 0 \\ 2 & 0 \end{vmatrix} + 1 \begin{vmatrix} 0 & 1 \\ -1 & 0 \end{vmatrix} \right) - \left(-1 \begin{vmatrix} 2 & 1 \\ 1 & 0 \end{vmatrix} - 2 \begin{vmatrix} 1 & 1 \\ 2 & 1 \end{vmatrix} \right)$

$= -2 \, (0) - 1 \, (1) + 1 \, (-1) + 2 \, (-1) = -4,$

$|D_x| = \begin{vmatrix} 0 & 1 & 1 & 1 \\ 0 & 0 & 0 & 1 \\ 0 & 1 & -1 & 0 \\ 1 & 0 & 2 & 0 \end{vmatrix} = -1 \begin{vmatrix} 1 & 1 & 1 \\ 0 & 0 & 1 \\ 1 & -1 & 0 \end{vmatrix} = -1 \, (-1) \begin{vmatrix} 1 & 1 \\ 1 & -1 \end{vmatrix} = -2,$

$|D_y| = \begin{vmatrix} 1 & 0 & 1 & 1 \\ 2 & 0 & 0 & 1 \\ 0 & 0 & -1 & 0 \\ 1 & 1 & 2 & 0 \end{vmatrix} = 1 \begin{vmatrix} 1 & 1 & 1 \\ 2 & 0 & 1 \\ 0 & -1 & 0 \end{vmatrix} = 1 \begin{vmatrix} 0 & 1 \\ -1 & 0 \end{vmatrix} - 2 \begin{vmatrix} 1 & 1 \\ -1 & 0 \end{vmatrix} = 1 - 2 \, (1) = -1,$

$|D_z| = \begin{vmatrix} 1 & 1 & 0 & 1 \\ 2 & 0 & 0 & 1 \\ 0 & 1 & 0 & 0 \\ 1 & 0 & 1 & 0 \end{vmatrix} = -1 \begin{vmatrix} 1 & 1 & 1 \\ 2 & 0 & 1 \\ 0 & 1 & 0 \end{vmatrix} = -1 \begin{vmatrix} 0 & 1 \\ 1 & 0 \end{vmatrix} + 2 \begin{vmatrix} 1 & 1 \\ 1 & 0 \end{vmatrix} = -1 \, (-1) + 2 \, (-1) = -1, \text{ and}$

$|D_w| = \begin{vmatrix} 1 & 1 & 1 & 0 \\ 2 & 0 & 0 & 0 \\ 0 & 1 & -1 & 0 \\ 1 & 0 & 2 & 1 \end{vmatrix} = 1 \begin{vmatrix} 1 & 1 & 1 \\ 2 & 0 & 0 \\ 0 & 1 & -1 \end{vmatrix} = -2 \begin{vmatrix} 1 & 1 \\ 1 & -1 \end{vmatrix} = -2 \, (-2) = 4. \text{ Hence, we have } x = \frac{|D_x|}{|D|} = \frac{-2}{-4} = \frac{1}{2},$

$y = \frac{|D_y|}{|D|} = \frac{-1}{-4} = \frac{1}{4}, z = \frac{|D_z|}{|D|} = \frac{-1}{-4} = \frac{1}{4}, \text{ and } w = \frac{|D_w|}{|D|} = \frac{4}{-4} = -1, \text{ and the solution is } \left(\frac{1}{2}, \frac{1}{4}, \frac{1}{4}, -1 \right).$

45.
$$\begin{vmatrix} a & 0 & 0 & 0 & 0 \\ 0 & b & 0 & 0 & 0 \\ 0 & 0 & c & 0 & 0 \\ 0 & 0 & 0 & d & 0 \\ 0 & 0 & 0 & 0 & e \end{vmatrix} = a \begin{vmatrix} b & 0 & 0 & 0 \\ 0 & c & 0 & 0 \\ 0 & 0 & d & 0 \\ 0 & 0 & 0 & e \end{vmatrix} = ab \begin{vmatrix} c & 0 & 0 \\ 0 & d & 0 \\ 0 & 0 & e \end{vmatrix} = abc \begin{vmatrix} d & 0 \\ 0 & e \end{vmatrix} = abcde$$

47.
$$\begin{vmatrix} x & 12 & 13 \\ 0 & x-1 & 23 \\ 0 & 0 & x-2 \end{vmatrix} = 0 \quad \Leftrightarrow \quad (x-2)\begin{vmatrix} x & 12 \\ 0 & x-1 \end{vmatrix} = 0 \quad \Leftrightarrow \quad (x-2) \cdot x (x-1) = 0 \quad \Leftrightarrow \quad x = 0, 1, \text{ or } 2$$

49.
$$\begin{vmatrix} 1 & 0 & x \\ x^2 & 1 & 0 \\ x & 0 & 1 \end{vmatrix} = 0 \quad \Leftrightarrow \quad 1 \begin{vmatrix} 1 & 0 \\ 0 & 1 \end{vmatrix} + x \begin{vmatrix} x^2 & 1 \\ x & 0 \end{vmatrix} = 0 \quad \Leftrightarrow \quad 1 - x^2 = 0 \quad \Leftrightarrow \quad x^2 = 1 \quad \Leftrightarrow \quad x = \pm 1$$

51. Area $= \pm\dfrac{1}{2} \begin{vmatrix} 0 & 0 & 1 \\ 6 & 2 & 1 \\ 3 & 8 & 1 \end{vmatrix} = \pm\dfrac{1}{2} \begin{vmatrix} 6 & 2 \\ 3 & 8 \end{vmatrix} = \pm\dfrac{1}{2}(48-6) = \dfrac{1}{2}(42) = 21$

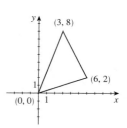

53. Area $= \pm\dfrac{1}{2} \begin{vmatrix} -1 & 3 & 1 \\ 2 & 9 & 1 \\ 5 & -6 & 1 \end{vmatrix} = \pm\dfrac{1}{2}\left[-1 \begin{vmatrix} 9 & 1 \\ -6 & 1 \end{vmatrix} - 3 \begin{vmatrix} 2 & 1 \\ 5 & 1 \end{vmatrix} + 1 \begin{vmatrix} 2 & 9 \\ 5 & -6 \end{vmatrix} \right]$

$$= \pm\dfrac{1}{2}\left[-1(9+6) - 3(2-5) + 1(-12-45) \right]$$

$$= \pm\dfrac{1}{2}\left[-15 - 3(-3) + (-57) \right] = \pm\dfrac{1}{2}(-63) = \dfrac{63}{2}$$

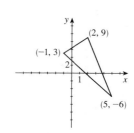

55.
$$\begin{vmatrix} 1 & x & x^2 \\ 1 & y & y^2 \\ 1 & z & z^2 \end{vmatrix} = 1 \begin{vmatrix} y & y^2 \\ z & z^2 \end{vmatrix} - 1 \begin{vmatrix} x & x^2 \\ z & z^2 \end{vmatrix} + 1 \begin{vmatrix} x & x^2 \\ y & y^2 \end{vmatrix} = yz^2 - y^2 z - \left(xz^2 - x^2 z \right) + \left(xy^2 - xy^2 \right)$$

$$= yz^2 - y^2 z - xz^2 - x^2 z + xy^2 - xy^2 + xyz - xyz = xyz - xz^2 - y^2 z + yz^2 - x^2 y + x^2 z + zy^2 - xyz$$

$$= z \left(xy - xz - y^2 + yz \right) - x \left(xy - xz - y^2 + yz \right) = (z - x) \left(xy - xz - y^2 + yz \right)$$

$$= (z - x) \left[x (y - z) - y (y - z) \right] = (z - x)(x - y)(y - z)$$

57. (a) Using the points $(10, 25)$, $(15, 33.75)$, and $(40, 40)$, we substitute for x and y and get the system

$$\begin{cases} 100a + 10b + c = 25 \\ 225a + 15b + c = 33.75 \\ 1600a + 40b + c = 40 \end{cases}$$

(b) $|D| = \begin{vmatrix} 100 & 10 & 1 \\ 225 & 15 & 1 \\ 1600 & 40 & 1 \end{vmatrix} = 1 \cdot \begin{vmatrix} 225 & 15 \\ 1600 & 40 \end{vmatrix} - 1 \cdot \begin{vmatrix} 100 & 10 \\ 1600 & 40 \end{vmatrix} + 1 \cdot \begin{vmatrix} 100 & 10 \\ 225 & 15 \end{vmatrix}$

$= (9000 - 24{,}000) - (4000 - 16{,}000) + (1500 - 2250) = -15{,}000 + 12{,}000 - 750 = -3750,$

$|D_a| = \begin{vmatrix} 25 & 10 & 1 \\ 33.75 & 15 & 1 \\ 40 & 40 & 1 \end{vmatrix} = 1 \cdot \begin{vmatrix} 33.75 & 15 \\ 40 & 40 \end{vmatrix} - 1 \cdot \begin{vmatrix} 25 & 10 \\ 40 & 40 \end{vmatrix} + 1 \cdot \begin{vmatrix} 25 & 10 \\ 33.75 & 15 \end{vmatrix}$

$= (1350 - 600) - (1000 - 400) + (375 - 337.5) = 750 - 600 + 37.5 = 187.5,$

$|D_b| = \begin{vmatrix} 100 & 25 & 1 \\ 225 & 33.75 & 1 \\ 1600 & 40 & 1 \end{vmatrix} = 1 \cdot \begin{vmatrix} 225 & 33.75 \\ 1600 & 40 \end{vmatrix} - 1 \cdot \begin{vmatrix} 100 & 25 \\ 1600 & 40 \end{vmatrix} + 1 \cdot \begin{vmatrix} 100 & 25 \\ 225 & 33.75 \end{vmatrix}$

$= (9000 - 54{,}000) - (4000 - 40{,}000) + (3375 - 5625) = -45{,}000 + 36{,}000 - 2250 = -11{,}250,$ and

$|D_c| = \begin{vmatrix} 100 & 10 & 25 \\ 225 & 15 & 33.75 \\ 1600 & 40 & 40 \end{vmatrix} = 25 \cdot \begin{vmatrix} 225 & 15 \\ 1600 & 40 \end{vmatrix} - 33.75 \cdot \begin{vmatrix} 100 & 10 \\ 1600 & 40 \end{vmatrix} + 40 \cdot \begin{vmatrix} 100 & 10 \\ 225 & 15 \end{vmatrix}$

$= 25 \cdot (9{,}000 - 24{,}000) - 33.75 \cdot (4{,}000 - 16{,}000) + 40 \cdot (1{,}500 - 2{,}250)$

$= 25 \cdot (-15{,}000) + 33.75 \cdot 12{,}000 + 40 \cdot (-750) = -375{,}000 + 405{,}000 - 30{,}000 = 0.$

Thus, $a = \dfrac{|D_a|}{|D|} = \dfrac{187.5}{-3750} = 0.05$, $b = \dfrac{|D_b|}{|D|} = \dfrac{-11{,}250}{-3{,}750} = 3$, and $c = \dfrac{|D_c|}{|D|} = \dfrac{0}{-3{,}750} = 0$. Thus, the model is

$y = 0.05x^2 + 3x.$

59. (a) The coordinates of the vertices of the surrounding rectangle are (a_1, b_1), (a_2, b_1), (a_2, b_3), and (a_1, b_3). The area of the surrounding rectangle is given by $(a_2 - a_1) \cdot (b_3 - b_1) = a_2 b_3 + a_1 b_1 - a_2 b_1 - a_1 b_3 = a_1 b_1 + a_2 b_3 - a_1 b_3 - a_2 b_1$.

(b) The area of the three blue triangles are as follows:

Area of $\triangle((a_1, b_1), (a_2, b_1), (a_2, b_2))$: $\frac{1}{2}(a_2 - a_1) \cdot (b_2 - b_1) = \frac{1}{2}(a_2 b_2 + a_1 b_1 - a_2 b_1 - a_1 b_2)$

Area of $\triangle((a_2, b_2), (a_2, b_3), (a_3, b_3))$: $\frac{1}{2}(a_2 - a_3) \cdot (b_3 - b_2) = \frac{1}{2}(a_2 b_3 + a_3 b_2 - a_2 b_2 - a_3 b_3)$

Area of $\triangle((a_1, b), (a_1, b_3), (a_3, b_3))$: $\frac{1}{2}(a_3 - a_1) \cdot (b_3 - b_1) = \frac{1}{2}(a_3 b_3 + a_1 b_1 - a_3 b_1 - a_1 b_3)$.

Thus the sum of the areas of the blue triangles, B, is

$B = \frac{1}{2}(a_2 b_2 + a_1 b_1 - a_2 b_1 - a_1 b_2) + \frac{1}{2}(a_2 b_3 + a_3 b_2 - a_2 b_2 - a_3 b_3) + \frac{1}{2}(a_3 b_3 + a_1 b_1 - a_3 b_1 - a_1 b_3)$

$= \frac{1}{2}(a_1 b_1 + a_1 b_1 + a_2 b_2 + a_2 b_3 + a_3 b_2 + a_3 b_3) - \frac{1}{2}(a_1 b_2 + a_1 b_3 + a_2 b_1 + a_2 b_2 + a_3 b_1 + a_3 b_3)$

$= a_1 b_1 + \frac{1}{2}(a_2 b_3 + a_3 b_2) - \frac{1}{2}(a_1 b_2 + a_1 b_3 + a_2 b_1 + a_3 b_1)$

So the area of the red triangle A is the area of the rectangle minus the sum of the areas of the blue triangles, that is,

$A = (a_1 b_1 + a_2 b_3 - a_1 b_3 - a_2 b_1) - \left[a_1 b_1 + \frac{1}{2}(a_2 b_3 + a_3 b_2) - \frac{1}{2}(a_1 b_2 + a_1 b_3 + a_2 b_1 + a_3 b_1) \right]$

$= a_1 b_1 + a_2 b_3 - a_1 b_3 - a_2 b_1 - a_1 b_1 - \frac{1}{2}(a_2 b_3 + a_3 b_2) + \frac{1}{2}(a_1 b_2 + a_1 b_3 + a_2 b_1 + a_3 b_1)$

$= \frac{1}{2}(a_1 b_2 + a_2 b_3 + a_3 b_1) - \frac{1}{2}(a_1 b_3 + a_2 b_1 + a_3 b_2)$

(c) We first find $Q = \begin{vmatrix} a_1 & b_1 & 1 \\ a_2 & b_2 & 1 \\ a_3 & b_3 & 1 \end{vmatrix}$ by expanding about the third column.

$$Q = 1\begin{vmatrix} a_2 & b_2 \\ a_3 & b_3 \end{vmatrix} - 1\begin{vmatrix} a_1 & b_1 \\ a_3 & b_3 \end{vmatrix} + 1\begin{vmatrix} a_1 & b_1 \\ a_2 & b_2 \end{vmatrix} = a_2b_3 - a_3b_2 - (a_1b_3 - a_3b_1) + a_1b_2 - a_2b_1$$

$$= a_1b_2 + a_2b_3 + a_3b_1 - a_1b_3 - a_2b_1 - a_3b_2$$

So $\frac{1}{2}Q = \frac{1}{2}(a_1b_2 + a_2b_3 + a_3b_1) - \frac{1}{2}(a_1b_3 - a_2b_1 - a_3b_2)$, the area of the red triangle. Since $\frac{1}{2}Q$ is not always positive, the area is $\pm\frac{1}{2}Q$.

61. (a) Let $|M| = \begin{vmatrix} x & y & 1 \\ x_1 & y_1 & 1 \\ x_2 & y_2 & 1 \end{vmatrix}$. Then, expanding by the third column,

$$|M| = \begin{vmatrix} x_1 & y_1 \\ x_2 & y_2 \end{vmatrix} - \begin{vmatrix} x & y \\ x_2 & y_2 \end{vmatrix} + \begin{vmatrix} x & y \\ x_1 & y_1 \end{vmatrix} = (x_1y_2 - x_2y_1) - (xy_2 - x_2y) + (xy_1 - x_1y)$$

$$= x_1y_2 - x_2y_1 - xy_2 + x_2y + xy_1 - x_1y = x_2y - x_1y - xy_2 + xy_1 + x_1y_2 - x_2y_1$$

$$= (x_2 - x_1)y - (y_2 - y_1)x + x_1y_2 - x_2y_1$$

So $|M| = 0 \iff (x_2 - x_1)y - (y_2 - y_1)x + x_1y_2 - x_2y_1 = 0 \iff (x_2 - x_1)y = (y_2 - y_1)x - x_1y_2 + x_2y_1 \iff$

$(x_2 - x_1)y = (y_2 - y_1)x - x_1y_2 + x_1y_1 - x_1y_1 + x_2y_1 \iff y = \dfrac{y_2 - y_1}{x_2 - x_1}x - \dfrac{x_1(y_2 - y_1)}{x_2 - x_1} + \dfrac{y_1(x_2 - x_1)}{x_2 - x_1}$

$\iff y = \dfrac{y_2 - y_1}{x_2 - x_1}(x - x_1) + y_1 \iff y - y_1 = \dfrac{y_2 - y_1}{x_2 - x_1}(x - x_1)$, which is the "two-point" form of the equation

for the line passing through the points (x_1, y_1) and (x_2, y_2).

(b) Using the result of part (a), the line has equation

$$\begin{vmatrix} x & y & 1 \\ 20 & 50 & 1 \\ -10 & 25 & 1 \end{vmatrix} = 0 \iff \begin{vmatrix} 20 & 50 \\ -10 & 25 \end{vmatrix} - \begin{vmatrix} x & y \\ -10 & 25 \end{vmatrix} + \begin{vmatrix} x & y \\ 20 & 50 \end{vmatrix} = 0 \iff$$

$(500 + 500) - (25x + 10y) + (50x - 20y) = 0 \iff 25x - 30y + 1000 = 0 \iff 5x - 6y + 200 = 0.$

63. Gaussian elimination is superior, since it takes much longer to evaluate six 5×5 determinants than it does to perform one five-equation Gaussian elimination.

10.8   Partial Fractions

1. $\dfrac{1}{(x-1)(x+2)} = \dfrac{A}{x-1} + \dfrac{B}{x+2}$

3. $\dfrac{x^2 - 3x + 5}{(x-2)^2(x+4)} = \dfrac{A}{x-2} + \dfrac{B}{(x-2)^2} + \dfrac{C}{x+4}$

5. $\dfrac{x^2}{(x-3)(x^2+4)} = \dfrac{A}{x-3} + \dfrac{Bx + C}{x^2+4}$

7. $\dfrac{x^3 - 4x^2 + 2}{(x^2+1)(x^2+2)} = \dfrac{Ax + B}{x^2+1} + \dfrac{Cx + D}{x^2+2}$

9. $\dfrac{x^3 + x + 1}{x(2x-5)^3(x^2+2x+5)^2} = \dfrac{A}{x} + \dfrac{B}{2x-5} + \dfrac{C}{(2x-5)^2} + \dfrac{D}{(2x-5)^3} + \dfrac{Ex + F}{x^2+2x+5} + \dfrac{Gx + H}{(x^2+2x+5)^2}$

11. $\dfrac{2}{(x-1)(x+1)} = \dfrac{A}{x-1} + \dfrac{B}{x+1}$. Multiplying by $(x-1)(x+1)$, we get $2 = A(x+1) + B(x-1) \quad \Leftrightarrow$

$2 = Ax + A + Bx - B$. Thus $\begin{cases} A+B=0 \\ A-B=2 \end{cases}$ Adding we get $2A = 2 \quad \Leftrightarrow \quad A = 1$. Now $A+B=0 \quad \Leftrightarrow$

$B = -A$, so $B = -1$. Thus, the required partial fraction decomposition is $\dfrac{2}{(x-1)(x+1)} = \dfrac{1}{x-1} - \dfrac{1}{x+1}$.

13. $\dfrac{5}{(x-1)(x+4)} = \dfrac{A}{x-1} + \dfrac{B}{x+4}$. Multiplying by $(x-1)(x+4)$, we get $5 = A(x+4) + B(x-1) \quad \Leftrightarrow$

$5 = Ax + 4A + Bx - B$. Thus $\begin{cases} A+B=0 \\ 4A-B=5 \end{cases}$ Now $A+B=0 \quad \Leftrightarrow \quad B = -A$, so substituting, we

get $4A - (-A) = 5 \quad \Leftrightarrow \quad 5A = 5 \Leftrightarrow A = 1$ and $B = -1$. The required partial fraction decomposition is

$\dfrac{5}{(x-1)(x+4)} = \dfrac{1}{x-1} - \dfrac{1}{x+4}$.

15. $\dfrac{12}{x^2-9} = \dfrac{12}{(x-3)(x+3)} = \dfrac{A}{x-3} + \dfrac{B}{x+3}$. Multiplying by $(x-3)(x+3)$, we get $12 = A(x+3) + B(x-3)$

$\Leftrightarrow \quad 12 = Ax + 3A + Bx - 3B$. Thus $\begin{cases} A+B=0 \\ 3A-3B=12 \end{cases} \Leftrightarrow \begin{cases} A+B=0 \\ A-B=4 \end{cases}$ Adding, we get $2A = 4 \quad \Leftrightarrow$

$A = 2$. So $2 + B = 0 \quad \Leftrightarrow \quad B = -2$. The required partial fraction decomposition is $\dfrac{12}{x^2-9} = \dfrac{2}{x-3} - \dfrac{2}{x+3}$.

17. $\dfrac{4}{x^2-4} = \dfrac{4}{(x-2)(x+2)} = \dfrac{A}{x-2} + \dfrac{B}{x+2}$. Multiplying by $x^2 - 4$, we get

$4 = A(x+2) + B(x-2) = (A+B)x + (2A-2B)$, and so $\begin{cases} A+B=0 \\ 2A-2B=4 \end{cases} \Leftrightarrow \begin{cases} A+B=0 \\ A-B=2 \end{cases}$ Adding we

get $2A = 2 \quad \Leftrightarrow \quad A = 1$, and $B = -1$. Therefore, $\dfrac{4}{x^2-4} = \dfrac{1}{x-2} - \dfrac{1}{x+2}$.

19. $\dfrac{x+14}{x^2-2x-8} = \dfrac{x+14}{(x-4)(x+2)} = \dfrac{A}{x-4} + \dfrac{B}{x+2}$. Hence, $x+14 = A(x+2) + B(x-4) = (A+B)x + (2A-4B)$,

and so $\begin{cases} A+B=1 \\ 2A-4B=14 \end{cases} \Leftrightarrow \begin{cases} 2A+2B=2 \\ A-2B=7 \end{cases}$ Adding, we get $3A = 9 \quad \Leftrightarrow \quad A = 3$. So $(3) + B = 1 \quad \Leftrightarrow$

$B = -2$. Therefore, $\dfrac{x+14}{x^2-2x-8} = \dfrac{3}{x-4} - \dfrac{2}{x+2}$.

21. $\dfrac{x}{8x^2-10x+3} = \dfrac{x}{(4x-3)(2x-1)} = \dfrac{A}{4x-3} + \dfrac{B}{2x-1}$. Hence,

$x = A(2x-1) + B(4x-3) = (2A+4B)x + (-A-3B)$, and so $\begin{cases} 2A+4B=1 \\ -A-3B=0 \end{cases} \Leftrightarrow \begin{cases} 2A+4B=1 \\ -2A-6B=0 \end{cases}$

Adding, we get $-2B = 1 \quad \Leftrightarrow \quad B = -\frac{1}{2}$, and $A = \frac{3}{2}$. Therefore, $\dfrac{x}{8x^2-10x+3} = \dfrac{\frac{3}{2}}{4x-3} - \dfrac{\frac{1}{2}}{2x-1}$.

23. $\dfrac{9x^2 - 9x + 6}{2x^3 - x^2 - 8x + 4} = \dfrac{9x^2 - 9x + 6}{(x - 2)(x + 2)(2x - 1)} = \dfrac{A}{x - 2} + \dfrac{B}{x + 2} + \dfrac{C}{2x - 1}$. Thus,

$$9x^2 - 9x + 6 = A(x + 2)(2x - 1) + B(x - 2)(2x - 1) + C(x - 2)(x + 2)$$

$$= A(2x^2 + 3x - 2) + B(2x^2 - 5x + 2) + C(x^2 - 4)$$

$$= (2A + 2B + C)x^2 + (3A - 5B)x + (-2A + 2B - 4C)$$

This leads to the system $\begin{cases} 2A + 2B + C = 9 & \text{Coefficients of } x^2 \\ 3A - 5B = -9 & \text{Coefficients of } x \\ -2A + 2B - 4C = 6 & \text{Constant terms} \end{cases} \Leftrightarrow \begin{cases} 2A + 2B + C = 9 \\ 16B + 3C = 45 \\ 4B - 3C = 15 \end{cases} \Leftrightarrow$

$\begin{cases} 2A + 2B + C = 9 \\ 16B + 3C = 45 \\ 15C = -15 \end{cases}$ Hence, $-15C = 15 \Leftrightarrow C = -1$; $16B - 3 = 45 \Leftrightarrow B = 3$; and $2A + 6 - 1 = 9$

$\Leftrightarrow A = 2$. Therefore, $\dfrac{9x^2 - 9x + 6}{2x^3 - x^2 - 8x + 4} = \dfrac{2}{x - 2} + \dfrac{3}{x + 2} - \dfrac{1}{2x - 1}$.

25. $\dfrac{x^2 + 1}{x^3 + x^2} = \dfrac{x^2 + 1}{x^2(x + 1)} = \dfrac{A}{x} + \dfrac{B}{x^2} + \dfrac{C}{x + 1}$. Hence,

$x^2 + 1 = Ax(x + 1) + B(x + 1) + Cx^2 = (A + C)x^2 + (A + B)x + B$, and so $B = 1$; $A + 1 = 0 \Leftrightarrow A = -1$;

and $-1 + C = 1 \Leftrightarrow C = 2$. Therefore, $\dfrac{x^2 + 1}{x^3 + x^2} = \dfrac{-1}{x} + \dfrac{1}{x^2} + \dfrac{2}{x + 1}$.

27. $\dfrac{2x}{4x^2 + 12x + 9} = \dfrac{2x}{(2x + 3)^2} = \dfrac{A}{2x + 3} + \dfrac{B}{(2x + 3)^2}$. Hence, $2x = A(2x + 3) + B = 2Ax + (3A + B)$. So $2A = 2$

$\Leftrightarrow A = 1$; and $3(1) + B = 0 \Leftrightarrow B = -3$. Therefore, $\dfrac{2x}{4x^2 + 12x + 9} = \dfrac{1}{2x + 3} - \dfrac{3}{(2x + 3)^2}$.

29. $\dfrac{4x^2 - x - 2}{x^4 + 2x^3} = \dfrac{4x^2 - x - 2}{x^3(x + 2)} = \dfrac{A}{x} + \dfrac{B}{x^2} + \dfrac{C}{x^3} + \dfrac{D}{x + 2}$. Hence,

$$4x^2 - x - 2 = Ax^2(x + 2) + Bx(x + 2) + C(x + 2) + Dx^3$$

$$= (A + D)x^3 + (2A + B)x^2 + (2B + C)x + 2C$$

So $2C = -2 \Leftrightarrow C = -1$; $2B - 1 = -1 \Leftrightarrow B = 0$; $2A + 0 = 4 \Leftrightarrow A = 2$; and $2 + D = 0 \Leftrightarrow$

$D = -2$. Therefore, $\dfrac{4x^2 - x - 2}{x^4 + 2x^3} = \dfrac{2}{x} - \dfrac{1}{x^3} - \dfrac{2}{x + 2}$.

31. $\dfrac{-10x^2 + 27x - 14}{(x-1)^3(x+2)} = \dfrac{A}{x+2} + \dfrac{B}{x-1} + \dfrac{C}{(x-1)^2} + \dfrac{D}{(x-1)^3}$. Thus,

$$-10x^2 + 27x - 14 = A(x-1)^3 + B(x+2)(x-1)^2 + C(x+2)(x-1) + D(x+2)$$

$$= A(x^3 - 3x^2 + 3x - 1) + B(x+2)(x^2 - 2x + 1) + C(x^2 + x - 2) + D(x+2)$$

$$= A(x^3 - 3x^2 + 3x - 1) + B(x^3 - 3x + 2) + C(x^2 + x - 2) + D(x+2)$$

$$= (A+B)x^3 + (-3A+C)x^2 + (3A - 3B + C + D)x + (-A + 2B - 2C + 2D)$$

which leads to the system

$$\begin{cases} A + B & = 0 \quad \text{Coefficients of } x^3 \\ -3A \quad\;\; + C & = -10 \quad \text{Coefficients of } x^2 \\ 3A - 3B + C + D = 27 \quad \text{Coefficients of } x \\ -A + 2B - 2C + 2D = -14 \quad \text{Constant terms} \end{cases} \Leftrightarrow \begin{cases} A + B & = 0 \\ 3B + C & = -10 \\ -3B + 2C + D = 17 \\ 3B - 5C + 7D = -15 \end{cases} \Leftrightarrow$$

$$\begin{cases} A + B & = 0 \\ 3B + C & = -10 \\ 3C + D = 7 \\ -3C + 8D = 2 \end{cases} \Leftrightarrow \begin{cases} A + B & = 0 \\ 3B + C & = -10 \\ 3C + D = 7 \\ 9D = 9 \end{cases}$$

Hence, $9D = 9 \Leftrightarrow D = 1, 3C + 1 = 7 \Leftrightarrow C = 2, 3B + 2 = -10 \Leftrightarrow B = -4$, and $A - 4 = 0 \Leftrightarrow$

$A = 4$. Therefore, $\dfrac{-10x^2 + 27x - 14}{(x-1)^3(x+2)} = \dfrac{4}{x+2} - \dfrac{4}{x-1} + \dfrac{2}{(x-1)^2} + \dfrac{1}{(x-1)^3}$.

33. $\dfrac{3x^3 + 22x^2 + 53x + 41}{(x+2)^2(x+3)^2} = \dfrac{A}{x+2} + \dfrac{B}{(x+2)^2} + \dfrac{C}{x+3} + \dfrac{D}{(x+3)^2}$. Thus,

$$3x^3 + 22x^2 + 53x + 41 = A(x+2)(x+3)^2 + B(x+3)^2 + C(x+2)^2(x+3) + D(x+2)^2$$

$$= A(x^3 + 8x^2 + 21x + 18) + B(x^2 + 6x + 9)$$

$$+ C(x^3 + 7x^2 + 16x + 12) + D(x^2 + 4x + 4)$$

$$= (A+C)x^3 + (8A + B + 7C + D)x^2$$

$$+ (21A + 6B + 16C + 4D)x + (18A + 9B + 12C + 4D)$$

so we must solve the system $\begin{cases} A \quad\quad + C & = 3 \quad \text{Coefficients of } x^3 \\ 8A + B + 7C + D = 22 \quad \text{Coefficients of } x^2 \\ 21A + 6B + 16C + 4D = 53 \quad \text{Coefficients of } x \\ 18A + 9B + 12C + 4D = 41 \quad \text{Constant terms} \end{cases} \Leftrightarrow$

$$\begin{cases} A + C & = 3 \\ B - C + D = -2 \\ 6B - 5C + 4D = -10 \\ 9B - 6C + 4D = -13 \end{cases} \Leftrightarrow \begin{cases} A + C & = 3 \\ B - C + D = -2 \\ C - 2D = 2 \\ 3C - 5D = 5 \end{cases} \Leftrightarrow \begin{cases} A + C & = 3 \\ B - C + D = -2 \\ C - 2D = 2 \\ D = -1 \end{cases} \text{ Hence,}$$

$D = -1, C + 2 = 2 \Leftrightarrow C = 0, B - 0 - 1 = -2 \Leftrightarrow B = -1$, and $A + 0 = 3 \Leftrightarrow A = 3$. Therefore,

$$\dfrac{3x^3 + 22x^2 + 53x + 41}{(x+2)^2(x+3)^2} = \dfrac{3}{x+2} - \dfrac{1}{(x+2)^2} - \dfrac{1}{(x+3)^2}.$$

35. $\dfrac{x-3}{x^3+3x} = \dfrac{x-3}{x\left(x^2+3\right)} = \dfrac{A}{x} + \dfrac{Bx+C}{x^2+3}$. Hence, $x-3 = A\left(x^2+3\right) + Bx^2 + Cx = (A+B)\,x^2 + Cx + 3A$. So

$3A = -3 \Leftrightarrow A = -1$; $C = 1$; and $-1 + B = 0 \quad \Leftrightarrow \quad B = 1$. Therefore, $\dfrac{x-3}{x^3+3x} = -\dfrac{1}{x} + \dfrac{x+1}{x^2+3}$.

37. $\dfrac{2x^3+7x+5}{\left(x^2+x+2\right)\left(x^2+1\right)} = \dfrac{Ax+B}{x^2+x+2} + \dfrac{Cx+D}{x^2+1}$. Thus,

$$
\begin{aligned}
2x^3 + 7x + 5 &= (Ax+B)\left(x^2+1\right) + (Cx+D)\left(x^2+x+2\right) \\
&= Ax^3 + Ax + Bx^2 + B + Cx^3 + Cx^2 + 2Cx + Dx^2 + Dx + 2D \\
&= (A+C)\,x^3 + (B+C+D)\,x^2 + (A+2C+D)\,x + (B+2D)
\end{aligned}
$$

We must solve the system

$\begin{cases} A + \quad\ C \qquad\quad = 2 \\ \quad\ B + C + D = 0 \\ A + \qquad 2C + D = 7 \\ \quad\ B \qquad\ + 2D = 5 \end{cases}$
$\begin{array}{l} \text{Coefficients of } x^3 \\ \text{Coefficients of } x^2 \\ \text{Coefficients of } x \\ \text{Constant terms} \end{array}$
$\Leftrightarrow$
$\begin{cases} A + \quad\ C \quad\ = 2 \\ \quad B + C + D = 0 \\ \qquad\ C + D = 5 \\ \qquad\ C - D = -5 \end{cases}$
$\Leftrightarrow$
$\begin{cases} A + \quad\ C \quad\ = 2 \\ \quad B + C + \ D = 0 \\ \qquad\ C + \ D = 5 \\ \qquad\qquad 2D = 10 \end{cases}$

Hence, $2D = 10 \quad \Leftrightarrow \quad D = 5$, $C + 5 = 5 \quad \Leftrightarrow \quad C = 0$, $B + 0 + 5 = 0 \quad \Leftrightarrow \quad B = -5$, and $A + 0 = 2 \quad \Leftrightarrow$

$A = 2$. Therefore, $\dfrac{2x^3+7x+5}{\left(x^2+x+2\right)\left(x^2+1\right)} = \dfrac{2x-5}{x^2+x+2} + \dfrac{5}{x^2+1}$.

39. $\dfrac{x^4+x^3+x^2-x+1}{x\left(x^2+1\right)^2} = \dfrac{A}{x} + \dfrac{Bx+C}{x^2+1} + \dfrac{Dx+E}{\left(x^2+1\right)^2}$. Hence,

$$
\begin{aligned}
x^4+x^3+x^2-x+1 &= A\left(x^2+1\right)^2 + (Bx+C)\,x\left(x^2+1\right) + x\,(Dx+E) \\
&= A\left(x^4+2x^2+1\right) + \left(Bx^2+Cx\right)\left(x^2+1\right) + Dx^2 + Ex \\
&= A\left(x^4+2x^2+1\right) + Bx^4 + Bx^2 + Cx^3 + Cx + Dx^2 + Ex \\
&= (A+B)\,x^4 + Cx^3 + (2A+B+D)\,x^2 + (C+E)\,x + A
\end{aligned}
$$

So $A = 1$, $1 + B = 1 \quad \Leftrightarrow \quad B = 0$; $C = 1$; $2 + 0 + D = 1 \quad \Leftrightarrow \quad D = -1$; and $1 + E = -1 \quad \Leftrightarrow \quad E = -2$.

Therefore, $\dfrac{x^4+x^3+x^2-x+1}{x\left(x^2+1\right)^2} = \dfrac{1}{x} + \dfrac{1}{x^2+1} - \dfrac{x+2}{\left(x^2+1\right)^2}$.

41. We must first get a proper rational function. Using long division, we find that $\dfrac{x^5-2x^4+x^3+x+5}{x^3-2x^2+x-2} = x^2 + $

$\dfrac{2x^2+x+5}{x^3-2x^2+x-2} = x^2 + \dfrac{2x^2+x+5}{(x-2)\left(x^2+1\right)} = x^2 + \dfrac{A}{x-2} + \dfrac{Bx+C}{x^2+1}$. Hence,

$$
\begin{aligned}
2x^2+x+5 &= A\left(x^2+1\right) + (Bx+C)(x-2) = Ax^2 + A + Bx^2 + Cx - 2Bx - 2C \\
&= (A+B)\,x^2 + (C-2B)\,x + (A-2C)
\end{aligned}
$$

Equating coefficients, we get the system

$\begin{cases} A + B \qquad\quad = 2 \\ \quad -2B + \ C = 1 \\ A \qquad\ - 2C = 5 \end{cases}$
$\begin{array}{l} \text{Coefficients of } x^2 \\ \text{Coefficients of } x \\ \text{Constant terms} \end{array}$
$\Leftrightarrow$
$\begin{cases} A + B \qquad = 2 \\ \quad -2B + \ C = 1 \\ \qquad\ B + 2C = -3 \end{cases}$
$\Leftrightarrow$
$\begin{cases} A + B \qquad = 2 \\ \quad -2B + C = 1 \\ \qquad\quad 5C = -5 \end{cases}$

Therefore, $5C = -5 \quad \Leftrightarrow \quad C = -1$, $-2B - 1 = 1 \quad \Leftrightarrow \quad B = -1$, and $A - 1 = 2 \quad \Leftrightarrow \quad A = 3$. Therefore,

$\dfrac{x^5-2x^4+x^3+x+5}{x^3-2x^2+x-2} = x^2 + \dfrac{3}{x-2} - \dfrac{x+1}{x^2+1}$.

43. $\dfrac{ax+b}{x^2-1} = \dfrac{A}{x-1} + \dfrac{B}{x+1}$. Hence, $ax+b = A(x+1) + B(x-1) = (A+B)x + (A-B)$.

So $\begin{cases} A+B = a \\ A-B = b \end{cases}$ Adding, we get $2A = a+b \Leftrightarrow A = \dfrac{a+b}{2}$.

Substituting, we get $B = a - A = \dfrac{2a}{2} - \dfrac{a+b}{2} = \dfrac{a-b}{2}$. Therefore, $A = \dfrac{a+b}{2}$ and $B = \dfrac{a-b}{2}$.

45. (a) The expression $\dfrac{x}{x^2+1} + \dfrac{1}{x+1}$ is already a partial fraction decomposition. The denominator in the first term is a quadratic which cannot be factored and the degree of the numerator is less than 2. The denominator of the second term is linear and the numerator is a constant.

(b) The term $\dfrac{x}{(x+1)^2}$ can be decomposed further, since the numerator and denominator both have linear factors.

$\dfrac{x}{(x+1)^2} = \dfrac{A}{x+1} + \dfrac{B}{(x+1)^2}$. Hence, $x = A(x+1) + B = Ax + (A+B)$. So $A = 1$, $B = -1$, and

$\dfrac{x}{(x+1)^2} = \dfrac{1}{x+1} + \dfrac{-1}{(x+1)^2}$.

(c) The expression $\dfrac{1}{x+1} + \dfrac{2}{(x+1)^2}$ is already a partial fraction decomposition, since each numerator is constant.

(d) The expression $\dfrac{x+2}{(x^2+1)^2}$ is already a partial fraction decomposition, since the denominator is the square of a quadratic which cannot be factored, and the degree of the numerator is less than 2.

10.9 Systems of Inequalities

1. $x < 3$

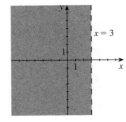

3. $y > x$

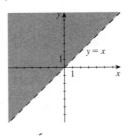

5. $y \le 2x + 2$

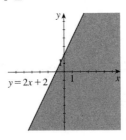

7. $2x - y \le 8$

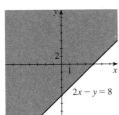

9. $4x + 5y < 20$

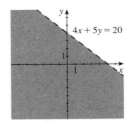

11. $y > x^2 + 1$

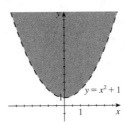

13. $x^2 + y^2 \le 25$

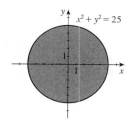

15. The boundary is a solid curve, so we have the inequality $y \le \frac{1}{2}x - 1$. We take the test point $(0, -2)$ and verify that it satisfies the inequality: $-2 \le \frac{1}{2}(0) - 1$.

17. The boundary is a broken curve, so we have the inequality $x^2 + y^2 > 4$. We take the test point $(0, 4)$ and verify that it satisfies the inequality: $0^2 + 4^2 > 4$.

19. $\begin{cases} x + y \le 4 \\ \quad y \ge x \end{cases}$ The vertices occur where $\begin{cases} x + y = 4 \\ \quad y = x \end{cases}$ Substituting, we have

$2x = 4 \quad \Leftrightarrow \quad x = 2$. Since $y = x$, the vertex is $(2, 2)$, and the solution set is not bounded.

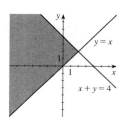

21. $\begin{cases} y < \frac{1}{4}x + 2 \\ y \ge 2x - 5 \end{cases}$ The vertex occurs where $\begin{cases} y = \frac{1}{4}x + 2 \\ y = 2x - 5 \end{cases}$ Substituting for y

gives $\frac{1}{4}x + 2 = 2x - 5 \quad \Leftrightarrow \quad \frac{7}{4}x = 7 \quad \Leftrightarrow \quad x = 4$, so $y = 3$. Hence, the vertex is $(4, 3)$, and the solution is not bounded.

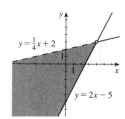

23. $\begin{cases} \quad\quad x \ge 0 \\ \quad\quad y \ge 0 \\ 3x + 5y \le 15 \\ 3x + 2y \le 9 \end{cases}$ From the graph, the points $(3, 0)$, $(0, 3)$ and $(0, 0)$ are

vertices, and the fourth vertex occurs where the lines $3x + 5y = 15$ and $3x + 2y = 9$ intersect. Subtracting these two equations gives $3y = 6 \quad \Leftrightarrow \quad y = 2$, and so $x = \frac{5}{3}$. Thus, the fourth vertex is $\left(\frac{5}{3}, 2\right)$, and the solution set is bounded.

25. $\begin{cases} y < 9 - x^2 \\ y \ge x + 3 \end{cases}$ The vertices occur where $\begin{cases} y = 9 - x^2 \\ y = x + 3 \end{cases}$ Substituting for y

gives $9 - x^2 = x + 3 \quad \Leftrightarrow \quad x^2 + x - 6 = 0 \quad \Leftrightarrow$

$(x - 2)(x + 3) = 0 \Rightarrow x = -3, x = 2$. Therefore, the vertices are $(-3, 0)$ and $(2, 5)$, and the solution set is bounded.

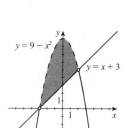

27. $\begin{cases} x^2 + y^2 \le 4 \\ x - y > 0 \end{cases}$ The vertices occur where $\begin{cases} x^2 + y^2 = 4 \\ x - y = 0 \end{cases}$ Since $x - y = 0$

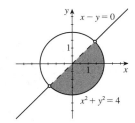

$\Leftrightarrow$ $x = y$, substituting for x gives $y^2 + y^2 = 4$ $\Leftrightarrow$ $y^2 = 2 \Rightarrow y = \pm\sqrt{2}$,

and $x = \pm\sqrt{2}$. Therefore, the vertices are $\left(-\sqrt{2}, -\sqrt{2}\right)$ and $\left(\sqrt{2}, \sqrt{2}\right)$, and the

solution set is bounded.

29. $\begin{cases} x^2 - y \le 0 \\ 2x^2 + y \le 12 \end{cases}$ The vertices occur where $\begin{cases} x^2 - y = 0 \\ 2x^2 + y = 12 \end{cases}$ $\Leftrightarrow$

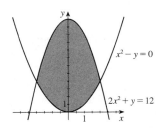

$\begin{cases} 2x^2 - 2y = 0 \\ 2x^2 + y = 12 \end{cases}$ Subtracting the equations gives $3y = 12$ $\Leftrightarrow$ $y = 4$, and

$x = \pm 2$. Thus, the vertices are $(2, 4)$ and $(-2, 4)$, and the solution set is bounded.

31. $\begin{cases} x + 2y \le 14 \\ 3x - y \ge 0 \\ x - y \ge 2 \end{cases}$ We find the vertices of the region by solving pairs of the

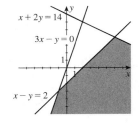

corresponding equations: $\begin{cases} x + 2y = 14 \\ x - y = 2 \end{cases}$ $\Leftrightarrow$ $\begin{cases} x + 2y = 14 \\ 3y = 12 \end{cases}$ $\Leftrightarrow y = 4$ and

$x = 6$. $\begin{cases} 3x - y = 0 \\ x - y = 2 \end{cases}$ $\Leftrightarrow$ $\begin{cases} 3x - y = 0 \\ 2x - y = -2 \end{cases}$ $\Leftrightarrow x = -1$ and $y = -3$. Therefore,

the vertices are $(6, 4)$ and $(-1, -3)$, and the solution set is not bounded.

33. $\begin{cases} x \ge 0, y \ge 0 \\ x \le 5, x + y \le 7 \end{cases}$ The points of intersection are $(0, 7)$, $(0, 0)$, $(7, 0)$, $(5, 2)$,

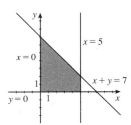

and $(5, 0)$. However, the point $(7, 0)$ is not in the solution set. Therefore, the

vertices are $(0, 7)$, $(0, 0)$, $(5, 0)$, and $(5, 2)$, and the solution set is bounded.

35. $\begin{cases} y > x + 1 \\ x + 2y \le 12 \\ x + 1 > 0 \end{cases}$ We find the vertices of the region by solving pairs of the

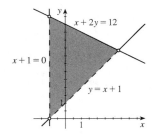

corresponding equations. Using $x = -1$ and substituting for x in the line

$y = x + 1$ gives the point $(-1, 0)$. Substituting for x in the line $x + 2y = 12$ gives

the point $\left(-1, \frac{13}{2}\right)$. $\begin{cases} y = x + 1 \\ x + 2y = 12 \end{cases}$ $\Leftrightarrow$ $x = y - 1$ and $y - 1 + 2y = 12$

$\Leftrightarrow$ $3y = 13$ $\Leftrightarrow$ $y = \frac{13}{3}$ and $x = \frac{10}{3}$. So the vertices are $(-1, 0)$, $\left(-1, \frac{13}{2}\right)$,

and $\left(\frac{10}{3}, \frac{13}{3}\right)$, and none of these vertices is in the solution set. The solution set is

bounded.

37. $\begin{cases} x^2 + y^2 \le 8 \\ x \ge 2,\, y \ge 0 \end{cases}$ The intersection points are $(2, \pm 2)$, $(2, 0)$, and $(2\sqrt{2}, 0)$.

However, since $(2, -2)$ is not part of the solution set, the vertices are $(2, 2)$, $(2, 0)$, and $(2\sqrt{2}, 0)$. The solution set is bounded.

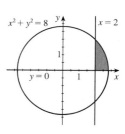

39. $\begin{cases} x^2 + y^2 < 9 \\ x + y > 0,\, x \le 0 \end{cases}$ Substituting $x = 0$ into the equations $x^2 + y^2 = 9$ and

$x + y = 0$ gives the vertices $(0, \pm 3)$ and $(0, 0)$. To find the points of intersection for the equations $x^2 + y^2 = 9$ and $x + y = 0$, we solve for $x = -y$ and substitute into the first equation. This gives $(-y)^2 + y^2 = 9 \Rightarrow y = \pm \frac{3\sqrt{2}}{2}$. The points $(0, -3)$ and $\left(\frac{3\sqrt{2}}{2}, -\frac{3\sqrt{2}}{2}\right)$ lie away from the solution set, so the vertices are $(0, 0)$, $(0, 3)$, and $\left(-\frac{3\sqrt{2}}{2}, \frac{3\sqrt{2}}{2}\right)$. Note that the vertices are not solutions in this case. The solution set is bounded.

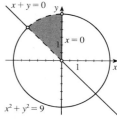

41. $\begin{cases} y \ge x - 3 \\ y \ge -2x + 6 \\ y \le 8 \end{cases}$ Using a graphing calculator, we find the region shown. The

vertices are $(3, 0)$, $(-1, 8)$, and $(11, 8)$.

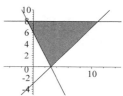

43. $\begin{cases} y \le 6x - x^2 \\ x + y \ge 4 \end{cases}$ Using a graphing calculator, we find the region shown.

The vertices are $(0.6, 3.4)$ and $(6.4, -2.4)$.

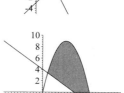

45. Let x be the number of fiction books published in a year and y the number of nonfiction books. Then the following system of inequalities holds:

$\begin{cases} x \ge 0,\, y \ge 0 \\ x + y \le 100 \\ y \ge 20,\, x \ge y \end{cases}$ From the graph, we see that the vertices are $(50, 50)$, $(80, 20)$

and $(20, 20)$.

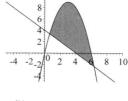

47. Let x be the number of Standard Blend packages and y be the number of Deluxe Blend packages. Since there are 16 ounces per pound, we get the following system

of inequalities holds: $\begin{cases} x \ge 0 \\ y \ge 0 \\ \frac{1}{4}x + \frac{5}{8}y \le 80 \\ \frac{3}{4}x + \frac{3}{8}y \le 90 \end{cases}$ From the graph, we see that the

vertices are $(0, 0)$, $(120, 0)$, $(70, 100)$ and $(0, 128)$.

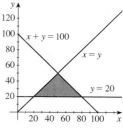

49. $x + 2y > 4$, $-x + y < 1$, $x + 3y \leq 9$, $x < 3$.

Method 1:   We shade the solution to each inequality with lines perpendicular to the boundary. As you can see, as the number of inequalities in the system increases, it gets harder to locate the region where *all* of the shaded parts overlap.

Method 2: Here, if a region is shaded then it fails to satisfy at least one inequality. As a result, the region that is left unshaded satisfies each inequality, and is the solution to the system of inequalities. In this case, this method makes it easier to identify the solution set.

To finish, we find the vertices of the solution set. The line $x = 3$ intersects the line $x + 2y = 4$ at $\left(3, \frac{1}{2}\right)$ and the line $x + 3y = 9$ at $(3, 2)$. To find where the lines $-x + y = 1$ and $x + 2y = 4$ intersect, we add the two equations, which gives $3y = 5$ $\Leftrightarrow$ $y = \frac{5}{3}$, and $x = \frac{2}{3}$. To find where the lines $-x + y = 1$ and $x + 3y = 9$ intersect, we add the two equations, which gives $4y = 10$ $\Leftrightarrow$ $y = \frac{10}{4} = \frac{5}{2}$, and $x = \frac{3}{2}$. The vertices are $\left(3, \frac{1}{2}\right)$, $(3, 2)$, $\left(\frac{2}{3}, \frac{5}{3}\right)$, and $\left(\frac{3}{2}, \frac{5}{2}\right)$, and the solution set is bounded.

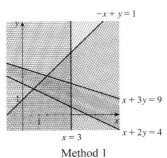

Method 1

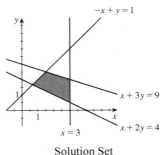

Method 2   Solution Set

Chapter 10 Review

1. $\begin{cases} 2x + 3y = 7 \\ x - 2y = 0 \end{cases}$ By inspection of the graph, it appears that $(2, 1)$ is the solution to the system. We check this in both

equations to verify that it is the solution. $2(2) + 3(1) = 4 + 3 = 7$ and $2 - 2(1) = 2 - 2 = 0$. Since both equations are satisfied, the solution is indeed $(2, 1)$.

3. $\begin{cases} x^2 + y = 2 \\ x^2 - 3x - y = 0 \end{cases}$ By inspection of the graph, it appears that $(2, -2)$ is a solution to the system, but is difficult

to get accurate values for the other point. Adding the equations, we get $2x^2 - 3x = 2 \Leftrightarrow 2x^2 - 3x - 2 = 0$ $\Leftrightarrow$ $(2x + 1)(x - 2) = 0$. So $2x + 1 = 0$ $\Leftrightarrow$ $x = -\frac{1}{2}$ or $x = 2$. If $x = -\frac{1}{2}$, then $\left(-\frac{1}{2}\right)^2 + y = 2 \Leftrightarrow y = \frac{7}{4}$. If $x = 2$, then $2^2 + y = 2$ $\Leftrightarrow$ $y = -2$. Thus, the solutions are $\left(-\frac{1}{2}, \frac{7}{4}\right)$ and $(2, -2)$.

5. $\begin{cases} 3x - y = 5 \\ 2x + y = 5 \end{cases}$ Adding, we get $5x = 10$ $\Leftrightarrow$ $x = 2$. So $2(2) + y = 5$ $\Leftrightarrow$

$y = 1$. Thus, the solution is $(2, 1)$.

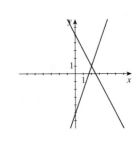

7. $\begin{cases} 2x - 7y = 28 \\ y = \frac{2}{7}x - 4 \end{cases}$ $\Leftrightarrow$ $\begin{cases} 2x - 7y = 28 \\ 2x - 7y = 28 \end{cases}$ Since these equations

represent the same line, any point on this line will satisfy the system. Thus the

solution are $\left(x, \frac{2}{7}x - 4\right)$, where x is any real number.

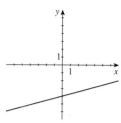

9. $\begin{cases} 2x - y = 1 \\ x + 3y = 10 \\ 3x + 4y = 15 \end{cases}$ Solving the first equation for y, we get $y = -2x + 1$.

Substituting into the second equation gives $x + 3(-2x + 1) = 10$ $\Leftrightarrow$

$-5x = 7$ $\Leftrightarrow$ $x = -\frac{7}{5}$. So $y = -\left(-\frac{7}{5}\right) + 1 = \frac{12}{5}$. Checking the point

$\left(-\frac{7}{5}, \frac{12}{5}\right)$ in the third equation we have $3\left(-\frac{7}{5}\right) + 4\left(\frac{12}{5}\right) \overset{?}{=} 15$ but

$-\frac{21}{5} + \frac{48}{5} \neq 15$. Thus, there is no solution, and the lines do not intersect at one

point.

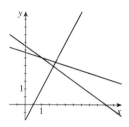

11. $\begin{cases} y = x^2 + 2x \\ y = 6 + x \end{cases}$ Substituting for y gives $6 + x = x^2 + 2x$ $\Leftrightarrow$ $x^2 + x - 6 = 0$. Factoring, we have

$(x - 2)(x + 3) = 0$. Thus $x = 2$ or -3. If $x = 2$, then $y = 8$, and if $x = -3$, then $y = 3$. Thus the solutions are $(-3, 3)$

and $(2, 8)$.

13. $\begin{cases} 3x + \dfrac{4}{y} = 6 \\ x - \dfrac{8}{y} = 4 \end{cases}$ Adding twice the first equation to the second gives $7x = 16$ $\Leftrightarrow$ $x = \frac{16}{7}$. So $\dfrac{16}{7} - \dfrac{8}{y} = 4$ $\Leftrightarrow$

$16y - 56 = 28y$ $\Leftrightarrow$ $-12y = 56$ $\Leftrightarrow$ $y = -\frac{14}{3}$. Thus, the solution is $\left(\frac{16}{7}, -\frac{14}{3}\right)$.

15. $\begin{cases} 0.32x + 0.43y = 0 \\ 7x - 12y = 341 \end{cases}$ $\Leftrightarrow$ $\begin{cases} y = -\dfrac{32x}{43} \\ y = \dfrac{7x - 341}{12} \end{cases}$

The solution is approximately $(21.41, -15.93)$.

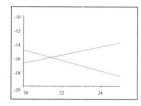

17. $\begin{cases} x - y^2 = 10 \\ x = \frac{1}{22}y + 12 \end{cases}$ $\Leftrightarrow$ $\begin{cases} y = \pm\sqrt{x - 10} \\ y = 22(x - 12) \end{cases}$

The solutions are $(11.94, -1.39)$ and $(12.07, 1.44)$.

19. (a) 2×3

   (b) Yes, this matrix is in row-echelon form.

   (c) No, this matrix is not in reduced row-echelon form,
   since the leading 1 in the second row does not have a 0
   above it.

   (d) $\begin{cases} x + 2y = -5 \\ y = 3 \end{cases}$

21. (a) 3×4

   (b) Yes, this matrix is in row-echelon form.

   (c) Yes, this matrix is in reduced row-echelon form.

   (d) $\begin{cases} x \quad + 8z = 0 \\ y + 5z = -1 \\ 0 = 0 \end{cases}$

23. (a) 3×4

(b) No, this matrix is not in row-echelon form. The leading 1 in the second row is not to the left of the one above it.

(c) No, this matrix is not in reduced row-echelon form.

(d) $\begin{cases} \phantom{x + {}} y - 3z = 4 \\ x + y \phantom{{}+ z} = 7 \\ x + 2y + z = 2 \end{cases}$

25. $\begin{cases} x + y + 2z = 6 \\ 2x \phantom{{}+ 2y} + 5z = 12 \\ x + 2y + 3z = 9 \end{cases} \Leftrightarrow \begin{cases} x + y + 2z = 6 \\ \phantom{x + {}} 2y - z = 0 \\ \phantom{x + {}} 4y + z = 6 \end{cases} \Leftrightarrow \begin{cases} x + y + 2z = 6 \\ \phantom{x + {}} 2y - z = 0 \\ \phantom{x + {}} 3z = 6 \end{cases}$ Therefore, $3z = 6 \quad \Leftrightarrow$

$z = 2, 2y - 2 = 0 \quad \Leftrightarrow \quad y = 1$, and $x + 1 + 2\,(2) = 6 \quad \Leftrightarrow \quad x = 1$. Hence, the solution is $(1, 1, 2)$.

27. $\begin{cases} x - 2y + 3z = 1 \\ 2x - y + z = 3 \\ 2x - 7y + 11z = 2 \end{cases} \Leftrightarrow \begin{cases} x - 2y + 3z = 1 \\ \phantom{x + {}} 3y - 5z = 1 \\ \phantom{x + {}} 6y - 10z = 1 \end{cases} \Leftrightarrow \begin{cases} x - 2y + 3z = 1 \\ \phantom{x + {}} 3y - 5z = 1 \\ \phantom{x + {}} 0 = -1 \end{cases}$ which is impossible.

Therefore, the system has no solution.

29. $\begin{bmatrix} 1 & 2 & 2 & 6 \\ 1 & -1 & 0 & -1 \\ 2 & 1 & 3 & 7 \end{bmatrix} \xrightarrow{R_2 \leftrightarrow R_1} \begin{bmatrix} 1 & -1 & 0 & -1 \\ 1 & 2 & 2 & 6 \\ 2 & 1 & 3 & 7 \end{bmatrix} \xrightarrow[R_3 - 2R_1 \to R_3]{R_2 - R_1 \to R_2} \begin{bmatrix} 1 & -1 & 0 & -1 \\ 0 & 3 & 2 & 7 \\ 0 & 3 & 3 & 9 \end{bmatrix} \xrightarrow{R_3 - R_2 \to R_3} \begin{bmatrix} 1 & -1 & 0 & -1 \\ 0 & 3 & 2 & 7 \\ 0 & 0 & 1 & 2 \end{bmatrix}.$

Thus, $z = 2$, $3y + 2\,(2) = 7 \quad \Leftrightarrow \quad 3y = 3 \Leftrightarrow y = 1$, and $x - (1) = -1 \quad \Leftrightarrow \quad x = 0$, and so the solution is $(0, 1, 2)$.

31. $\begin{bmatrix} 1 & -2 & 3 & -2 \\ 2 & -1 & 1 & 2 \\ 2 & -7 & 11 & -9 \end{bmatrix} \xrightarrow[R_3 - 2R_1 \to R_3]{R_2 - 2R_1 \to R_2} \begin{bmatrix} 1 & -2 & 3 & -2 \\ 0 & 3 & -5 & 6 \\ 0 & -3 & 5 & -5 \end{bmatrix} \xrightarrow{R_3 + R_2 \to R_3} \begin{bmatrix} 1 & -2 & 3 & -2 \\ 0 & 3 & -5 & 6 \\ 0 & 0 & 0 & 1 \end{bmatrix}.$

The last row corresponds to the equation $0 = 1$, which is always false. Thus, there is no solution.

33. $\begin{bmatrix} 1 & 1 & 1 & 1 & 0 \\ 1 & -1 & -4 & -1 & -1 \\ 1 & -2 & 0 & 4 & -7 \\ 2 & 2 & 3 & 4 & -3 \end{bmatrix} \xrightarrow[\substack{R_3 - R_1 \to R_3 \\ R_4 - 2R_1 \to R_4}]{R_2 - R_1 \to R_2} \begin{bmatrix} 1 & 1 & 1 & 1 & 0 \\ 0 & -2 & -5 & -2 & -1 \\ 0 & -3 & -1 & 3 & -7 \\ 0 & 0 & 1 & 2 & -3 \end{bmatrix} \xrightarrow{-R_3 + R_2 \to R_3} \begin{bmatrix} 1 & 1 & 1 & 1 & 0 \\ 0 & 1 & -4 & -5 & 6 \\ 0 & -3 & -1 & 3 & -7 \\ 0 & 0 & 1 & 2 & -3 \end{bmatrix}$

$\xrightarrow{R_3 + 3R_2 \to R_3} \begin{bmatrix} 1 & 1 & 1 & 1 & 0 \\ 0 & 1 & -4 & -5 & 6 \\ 0 & 0 & -13 & -12 & 11 \\ 0 & 0 & 1 & 2 & -3 \end{bmatrix} \xrightarrow{R_3 \leftrightarrow R_4} \begin{bmatrix} 1 & 1 & 1 & 1 & 0 \\ 0 & 1 & -4 & -5 & 6 \\ 0 & 0 & 1 & 2 & -3 \\ 0 & 0 & -13 & -12 & 11 \end{bmatrix} \xrightarrow{R_4 + 13R_3 \to R_4} \begin{bmatrix} 1 & 1 & 1 & 1 & 0 \\ 0 & 1 & -4 & -5 & 6 \\ 0 & 0 & 1 & 2 & -3 \\ 0 & 0 & 0 & 14 & -28 \end{bmatrix}.$

Therefore, $14w = -28 \quad \Leftrightarrow \quad w = -2$, $z + 2\,(-2) = -3 \quad \Leftrightarrow \quad z = 1$, $y - 4\,(1) - 5\,(-2) = 6 \quad \Leftrightarrow \quad y = 0$, and $x + 0 + 1 + (-2) = 0 \quad \Leftrightarrow \quad x = 1$. So the solution is $(1, 0, 1, -2)$.

35. $\begin{cases} x - 3y + z = 4 \\ 4x - y + 15z = 5 \end{cases} \Leftrightarrow \begin{cases} x - 3y + z = 4 \\ \phantom{x + {}} y + z = -1 \end{cases}$ Thus, the system has infinitely many solutions given by

$z = t$, $y + t = -1 \quad \Leftrightarrow \quad y = -1 - t$, and $x + 3\,(1 + t) + t = 4 \quad \Leftrightarrow \quad x = 1 - 4t$. Therefore, the solutions are $(1 - 4t, -1 - t, t)$, where t is any real number.

37. $\begin{cases} -x + 4y + z = 8 \\ 2x - 6y + z = -9 \\ x - 6y - 4z = -15 \end{cases} \Leftrightarrow \begin{cases} -x + 4y + z = 8 \\ 2y + 3z = 7 \\ 6y + 9z = 21 \end{cases} \Leftrightarrow \begin{cases} -x + 4y + z = 8 \\ 2y + 3z = 7 \\ 0 = 0 \end{cases}$

Thus, the system has infinitely many solutions. Letting $z = t$, we find $2y + 3t = 7 \Leftrightarrow y = \frac{7}{2} - \frac{3}{2}t$, and $-x + 4\left(\frac{7}{2} - \frac{3}{2}t\right) + t = 8 \Leftrightarrow x = 6 - 5t$. Therefore, the solutions are $\left(6 - 5t, \frac{7}{2} - \frac{3}{2}t, t\right)$, where t is any real number.

39. $\begin{bmatrix} 1 & -1 & 3 & 2 \\ 2 & 1 & 1 & 2 \\ 3 & 0 & 4 & 4 \end{bmatrix} \xrightarrow[R_3 - 3R_1 \to R_3]{R_2 - 2R_1 \to R_2} \begin{bmatrix} 1 & -1 & 3 & 2 \\ 0 & 3 & -5 & -2 \\ 0 & 3 & -5 & -2 \end{bmatrix} \xrightarrow{R_3 - R_2 \to R_3} \begin{bmatrix} 1 & -1 & 3 & 2 \\ 0 & 3 & -5 & -2 \\ 0 & 0 & 0 & 0 \end{bmatrix} \xrightarrow{\frac{1}{3}R_2}$

$\begin{bmatrix} 1 & -1 & 3 & 2 \\ 0 & 1 & -\frac{5}{3} & -\frac{2}{3} \\ 0 & 0 & 0 & 0 \end{bmatrix} \xrightarrow{R_1 + R_2 \to R_1} \begin{bmatrix} 1 & 0 & \frac{4}{3} & \frac{4}{3} \\ 0 & 1 & -\frac{5}{3} & -\frac{2}{3} \\ 0 & 0 & 0 & 0 \end{bmatrix}$. The system is dependent, so let $z = t$: $y - \frac{5}{3}t = -\frac{2}{3} \Leftrightarrow$

$y = \frac{5}{3}t - \frac{2}{3}$ and $x + \frac{4}{3}t = \frac{4}{3} \Leftrightarrow x = -\frac{4}{3}t + \frac{4}{3}$. So the solution is $\left(-\frac{4}{3}t + \frac{4}{3}, \frac{5}{3}t - \frac{2}{3}, t\right)$, where t is any real number.

41. $\begin{bmatrix} 1 & -1 & 1 & -1 & 0 \\ 3 & -1 & -1 & -1 & 2 \end{bmatrix} \xrightarrow{R_2 - 3R_1 \to R_2} \begin{bmatrix} 1 & -1 & 1 & -1 & 0 \\ 0 & 2 & -4 & 2 & 2 \end{bmatrix} \xrightarrow{\frac{1}{2}R_4} \begin{bmatrix} 1 & -1 & 1 & -1 & 0 \\ 0 & 1 & -2 & 1 & 1 \end{bmatrix} \xrightarrow{R_1 + R_2 \to R_1}$

$\begin{bmatrix} 1 & 0 & -1 & 0 & 1 \\ 0 & 1 & -2 & 1 & 1 \end{bmatrix}$. Since the system is dependent, Let $z = s$ and $w = t$. Then $y - 2s + t = 1 \Leftrightarrow y = 2s - t + 1$

and $x - s = 1 \Leftrightarrow x = s + 1$. So the solution is $(s + 1, 2s - t + 1, s, t)$, where s and t are any real numbers.

43. $\begin{bmatrix} 1 & -1 & 1 & 0 \\ 3 & 2 & -1 & 6 \\ 1 & 4 & -3 & 3 \end{bmatrix} \xrightarrow[R_3 - R_1 \to R_3]{R_2 - 3R_1 \to R_2} \begin{bmatrix} 1 & -1 & 1 & 0 \\ 0 & 5 & -4 & 6 \\ 0 & 5 & -4 & 3 \end{bmatrix} \xrightarrow{R_3 - R_2 \to R_3} \begin{bmatrix} 1 & -1 & 1 & 0 \\ 0 & 5 & -4 & 6 \\ 0 & 0 & 0 & 3 \end{bmatrix}$. The last row of this

matrix corresponds to the equation $0 = 3$, which is always false. Hence there is no solution.

45. $\begin{bmatrix} 1 & 1 & -1 & -1 & 2 \\ 1 & -1 & 1 & -1 & 0 \\ 2 & 0 & 0 & 2 & 2 \\ 2 & 4 & -4 & -2 & 6 \end{bmatrix} \xrightarrow{R_1 \leftrightarrow \frac{1}{2}R_3} \begin{bmatrix} 1 & 0 & 0 & 1 & 1 \\ 1 & -1 & 1 & -1 & 0 \\ 1 & 1 & -1 & -1 & 2 \\ 2 & 4 & -4 & -2 & 6 \end{bmatrix} \xrightarrow[R_4 - 2R_1 \to R_4]{\substack{R_2 - R_1 \to R_2 \\ R_3 - R_1 \to R_3}} \begin{bmatrix} 1 & 0 & 0 & 1 & 1 \\ 0 & -1 & 1 & -2 & -1 \\ 0 & 1 & -1 & -2 & 1 \\ 0 & 4 & -4 & -4 & 4 \end{bmatrix}$

$\xrightarrow[R_4 + 4R_2 \to R_4]{R_3 + R_2 \to R_3} \begin{bmatrix} 1 & 0 & 0 & 1 & 1 \\ 0 & -1 & 1 & -2 & -1 \\ 0 & 0 & 0 & -4 & 0 \\ 0 & 0 & 0 & -12 & 0 \end{bmatrix} \xrightarrow[-\frac{1}{12}R_4]{-\frac{1}{4}R_3} \begin{bmatrix} 1 & 0 & 0 & 1 & 1 \\ 0 & -1 & 1 & -2 & -1 \\ 0 & 0 & 0 & 0 & 0 \\ 0 & 0 & 0 & 1 & 0 \end{bmatrix} \xrightarrow[\substack{R_2 + 2R_3 \to R_2 \\ R_4 - R_3 \to R_4}]{R_1 - R_3 \to R_1} \begin{bmatrix} 1 & 0 & 0 & 0 & 1 \\ 0 & -1 & 1 & 0 & -1 \\ 0 & 0 & 0 & 1 & 0 \\ 0 & 0 & 0 & 0 & 0 \end{bmatrix}$.

This system is dependent. Let $z = t$, so $-y + t = -1 \Leftrightarrow y = t + 1$; $x = 1 \Leftrightarrow x = 1$. So the solution is $(1, t + 1, t, 0)$, where t is any real number.

47. Let x be the amount in the 6% account and y the amount in the 7% account. The system is $\begin{cases} y = 2x \\ 0.06x + 0.07y = 600 \end{cases}$

Substituting gives $0.06x + 0.07(2x) = 600 \Leftrightarrow 0.2x = 600 \Leftrightarrow x = 3000$, so $y = 2(3000) = 6000$. Hence, the man has \$3,000 invested at 6% and \$6,000 invested at 7%.

49. Let x be the amount invested in Bank A, y the amount invested in Bank B, and z the amount invested in Bank C.

We get the following system:
$$\begin{cases} x + y + z = 60{,}000 \\ 0.02x + 0.025y + 0.03z = 1575 \\ 2x + 2z = y \end{cases} \Leftrightarrow \begin{cases} x + y + z = 60{,}000 \\ 2x + 2.5y + 3z = 157{,}500 \\ 2x - y + 2z = 0 \end{cases}$$

which has matrix representation
$$\begin{bmatrix} 1 & 1 & 1 & 60{,}000 \\ 2 & 2.5 & 3 & 157{,}500 \\ 2 & -1 & 2 & 0 \end{bmatrix} \xrightarrow[R_3 - 2R_1 \to R_3]{R_2 - 2R_1 \to R_2} \begin{bmatrix} 1 & 1 & 1 & 60{,}000 \\ 0 & 0.5 & 1 & 37{,}500 \\ 0 & -3 & 0 & -120{,}000 \end{bmatrix} \xrightarrow{R_2 \leftrightarrow -\frac{1}{3}R_3}$$

$$\begin{bmatrix} 1 & 1 & 1 & 60{,}000 \\ 0 & 1 & 0 & 40{,}000 \\ 0 & 0.5 & 1 & 37{,}500 \end{bmatrix} \xrightarrow[R_3 - 0.5R_2 \to R_3]{R_1 - R_2 \to R_1} \begin{bmatrix} 1 & 0 & 1 & 20{,}000 \\ 0 & 1 & 0 & 40{,}000 \\ 0 & 0 & 1 & 17{,}500 \end{bmatrix} \xrightarrow{R_1 - R_3 \to R_1} \begin{bmatrix} 1 & 0 & 1 & 2{,}500 \\ 0 & 1 & 0 & 40{,}000 \\ 0 & 0 & 1 & 17{,}500 \end{bmatrix}.$$ Thus, she invests

$2,500 in Bank A, $40,000 in Bank B, and $17,500 in Bank C.

In Solutions 51–61, the matrices A, B, C, D, E, F, and G are defined as follows:

$$A = \begin{bmatrix} 2 & 0 & -1 \end{bmatrix} \qquad B = \begin{bmatrix} 1 & 2 & 4 \\ -2 & 1 & 0 \end{bmatrix} \qquad C = \begin{bmatrix} \frac{1}{2} & 3 \\ 2 & \frac{3}{2} \\ -2 & 1 \end{bmatrix}$$

$$D = \begin{bmatrix} 1 & 4 \\ 0 & -1 \\ 2 & 0 \end{bmatrix} \qquad E = \begin{bmatrix} 2 & -1 \\ -\frac{1}{2} & 1 \end{bmatrix} \qquad F = \begin{bmatrix} 4 & 0 & 2 \\ -1 & 1 & 0 \\ 7 & 5 & 0 \end{bmatrix} \qquad G = \begin{bmatrix} 5 \end{bmatrix}$$

51. $A + B$ is not defined because the matrix dimensions 1×3 and 2×3 are not compatible.

53. $2C + 3D = 2\begin{bmatrix} \frac{1}{2} & 3 \\ 2 & \frac{3}{2} \\ -2 & 1 \end{bmatrix} + 3\begin{bmatrix} 1 & 4 \\ 0 & -1 \\ 2 & 0 \end{bmatrix} = \begin{bmatrix} 1 & 6 \\ 4 & 3 \\ -4 & 2 \end{bmatrix} + \begin{bmatrix} 3 & 12 \\ 0 & -3 \\ 6 & 0 \end{bmatrix} = \begin{bmatrix} 4 & 18 \\ 4 & 0 \\ 2 & 2 \end{bmatrix}$

55. $GA = \begin{bmatrix} 5 \end{bmatrix}\begin{bmatrix} 2 & 0 & -1 \end{bmatrix} = \begin{bmatrix} 10 & 0 & -5 \end{bmatrix}$

57. $BC = \begin{bmatrix} 1 & 2 & 4 \\ -2 & 1 & 0 \end{bmatrix}\begin{bmatrix} \frac{1}{2} & 3 \\ 2 & \frac{3}{2} \\ -2 & 1 \end{bmatrix} = \begin{bmatrix} -\frac{7}{2} & 10 \\ 1 & -\frac{9}{2} \end{bmatrix}$

59. $BF = \begin{bmatrix} 1 & 2 & 4 \\ -2 & 1 & 0 \end{bmatrix}\begin{bmatrix} 4 & 0 & 2 \\ -1 & 1 & 0 \\ 7 & 5 & 0 \end{bmatrix} = \begin{bmatrix} 30 & 22 & 2 \\ -9 & 1 & -4 \end{bmatrix}$

61. $(C + D)E = \left(\begin{bmatrix} \frac{1}{2} & 3 \\ 2 & \frac{3}{2} \\ -2 & 1 \end{bmatrix} + \begin{bmatrix} 1 & 4 \\ 0 & -1 \\ 2 & 0 \end{bmatrix}\right)\begin{bmatrix} 2 & -1 \\ -\frac{1}{2} & 1 \end{bmatrix} = \begin{bmatrix} \frac{3}{2} & 7 \\ 2 & \frac{1}{2} \\ 0 & 1 \end{bmatrix}\begin{bmatrix} 2 & -1 \\ -\frac{1}{2} & 1 \end{bmatrix} = \begin{bmatrix} -\frac{1}{2} & \frac{11}{2} \\ \frac{15}{4} & -\frac{3}{2} \\ -\frac{1}{2} & 1 \end{bmatrix}$

63. $AB = \begin{bmatrix} 2 & -5 \\ -2 & 6 \end{bmatrix}\begin{bmatrix} 3 & \frac{5}{2} \\ 1 & 1 \end{bmatrix} = \begin{bmatrix} 1 & 0 \\ 0 & 1 \end{bmatrix}$ and $BA = \begin{bmatrix} 3 & \frac{5}{2} \\ 1 & 1 \end{bmatrix}\begin{bmatrix} 2 & -5 \\ -2 & 6 \end{bmatrix} = \begin{bmatrix} 1 & 0 \\ 0 & 1 \end{bmatrix}.$

In Solutions 65–69, $A = \begin{bmatrix} 2 & 1 \\ 3 & 2 \end{bmatrix}$, $B = \begin{bmatrix} 1 & -2 \\ -2 & 4 \end{bmatrix}$, **and** $C = \begin{bmatrix} 0 & 1 & 3 \\ -2 & 4 & 0 \end{bmatrix}$.

65. $A + 3X = B \Leftrightarrow 3X = B - A \Leftrightarrow X = \frac{1}{3}(B - A)$. Thus,

$$X = \frac{1}{3}\left(\begin{bmatrix} 1 & -2 \\ -2 & 4 \end{bmatrix} - \begin{bmatrix} 2 & 1 \\ 3 & 2 \end{bmatrix} \right) = \frac{1}{3} \begin{bmatrix} -1 & -3 \\ -5 & 2 \end{bmatrix}.$$

67. $2(X - A) = 3B \Leftrightarrow X - A = \frac{3}{2}B \Leftrightarrow X = A + \frac{3}{2}B$. Thus,

$$X = \begin{bmatrix} 2 & 1 \\ 3 & 2 \end{bmatrix} + \frac{3}{2} \begin{bmatrix} 1 & -2 \\ -2 & 4 \end{bmatrix} = \begin{bmatrix} 2 & 1 \\ 3 & 2 \end{bmatrix} + \begin{bmatrix} \frac{3}{2} & -3 \\ -3 & 6 \end{bmatrix} = \begin{bmatrix} \frac{7}{2} & -2 \\ 0 & 8 \end{bmatrix}.$$

69. $AX = C \Leftrightarrow A^{-1}AX = X = A^{-1}C$. Now

$$A^{-1} = \frac{1}{4-3} \begin{bmatrix} 2 & -1 \\ -3 & 2 \end{bmatrix} = \begin{bmatrix} 2 & -1 \\ -3 & 2 \end{bmatrix}.$$ Thus, $X = A^{-1}C = \begin{bmatrix} 2 & -1 \\ -3 & 2 \end{bmatrix} \begin{bmatrix} 0 & 1 & 3 \\ -2 & 4 & 0 \end{bmatrix} = \begin{bmatrix} 2 & -2 & 6 \\ -4 & 5 & -9 \end{bmatrix}.$

71. $D = \begin{bmatrix} 1 & 4 \\ 2 & 9 \end{bmatrix}$. Then $|D| = 1(9) - 2(4) = 1$, and so $D^{-1} = \begin{bmatrix} 9 & -4 \\ -2 & 1 \end{bmatrix}$.

73. $D = \begin{bmatrix} 4 & -12 \\ -2 & 6 \end{bmatrix}$. Then $|D| = 4(6) - 2(12) = 0$, and so D has no inverse.

75. $D = \begin{bmatrix} 3 & 0 & 1 \\ 2 & -3 & 0 \\ 4 & -2 & 1 \end{bmatrix}$. Then, $|D| = 1 \begin{vmatrix} 2 & -3 \\ 4 & -2 \end{vmatrix} + 1 \begin{vmatrix} 3 & 0 \\ 2 & -3 \end{vmatrix} = -4 + 12 - 9 = -1$. So D^{-1} exists.

$$\begin{bmatrix} 3 & 0 & 1 & 1 & 0 & 0 \\ 2 & -3 & 0 & 0 & 1 & 0 \\ 4 & -2 & 1 & 0 & 0 & 1 \end{bmatrix} \xrightarrow{R_1 - R_2 \to R_1} \begin{bmatrix} 1 & 3 & 1 & 1 & -1 & 0 \\ 2 & -3 & 0 & 0 & 1 & 0 \\ 4 & -2 & 1 & 0 & 0 & 1 \end{bmatrix} \xrightarrow[R_3 - 4R_1 \to R_3]{R_2 - 2R_1 \to R_2} \begin{bmatrix} 1 & 3 & 1 & 1 & -1 & 0 \\ 0 & -9 & -2 & -2 & 3 & 0 \\ 0 & -14 & -3 & -4 & 4 & 1 \end{bmatrix} \xrightarrow[-2R_3]{-3R_2}$$

$$\begin{bmatrix} 1 & 3 & 1 & 1 & -1 & 0 \\ 0 & 27 & 6 & 6 & -9 & 0 \\ 0 & 28 & 6 & 8 & -8 & -2 \end{bmatrix} \xrightarrow{R_3 - R_2 \to R_3} \begin{bmatrix} 1 & 3 & 1 & 1 & -1 & 0 \\ 0 & 27 & 6 & 6 & -9 & 0 \\ 0 & 1 & 0 & 2 & 1 & -2 \end{bmatrix} \xrightarrow[\frac{1}{3}R_3]{R_3 \leftrightarrow R_2} \begin{bmatrix} 1 & 3 & 1 & 1 & -1 & 0 \\ 0 & 1 & 0 & 2 & 1 & -2 \\ 0 & 9 & 2 & 2 & -3 & 0 \end{bmatrix} \xrightarrow[R_1 - 3R_2 \to R_1]{R_3 - 9R_2 \to R_3}$$

$$\begin{bmatrix} 1 & 0 & 1 & -5 & -4 & 6 \\ 0 & 1 & 0 & 2 & 1 & -2 \\ 0 & 0 & 2 & -16 & -12 & 18 \end{bmatrix} \xrightarrow[R_1 - R_3 \to R_1]{\frac{1}{2}R_3} \begin{bmatrix} 1 & 0 & 0 & 3 & 2 & -3 \\ 0 & 1 & 0 & 2 & 1 & -2 \\ 0 & 0 & 1 & -8 & -6 & 9 \end{bmatrix}.$$ Thus, $D^{-1} = \begin{bmatrix} 3 & 2 & -3 \\ 2 & 1 & -2 \\ -8 & -6 & 9 \end{bmatrix}.$

77. $D = \begin{bmatrix} 1 & 0 & 0 & 1 \\ 0 & 2 & 0 & 2 \\ 0 & 0 & 3 & 3 \\ 0 & 0 & 0 & 4 \end{bmatrix}$. Thus, $|D| = \begin{vmatrix} 2 & 0 & 2 \\ 0 & 3 & 3 \\ 0 & 0 & 4 \end{vmatrix} = 2 \begin{vmatrix} 3 & 3 \\ 0 & 4 \end{vmatrix} = 24$ and D^{-1} exists.

$$\begin{bmatrix} 1 & 0 & 0 & 1 & 1 & 0 & 0 & 0 \\ 0 & 2 & 0 & 2 & 0 & 1 & 0 & 0 \\ 0 & 0 & 3 & 3 & 0 & 0 & 1 & 0 \\ 0 & 0 & 0 & 4 & 0 & 0 & 0 & 1 \end{bmatrix} \xrightarrow[\frac{1}{4}R_4]{\substack{\frac{1}{2}R_2 \\ \frac{1}{3}R_3}}$$

$$\begin{bmatrix} 1 & 0 & 0 & 1 & 1 & 0 & 0 & 0 \\ 0 & 1 & 0 & 1 & 0 & \frac{1}{2} & 0 & 0 \\ 0 & 0 & 1 & 1 & 0 & 0 & \frac{1}{3} & 0 \\ 0 & 0 & 0 & 1 & 0 & 0 & 0 & \frac{1}{4} \end{bmatrix} \xrightarrow[R_3 - R_4 \to R_3]{\substack{R_1 - R_4 \to R_1 \\ R_2 - R_4 \to R_2}} \begin{bmatrix} 1 & 0 & 0 & 0 & 1 & 0 & 0 & -\frac{1}{4} \\ 0 & 1 & 0 & 0 & 0 & \frac{1}{2} & 0 & -\frac{1}{4} \\ 0 & 0 & 1 & 0 & 0 & 0 & \frac{1}{3} & -\frac{1}{4} \\ 0 & 0 & 0 & 1 & 0 & 0 & 0 & \frac{1}{4} \end{bmatrix}.$$ Therefore, $D^{-1} = \begin{bmatrix} 1 & 0 & 0 & -\frac{1}{4} \\ 0 & \frac{1}{2} & 0 & -\frac{1}{4} \\ 0 & 0 & \frac{1}{3} & -\frac{1}{4} \\ 0 & 0 & 0 & \frac{1}{4} \end{bmatrix}.$

79. $\begin{bmatrix} 12 & -5 \\ 5 & -2 \end{bmatrix} \begin{bmatrix} x \\ y \end{bmatrix} = \begin{bmatrix} 10 \\ 17 \end{bmatrix}$. If we let $A = \begin{bmatrix} 12 & -5 \\ 5 & -2 \end{bmatrix}$, then $A^{-1} = \dfrac{1}{-24+25} \begin{bmatrix} -2 & 5 \\ -5 & 12 \end{bmatrix} = \begin{bmatrix} -2 & 5 \\ -5 & 12 \end{bmatrix}$, and so

$\begin{bmatrix} x \\ y \end{bmatrix} = \begin{bmatrix} -2 & 5 \\ -5 & 12 \end{bmatrix} \begin{bmatrix} 10 \\ 17 \end{bmatrix} = \begin{bmatrix} 65 \\ 154 \end{bmatrix}$. Therefore, the solution is $(65, 154)$.

81. $\begin{bmatrix} 2 & 1 & 5 \\ 1 & 2 & 2 \\ 1 & 0 & 3 \end{bmatrix} \begin{bmatrix} x \\ y \\ z \end{bmatrix} = \begin{bmatrix} \frac{1}{3} \\ \frac{1}{4} \\ \frac{1}{6} \end{bmatrix}$. Let $A = \begin{bmatrix} 2 & 1 & 5 \\ 1 & 2 & 2 \\ 1 & 0 & 3 \end{bmatrix}$. Then $\begin{bmatrix} 2 & 1 & 5 & 1 & 0 & 0 \\ 1 & 2 & 2 & 0 & 1 & 0 \\ 1 & 0 & 3 & 0 & 0 & 1 \end{bmatrix} \xrightarrow{R_1 \leftrightarrow R_2} \begin{bmatrix} 1 & 2 & 2 & 0 & 1 & 0 \\ 2 & 1 & 5 & 1 & 0 & 0 \\ 1 & 0 & 3 & 0 & 0 & 1 \end{bmatrix}$

$\xrightarrow[R_3 - R_1 \to R_3]{R_2 - 2R_1 \to R_2} \begin{bmatrix} 1 & 2 & 2 & 0 & 1 & 0 \\ 0 & -3 & 1 & 1 & -2 & 0 \\ 0 & -2 & 1 & 0 & -1 & 1 \end{bmatrix} \xrightarrow{R_2 - 2R_3 \to R_2} \begin{bmatrix} 1 & 2 & 2 & 0 & 1 & 0 \\ 0 & 1 & -1 & 1 & 0 & -2 \\ 0 & -2 & 1 & 0 & -1 & 1 \end{bmatrix} \xrightarrow[R_3 \to R_3 + 2R_2]{R_1 - 2R_2 \to R_1}$

$\begin{bmatrix} 1 & 0 & 4 & -2 & 1 & 4 \\ 0 & 1 & -1 & 1 & 0 & -2 \\ 0 & 0 & -1 & 2 & -1 & -3 \end{bmatrix} \xrightarrow{-R_3} \begin{bmatrix} 1 & 0 & 4 & -2 & 1 & 4 \\ 0 & 1 & -1 & 1 & 0 & -2 \\ 0 & 0 & 1 & -2 & 1 & 3 \end{bmatrix} \xrightarrow[R_2 + R_3 \to R_2]{R_1 - 4R_3 \to R_1} \begin{bmatrix} 1 & 0 & 0 & 6 & -3 & -8 \\ 0 & 1 & 0 & -1 & 1 & 1 \\ 0 & 0 & 1 & -2 & 1 & 3 \end{bmatrix}$. Hence,

$A^{-1} = \begin{bmatrix} 6 & -3 & -8 \\ -1 & 1 & 1 \\ -2 & 1 & 3 \end{bmatrix}$ and $\begin{bmatrix} x \\ y \\ z \end{bmatrix} = \begin{bmatrix} 6 & -3 & -8 \\ -1 & 1 & 1 \\ -2 & 1 & 3 \end{bmatrix} \begin{bmatrix} \frac{1}{3} \\ \frac{1}{4} \\ \frac{1}{6} \end{bmatrix} = \begin{bmatrix} -\frac{1}{12} \\ \frac{1}{12} \\ \frac{1}{12} \end{bmatrix}$, and so the solution is $\left(-\frac{1}{12}, \frac{1}{12}, \frac{1}{12} \right)$.

83. $|D| = \begin{vmatrix} 2 & 7 \\ 6 & 16 \end{vmatrix} = 32 - 42 = -10$, $|D_x| = \begin{vmatrix} 13 & 7 \\ 30 & 16 \end{vmatrix} = 208 - 210 = -2$, and $|D_y| = \begin{vmatrix} 2 & 13 \\ 6 & 30 \end{vmatrix} = 60 - 78 = -18$.

Therefore, $x = \frac{-2}{-10} = \frac{1}{5}$ and $y = \frac{-18}{-10} = \frac{9}{5}$, and so the solution is $\left(\frac{1}{5}, \frac{9}{5} \right)$.

85. $|D| = \begin{vmatrix} 2 & -1 & 5 \\ -1 & 7 & 0 \\ 5 & 4 & 3 \end{vmatrix} = 5 \begin{vmatrix} -1 & 7 \\ 5 & 4 \end{vmatrix} + 3 \begin{vmatrix} 2 & -1 \\ -1 & 7 \end{vmatrix} = -195 + 39 = -156$,

$|D_x| = \begin{vmatrix} 0 & -1 & 5 \\ 9 & 7 & 0 \\ -9 & 4 & 3 \end{vmatrix} = 5 \begin{vmatrix} 9 & 7 \\ -9 & 4 \end{vmatrix} + 3 \begin{vmatrix} 0 & -1 \\ 9 & 7 \end{vmatrix} = 495 + 27 = 522$,

$|D_y| = \begin{vmatrix} 2 & 0 & 5 \\ -1 & 9 & 0 \\ 5 & -9 & 3 \end{vmatrix} = 5 \begin{vmatrix} -1 & 9 \\ 5 & -9 \end{vmatrix} + 3 \begin{vmatrix} 2 & 0 \\ -1 & 9 \end{vmatrix} = -180 + 54 = -126$, and

$|D_z| = \begin{vmatrix} 2 & -1 & 0 \\ -1 & 7 & 9 \\ 5 & 4 & -9 \end{vmatrix} = -9 \begin{vmatrix} 2 & -1 \\ 5 & 4 \end{vmatrix} - 9 \begin{vmatrix} 2 & -1 \\ -1 & 7 \end{vmatrix} = -117 - 117 = -234$.

Therefore, $x = \frac{522}{-156} = -\frac{87}{26}$, $y = \frac{-126}{-156} = \frac{21}{26}$, and $z = \frac{-234}{-156} = \frac{3}{2}$, and so the solution is $\left(-\frac{87}{26}, \frac{21}{26}, \frac{3}{2} \right)$.

87. The area is $\pm \dfrac{1}{2} \begin{vmatrix} -1 & 3 & 1 \\ 3 & 1 & 1 \\ -2 & -2 & 1 \end{vmatrix} = \pm \dfrac{1}{2} \left(\begin{vmatrix} 3 & 1 \\ -2 & -2 \end{vmatrix} - \begin{vmatrix} -1 & 3 \\ -2 & -2 \end{vmatrix} + \begin{vmatrix} -1 & 3 \\ 3 & 1 \end{vmatrix} \right) = \pm \dfrac{1}{2} (-4 - 8 - 10) = 11$.

89. $\dfrac{3x+1}{x^2-2x-15} = \dfrac{3x+1}{(x-5)(x+3)} = \dfrac{A}{x-5} + \dfrac{B}{x+3}$. Thus, $3x+1 = A(x+3)+B(x-5) = x(A+B)+(3A-5B)$,

and so $\begin{cases} A+B=3 \\ 3A-5B=1 \end{cases} \Leftrightarrow \begin{cases} -3A-3B=-9 \\ 3A-5B=1 \end{cases}$ Adding, we have $-8B=-8 \quad \Leftrightarrow \quad B=1$, and $A=2$.

Hence, $\dfrac{3x+1}{x^2-2x-15} = \dfrac{2}{x-5} + \dfrac{1}{x+3}$.

91. $\dfrac{2x-4}{x(x-1)^2} = \dfrac{A}{x} + \dfrac{B}{x-1} + \dfrac{C}{(x-1)^2}$. Then $2x-4 = A(x-1)^2 + Bx(x-1) + Cx = Ax^2 - 2Ax + A + Bx^2 - Bx + Cx = x^2(A+B) + x(-2A-B+C) + A$. So $A=-4$, $-4+B=0 \quad \Leftrightarrow \quad B=4$, and $8-4+C=2 \quad \Leftrightarrow$ $C=-2$. Therefore, $\dfrac{2x-4}{x(x-1)^2} = -\dfrac{4}{x} + \dfrac{4}{x-1} - \dfrac{2}{(x-1)^2}$.

93. $\dfrac{2x-1}{x^3+x} = \dfrac{2x-1}{x(x^2+1)} = \dfrac{A}{x} + \dfrac{Bx+C}{x^2+1}$. Then $2x-1 = A(x^2+1) + (Bx+C)x = Ax^2 + A + Bx^2 + Cx = (A+B)x^2 + Cx + A$. So $A=-1$, $C=2$, and $A+B=0$ gives us $B=1$. Thus $\dfrac{2x-1}{x^3+x} = -\dfrac{1}{x} + \dfrac{x+2}{x^2+1}$.

95. The boundary is a solid curve, so we have the inequality $x+y^2 \le 4$. We take the test point $(0,0)$ and verify that it satisfies the inequality: $0+0^2 \le 4$.

97. $3x+y \le 6$

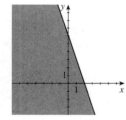

99. $x^2+y^2>9$

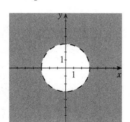

101. $\begin{cases} y \ge x^2 - 3x \\ y \le \frac{1}{3}x - 1 \end{cases}$

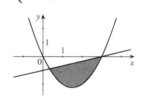

103. $\begin{cases} x+y \ge 2 \\ y-x \le 2 \\ x \le 3 \end{cases}$

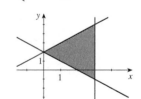

105. $\begin{cases} x^2+y^2<9 \\ x+y<0 \end{cases}$ The vertices occur where $y=-x$. By substitution,

$x^2+x^2=9 \quad \Leftrightarrow \quad x=\pm\dfrac{3}{\sqrt{2}}$, and so $y=\mp\dfrac{3}{\sqrt{2}}$. Therefore, the vertices are

$\left(\dfrac{3}{\sqrt{2}}, -\dfrac{3}{\sqrt{2}}\right)$ and $\left(-\dfrac{3}{\sqrt{2}}, \dfrac{3}{\sqrt{2}}\right)$ and the solution set is bounded.

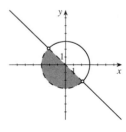

107. $\begin{cases} x \ge 0,\, y \ge 0 \\ x+2y \le 12 \\ y \le x+4 \end{cases}$ The intersection points are $(-4,0)$, $(0,4)$, $\left(\frac{4}{3}, \frac{16}{3}\right)$, $(0,6)$,

$(0,0)$, and $(12,0)$. Since the points $(-4,0)$ and $(0,6)$ are not in the solution set, the vertices are $(0,4)$, $\left(\frac{4}{3}, \frac{16}{3}\right)$, $(12,0)$, and $(0,0)$. The solution set is bounded.

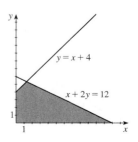

109. $\begin{cases} -x + y + z = a \\ x - y + z = b \\ x + y - z = c \end{cases}$ $\Leftrightarrow$ $\begin{cases} -x + y + z = a \\ \quad\quad 2z = a + b \\ \quad 2y \quad = a + c \end{cases}$ Thus, $y = \dfrac{a+c}{2}$, $z = \dfrac{a+b}{2}$, and $-x + \dfrac{a+c}{2} + \dfrac{a+b}{2} = a$

$\Leftrightarrow$ $x = \dfrac{b+c}{2}$. The solution is $\left(\dfrac{b+c}{2}, \dfrac{a+c}{2}, \dfrac{a+b}{2} \right)$.

111. Solving the second equation for y, we have $y = kx$. Substituting for y in the first equation gives us

$x + kx = 12 \Leftrightarrow (1+k)\,x = 12 \quad\Leftrightarrow\quad x = \dfrac{12}{k+1}$. Substituting for y in the third equation gives us

$kx - x = 2k \quad\Leftrightarrow\quad (k-1)\,x = 2k \quad\Leftrightarrow\quad x = \dfrac{2k}{k-1}$. These points of intersection are the same when the

x-values are equal. Thus, $\dfrac{12}{k+1} = \dfrac{2k}{k-1} \quad\Leftrightarrow\quad 12\,(k-1) = 2k\,(k+1) \quad\Leftrightarrow\quad 12k - 12 = 2k^2 + 2k \quad\Leftrightarrow$

$0 = 2k^2 - 10k + 12 = 2\,(k^2 - 5k + 6) = 2\,(k-3)\,(k+2)$. Hence, $k = 2$ or $k = 3$.

Chapter 10 Test

1. (a) The system is linear.

(b) $\begin{cases} x + 3y = 7 \\ 5x + 2y = -4 \end{cases}$ Multiplying the first equation by -5 and then adding gives $-13y = -39 \quad\Leftrightarrow\quad y = 3$. So

$x + 3\,(3) = 7 \quad\Leftrightarrow\quad x = -2$. Thus, the solution is $(-2, 3)$.

2. (a) The system is nonlinear.

(b) $\begin{cases} 6x + y^2 = 10 \\ 3x - y = 5 \end{cases}$ $\Leftrightarrow$ $\begin{cases} 6x + y^2 = 10 \\ y^2 + 2y = 0 \end{cases}$ Thus $y^2 + 2y = y\,(y+2) = 0$, so either $y = 0$ or $y = -2$. If $y = 0$,

then $3x = 5 \quad\Leftrightarrow\quad x = \frac{5}{3}$ and if $y = -2$, then $3x - (-2) = 5 \quad\Leftrightarrow\quad 3x = 3 \quad\Leftrightarrow\quad x = 1$. Thus the solutions

are $\left(\frac{5}{3}, 0 \right)$ and $(1, -2)$.

3. $\begin{cases} x - 2y = 1 \\ \quad y = x^3 - 2x^2 \end{cases}$

The solutions are approximately $(-0.55, -0.78)$, $(0.43, -0.29)$, and $(2.12, 0.56)$.

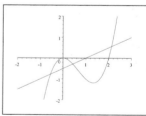

4. Let w be the speed of the wind and a the speed of the airplane in still air, in kilometers per hour. Then the speed of the of the plane flying against the wind is $a - w$ and the speed of the plane flying with the wind is $a + w$. Using distance $=$ rate $\times$ time,

we get the system $\begin{cases} 600 = 2.5\,(a - w) \\ 300 = \frac{50}{60}\,(a + w) \end{cases}$ $\Leftrightarrow$ $\begin{cases} 240 = a - w \\ 360 = a + w \end{cases}$ Adding the two equations, we get $600 = 2a \quad\Leftrightarrow$

$a = 300$. So $360 = 300 + w \quad\Leftrightarrow\quad w = 60$. Thus the speed of the airplane in still air is 300 km/h and the speed of the wind is 60 km/h.

5. (a) This matrix is in row-echelon form, but not in reduced row-echelon form, since the leading 1 in the second row does not have a 0 above it.

(b) This matrix is in reduced row-echelon form.

(c) This matrix is in neither row-echelon form nor reduced row-echelon form.

6. (a) $\begin{cases} x - y + 2z = 0 \\ 2x - 4y + 5z = -5 \\ 2y - 3z = 5 \end{cases}$ has the matrix representation $\begin{bmatrix} 1 & -1 & 2 & 0 \\ 2 & -4 & 5 & -5 \\ 0 & 2 & -3 & 5 \end{bmatrix}$ $\xrightarrow{R_2 - R_1 \to R_2}$ $\begin{bmatrix} 1 & -1 & 2 & 0 \\ 0 & -2 & 1 & -5 \\ 0 & 2 & -3 & 5 \end{bmatrix}$

$\xrightarrow{R_3 + R_2 \to R_3}$ $\begin{bmatrix} 1 & -1 & 2 & 0 \\ 0 & -2 & 1 & -5 \\ 0 & 0 & -2 & 0 \end{bmatrix}$ $\xrightarrow{-\frac{1}{2}R_3}$ $\begin{bmatrix} 1 & -1 & 2 & 0 \\ 0 & -2 & 1 & -5 \\ 0 & 0 & 1 & 0 \end{bmatrix}$. Thus $z = 0$, $-2y + 0 = -5$ $\Leftrightarrow$ $y = \frac{5}{2}$, and

$x - \frac{5}{2} + 2(0) = 0$ $\Leftrightarrow$ $x = \frac{5}{2}$. Thus, the solution is $\left(\frac{5}{2}, \frac{5}{2}, 0\right)$.

(b) $\begin{cases} 2x - 3y + z = 3 \\ x + 2y + 2z = -1 \\ 4x + y + 5z = 4 \end{cases}$ has the matrix representation $\begin{bmatrix} 2 & -3 & 1 & 3 \\ 1 & 2 & 2 & -1 \\ 4 & 1 & 5 & 4 \end{bmatrix}$ $\xrightarrow{R_1 \leftrightarrow R_2}$ $\begin{bmatrix} 1 & 2 & 2 & -1 \\ 2 & -3 & 1 & 3 \\ 4 & 1 & 5 & 4 \end{bmatrix}$

$\xrightarrow[R_3 - 4R_1 \to R_3]{R_2 - 2R_1 \to R_2}$ $\begin{bmatrix} 1 & 2 & 2 & -1 \\ 0 & -7 & -3 & 5 \\ 0 & -7 & -3 & 8 \end{bmatrix}$ $\xrightarrow{R_3 - R_2 \to R_3}$ $\begin{bmatrix} 1 & 2 & 2 & -1 \\ 0 & -7 & -3 & 5 \\ 0 & 0 & 0 & 3 \end{bmatrix}$.

Since the last row corresponds to the equation $0 = 3$, this system has no solution.

7. $\begin{cases} x + 3y - z = 0 \\ 3x + 4y - 2z = -1 \\ -x + 2y = 1 \end{cases}$ has the matrix representation $\begin{bmatrix} 1 & 3 & -1 & 0 \\ 3 & 4 & -2 & -1 \\ -1 & 2 & 0 & 1 \end{bmatrix}$ $\xrightarrow[R_3 + R_1 \to R_3]{R_2 - 3R_1 \to R_2}$ $\begin{bmatrix} 1 & 3 & -1 & 0 \\ 0 & -5 & 1 & -1 \\ 0 & 5 & -1 & 1 \end{bmatrix}$

$\xrightarrow{R_3 - R_2 \to R_3}$ $\begin{bmatrix} 1 & 3 & -1 & 0 \\ 0 & -5 & 1 & -1 \\ 0 & 0 & 0 & 0 \end{bmatrix}$ $\xrightarrow{-\frac{1}{5}R_2}$ $\begin{bmatrix} 1 & 3 & -1 & 0 \\ 0 & 1 & -\frac{1}{5} & \frac{1}{5} \\ 0 & 0 & 0 & 0 \end{bmatrix}$ $\xrightarrow{R_1 - 3R_2 \to R_1}$ $\begin{bmatrix} 1 & 0 & -\frac{2}{5} & -\frac{3}{5} \\ 0 & 1 & -\frac{1}{5} & \frac{1}{5} \\ 0 & 0 & 0 & 0 \end{bmatrix}$.

Since this system is dependent, let $z = t$. Then $y - \frac{1}{5}t = \frac{1}{5}$ $\Leftrightarrow$ $y = \frac{1}{5}t + \frac{1}{5}$ and $x - \frac{2}{5}t = -\frac{3}{5}$ $\Leftrightarrow$ $x = \frac{2}{5}t - \frac{3}{5}$.
Thus, the solution is $\left(\frac{2}{5}t - \frac{3}{5}, \frac{1}{5}t + \frac{1}{5}, t\right)$.

8. Let x, y, and z represent the price in dollars for coffee, juice, and donuts respectively. Then the system of equations is

$\begin{cases} 2x + y + 2z = 6.25 \quad \text{Anne} \\ x + 3z = 3.75 \quad \text{Barry} \\ 3x + y + 4z = 9.25 \quad \text{Cathy} \end{cases}$ $\Leftrightarrow$ $\begin{cases} 2x + y + 2z = 6.25 \\ y - 4z = -1.25 \\ y - 5z = -2.00 \end{cases}$ $\Leftrightarrow$ $\begin{cases} 2x + y + 2z = 6.25 \\ y - 4z = -1.25 \\ z = 0.75 \end{cases}$

Thus, $z = 0.75$, $y - 4(0.75) = -1.25$ $\Leftrightarrow$ $y = 1.75$, and $2x + 1.75 + 2(0.75) = 6.25$ $\Leftrightarrow$ $x = 1.5$. Thus coffee costs \$1.50, juice costs \$1.75, and donuts cost \$0.75.

9. $A = \begin{bmatrix} 2 & 3 \\ 2 & 4 \end{bmatrix}$, $B = \begin{bmatrix} 2 & 4 \\ -1 & 1 \\ 3 & 0 \end{bmatrix}$, and $C = \begin{bmatrix} 1 & 0 & 4 \\ -1 & 1 & 2 \\ 0 & 1 & 3 \end{bmatrix}$.

(a) $A + B$ is undefined because A is 2×2 and B is 3×2, so they have incompatible dimensions.

(b) AB is undefined because A is 2×2 and B is 3×2, so they have incompatible dimensions.

(c) $BA - 3B = \begin{bmatrix} 2 & 4 \\ -1 & 1 \\ 3 & 0 \end{bmatrix} \begin{bmatrix} 2 & 3 \\ 2 & 4 \end{bmatrix} - 3 \begin{bmatrix} 2 & 4 \\ -1 & 1 \\ 3 & 0 \end{bmatrix} = \begin{bmatrix} 12 & 22 \\ 0 & 1 \\ 6 & 9 \end{bmatrix} - \begin{bmatrix} 6 & 12 \\ -3 & 3 \\ 9 & 0 \end{bmatrix} = \begin{bmatrix} 6 & 10 \\ 3 & -2 \\ -3 & 9 \end{bmatrix}$

(d) $CBA = \begin{bmatrix} 1 & 0 & 4 \\ -1 & 1 & 2 \\ 0 & 1 & 3 \end{bmatrix} \begin{bmatrix} 2 & 4 \\ -1 & 1 \\ 3 & 0 \end{bmatrix} \begin{bmatrix} 2 & 3 \\ 2 & 4 \end{bmatrix} = \begin{bmatrix} 14 & 4 \\ 3 & -3 \\ 8 & 1 \end{bmatrix} \begin{bmatrix} 2 & 3 \\ 2 & 4 \end{bmatrix} = \begin{bmatrix} 36 & 58 \\ 0 & -3 \\ 18 & 28 \end{bmatrix}$

(e) $A = \begin{bmatrix} 2 & 3 \\ 2 & 4 \end{bmatrix} \quad \Leftrightarrow \quad A^{-1} = \frac{1}{8 - 6} \begin{bmatrix} 4 & -3 \\ -2 & 2 \end{bmatrix} = \begin{bmatrix} 2 & -\frac{3}{2} \\ -1 & 1 \end{bmatrix}$

(f) B^{-1} does not exist because B is not a square matrix.   **(g)** $\det(B)$ is not defined because B is not a square matrix.

(h) $\det(C) = \begin{vmatrix} 1 & 0 & 4 \\ -1 & 1 & 2 \\ 0 & 1 & 3 \end{vmatrix} = 1 \begin{vmatrix} 1 & 2 \\ 1 & 3 \end{vmatrix} + 4 \begin{vmatrix} -1 & 1 \\ 0 & 1 \end{vmatrix} = 1 - 4 = -3$

10. (a) The system $\begin{cases} 4x - 3y = 10 \\ 3x - 2y = 30 \end{cases}$ is equivalent to the matrix equation $\begin{bmatrix} 4 & -3 \\ 3 & -2 \end{bmatrix} \begin{bmatrix} x \\ y \end{bmatrix} = \begin{bmatrix} 10 \\ 30 \end{bmatrix}$.

(b) We have $|D| = \begin{vmatrix} 4 & -3 \\ 3 & -2 \end{vmatrix} = 4(-2) - 3(-3) = 1$. So $D^{-1} = \begin{bmatrix} -2 & 3 \\ -3 & 4 \end{bmatrix}$ and $\begin{bmatrix} x \\ y \end{bmatrix} = \begin{bmatrix} -2 & 3 \\ -3 & 4 \end{bmatrix} \begin{bmatrix} 10 \\ 30 \end{bmatrix} = \begin{bmatrix} 70 \\ 90 \end{bmatrix}$.

Therefore, $x = 70$ and $y = 90$.

11. $|A| = \begin{vmatrix} 1 & 4 & 1 \\ 0 & 2 & 0 \\ 1 & 0 & 1 \end{vmatrix} = 2 \begin{vmatrix} 1 & 1 \\ 1 & 1 \end{vmatrix} = 0$, $|B| = \begin{vmatrix} 1 & 4 & 0 \\ 0 & 2 & 0 \\ -3 & 0 & 1 \end{vmatrix} = 2 \begin{vmatrix} 1 & 0 \\ -3 & 1 \end{vmatrix} = 2$. Since $|A| = 0$, A does not have an inverse, and

since $|B| \neq 0$, B does have an inverse. $\begin{bmatrix} 1 & 4 & 0 & 1 & 0 & 0 \\ 0 & 2 & 0 & 0 & 1 & 0 \\ -3 & 0 & 1 & 0 & 0 & 1 \end{bmatrix} \xrightarrow{R_3 + 3R_1 \to R_3} \begin{bmatrix} 1 & 4 & 0 & 1 & 0 & 0 \\ 0 & 2 & 0 & 0 & 1 & 0 \\ 0 & 12 & 1 & 3 & 0 & 1 \end{bmatrix} \xrightarrow[R_3 - 6R_2 \to R_3]{R_1 - 2R_2 \to R_1}$

$\begin{bmatrix} 1 & 0 & 0 & 1 & -2 & 0 \\ 0 & 2 & 0 & 0 & 1 & 0 \\ 0 & 0 & 1 & 3 & -6 & 1 \end{bmatrix} \xrightarrow{\frac{1}{2}R_2} \begin{bmatrix} 1 & 0 & 0 & 1 & -2 & 0 \\ 0 & 1 & 0 & 0 & \frac{1}{2} & 0 \\ 0 & 0 & 1 & 3 & -6 & 1 \end{bmatrix}$. Therefore, $B^{-1} = \begin{bmatrix} 1 & -2 & 0 \\ 0 & \frac{1}{2} & 0 \\ 3 & -6 & 1 \end{bmatrix}$.

12. $\begin{cases} 2x & - z = 14 \\ 3x - y + 5z = 0 \\ 4x + 2y + 3z = -2 \end{cases}$ Then $|D| = \begin{vmatrix} 2 & 0 & -1 \\ 3 & -1 & 5 \\ 4 & 2 & 3 \end{vmatrix} = 2 \begin{vmatrix} -1 & 5 \\ 2 & 3 \end{vmatrix} - 1 \begin{vmatrix} 3 & -1 \\ 4 & 2 \end{vmatrix} = -26 - 10 = -36$,

$|D_x| = \begin{vmatrix} 14 & 0 & -1 \\ 0 & -1 & 5 \\ -2 & 2 & 3 \end{vmatrix} = 14 \begin{vmatrix} -1 & 5 \\ 2 & 3 \end{vmatrix} - 1 \begin{vmatrix} 0 & -1 \\ 4 & 2 \end{vmatrix} = -182 + 2 = -180$,

$|D_y| = \begin{vmatrix} 2 & 14 & -1 \\ 3 & 0 & 5 \\ 4 & -2 & 3 \end{vmatrix} = -3 \begin{vmatrix} 14 & -1 \\ -2 & 3 \end{vmatrix} - 5 \begin{vmatrix} 2 & 14 \\ 4 & -2 \end{vmatrix} = -120 + 300 = 180$, and

$|D_z| = \begin{vmatrix} 2 & 0 & 14 \\ 3 & -1 & 0 \\ 4 & 2 & -2 \end{vmatrix} = 2 \begin{vmatrix} -1 & 0 \\ 2 & -2 \end{vmatrix} + 14 \begin{vmatrix} 3 & -1 \\ 2 & 2 \end{vmatrix} = 4 + 140 = 144$.

Therefore, $x = \frac{-180}{-36} = 5$, $y = \frac{180}{-36} = -5$, $z = \frac{144}{-36} = -4$, and so the solution is $(5, -5, -4)$.

13. (a) $\dfrac{4x - 1}{(x - 1)^2 (x + 2)} = \dfrac{A}{x - 1} + \dfrac{B}{(x - 1)^2} + \dfrac{C}{x + 2}$. Thus,

$$4x - 1 = A(x - 1)(x + 2) + B(x + 2) + C(x - 1)^2 = A(x^2 + x - 2) + B(x + 2) + C(x^2 - 2x + 1)$$
$$= (A + C)x^2 + (A + B - 2C)x + (-2A + 2B + C)$$

which leads to the system of equations

$$\begin{cases} A + C = 0 \\ A + B - 2C = 4 \\ -2A + 2B + C = -1 \end{cases} \Leftrightarrow \begin{cases} A + C = 0 \\ B - 3C = 4 \\ 2B + 3C = -1 \end{cases} \Leftrightarrow \begin{cases} A + C = 0 \\ B - 3C = 4 \\ 9C = -9 \end{cases}$$

Therefore, $9C = -9 \Leftrightarrow C = -1$, $B - 3(-1) = 4 \Leftrightarrow B = 1$, and $A + (-1) = 0 \Leftrightarrow A = 1$.

Therefore, $\dfrac{4x - 1}{(x - 1)^2 (x + 2)} = \dfrac{1}{x - 1} + \dfrac{1}{(x - 1)^2} - \dfrac{1}{x + 2}$.

(b) $\dfrac{2x - 3}{x^3 + 3x} = \dfrac{2x - 3}{x(x^2 + 3)} = \dfrac{A}{x} + \dfrac{Bx + C}{x^2 + 3}$. Then

$$2x - 3 = A(x^2 + 3) + (Bx + C)x = Ax^2 + 3A + Bx^2 + Cx = (A + B)x^2 + Cx + 3A.$$

So $3A = -3 \Leftrightarrow A = -1$, $C = 2$ and $A + B = 0$ gives us $B = 1$. Thus $\dfrac{2x - 3}{x^3 + x} = -\dfrac{1}{x} + \dfrac{x + 2}{x^2 + 3}$.

14. (a) $\begin{cases} 2x + y \le 8 \\ x - y \ge -2 \\ x + 2y \ge 4 \end{cases}$ From the graph, the points $(4, 0)$ and $(0, 2)$ are vertices. The

third vertex occurs where the lines $2x + y = 8$ and $x - y = -2$ intersect. Adding these two equations gives $3x = 6 \Leftrightarrow x = 2$, and so $y = 8 - 2(2) = 4$. Thus, the third vertex is $(2, 4)$.

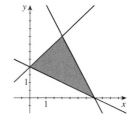

(b) $\begin{cases} x^2 - 5 + y \le 0 \\ y \le 5 + 2x \end{cases}$ Substituting $y = 5 + 2x$ into the first equation gives

$x^2 - 5 + (5 + 2x) = 0 \Leftrightarrow x^2 + 2x = 0 \Leftrightarrow x(x + 2) = 0 \Leftrightarrow x = 0$ or $x = -2$. If $x = 0$, then $y = 5 + 2(0) = 5$, and if $x = -2$, then $y = 5 + 2(-2) = 1$. Thus, the vertices are $(0, 5)$ and $(-2, 1)$.

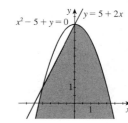

Focus on Modeling: Linear Programming

1.

Vertex	$M = 200 - x - y$
$(0, 2)$	$200 - (0) - (2) = 198$
$(0, 5)$	$200 - (0) - (5) = 195$
$(4, 0)$	$200 - (4) - (0) = 196$

Thus, the maximum value is 198 and the minimum value is 195.

3. $\begin{cases} x \geq 0, y \geq 0 \\ 2x + y \leq 10 \\ 2x + 4y \leq 28 \end{cases}$ The objective function is $P = 140 - x + 3y$. From the graph,

the vertices are $(0, 0)$, $(5, 0)$, $(2, 6)$, and $(0, 7)$.

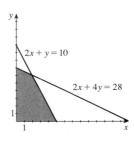

Vertex	$P = 140 - x + 3y$
$(0, 0)$	$140 - (0) + 3 (0) = 140$
$(5, 0)$	$140 - (5) + 3 (0) = 135$
$(2, 6)$	$140 - (2) + 3 (6) = 156$
$(0, 7)$	$140 - (0) + 3 (7) = 161$

Thus the maximum value is 161, and the minimum value is 135.

5. Let t be the number of tables made daily and c be the number of chairs made daily. Then the data given can be summarized by the following table:

	Tables t	Chairs c	Available time
Carpentry	2 h	3 h	108 h
Finishing	1 h	$\frac{1}{2}$ h	20 h
Profit	$35	$20	

Thus we wish to maximize the total profit $P = 35t + 20c$ subject to the constraints $\begin{cases} 2t + 3c \leq 108 \\ t + \frac{1}{2}c \leq 20 \\ t \geq 0, c \geq 0 \end{cases}$

From the graph, the vertices occur at $(0, 0)$, $(20, 0)$, $(0, 36)$, and $(3, 34)$.

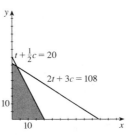

Vertex	$P = 35t + 20c$
$(0, 0)$	$35 (0) + 20 (0) = 0$
$(20, 0)$	$35 (20) + 20 (0) = 700$
$(0, 36)$	$35 (0) + 20 (36) = 720$
$(3, 34)$	$35 (3) + 20 (34) = 785$

Hence, 3 tables and 34 chairs should be produced daily for a maximum profit of $785.

7. Let x be the number of crates of oranges and y the number of crates of grapefruit. Then the data given can be summarized by the following table:

	Oranges	Grapefruit	Available
Volume	4 ft^3	6 ft^3	300 ft^3
Weight	80 lb	100 lb	5600 lb
Profit	$2.50	$4.00	

In addition, $x \geq y$. Thus we wish to maximize the total profit $P = 2.5x + 4y$ subject to the constraints

$$\begin{cases} x \geq 0,\ y \geq 0,\ x \geq y \\ 4x + 6y \leq 300 \\ 80x + 100y \leq 5600 \end{cases}$$

From the graph, the vertices occur at $(0, 0)$, $(30, 30)$, $(45, 20)$, and $(70, 0)$.

Vertex	$P = 2.5x + 4y$
$(0, 0)$	$2.5\,(0) + 4\,(0) = \quad 0$
$(30, 30)$	$2.5\,(30) + 4\,(30) = 195$
$(45, 20)$	$2.5\,(45) + 4\,(20) = 192.5$
$(70, 0)$	$2.5\,(70) + 4\,(0) = 175$

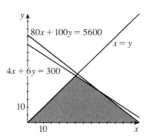

Thus, she should carry 30 crates of oranges and 30 crates of grapefruit for a maximum profit of $195.

9. Let x be the number of stereo sets shipped from Long Beach to Santa Monica and y the number of stereo sets shipped from Long Beach to El Toro. Thus, $15 - x$ sets must be shipped to Santa Monica from Pasadena and $19 - y$ sets to El Toro from Pasadena. Thus, $x \geq 0$, $y \geq 0$, $15 - x \geq 0$, $19 - y \geq 0$, $x + y \leq 24$, and $(15 - x) + (19 - y) \leq 18$. Simplifying, we get

the constraints $\begin{cases} x \geq 0,\ y \geq 0 \\ x \leq 15,\ y \leq 19 \\ x + y \leq 24 \\ x + y \geq 16 \end{cases}$

The objective function is the cost $C = 5x + 6y + 4\,(15 - x) + 5.5\,(19 - y) = x + 0.5y + 164.5$, which we wish to minimize. From the graph, the vertices occur at $(0, 16)$, $(0, 19)$, $(5, 19)$, $(15, 9)$, and $(15, 1)$.

Vertex	$C = x + 0.5y + 164.5$
$(0, 16)$	$(0) + 0.5\,(16) + 164.5 = 172.5$
$(0, 19)$	$(0) + 0.5\,(19) + 164.5 = 174$
$(5, 19)$	$(5) + 0.5\,(19) + 164.5 = 179$
$(15, 9)$	$(15) + 0.5\,(9) + 164.5 = 184$
$(15, 1)$	$(15) + 0.5\,(1) + 164.5 = 180$

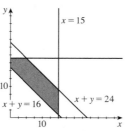

The minimum cost is $172.50 and occurs when $x = 0$ and $y = 16$. Hence, no stereo should be shipped from Long Beach to Santa Monica, 16 from Long Beach to El Toro, 15 from Pasadena to Santa Monica, and 3 from Pasadena to El Toro.

11. Let x be the number of bags of standard mixtures and y be the number of bags of deluxe mixtures. Then the data can be summarized by the following table:

	Standard	Deluxe	Available
Cashews	100 g	150 g	15 kg
Peanuts	200 g	50 g	20 kg
Selling price	$1.95	$2.20	

Thus the total revenue, which we want to maximize, is given by $R = 1.95x + 2.25y$. We have the constraints

$$\begin{cases} x \geq 0, y \geq 0, x \geq y \\ 0.1x + 0.15y \leq 15 \\ 0.2x + 0.05y \leq 20 \end{cases} \Leftrightarrow \begin{cases} x \geq 0, y \geq 0, x \geq y \\ 10x + 15y \leq 1500 \\ 20x + 5y \leq 2000 \end{cases}$$

From the graph, the vertices occur at $(0,0)$, $(60, 60)$, $(90, 40)$, and $(100, 0)$.

Vertex	$R = 1.95x + 2.25y$
$(0, 0)$	$1.96\,(0) + 2.25\,(0) = 0$
$(60, 60)$	$1.95\,(60) + 2.25\,(60) = 252$
$(90, 40)$	$1.95\,(90) + 2.25\,(40) = 265.5$
$(100, 0)$	$1.95\,(100) + 2.25\,(0) = 195$

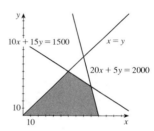

Hence, he should pack 90 bags of standard and 40 bags of deluxe mixture for a maximum revenue of $265.50.

13. Let x be the amount in municipal bonds and y the amount in bank certificates, both in dollars. Then $12000 - x - y$ is the amount in high-risk bonds. So our constraints can be stated as

$$\begin{cases} x \geq 0, y \geq 0, x \geq 3y \\ 12{,}000 - x - y \geq 0 \\ 12{,}000 - x - y \leq 2000 \end{cases} \Leftrightarrow \begin{cases} x \geq 0, y \geq 0, x \geq 3y \\ x + y \leq 12{,}000 \\ x + y \geq 10{,}000 \end{cases}$$

From the graph, the vertices occur at $(7500, 2500)$, $(10000, 0)$, $(12000, 0)$, and $(9000, 3000)$. The objective function is $P = 0.07x + 0.08y + 0.12\,(12000 - x - y) = 1440 - 0.05x - 0.04y$, which we wish to maximize.

Vertex	$P = 1440 - 0.05x - 0.04y$
$(7500, 2500)$	$1440 - 0.05\,(7500) - 0.04\,(2500) = 965$
$(10000, 0)$	$1440 - 0.05\,(10{,}000) - 0.04\,(0) = 940$
$(12000, 0)$	$1440 - 0.05\,(12{,}000) - 0.04\,(0) = 840$
$(9000, 3000)$	$1440 - 0.05\,(9000) - 0.04\,(3000) = 870$

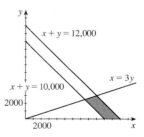

Hence, she should invest $7500 in municipal bonds, $2500 in bank certificates, and the remaining $2000 in high-risk bonds for a maximum yield of $965.

15. Let g be the number of games published and e be the number of educational programs published. Then the number of utility programs published is $36 - g - e$. Hence we wish to maximize profit, $P = 5000g + 8000e + 6000(36 - g - e) = 216{,}000 - 1000g + 2000e$, subject to the constraints

$$\begin{cases} g \geq 4, e \geq 0 \\ 36 - g - e \geq 0 \\ 36 - g - e \leq 2e \end{cases} \Leftrightarrow \begin{cases} g \geq 4, e \geq 0 \\ g + e \leq 36 \\ g + 3e \geq 36. \end{cases}$$

From the graph, the vertices are at $\left(4, \frac{32}{3}\right)$, $(4, 32)$, and $(36, 0)$. The objective function is $P = 216{,}000 - 1000g + 2000e$.

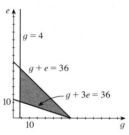

Vertex	$P = 216{,}000 - 1000g + 2000e$
$\left(4, \frac{32}{3}\right)$	$216{,}000 - 1000\,(4) + 2000\left(\frac{32}{3}\right) = 233{,}333.33$
$(4, 32)$	$216{,}000 - 1000\,(4) + 2000\,(32) = 276{,}000$
$(36, 0)$	$216{,}000 - 1000\,(36) + 2000\,(0) = 180{,}000$

So, they should publish 4 games, 32 educational programs, and no utility program for a maximum profit of $276,000 annually.

11 Analytic Geometry

11.1 Parabolas

1. $y^2 = 2x$ is Graph III, which opens to the right and is not as wide as the graph for Exercise 5.

3. $x^2 = -6y$ is Graph II, which opens downward and is narrower than the graph for Exercise 6.

5. $y^2 - 8x = 0$ is Graph VI, which opens to the right and is wider than the graph for Exercise 1.

7. $y^2 = 4x$. Then $4p = 4 \Leftrightarrow p = 1$. The focus is $(1, 0)$, the directrix is $x = -1$, and the focal diameter is 4.

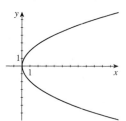

9. $x^2 = 9y$. Then $4p = 9 \Leftrightarrow p = \frac{9}{4}$. The focus is $\left(0, \frac{9}{4}\right)$, the directrix is $y = -\frac{9}{4}$, and the focal diameter is 9.

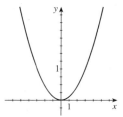

11. $y = 5x^2 \Leftrightarrow x^2 = \frac{1}{5}y$. Then $4p = \frac{1}{5} \Leftrightarrow p = \frac{1}{20}$. The focus is $\left(0, \frac{1}{20}\right)$, the directrix is $y = -\frac{1}{20}$, and the focal diameter is $\frac{1}{5}$.

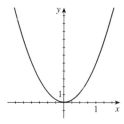

13. $x = -8y^2 \Leftrightarrow y^2 = -\frac{1}{8}x$. Then $4p = -\frac{1}{8} \Leftrightarrow p = -\frac{1}{32}$. The focus is $\left(-\frac{1}{32}, 0\right)$, the directrix is $x = \frac{1}{32}$, and the focal diameter is $\frac{1}{8}$.

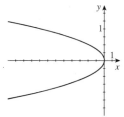

15. $x^2 + 6y = 0 \Leftrightarrow x^2 = -6y$. Then $4p = -6 \Leftrightarrow p = -\frac{3}{2}$. The focus is $\left(0, -\frac{3}{2}\right)$, the directrix is $y = \frac{3}{2}$, and the focal diameter is 6.

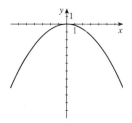

17. $5x + 3y^2 = 0 \Leftrightarrow y^2 = -\frac{5}{3}x$. Then $4p = -\frac{5}{3} \Leftrightarrow p = -\frac{5}{12}$. The focus is $\left(-\frac{5}{12}, 0\right)$, the directrix is $x = \frac{5}{12}$, and the focal diameter is $\frac{5}{3}$.

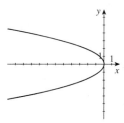

19. $x^2 = 16y$

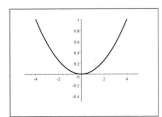

21. $y^2 = -\frac{1}{3}x$

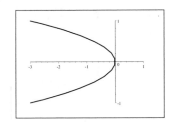

23. $4x + y^2 = 0$

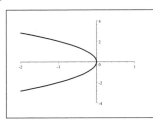

25. Since the focus is $(0, 2)$, $p = 2$ $\Leftrightarrow$ $4p = 8$. Hence, an equation of the parabola is $x^2 = 8y$.

27. Since the focus is $(-8, 0)$, $p = -8$ $\Leftrightarrow$ $4p = -32$. Hence, an equation of the parabola is $y^2 = -32x$.

29. Since the directrix is $x = 2$, $p = -2$ $\Leftrightarrow$ $4p = -8$. Hence, an equation of the parabola is $y^2 = -8x$.

31. Since the directrix is $y = -10$, $p = 10$ $\Leftrightarrow$ $4p = 40$. Hence, an equation of the parabola is $x^2 = 40y$.

33. The focus is on the positive x-axis, so the parabola opens horizontally with $2p = 2$ $\Leftrightarrow$ $4p = 4$. So an equation of the parabola is $y^2 = 4x$.

35. Since the parabola opens upward with focus 5 units from the vertex, the focus is $(5, 0)$. So $p = 5$ $\Leftrightarrow$ $4p = 20$. Thus an equation of the parabola is $x^2 = 20y$.

37. $p = 2$ $\Leftrightarrow$ $4p = 8$. Since the parabola opens upward, its equation is $x^2 = 8y$.

39. $p = 4$ $\Leftrightarrow$ $4p = 16$. Since the parabola opens to the left, its equation is $y^2 = -16x$.

41. The focal diameter is $4p = \frac{3}{2} + \frac{3}{2} = 3$. Since the parabola opens to the left, its equation is $y^2 = -3x$.

43. The equation of the parabola has the form $y^2 = 4px$. Since the parabola passes through the point $(4, -2)$, $(-2)^2 = 4p(4)$ $\Leftrightarrow$ $4p = 1$, and so an equation is $y^2 = x$.

45. The area of the shaded region is width $\times$ height $= 4p \cdot p = 8$, and so $p^2 = 2$ $\Leftrightarrow$ $p = -\sqrt{2}$ (because the parabola opens downward). Therefore, an equation is $x^2 = 4py = -4\sqrt{2}y$ $\Leftrightarrow$ $x^2 = -4\sqrt{2}y$.

47. (a) A parabola with directrix $y = -p$ has equation $x^2 = 4py$. If the directrix is $y = \frac{1}{2}$, then $p = -\frac{1}{2}$, so an equation is $x^2 = 4\left(-\frac{1}{2}\right)y$ $\Leftrightarrow$ $x^2 = -2y$. If the directrix is $y = 1$, then $p = -1$, so an equation is $x^2 = 4(-1)y$ $\Leftrightarrow$ $x^2 = -4y$. If the directrix is $y = 4$, then $p = -4$, so an equation is $x^2 = 4(-4)y$ $\Leftrightarrow$ $x^2 = -16y$. If the directrix is $y = 8$, then $p = -8$, so an equation is $x^2 = 4(-8)y$ $\Leftrightarrow$ $x^2 = -32y$.

(b)

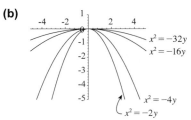

As the directrix moves further from the vertex, the parabolas get flatter.

49. (a) Since the focal diameter is 12 cm, $4p = 12$. Hence, the parabola has equation $y^2 = 12x$.

(b) At a point 20 cm horizontally from the vertex, the parabola passes through the point $(20, y)$, and hence from part (a), $y^2 = 12(20)$ $\Leftrightarrow$ $y^2 = 240$ $\Leftrightarrow$ $y = \pm 4\sqrt{15}$. Thus, $|CD| = 8\sqrt{15} \approx 31$ cm.

51. With the vertex at the origin, the top of one tower will be at the point $(300, 150)$. Inserting this point into the equation $x^2 = 4py$ gives $(300)^2 = 4p(150)$ $\Leftrightarrow$ $90000 = 600p$ $\Leftrightarrow$ $p = 150$. So an equation of the parabolic part of the cables is $x^2 = 4(150)y$ $\Leftrightarrow$ $x^2 = 600y$.

53. Many answers are possible: satellite dish TV antennas, sound surveillance equipment, solar collectors for hot water heating or electricity generation, bridge pillars, etc.

11.2 **Ellipses**

1. $\dfrac{x^2}{16} + \dfrac{y^2}{4} = 1$ is Graph II. The major axis is horizontal and the vertices are $(\pm 4, 0)$.

3. $4x^2 + y^2 = 4$ is Graph I. The major axis is vertical and the vertices are $(0, \pm 2)$.

5. $\dfrac{x^2}{25} + \dfrac{y^2}{9} = 1$. This ellipse has $a = 5$, $b = 3$, and so $c^2 = a^2 - b^2 = 16$ $\Leftrightarrow$

$c = 4$. The vertices are $(\pm 5, 0)$, the foci are $(\pm 4, 0)$, the eccentricity is

$e = \dfrac{c}{a} = \dfrac{4}{5} = 0.8$, the length of the major axis is $2a = 10$, and the length of the

minor axis is $2b = 6$.

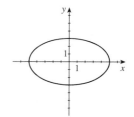

7. $9x^2 + 4y^2 = 36$ $\Leftrightarrow$ $\dfrac{x^2}{4} + \dfrac{y^2}{9} = 1$. This ellipse has $a = 3$, $b = 2$, and so

$c^2 = 9 - 4 = 5$ $\Leftrightarrow$ $c = \sqrt{5}$. The vertices are $(0, \pm 3)$, the foci are $\left(0, \pm\sqrt{5}\right)$,

the eccentricity is $e = \dfrac{c}{a} = \dfrac{\sqrt{5}}{3}$, the length of the major axis is $2a = 6$, and the

length of the minor axis is $2b = 4$.

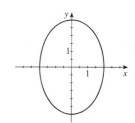

9. $x^2 + 4y^2 = 16$ $\Leftrightarrow$ $\dfrac{x^2}{16} + \dfrac{y^2}{4} = 1$. This ellipse has $a = 4$, $b = 2$, and so

$c^2 = 16 - 4 = 12$ $\Leftrightarrow$ $c = 2\sqrt{3}$. The vertices are $(\pm 4, 0)$, the foci are

$\left(\pm 2\sqrt{3}, 0\right)$, the eccentricity is $e = \dfrac{c}{a} = \dfrac{2\sqrt{3}}{4} = \dfrac{\sqrt{3}}{2}$, the length of the major axis

is $2a = 8$, and the length of the minor axis is $2b = 4$.

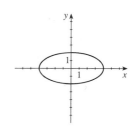

11. $2x^2 + y^2 = 3$ $\Leftrightarrow$ $\dfrac{x^2}{\frac{3}{2}} + \dfrac{y^2}{3} = 1$. This ellipse has $a = \sqrt{3}$, $b = \sqrt{\dfrac{3}{2}}$, and so

$c^2 = 3 - \dfrac{3}{2} = \dfrac{3}{2}$ $\Leftrightarrow$ $c = \sqrt{\dfrac{3}{2}} = \dfrac{\sqrt{6}}{2}$. The vertices are $(0, \pm\sqrt{3})$, the foci are

$\left(0, \pm\dfrac{\sqrt{6}}{2}\right)$, the eccentricity is $e = \dfrac{c}{a} = \dfrac{\frac{\sqrt{6}}{2}}{\sqrt{3}} = \dfrac{\sqrt{2}}{2}$, the length of the major axis is

$2a = 2\sqrt{3}$, and the length of the minor axis is $2b = 2 \cdot \dfrac{\sqrt{6}}{2} = \sqrt{6}$.

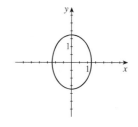

13. $x^2 + 4y^2 = 1 \Leftrightarrow \dfrac{x^2}{1} + \dfrac{y^2}{\frac{1}{4}} = 1$. This ellipse has $a = 1$, $b = \frac{1}{2}$, and so

$c^2 = 1 - \frac{1}{4} = \frac{3}{4} \Leftrightarrow c = \frac{\sqrt{3}}{2}$. The vertices are $(\pm 1, 0)$, the foci are $\left(\pm\frac{\sqrt{3}}{2}, 0\right)$,

the eccentricity is $e = \dfrac{c}{a} = \dfrac{\sqrt{3}/2}{1} = \dfrac{\sqrt{3}}{2}$, the length of the major axis is $2a = 2$,

and the length of the minor axis is $2b = 1$.

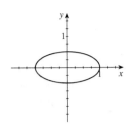

15. $\frac{1}{2}x^2 + \frac{1}{8}y^2 = \frac{1}{4} \Leftrightarrow 2x^2 + \frac{1}{2}y^2 = 1 \Leftrightarrow \dfrac{x^2}{\frac{1}{2}} + \dfrac{y^2}{2} = 1$. This ellipse has

$a = \sqrt{2}$, $b = \frac{1}{\sqrt{2}}$, and so $c^2 = 2 - \frac{1}{2} = \frac{3}{2} \Leftrightarrow c = \sqrt{\frac{3}{2}} = \frac{\sqrt{6}}{2}$. The vertices are

$(0, \pm\sqrt{2})$, the foci are $\left(0, \pm\frac{\sqrt{6}}{2}\right)$, the eccentricity is $e = \dfrac{c}{a} = \dfrac{\frac{\sqrt{6}}{2}}{\sqrt{2}} = \dfrac{\sqrt{3}}{2}$, the

length of the major axis is $2a = 2\sqrt{2}$, and length of the minor axis is $2b = \sqrt{2}$.

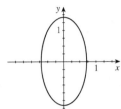

17. $y^2 = 1 - 2x^2 \Leftrightarrow 2x^2 + y^2 = 1 \Leftrightarrow \dfrac{x^2}{\frac{1}{2}} + \dfrac{y^2}{1} = 1$. This ellipse has $a = 1$,

$b = \frac{\sqrt{2}}{2}$, and so $c^2 = 1 - \frac{1}{2} = \frac{1}{2} \Leftrightarrow c = \frac{\sqrt{2}}{2}$. The vertices are $(0, \pm 1)$, the foci

are $\left(0, \pm\frac{\sqrt{2}}{2}\right)$, the eccentricity is $e = \dfrac{c}{a} = \dfrac{1/\sqrt{2}}{1} = \dfrac{\sqrt{2}}{2}$, the length of the major

axis is $2a = 2$, and the length of the minor axis is $2b = \sqrt{2}$.

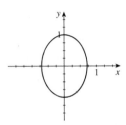

19. This ellipse has a horizontal major axis with $a = 5$ and $b = 4$, so an equation is $\dfrac{x^2}{(5)^2} + \dfrac{y^2}{(4)^2} = 1 \Leftrightarrow \dfrac{x^2}{25} + \dfrac{y^2}{16} = 1$.

21. This ellipse has a vertical major axis with $c = 2$ and $b = 2$. So $a^2 = c^2 + b^2 = 2^2 + 2^2 = 8 \Leftrightarrow a = 2\sqrt{2}$. So an

equation is $\dfrac{x^2}{(2)^2} + \dfrac{y^2}{(2\sqrt{2})^2} = 1 \Leftrightarrow \dfrac{x^2}{4} + \dfrac{y^2}{8} = 1$.

23. This ellipse has a horizontal major axis with $a = 16$, so an equation of the ellipse is of the form $\dfrac{x^2}{16^2} + \dfrac{y^2}{b^2} = 1$. Substituting

the point $(8, 6)$ into the equation, we get $\frac{64}{256} + \frac{36}{b^2} = 1 \Leftrightarrow \frac{36}{b^2} = 1 - \frac{1}{4} \Leftrightarrow \frac{36}{b^2} = \frac{3}{4} \Leftrightarrow b^2 = \frac{4\,(36)}{3} = 48$.

Thus, an equation of the ellipse is $\dfrac{x^2}{256} + \dfrac{y^2}{48} = 1$.

25. $\dfrac{x^2}{25} + \dfrac{y^2}{20} = 1 \Leftrightarrow \dfrac{y^2}{20} = 1 - \dfrac{x^2}{25} \Leftrightarrow$

$y^2 = 20 - \dfrac{4x^2}{5} \Rightarrow y = \pm\sqrt{20 - \dfrac{4x^2}{5}}$.

27. $6x^2 + y^2 = 36 \Leftrightarrow y^2 = 36 - 6x^2 \Rightarrow$
$y = \pm\sqrt{36 - 6x^2}$.

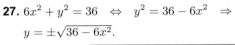

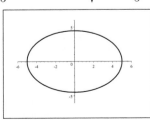

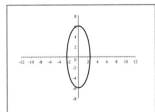

29. The foci are $(\pm 4, 0)$, and the vertices are $(\pm 5, 0)$. Thus, $c = 4$ and $a = 5$, and so $b^2 = 25 - 16 = 9$. Therefore, an equation

of the ellipse is $\dfrac{x^2}{25} + \dfrac{y^2}{9} = 1$.

31. The length of the major axis is $2a = 4 \Leftrightarrow a = 2$, the length of the minor axis is $2b = 2 \Leftrightarrow b = 1$, and the foci are on the y-axis. Therefore, an equation of the ellipse is $x^2 + \dfrac{y^2}{4} = 1$.

33. The foci are $(0, \pm 2)$, and the length of the minor axis is $2b = 6 \Leftrightarrow b = 3$. Thus, $a^2 = 4 + 9 = 13$. Since the foci are on the y-axis, an equation is $\dfrac{x^2}{9} + \dfrac{y^2}{13} = 1$.

35. The endpoints of the major axis are $(\pm 10, 0) \Leftrightarrow a = 10$, and the distance between the foci is $2c = 6 \Leftrightarrow c = 3$. Therefore, $b^2 = 100 - 9 = 91$, and so an equation of the ellipse is $\dfrac{x^2}{100} + \dfrac{y^2}{91} = 1$.

37. The length of the major axis is 10, so $2a = 10 \Leftrightarrow a = 5$, and the foci are on the x-axis, so the form of the equation is $\dfrac{x^2}{25} + \dfrac{y^2}{b^2} = 1$. Since the ellipse passes through $(\sqrt{5}, 2)$, we know that $\dfrac{(\sqrt{5})^2}{25} + \dfrac{(2)^2}{b^2} = 1 \Leftrightarrow \dfrac{5}{25} + \dfrac{4}{b^2} = 1 \Leftrightarrow \dfrac{4}{b^2} = \dfrac{4}{5} \Leftrightarrow b^2 = 5$, and so an equation is $\dfrac{x^2}{25} + \dfrac{y^2}{5} = 1$.

39. Since the foci are $(\pm 1.5, 0)$, we have $c = \dfrac{3}{2}$. Since the eccentricity is $0.8 = \dfrac{c}{a}$, we have $a = \dfrac{\frac{3}{2}}{\frac{4}{5}} = \dfrac{15}{8}$, and so $b^2 = \dfrac{225}{64} - \dfrac{9}{4} = \dfrac{225 - 16 \cdot 9}{64} = \dfrac{81}{64}$. Therefore, an equation of the ellipse is $\dfrac{x^2}{(15/8)^2} + \dfrac{y^2}{81/64} = 1 \Leftrightarrow \dfrac{64x^2}{225} + \dfrac{64y^2}{81} = 1$.

41. $\begin{cases} 4x^2 + y^2 = 4 \\ 4x^2 + 9y^2 = 36 \end{cases}$ Subtracting the first equation from the second gives

$8y^2 = 32 \Leftrightarrow y^2 = 4 \Leftrightarrow y = \pm 2$. Substituting $y = \pm 2$ in the first equation gives $4x^2 + (\pm 2)^2 = 4 \Leftrightarrow x = 0$, and so the points of intersection are $(0, \pm 2)$.

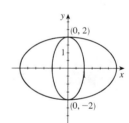

43. $\begin{cases} 100x^2 + 25y^2 = 100 \\ x^2 + \dfrac{y^2}{9} = 1 \end{cases}$ Dividing the first equation by 100 gives $x^2 + \dfrac{y^2}{4} = 1$.

Subtracting this equation from the second equation gives $\dfrac{y^2}{9} - \dfrac{y^2}{4} = 0 \Leftrightarrow \left(\dfrac{1}{9} - \dfrac{1}{4}\right) y^2 = 0 \Leftrightarrow y = 0$. Substituting $y = 0$ in the second equation gives $x^2 + (0)^2 = 1 \Leftrightarrow x = \pm 1$, and so the points of intersection are $(\pm 1, 0)$.

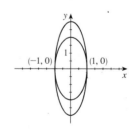

45. (a) $x^2 + ky^2 = 100 \Leftrightarrow ky^2 = 100 - x^2 \Leftrightarrow y = \pm \dfrac{1}{k}\sqrt{100 - x^2}$.

For the top half, we graph $y = \dfrac{1}{k}\sqrt{100 - x^2}$ for $k = 4, 10, 25$, and 50.

(b) This family of ellipses have common major axes and vertices, and the eccentricity increases as k increases.

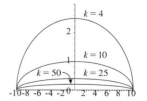

47. Using the perihelion, $a - c = 147,000,000$, while using the aphelion, $a + c = 153,000,000$. Adding, we have $2a = 300,000,000 \Leftrightarrow a = 150,000,000$. So $b^2 = a^2 - c^2 = (150 \times 10^6)^2 - (3 \times 10^6)^2 = 22{,}491 \times 10^{12} = 2.2491 \times 10^{16}$. Thus, an equation of the orbit is $\dfrac{x^2}{2.2500 \times 10^{16}} + \dfrac{y^2}{2.2491 \times 10^{16}} = 1$.

49. Using the perilune, $a - c = 1075 + 68 = 1143$, and using the apolune, $a + c = 1075 + 195 = 1270$. Adding, we get

$2a = 2413 \quad \Leftrightarrow \quad a = 1206.5$. So $c = 1270 - 1206.5 \quad \Leftrightarrow \quad c = 63.5$. Therefore, $b^2 = (1206.5)^2 - (63.5)^2 = 1{,}451{,}610$.

Since $a^2 \approx 1{,}455{,}642$, an equation of Apollo 11's orbit is $\dfrac{x^2}{1{,}455{,}642} + \dfrac{y^2}{1{,}451{,}610} = 1$.

51. From the diagram, $a = 40$ and $b = 20$, and so an equation of the ellipse whose top half is the window is $\dfrac{x^2}{1600} + \dfrac{y^2}{400} = 1$.

Since the ellipse passes through the point $(25, h)$, by substituting, we have $\dfrac{25^2}{1600} + \dfrac{h^2}{400} = 1 \quad \Leftrightarrow \quad 625 + 4y^2 = 1600$

$\Leftrightarrow \quad y = \dfrac{\sqrt{975}}{2} = \dfrac{5\sqrt{39}}{2} \approx 15.61$ in. Therefore, the window is approximately 15.6 inches high at the specified point.

53. We start with the flashlight perpendicular to the wall; this shape is a circle. As the angle of elevation increases, the shape of the light changes to an ellipse. When the flashlight is angled so that the outer edge of the light cone is parallel to the wall, the shape of the light is a parabola. Finally, as the angle of elevation increases further, the shape of the light is hyperbolic.

55. The shape drawn on the paper is almost, but not quite, an ellipse. For example, when the bottle has radius 1 unit and the compass legs are set 1 unit apart, then it can be shown that an equation of the resulting curve is $1 + y^2 = 2 \cos x$. The graph of this curve differs very slightly from the ellipse with the same major and minor axis. This example shows that in mathematics, things are not always as they appear to be.

11.3   Hyperbolas

1. $\dfrac{x^2}{4} - y^2 = 1$ is Graph III, which opens horizontally and has vertices at $(\pm 2, 0)$.

3. $16y^2 - x^2 = 144$ is Graph II, which pens vertically and has vertices at $(0, \pm 3)$.

5. The hyperbola $\dfrac{x^2}{4} - \dfrac{y^2}{16} = 1$ has $a = 2$, $b = 4$, and $c^2 = 16 + 4 \quad \Rightarrow \quad c = 2\sqrt{5}$. The vertices are $(\pm 2, 0)$, the foci are $\left(\pm 2\sqrt{5}, 0\right)$, and the asymptotes are $y = \pm \frac{4}{2}x$ $\Leftrightarrow \quad y = \pm 2x$.

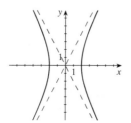

7. The hyperbola $\dfrac{y^2}{1} - \dfrac{x^2}{25} = 1$ has $a = 1$, $b = 5$, and $c^2 = 1 + 25 = 26 \quad \Rightarrow \quad c = \sqrt{26}$. The vertices are $(0, \pm 1)$, the foci are $\left(0, \pm\sqrt{26}\right)$, and the asymptotes are $y = \pm \frac{1}{5}x$.

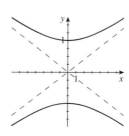

9. The hyperbola $x^2 - y^2 = 1$ has $a = 1$, $b = 1$, and
$c^2 = 1 + 1 = 2 \implies c = \sqrt{2}$. The vertices are $(\pm 1, 0)$,
the foci are $\left(\pm\sqrt{2}, 0\right)$, and the asymptotes are $y = \pm x$.

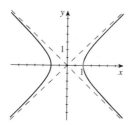

11. The hyperbola $25y^2 - 9x^2 = 225 \iff \dfrac{y^2}{9} - \dfrac{x^2}{25} = 1$
has $a = 3$, $b = 5$, and $c^2 = 25 + 9 = 34 \implies$
$c = \sqrt{34}$. The vertices are $(0, \pm 3)$, the foci are
$\left(0, \pm\sqrt{34}\right)$, and the asymptotes are $y = \pm\frac{3}{5}x$.

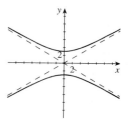

13. The hyperbola $x^2 - 4y^2 - 8 = 0 \iff \dfrac{x^2}{8} - \dfrac{y^2}{2} = 1$
has $a = \sqrt{8}$, $b = \sqrt{2}$, and $c^2 = 8 + 2 = 10 \implies$
$c = \sqrt{10}$. The vertices are $\left(\pm 2\sqrt{2}, 0\right)$, the foci are
$\left(\pm\sqrt{10}, 0\right)$, and the asymptotes are $y = \pm\frac{\sqrt{2}}{\sqrt{8}}x = \pm\frac{1}{2}x$.

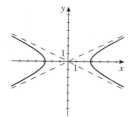

15. The hyperbola $4y^2 - x^2 = 1 \iff \dfrac{y^2}{\frac{1}{4}} - x^2 = 1$ has
$a = \frac{1}{2}$, $b = 1$, and $c^2 = \frac{1}{4} + 1 = \frac{5}{4} \implies c = \frac{\sqrt{5}}{2}$. The
vertices are $\left(0, \pm\frac{1}{2}\right)$, the foci are $\left(0, \pm\frac{\sqrt{5}}{2}\right)$, and the
asymptotes are $y = \pm\frac{1/2}{1}x = \pm\frac{1}{2}x$.

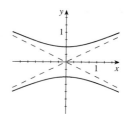

17. From the graph, the foci are $(\pm 4, 0)$, and the vertices are $(\pm 2, 0)$, so $c = 4$ and $a = 2$. Thus, $b^2 = 16 - 4 = 12$, and since
the vertices are on the x-axis, an equation of the hyperbola is $\dfrac{x^2}{4} - \dfrac{y^2}{12} = 1$.

19. From the graph, the vertices are $(0, \pm 4)$, the foci are on the y-axis, and the hyperbola passes through the point $(3, -5)$. So
the equation is of the form $\dfrac{y^2}{16} - \dfrac{x^2}{b^2} = 1$. Substituting the point $(3, -5)$, we have $\dfrac{(-5)^2}{16} - \dfrac{(3)^2}{b^2} = 1 \iff \dfrac{25}{16} - 1 = \dfrac{9}{b^2}$
$\iff \dfrac{9}{16} = \dfrac{9}{b^2} \iff b^2 = 16$. Thus, an equation of the hyperbola is $\dfrac{y^2}{16} - \dfrac{x^2}{16} = 1$.

21. The vertices are $(\pm 3, 0)$, so $a = 3$. Since the asymptotes are $y = \pm\frac{1}{2}x = \pm\dfrac{b}{a}x$, we have $\dfrac{b}{3} = \dfrac{1}{2} \iff b = \dfrac{3}{2}$. Since the
vertices are on the x-axis, an equation is $\dfrac{x^2}{3^2} - \dfrac{y^2}{(3/2)^2} = 1 \iff \dfrac{x^2}{9} - \dfrac{4y^2}{9} = 1$.

23. $x^2 - 2y^2 = 8$ $\Leftrightarrow$ $2y^2 = x^2 - 8 \Leftrightarrow y^2 = \frac{1}{2}x^2 - 4$ **25.** $\dfrac{y^2}{2} - \dfrac{x^2}{6} = 1$ $\Leftrightarrow$ $\dfrac{y^2}{2} = \dfrac{x^2}{6} + 1 \Leftrightarrow y^2 = \dfrac{x^2}{3} + 2$ $\Rightarrow$

$\Rightarrow y = \pm\sqrt{\frac{1}{2}x^2 - 4}$

$$y = \pm\sqrt{\frac{x^2}{3} + 2}$$

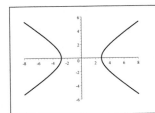   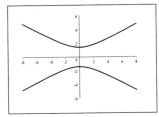

27. The foci are $(\pm 5, 0)$ and the vertices are $(\pm 3, 0)$, so $c = 5$ and $a = 3$. Then $b^2 = 25 - 9 = 16$, and since the vertices are on the x-axis, an equation of the hyperbola is $\dfrac{x^2}{9} - \dfrac{y^2}{16} = 1$.

29. The foci are $(0, \pm 2)$ and the vertices are $(0, \pm 1)$, so $c = 2$ and $a = 1$. Then $b^2 = 4 - 1 = 3$, and since the vertices are on the y-axis, an equation is $y^2 - \dfrac{x^2}{3} = 1$.

31. The vertices are $(\pm 1, 0)$ and the asymptotes are $y = \pm 5x$, so $a = 1$. The asymptotes are $y = \pm\dfrac{b}{a}x$, so $\dfrac{b}{1} = 5$ $\Leftrightarrow$ $b = 5$. Therefore, an equation of the hyperbola is $x^2 - \dfrac{y^2}{25} = 1$.

33. The foci are $(0, \pm 8)$, and the asymptotes are $y = \pm\frac{1}{2}x$, so $c = 8$. The asymptotes are $y = \pm\dfrac{a}{b}x$, so $\dfrac{a}{b} = \dfrac{1}{2}$ and $b = 2a$. Since $a^2 + b^2 = c^2 = 64$, we have $a^2 + 4a^2 = 64$ $\Leftrightarrow$ $a^2 = \frac{64}{5}$ and $b^2 = 4a^2 = \frac{256}{5}$. Thus, an equation of the hyperbola is $\dfrac{y^2}{64/5} - \dfrac{x^2}{256/5} = 1$ $\Leftrightarrow$ $\dfrac{5y^2}{64} - \dfrac{5x^2}{256} = 1$.

35. The asymptotes of the hyperbola are $y = \pm x$, so $b = a$. Since the hyperbola passes through the point $(5, 3)$, its foci are on the x-axis, and an equation is of the form, $\dfrac{x^2}{a^2} - \dfrac{y^2}{a^2} = 1$, so it follows that $\dfrac{25}{a^2} - \dfrac{9}{a^2} = 1$ $\Leftrightarrow$ $a^2 = 16 = b^2$. Therefore, an equation of the hyperbola is $\dfrac{x^2}{16} - \dfrac{y^2}{16} = 1$.

37. The foci are $(\pm 5, 0)$, and the length of the transverse axis is 6, so $c = 5$ and $2a = 6$ $\Leftrightarrow$ $a = 3$. Thus, $b^2 = 25 - 9 = 16$, and an equation is $\dfrac{x^2}{9} - \dfrac{y^2}{16} = 1$.

39. (a) The hyperbola $x^2 - y^2 = 5$ $\Leftrightarrow$ $\dfrac{x^2}{5} - \dfrac{y^2}{5} = 1$ has $a = \sqrt{5}$ and $b = \sqrt{5}$. Thus, the asymptotes are $y = \pm x$, and their slopes are $m_1 = 1$ and $m_2 = -1$. Since $m_1 \cdot m_2 = -1$, the asymptotes are perpendicular.

(b) Since the asymptotes are perpendicular, they must have slopes ± 1, so $a = b$. Therefore, $c^2 = 2a^2$ $\Leftrightarrow$ $a^2 = \dfrac{c^2}{2}$, and since the vertices are on the x-axis, an equation is $\dfrac{x^2}{\frac{1}{2}c^2} - \dfrac{y^2}{\frac{1}{2}c^2} = 1$ $\Leftrightarrow$ $x^2 - y^2 = \dfrac{c^2}{2}$.

41. $\sqrt{(x+c)^2 + y^2} - \sqrt{(x-c)^2 + y^2} = \pm 2a$. Let us consider the positive case only. Then

$\sqrt{(x+c)^2 + y^2} = 2a + \sqrt{(x-c)^2 + y^2}$, and squaring both sides gives $x^2 + 2cx + c^2 + y^2 = 4a^2 + 4a\sqrt{(x-c)^2 + y^2} +$

$x^2 - 2cx + c^2 + y^2 \quad \Leftrightarrow \quad 4a\sqrt{(x-c)^2 + y^2} = 4cx - 4a^2$. Dividing by 4 and squaring both sides gives

$a^2\left(x^2 - 2cx + c^2 + y^2\right) = c^2x^2 - 2a^2cx + a^4 \quad \Leftrightarrow \quad a^2x^2 - 2a^2cx + a^2c^2 + a^2y^2 = c^2x^2 - 2a^2cx + a^4$

$\Leftrightarrow \quad a^2x^2 + a^2c^2 + a^2y^2 = c^2x^2 + a^4$. Rearranging the order, we have $c^2x^2 - a^2x^2 - a^2y^2 = a^2c^2 - a^4 \quad \Leftrightarrow$

$\left(c^2 - a^2\right)x^2 - a^2y^2 = a^2\left(c^2 - a^2\right)$. The negative case gives the same result.

43. (a) From the equation, we have $a^2 = k$ and $b^2 = 16 - k$. Thus, $c^2 = a^2 + b^2 = k + 16 - k = 16 \quad \Rightarrow \quad c = \pm 4$. Thus the foci of the family of hyperbolas are $(0, \pm 4)$.

(b) $\dfrac{y^2}{k} - \dfrac{x^2}{16 - k} = 1 \quad \Leftrightarrow \quad y^2 = k\left(1 + \dfrac{x^2}{16 - k}\right) \quad \Rightarrow$

$y = \pm\sqrt{k + \dfrac{kx^2}{16 - k}}$. For the top branch, we graph $y = \sqrt{k + \dfrac{kx^2}{16 - k}}$,

$k = 1, 4, 8, 12$. As k increases, the asymptotes get steeper and the vertices move further apart.

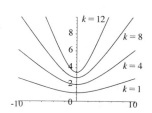

45. Since the asymptotes are perpendicular, $a = b$. Also, since the sun is a focus and the closest distance is 2×10^9, it follows that $c - a = 2 \times 10^9$. Now $c^2 = a^2 + b^2 = 2a^2$, and so $c = \sqrt{2}a$. Thus, $\sqrt{2}a - a = 2 \times 10^9 \quad \Rightarrow \quad a = \dfrac{2 \times 10^9}{\sqrt{2} - 1}$

and $a^2 = b^2 = \dfrac{4 \times 10^{18}}{3 - 2\sqrt{2}} \approx 2.3 \times 10^{19}$. Therefore, an equation of the hyperbola is $\dfrac{x^2}{2.3 \times 10^{19}} - \dfrac{y^2}{2.3 \times 10^{19}} = 1 \quad \Leftrightarrow$

$x^2 - y^2 = 2.3 \times 10^{19}$.

47. Some possible answers are: as cross-sections of nuclear power plant cooling towers, or as reflectors for camouflaging the location of secret installations.

11.4 Shifted Conics

1. The ellipse $\dfrac{(x - 2)^2}{9} + \dfrac{(y - 1)^2}{4} = 1$ is obtained from the ellipse $\dfrac{x^2}{9} + \dfrac{y^2}{4} = 1$

by shifting it 2 units to the right and 1 unit upward. So $a = 3$, $b = 2$, and

$c = \sqrt{9 - 4} = \sqrt{5}$. The center is $(2, 1)$, the foci are $(2 \pm \sqrt{5}, 1)$, the vertices are

$(2 \pm 3, 1) = (-1, 1)$ and $(5, 1)$, the length of the major axis is $2a = 6$, and the

length of the minor axis is $2b = 4$.

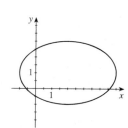

3. The ellipse $\dfrac{x^2}{9} + \dfrac{(y + 5)^2}{25} = 1$ is obtained from the ellipse $\dfrac{x^2}{9} + \dfrac{y^2}{25} = 1$ by

shifting it 5 units downward. So $a = 5$, $b = 3$, and $c = \sqrt{25 - 9} = 4$. The center

is $(0, -5)$, the foci are $(0, -5 \pm 4) = (0, -9)$ and $(0, -1)$, the vertices are

$(0, -5 \pm 5) = (0, -10)$ and $(0, 0)$, the length of the major axis is $2a = 10$, and

the length of the minor axis is $2b = 6$.

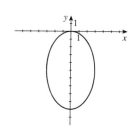

5. The parabola $(x-3)^2 = 8(y+1)$ is obtained from the parabola $x^2 = 8y$ by shifting it 3 units to the right and 1 unit down. So $4p = 8 \Leftrightarrow p = 2$. The vertex is $(3, -1)$, the focus is $(3, -1+2) = (3, 1)$, and the directrix is $y = -1 - 2 = -3$.

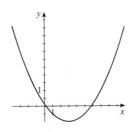

7. The parabola $-4\left(x+\frac{1}{2}\right)^2 = y \Leftrightarrow \left(x+\frac{1}{2}\right)^2 = -\frac{1}{4}y$ is obtained from the parabola $x^2 = -\frac{1}{4}y$ by shifting it $\frac{1}{2}$ unit to the left. So $4p = -\frac{1}{4} \Leftrightarrow p = -\frac{1}{16}$. The vertex is $\left(-\frac{1}{2}, 0\right)$, the focus is $\left(-\frac{1}{2}, 0 - \frac{1}{16}\right) = \left(-\frac{1}{2}, -\frac{1}{16}\right)$, and the directrix is $y = 0 + \frac{1}{16} = \frac{1}{16}$.

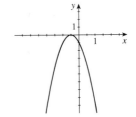

9. The hyperbola $\dfrac{(x+1)^2}{9} - \dfrac{(y-3)^2}{16} = 1$ is obtained from the hyperbola $\dfrac{x^2}{9} - \dfrac{y^2}{16} = 1$ by shifting it 1 unit to the left and 3 units up. So $a = 3$, $b = 4$, and $c = \sqrt{9+16} = 5$. The center is $(-1, 3)$, the foci are $(-1 \pm 5, 3) = (-6, 3)$ and $(4, 3)$, the vertices are $(-1 \pm 3, 3) = (-4, 3)$ and $(2, 3)$, and the asymptotes are $(y-3) = \pm\frac{4}{3}(x+1) \Leftrightarrow y = \pm\frac{4}{3}(x+1) + 3 \Leftrightarrow 3y = 4x + 13$ and $3y = -4x + 5$.

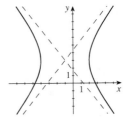

11. The hyperbola $y^2 - \dfrac{(x+1)^2}{4} = 1$ is obtained from the hyperbola $y^2 - \dfrac{x^2}{4} = 1$ by shifting it 1 unit to the left. So $a = 1$, $b = 2$, and $c = \sqrt{1+4} = \sqrt{5}$. The center is $(-1, 0)$, the foci are $\left(-1, \pm\sqrt{5}\right) = \left(-1, -\sqrt{5}\right)$ and $\left(-1, \sqrt{5}\right)$, the vertices are $(-1, \pm 1) = (-1, -1)$ and $(-1, 1)$, and the asymptotes are $y = \pm\frac{1}{2}(x+1) \Leftrightarrow y = \frac{1}{2}(x+1)$ and $y = -\frac{1}{2}(x+1)$.

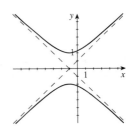

13. This is a parabola that opens down with its vertex at $(0, 4)$, so its equation is of the form $x^2 = a(y-4)$. Since $(1, 0)$ is a point on this parabola, we have $(1)^2 = a(0-4) \Leftrightarrow 1 = -4a \Leftrightarrow a = -\frac{1}{4}$. Thus, an equation is $x^2 = -\frac{1}{4}(y-4)$.

15. This is an ellipse with the major axis parallel to the x-axis, with one vertex at $(0, 0)$, the other vertex at $(10, 0)$, and one focus at $(8, 0)$. The center is at $\left(\frac{0+10}{2}, 0\right) = (5, 0)$, $a = 5$, and $c = 3$ (the distance from one focus to the center). So $b^2 = a^2 - c^2 = 25 - 9 = 16$. Thus, an equation is $\dfrac{(x-5)^2}{25} + \dfrac{y^2}{16} = 1$.

17. This is a hyperbola with center $(0, 1)$ and vertices $(0, 0)$ and $(0, 2)$. Since a is the distance form the center to a vertex, we have $a = 1$. The slope of the given asymptote is 1, so $\dfrac{a}{b} = 1 \Leftrightarrow b = 1$. Thus, an equation of the hyperbola is $(y-1)^2 - x^2 = 1$.

19. $9x^2 - 36x + 4y^2 = 0$ $\Leftrightarrow$ $9\left(x^2 - 4x + 4\right) - 36 + 4y^2 = 0$ $\Leftrightarrow$

$9\left(x - 2\right)^2 + 4y^2 = 36$ $\Leftrightarrow$ $\dfrac{(x-2)^2}{4} + \dfrac{y^2}{9} = 1$. This is an ellipse with $a = 3$,

$b = 2$, and $c = \sqrt{9 - 4} = \sqrt{5}$. The center is $(2, 0)$, the foci are $\left(2, \pm\sqrt{5}\right)$, the

vertices are $(2, \pm 3)$, the length of the major axis is $2a = 6$, and the length of the

minor axis is $2b = 4$.

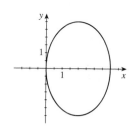

21. $x^2 - 4y^2 - 2x + 16y = 20$ $\Leftrightarrow$

$(x^2 - 2x + 1) - 4(y^2 - 4y + 4) = 20 + 1 - 16$ $\Leftrightarrow$ $(x - 1)^2 - 4(y - 2)^2 = 5$

$\Leftrightarrow$ $\dfrac{(x-1)^2}{5} - \dfrac{(y-2)^2}{\frac{5}{4}} = 1$. This is a hyperbola with $a = \sqrt{5}$, $b = \frac{1}{2}\sqrt{5}$, and

$c = \sqrt{5 + \frac{5}{4}} = \frac{5}{2}$. The center is $(1, 2)$, the foci are $\left(1 \pm \frac{5}{2}, 2\right) = \left(-\frac{3}{2}, 2\right)$ and

$\left(\frac{7}{2}, 2\right)$, the vertices are $\left(1 \pm \sqrt{5}, 2\right)$, and the asymptotes are $y - 2 = \pm\frac{1}{2}\left(x - 1\right)$

$\Leftrightarrow$ $y = \pm\frac{1}{2}\left(x - 1\right) + 2$ $\Leftrightarrow$ $y = \frac{1}{2}x + \frac{3}{2}$ and $y = -\frac{1}{2}x + \frac{5}{2}$.

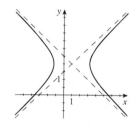

23. $4x^2 + 25y^2 - 24x + 250y + 561 = 0$ $\Leftrightarrow$ $4\left(x^2 - 6x + 9\right) +$

$25\left(y^2 + 10y + 25\right) = -561 + 36 + 625 \Leftrightarrow$ $4\left(x - 3\right)^2 + 25\left(y + 5\right)^2 = 100$

$\Leftrightarrow$ $\dfrac{(x-3)^2}{25} + \dfrac{(y+5)^2}{4} = 1$. This is an ellipse with $a = 5$, $b = 2$, and

$c = \sqrt{25 - 4} = \sqrt{21}$. The center is $(3, -5)$, the foci are $\left(3 \pm \sqrt{21}, -5\right)$, the

vertices are $(3 \pm 5, -5) = (-2, -5)$ and $(8, -5)$, the length of the major axis is

$2a = 10$, and the length of the minor axis is $2b = 4$.

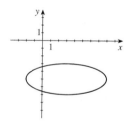

25. $16x^2 - 9y^2 - 96x + 288 = 0$ $\Leftrightarrow$ $16\left(x^2 - 6x\right) - 9y^2 + 288 = 0$ $\Leftrightarrow$

$16\left(x^2 - 6x + 9\right) - 9y^2 = 144 - 288$ $\Leftrightarrow$ $16\left(x - 3\right)^2 - 9y^2 = -144$ $\Leftrightarrow$

$\dfrac{y^2}{16} - \dfrac{(x-3)^2}{9} = 1$. This is a hyperbola with $a = 4$, $b = 3$, and

$c = \sqrt{16 + 9} = 5$. The center is $(3, 0)$, the foci are $(3, \pm 5)$, the vertices are

$(3, \pm 4)$, and the asymptotes are $y = \pm\frac{4}{3}\left(x - 3\right)$ $\Leftrightarrow$ $y = \frac{4}{3}x - 4$ and

$y = 4 - \frac{4}{3}x$.

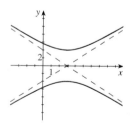

27. $x^2 + 16 = 4\left(y^2 + 2x\right)$ $\Leftrightarrow$ $x^2 - 8x - 4y^2 + 16 = 0$ $\Leftrightarrow$

$(x^2 - 8x + 16) - 4y^2 = -16 + 16$ $\Leftrightarrow$ $4y^2 = (x - 4)^2$ $\Leftrightarrow$

$y = \pm\frac{1}{2}\left(x - 4\right)$. Thus, the conic is degenerate, and its graph is the pair of lines

$y = \frac{1}{2}\left(x - 4\right)$ and $y = -\frac{1}{2}\left(x - 4\right)$.

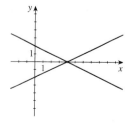

29. $3x^2 + 4y^2 - 6x - 24y + 39 = 0 \;\Leftrightarrow\; 3\left(x^2 - 2x\right) + 4\left(y^2 - 6y\right) = -39 \;\Leftrightarrow$

$3\left(x^2 - 2x + 1\right) + 4\left(y^2 - 6y + 9\right) = -39 + 3 + 36 \;\Leftrightarrow$

$3\left(x - 1\right)^2 + 4\left(y - 3\right)^2 = 0 \;\Leftrightarrow\; x = 1$ and $y = 3$. This is a degenerate conic

whose graph is the point $(1, 3)$.

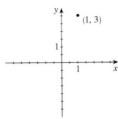

31. $2x^2 - 4x + y + 5 = 0 \;\Leftrightarrow\; y = -2x^2 + 4x - 5.$

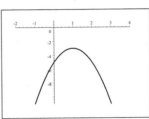

33. $9x^2 + 36 = y^2 + 36x + 6y \;\Leftrightarrow\; x^2 - 36x + 36 = y^2 + 6y \;\Leftrightarrow$

$9x^2 - 36x + 45 = y^2 + 6y + 9 \;\Leftrightarrow\; 9\left(x^2 - 4x + 5\right) = \left(y + 3\right)^2 \;\Leftrightarrow$

$y + 3 = \pm\sqrt{9\left(x^2 - 4x + 5\right)} \;\Leftrightarrow\; y = -3 \pm 3\sqrt{x^2 - 4x + 5}$

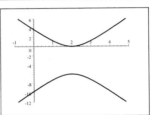

35. $4x^2 + y^2 + 4\left(x - 2y\right) + F = 0 \;\Leftrightarrow\; 4\left(x^2 + x\right) + \left(y^2 - 8y\right) = -F \;\Leftrightarrow$

$4\left(x^2 + x + \frac{1}{4}\right) + \left(y^2 - 8y + 16\right) = 16 + 1 - F \;\Leftrightarrow\; 4\left(x + \frac{1}{2}\right)^2 + \left(y - 1\right)^2 = 17 - F$

(a) For an ellipse, $17 - F > 0 \;\Leftrightarrow\; F < 17.$

(b) For a single point, $17 - F = 0 \;\Leftrightarrow\; F = 17.$

(c) For the empty set, $17 - F < 0 \;\Leftrightarrow\; F > 17.$

37. (a) $x^2 = 4p\left(y + p\right)$, for

$p = -2, -\frac{3}{2}, -1, -\frac{1}{2}, \frac{1}{2}, 1, \frac{3}{2}, 2.$

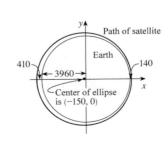

(b) The graph of $x^2 = 4p\left(y + p\right)$ is obtained by shifting the graph of $x^2 = 4py$ vertically $-p$ units so that the vertex is at $(0, -p)$. The focus of $x^2 = 4py$ is at $(0, p)$, so this point is also shifted $-p$ units vertically to the point $(0, p - p) = (0, 0)$. Thus, the focus is located at the origin.

(c) The parabolas become narrower as the vertex moves toward the origin.

39. Since the height of the satellite above the earth varies between 140 and 440, the length of the major axis is $2a = 140 + 2\left(3960\right) + 440 = 8500 \;\Leftrightarrow\; a = 4250.$

Since the center of the earth is at one focus, we have

$a - c = \text{(earth radius)} + 140 = 3960 + 140 = 4100 \;\Leftrightarrow$

$c = a - 4100 = 4250 - 4100 = 150.$ Thus, the center of the ellipse is $(-150, 0).$

So $b^2 = a^2 - c^2 = 4250^2 - 150^2 = 18{,}062{,}500 - 22500 = 18{,}040{,}000.$ Hence,

an equation is $\dfrac{\left(x + 150\right)^2}{18{,}062{,}500} + \dfrac{y^2}{18{,}040{,}000} = 1.$

11.5 Rotation of Axes

1. $(x, y) = (1, 1)$, $\phi = 45°$. Then $X = x \cos \phi + y \sin \phi = 1 \cdot \frac{1}{\sqrt{2}} + 1 \cdot \frac{1}{\sqrt{2}} = \sqrt{2}$ and
$Y = -x \sin \phi + y \cos \phi = -1 \cdot \frac{1}{\sqrt{2}} + 1 \cdot \frac{1}{\sqrt{2}} = 0$. Therefore, the XY-coordinates of the given point are $(X, Y) = \left(\sqrt{2}, 0 \right)$.

3. $(x, y) = \left(3, -\sqrt{3} \right)$, $\phi = 60°$. Then $X = x \cos \phi + y \sin \phi = 3 \cdot \frac{1}{2} - \sqrt{3} \cdot \frac{\sqrt{3}}{2} = 0$ and
$Y = -x \sin \phi + y \cos \phi = -3 \cdot \frac{\sqrt{3}}{2} - \sqrt{3} \cdot \frac{1}{2} = -2\sqrt{3}$. Therefore, the XY-coordinates of the given point are
$(X, Y) = \left(0, -2\sqrt{3} \right)$.

5. $(x, y) = (0, 2)$, $\phi = 55°$. Then $X = x \cos \phi + y \sin \phi = 0 \cos 55° + 2 \sin 55° \approx 1.6383$ and
$Y = -x \sin \phi + y \cos \phi = -0 \sin 55° + 2 \cos 55° \approx 1.1472$. Therefore, the XY-coordinates of the given point are
approximately $(X, Y) = (1.6383, 1.1472)$.

7. $x^2 - 3y^2 = 4$, $\phi = 60°$. Then $x = X \cos 60° - Y \sin 60° = \frac{1}{2} X - \frac{\sqrt{3}}{2} Y$ and $y = X \sin 60° + Y \cos 60° = \frac{\sqrt{3}}{2} X + \frac{1}{2} Y$.
Substituting these values into the equation, we get $\left(\frac{1}{2} X - \frac{\sqrt{3}}{2} Y \right)^2 - 3 \left(\frac{\sqrt{3}}{2} X + \frac{1}{2} Y \right)^2 = 4$ $\Leftrightarrow$
$\frac{X^2}{4} - \frac{\sqrt{3} XY}{2} + \frac{3Y^2}{4} - 3 \left(\frac{3X^2}{4} + \frac{\sqrt{3} XY}{2} + \frac{Y^2}{4} \right) = 4$ $\Leftrightarrow$ $\frac{X^2}{4} - \frac{9}{4} X^2 + \frac{3Y^2}{4} - \frac{3Y^2}{4} - \frac{\sqrt{3} XY}{2} - \frac{3\sqrt{3} XY}{2} = 4$
$\Leftrightarrow$ $-2X^2 - 2\sqrt{3} XY = 4$ $\Leftrightarrow$ $X^2 + \sqrt{3} XY = -2$.

9. $x^2 - y^2 = 2y$, $\phi = \cos^{-1} \left(\frac{3}{5} \right)$. So $\cos \phi = \frac{3}{5}$ and $\sin \phi = \frac{4}{5}$. Then
$(X \cos \phi - Y \sin \phi)^2 - (X \sin \phi + Y \cos \phi)^2 = 2 (X \sin \phi + Y \cos \phi)$ $\Leftrightarrow$ $\left(\frac{3}{5} X - \frac{4}{5} Y \right)^2 -$
$\left(\frac{4}{5} X + \frac{3}{5} Y \right)^2 = 2 \left(\frac{4}{5} X + \frac{3}{5} Y \right)$ $\Leftrightarrow$ $\frac{9X^2}{25} - \frac{24XY}{25} + \frac{16Y^2}{25} - \frac{16X^2}{25} - \frac{24XY}{25} - \frac{9Y^2}{25} = \frac{8X}{5} + \frac{6Y}{5}$ $\Leftrightarrow$
$-\frac{7X^2}{25} - \frac{48XY}{25} + \frac{7Y^2}{25} - \frac{8X}{5} - \frac{6Y}{5} = 0$ $\Leftrightarrow$ $7Y^2 - 48XY - 7X^2 - 40X - 30Y = 0$.

11. $x^2 + 2\sqrt{3} xy - y^2 = 4$, $\phi = 30°$. Then $x = X \cos 30° - Y \sin 30° = \frac{\sqrt{3}}{2} X - \frac{1}{2} Y = \frac{1}{2} \left(\sqrt{3} X - Y \right)$
and $y = X \sin 30° + Y \cos 30° = \frac{1}{2} X + \frac{\sqrt{3}}{2} Y = \frac{1}{2} \left(X + \sqrt{3} Y \right)$. Substituting these values into the
equation, we get $\left[\frac{1}{2} \left(\sqrt{3} X - Y \right) \right]^2 + 2\sqrt{3} \left[\frac{1}{2} \left(\sqrt{3} X - Y \right) \right] \left[\frac{1}{2} \left(X + \sqrt{3} Y \right) \right] - \left[\frac{1}{2} \left(X + \sqrt{3} Y \right) \right]^2 = 4$ $\Leftrightarrow$
$\left(\sqrt{3} X - Y \right)^2 + 2\sqrt{3} \left(\sqrt{3} X - Y \right) \left(X + \sqrt{3} Y \right) - \left(X + \sqrt{3} Y \right)^2 = 16$ $\Leftrightarrow$
$\left(3X^2 - 2\sqrt{3} XY + Y^2 \right) + \left(6X^2 + 4\sqrt{3} XY - 6Y^2 \right) - \left(X^2 + 2\sqrt{3} XY + 3Y^2 \right) = 16$ $\Leftrightarrow$ $8X^2 - 8Y^2 = 16$ $\Leftrightarrow$
$\frac{X^2}{2} - \frac{Y^2}{2} = 1$. This is a hyperbola.

13. (a) $xy = 8$ $\Leftrightarrow$ $0x^2 + xy + 0y^2 = 8$. So $A = 0$, $B = 1$, and $C = 0$, and
so the discriminant is $B^2 - 4AC = 1^2 - 4 (0) (0) = 1$. Since the
discriminant is positive, the equation represents a hyperbola.

(b) $\cot 2\phi = \frac{A - C}{B} = 0$ $\Rightarrow$ $2\phi = 90°$ $\Leftrightarrow$ $\phi = 45°$. Therefore,

$x = \frac{\sqrt{2}}{2} X - \frac{\sqrt{2}}{2} Y$ and $y = \frac{\sqrt{2}}{2} X + \frac{\sqrt{2}}{2} Y$. After substitution, the original

equation becomes $\left(\frac{\sqrt{2}}{2} X - \frac{\sqrt{2}}{2} Y \right) \left(\frac{\sqrt{2}}{2} X + \frac{\sqrt{2}}{2} Y \right) = 8$ $\Leftrightarrow$

$\frac{(X - Y)(X + Y)}{2} = 8$ $\Leftrightarrow$ $\frac{X^2}{16} - \frac{Y^2}{16} = 1$. This is a hyperbola with $a = 4$, $b = 4$, and $c = 4\sqrt{2}$. Hence, the

vertices are $V (\pm 4, 0)$ and the foci are $F \left(\pm 4\sqrt{2}, 0 \right)$.

(c)

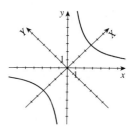

15. (a) $x^2 + 2xy + y^2 + x - y = 0$. So $A = 1$, $B = 2$, and $C = 1$, and so the discriminant is $B^2 - 4AC = 2^2 - 4(1)(1) = 0$. Since the discriminant is zero, the equation represents a parabola.

(c)

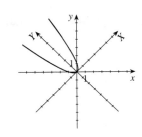

(b) $\cot 2\phi = \dfrac{A - C}{B} = 0 \quad \Rightarrow \quad 2\phi = 90° \quad \Leftrightarrow \quad \phi = 45°$. Therefore,

$x = \frac{\sqrt{2}}{2}X - \frac{\sqrt{2}}{2}Y$ and $y = \frac{\sqrt{2}}{2}X + \frac{\sqrt{2}}{2}Y$. After substitution, the original equation becomes

$$\left(\tfrac{\sqrt{2}}{2}X - \tfrac{\sqrt{2}}{2}Y\right)^2 + 2\left(\tfrac{\sqrt{2}}{2}X - \tfrac{\sqrt{2}}{2}Y\right)\left(\tfrac{\sqrt{2}}{2}X + \tfrac{\sqrt{2}}{2}Y\right)$$

$$+ \left(\tfrac{\sqrt{2}}{2}X + \tfrac{\sqrt{2}}{2}Y\right)^2 + \left(\tfrac{\sqrt{2}}{2}X - \tfrac{\sqrt{2}}{2}Y\right) - \left(\tfrac{\sqrt{2}}{2}X + \tfrac{\sqrt{2}}{2}Y\right) = 0 \quad \Leftrightarrow$$

$$\tfrac{1}{2}X^2 - XY + \tfrac{1}{2}Y^2 + X^2 - Y^2 + \tfrac{1}{2}X^2 + XY + Y^2 - \sqrt{2}Y = 0 \quad \Leftrightarrow \quad 2X^2 - \sqrt{2}Y = 0 \quad \Leftrightarrow \quad X^2 = \tfrac{\sqrt{2}}{2}Y.$$

This is a parabola with $4p = \frac{1}{\sqrt{2}}$ and hence the focus is $F\left(0, \frac{1}{4\sqrt{2}}\right)$.

17. (a) $x^2 + 2\sqrt{3}xy - y^2 + 2 = 0$. So $A = 1$, $B = 2\sqrt{3}$, and $C = -1$, and so the discriminant is $B^2 - 4AC = \left(2\sqrt{3}\right)^2 - 4(1)(-1) > 0$. Since the discriminant is positive, the equation represents a hyperbola.

(c)

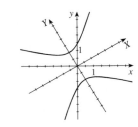

(b) $\cot 2\phi = \dfrac{A - C}{B} = \dfrac{1 + 1}{2\sqrt{3}} = \dfrac{1}{\sqrt{3}} \quad \Rightarrow \quad 2\phi = 60° \quad \Leftrightarrow \quad \phi = 30°$.

Therefore, $x = \frac{\sqrt{3}}{2}X - \frac{1}{2}Y$ and $y = \frac{1}{2}X + \frac{\sqrt{3}}{2}Y$. After substitution, the original equation becomes

$$\left(\tfrac{\sqrt{3}}{2}X - \tfrac{1}{2}Y\right)^2 + 2\sqrt{3}\left(\tfrac{\sqrt{3}}{2}X - \tfrac{1}{2}Y\right)\left(\tfrac{1}{2}X + \tfrac{\sqrt{3}}{2}Y\right) - \left(\tfrac{1}{2}X + \tfrac{\sqrt{3}}{2}Y\right)^2 + 2 = 0 \quad \Leftrightarrow$$

$$\tfrac{3}{4}X^2 - \tfrac{\sqrt{3}}{2}XY + \tfrac{1}{4}Y^2 + \tfrac{\sqrt{3}}{2}\left(\sqrt{3}X^2 + 2XY - \sqrt{3}Y^2\right) - \tfrac{1}{4}X^2 - \tfrac{\sqrt{3}}{2}XY - \tfrac{3}{4}Y^2 + 2 = 0 \quad \Leftrightarrow$$

$$X^2\left(\tfrac{3}{4} + \tfrac{3}{2} - \tfrac{1}{4}\right) + XY\left(-\tfrac{\sqrt{3}}{2} + \sqrt{3} - \tfrac{\sqrt{3}}{2}\right) + Y^2\left(\tfrac{1}{4} - \tfrac{3}{2} - \tfrac{3}{4}\right) = -2 \quad \Leftrightarrow \quad 2X^2 - 2Y^2 = -2 \quad \Leftrightarrow$$

$$Y^2 - X^2 = 1.$$

19. (a) $11x^2 - 24xy + 4y^2 + 20 = 0$. So $A = 11$, $B = -24$, and $C = 4$, and so the discriminant is $B^2 - 4AC = (-24)^2 - 4(11)(4) > 0$. Since the discriminant is positive, the equation represents a hyperbola.

(c)

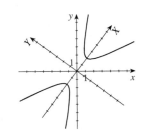

(b) $\cot 2\phi = \dfrac{A - C}{B} = \dfrac{11 - 4}{-24} = -\dfrac{7}{24} \quad \Rightarrow \quad \cos 2\phi = -\dfrac{7}{25}$. Therefore,

$\cos\phi = \sqrt{\dfrac{1 + (-7/25)}{2}} = \dfrac{3}{5}$ and $\sin\phi = \sqrt{\dfrac{1 - (-7/25)}{2}} = \dfrac{4}{5}$. Hence,

$x = \dfrac{3X}{5} - \dfrac{4}{5}Y$ and $y = \dfrac{4}{5}X + \dfrac{3}{5}Y$. After substitution, the original equation becomes

Since $\cos 2\phi = -\frac{7}{25}$, we have $2\phi \approx 106.26°$, so $\phi \approx 53°$.

$$11\left(\tfrac{3}{5}X - \tfrac{4}{5}Y\right)^2 - 24\left(\tfrac{3}{5}X - \tfrac{4}{5}Y\right)\left(\tfrac{4}{5}X + \tfrac{3}{5}Y\right) + 4\left(\tfrac{4}{5}X + \tfrac{3}{5}Y\right)^2 + 20 = 0 \quad \Leftrightarrow$$

$$\tfrac{11}{25}\left(9X^2 - 24XY + 16Y^2\right) - \tfrac{24}{25}\left(12X^2 - 7XY - 12Y^2\right) + \tfrac{4}{25}\left(16X^2 + 24XY + 9Y^2\right) + 20 = 0 \quad \Leftrightarrow$$

$$X^2\left(99 - 288 + 64\right) + XY\left(-264 + 168 + 96\right) + Y^2\left(176 + 288 + 36\right) = -500 \quad \Leftrightarrow$$

$$-125X^2 + 500Y^2 = -500 \quad \Leftrightarrow \quad \tfrac{1}{4}X^2 - Y^2 = 1.$$

21. (a) $\sqrt{3}x^2 + 3xy = 3$. So $A = \sqrt{3}$, $B = 3$, and $C = 0$, and so the
discriminant is $B^2 - 4AC = (3)^2 - 4\left(\sqrt{3}\right)(0) = 9$. Since the
discriminant is positive, the equation represents a hyperbola.

(c)

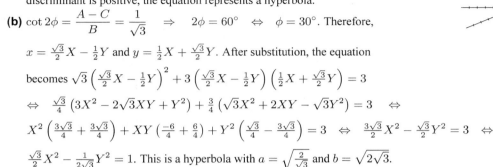

(b) $\cot 2\phi = \dfrac{A - C}{B} = \dfrac{1}{\sqrt{3}} \quad \Rightarrow \quad 2\phi = 60° \quad \Leftrightarrow \quad \phi = 30°$. Therefore,

$x = \frac{\sqrt{3}}{2}X - \frac{1}{2}Y$ and $y = \frac{1}{2}X + \frac{\sqrt{3}}{2}Y$. After substitution, the equation

becomes $\sqrt{3}\left(\frac{\sqrt{3}}{2}X - \frac{1}{2}Y\right)^2 + 3\left(\frac{\sqrt{3}}{2}X - \frac{1}{2}Y\right)\left(\frac{1}{2}X + \frac{\sqrt{3}}{2}Y\right) = 3$

$\Leftrightarrow \quad \frac{\sqrt{3}}{4}\left(3X^2 - 2\sqrt{3}XY + Y^2\right) + \frac{3}{4}\left(\sqrt{3}X^2 + 2XY - \sqrt{3}Y^2\right) = 3 \quad \Leftrightarrow$

$X^2\left(\frac{3\sqrt{3}}{4} + \frac{3\sqrt{3}}{4}\right) + XY\left(\frac{-6}{4} + \frac{6}{4}\right) + Y^2\left(\frac{\sqrt{3}}{4} - \frac{3\sqrt{3}}{4}\right) = 3 \quad \Leftrightarrow \quad \frac{3\sqrt{3}}{2}X^2 - \frac{\sqrt{3}}{2}Y^2 = 3 \quad \Leftrightarrow$

$\frac{\sqrt{3}}{2}X^2 - \frac{1}{2\sqrt{3}}Y^2 = 1$. This is a hyperbola with $a = \sqrt{\frac{2}{\sqrt{3}}}$ and $b = \sqrt{2\sqrt{3}}$.

23. (a) $2\sqrt{3}x^2 - 6xy + \sqrt{3}x + 3y = 0$. So $A = 2\sqrt{3}$, $B = -6$, and $C = 0$, and
so the discriminant is $B^2 - 4AC = (-6)^2 - 4\left(2\sqrt{3}\right)(0) = 36$. Since
the discriminant is positive, the equation represents a hyperbola.

(c)

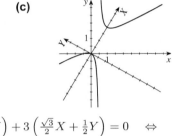

(b) $\cot 2\phi = \dfrac{A - C}{B} = \dfrac{2\sqrt{3}}{-6} = -\dfrac{1}{\sqrt{3}} \quad \Rightarrow \quad 2\phi = 120° \quad \Leftrightarrow \quad \phi = 60°$.

Therefore, $x = \frac{1}{2}X - \frac{\sqrt{3}}{2}Y$ and $y = \frac{\sqrt{3}}{2}X + \frac{1}{2}Y$, and substituting gives

$2\sqrt{3}\left(\frac{1}{2}X - \frac{\sqrt{3}}{2}Y\right)^2 - 6\left(\frac{1}{2}X - \frac{\sqrt{3}}{2}Y\right)\left(\frac{\sqrt{3}}{2}X + \frac{1}{2}Y\right) + \sqrt{3}\left(\frac{1}{2}X - \frac{\sqrt{3}}{2}Y\right) + 3\left(\frac{\sqrt{3}}{2}X + \frac{1}{2}Y\right) = 0 \quad \Leftrightarrow$

$\frac{\sqrt{3}}{2}\left(X^2 - 2\sqrt{3}XY + 3Y^2\right) - \frac{3}{2}\left(\sqrt{3}X^2 - 2XY - \sqrt{3}Y^2\right) + \frac{\sqrt{3}}{2}\left(X - \sqrt{3}Y\right) + \frac{3}{2}\left(\sqrt{3}X + Y\right) = 0 \quad \Leftrightarrow$

$X^2\left(\frac{\sqrt{3}}{2} - \frac{3\sqrt{3}}{2}\right) + X\left(\frac{\sqrt{3}}{2} + \frac{3\sqrt{3}}{2}\right) + XY\left(-3 + 3\right) + Y^2\left(\frac{3\sqrt{3}}{2} + \frac{3\sqrt{3}}{2}\right) + Y\left(-\frac{3}{2} + \frac{3}{2}\right) = 0 \quad \Leftrightarrow$

$-\sqrt{3}X^2 + 2\sqrt{3}X + 3\sqrt{3}Y^2 = 0 \quad \Leftrightarrow \quad -X^2 + 2X + 3Y^2 = 0 \quad \Leftrightarrow \quad 3Y^2 - \left(X^2 - 2X + 1\right) = -1 \quad \Leftrightarrow$

$(X - 1)^2 - 3Y^2 = 1$. This is a hyperbola with $a = 1$, $b = \frac{\sqrt{3}}{3}$, $c = \sqrt{1 + \frac{1}{3}} = \frac{2}{\sqrt{3}}$, and $C(1, 0)$.

25. (a) $52x^2 + 72xy + 73y^2 = 40x - 30y + 75$. So $A = 52$, $B = 72$, and $C = 73$, and so the discriminant is
$B^2 - 4AC = (72)^2 - 4(52)(73) = -10{,}000$. Since the discriminant is decidedly negative, the equation represents an
ellipse.

(b) $\cot 2\phi = \dfrac{A - C}{B} = \dfrac{52 - 73}{72} = -\dfrac{7}{24}$. Therefore, as in Exercise 19(b), we get $\cos\phi = \frac{3}{5}$, $\sin\phi = \frac{4}{5}$, and

$x = \frac{3}{5}X - \frac{4}{5}Y$, $y = \frac{4}{5}X + \frac{3}{5}Y$. By substitution,

$52\left(\frac{3}{5}X - \frac{4}{5}Y\right)^2 + 72\left(\frac{3}{5}X - \frac{4}{5}Y\right)\left(\frac{4}{5}X + \frac{3}{5}Y\right) + 73\left(\frac{4}{5}X + \frac{3}{5}Y\right)^2$

$\qquad\qquad = 40\left(\frac{3}{5}X - \frac{4}{5}Y\right) - 30\left(\frac{4}{5}X + \frac{3}{5}Y\right) + 75 \quad \Leftrightarrow$

(c)

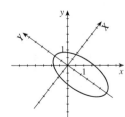

$\frac{52}{25}\left(9X^2 - 24XY + 16Y^2\right) + \frac{72}{25}\left(12X^2 - 7XY - 12Y^2\right)$

$+ \frac{73}{25}\left(16X^2 + 24XY + 9Y^2\right) = 24X - 32Y - 24X - 18Y + 75 \quad \Leftrightarrow$

$468X^2 + 832Y^2 + 864X^2 - 864Y^2 + 1168X^2 + 657Y^2 = -1250Y + 1875$

$\Leftrightarrow \quad 2500X^2 + 625Y^2 + 1250Y = 1875 \quad \Leftrightarrow$

$100X^2 + 25Y^2 + 50Y = 75 \quad \Leftrightarrow \quad X^2 + \frac{1}{4}(Y + 1)^2 = 1$. This is an

ellipse with $a = 2$, $b = 1$, $c = \sqrt{4 - 1} = \sqrt{3}$, and center $C(0, -1)$.

Since $\cos 2\phi = -\frac{7}{25}$, we have

$2\phi = \cos^{-1}\left(-\frac{7}{25}\right) \approx 106.26°$

and so $\phi \approx 53°$.

27. (a) The discriminant is $B^2 - 4AC = (-4)^2 + 4(2)(2) = 0$. Since the discriminant is 0, the equation represents a parabola.

(b) $2x^2 - 4xy + 2y^2 - 5x - 5 = 0 \quad \Leftrightarrow \quad 2y^2 - 4xy = -2x^2 + 5x + 5$

$\Leftrightarrow \quad 2\left(y^2 - 2xy\right) = -2x^2 + 5x + 5 \quad \Leftrightarrow$

$2\left(y^2 - 2xy + x^2\right) = -2x^2 + 5x + 5 + 2x^2 \quad \Leftrightarrow$

$2\left(y - x\right)^2 = 5x + 5 \quad \Leftrightarrow$

$\left(y - x\right)^2 = \frac{5}{2}x + \frac{5}{2} \quad \Rightarrow y - x = \pm\sqrt{\frac{5}{2}x + \frac{5}{2}} \quad \Leftrightarrow$

$y = x \pm \sqrt{\frac{5}{2}x + \frac{5}{2}}$

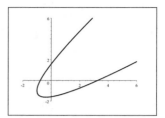

29. (a) The discriminant is $B^2 - 4AC = 10^2 + 4(6)(3) = 28 > 0$. Since the discriminant is positive, the equation represents a hyperbola.

(b) $6x^2 + 10xy + 3y^2 - 6y = 36 \quad \Leftrightarrow \quad 3y^2 + 10xy - 6y = 36 - 6x^2 \quad \Leftrightarrow$

$3y^2 + 2(5x - 3)y = 36 - 6x^2 \quad \Leftrightarrow \quad y^2 + 2\left(\frac{5}{3}x - 1\right)y = 12 - 2x^2$

$\Leftrightarrow \quad y^2 + 2\left(\frac{5}{3}x - 1\right)y + \left(\frac{5}{3}x - 1\right)^2 = \left(\frac{5}{3}x - 1\right)^2 + 12 - 2x^2 \quad \Leftrightarrow$

$\left[y + \left(\frac{5}{3}x - 1\right)\right]^2 = \frac{25}{9}x^2 - \frac{10}{3}x + 1 + 12 - 2x^2 \quad \Leftrightarrow$

$\left[y + \left(\frac{5}{3}x - 1\right)\right]^2 = \frac{7}{9}x^2 - \frac{10}{3}x + 13 \quad \Leftrightarrow$

$y + \left(\frac{5}{3}x - 1\right) = \pm\sqrt{\frac{7}{9}x^2 - \frac{10}{3}x + 13} \quad \Leftrightarrow$

$y = -\frac{5}{3}x + 1 \pm \sqrt{\frac{7}{9}x^2 - \frac{10}{3}x + 13}$

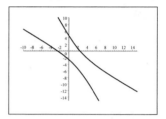

31. (a) $7x^2 + 48xy - 7y^2 - 200x - 150y + 600 = 0$. Then $A = 7$, $B = 48$, and $C = -7$, and so the discriminant is $B^2 - 4AC = (48)^2 - 4(7)(7) > 0$. Since the discriminant is positive, the equation represents a hyperbola. We now find the equation in terms of XY-coordinates. We have $\cot 2\phi = \dfrac{A - C}{B} = \dfrac{7}{24} \quad \Rightarrow \quad \cos\phi = \frac{4}{5}$ and $\sin\phi = \frac{3}{5}$.

Therefore, $x = \frac{4}{5}X - \frac{3}{5}Y$ and $y = \frac{3}{5}X + \frac{4}{5}Y$, and substitution gives

$7\left(\frac{4}{5}X - \frac{3}{5}Y\right)^2 + 48\left(\frac{4}{5}X - \frac{3}{5}Y\right)\left(\frac{3}{5}X + \frac{4}{5}Y\right) - 7\left(\frac{3}{5}X + \frac{4}{5}Y\right)^2 - 200\left(\frac{4}{5}X - \frac{3}{5}Y\right) - 150\left(\frac{3}{5}X + \frac{4}{5}Y\right) + 600 = 0$

$\Leftrightarrow \quad \frac{7}{25}\left(16X^2 - 24XY + 9Y^2\right) + \frac{48}{25}\left(12X^2 + 7XY - 12Y^2\right) - \frac{7}{25}\left(9X^2 + 24XY + 16Y^2\right)$

$\qquad\qquad\qquad\qquad\qquad\qquad - 160X + 120Y - 90X - 120Y + 600 = 0$

$\Leftrightarrow \quad 112X^2 - 168XY + 63Y^2 + 576X^2 + 336XY - 576Y^2 - 63X^2 - 168XY - 112Y^2 - 6250X + 15{,}000 = 0$

$\Leftrightarrow \quad 25X^2 - 25Y^2 - 250X + 600 = 0 \quad \Leftrightarrow \quad 25\left(X^2 - 10X + 25\right) - 25Y^2 = -600 + 625 \quad \Leftrightarrow$

$(X - 5)^2 - Y^2 = 1$. This is a hyperbola with $a = 1$, $b = 1$, $c = \sqrt{1 + 1} = \sqrt{2}$, and center $C(5, 0)$.

(b) In the XY-plane, the center is $C(5, 0)$, the vertices are $V(5 \pm 1, 0) = V_1(4, 0)$ and $V_2(6, 0)$, and the foci are $F\left(5 \pm \sqrt{2}, 0\right)$. In the xy-plane, the center is $C\left(\frac{4}{5} \cdot 5 - \frac{3}{5} \cdot 0, \frac{3}{5} \cdot 5 + \frac{4}{5} \cdot 0\right) = C(4, 3)$, the vertices are $V_1\left(\frac{4}{5} \cdot 4 - \frac{3}{5} \cdot 0, \frac{3}{5} \cdot 4 + \frac{4}{5} \cdot 0\right) = V_1\left(\frac{16}{5}, \frac{12}{5}\right)$ and $V_2\left(\frac{4}{5} \cdot 6 - \frac{3}{5} \cdot 0, \frac{3}{5} \cdot 6 + \frac{4}{5} \cdot 0\right) = V_2\left(\frac{24}{5}, \frac{18}{5}\right)$, and the foci are $F_1\left(4 + \frac{4}{5}\sqrt{2}, 3 + \frac{3}{5}\sqrt{2}\right)$ and $F_2\left(4 - \frac{4}{5}\sqrt{2}, 3 - \frac{3}{5}\sqrt{2}\right)$.

(c) In the XY-plane, the equations of the asymptotes are $Y = X - 5$ and $Y = -X + 5$. In the xy-plane, these equations become $-x \cdot \frac{3}{5} + y \cdot \frac{4}{5} = x \cdot \frac{4}{5} + y \cdot \frac{3}{5} - 5 \quad \Leftrightarrow \quad 7x - y - 25 = 0$. Similarly, $-x \cdot \frac{3}{5} + y \cdot \frac{4}{5} = -x \cdot \frac{4}{5} - y \cdot \frac{3}{5} + 5$ $\Leftrightarrow \quad x + 7y - 25 = 0$.

33. We use the hint and eliminate Y by adding: $x = X \cos\phi - Y \sin\phi$ $\Leftrightarrow$ $x \cos\phi = X \cos^2\phi - Y \sin\phi \cos\phi$ and $y = X \sin\phi + Y \cos\phi$ $\Leftrightarrow$ $y \sin\phi = X \sin^2\phi + Y \sin\phi \cos\phi$, and adding these two equations gives $x \cos\phi + y \sin\phi = X \left(\cos^2\phi + \sin^2\phi\right)$ $\Leftrightarrow$ $x \cos\phi + y \sin\phi = X$. In a similar manner, we eliminate X by subtracting: $x = X \cos\phi - Y \sin\phi$ $\Leftrightarrow$ $-x \sin\phi = -X \cos\phi \sin\phi + Y \sin^2\phi$ and $y = X \sin\phi + Y \cos\phi$ $\Leftrightarrow$ $y \cos\phi = X \sin\phi \cos\phi + Y \cos^2\phi$, so $-x \sin\phi + y \cos\phi = Y \left(\cos^2\phi + \sin^2\phi\right)$ $\Leftrightarrow$ $-x \sin\phi + y \cos\phi = Y$. Thus, $X = x \cos\phi + y \sin\phi$ and $Y = -x \sin\phi + y \cos\phi$.

35. $Z = \begin{bmatrix} x \\ y \end{bmatrix}$, $Z' = \begin{bmatrix} X \\ Y \end{bmatrix}$, and $R = \begin{bmatrix} \cos\phi & -\sin\phi \\ \sin\phi & \cos\phi \end{bmatrix}$.

Thus $Z = RZ'$ $\Leftrightarrow$ $\begin{bmatrix} x \\ y \end{bmatrix} = \begin{bmatrix} \cos\phi & -\sin\phi \\ \sin\phi & \cos\phi \end{bmatrix}\begin{bmatrix} X \\ Y \end{bmatrix} = \begin{bmatrix} X\cos\phi - Y\sin\phi Y \\ X\sin\phi + Y\cos\phi \end{bmatrix}$. Equating the entries in this matrix equation gives the first pair of rotation of axes formulas. Now

$R^{-1} = \dfrac{1}{\cos^2\phi + \sin^2\phi}\begin{bmatrix} \cos\phi & \sin\phi \\ -\sin\phi & \cos\phi \end{bmatrix} = \begin{bmatrix} \cos\phi & \sin\phi \\ -\sin\phi & \cos\phi \end{bmatrix}$ and so $Z' = R^{-1}Z$ $\Leftrightarrow$

$\begin{bmatrix} X \\ Y \end{bmatrix} = \begin{bmatrix} \cos\phi & \sin\phi \\ -\sin\phi & \cos\phi \end{bmatrix}\begin{bmatrix} x \\ y \end{bmatrix} = \begin{bmatrix} x\cos\phi + y\sin\phi \\ -x\sin\phi + y\cos\phi \end{bmatrix}$. Equating the entries in this matrix equation gives the second pair of rotation of axes formulas.

37. Let P be the point (x_1, y_1) and Q be the point (x_2, y_2) and let $P'(X_1, Y_1)$ and $Q'(X_2, Y_2)$ be the images of P and Q under the rotation of ϕ. So $X_1 = x_1 \cos\phi + y_1 \sin\phi$, $Y_1 = -x_1 \sin\phi + y_1 \cos\phi$, $X_2 = x_2 \cos\phi + y_2 \sin\phi$, and $Y_2 = -x_2 \sin\phi + y_2 \cos\phi$. Thus $d(P', Q') = \sqrt{(X_2 - X_1)^2 + (Y_2 - Y_1)^2}$, where

$$
\begin{aligned}
(X_2 - X_1)^2 &= [(x_2 \cos\phi + y_2 \sin\phi) - (x_1 \cos\phi + y_1 \sin\phi)]^2 = [(x_2 - x_1)\cos\phi + (y_2 - y_1)\sin\phi]^2 \\
&= (x_2 - x_1)^2 \cos^2\phi + (x_2 - x_1)(y_2 - y_1)\sin\phi\cos\phi + (y_2 - y_1)^2 \sin^2\phi
\end{aligned}
$$

and

$$
\begin{aligned}
(Y_2 - Y_1)^2 &= [(-x_2 \sin\phi + y_2 \cos\phi) - (-x_1 \sin\phi + y_1 \cos\phi)]^2 = [-(x_2 - x_1)\sin\phi + (y_2 - y_1)\cos\phi]^2 \\
&= (x_2 - x_1)^2 \sin^2\phi - (x_2 - x_1)(y_2 - y_1)\sin\phi\cos\phi + (y_2 - y_1)^2 \cos^2\phi
\end{aligned}
$$

So

$$
\begin{aligned}
(X_2 - X_1)^2 + (Y_2 - Y_1)^2 &= (x_2 - x_1)^2 \cos^2\phi + (x_2 - x_1)(y_2 - y_1)\sin\phi\cos\phi + (y_2 - y_1)^2 \sin^2\phi \\
&\quad + (x_2 - x_1)^2 \sin^2\phi - (x_2 - x_1)(y_2 - y_1)\sin\phi\cos\phi + (y_2 - y_1)^2 \cos^2\phi \\
&= (x_2 - x_1)^2 \cos^2\phi + (y_2 - y_1)^2 \sin^2\phi + (x_2 - x_1)^2 \sin^2\phi + (y_2 - y_1)^2 \cos^2\phi \\
&= (x_2 - x_1)^2 \left(\cos^2\phi + \sin^2\phi\right) + (y_2 - y_1)^2 \left(\sin^2\phi + \cos^2\phi\right) = (x_2 - x_1)^2 + (y_2 - y_1)^2
\end{aligned}
$$

Thus, $d(P', Q') = \sqrt{(X_2 - X_1)^2 + (Y_2 - Y_1)^2} = \sqrt{(x_2 - x_1)^2 + (y_2 - y_1)^2} = d(P, Q)$.

11.6 Polar Equations of Conics

1. Substituting $e = \frac{2}{3}$ and $d = 3$ into the equation of a conic with vertical directrix, $r = \dfrac{\frac{2}{3}\cdot 3}{1 + \frac{2}{3}\cos\theta}$ $\Leftrightarrow$ $r = \dfrac{6}{3 + 2\cos\theta}$.

3. Substituting $e = 1$ and $d = 2$ into the equation of a conic with horizontal directrix, $r = \dfrac{1\cdot 2}{1 + \sin\theta}$ $\Leftrightarrow$ $r = \dfrac{2}{1 + \sin\theta}$.

5. $r = 5\sec\theta \iff r\cos\theta = 5 \iff x = 5$. So $d = 5$ and $e = 4$ gives $r = \dfrac{4 \cdot 5}{1 + 4\cos\theta} \iff r = \dfrac{20}{1 + 4\cos\theta}$.

7. Since this is a parabola whose focus is at the origin and vertex at $(5, \pi/2)$, the directrix must be $y = 10$. So $d = 10$ and $e = 1$ gives $r = \dfrac{1 \cdot 10}{1 + \sin\theta} = \dfrac{10}{1 + \sin\theta}$.

9. $r = \dfrac{6}{1 + \cos\theta}$ is Graph II. The eccentricity is 1, so this is a parabola. When $\theta = 0$, we have $r = 3$ and when $\theta = \frac{\pi}{2}$, we have $r = 6$.

11. $r = \dfrac{3}{1 - 2\sin\theta}$ is Graph VI. $e = 2$, so this is a hyperbola. When $\theta = 0$, $r = 3$, and when $\theta = \pi$, $r = 3$.

13. $r = \dfrac{12}{3 + 2\sin\theta}$ is Graph IV. $r = \dfrac{4}{1 + \frac{2}{3}\sin\theta}$, so $e = \frac{2}{3}$ and this is an ellipse. When $\theta = 0$, $r = 4$, and when $\theta = \pi$, $r = 4$.

15. (a) $r = \dfrac{4}{1 + 3\cos\theta} \Rightarrow e = 3$, so the conic is a hyperbola.

    (b) The vertices occur where $\theta = 0$ and $\theta = \pi$. Now $\theta = 0 \Rightarrow$
$r = \dfrac{4}{1 + 3\cos 0} = 1$, and $\theta = \pi \Rightarrow r = \dfrac{4}{1 + 3\cos\pi} = \dfrac{4}{-2} = -2$. Thus the
vertices are $(1, 0)$ and $(-2, \pi)$.

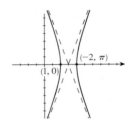

17. (a) $r = \dfrac{2}{1 - \cos\theta} \Rightarrow e = 1$, so the conic is a parabola.

    (b) Substituting $\theta = \pi$, we have $r = \dfrac{2}{1 - \cos\pi} = \dfrac{2}{2} = 1$. Thus the vertex is $(1, \pi)$.

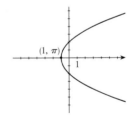

19. (a) $r = \dfrac{6}{2 + \sin\theta} \iff r = \dfrac{\frac{1}{2} \cdot 6}{1 + \frac{1}{2}\sin\theta} \Rightarrow e = \frac{1}{2}$, so the conic is an ellipse.

    (b) The vertices occur where $\theta = \frac{\pi}{2}$ and $\theta = \frac{3\pi}{2}$. Now $\theta = \frac{\pi}{2} \Rightarrow$
$r = \dfrac{6}{2 + \sin\frac{\pi}{2}} = \dfrac{6}{3} = 2$ and $\theta = \frac{3\pi}{2} \Rightarrow r = \dfrac{6}{2 + \sin\frac{3\pi}{2}} = \dfrac{6}{1} = 6$. Thus, the
vertices are $\left(2, \frac{\pi}{2}\right)$ and $\left(6, \frac{3\pi}{2}\right)$.

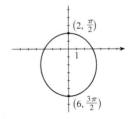

21. (a) $r = \dfrac{7}{2 - 5\sin\theta} \iff r = \dfrac{\frac{7}{2}}{1 - \frac{5}{2}\sin\theta} \Rightarrow e = \frac{5}{2}$, so the conic is a
hyperbola.

    (b) The vertices occur where $\theta = \frac{\pi}{2}$ and $\theta = \frac{3\pi}{2}$. $r = \dfrac{7}{2 - 5\sin\frac{\pi}{2}} = \dfrac{7}{-3} = -\frac{7}{3}$ and

    $\theta = \frac{3\pi}{2} \Rightarrow r = \dfrac{7}{2 - 5\sin\frac{3\pi}{2}} = \frac{7}{7} = 1$. Thus, the vertices are $\left(-\frac{7}{3}, \frac{\pi}{2}\right)$ and
$\left(1, \frac{3\pi}{2}\right)$.

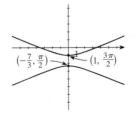

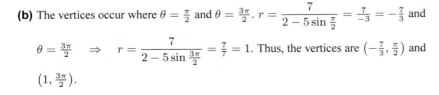

23. (a) $r = \dfrac{1}{4 - 3\cos\theta}$ $\Leftrightarrow$ $r = \dfrac{\frac{1}{4}}{1 - \frac{3}{4}\cos\theta}$ $\Rightarrow$ $e = \frac{3}{4}$. Since $ed = \frac{1}{4}$ we

have $d = \dfrac{\frac{1}{4}}{e} = \frac{4}{3} \cdot \frac{1}{4} = \frac{1}{3}$. Therefore the eccentricity is $\frac{3}{4}$ and the directrix is

$x = -\frac{1}{3}$ $\Leftrightarrow$ $r = -\frac{1}{3}\sec\theta$.

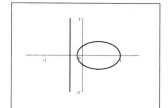

(b) We replace θ by $\theta - \frac{\pi}{3}$ to get $r = \dfrac{1}{4 - 3\cos\left(\theta - \frac{\pi}{3}\right)}$.

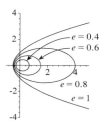

25. The ellipse is nearly circular when e is close to 0 and becomes more elongated as
$e \to 1^-$. At $e = 1$, the curve becomes a parabola.

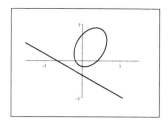

27. (a) Since the polar form of an ellipse with directrix $x = -d$ is $r = \dfrac{ed}{1 - e\cos\theta}$ we need to show that $ed = a\left(1 - e^2\right)$.

From the proof of the Equivalent Description of Conics we have $a^2 = \dfrac{e^2 d^2}{\left(1 - e^2\right)^2}$. Since the conic is an ellipse, $e < 1$

and so the quantities a, d, and $\left(1 - e^2\right)$ are all positive. Thus we can take the square roots of both sides and maintain

equality. Thus $a^2 = \dfrac{e^2 d^2}{\left(1 - e^2\right)^2}$ $\Leftrightarrow$ $a = \dfrac{ed}{1 - e^2}$ $\Leftrightarrow$ $ed = a\left(1 - e^2\right)$. As a result, $r = \dfrac{ed}{1 - e\cos\theta}$ $\Leftrightarrow$

$r = \dfrac{a\left(1 - e^2\right)}{1 - e\cos\theta}$.

(b) Since $2a = 2.99 \times 10^8$ we have $a = 1.495 \times 10^8$, so a polar equation for the earth's orbit (using $e \approx 0.017$) is

$r = \dfrac{1.495 \times 10^8 \left[1 - (0.017)^2\right]}{1 - 0.017\cos\theta} \approx \dfrac{1.49 \times 10^8}{1 - 0.017\cos\theta}$.

29. From Exercise 28, we know that at perihelion $r = 4.43 \times 10^9 = a\left(1 - e\right)$ and at aphelion $r = 7.37 \times 10^9 = a\left(1 + e\right)$.

Dividing these equations gives $\dfrac{7.37 \times 10^9}{4.43 \times 10^9} = \dfrac{a\left(1 + e\right)}{a\left(1 - e\right)}$ $\Leftrightarrow$ $1.664 = \dfrac{1 + e}{1 - e}$ $\Leftrightarrow$ $1.664\left(1 - e\right) = 1 + e$ $\Leftrightarrow$

$1.664 - 1 = e + 1.664$ $\Leftrightarrow$ $0.664 = 2.664e$ $\Leftrightarrow$ $e = \dfrac{0.664}{2.664} \approx 0.25$.

31. The r-coordinate of the satellite will be its distance from the focus (the center of the earth). From the r-coordinate we can
easily calculate the height of the satellite.

11.7   Plane Curves and Parametric Equations

1. (a) $x = 2t$, $y = t + 6$

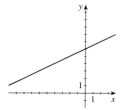

(b) Since $x = 2t$, $t = \dfrac{x}{2}$ and so $y = \dfrac{x}{2} + 6 \Leftrightarrow$
$x - 2y + 12 = 0$.

3. (a) $x = t^2$, $y = t - 2$, $2 \le t \le 4$

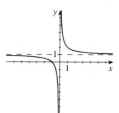

(b) Since $y = t - 2 \quad \Leftrightarrow \quad t = y + 2$, we have $x = t^2$
$\Leftrightarrow \quad x = (y + 2)^2$, and since $2 \le t \le 4$, we have
$4 \le x \le 16$.

5. (a) $x = \sqrt{t}$, $y = 1 - t \quad \Rightarrow \quad t \ge 0$

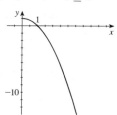

(b) Since $x = \sqrt{t}$, we have $x^2 = t$, and so $y = 1 - x^2$
with $x \ge 0$.

7. (a) $x = \dfrac{1}{t}$, $y = t + 1$

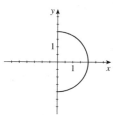

(b) Since $x = \dfrac{1}{t}$ we have $t = \dfrac{1}{x}$ and so $y = \dfrac{1}{x} + 1$.

9. (a) $x = 4t^2$, $y = 8t^3$

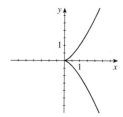

(b) Since $y = 8t^3 \quad \Leftrightarrow \quad y^2 = 64t^6 = \left(4t^2\right)^3 = x^3$, we
have $y^2 = x^3$.

11. (a) $x = 2 \sin t$, $y = 2 \cos t$, $0 \le t \le \pi$

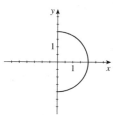

(b) $x^2 = (2 \sin t)^2 = 4 \sin^2 t$ and $y^2 = 4 \cos^2 t$. Hence,
$x^2 + y^2 = 4 \sin^2 t + 4 \cos^2 t = 4 \quad \Leftrightarrow$
$x^2 + y^2 = 4$, where $x \ge 0$.

13. (a) $x = \sin^2 t$, $y = \sin^4 t$

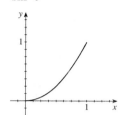

(b) Since $x = \sin^2 t$ we have $x^2 = \sin^4 t$ and so $y = x^2$.
But since $0 \le \sin^2 t \le 1$ we only get the part of this
parabola for which $0 \le x \le 1$.

15. (a) $x = \cos t$, $y = \cos 2t$

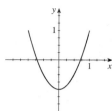

(b) Since $x = \cos t$ we have $x^2 = \cos^2 t$, so
$2x^2 - 1 = 2\cos^2 t - 1 = \cos 2t = y$. Hence, the
rectangular equation is $y = 2x^2 - 1$, $-1 \le x \le 1$.

17. (a) $x = \sec t$, $y = \tan t$, $0 \le t < \frac{\pi}{2}$ $\Rightarrow$ $x \ge 1$ and
$y \ge 0$.

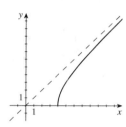

(b) $x^2 = \sec^2 t$, $y^2 = \tan^2 t$, and
$y^2 + 1 = \tan^2 t + 1 = \sec^2 t = x^2$. Therefore,
$y^2 + 1 = x^2$ $\Leftrightarrow$ $x^2 - y^2 = 1$, $x \ge 1$, $y \ge 0$.

19. (a) $x = e^t$, $y = e^{-t}$ $\Rightarrow$ $x > 0, y > 0$.

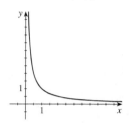

(b) $xy = e^t \cdot e^{-t} = e^0 = 1$. Hence, the equation is
$xy = 1$, with $x > 0$, $y > 0$.

21. (a) $x = \cos^2 t$, $y = \sin^2 t$

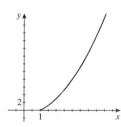

(b) $x + y = \cos^2 t + \sin^2 t = 1$. Hence, the equation is
$x + y = 1$ with $0 \le x, y \le 1$.

23. Since the line passes through the point $(4, -1)$ and has
slope $\frac{1}{2}$, parametric equations for the line are $x = 4 + t$,
$y = -1 + \frac{1}{2}t$.

25. Since the line passes through the points $(6, 7)$ and $(7, 8)$, its slope is $\dfrac{8 - 7}{7 - 6} = 1$. Thus, parametric equations for the line are
$x = 6 + t$, $y = 7 + t$.

27. Since $\cos^2 t + \sin^2 t = 1$, we have $a^2 \cos^2 t + a^2 \sin^2 t = a^2$. If we let $x = a \cos t$ and $y = a \sin t$, then $x^2 + y^2 = a^2$.
Hence, parametric equations for the circle are $x = a \cos t$, $y = a \sin t$.

29. $x = a \tan \theta$ $\Leftrightarrow$ $\tan \theta = \dfrac{x}{a}$ $\Rightarrow$ $\tan^2 \theta = \dfrac{x^2}{a^2}$. Also, $y = b \sec \theta$ $\Leftrightarrow$ $\sec \theta = \dfrac{y}{b}$ $\Rightarrow$ $\sec^2 \theta = \dfrac{y^2}{b^2}$. Since
$\tan^2 \theta = \sec^2 \theta - 1$, we have $\dfrac{x^2}{a^2} = \dfrac{y^2}{b^2} - 1$ $\Leftrightarrow$ $\dfrac{y^2}{b^2} - \dfrac{x^2}{a^2} = 1$, which is the equation of a hyperbola.

31. $x = t \cos t$, $y = t \sin t$, $t \geq 0$

t	x	y
0	0	0
$\frac{\pi}{4}$	$\frac{\pi\sqrt{2}}{8}$	$\frac{\pi\sqrt{2}}{8}$
$\frac{\pi}{2}$	0	$\frac{\pi}{2}$
$\frac{3\pi}{4}$	$-\frac{3\pi\sqrt{2}}{8}$	$\frac{3\pi\sqrt{2}}{8}$
π	$-\pi$	0

t	x	y
$\frac{5\pi}{4}$	$-\frac{5\pi\sqrt{2}}{8}$	$-\frac{5\pi\sqrt{2}}{8}$
$\frac{3\pi}{2}$	0	$-\frac{3\pi}{2}$
$\frac{7\pi}{4}$	$\frac{7\pi\sqrt{2}}{8}$	$-\frac{7\pi\sqrt{2}}{8}$
2π	2π	0

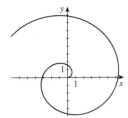

33. $x = \dfrac{3t}{1+t^3}$, $y = \dfrac{3t^2}{1+t^3}$, $t \neq -1$

t	x	y
-0.9	-9.96	8.97
-0.75	-3.89	2.92
-0.5	-1.71	0.86
0	0	0
0.5	1.33	0.67
1	1.5	1.5
1.5	1.03	1.54

t	x	y
2	0.67	1.33
2.5	0.45	1.13
3	0.32	0.96
4	0.18	0.74
5	0.12	0.60
6	0.08	0.50

t	x	y
-1.1	9.97	-10.97
-1.25	3.93	-4.92
-1.5	1.89	-2.84
-2	0.86	-1.71
-2.5	0.51	-1.28
-3	0.35	-1.04
-3.5	0.25	-0.88

t	x	y
-4	0.19	-0.76
-4.5	0.15	-0.67
-5	0.12	-0.60
-6	0.08	-0.50
-7	0.06	-0.43
-8	0.05	-0.38

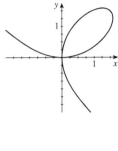

As $t \to -1^-$ we have $x \to -\infty$ and $y \to \infty$. As $t \to -1^+$ we have $x \to \infty$ and $y \to -\infty$. As $t \to \infty$ we have $x \to 0^+$ and $y \to 0^+$. As $t \to -\infty$ we have $x \to 0^+$ and $y \to 0^-$.

35. $x = (v_0 \cos \alpha)\, t$, $y = (v_0 \sin \alpha)\, t - 16t^2$. From the equation for x, $t = \dfrac{x}{v_0 \cos \alpha}$. Substituting into the equation for y gives

$$y = (v_0 \sin \alpha)\, \frac{x}{v_0 \cos \alpha} - 16 \left(\frac{x}{v_0 \cos \alpha} \right)^2 = x \tan \alpha - \frac{16x^2}{v_0^2 \cos^2 \alpha}. \quad \text{Thus the equation is of the form } y = c_1 x - c_2 x^2,$$

where c_1 and c_2 are constants, so its graph is a parabola.

37. $x = \sin t$, $y = 2 \cos 3t$

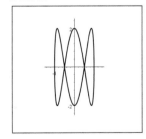

39. $x = 3 \sin 5t$, $y = 5 \cos 3t$

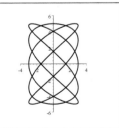

41. $x = \sin (\cos t)$, $y = \cos t^{3/2}$, $0 \leq t \leq 2\pi$

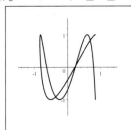

43. (a) $r = e^{\theta/12}$, $0 \leq \theta \leq 4\pi$ $\Rightarrow$

$x = e^{t/12} \cos t$, $y = e^{t/12} \sin t$

(b)

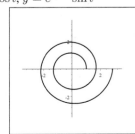

45. (a) $r = \dfrac{4}{2 - \cos\theta}$ $\Leftrightarrow$ $x = \dfrac{4\cos t}{2 - \cos t}, y = \dfrac{4\sin t}{2 - \cos t}$

(b)

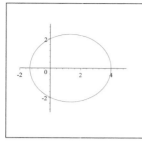

47. $x = t^3 - 2t, y = t^2 - t$ is Graph III, since

$$y = t^2 - t = \left(t^2 - t + \tfrac{1}{4}\right) - \tfrac{1}{4} = \left(t - \tfrac{1}{2}\right)^2 - \tfrac{1}{4}, \text{ and so}$$

$y \geq -\tfrac{1}{4}$ on this curve, while x is unbounded.

49. $x = t + \sin 2t, y = t + \sin 3t$ is Graph II, since the values of x and y oscillate about their values on the line $x = t, y = t$
   $\Leftrightarrow$ $y = x$.

51. (a) If we modify Figure 8 so that $|PC| = b$, then by the
same reasoning as in Example 6, we see that
$x = |OT| - |PQ| = a\theta - b\sin\theta$ and
$y = |TC| - |CQ| = a - b\cos\theta$.
We graph the case where $a = 3$ and $b = 2$.

(b)

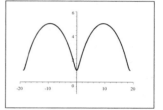

53. (a) We first note that the center of circle C (the small circle) has coordinates
$([a - b]\cos\theta, [a - b]\sin\theta)$. Now the arc PQ has the same length as the arc
$P'Q$, so $b\phi = a\theta$ $\Leftrightarrow$ $\phi = \dfrac{a}{b}\theta$, and so $\phi - \theta = \dfrac{a}{b}\theta - \theta = \dfrac{a - b}{b}\theta$. Thus
the x-coordinate of P is the x-coordinate of the center of circle C plus
$b\cos(\phi - \theta) = b\cos\left(\dfrac{a - b}{b}\theta\right)$, and the y-coordinate of P is the
y-coordinate of the center of circle C minus
$b \cdot \sin(\phi - \theta) = b\sin\left(\dfrac{a - b}{b}\theta\right)$. So $x = (a - b)\cos\theta + b\cos\left(\dfrac{a - b}{b}\theta\right)$
and $y = (a - b)\sin\theta - b\sin\left(\dfrac{a - b}{b}\theta\right)$.

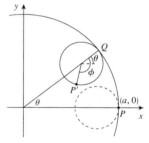

(b) If $a = 4b, b = \dfrac{a}{4}$, and $x = \tfrac{3}{4}a\cos\theta + \tfrac{1}{4}a\cos 3\theta, y = \tfrac{3}{4}a\sin\theta - \tfrac{1}{4}a\sin 3\theta$.
From Example 2 in Section 7.3, $\cos 3\theta = 4\cos^3\theta - 3\cos\theta$. Similarly, one
can prove that $\sin 3\theta = 3\sin\theta - 4\sin^3\theta$. Substituting, we get
$x = \tfrac{3}{4}a\cos\theta + \tfrac{1}{4}a\left(4\cos^3\theta - 3\cos\theta\right) = a\cos^3\theta$
$y = \tfrac{3}{4}a\sin\theta - \tfrac{1}{4}a\left(3\sin\theta - 4\sin^3\theta\right) = a\sin^3\theta$. Thus,
$x^{2/3} + y^{2/3} = a^{2/3}\cos^2\theta + a^{2/3}\sin^2\theta = a^{2/3}$, so $x^{2/3} + y^{2/3} = a^{2/3}$.

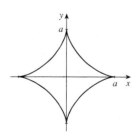

55. A polar equation for the circle is $r = 2a\sin\theta$. Thus the coordinates of Q are $x = r\cos\theta = 2a\sin\theta\cos\theta$ and
$y = r\sin\theta = 2a\sin^2\theta$. The coordinates of R are $x = 2a\cot\theta$ and $y = 2a$. Since P is the midpoint of QR, we use the
midpoint formula to get $x = a(\sin\theta\cos\theta + \cot\theta)$ and $y = a(1 + \sin^2\theta)$.

57. (a) A has coordinates $(a\cos\theta, a\sin\theta)$. Since OA is perpendicular to AB, $\triangle OAB$ is a right triangle and B has coordinates $(a\sec\theta, 0)$. It follows that P has coordinates $(a\sec\theta, b\sin\theta)$. Thus, the parametric equations are $x = a\sec\theta$, $y = b\sin\theta$.

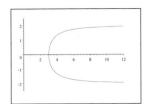

(b) The right half of the curve is graphed with $a = 3$ and $b = 2$.

59. We use the equation for y from Example 6 and solve for θ. Thus for $0 \le \theta \le \pi$, $y = a(1 - \cos\theta) \Leftrightarrow \dfrac{a-y}{a} = \cos\theta$

$\Leftrightarrow \quad \theta = \cos^{-1}\left(\dfrac{a-y}{a}\right)$. Substituting into the equation for x, we get $x = a\left[\cos^{-1}\left(\dfrac{a-y}{a}\right) - \sin\left(\cos^{-1}\left(\dfrac{a-y}{a}\right)\right)\right]$.

However, $\sin\left(\cos^{-1}\left(\dfrac{a-y}{a}\right)\right) = \sqrt{1 - \left(\dfrac{a-y}{a}\right)^2} = \dfrac{\sqrt{2ay - y^2}}{a}$. Thus, $x = a\left[\cos^{-1}\left(\dfrac{a-y}{a}\right) - \dfrac{\sqrt{2ay - y^2}}{a}\right]$,

and we have $\dfrac{\sqrt{2ay - y^2} + x}{a} = \cos^{-1}\left(\dfrac{a-y}{a}\right) \quad \Rightarrow \quad 1 - \dfrac{y}{a} = \cos\left(\dfrac{\sqrt{2ay - y^2} + x}{a}\right) \quad \Rightarrow$

$y = a\left[1 - \cos\left(\dfrac{\sqrt{2ay - y^2} + x}{a}\right)\right]$.

61. (a) In the figure, since OQ and QT are perpendicular and OT and TD are perpendicular, the angles formed by their intersections are equal, that is, $\theta = \angle DTQ$. Now the coordinates of T are $(\cos\theta, \sin\theta)$. Since $|TD|$ is the length of the string that has been unwound from the circle, it must also have arc length θ, so $|TD| = \theta$. Thus the x-displacement from T to D is $\theta \cdot \sin\theta$ while the y-displacement from T to D is $\theta \cdot \cos\theta$. So the coordinates of D are $x = \cos\theta + \theta\sin\theta$ and $y = \sin\theta - \theta\cos\theta$.

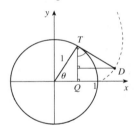

(b)

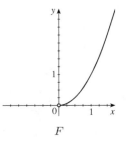

63. C: $x = t$, $y = t^2$; $\quad D$: $x = \sqrt{t}$, $y = t$, $t \ge 0$ $\quad E$: $x = \sin t$, $y = 1 - \cos^2 t$ $\quad F$: $x = e^t$, $y = e^{2t}$

(a) For C, $x = t$, $y = t^2 \quad \Rightarrow \quad y = x^2$.

For D, $x = \sqrt{t}$, $y = t \quad \Rightarrow \quad y = x^2$.

For E, $x = \sin t \quad \Rightarrow \quad x^2 = \sin^2 t = 1 - \cos^2 t = y$ and so $y = x^2$.

For F, $x = e^t \quad \Rightarrow \quad x^2 = e^{2t} = y$ and so $y = x^2$. Therefore, the points on all four curves satisfy the same rectangular equation.

(b) Curve C is the entire parabola $y = x^2$. Curve D is the right half of the parabola because $t \ge 0$, so $x \ge 0$. Curve E is the portion of the parabola for $-1 \le x \le 1$. Curve F is the portion of the parabola where $x > 0$, since $e^t > 0$ for all t.

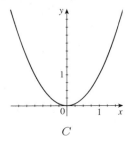

C

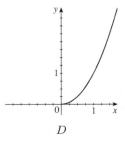

D

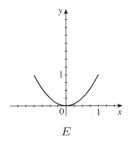

E

F

Chapter 11 Review

1. $y^2 = 4x$. This is a parabola with $4p = 4 \iff p = 1$. The vertex is $(0,0)$, the focus is $(1,0)$, and the directrix is $x = -1$.

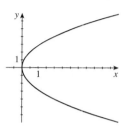

3. $x^2 + 8y = 0 \iff x^2 = -8y$. This is a parabola with $4p = -8 \iff p = -2$. The vertex is $(0,0)$, the focus is $(0,-2)$, and the directrix is $y = 2$.

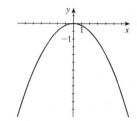

5. $x - y^2 + 4y - 2 = 0 \iff x - (y^2 - 4y + 4) - 2 = -4 \iff$
$x - (y-2)^2 = -2 \iff (y-2)^2 = x + 2$. This is a parabola with $4p = 1 \iff$
$p = \frac{1}{4}$. The vertex is $(-2, 2)$, the focus is $\left(-2 + \frac{1}{4}, 2\right) = \left(-\frac{7}{4}, 2\right)$, and the
directrix is $x = -2 - \frac{1}{4} = -\frac{9}{4}$.

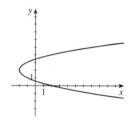

7. $\frac{1}{2}x^2 + 2x = 2y + 4 \iff x^2 + 4x = 4y + 8 \iff x^2 + 4x + 4 = 4y + 8$
$\iff (x+2)^2 = 4(y+3)$. This is a parabola with $4p = 4 \iff p = 1$. The
vertex is $(-2, -3)$, the focus is $(-2, -3+1) = (-2, -2)$, and the directrix is
$y = -3 - 1 = -4$.

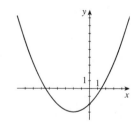

9. $\dfrac{x^2}{9} + \dfrac{y^2}{25} = 1$. This is an ellipse with $a = 5$, $b = 3$, and $c = \sqrt{25 - 9} = 4$. The
center is $(0,0)$, the vertices are $(0, \pm 5)$, the foci are $(0, \pm 4)$, the length of the
major axis is $2a = 10$, and the length of the minor axis is $2b = 6$.

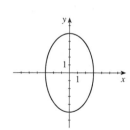

11. $x^2 + 4y^2 = 16 \iff \dfrac{x^2}{16} + \dfrac{y^2}{4} = 1$. This is an ellipse with $a = 4$, $b = 2$, and
$c = \sqrt{16 - 4} = 2\sqrt{3}$. The center is $(0,0)$, the vertices are $(\pm 4, 0)$, the foci are
$\left(\pm 2\sqrt{3}, 0\right)$, the length of the major axis is $2a = 8$, and the length of the minor axis
is $2b = 4$.

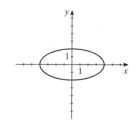

13. $\dfrac{(x-3)^2}{9} + \dfrac{y^2}{16} = 1$. This is an ellipse with $a = 4$, $b = 3$, and $c = \sqrt{16-9} = \sqrt{7}$.

The center is $(3, 0)$, the vertices are $(3, \pm 4)$, the foci are $(3, \pm\sqrt{7})$, the length of the major axis is $2a = 8$, and the length of the minor axis is $2b = 6$.

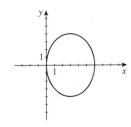

15. $4x^2 + 9y^2 = 36y$ $\Leftrightarrow$ $4x^2 + 9\left(y^2 - 4y + 4\right) = 36$ $\Leftrightarrow$

$4x^2 + 9\left(y-2\right)^2 = 36$ $\Leftrightarrow$ $\dfrac{x^2}{9} + \dfrac{(y-2)^2}{4} = 1$. This is an ellipse with $a = 3$,

$b = 2$, and $c = \sqrt{9-4} = \sqrt{5}$. The center is $(0, 2)$, the vertices are $(\pm 3, 2)$, the

foci are $(\pm\sqrt{5}, 2)$, the length of the major axis is $2a = 6$, and the length of the

minor axis is $2b = 4$.

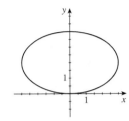

17. $-\dfrac{x^2}{9} + \dfrac{y^2}{16} = 1$ $\Leftrightarrow$ $\dfrac{y^2}{16} - \dfrac{x^2}{9} = 0$. This is a hyperbola with $a = 4$, $b = 3$, and

$c = \sqrt{16+9} = \sqrt{25} = 5$. The center is $(0, 0)$, the vertices are $(0, \pm 4)$, the foci

are $(0, \pm 5)$, and the asymptotes are $y = \pm\frac{4}{3}x$.

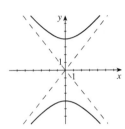

19. $x^2 - 2y^2 = 16$ $\Leftrightarrow$ $\dfrac{x^2}{16} - \dfrac{y^2}{8} = 1$. This is a hyperbola with $a = 4$, $b = 2\sqrt{2}$,

and $c = \sqrt{16+8} = \sqrt{24} = 2\sqrt{6}$. The center is $(0, 0)$, the vertices are $(\pm 4, 0)$,

the foci are $(\pm 2\sqrt{6}, 0)$, and the asymptotes are $y = \pm\frac{2\sqrt{2}}{4}x$ $\Leftrightarrow$ $y = \pm\frac{1}{\sqrt{2}}x$.

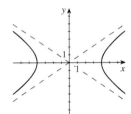

21. $\dfrac{(x+4)^2}{16} - \dfrac{y^2}{16} = 1$. This is a hyperbola with $a = 4$, $b = 4$ and

$c = \sqrt{16+16} = 4\sqrt{2}$. The center is $(-4, 0)$, the vertices are $(-4 \pm 4, 0)$ which

are $(-8, 0)$ and $(0, 0)$, the foci are $(-4 \pm 4\sqrt{2}, 0)$, and the asymptotes are

$y = \pm(x+4)$.

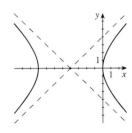

23. $9y^2 + 18y = x^2 + 6x + 18$ $\Leftrightarrow$

$9\left(y^2 + 2y + 1\right) = \left(x^2 + 6x + 9\right) + 9 - 9 + 18$ $\Leftrightarrow$

$9(y+1)^2 - (x+3)^2 = 18$ $\Leftrightarrow$ $\dfrac{(y+1)^2}{2} - \dfrac{(x+3)^2}{18} = 1$. This is a

hyperbola with $a = \sqrt{2}$, $b = 3\sqrt{2}$, and $c = \sqrt{2+18} = 2\sqrt{5}$. The center is

$(-3, -1)$, the vertices are $(-3, -1 \pm \sqrt{2})$, the foci are $(-3, -1 \pm 2\sqrt{5})$, and the

asymptotes are $y + 1 = \pm\frac{1}{3}(x+3)$ $\Leftrightarrow$ $y = \frac{1}{3}x$ and $y = -\frac{1}{3}x - 2$.

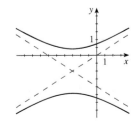

25. This is a parabola that opens to the right with its vertex at $(0,0)$ and the focus at $(2,0)$. So $p = 2$, and the equation is $y^2 = 4(2)x \iff y^2 = 8x$.

27. From the graph, the center is $(0,0)$, and the vertices are $(0,-4)$ and $(0,4)$. Since a is the distance from the center to a vertex, we have $a = 4$. Because one focus is $(0,5)$, we have $c = 5$, and since $c^2 = a^2 + b^2$, we have $25 = 16 + b^2 \iff b^2 = 9$. Thus an equation of the hyperbola is $\dfrac{y^2}{16} - \dfrac{x^2}{9} = 1$.

29. From the graph, the center of the ellipse is $(4,2)$, and so $a = 4$ and $b = 2$. The equation is $\dfrac{(x-4)^2}{4^2} + \dfrac{(y-2)^2}{2^2} = 1 \iff \dfrac{(x-4)^2}{16} + \dfrac{(y-2)^2}{4} = 1$.

31. $\dfrac{x^2}{12} + y = 1 \iff \dfrac{x^2}{12} = -(y-1) \iff x^2 = -12(y-1)$. This is a parabola with $4p = -12 \iff p = -3$. The vertex is $(0,1)$ and the focus is $(0, 1-3) = (0,-2)$.

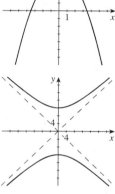

33. $x^2 - y^2 + 144 = 0 \iff \dfrac{y^2}{144} - \dfrac{x^2}{144} = 1$. This is a hyperbola with $a = 12$, $b = 12$, and $c = \sqrt{144 + 144} = 12\sqrt{2}$. The vertices are $(0, \pm 12)$ and the foci are $\left(0, \pm 12\sqrt{2}\right)$.

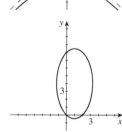

35. $4x^2 + y^2 = 8(x+y) \iff 4(x^2 - 2x) + (y^2 - 8y) = 0 \iff 4(x^2 - 2x + 1) + (y^2 - 8y + 16) = 4 + 16 \iff 4(x-1)^2 + (y-4)^2 = 20$
$\iff \dfrac{(x-1)^2}{5} + \dfrac{(y-4)^2}{20} = 1$. This is an ellipse with $a = 2\sqrt{5}$, $b = \sqrt{5}$, and $c = \sqrt{20 - 5} = \sqrt{15}$. The vertices are $\left(1, 4 \pm 2\sqrt{5}\right)$ and the foci are $\left(1, 4 \pm \sqrt{15}\right)$.

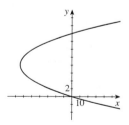

37. $x = y^2 - 16y \iff x + 64 = y^2 - 16y + 64 \iff (y-8)^2 = x + 64$. This is a parabola with $4p = 1 \iff p = \frac{1}{4}$. The vertex is $(-64, 8)$ and the focus is $\left(-64 + \frac{1}{4}, 8\right) = \left(-\frac{255}{4}, 8\right)$.

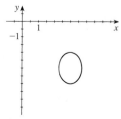

39. $2x^2 - 12x + y^2 + 6y + 26 = 0 \iff 2(x^2 - 6x) + (y^2 + 6y) = -26 \iff 2(x^2 - 6x + 9) + (y^2 + 6y + 9) = -26 + 18 + 9 \iff 2(x-3)^2 + (y+3)^2 = 1 \iff \dfrac{(x-3)^2}{\frac{1}{2}} + (y+3)^2 = 1$. This is an ellipse with $a = 1$, $b = \frac{\sqrt{2}}{2}$, and $c = \sqrt{1 - \frac{1}{2}} = \frac{\sqrt{2}}{2}$. The vertices are $(3, -3 \pm 1) = (3, -4)$ and $(3, -2)$, and the foci are $\left(3, -3 \pm \frac{\sqrt{2}}{2}\right)$.

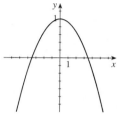

41. $9x^2 + 8y^2 - 15x + 8y + 27 = 0 \Leftrightarrow 9\left(x^2 - \frac{5}{3}x + \frac{25}{36}\right) + 8\left(y^2 + y + \frac{1}{4}\right) = -27 + \frac{25}{4} + 2 \Leftrightarrow$

$9\left(x - \frac{5}{6}\right)^2 + 8\left(y + \frac{1}{2}\right)^2 = -\frac{75}{4}$. However, since the left-hand side of the equation is greater than or equal to 0, there is no point that satisfies this equation. The graph is empty.

43. The parabola has focus $(0, 1)$ and directrix $y = -1$. Therefore, $p = 1$ and so $4p = 4$. Since the focus is on the y-axis and the vertex is $(0, 0)$, an equation of the parabola is $x^2 = 4y$.

45. The hyperbola has vertices $(0, \pm 2)$ and asymptotes $y = \pm\frac{1}{2}x$. Therefore, $a = 2$, and the foci are on the y-axis. Since the slopes of the asymptotes are $\pm\frac{1}{2} = \pm\frac{a}{b} \Leftrightarrow b = 2a = 4$, an equation of the hyperbola is $\dfrac{y^2}{4} - \dfrac{x^2}{16} = 1$.

47. The ellipse has foci $F_1(1, 1)$ and $F_2(1, 3)$, and one vertex is on the x-axis. Thus, $2c = 3 - 1 = 2 \Leftrightarrow c = 1$, and so the center of the ellipse is $C(1, 2)$. Also, since one vertex is on the x-axis, $a = 2 - 0 = 2$, and thus $b^2 = 4 - 1 = 3$. So an equation of the ellipse is $\dfrac{(x - 1)^2}{3} + \dfrac{(y - 2)^2}{4} = 1$.

49. The ellipse has vertices $V_1(7, 12)$ and $V_2(7, -8)$ and passes through the point $P(1, 8)$. Thus, $2a = 12 - (-8) = 20 \Leftrightarrow$ $a = 10$, and the center is $\left(7, \dfrac{-8 + 12}{2}\right) = (7, 2)$. Thus an equation of the ellipse has the form $\dfrac{(x - 7)^2}{b^2} + \dfrac{(y - 2)^2}{100} = 1$.

Since the point $P(1, 8)$ is on the ellipse, $\dfrac{(1 - 7)^2}{b^2} + \dfrac{(8 - 2)^2}{100} = 1 \Leftrightarrow 3600 + 36b^2 = 100b^2 \Leftrightarrow 64b^2 = 3600 \Leftrightarrow$ $b^2 = \frac{225}{4}$. Therefore, an equation of the ellipse is $\dfrac{(x - 7)^2}{225/4} + \dfrac{(y - 2)^2}{100} = 1 \Leftrightarrow \dfrac{4(x - 7)^2}{225} + \dfrac{(y - 2)^2}{100} = 1$.

51. The length of the major axis is $2a = 186{,}000{,}000 \Leftrightarrow a = 93{,}000{,}000$. The eccentricity is $e = c/a = 0.017$, and so $c = 0.017\,(93{,}000{,}000) = 1{,}581{,}000$.

(a) The earth is closest to the sun when the distance is $a - c = 93{,}000{,}000 - 1{,}581{,}000 = 91{,}419{,}000$.

(b) The earth is furthest from the sun when the distance is $a + c = 93{,}000{,}000 + 1{,}581{,}000 = 94{,}581{,}000$.

53. (a) The graphs of $\dfrac{x^2}{16 + k^2} + \dfrac{y^2}{k^2} = 1$ for $k = 1, 2, 4,$ and 8 are shown in the figure.

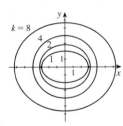

(b) $c^2 = \left(16 + k^2\right) - k^2 = 16 \Rightarrow c = \pm 4$. Since the center is $(0, 0)$, the foci of each of the ellipses are $(\pm 4, 0)$.

55. (a) $x^2 + 4xy + y^2 = 1$. Then $A = 1$, $B = 4$, and $C = 1$, so the discriminant is $4^2 - 4(1)(1) = 12$. Since the discriminant is positive, the equation represents a hyperbola.

(b) $\cot 2\phi = \dfrac{A - C}{B} = \dfrac{1 - 1}{4} = 0 \Rightarrow 2\phi = 90° \Leftrightarrow \phi = 45°$. Therefore, $x = \frac{\sqrt{2}}{2}X - \frac{\sqrt{2}}{2}Y$ and $y = \frac{\sqrt{2}}{2}X + \frac{\sqrt{2}}{2}Y$. Substituting into the original equation gives

$$\left(\tfrac{\sqrt{2}}{2}X - \tfrac{\sqrt{2}}{2}Y\right)^2 + 4\left(\tfrac{\sqrt{2}}{2}X - \tfrac{\sqrt{2}}{2}Y\right)\left(\tfrac{\sqrt{2}}{2}X + \tfrac{\sqrt{2}}{2}Y\right) + \left(\tfrac{\sqrt{2}}{2}X + \tfrac{\sqrt{2}}{2}Y\right)^2 = 1 \Leftrightarrow$$

$$\tfrac{1}{2}\left(X^2 - 2XY + Y^2\right) + 2\left(X^2 + XY - XY - Y^2\right) + \tfrac{1}{2}\left(X^2 + 2XY + Y^2\right) = 1 \qquad \textbf{(c)}$$

$\Leftrightarrow 3X^2 - Y^2 = 1 \Leftrightarrow 3X^2 - Y^2 = 1$. This is a hyperbola with $a = \frac{1}{\sqrt{3}}$, $b = 1$, and

$c = \sqrt{\frac{1}{3} + 1} = \frac{2}{\sqrt{3}}$. Therefore, the hyperbola has vertices $V\left(\pm\frac{1}{\sqrt{3}}, 0\right)$ and foci

$F\left(\pm\frac{2}{\sqrt{3}}, 0\right)$, in XY-coordinates.

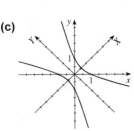

57. (a) $7x^2 - 6\sqrt{3}xy + 13y^2 - 4\sqrt{3}x - 4y = 0$. Then $A = 7$, $B = -6\sqrt{3}$, and $C = 13$, so the discriminant is

$\left(-6\sqrt{3}\right)^2 - 4\,(7)\,(13) = -256$. Since the discriminant is negative, the equation represents an ellipse.

(b) $\cot 2\phi = \dfrac{A - C}{B} = \dfrac{7 - 13}{-6\sqrt{3}} = \dfrac{1}{\sqrt{3}}$ $\Rightarrow$ $2\phi = 60°$ $\Leftrightarrow$ $\phi = 30°$. Therefore, $x = \frac{\sqrt{3}}{2}X - \frac{1}{2}Y$ and

$y = \frac{1}{2}X + \frac{\sqrt{3}}{2}Y$. Substituting into the original equation gives

$7\left(\frac{\sqrt{3}}{2}X - \frac{1}{2}Y\right)^2 - 6\sqrt{3}\left(\frac{\sqrt{3}}{2}X - \frac{1}{2}Y\right)\left(\frac{1}{2}X + \frac{\sqrt{3}}{2}Y\right)$

$$+ 13\left(\frac{1}{2}X + \frac{\sqrt{3}}{2}Y\right)^2 - 4\sqrt{3}\left(\frac{\sqrt{3}}{2}X - \frac{1}{2}Y\right) - 4\left(\frac{1}{2}X + \frac{\sqrt{3}}{2}Y\right) = 0 \quad \Leftrightarrow$$

$\frac{7}{4}\left(3X^2 - 2\sqrt{3}XY + Y^2\right) - \frac{3\sqrt{3}}{2}\left(\sqrt{3}X^2 + 3XY - XY - \sqrt{3}Y^2\right)$

$$+ \frac{13}{4}\left(X^2 + 2\sqrt{3}XY + 3Y^2\right) - 6X + 2\sqrt{3}Y - 2X - 2\sqrt{3}Y = 0 \quad \Leftrightarrow$$

$X^2\left(\frac{21}{4} - \frac{9}{2} + \frac{13}{4}\right) - 8X + Y^2\left(\frac{7}{4} + \frac{9}{2} + \frac{39}{4}\right) = 0$ $\Leftrightarrow$ **(c)**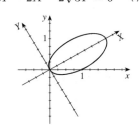

$4X^2 - 8X + 16Y^2 = 0$ $\Leftrightarrow$ $4\left(X^2 - 2X + 1\right) + 16Y^2 = 4$ $\Leftrightarrow$

$(X - 1)^2 + 4Y^2 = 1$. This ellipse has $a = 1$, $b = \frac{1}{2}$, and $c = \sqrt{1 - \frac{1}{4}} = \frac{1}{2}\sqrt{3}$.

Therefore, the vertices are $V\,(1 \pm 1, 0) = V_1\,(0, 0)$ and $V_2\,(2, 0)$ and the foci are

$F\left(1 \pm \frac{1}{2}\sqrt{3}, 0\right)$.

59. $5x^2 + 3y^2 = 60$ $\Leftrightarrow$ $3y^2 = 60 - 5x^2$ $\Leftrightarrow$ $y^2 = 20 - \frac{5}{3}x^2$. This conic is
an ellipse.

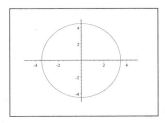

61. $6x + y^2 - 12y = 30$ $\Leftrightarrow$ $y^2 - 12y = 30 - 6x$ $\Leftrightarrow$

$y^2 - 12y + 36 = 66 - 6x$ $\Leftrightarrow$ $(y - 6)^2 = 66 - 6x$ $\Leftrightarrow$

$y - 6 = \pm\sqrt{66 - 6x}$ $\Leftrightarrow$ $y = 6 \pm \sqrt{66 - 6x}$. This conic is a parabola.

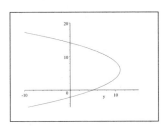

63. (a) $r = \dfrac{1}{1 - \cos\theta}$ $\Rightarrow$ $e = 1$. Therefore, this is a

parabola.

(b)

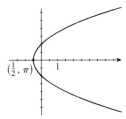

65. (a) $r = \dfrac{4}{1 + 2\sin\theta}$ $\Rightarrow$ $e = 2$. Therefore this is a

hyperbola.

(b)

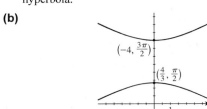

67. (a)

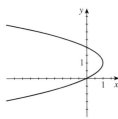

(b) $x = 1 - t^2, y = 1 + t \quad \Leftrightarrow \quad t = y - 1.$
Substituting for t gives $x = 1 - (y - 1)^2 \quad \Leftrightarrow$
$(y - 1)^2 = 1 - x$ which is the rectangular coordinate equation.

69. (a)

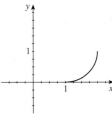

(b) $x = 1 + \cos t \quad \Leftrightarrow \quad \cos t = x - 1$, and
$y = 1 - \sin t \quad \Leftrightarrow \quad \sin t = 1 - y.$ Since
$\cos^2 t + \sin^2 t = 1$, it follows that
$(x - 1)^2 + (1 - y)^2 = 1 \quad \Leftrightarrow$
$(x - 1)^2 + (y - 1)^2 = 1.$ Since t is restricted by
$0 \le t \le \frac{\pi}{2}, 1 + \cos 0 \le x \le 1 + \cos \frac{\pi}{2} \quad \Leftrightarrow$
$1 \le x \le 2$, and similarly, $0 \le y \le 1$. (This is the lower right quarter of the circle.)

71. $x = \cos 2t, y = \sin 3t$

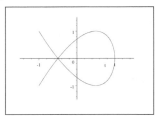

73. The coordinates of Q are $x = \cos \theta$ and $y = \sin \theta$. The coordinates of R are $x = 1$ and $y = \tan \theta$. Hence, the midpoint P is $\left(\dfrac{1 + \cos \theta}{2}, \dfrac{\sin \theta + \tan \theta}{2} \right)$, so parametric equations for the curve are $x = \dfrac{1 + \cos \theta}{2}$ and $y = \dfrac{\sin \theta + \tan \theta}{2}.$

Chapter 11 Test

1. $x^2 = -12y$. This is a parabola with $4p = -12 \quad \Leftrightarrow \quad p = -3$. The focus is $(0, -3)$ and the directrix is $y = 3$.

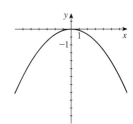

2. $\dfrac{x^2}{16} + \dfrac{y^2}{4} = 1$. This is an ellipse with $a = 4, b = 2$, and $c = \sqrt{16 - 4} = 2\sqrt{3}$.
The vertices are $(\pm 4, 0)$, the foci are $(\pm 2\sqrt{3}, 0)$, the length of the major axis is $2a = 8$, and the length of the minor axis is $2b = 4$.

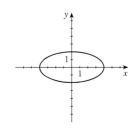

3. $\dfrac{y^2}{9} - \dfrac{x^2}{16} = 1$. This is a hyperbola with $a = 3$, $b = 4$, and $c = \sqrt{9 + 16} = 5$. The

vertices are $(0, \pm 3)$, the foci are $(0, \pm 5)$, and the asymptotes are $y = \pm \frac{3}{4}x$.

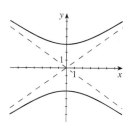

4. This is a parabola that opens to the left with its vertex at $(0, 0)$. So its equation is of the form $y^2 = 4px$ with $p < 0$.
Substituting the point $(-4, 2)$, we have $2^2 = 4p(-4)$ $\Leftrightarrow$ $4 = -16p$ $\Leftrightarrow$ $p = -\frac{1}{4}$. So an equation is
$y^2 = 4\left(-\frac{1}{4}\right)x$ $\Leftrightarrow$ $y^2 = -x$.

5. This is an ellipse tangent to the x-axis at $(0, 0)$ and with one vertex at the point $(4, 3)$. The center is $(0, 3)$, and $a = 4$ and
$b = 3$. Thus the equation is $\dfrac{x^2}{16} + \dfrac{(y - 3)^2}{9} = 1$.

6. This a hyperbola with a horizontal transverse axis, vertices at $(1, 0)$ and $(3, 0)$, and foci at $(0, 0)$ and $(4, 0)$. Thus the center
is $(2, 0)$, and $a = 3 - 2 = 1$ and $c = 4 - 2 = 2$. Thus $b^2 = 2^2 - 1^2 = 3$. So an equation is $\dfrac{(x - 2)^2}{1^2} - \dfrac{y^2}{3} = 1$ $\Leftrightarrow$
$(x - 2)^2 - \dfrac{y^2}{3} = 1$.

7. $16x^2 + 36y^2 - 96x + 36y + 9 = 0$ $\Leftrightarrow$ $16\left(x^2 - 6x\right) + 36\left(y^2 + y\right) = -9$

$\Leftrightarrow$ $16\left(x^2 - 6x + 9\right) + 36\left(y^2 + y + \frac{1}{4}\right) = -9 + 144 + 9$ $\Leftrightarrow$

$16\left(x - 3\right)^2 + 36\left(y + \frac{1}{2}\right)^2 = 144$ $\Leftrightarrow$ $\dfrac{(x - 3)^2}{9} + \dfrac{\left(y + \frac{1}{2}\right)^2}{4} = 1$. This is an

ellipse with $a = 3$, $b = 2$, and $c = \sqrt{9 - 4} = \sqrt{5}$. The center is $\left(3, -\frac{1}{2}\right)$, the

vertices are $\left(3 \pm 3, -\frac{1}{2}\right) = \left(0, -\frac{1}{2}\right)$ and $\left(6, -\frac{1}{2}\right)$, and the foci are

$(h \pm c, k) = \left(3 \pm \sqrt{5}, -\frac{1}{2}\right) = \left(3 + \sqrt{5}, -\frac{1}{2}\right)$ and $\left(3 - \sqrt{5}, -\frac{1}{2}\right)$.

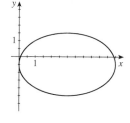

8. $9x^2 - 8y^2 + 36x + 64y = 164$ $\Leftrightarrow$ $9\left(x^2 + 4x\right) - 8\left(y^2 - 8y\right) = 164$ $\Leftrightarrow$
$9\left(x^2 + 4x + 4\right) - 8\left(y^2 - 8y + 16\right) = 164 + 36 - 128$ $\Leftrightarrow$

$9\left(x + 2\right)^2 - 8\left(y - 4\right)^2 = 72$ $\Leftrightarrow$ $\dfrac{(x + 2)^2}{8} - \dfrac{(y - 4)^2}{9} = 1$. This conic is a

hyperbola.

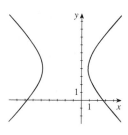

9. $2x + y^2 + 8y + 8 = 0$ $\Leftrightarrow$ $y^2 + 8y + 16 = -2x - 8 + 16$ $\Leftrightarrow$
$(y + 4)^2 = -2(x - 4)$. This is a parabola with $4p = -2$ $\Leftrightarrow$ $p = -\frac{1}{2}$. The
vertex is $(4, -4)$ and the focus is $\left(4 - \frac{1}{2}, -4\right) = \left(\frac{7}{2}, -4\right)$.

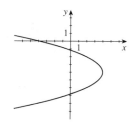

10. The hyperbola has foci $(0, \pm 5)$ and asymptotes $y = \pm\frac{3}{4}x$. Since the foci are $(0, \pm 5)$, $c = 5$, the foci are on the y-axis, and the center is $(0, 0)$. Also, since $y = \pm\frac{3}{4}x = \pm\frac{a}{b}x$, it follows that $\frac{a}{b} = \frac{3}{4}$ $\Leftrightarrow$ $a = \frac{3}{4}b$. Then $c^2 = 5^2 = 25 = a^2 + b^2 = \left(\frac{3}{4}b\right)^2 + b^2 = \frac{25}{16}b^2$ $\Leftrightarrow$ $b^2 = 16$, and by substitution, $a = \frac{3}{4}(4) = 3$. Therefore, an equation of the hyperbola is $\dfrac{y^2}{9} - \dfrac{x^2}{16} = 1$.

11. The parabola has focus $(2, 4)$ and directrix the x-axis ($y = 0$). Therefore, $2p = 4 - 0 = 4$ $\Leftrightarrow$ $p = 2$ $\Leftrightarrow$ $4p = 8$, and the vertex is $(2, 4 - p) = (2, 2)$. Hence, an equation of the parabola is $(x - 2)^2 = 8(y - 2)$ $\Leftrightarrow$ $x^2 - 4x + 4 = 8y - 16$ $\Leftrightarrow$ $x^2 - 4x - 8y + 20 = 0$.

12. We place the vertex of the parabola at the origin, so the parabola contains the points $(3, \pm 3)$, and the equation is of the form $y^2 = 4px$. Substituting the point $(3, 3)$, we get $3^2 = 4p(3)$ $\Leftrightarrow$ $9 = 12p$ $\Leftrightarrow$ $p = \frac{3}{4}$. So the focus is $\left(\frac{3}{4}, 0\right)$, and we should place the light bulb $\frac{3}{4}$ inch from the vertex.

13. (a) $5x^2 + 4xy + 2y^2 = 18$. Then $A = 5$, $B = 4$, and $C = 2$, so the discriminant is $(4)^2 - 4(5)(2) = -24$. Since the discriminant is negative, the equation represents an ellipse.

(b) $\cot 2\phi = \dfrac{A - C}{B} = \dfrac{5 - 2}{4} = \dfrac{3}{4}$. Thus, $\cos 2\phi = \frac{3}{5}$ and so $\cos \phi = \sqrt{\dfrac{1 + (3/5)}{2}} = \dfrac{2\sqrt{5}}{5}$, $\sin\phi = \sqrt{\dfrac{1 - (3/5)}{2}} = \dfrac{\sqrt{5}}{5}$. It follows that $x = \dfrac{2\sqrt{5}}{5}X - \dfrac{\sqrt{5}}{5}Y$ and $y = \dfrac{\sqrt{5}}{5}X + \dfrac{2\sqrt{5}}{5}Y$. By substitution, $5\left(\dfrac{2\sqrt{5}}{5}X - \dfrac{\sqrt{5}}{5}Y\right)^2 + 4\left(\dfrac{2\sqrt{5}}{5}X - \dfrac{\sqrt{5}}{5}Y\right)\left(\dfrac{\sqrt{5}}{5}X + \dfrac{2\sqrt{5}}{5}Y\right) + 2\left(\dfrac{\sqrt{5}}{5}X + \dfrac{2\sqrt{5}}{5}Y\right)^2 = 18$ $\Leftrightarrow$ $4X^2 - 4XY + Y^2 + \frac{4}{5}\left(2X^2 + 4XY - XY - 2Y^2\right) + \frac{2}{5}\left(X^2 + 4XY + 4Y^2\right) = 18$ $\Leftrightarrow$ $X^2\left(4 + \frac{8}{5} + \frac{2}{5}\right) + XY\left(-4 + \frac{12}{5} + \frac{8}{5}\right) + Y^2\left(1 - \frac{8}{5} + \frac{4}{5}\right) = 18$ $\Leftrightarrow$ $6X^2 + Y^2 = 18$ $\Leftrightarrow$ $\dfrac{X^2}{3} + \dfrac{Y^2}{18} = 1$. This is an ellipse with $a = 3\sqrt{2}$ and $b = \sqrt{3}$.

(c)

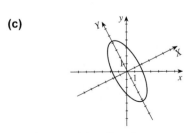

Since $\cos 2\phi = \frac{3}{5}$ we have $2\phi = \cos^{-1}\frac{3}{5} \approx 53.13°$, so $\phi \approx 27°$.

(d) In XY-coordinates, the vertices are $V\left(0, \pm 3\sqrt{2}\right)$. Therefore, in xy-coordinates, the vertices are $x = -\dfrac{3\sqrt{2}}{\sqrt{5}}$ and $y = \dfrac{6\sqrt{2}}{\sqrt{5}}$ $\Rightarrow$ $V_1\left(-\dfrac{3\sqrt{2}}{\sqrt{5}}, \dfrac{6\sqrt{2}}{\sqrt{5}}\right)$, and $x = \dfrac{3\sqrt{2}}{\sqrt{5}}$ and $y = -\dfrac{6\sqrt{2}}{\sqrt{5}}$ $\Rightarrow$ $V_2\left(\dfrac{3\sqrt{2}}{\sqrt{5}}, \dfrac{-6\sqrt{2}}{\sqrt{5}}\right)$.

14. (a) Since the focus of this conic is the origin and the directrix is $x = 2$, the equation has the form $r = \dfrac{ed}{1 + e \cos \theta}$. Subsituting $e = \frac{1}{2}$ and $d = 2$ we get $r = \dfrac{1}{1 + \frac{1}{2} \cos \theta}$ $\Leftrightarrow$ $r = \dfrac{2}{2 + \cos \theta}$.

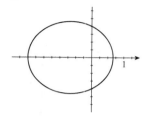

(b) $r = \dfrac{3}{2 - \sin \theta}$ $\Leftrightarrow$ $r = \dfrac{\frac{3}{2}}{1 - \frac{1}{2} \sin \theta}$. So $e = \frac{1}{2}$ and the conic is an ellipse.

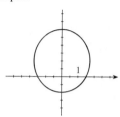

15. (a) $x = 3 \sin \theta + 3$, $y = 2 \cos \theta$, $0 \le \theta \le \pi$. From the work of part (b), we see that this is the half-ellipse shown.

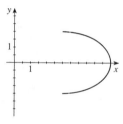

(b) $x = 3 \sin \theta + 3 \Leftrightarrow x - 3 = 3 \sin \theta \Leftrightarrow \dfrac{x - 3}{3} = \sin \theta$. Squaring both sides gives $\dfrac{(x - 3)^2}{9} = \sin^2 \theta$. Similarly, $y = 2 \cos \theta \Leftrightarrow \dfrac{y}{2} = \cos \theta$, and squaring both sides gives $\dfrac{y^2}{4} = \cos^2 \theta$. Since $\sin^2 \theta + \cos^2 \theta = 1$, it follows that $\dfrac{(x - 3)^2}{9} + \dfrac{y^2}{4} = 1$. Since $0 \le \theta \le \pi$, $\sin \theta \ge 0$, so $3 \sin \theta \ge 0 \Rightarrow 3 \sin \theta + 3 \ge 3$, and so $x \ge 3$. Thus the curve consists of only the right half of the ellipse.

Focus on Modeling: The Path of a Projectile

1. From $x = (v_0 \cos \theta) t$, we get $t = \dfrac{x}{v_0 \cos \theta}$. Substituting this value for t into the equation for y, we get

$$y = (v_0 \sin \theta) t - \tfrac{1}{2} g t^2 \quad \Leftrightarrow \quad y = (v_0 \sin \theta) \left(\frac{x}{v_0 \cos \theta} \right) - \tfrac{1}{2} g \left(\frac{x}{v_0 \cos \theta} \right)^2 \quad \Leftrightarrow \quad y = (\tan \theta) x - \frac{g}{2 v_0^2 \cos^2 \theta} x^2.$$

This shows that y is a quadratic function of x, so its graph is a parabola as long as $\theta \neq 90°$. When $\theta = 90°$, the path of the projectile is a straight line up (then down).

3. (a) We use the equation $t = \dfrac{2 v_0 \sin \theta}{g}$. Substituting $g \approx 32$ ft/s^2, $\theta = 5°$, and $v_0 = 1000$, we get

$$t = \frac{2 \cdot 1000 \cdot \sin 5°}{32} \approx 5.447 \text{ seconds.}$$

(b) Substituting the given values into $y = (v_0 \sin \theta) t - \tfrac{1}{2} g t^2$, we get

$y = 87.2t - 16t^2$. The maximum value of y is attained at the vertex of the parabola; thus $y = 87.2t - 16t^2 = -16 \left(t^2 - 5.45t \right) \quad \Leftrightarrow$

$y = -16 \left[t^2 - 2 (2.725) t + 7.425625 \right] + 118.7$. Thus the greatest height is 118.7 ft.

(c) The rocket hits the ground after 5.447 s, so substituting this into the expression for the horizontal distance gives

$x = (1000 \cos 5°) \, 5.447 = 5426$ ft.

(d)

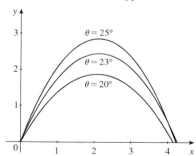

5. We use the equation of the parabola from Exercise 1 and find its vertex: $y = (\tan \theta) x - \dfrac{g}{2 v_0^2 \cos^2 \theta} x^2 \quad \Leftrightarrow$

$$y = -\frac{g}{2 v_0^2 \cos^2 \theta} \left[x^2 - \frac{2 v_0^2 \sin \theta \cos \theta x}{g} \right] \quad \Leftrightarrow \quad y = -\frac{g}{2 v_0^2 \cos^2 \theta} \left[x^2 - \frac{2 v_0^2 \sin \theta \cos \theta x}{g} + \left(\frac{v_0^2 \sin \theta \cos \theta}{g} \right)^2 \right] +$$

$$\frac{g}{2 v_0^2 \cos^2 \theta} \cdot \left(\frac{v_0^2 \sin \theta \cos \theta}{g} \right)^2 \quad \Leftrightarrow \quad y = -\frac{g}{2 v_0^2 \cos^2 \theta} \left[x - \frac{v_0^2 \sin \theta \cos \theta}{g} \right]^2 + \frac{v_0^2 \sin^2 \theta}{2g}. \text{ Thus the vertex is at}$$

$\left(\dfrac{v_0^2 \sin \theta \cos \theta}{g}, \dfrac{v_0^2 \sin^2 \theta}{2g} \right)$, so the maximum height is $\dfrac{v_0^2 \sin^2 \theta}{2g}$.

7. In Exercise 6 we derived the equations $x = (v_0 \cos \theta - w) t$, $y = (v_0 \sin \theta) t - \tfrac{1}{2} g t^2$. We plot the graphs for the given values of v_0, w, and θ in the first figure. We see that the optimal firing angle appears to be between $15°$ and $30°$. The projectile will be blown backwards if the horizontal component of its velocity is less than the speed of the wind, that is, $32 \cos \theta < 24 \quad \Leftrightarrow \quad \cos \theta < \tfrac{3}{4} \quad \Rightarrow \quad \theta > 41.4°$.

In the second figure, we graph the trajectory for $\theta = 20°$, $\theta = 23°$, and $\theta = 25°$. The solution appears to be close to $23°$.

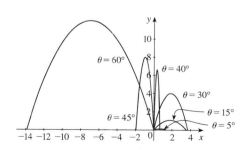

12 Sequences and Series

12.1 Sequences and Summation Notation

1. $a_n = n + 1$. Then $a_1 = 1 + 1 = 2$, $a_2 = 2 + 1 = 3$, $a_3 = 3 + 1 = 4$, $a_4 = 4 + 1 = 5$, and $a_{100} = 100 + 1 = 101$.

3. $a_n = \dfrac{1}{n + 1}$. Then $a_1 = \dfrac{1}{1 + 1} = \dfrac{1}{2}$, $a_2 = \dfrac{1}{2 + 1} = \dfrac{1}{3}$, $a_3 = \dfrac{1}{3 + 1} = \dfrac{1}{4}$, $a_4 = \dfrac{1}{4 + 1} = \dfrac{1}{5}$, and $a_{100} = \dfrac{1}{100 + 1} = \dfrac{1}{101}$.

5. $a_n = \dfrac{(-1)^n}{n^2}$. Then $a_1 = \dfrac{(-1)^1}{1^2} = -1$, $a_2 = \dfrac{(-1)^2}{2^2} = \dfrac{1}{4}$, $a_3 = \dfrac{(-1)^3}{3^2} = -\dfrac{1}{9}$, $a_4 = \dfrac{(-1)^4}{4^2} = \dfrac{1}{16}$, and $a_{100} = \dfrac{(-1)^{100}}{100^2} = \dfrac{1}{10,000}$.

7. $a_n = 1 + (-1)^n$. Then $a_1 = 1 + (-1)^1 = 0$, $a_2 = 1 + (-1)^2 = 2$, $a_3 = 1 + (-1)^3 = 0$, $a_4 = 1 + (-1)^4 = 2$, and $a_{100} = 1 + (-1)^{100} = 2$.

9. $a_n = n^n$. Then $a_1 = 1^1 = 1$, $a_2 = 2^2 = 4$, $a_3 = 3^3 = 27$, $a_4 = 4^4 = 256$, and $a_{100} = 100^{100} = 10^{200}$.

11. $a_n = 2(a_{n-1} - 2)$ and $a_1 = 3$. Then $a_2 = 2[(3) - 2] = 2$, $a_3 = 2[(2) - 2] = 0$, $a_4 = 2[(0) - 2] = -4$, and $a_5 = 2[(-4) - 2] = -12$.

13. $a_n = 2a_{n-1} + 1$ and $a_1 = 1$. Then $a_2 = 2(1) + 1 = 3$, $a_3 = 2(3) + 1 = 7$, $a_4 = 2(7) + 1 = 15$, and $a_5 = 2(15) + 1 = 31$.

15. $a_n = a_{n-1} + a_{n-2}$, $a_1 = 1$, and $a_2 = 2$. Then $a_3 = 2 + 1 = 3$, $a_4 = 3 + 2 = 5$, and $a_5 = 5 + 3 = 8$.

17. (a) $a_1 = 7$, $a_2 = 11$, $a_3 = 15$, $a_4 = 19$, $a_5 = 23$, $a_6 = 27$, $a_7 = 31$, $a_8 = 35$, $a_9 = 39$, $a_{10} = 43$

(b)

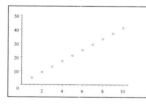

19. (a) $a_1 = \dfrac{12}{1} = 12$, $a_2 = \dfrac{12}{2} = 6$, $a_3 = \dfrac{12}{3} = 4$, $a_4 = \dfrac{12}{4} = 3$, $a_5 = \dfrac{12}{5}$, $a_6 = \dfrac{12}{6} = 2$, $a_7 = \dfrac{12}{7}$, $a_8 = \dfrac{12}{8} = \dfrac{3}{2}$, $a_9 = \dfrac{12}{9} = \dfrac{4}{3}$, $a_{10} = \dfrac{12}{10} = \dfrac{6}{5}$

(b)
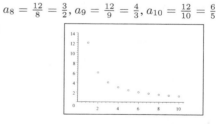

21. (a) $a_1 = 2$, $a_2 = 0.5$, $a_3 = 2$, $a_4 = 0.5$, $a_5 = 2$, $a_6 = 0.5$, $a_7 = 2$, $a_8 = 0.5$, $a_9 = 2$, $a_{10} = 0.5$

(b)

23. $2, 4, 8, 16, \ldots$. All are powers of 2, so $a_1 = 2$, $a_2 = 2^2$, $a_3 = 2^3$, $a_4 = 2^4$, $\ldots$ Thus $a_n = 2^n$.

25. $1, 4, 7, 10, \ldots$. The difference between any two consecutive terms is 3, so $a_1 = 3(1) - 2$, $a_2 = 3(2) - 2$, $a_3 = 3(3) - 2$, $a_4 = 3(4) - 2$, $\ldots$ Thus $a_n = 3n - 2$.

27. $1, \frac{3}{4}, \frac{5}{9}, \frac{7}{16}, \frac{9}{25}, \dots$. We consider the numerator separately from the denominator. The numerators of the terms differ by 2, and the denominators are perfect squares. So $a_1 = \dfrac{2(1) - 1}{1^2}$, $a_2 = \dfrac{2(2) - 1}{2^2}$, $a_3 = \dfrac{2(3) - 1}{3^2}$, $a_4 = \dfrac{2(4) - 1}{4^2}$, $a_5 = \dfrac{2(5) - 1}{5^2}, \dots$. Thus $a_n = \dfrac{2n - 1}{n^2}$.

29. $0, 2, 0, 2, 0, 2, \dots$. These terms alternate between 0 and 2. So $a_1 = 1 - 1$, $a_2 = 1 + 1$, $a_3 = 1 - 1$, $a_4 = 1 + 1$, $a_5 = 1 - 1$, $a_6 = 1 + 1, \dots$ Thus $a_n = 1 + (-1)^n$.

31. $a_1 = 1$, $a_2 = 3$, $a_3 = 5$, $a_4 = 7, \dots$. Therefore, $a_n = 2n - 1$. So $S_1 = 1$, $S_2 = 1 + 3 = 4$, $S_3 = 1 + 3 + 5 + = 9$, $S_4 = 1 + 3 + 5 + 7 = 16$, $S_5 = 1 + 3 + 5 + 7 + 9 = 25$, and $S_6 = 1 + 3 + 5 + 7 + 9 + 11 = 36$.

33. $a_1 = \frac{1}{3}$, $a_2 = \dfrac{1}{3^2}$, $a_3 = \dfrac{1}{3^3}$, $a_4 = \dfrac{1}{3^4}, \dots$. Therefore, $a_n = \dfrac{1}{3^n}$. So $S_1 = \frac{1}{3}$, $S_2 = \frac{1}{3} + \dfrac{1}{3^2} = \frac{4}{9}$, $S_3 = \frac{1}{3} + \dfrac{1}{3^2} + \dfrac{1}{3^3} = \frac{13}{27}$, $S_4 = \frac{1}{3} + \dfrac{1}{3^2} + \dfrac{1}{3^3} + \dfrac{1}{3^4} = \frac{40}{81}$, and $S_5 = \frac{1}{3} + \dfrac{1}{3^2} + \dfrac{1}{3^3} + \dfrac{1}{3^4} + \dfrac{1}{3^5} = \frac{121}{243}$, $S_6 = \frac{1}{3} + \dfrac{1}{3^2} + \dfrac{1}{3^3} + \dfrac{1}{3^4} + \dfrac{1}{3^5} + \dfrac{1}{3^6} = \frac{364}{729}$.

35. $a_n = \dfrac{2}{3^n}$. So $S_1 = \frac{2}{3}$, $S_2 = \frac{2}{3} + \dfrac{2}{3^2} = \frac{8}{9}$, $S_3 = \frac{2}{3} + \dfrac{2}{3^2} + \dfrac{2}{3^3} = \frac{26}{27}$, and $S_4 = \frac{2}{3} + \dfrac{2}{3^2} + \dfrac{2}{3^3} + \dfrac{2}{3^4} = \frac{80}{81}$. Therefore,

$S_n = \dfrac{3^n - 1}{3^n}$.

37. $a_n = \sqrt{n} - \sqrt{n+1}$. So $S_1 = \sqrt{1} - \sqrt{2} = 1 - \sqrt{2}$, $S_2 = \left(\sqrt{1} - \sqrt{2}\right) + \left(\sqrt{2} - \sqrt{3}\right) = 1 + \left(-\sqrt{2} + \sqrt{2}\right) - \sqrt{3} = 1 - \sqrt{3}$,

$S_3 = \left(\sqrt{1} - \sqrt{2}\right) + \left(\sqrt{2} - \sqrt{3}\right) + \left(\sqrt{3} - \sqrt{4}\right) = 1 + \left(-\sqrt{2} + \sqrt{2}\right) + \left(-\sqrt{3} + \sqrt{3}\right) - \sqrt{4} = 1 - \sqrt{4}$,

$S_4 = \left(\sqrt{1} - \sqrt{2}\right) + \left(\sqrt{2} - \sqrt{3}\right) + \left(\sqrt{3} - \sqrt{4}\right) + \left(\sqrt{4} - \sqrt{5}\right)$

$\quad = 1 + \left(-\sqrt{2} + \sqrt{2}\right) + \left(-\sqrt{3} + \sqrt{3}\right) + \left(-\sqrt{4} + \sqrt{4}\right) - \sqrt{5} = 1 - \sqrt{5}$

Therefore,

$S_n = \left(\sqrt{1} - \sqrt{2}\right) + \left(\sqrt{2} - \sqrt{3}\right) + \cdots + \left(\sqrt{n} - \sqrt{n+1}\right)$

$\quad = 1 + \left(-\sqrt{2} + \sqrt{2}\right) + \left(-\sqrt{3} + \sqrt{3}\right) + \cdots + \left(-\sqrt{n} + \sqrt{n}\right) - \sqrt{n+1} = 1 - \sqrt{n+1}$

39. $\sum_{k=1}^{4} k = 1 + 2 + 3 + 4 = 10$

41. $\sum_{k=1}^{3} \dfrac{1}{k} = 1 + \frac{1}{2} + \frac{1}{3} = \frac{6}{6} + \frac{3}{6} + \frac{2}{6} = \frac{11}{6}$

43. $\sum_{i=1}^{8} \left[1 + (-1)^i\right] = 0 + 2 + 0 + 2 + 0 + 2 + 0 + 2 = 8$

45. $\sum_{k=1}^{5} 2^{k-1} = 2^0 + 2^1 + 2^2 + 2^3 + 2^4 = 1 + 2 + 4 + 8 + 16 = 31$

47. 385   **49.** 46,438   **51.** 22

53. $\sum_{k=1}^{5} \sqrt{k} = \sqrt{1} + \sqrt{2} + \sqrt{3} + \sqrt{4} + \sqrt{5}$

55. $\sum_{k=0}^{6} \sqrt{k+4} = \sqrt{4} + \sqrt{5} + \sqrt{6} + \sqrt{7} + \sqrt{8} + \sqrt{9} + \sqrt{10}$

57. $\sum_{k=3}^{100} x^k = x^3 + x^4 + x^5 + \cdots + x^{100}$

59. $1 + 2 + 3 + 4 + \cdots + 100 = \sum_{k=1}^{100} k$

61. $1^2 + 2^2 + 3^2 + \cdots + 10^2 = \sum_{k=1}^{10} k^2$

63. $\dfrac{1}{1 \cdot 2} + \dfrac{1}{2 \cdot 3} + \dfrac{1}{3 \cdot 4} + \cdots + \dfrac{1}{999 \cdot 1000} = \sum_{k=1}^{999} \dfrac{1}{k(k+1)}$

65. $1 + x + x^2 + x^3 + \cdots + x^{100} = \sum_{k=0}^{100} x^k$

67. $\sqrt{2}, \sqrt{2\sqrt{2}}, \sqrt{2\sqrt{2\sqrt{2}}}, \sqrt{2\sqrt{2\sqrt{2\sqrt{2}}}}, \dots$. We simplify each term in an attempt to determine a formula for a_n. So $a_1 = 2^{1/2}$, $a_2 = \sqrt{2 \cdot 2^{1/2}} = \sqrt{2^{3/2}} = 2^{3/4}$, $a_3 = \sqrt{2 \cdot 2^{3/4}} = \sqrt{2^{7/4}} = 2^{7/8}$, $a_4 = \sqrt{2 \cdot 2^{7/8}} = \sqrt{2^{15/8}} = 2^{15/16}, \dots$. Thus $a_n = 2^{(2^n - 1)/2^n}$.

69. (a) $A_1 = \$2004$, $A_2 = \$2008.01$, $A_3 = \$2012.02$, $A_4 = \$2016.05$, $A_5 = \$2020.08$, $A_6 = \$2024.12$

(b) Since 3 years is 36 months, we get $A_{36} = \$2149.16$.

71. (a) $P_1 = 35{,}700$, $P_2 = 36{,}414$, $P_3 = 37{,}142$, $P_4 = 37{,}885$, $P_5 = 38{,}643$

(b) Since 2014 is 10 years after 2004, $P_{10} = 42{,}665$.

73. (a) The number of catfish at the end of the month, P_n, is the population at the start of the month, P_{n-1}, plus the increase in population, $0.08P_{n-1}$, minus the 300 catfish harvested. Thus $P_n = P_{n-1} + 0.08P_{n-1} - 300 \quad \Leftrightarrow \quad P_n = 1.08P_{n-1} - 300$.

(b) $P_1 = 5100$, $P_2 = 5208$, $P_3 = 5325$, $P_4 = 5451$, $P_5 = 5587$, $P_6 = 5734$, $P_7 = 5892$, $P_8 = 6064$, $P_9 = 6249$, $P_{10} = 6449$, $P_{11} = 6665$, $P_{12} = 6898$. Thus there should be 6898 catfish in the pond at the end of 12 months.

75. (a) Let S_n be his salary in the nth year. Then $S_1 = \$30{,}000$. Since his salary increase by 2000 each year, $S_n = S_{n-1} + 2000$. Thus $S_1 = \$30{,}000$ and $S_n = S_{n-1} + 2000$.

(b) $S_5 = S_4 + 2000 = (S_3 + 2000) + 2000 = (S_2 + 2000) + 4000 = (S_1 + 2000) + 6000 = \$38{,}000$.

77. Let F_n be the number of pairs of rabbits in the nth month. Clearly $F_1 = F_2 = 1$. In the nth month each pair that is two or more months old (that is, F_{n-2} pairs) will add a pair of offspring to the F_{n-1} pairs already present. Thus $F_n = F_{n-1} + F_{n-2}$. So F_n is the Fibonacci sequence.

79. $a_{n+1} = \begin{cases} \dfrac{a_n}{2} & \text{if } a_n \text{ is even} \\ 3a_n + 1 & \text{if } a_n \text{ is odd} \end{cases}$ With $a_1 = 11$, we have $a_2 = 34$, $a_3 = 17$, $a_4 = 52$, $a_5 = 26$, $a_6 = 13$, $a_7 = 40$, $a_8 = 20$, $a_9 = 10$, $a_{10} = 5$, $a_{11} = 16$, $a_{12} = 8$, $a_{13} = 4$, $a_{14} = 2$, $a_{15} = 1$, $a_{16} = 4$, $a_{17} = 2$, $a_{18} = 1$, ... (with 4, 2, 1 repeating). So $a_{3n+1} = 4$, $a_{3n+2} = 2$, and $a_{3n} = 1$, for $n \geq 5$. With $a_1 = 25$, we have $a_2 = 76$, $a_3 = 38$, $a_4 = 19$, $a_5 = 58$, $a_6 = 29$, $a_7 = 88$, $a_8 = 44$, $a_9 = 22$, $a_{10} = 11$, $a_{11} = 34$, $a_{12} = 17$, $a_{13} = 52$, $a_{14} = 26$, $a_{15} = 13$, $a_{16} = 40$, $a_{17} = 20$, $a_{18} = 10$, $a_{19} = 5$, $a_{20} = 16$, $a_{21} = 8$, $a_{22} = 4$, $a_{23} = 2$, $a_{24} = 1$, $a_{25} = 4$, $a_{26} = 2$, $a_{27} = 1$, ... (with 4, 2, 1 repeating). So $a_{3n+1} = 4$, $a_{3n+2} = 2$, and $a_{3n+3} = 1$ for $n \geq 7$.
We conjecture that the sequence will always return to the numbers 4, 2, 1 repeating.

12.2 Arithmetic Sequences

1. (a) $a_1 = 5 + 2(1 - 1) = 5$,
$a_2 = 5 + 2(2 - 1) = 5 + 2 = 7$,
$a_3 = 5 + 2(3 - 1) = 5 + 4 = 9$,
$a_4 = 5 + 2(4 - 1) = 5 + 6 = 11$,
$a_5 = 5 + 2(5 - 1) = 5 + 8 = 13$

(b) The common difference is 2.

(c)

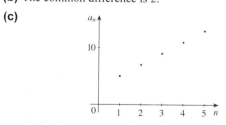

3. (a) $a_1 = \frac{5}{2} - (1 - 1) = \frac{5}{2}$, $a_2 = \frac{5}{2} - (2 - 1) = \frac{3}{2}$,
$a_3 = \frac{5}{2} - (3 - 1) = \frac{1}{2}$, $a_4 = \frac{5}{2} - (4 - 1) = -\frac{1}{2}$,
$a_5 = \frac{5}{2} - (5 - 1) = -\frac{3}{2}$

(b) The common difference is -1.

(c)

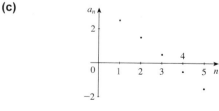

5. $a = 3$, $d = 5$, $a_n = a + d(n - 1) = 3 + 5(n - 1)$. So $a_{10} = 3 + 5(10 - 1) = 48$.

7. $a = \frac{5}{2}$, $d = -\frac{1}{2}$, $a_n = a + d(n - 1) = \frac{5}{2} - \frac{1}{2}(n - 1)$. So $a_{10} = \frac{5}{2} - \frac{1}{2}(10 - 1) = -2$.

9. $a_4 - a_3 = 14 - 11 = 3$, $a_3 - a_2 = 11 - 8 = 3$, $a_2 - a_1 = 8 - 5 = 3$. This sequence is arithmetic with common difference 3.

11. Since $a_2 - a_1 = 4 - 2 = 2$ and $a_4 - a_3 = 16 - 8 = 8$, the terms of the sequence do not have a common difference. This sequence is not arithmetic.

13. $a_4 - a_3 = -\frac{3}{2} - 0 = -\frac{3}{2}$, $a_3 - a_2 = 0 - \frac{3}{2} = -\frac{3}{2}$, $a_2 - a_1 = \frac{3}{2} - 3 = -\frac{3}{2}$. This sequence is arithmetic with common difference $-\frac{3}{2}$.

15. $a_4 - a_3 = 7.7 - 6.0 = 1.7$, $a_3 - a_2 = 6.0 - 4.3 = 1.7$, $4. - a_1 = 4.3 - 2.6 = 1.7$. This sequence is arithmetic with common difference 1.7.

17. $a_1 = 4 + 7(1) = 11$, $a_2 = 4 + 7(2) = 18$, $a_3 = 4 + 7(3) = 25$, $a_4 = 4 + 7(4) = 32$, $a_5 = 4 + 7(5) = 39$. This sequence is arithmetic, the common difference is $d = 7$ and $a_n = 4 + 7n = 4 + 7n - 7 + 7 = 11 + 7(n-1)$.

19. $a_1 = \frac{1}{1 + 2(1)} = \frac{1}{3}$, $a_2 = \frac{1}{1 + 2(2)} = \frac{1}{5}$, $a_3 = \frac{1}{1 + 2(3)} = \frac{1}{7}$, $a_4 = \frac{1}{1 + 2(4)} = \frac{1}{9}$, $a_5 = \frac{1}{1 + 2(5)} = \frac{1}{11}$. Since $a_4 - a_3 = \frac{1}{9} - \frac{1}{7} = -\frac{2}{63}$ and $a_3 - a_2 = \frac{1}{7} - \frac{1}{3} = -\frac{2}{21}$, the terms of the sequence do not have a common difference. This sequence is not arithmetic.

21. $a_1 = 6(1) - 10 = -4$, $a_2 = 6(2) - 10 = 2$, $a_3 = 6(3) - 10 = 8$, $a_4 = 6(4) - 10 = 14$, $a_5 = 6(5) - 10 = 20$. This sequence is arithmetic, the common difference is $d = 6$ and $a_n = 6n - 10 = 6n - 6 + 6 - 10 = -4 + 6(n-1)$.

23. $2, 5, 8, 11, \ldots$. Then $d = a_2 - a_1 = 5 - 2 = 3$, $a_5 = a_4 + 3 = 11 + 3 = 14$, $a_n = 2 + 3(n-1)$, and $a_{100} = 2 + 3(99) = 299$.

25. $4, 9, 14, 19, \ldots$. Then $d = a_2 - a_1 = 9 - 4 = 5$, $a_5 = a_4 + 5 = 19 + 5 = 24$, $a_n = 4 + 5(n-1)$, and $a_{100} = 4 + 5(99) = 499$.

27. $-12, -8, -4, 0, \ldots$. Then $d = a_2 - a_1 = -8 - (-12) = 4$, $a_5 = a_4 + 4 = 0 + 4 = 4$, $a_n = -12 + 4(n-1)$, and $a_{100} = -12 + 4(99) = 384$.

29. $25, 26.5, 28, 29.5, \ldots$. Then $d = a_2 - a_1 = 26.5 - 25 = 1.5$, $a_5 = a_4 + 1.5 = 29.5 + 1.5 = 31$, $a_n = 25 + 1.5(n-1)$, $a_{100} = 25 + 1.5(99) = 173.5$.

31. $2, 2 + s, 2 + 2s, 2 + 3s, \ldots$. Then $d = a_2 - a_1 = 2 + s - 2 = s$, $a_5 = a_4 + s = 2 + 3s + s = 2 + 4s$, $a_n = 2 + (n-1)s$, and $a_{100} = 2 + 99s$.

33. $a_{10} = \frac{55}{2}$, $a_2 = \frac{7}{2}$, and $a_n = a + d(n-1)$. Then $a_2 = a + d = \frac{7}{2} \Leftrightarrow d = \frac{7}{2} - a$. Substituting into $a_{10} = a + 9d = \frac{55}{2}$ gives $a + 9\left(\frac{7}{2} - a\right) = \frac{55}{2} \Leftrightarrow a = \frac{1}{2}$. Thus, the first term is $a_1 = \frac{1}{2}$.

35. $a_{100} = 98$ and $d = 2$. Note that $a_{100} = a + 99d = a + 99(2) = a + 198$. Since $a_{100} = 98$, we have $a + 198 = a_{100} = 98 \Leftrightarrow a = -100$. Hence, $a_1 = -100$, $a_2 = -100 + 2 = -98$, and $a_3 = -100 + 4 = -96$.

37. The arithmetic sequence is $1, 4, 7, \ldots$. So $d = 4 - 1 = 3$ and $a_n = 1 + 3(n-1)$. Then $a_n = 88 \Leftrightarrow 1 + 3(n-1) = 88 \Leftrightarrow 3(n-1) = 87 \Leftrightarrow n - 1 = 29 \Leftrightarrow n = 30$. So 88 is the 30th term.

39. $a = 1$, $d = 2$, $n = 10$. Then $S_{10} = \frac{10}{2}[2a + (10-1)d] = \frac{10}{2}[2 \cdot 1 + 9 \cdot 2] = 100$.

41. $a = 4$, $d = 2$, $n = 20$. Then $S_{20} = \frac{20}{2}[2a + (20-1)d] = \frac{20}{2}[2 \cdot 4 + 19 \cdot 2] = 460$.

43. $a_1 = 55$, $d = 12$, $n = 10$. Then $S_{10} = \frac{10}{2}[2a + (10-1)d] = \frac{10}{2}[2 \cdot 55 + 9 \cdot 12] = 1090$.

45. $1 + 5 + 9 + \cdots + 401$ is a partial sum of an arithmetic series, where $a = 1$ and $d = 5 - 1 = 4$. The last term is $401 = a_n = 1 + 4(n-1)$, so $n - 1 = 100 \Leftrightarrow n = 101$. So the partial sum is $S_{101} = \frac{101}{2}(1 + 401) = 101 \cdot 201 = 20{,}301$.

47. $0.7 + 2.7 + 4.7 + \cdots + 56.7$ is a partial sum of an arithmetic series, where $a = 0.7$ and $d = 2.7 - 0.7 = 2$. The last term is $56.7 = a_n = 0.7 + 2(n-1) \Leftrightarrow 28 = n - 1 \Leftrightarrow n = 29$. So the partial sum is $S_{29} = \frac{29}{2}(0.7 + 56.7) = 832.3$.

49. $\sum_{k=0}^{10}(3 + 0.25k)$ is a partial sum of an arithmetic series where $a = 3 + 0.25 \cdot 0 = 3$ and $d = 0.25$. The last term is $a_{11} = 3 + 0.25 \cdot 10 = 5.5$. So the partial sum is $S_{11} = \frac{11}{2}(3 + 5.5) = 46.75$.

51. Let x denote the length of the side between the length of the other two sides. Then the lengths of the three sides of the triangle are $x - a$, x, and $x + a$, for some $a > 0$. Since $x + a$ is the longest side, it is the hypotenuse, and by the Pythagorean Theorem, we know that $(x - a)^2 + x^2 = (x + a)^2 \Leftrightarrow x^2 - 2ax + a^2 + x^2 = x^2 + 2ax + a^2 \Leftrightarrow x^2 - 4ax = 0 \Leftrightarrow x(x - 4a) = 0 \Rightarrow x = 4a$ ($x = 0$ is not a possible solution). Thus, the lengths of the three sides are $x - a = 4a - a = 3a$, $x = 4a$, and $x + a = 4a + a = 5a$. The lengths $3a$, $4a$, $5a$ are proportional to 3, 4, 5, and so the triangle is similar to a 3-4-5 triangle.

53. The sequence $1, \frac{3}{5}, \frac{3}{7}, \frac{1}{3}, \ldots$ is harmonic if $1, \frac{5}{3}, \frac{7}{3}, 3, \ldots$ forms an arithmetic sequence. Since $\frac{5}{3} - 1 = \frac{7}{3} - \frac{5}{3} = 3 - \frac{7}{3} = \frac{2}{3}$, the sequence of reciprocals is arithmetic and thus the original sequence is harmonic.

55. We have an arithmetic sequence with $a = 5$ and $d = 2$. We seek n such that $2700 = S_n = \frac{n}{2}\left[2a + (n-1)d\right]$.

Solving for n, we have $2700 = \frac{n}{2}\left[10 + 2(n-1)\right] \Leftrightarrow 5400 = 10n + 2n^2 - 2n \Leftrightarrow n^2 + 4n - 2700 = 0 \Leftrightarrow$ $(n - 50)(n + 54) = 0 \Leftrightarrow n = 50$ or $n = -54$. Since n is a positive integer, 50 terms of the sequence must be added to get 2700.

57. The diminishing values of the computer form an arithmetic sequence with $a_1 = 12500$ and common difference $d = -1875$. Thus the value of the computer after 6 years is $a_7 = 12500 + (7 - 1)(-1875) = \$1250$.

59. The increasing values of the man's salary form an arithmetic sequence with $a_1 = 30,000$ and common difference $d = 2300$. Then his total earnings for a ten-year period are $S_{10} = \frac{10}{2}\left[2(30,000) + 9(2300)\right] = 403,500$. Thus his total earnings for the 10 year period are $\$403,500$.

61. The number of seats in the nth row is given by the nth term of an arithmetic sequence with $a_1 = 15$ and common difference $d = 3$. We need to find n such that $S_n = 870$. So we solve $870 = S_n = \frac{n}{2}\left[2(15) + (n-1)3\right]$ for n. We have $870 = \frac{n}{2}(27 + 3n) \Leftrightarrow 1740 = 3n^2 + 27n \Leftrightarrow 3n^2 + 27n - 1740 = 0 \Leftrightarrow n^2 + 9n - 580 = 0 \Leftrightarrow$ $(x - 20)(x + 29) = 0 \Rightarrow n = 20$ or $n = -29$. Since the number of rows is positive, the theater must have 20 rows.

63. The number of gifts on the 12th day is $1 + 2 + 3 + 4 + \cdots + 12$. Since $a_2 - a_1 = a_3 - a_2 = a_4 - a_3 = \cdots = 1$, the number of gifts on the 12th day is the partial sum of an arithmetic sequence with $a = 1$ and $d = 1$. So the sum is $S_{12} = 12\left(\frac{1 + 12}{2}\right) = 6 \cdot 13 = 78$.

12.3   Geometric Sequences

1. (a) $a_1 = 5(2)^0 = 5$, $a_2 = 5(2)^1 = 10$,
$a_3 = 5(2)^2 = 20$, $a_4 = 5(2)^3 = 40$,
$a_5 = 5(2)^4 = 80$

(b) The common ratio is 2.

(c)

3. (a) $a_1 = \frac{5}{2}\left(-\frac{1}{2}\right)^0 = \frac{5}{2}$, $a_2 = \frac{5}{2}\left(-\frac{1}{2}\right)^1 = -\frac{5}{4}$,
$a_3 = \frac{5}{2}\left(-\frac{1}{2}\right)^2 = \frac{5}{8}$, $a_4 = \frac{5}{2}\left(-\frac{1}{2}\right)^3 = -\frac{5}{16}$,
$a_5 = \frac{5}{2}\left(-\frac{1}{2}\right)^4 = \frac{5}{32}$

(b) The common ratio is $-\frac{1}{2}$.

(c)

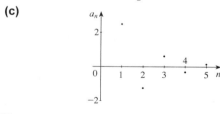

5. $a = 3$, $r = 5$. So $a_n = ar^{n-1} = 3(5)^{n-1}$ and $a_4 = 3 \cdot 5^3 = 375$.

7. $a = \frac{5}{2}$, $r = -\frac{1}{2}$. So $a_n = ar^{n-1} = \frac{5}{2}\left(-\frac{1}{2}\right)^{n-1}$ and $a_4 = \frac{5}{2} \cdot \left(-\frac{1}{2}\right)^3 = -\frac{5}{16}$.

9. $\frac{a_2}{a_1} = \frac{4}{2} = 2$, $\frac{a_3}{a_2} = \frac{8}{4} = 2$, $\frac{a_4}{a_3} = \frac{16}{8} = 2$. Since these ratios are the same, the sequence is geometric with the common ratio 2.

11. $\frac{a_2}{a_1} = \frac{3/2}{3} = \frac{1}{2}$, $\frac{a_3}{a_2} = \frac{3/4}{3/2} = \frac{1}{2}$, $\frac{a_4}{a_3} = \frac{3/8}{3/4} = \frac{1}{2}$. Since these ratios are the same, the sequence is geometric with the common ratio $\frac{1}{2}$.

13. $\frac{a_2}{a_1} = \frac{1/3}{1/2} = \frac{2}{3}$, $\frac{a_4}{a_3} = \frac{1/5}{1/4} = \frac{4}{5}$. Since these ratios are not the same, this is not a geometric sequence.

15. $\frac{a_2}{a_1} = \frac{1.1}{1.0} = 1.1$, $\frac{a_3}{a_2} = \frac{1.21}{1.1} = 1.1$, $\frac{a_4}{a_3} = \frac{1.331}{1.21} = 1.1$. Since these ratios are the same, the sequence is geometric with the common ratio 1.1.

17. $a_1 = 2\,(3)^1 = 6$, $a_2 = 2\,(3)^2 = 18$, $a_3 = 2\,(3)^3 = 54$, $a_4 = 2\,(3)^4 = 162$, $a_5 = 2\,(3)^5 = 486$. This sequence is geometric, the common ratio is $r = 3$ and $a_n = a_1 r^{n-1} = 6\,(3)^{n-1}$.

19. $a_1 = \frac{1}{4}$, $a_2 = \frac{1}{4^2} = \frac{1}{16}$, $a_3 = \frac{1}{4^3} = \frac{1}{64}$, $a_4 = \frac{1}{4^4} = \frac{1}{256}$, $a_5 = \frac{1}{4^5} = \frac{1}{1024}$. This sequence is geometric, the common ratio is $r = \frac{1}{4}$ and $a_n = a_1 r^{n-1} = \frac{1}{4}\left(\frac{1}{4}\right)^{n-1}$.

21. Since $\ln a^b = b \ln a$, we have $a_1 = \ln\left(5^0\right) = \ln 1 = 0$, $a_2 = \ln\left(5^1\right) = \ln 5$, $a_3 = \ln\left(5^2\right) = 2 \ln 5$, $a_4 = \ln\left(5^3\right) = 3 \ln 5$, $a_5 = \ln\left(5^4\right) = 4 \ln 5$. Since $a_1 = 0$ and $a_2 \neq 0$, this sequence is not geometric.

23. $2, 6, 18, 54, \ldots$. Then $r = \frac{a_2}{a_1} = \frac{6}{2} = 3$, $a_5 = a_4 \cdot 3 = 54\,(3) = 162$, and $a_n = 2 \cdot 3^{n-1}$.

25. $0.3, -0.09, 0.027, -0.0081, \ldots$. Then $r = \frac{a_2}{a_1} = \frac{-0.09}{0.3} = -0.3$, $a_5 = a_4 \cdot (-0.3) = -0.0081\,(-0.3) = 0.00243$, and $a_n = 0.3\,(-0.3)^{n-1}$.

27. $144, -12, 1, -\frac{1}{12}, \ldots$. Then $r = \frac{a_2}{a_1} = \frac{-12}{144} = -\frac{1}{12}$, $a_5 = a_4 \cdot \left(-\frac{1}{12}\right) = -\frac{1}{12}\left(-\frac{1}{12}\right) = \frac{1}{144}$, $a_n = 144\left(-\frac{1}{12}\right)^{n-1}$.

29. $3, 3^{5/3}, 3^{7/3}, 27, \ldots$. Then $r = \frac{a_2}{a_1} = \frac{3^{5/3}}{3} = 3^{2/3}$, $a_5 = a_4 \cdot \left(3^{2/3}\right) = 27 \cdot 3^{2/3} = 3^{11/3}$, and $a_n = 3\left(3^{2/3}\right)^{n-1} = 3 \cdot 3^{(2n-2)/3} = 3^{(2n+1)/3}$.

31. $1, s^{2/7}, s^{4/7}, s^{6/7}, \ldots$. Then $r = \frac{a_2}{a_1} = \frac{s^{2/7}}{1} = s^{2/7}$, $a_5 = a_4 \cdot s^{2/7} = s^{6/7} \cdot s^{2/7} = s^{8/7}$, and $a_n = \left(s^{2/7}\right)^{n-1} = s^{(2n-2)/7}$.

33. $a_1 = 8$, $a_2 = 4$. Thus $r = \frac{a_2}{a_1} = \frac{4}{8} = \frac{1}{2}$ and $a_5 = a_1 r^{5-1} = 8\left(\frac{1}{2}\right)^4 = \frac{8}{16} = \frac{1}{2}$.

35. $r = \frac{2}{5}$, $a_4 = \frac{5}{2}$. Since $r = \frac{a_4}{a_3}$, we have $a_3 = \frac{a_4}{r} = \frac{5/2}{2/5} = \frac{25}{4}$.

37. The geometric sequence is $2, 6, 18, \ldots$. Thus $r = \frac{a_2}{a_1} = \frac{6}{2} = 3$. We need to find n so that $a_n = 2 \cdot 3^{n-1} = 118{,}098 \Leftrightarrow 3^{n-1} = 59{,}049 \Leftrightarrow n - 1 = \log_3 59{,}049 = 10 \Leftrightarrow n = 11$. Therefore, $118{,}098$ is the 11th term of the geometric sequence.

39. $a = 5$, $r = 2$, $n = 6$. Then $S_6 = 5\frac{1 - 2^6}{1 - 2} = (-5)(-63) = 315$.

41. $a_3 = 28$, $a_6 = 224$, $n = 6$. So $\frac{a_6}{a_3} = \frac{ar^5}{ar^2} = r^3$. So we have $r^3 = \frac{a_6}{a_3} = \frac{224}{28} = 8$, and hence $r = 2$. Since $a_3 = a \cdot r^2$, we get $a = \frac{a_3}{r^2} = \frac{28}{2^2} = 7$. So $S_6 = 7\frac{1 - 2^6}{1 - 2} = (-7)(-63) = 441$.

43. $1 + 3 + 9 + \cdots + 2187$ is a partial sum of a geometric sequence, where $a = 1$ and $r = \frac{a_2}{a_1} = \frac{3}{1} = 3$. Then the last term is $2187 = a_n = 1 \cdot 3^{n-1} \Leftrightarrow n - 1 = \log_3 2187 = 7 \Leftrightarrow n = 8$. So the partial sum is $S_8 = (1)\frac{1 - 3^8}{1 - 3} = 3280$.

45. $\sum_{k=0}^{10} 3\left(\frac{1}{2}\right)^k$ is a partial sum of a geometric sequence, where $a = 3$, $r = \frac{1}{2}$, and $n = 11$. So the partial sum is $S_{11} = (3)\frac{1 - \left(\frac{1}{2}\right)^{11}}{1 - \left(\frac{1}{2}\right)} = 6\left[1 - \left(\frac{1}{2}\right)^{11}\right] = \frac{6141}{1024} \approx 5.997070313$.

47. $1 + \frac{1}{3} + \frac{1}{9} + \frac{1}{27} + \cdots$ is an infinite geometric series with $a = 1$ and $r = \frac{1}{3}$. Therefore, the sum of the series is $S = \frac{a}{1 - r} = \frac{1}{1 - \left(\frac{1}{3}\right)} = \frac{3}{2}$.

49. $1 - \frac{1}{3} + \frac{1}{9} - \frac{1}{27} + \cdots$ is an infinite geometric series with $a = 1$ and $r = -\frac{1}{3}$. Therefore, the sum of the series is

$$S = \frac{a}{1-r} = \frac{1}{1 - \left(-\frac{1}{3}\right)} = \frac{3}{4}.$$

51. $\frac{1}{3^6} + \frac{1}{3^8} + \frac{1}{3^{10}} + \frac{1}{3^{12}} + \cdots$ is an infinite geometric series with $a = \frac{1}{3^6}$ and $r = \frac{1}{3^2} = \frac{1}{9}$. Therefore, the sum of the series

is $S = \frac{a}{1-r} = \frac{\frac{1}{3^6}}{1 - \left(\frac{1}{9}\right)} = \frac{1}{3^6} \cdot \frac{9}{8} = \frac{1}{648}.$

53. $-\frac{100}{9} + \frac{10}{3} - 1 + \frac{3}{10} - \cdots$ is an infinite geometric series with $a = -\frac{100}{9}$ and $r = \frac{\frac{10}{3}}{-\frac{100}{9}} = \frac{3}{10}$. Therefore, the sum of the

series is $S = \frac{a}{1-r} = \frac{-\frac{100}{9}}{1 - \left(-\frac{3}{10}\right)} = \frac{-\frac{100}{9}}{\frac{13}{10}} = -\frac{100}{9} \cdot \frac{10}{13} = -\frac{1000}{117}.$

55. $0.777\ldots = \frac{7}{10} + \frac{7}{100} + \frac{7}{1000} + \cdots$ is an infinite geometric series with $a = \frac{7}{10}$ and $r = \frac{1}{10}$. Thus

$0.777\ldots = \frac{a}{1-r} = \frac{\frac{7}{10}}{1 - \frac{1}{10}} = \frac{7}{9}.$

57. $0.030303\ldots = \frac{3}{100} + \frac{3}{10,000} + \frac{3}{1,000,000} + \cdots$ is an infinite geometric series with $a = \frac{3}{100}$ and $r = \frac{1}{100}$. Thus

$0.030303\ldots = \frac{a}{1-r} = \frac{\frac{3}{100}}{1 - \frac{1}{100}} = \frac{3}{99} = \frac{1}{33}.$

59. $0.\overline{112} = 0.112112112\ldots = \frac{112}{1000} + \frac{112}{1,000,000} + \frac{112}{1,000,000,000} + \cdots$ is an infinite geometric series with $a = \frac{112}{1000}$ and

$r = \frac{1}{1000}$. Thus $0.112112112\ldots = \frac{a}{1-r} = \frac{\frac{112}{1000}}{1 - \frac{1}{1000}} = \frac{112}{999}.$

61. Since we have 5 terms, let us denote $a_1 = 5$ and $a_5 = 80$. Also, $\frac{a_5}{a_1} = r^4$ because the sequence is geometric, and so

$r^4 = \frac{80}{5} = 16 \quad \Leftrightarrow \quad r = \pm 2$. If $r = 2$, the three geometric means are $a_2 = 10$, $a_3 = 20$, and $a_4 = 40$. (If $r = -2$, the

three geometric means are $a_2 = -10$, $a_3 = 20$, and $a_4 = -40$, but these are not between 5 and 80.)

63. (a) The value at the end of the year is equal to the the value at beginning less the depreciation, so

$V_n = V_{n-1} - 0.2V_{n-1} = 0.8V_{n-1}$ with $V_1 = 160,000$. Thus $V_n = 160,000 \cdot 0.8^{n-1}$.

(b) $V_n < 100,000 \quad \Leftrightarrow \quad 0.8^{n-1} \cdot 160,000 < 100,000 \quad \Leftrightarrow \quad 0.8^{n-1} < 0.625 \quad \Leftrightarrow \quad (n-1)\log 0.8 < \log 0.625 \quad \Leftrightarrow$

$n - 1 > \frac{\log 0.625}{\log 0.8} = 2.11$. Thus it will depreciate to below $100,000 during the fourth year.

65. Since the ball is dropped from a height of 80 feet, $a = 80$. Also since the ball rebounds three-fourths of the distance

fallen, $r = \frac{3}{4}$. So on the nth bounce, the ball attains a height of $a_n = 80\left(\frac{3}{4}\right)^n$. Hence, on the fifth bounce, the ball goes

$a_5 = 80\left(\frac{3}{4}\right)^5 = \frac{80 \cdot 243}{1024} \approx 19$ ft high.

67. Let a_n be the amount of water remaining at the nth stage. We start with 5 gallons, so $a = 5$. When 1 gallon (that is, $\frac{1}{5}$ of the

mixture) is removed, $\frac{4}{5}$ of the mixture (and hence $\frac{4}{5}$ of the water in the mixture) remains. Thus, $a_1 = 5 \cdot \frac{4}{5}$, $a_2 = 5 \cdot \frac{4}{5} \cdot \frac{4}{5}, \ldots$,

and in general, $a_n = 5\left(\frac{4}{5}\right)^n$. The amount of water remaining after 3 repetitions is $a_3 = 5\left(\frac{4}{5}\right)^3 = \frac{64}{25}$, and after

5 repetitions it is $a_5 = 5\left(\frac{4}{5}\right)^5 = \frac{1024}{625}.$

69. Let a_n be the height the ball reaches on the nth bounce. From the given information, a_n is the geometric sequence $a_n = 9 \cdot \left(\frac{1}{3}\right)^n$. (Notice that the ball hits the ground for the fifth time after the fourth bounce.)

(a) $a_0 = 9$, $a_1 = 9 \cdot \frac{1}{3} = 3$, $a_2 = 9 \cdot \left(\frac{1}{3}\right)^2 = 1$, $a_3 = 9 \cdot \left(\frac{1}{3}\right)^3 = \frac{1}{3}$, and $a_4 = 9 \cdot \left(\frac{1}{3}\right)^4 = \frac{1}{9}$. The total distance traveled is
$a_0 + 2a_1 + 2a_2 + 2a_3 + 2a_4 = 9 + 2 \cdot 3 + 2 \cdot 1 + 2 \cdot \frac{1}{3} + 2 \cdot \frac{1}{9} = \frac{161}{9} = 17\frac{8}{9}$ ft.

(b) The total distance traveled at the instant the ball hits the ground for the nth time is
$$
\begin{aligned}
D_n &= 9 + 2 \cdot 9 \cdot \frac{1}{3} + 2 \cdot 9 \cdot \left(\frac{1}{3}\right)^2 + 2 \cdot 9 \cdot \left(\frac{1}{3}\right)^3 + 2 \cdot 9 \cdot \left(\frac{1}{3}\right)^4 + \cdots + 2 \cdot 9 \cdot \left(\frac{1}{3}\right)^{n-1} \\
&= 2\left[9 + 9 \cdot \frac{1}{3} + 9 \cdot \left(\frac{1}{3}\right)^2 + 9 \cdot \left(\frac{1}{3}\right)^3 + 9 \cdot \left(\frac{1}{3}\right)^4 + \cdots + 9 \cdot \left(\frac{1}{3}\right)^{n-1}\right] - 9 \\
&= 2\left[9 \cdot \frac{1 - \left(\frac{1}{3}\right)^n}{1 - \frac{1}{3}}\right] - 9 = 27\left[1 - \left(\frac{1}{3}\right)^n\right] - 9 = 18 - \left(\frac{1}{3}\right)^{n-3}
\end{aligned}
$$

71. Let $a_1 = 1$ be the man with 7 wives. Also, let $a_2 = 7$ (the wives), $a_3 = 7a_2 = 7^2$ (the sacks), $a_4 = 7a_3 = 7^3$ (the cats), and $a_5 = 7a_4 = 7^4$ (the kits). The total is $a_1 + a_2 + a_3 + a_4 + a_5 = 1 + 7 + 7^2 + 7^3 + 7^4$, which is a partial sum of a geometric sequence with $a = 1$ and $r = 7$. Thus, the number in the party is $S_5 = 1 \cdot \frac{1 - 7^5}{1 - 7} = 2801$.

73. Let a_n be the height the ball reaches on the nth bounce. We have $a_0 = 1$ and $a_n = \frac{1}{2}a_{n-1}$. Since the total distance d traveled includes the bounce up as well and the distance down, we have
$$
\begin{aligned}
d &= a_0 + 2 \cdot a_1 + 2 \cdot a_2 + \cdots = 1 + 2\left(\frac{1}{2}\right) + 2\left(\frac{1}{2}\right)^2 + 2\left(\frac{1}{2}\right)^3 + 2\left(\frac{1}{2}\right)^4 + \cdots \\
&= 1 + 1 + \frac{1}{2} + \left(\frac{1}{2}\right)^2 + \left(\frac{1}{2}\right)^3 + \cdots = 1 + \sum_{i=0}^{\infty} \left(\frac{1}{2}\right)^i = 1 + \frac{1}{1 - \frac{1}{2}} = 3
\end{aligned}
$$
Thus the total distance traveled is about 3 m.

75. **(a)** If a square has side x, then by the Pythagorean Theorem the length of the side of the square formed by joining the midpoints is, $\sqrt{\left(\frac{x}{2}\right)^2 + \left(\frac{x}{2}\right)^2} = \sqrt{\frac{x^2}{4} + \frac{x^2}{4}} = \frac{x}{\sqrt{2}}$. In our case, $x = 1$ and the side of the first inscribed square is $\frac{1}{\sqrt{2}}$, the side of the second inscribed square is $\frac{1}{\sqrt{2}} \cdot \frac{1}{\sqrt{2}} = \left(\frac{1}{\sqrt{2}}\right)^2$, the side of the third inscribed square is $\left(\frac{1}{\sqrt{2}}\right)^3$, and so on. Since this pattern continues, the total area of all the squares is
$$A = 1^2 + \left(\frac{1}{\sqrt{2}}\right)^2 + \left(\frac{1}{\sqrt{2}}\right)^4 + \left(\frac{1}{\sqrt{2}}\right)^6 + \cdots = 1 + \frac{1}{2} + \left(\frac{1}{2}\right)^2 + \left(\frac{1}{2}\right)^3 + \cdots = \frac{1}{1 - \frac{1}{2}} = 2.$$

(b) As in part (a), the sides of the squares are $1, \frac{1}{\sqrt{2}}, \left(\frac{1}{\sqrt{2}}\right)^2, \left(\frac{1}{\sqrt{2}}\right)^3, \ldots$. Thus the sum of the perimeters is
$$S = 4 \cdot 1 + 4 \cdot \frac{1}{\sqrt{2}} + 4 \cdot \left(\frac{1}{\sqrt{2}}\right)^2 + 4 \cdot \left(\frac{1}{\sqrt{2}}\right)^3 + \cdots,$$ which is an infinite geometric series with $a = 4$ and $r = \frac{1}{\sqrt{2}}$.
Thus the sum of the perimeters is $S = \frac{4}{1 - \frac{1}{\sqrt{2}}} = \frac{4\sqrt{2}}{\sqrt{2} - 1} = \frac{4\sqrt{2}}{\sqrt{2} - 1} \cdot \frac{\sqrt{2} + 1}{\sqrt{2} + 1} = \frac{4 \cdot 2 + 4\sqrt{2}}{2 - 1} = 8 + 4\sqrt{2}$.

77. Let a_n denote the area colored blue at nth stage. Since only the middle squares are colored blue, $a_n = \frac{1}{9} \times$ (area remaining yellow at the $(n-1)$th stage). Also, the area remaining yellow at the nth stage is $\frac{8}{9}$ of the area remaining yellow at the preceding stage. So $a_1 = \frac{1}{9}$, $a_2 = \frac{1}{9}\left(\frac{8}{9}\right)$, $a_3 = \frac{1}{9}\left(\frac{8}{9}\right)^2$, $a_4 = \frac{1}{9}\left(\frac{8}{9}\right)^3, \ldots$. Thus the total area colored blue $A = \frac{1}{9} + \frac{1}{9}\left(\frac{8}{9}\right) + \frac{1}{9}\left(\frac{8}{9}\right)^2 + \frac{1}{9}\left(\frac{8}{9}\right)^3 + \cdots$ is an infinite geometric series with $a = \frac{1}{9}$ and $r = \frac{8}{9}$. So the total area is $A = \frac{\frac{1}{9}}{1 - \frac{8}{9}} = 1$.

79. Let $a_1, a_2, a_3, \ldots$ be a geometric sequence with common ratio r. Thus $a_2 = a_1 r$, $a_3 = a_1 \cdot r^2, \ldots, a_n = a_1 \cdot r^{n-1}$. Hence,

$$\frac{1}{a_2} = \frac{1}{a_1 \cdot r} = \frac{1}{a_1} \cdot \frac{1}{r}, \frac{1}{a_3} = \frac{1}{a_1 \cdot r^2} = \frac{1}{a_1} \cdot \frac{1}{r^2} = \frac{1}{a_1} \left(\frac{1}{r}\right)^2, \ldots \frac{1}{a_n} = \frac{1}{a_1 \cdot r^{n-1}} = \frac{1}{a_1} \cdot \frac{1}{r^{n-1}} = \frac{1}{a_1} \left(\frac{1}{r}\right)^{n-1}, \text{ and }$$

so $\dfrac{1}{a_1}, \dfrac{1}{a_2}, \dfrac{1}{a_3}, \ldots$ is a geometric sequence with common ratio $\dfrac{1}{r}$.

81. Since $a_1, a_2, a_3, \ldots$ is an arithmetic sequence with common difference d, the terms can be expressed as $a_2 = a_1 + d$, $a_3 = a_1 + 2d, \ldots, a_n = a_1 + (n-1)d$. So $10^{a_2} = 10^{a_1 + d} = 10^{a_1} \cdot 10^d$, $10^{a_3} = 10^{a_1 + 2d} = 10^{a_1} \cdot \left(10^d\right)^2, \ldots$, $10^{a_n} = 10^{a_1 + (n-1)d} = 10^{a_1} \cdot \left(10^d\right)^{n-1}$, and so $10^{a_1}, 10^{a_2}, 10^{a_3}, \ldots$ is a geometric sequence with common ratio $r = 10^d$.

12.4   Mathematics of Finance

1. $n = 10$, $R = \$1000$, $i = 0.06$. So $A_f = R\dfrac{(1+i)^n - 1}{i} = 1000\dfrac{(1+0.06)^{10} - 1}{0.06} = \$13{,}180.79$.

3. $n = 20$, $R = \$5000$, $i = 0.12$. So $A_f = R\dfrac{(1+i)^n - 1}{i} = 5000\dfrac{(1+0.12)^{20} - 1}{0.12} = \$360{,}262.21$.

5. $n = 16$, $R = \$300$, $i = \dfrac{0.08}{4} = 0.02$. So $A_f = R\dfrac{(1+i)^n - 1}{i} = 300\dfrac{(1+0.02)^{16} - 1}{0.02} = \$5{,}591.79$.

7. $A_f = \$2000$, $i = \dfrac{0.06}{12} = 0.005$, $n = 8$. Then $R = \dfrac{iA_f}{(1+i)^n - 1} = \dfrac{(0.005)(2000)}{(1+0.005)^8 - 1} = \$245.66$.

9. $R = \$200$, $n = 20$, $i = \dfrac{0.09}{2} = 0.045$. So $A_p = R\dfrac{1-(1+i)^{-n}}{i} = (200)\dfrac{1-(1+0.045)^{-20}}{0.045} = \$2601.59$.

11. $A_p = \$12{,}000$, $i = \dfrac{0.105}{12} = 0.00875$, $n = 48$. Then $R = \dfrac{iA_p}{1-(1+i)^{-n}} = \dfrac{(0.00875)(12000)}{1-(1+0.00875)^{-48}} = \$307.24$.

13. $A_p = \$100{,}000$, $i = \dfrac{0.08}{12} \approx 0.006667$, $n = 360$. Then $R = \dfrac{iA_p}{1-(1+i)^{-n}} = \dfrac{(0.006667)(100{,}000)}{1-(1+0.006667)^{-360}} = \$733.76$.

Therefore, the total amount paid on this loan over the 30 year period is $(360)(733.76) = \$264{,}153.60$.

15. $A_p = 100{,}000$, $n = 360$, $i = \dfrac{0.0975}{12} = 0.008125$.

(a) $R = \dfrac{iA_p}{1-(1+i)^{-n}} = \dfrac{(0.008125)(100{,}000)}{1-(1+0.008125)^{-360}} = \$859.15$.

(b) The total amount that will be paid over the 30 year period is $(360)(859.15) = \$309{,}294.00$.

(c) $R = \$859.15$, $i = \dfrac{0.0975}{12} = 0.008125$, $n = 360$. So $A_f = 859.15\dfrac{(1+0.008125)^{360} - 1}{0.008125} = \$1{,}841{,}519.29$.

17. $R = \$30$, $i = \dfrac{0.10}{12} \approx 0.008333$, $n = 12$. Then $A_p = R\dfrac{1-(1+i)^{-n}}{i} = 30\dfrac{1-(1+0.008333)^{-12}}{0.008333} = \$341.24$.

19. $A_p = \$640$, $R = \$32$, $n = 24$. We want to solve the equation $R = \dfrac{iA_p}{1-(1+i)^n}$

for the interest rate i. Let x be the interest rate, then $i = \dfrac{x}{12}$. So we can express R

as a function of x by $R(x) = \dfrac{\dfrac{x}{12} \cdot 640}{1 - \left(1 + \dfrac{x}{12}\right)^{-24}}$. We graph $R(x)$ and $y = 32$ in

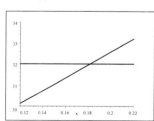

the rectangle $[0.12, 0.22] \times [30, 34]$. The x-coordinate of the intersection is about 0.1816, which corresponds to an interest rate of 18.16%.

21. $A_p = \$189.99$, $R = \$10.50$, $n = 20$. We want to solve the equation

$R = \dfrac{iA_p}{1 - (1 + i)^n}$ for the interest rate i. Let x be the interest rate, then $i = \dfrac{x}{12}$.

So we can express R as a function of x by $R(x) = \dfrac{\dfrac{x}{12} \cdot 189.99}{1 - \left(1 + \dfrac{x}{12}\right)^{-20}}$. We graph

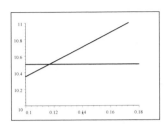

$R(x)$ and $y = 10.50$ in the rectangle $[0.10, 0.18] \times [10, 11]$. The x-coordinate of the intersection is about 0.1168, which corresponds to an interest rate of 11.68%.

23. (a) The present value of the kth payment is $PV = \dfrac{R}{(1 + i)^k}$. The amount of money to be invested now (A_p) to ensure an annuity in perpetuity is the (infinite) sum of the present values of each of the payments, as shown in the time line.

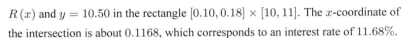

(b) $A_p = \dfrac{R}{1 + i} + \dfrac{R}{(1 + i)^2} + \dfrac{R}{(1 + i)^3} + \cdots + \dfrac{R}{(1 + i)^n} + \cdots$. This is an infinite geometric series with $a = \dfrac{R}{1 + i}$ and

$r = \dfrac{1}{1 + i}$. Therefore, $A_p = \dfrac{\dfrac{R}{1 + i}}{1 - \dfrac{1}{1 + i}} = \dfrac{R}{1 + i} \cdot \dfrac{1 + i}{i} = \dfrac{R}{i}$.

(c) Using the result from part (b), we have $R = 5000$ and $i = 0.10$. Then $A_p = \dfrac{R}{i} = \dfrac{5000}{0.10} = \$50,000$.

(d) We are given two different time periods: the interest is compounded quarterly, while the annuity is paid yearly. In order to use the formula in part (b), we need to find the effective annual interest rate produced by 8% interest compounded quarterly, which is $(1 + i/n)^n - 1$. Now $i = 8\%$ and it is compounded quarterly, $n = 4$, so the effective annual yield is $(1 + 0.02)^4 - 1 \approx 0.08243216$. Thus by the formula in part (b), the amount that must be invested is $A_p = \frac{3000}{0.08243216} = \$36,393.56$.

12.5   Mathematical Induction

1. Let $P(n)$ denote the statement $2 + 4 + 6 + \cdots + 2n = n(n + 1)$.

Step 1: $P(1)$ is the statement that $2 = 1(1 + 1)$, which is true.

Step 2: Assume that $P(k)$ is true; that is, $2 + 4 + 6 + \cdots + 2k = k(k + 1)$. We want to use this to show that $P(k + 1)$ is true. Now

$$\begin{aligned} 2 + 4 + 6 + \cdots + 2k + 2(k + 1) &= k(k + 1) + 2(k + 1) && \text{induction hypothesis} \\ &= (k + 1)(k + 2) = (k + 1)[(k + 1) + 1] \end{aligned}$$

Thus, $P(k + 1)$ follows from $P(k)$. So by the Principle of Mathematical Induction, $P(n)$ is true for all n.

3. Let $P(n)$ denote the statement $5 + 8 + 11 + \cdots + (3n + 2) = \dfrac{n(3n + 7)}{2}$.

Step 1: We need to show that $P(1)$ is true. But $P(1)$ says that $5 = \dfrac{1 \cdot (3 \cdot 1 + 7)}{2}$, which is true.

Step 2: Assume that $P(k)$ is true; that is, $5 + 8 + 11 + \cdots + (3k + 2) = \dfrac{k(3k + 7)}{2}$. We want to use this to show that $P(k + 1)$ is true. Now

$$
\begin{aligned}
5 + 8 + 11 + \cdots + (3k + 2) + [3(k + 1) + 2] &= \frac{k(3k + 7)}{2} + (3k + 5) \qquad \text{induction hypothesis} \\
&= \frac{3k^2 + 7k}{2} + \frac{6k + 10}{2} = \frac{3k^2 + 13k + 10}{2} \\
&= \frac{(3k + 10)(k + 1)}{2} = \frac{(k + 1)[3(k + 1) + 7]}{2}
\end{aligned}
$$

Thus, $P(k + 1)$ follows from $P(k)$. So by the Principle of Mathematical Induction, $P(n)$ is true for all n.

5. Let $P(n)$ denote the statement $1 \cdot 2 + 2 \cdot 3 + 3 \cdot 4 + \cdots + n(n + 1) = \dfrac{n(n + 1)(n + 2)}{3}$.

Step 1: $P(1)$ is the statement that $1 \cdot 2 = \dfrac{1 \cdot (1 + 1) \cdot (1 + 2)}{3}$, which is true.

Step 2: Assume that $P(k)$ is true; that is, $1 \cdot 2 + 2 \cdot 3 + 3 \cdot 4 + \cdots + k(k + 1) = \dfrac{k(k + 1)(k + 2)}{3}$. We want to use this to show that $P(k + 1)$ is true. Now

$$
\begin{aligned}
1 \cdot 2 + 2 \cdot 3 + 3 \cdot 4 + \cdots + k(k + 1) &+ (k + 1)[(k + 1) + 1] \\
&= \frac{k(k + 1)(k + 2)}{3} + (k + 1)(k + 2) \qquad \text{induction hypothesis} \\
&= \frac{k(k + 1)(k + 2)}{3} + \frac{3(k + 1)(k + 2)}{3} = \frac{(k + 1)(k + 2)(k + 3)}{3}
\end{aligned}
$$

Thus, $P(k + 1)$ follows from $P(k)$. So by the Principle of Mathematical Induction, $P(n)$ is true for all n.

7. Let $P(n)$ denote the statement $1^3 + 2^3 + 3^3 + \cdots + n^3 = \dfrac{n^2(n + 1)^2}{4}$.

Step 1: $P(1)$ is the statement that $1^3 = \dfrac{1^2 \cdot (1 + 1)^2}{4}$, which is clearly true.

Step 2: Assume that $P(k)$ is true; that is, $1^3 + 2^3 + 3^3 + \cdots + k^3 = \dfrac{k^2(k + 1)^2}{4}$. We want to use this to show that $P(k + 1)$ is true. Now

$$
\begin{aligned}
1^3 + 2^3 + 3^3 + \cdots + k^3 + (k + 1)^3 &= \frac{k^2(k + 1)^2}{4} + (k + 1)^3 \qquad \text{induction hypothesis} \\
&= \frac{(k + 1)^2[k^2 + 4(k + 1)]}{4} = \frac{(k + 1)^2[k^2 + 4k + 4]}{4} \\
&= \frac{(k + 1)^2(k + 2)^2}{4} = \frac{(k + 1)^2[(k + 1) + 1]^2}{4}
\end{aligned}
$$

Thus, $P(k + 1)$ follows from $P(k)$. So by the Principle of Mathematical Induction, $P(n)$ is true for all n.

9. Let $P(n)$ denote the statement $2^3 + 4^3 + 6^3 + \cdots + (2n)^3 = 2n^2(n+1)^2$.

Step 1: $P(1)$ is true since $2^3 = 2(1)^2(1+1)^2 = 2 \cdot 4 = 8$.

Step 2: Assume that $P(k)$ is true; that is, $2^3 + 4^3 + 6^3 + \cdots + (2k)^3 = 2k^2(k+1)^2$. We want to use this to show that $P(k+1)$ is true. Now

$$
\begin{aligned}
2^3 + 4^3 + 6^3 + \cdots + (2k)^3 + [2(k+1)]^3 &= 2k^2(k+1)^2 + [2(k+1)]^3 &\text{induction hypothesis}\\
&= 2k^2(k+1)^2 + 8(k+1)(k+1)^2 = (k+1)^2(2k^2 + 8k + 8)\\
&= 2(k+1)^2(k+2)^2 = 2(k+1)^2[(k+1)+1]^2
\end{aligned}
$$

Thus, $P(k+1)$ follows from $P(k)$. So by the Principle of Mathematical Induction, $P(n)$ is true for all n.

11. Let $P(n)$ denote the statement $1 \cdot 2 + 2 \cdot 2^2 + 3 \cdot 2^3 + 4 \cdot 2^4 + \cdots + n \cdot 2^n = 2[1 + (n-1)2^n]$.

Step 1: $P(1)$ is the statement that $1 \cdot 2 = 2[1+0]$, which is clearly true.

Step 2: Assume that $P(k)$ is true; that is, $1 \cdot 2 + 2 \cdot 2^2 + 3 \cdot 2^3 + 4 \cdot 2^4 + \cdots + k \cdot 2^k = 2[1 + (k-1)2^k]$. We want to use this to show that $P(k+1)$ is true. Now

$$1 \cdot 2 + 2 \cdot 2^2 + 3 \cdot 2^3 + 4 \cdot 2^4 + \cdots + k \cdot 2^k + (k+1) \cdot 2^{(k+1)}$$

$$
\begin{aligned}
&= 2\left[1 + (k-1)2^k\right] + (k+1) \cdot 2^{k+1} &\text{induction hypothesis}\\
&= 2\left[1 + (k-1) \cdot 2^k + (k+1) \cdot 2^k\right] = 2\left[1 + 2k \cdot 2^k\right]\\
&= 2\left[1 + k \cdot 2^{k+1}\right] = 2\left\{1 + [(k+1) - 1]2^{k+1}\right\}
\end{aligned}
$$

Thus $P(k+1)$ follows from $P(k)$. So by the Principle of Mathematical Induction, $P(n)$ is true for all n.

13. Let $P(n)$ denote the statement $n^2 + n$ is divisible by 2.

Step 1: $P(1)$ is the statement that $1^2 + 1 = 2$ is divisible by 2, which is clearly true.

Step 2: Assume that $P(k)$ is true; that is, $k^2 + k$ is divisible by 2. Now $(k+1)^2 + (k+1) = k^2 + 2k + 1 + k + 1 = (k^2 + k) + 2k + 2 = (k^2 + k) + 2(k+1)$. By the induction hypothesis, $k^2 + k$ is divisible by 2, and clearly $2(k+1)$ is divisible by 2. Thus, the sum is divisible by 2, so $P(k+1)$ is true. Therefore, $P(k+1)$ follows from $P(k)$. So by the Principle of Mathematical Induction, $P(n)$ is true for all n.

15. Let $P(n)$ denote the statement that $n^2 - n + 41$ is odd.

Step 1: $P(1)$ is the statement that $1^2 - 1 + 41 = 41$ is odd, which is clearly true.

Step 2: Assume that $P(k)$ is true; that is, $k^2 - k + 41$ is odd. We want to use this to show that $P(k+1)$ is true. Now, $(k+1)^2 - (k+1) + 41 = k^2 + 2k + 1 - k - 1 + 41 = (k^2 - k + 41) + 2k$, which is also odd because $k^2 - k + 41$ is odd by the induction hypothesis, $2k$ is always even, and an odd number plus an even number is always odd. Therefore, $P(k+1)$ follows from $P(k)$. So by the Principle of Mathematical Induction, $P(n)$ is true for all n.

17. Let $P(n)$ denote the statement that $8^n - 3^n$ is divisible by 5.

Step 1: $P(1)$ is the statement that $8^1 - 3^1 = 5$ is divisible by 5, which is clearly true.

Step 2: Assume that $P(k)$ is true; that is, $8^k - 3^k$ is divisible by 5. We want to use this to show that $P(k+1)$ is true. Now, $8^{k+1} - 3^{k+1} = 8 \cdot 8^k - 3 \cdot 3^k = 8 \cdot 8^k - (8-5) \cdot 3^k = 8 \cdot (8^k - 3^k) + 5 \cdot 3^k$, which is divisible by 5 because $8^k - 3^k$ is divisible by 5 by our induction hypothesis, and $5 \cdot 3^k$ is divisible by 5. Thus $P(k+1)$ follows from $P(k)$. So by the Principle of Mathematical Induction, $P(n)$ is true for all n.

19. Let $P(n)$ denote the statement $n < 2^n$.

Step 1: $P(1)$ is the statement that $1 < 2^1 = 2$, which is clearly true.

Step 2: Assume that $P(k)$ is true; that is, $k < 2^k$. We want to use this to show that $P(k+1)$ is true. Adding 1 to both sides of $P(k)$ we have $k + 1 < 2^k + 1$. Since $1 < 2^k$ for $k \geq 1$, we have $2^k + 1 < 2^k + 2^k = 2 \cdot 2^k = 2^{k+1}$. Thus $k + 1 < 2^{k+1}$, which is exactly $P(k+1)$. Therefore, $P(k+1)$ follows from $P(k)$. So by the Principle of Mathematical Induction, $P(n)$ is true for all n.

21. Let $P(n)$ denote the statement $(1 + x)^n \geq 1 + nx$, if $x > -1$.

Step 1: $P(1)$ is the statement that $(1 + x)^1 \geq 1 + 1x$, which is clearly true.

Step 2: Assume that $P(k)$ is true; that is, $(1 + x)^k \geq 1 + kx$. Now, $(1 + x)^{k+1} = (1 + x)(1 + x)^k \geq (1 + x)(1 + kx)$, by the induction hypothesis. Since $(1 + x)(1 + kx) = 1 + (k + 1)x + kx^2 \geq 1 + (k + 1)x$ (since $kx^2 \geq 0$), we have $(1 + x)^{k+1} \geq 1 + (k + 1)x$, which is $P(k+1)$. Thus $P(k+1)$ follows from $P(k)$. So the Principle of Mathematical Induction, $P(n)$ is true for all n.

23. Let $P(n)$ be the statement that $a_n = 5 \cdot 3^{n-1}$.

Step 1: $P(1)$ is the statement that $a_1 = 5 \cdot 3^0 = 5$, which is true.

Step 2: Assume that $P(k)$ is true; that is, $a_k = 5 \cdot 3^{k-1}$. We want to use this to show that $P(k+1)$ is true. Now, $a_{k+1} = 3a_k = 3 \cdot (5 \cdot 3^{k-1})$, by the induction hypothesis. Therefore, $a_{k+1} = 3 \cdot (5 \cdot 3^{k-1}) = 5 \cdot 3^k$, which is exactly $P(k+1)$. Thus, $P(k+1)$ follows from $P(k)$. So by the Principle of Mathematical Induction, $P(n)$ is true for all n.

25. Let $P(n)$ be the statement that $x - y$ is a factor of $x^n - y^n$ for all natural numbers n.

Step 1: $P(1)$ is the statement that $x - y$ is a factor of $x^1 - y^1$, which is clearly true.

Step 2: Assume that $P(k)$ is true; that is, $x - y$ is a factor of $x^k - y^k$. We want to use this to show that $P(k+1)$ is true. Now, $x^{k+1} - y^{k+1} = x^{k+1} - x^k y + x^k y - y^{k+1} = x^k (x - y) + (x^k - y^k) y$, for which $x - y$ is a factor because $x - y$ is a factor of $x^k (x - y)$, and $x - y$ is a factor of $(x^k - y^k) y$, by the induction hypothesis. Thus $P(k+1)$ follows from $P(k)$. So by the Principle of Mathematical Induction, $P(n)$ is true for all n.

27. Let $P(n)$ denote the statement that F_{3n} is even for all natural numbers n.

Step 1: $P(1)$ is the statement that F_3 is even. Since $F_3 = F_2 + F_1 = 1 + 1 = 2$, this statement is true.

Step 2: Assume that $P(k)$ is true; that is, F_{3k} is even. We want to use this to show that $P(k+1)$ is true. Now, $F_{3(k+1)} = F_{3k+3} = F_{3k+2} + F_{3k+1} = F_{3k+1} + F_{3k} + F_{3k+1} = F_{3k} + 2 \cdot F_{3k+1}$, which is even because F_{3k} is even by the induction hypothesis, and $2 \cdot F_{3k+1}$ is even. Thus $P(k+1)$ follows from $P(k)$. So by the Principle of Mathematical Induction, $P(n)$ is true for all n.

29. Let $P(n)$ denote the statement that $F_1^2 + F_2^2 + F_3^2 + \cdots + F_n^2 = F_n \cdot F_{n+1}$.

Step 1: $P(1)$ is the statement that $F_1^2 = F_1 \cdot F_2$ or $1^2 = 1 \cdot 1$, which is true.

Step 2: Assume that $P(k)$ is true, that is, $F_1^2 + F_2^2 + F_3^2 + \cdots + F_k^2 = F_k \cdot F_{k+1}$. We want to use this to show that $P(k+1)$ is true. Now

$$
\begin{aligned}
F_1^2 + F_2^2 + F_3^2 + \cdots + F_k^2 + F_{k+1}^2 &= F_k \cdot F_{k+1} + F_{k+1}^2 && \text{induction hypothesis} \\
&= F_{k+1}(F_k + F_{k+1}) \\
&= F_{k+1} \cdot F_{k+2} && \text{by definition of the Fibonacci sequence}
\end{aligned}
$$

Thus $P(k+1)$ follows from $P(k)$. So by the Principle of Mathematical Induction, $P(n)$ is true for all n.

31. Let $P(n)$ denote the statement $\begin{bmatrix} 1 & 1 \\ 1 & 0 \end{bmatrix}^n = \begin{bmatrix} F_{n+1} & F_n \\ F_n & F_{n-1} \end{bmatrix}$.

Step 1: Since $\begin{bmatrix} 1 & 1 \\ 1 & 0 \end{bmatrix}^2 = \begin{bmatrix} 1 & 1 \\ 1 & 0 \end{bmatrix}\begin{bmatrix} 1 & 1 \\ 1 & 0 \end{bmatrix} = \begin{bmatrix} 2 & 1 \\ 1 & 1 \end{bmatrix} = \begin{bmatrix} F_3 & F_2 \\ F_2 & F_1 \end{bmatrix}$, it follows that $P(2)$ is true.

Step 2: Assume that $P(k)$ is true; that is, $\begin{bmatrix} 1 & 1 \\ 1 & 0 \end{bmatrix}^k = \begin{bmatrix} F_{k+1} & F_k \\ F_k & F_{k-1} \end{bmatrix}$. We show that $P(k+1)$ follows from this. Now,

$$\begin{bmatrix} 1 & 1 \\ 1 & 0 \end{bmatrix}^{k+1} = \begin{bmatrix} 1 & 1 \\ 1 & 0 \end{bmatrix}^k\begin{bmatrix} 1 & 1 \\ 1 & 0 \end{bmatrix} = \begin{bmatrix} F_{k+1} & F_k \\ F_k & F_{k-1} \end{bmatrix}\begin{bmatrix} 1 & 1 \\ 1 & 0 \end{bmatrix} \quad \text{induction hypothesis}$$

$$= \begin{bmatrix} F_{k+1} + F_k & F_{k+1} \\ F_k + F_{k-1} & F_k \end{bmatrix} = \begin{bmatrix} F_{k+2} & F_{k+1} \\ F_{k+1} & F_k \end{bmatrix} \quad \text{by definition of the Fibonacci sequence}$$

Thus $P(k+1)$ follows from $P(k)$. So by the Principle of Mathematical Induction, $P(n)$ is true for all $n \geq 2$.

33. Since $F_1 = 1$, $F_2 = 1$, $F_3 = 2$, $F_4 = 3$, $F_5 = 5$, $F_6 = 8$, $F_7 = 13$, ... our conjecture is that $F_n \geq n$, for all $n \geq 5$. Let $P(n)$ denote the statement that $F_n \geq n$.

Step 1: $P(5)$ is the statement that $F_5 = 5 \geq 5$, which is clearly true.

Step 2: Assume that $P(k)$ is true; that is, $F_k \geq k$, for some $k \geq 5$. We want to use this to show that $P(k+1)$ is true. Now, $F_{k+1} = F_k + F_{k-1} \geq k + F_{k-1}$ (by the induction hypothesis)$\geq k+1$ (because $F_{k-1} \geq 1$). Thus $P(k+1)$ follows from $P(k)$. So by the Principle of Mathematical Induction, $P(n)$ is true for all $n \geq 5$.

35. (a) $P(n) = n^2 - n + 11$ is prime for all n. This is false as the case for $n = 11$ demonstrates: $P(11) = 11^2 - 11 + 11 = 121$, which is not prime since $11^2 = 121$.

(b) $n^2 > n$, for all $n \geq 2$. This is true. Let $P(n)$ denote the statement that $n^2 > n$.

Step 1: $P(2)$ is the statement that $2^2 = 4 > 2$, which is clearly true.

Step 2: Assume that $P(k)$ is true; that is, $k^2 > k$. We want to use this to show that $P(k+1)$ is true. Now $(k+1)^2 = k^2 + 2k + 1$. Using the induction hypothesis (to replace k^2), we have $k^2 + 2k + 1 > k + 2k + 1 = 3k + 1 > k + 1$, since $k \geq 2$. Therefore, $(k+1)^2 > k+1$, which is exactly $P(k+1)$. Thus $P(k+1)$ follows from $P(k)$. So by the Principle of Mathematical Induction, $P(n)$ is true for all n.

(c) $2^{2n+1} + 1$ is divisible by 3, for all $n \geq 1$. This is true. Let $P(n)$ denote the statement that $2^{2n+1} + 1$ is divisible by 3.

Step 1: $P(1)$ is the statement that $2^3 + 1 = 9$ is divisible by 3, which is clearly true.

Step 2: Assume that $P(k)$ is true; that is, $2^{2k+1} + 1$ is divisible by 3. We want to use this to show that $P(k+1)$ is true. Now, $2^{2(k+1)+1} + 1 = 2^{2k+3} + 1 = 4 \cdot 2^{2k+1} + 1 = (3+1)2^{2k+1} + 1 = 3 \cdot 2^{2k+1} + (2^{2k+1} + 1)$, which is divisible by 3 since $2^{2k+1} + 1$ is divisible by 3 by the induction hypothesis, and $3 \cdot 2^{2k+1}$ is clearly divisible by 3. Thus $P(k+1)$ follows from $P(k)$. So by the Principle of Mathematical Induction, $P(n)$ is true for all n.

(d) The statement $n^3 \geq (n+1)^2$ for all $n \geq 2$ is false. The statement fails when $n = 2$: $2^3 = 8 < (2+1)^2 = 9$.

(e) $n^3 - n$ is divisible by 3, for all $n \geq 2$. This is true. Let $P(n)$ denote the statement that $n^3 - n$ is divisible by 3.

Step 1: $P(2)$ is the statement that $2^3 - 2 = 6$ is divisible by 3, which is clearly true.

Step 2: Assume that $P(k)$ is true; that is, $k^3 - k$ is divisible by 3. We want to use this to show that $P(k+1)$ is true. Now $(k+1)^3 - (k+1) = k^3 + 3k^2 + 3k + 1 - (k+1) = k^3 + 3k^2 + 2k = k^3 - k + 3k^2 + 2k + k = (k^3 - k) + 3(k^2 + k)$. The term $k^3 - k$ is divisible by 3 by our induction hypothesis, and the term $3(k^2 + k)$ is clearly divisible by 3. Thus $(k+1)^3 - (k+1)$ is divisible by 3, which is exactly $P(k+1)$. So by the Principle of Mathematical Induction, $P(n)$ is true for all n.

(f) $n^3 - 6n^2 + 11n$ is divisible by 6, for all $n \geq 1$. This is true. Let $P(n)$ denote the statement that $n^3 - 6n^2 + 11n$ is divisible by 6.

Step 1: $P(1)$ is the statement that $(1)^3 - 6(1)^2 + 11(1) = 6$ is divisible by 6, which is clearly true.

Step 2: Assume that $P(k)$ is true; that is, $k^3 - 6k^2 + 11k$ is divisible by 6. We show that $P(k+1)$ is then also true. Now

$$\begin{aligned}
(k+1)^3 - 6(k+1)^2 + 11(k+1) &= k^3 + 3k^2 + 3k + 1 - 6k^2 - 12k - 6 + 11k + 11 \\
&= k^3 - 3k^2 + 2k + 6 = k^3 - 6k^2 + 11k + (3k^2 - 9k + 6) \\
&= (k^3 - 6k^2 + 11k) + 3(k^2 - 3k + 2) = (k^3 - 6k^2 + 11k) + 3(k-1)(k-2)
\end{aligned}$$

In this last expression, the first term is divisible by 6 by our induction hypothesis. The second term is also divisible by 6. To see this, notice that $k - 1$ and $k - 2$ are consecutive natural numbers, and so one of them must be even (divisible by 2). Since 3 also appears in this second term, it follows that this term is divisible by 2 and 3 and so is divisible by 6. Thus $P(k+1)$ follows from $P(k)$. So by the Principle of Mathematical Induction, $P(n)$ is true for all n.

12.6 The Binomial Theorem

1. $(x+y)^6 = x^6 + 6x^5y + 15x^4y^2 + 20x^3y^3 + 15x^2y^4 + 6xy^5 + y^6$

3. $\left(x + \dfrac{1}{x}\right)^4 = x^4 + 4x^3 \cdot \dfrac{1}{x} + 6x^2 \left(\dfrac{1}{x}\right)^2 + 4x \left(\dfrac{1}{x}\right)^3 + \left(\dfrac{1}{x}\right)^4 = x^4 + 4x^2 + 6 + \dfrac{4}{x^2} + \dfrac{1}{x^4}$

5. $(x-1)^5 = x^5 - 5x^4 + 10x^3 - 10x^2 + 5x - 1$

7. $(x^2y - 1)^5 = (x^2y)^5 - 5(x^2y)^4 + 10(x^2y)^3 - 10(x^2y)^2 + 5x^2y - 1 = x^{10}y^5 - 5x^8y^4 + 10x^6y^3 - 10x^4y^2 + 5x^2y - 1$

9. $(2x - 3y)^3 = (2x)^3 - 3(2x)^2 3y + 3 \cdot 2x(3y)^2 - (3y)^3 = 8x^3 - 36x^2y + 54xy^2 - 27y^3$

11. $\left(\dfrac{1}{x} - \sqrt{x}\right)^5 = \left(\dfrac{1}{x}\right)^5 - 5\left(\dfrac{1}{x}\right)^4 \sqrt{x} + 10\left(\dfrac{1}{x}\right)^3 x - 10\left(\dfrac{1}{x}\right)^2 x\sqrt{x} + 5\left(\dfrac{1}{x}\right)x^2 - x^2\sqrt{x}$

$$= \dfrac{1}{x^5} - \dfrac{5}{x^{7/2}} + \dfrac{10}{x^2} - \dfrac{10}{x^{1/2}} + 5x - x^{5/2}$$

13. $\dbinom{6}{4} = \dfrac{6!}{4! \, 2!} = \dfrac{6 \cdot 5 \cdot 4!}{2 \cdot 1 \cdot 4!} = 15$

15. $\dbinom{100}{98} = \dfrac{100!}{98! \, 2!} = \dfrac{100 \cdot 99 \cdot 98!}{98! \cdot 2 \cdot 1} = 4950$

17. $\dbinom{3}{1}\dbinom{4}{2} = \dfrac{3!}{1! \, 2!} \dfrac{4!}{2! \, 2!} = \dfrac{3 \cdot 2! \cdot 4 \cdot 3 \cdot 2!}{1 \cdot 2! \cdot 2 \cdot 1 \cdot 2!} = 18$

19. $\binom{5}{0} + \binom{5}{1} + \binom{5}{2} + \binom{5}{3} + \binom{5}{4} + \binom{5}{5} = (1+1)^5 = 2^5 = 32$

21. $(x + 2y)^4 = \binom{4}{0}x^4 + \binom{4}{1}x^3 \cdot 2y + \binom{4}{2}x^2 \cdot 4y^2 + \binom{4}{3}x \cdot 8y^3 + \binom{4}{4}16y^4 = x^4 + 8x^3y + 24x^2y^2 + 32xy^3 + 16y^4$

23. $\left(1 + \dfrac{1}{x}\right)^6 = \binom{6}{0}1^6 + \binom{6}{1}1^5 \left(\dfrac{1}{x}\right) + \binom{6}{2}1^4 \left(\dfrac{1}{x}\right)^2 + \binom{6}{3}1^3 \left(\dfrac{1}{x}\right)^3 + \binom{6}{4}1^2 \left(\dfrac{1}{x}\right)^4 + \binom{6}{5}1 \left(\dfrac{1}{x}\right)^5 + \binom{6}{6}\left(\dfrac{1}{x}\right)^6$

$$= 1 + \dfrac{6}{x} + \dfrac{15}{x^2} + \dfrac{20}{x^3} + \dfrac{15}{x^4} + \dfrac{6}{x^5} + \dfrac{1}{x^6}$$

25. The first three terms in the expansion of $(x + 2y)^{20}$ are $\binom{20}{0}x^{20} = x^{20}$, $\binom{20}{1}x^{19} \cdot 2y = 40x^{19}y$, and $\binom{20}{2}x^{18} \cdot (2y)^2 = 760x^{18}y^2$.

27. The last two terms in the expansion of $\left(a^{2/3} + a^{1/3}\right)^{25}$ are $\binom{25}{24}a^{2/3} \cdot \left(a^{1/3}\right)^{24} = 25a^{26/3}$, and $\binom{25}{25}a^{25/3} = a^{25/3}$.

29. The middle term in the expansion of $(x^2 + 1)^{18}$ occurs when both terms are raised to the 9th power. So this term is $\binom{18}{9}(x^2)^9 1^9 = 48{,}620x^{18}$.

31. The 24th term in the expansion of $(a + b)^{25}$ is $\binom{25}{23}a^2b^{23} = 300a^2b^{23}$.

33. The 100th term in the expansion of $(1 + y)^{100}$ is $\binom{100}{99}1^1 \cdot y^{99} = 100y^{99}$.

35. The term that contains x^4 in the expansion of $(x + 2y)^{10}$ has exponent $r = 4$. So this term is $\binom{10}{4}x^4 \cdot (2y)^{10-4} = 13{,}440x^4y^6$.

37. The rth term is $\binom{12}{r}a^r (b^2)^{12-r} = \binom{12}{r}a^r b^{24-2r}$. Thus the term that contains b^8 occurs where $24 - 2r = 8 \iff r = 8$. So the term is $\binom{12}{8}a^8b^8 = 495a^8b^8$.

39. $x^4 + 4x^3y + 6x^2y^2 + 4xy^3 + y^4 = (x + y)^4$

41. $8a^3 + 12a^2b + 6ab^2 + b^3 = \binom{3}{0}(2a)^3 + \binom{3}{1}(2a)^2 b + \binom{3}{2}2ab^2 + \binom{3}{3}b^3 = (2a + b)^3$

43. $\dfrac{(x + h)^3 - x^3}{h} = \dfrac{x^3 + 3x^2h + 3xh^2 + h^3 - x^3}{h} = \dfrac{3x^2h + 3xh^2 + h^3}{h} = \dfrac{h\left(3x^2 + 3xh + h^2\right)}{h} = 3x^2 + 3xh + h^2$

45. $(1.01)^{100} = (1 + 0.01)^{100}$. Now the first term in the expansion is $\binom{100}{0}1^{100} = 1$, the second term is $\binom{100}{1}1^{99}(0.01) = 1$, and the third term is $\binom{100}{2}1^{98}(0.01)^2 = 0.495$. Now each term is nonnegative, so $(1.01)^{100} = (1 + 0.01)^{100} > 1 + 1 + .0.495 > 2$. Thus $(1.01)^{100} > 2$.

47. $\binom{n}{1} = \dfrac{n!}{1!\,(n - 1)!} = \dfrac{n\,(n - 1)!}{1\,(n - 1)!} = \dfrac{n}{1} = n$. $\binom{n}{n - 1} = \dfrac{n!}{(n - 1)!\,1!} = \dfrac{n\,(n - 1)!}{(n - 1)!\,1} = n$. Therefore, $\binom{n}{1} = \binom{n}{n - 1} = n$.

49. (a) $\binom{n}{r - 1} + \binom{n}{r} = \dfrac{n!}{(r - 1)!\,[n - (r - 1)]!} + \dfrac{n!}{r!\,(n - r)!}$.

(b) $\dfrac{n!}{(r - 1)!\,[n - (r - 1)]!} + \dfrac{n!}{r!\,(n - r)!} = \dfrac{r \cdot n!}{r \cdot (r - 1)!\,(n - r + 1)!} + \dfrac{(n - r + 1) \cdot n!}{r!\,(n - r + 1)\,(n - r)!}$

$= \dfrac{r \cdot n!}{r!\,(n - r + 1)!} + \dfrac{(n - r + 1) \cdot n!}{r!\,(n - r + 1)!}$

Thus a common denominator is $r!\,(n - r + 1)!$.

(c) Therefore, using the results of parts (a) and (b),

$\binom{n}{r - 1} + \binom{n}{r} = \dfrac{n!}{(r - 1)!\,[n - (r - 1)]!} + \dfrac{n!}{r!\,(n - r)!} = \dfrac{r \cdot n!}{r!\,(n - r + 1)!} + \dfrac{(n - r + 1) \cdot n!}{r!\,(n - r + 1)!}$

$= \dfrac{r \cdot n! + (n - r + 1) \cdot n!}{r!\,(n - r + 1)!} = \dfrac{n!\,(r + n - r + 1)}{r!\,(n - r + 1)!} = \dfrac{n!\,(n + 1)}{r!\,(n + 1 - r)!} = \dfrac{(n + 1)!}{r!\,(n + 1 - r)!} = \binom{n + 1}{r}$

51. Notice that $(100!)^{101} = (100!)^{100} \cdot 100!$ and $(101!)^{100} = (101 \cdot 100!)^{100} = 101^{100} \cdot (100!)^{100}$. Now $100! = 1 \cdot 2 \cdot 3 \cdot 4 \cdots 99 \cdot 100$ and $101^{100} = 101 \cdot 101 \cdot 101 \cdots 101$. Thus each of these last two expressions consists of 100 factors multiplied together, and since each factor in the product for 101^{100} is larger than each factor in the product for $100!$, it follows that $100! < 101^{100}$. Thus $(100!)^{100} \cdot 100! < (100!)^{100} \cdot 101^{100}$. So $(100!)^{101} < (101!)^{100}$.

53. $0 = 0^n = (-1 + 1)^n = \binom{n}{0}(-1)^0 (1)^n + \binom{n}{1}(-1)^1 (1)^{n-1} + \binom{n}{2}(-1)^2 (1)^{n-2} + \cdots + \binom{n}{n}(-1)^n (1)^0$

$= \binom{n}{0} - \binom{n}{1} + \binom{n}{2} - \cdots + (-1)^k \binom{n}{k} + \cdots + (-1)^n \binom{n}{n}$

Chapter 12 Review

1. $a_n = \dfrac{n^2}{n+1}$. Then $a_1 = \dfrac{1^2}{1+1} = \dfrac{1}{2}$, $a_2 = \dfrac{2^2}{2+1} = \dfrac{4}{3}$, $a_3 = \dfrac{3^2}{3+1} = \dfrac{9}{4}$, $a_4 = \dfrac{4^2}{4+1} = \dfrac{16}{5}$, and

$a_{10} = \dfrac{10^2}{10+1} = \dfrac{100}{11}$.

3. $a_n = \dfrac{(-1)^n + 1}{n^3}$. Then $a_1 = \dfrac{(-1)^1 + 1}{1^3} = 0$, $a_2 = \dfrac{(-1)^2 + 1}{2^3} = \dfrac{2}{8} = \dfrac{1}{4}$, $a_3 = \dfrac{(-1)^3 + 1}{3^3} = 0$,

$a_4 = \dfrac{(-1)^4 + 1}{4^3} = \dfrac{2}{64} = \dfrac{1}{32}$, and $a_{10} = \dfrac{(-1)^{10} + 1}{10^3} = \dfrac{1}{500}$.

5. $a_n = \dfrac{(2n)!}{2^n n!}$. Then $a_1 = \dfrac{(2 \cdot 1)!}{2^1 \cdot 1!} = 1$, $a_2 = \dfrac{(2 \cdot 2)!}{2^2 \cdot 2!} = 3$, $a_3 = \dfrac{(2 \cdot 3)!}{2^3 \cdot 3!} = \dfrac{6 \cdot 5 \cdot 4}{8} = 15$,

$a_4 = \dfrac{(2 \cdot 4)!}{2^4 \cdot 4!} = \dfrac{8 \cdot 7 \cdot 6 \cdot 5}{16} = 105$, and $a_{10} = \dfrac{(2 \cdot 10)!}{2^{10} \cdot 10!} = 654{,}729{,}075$.

7. $a_n = a_{n-1} + 2n - 1$ and $a_1 = 1$. Then $a_2 = a_1 + 4 - 1 = 4$, $a_3 = a_2 + 6 - 1 = 9$, $a_4 = a_3 + 8 - 1 = 16$, $a_5 = a_4 + 10 - 1 = 25$, $a_6 = a_5 + 12 - 1 = 36$, and $a_7 = a_6 + 14 - 1 = 49$.

9. $a_n = a_{n-1} + 2a_{n-2}$, $a_1 = 1$ and $a_2 = 3$. Then $a_3 = a_2 + 2a_1 = 5$, $a_4 = a_3 + 2a_2 = 11$, $a_5 = a_4 + 2a_3 = 21$, $a_6 = a_5 + 2a_4 = 43$, and $a_7 = a_6 + 2a_5 = 85$.

11. (a) $a_1 = 2(1) + 5 = 7$, $a_2 = 2(2) + 5 = 9$,
$a_3 = 2(3) + 5 = 11$, $a_4 = 2(4) + 5 = 13$,
$a_5 = 2(5) + 5 = 15$

(b)

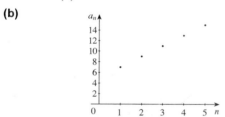

(c) This sequence is arithmetic with common difference 2.

13. (a) $a_1 = \dfrac{3^1}{2^2} = \dfrac{3}{4}$, $a_2 = \dfrac{3^2}{2^3} = \dfrac{9}{8}$, $a_3 = \dfrac{3^3}{2^4} = \dfrac{27}{16}$,
$a_4 = \dfrac{3^4}{2^5} = \dfrac{81}{32}$, $a_5 = \dfrac{3^5}{2^6} = \dfrac{243}{64}$

(b)

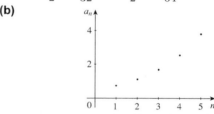

(c) This sequence is geometric with common ratio $\dfrac{3}{2}$.

15. $5, 5.5, 6, 6.5, \ldots$. Since $5.5 - 5 = 6 - 5.5 = 6.5 - 6 = 0.5$, this is an arithmetic sequence with $a_1 = 5$ and $d = 0.5$. Then $a_5 = a_4 + 0.5 = 7$.

17. $\sqrt{2}, 2\sqrt{2}, 3\sqrt{2}, 4\sqrt{2}, \ldots$. Since $2\sqrt{2} - \sqrt{2} = 3\sqrt{2} - 2\sqrt{2} = 4\sqrt{2} - 3\sqrt{2} = \sqrt{2}$, this is an arithmetic sequence with $a_1 = \sqrt{2}$ and $d = \sqrt{2}$. Then $a_5 = a_4 + \sqrt{2} = 4\sqrt{2} + \sqrt{2} = 5\sqrt{2}$.

19. $t - 3, t - 2, t - 1, t, \ldots$. Since $(t - 2) - (t - 3) = (t - 1) - (t - 2) = t - (t - 1) = 1$, this is an arithmetic sequence with $a_1 = t - 3$ and $d = 1$. Then $a_5 = a_4 + 1 = t + 1$.

21. $\dfrac{3}{4}, \dfrac{1}{2}, \dfrac{1}{3}, \dfrac{2}{9}, \ldots$. Since $\dfrac{\frac{1}{2}}{\frac{3}{4}} = \dfrac{\frac{1}{3}}{\frac{1}{2}} = \dfrac{\frac{2}{9}}{\frac{1}{3}} = \dfrac{2}{3}$, this is a geometric sequence with $a_1 = \dfrac{3}{4}$ and $r = \dfrac{2}{3}$. Then
$a_5 = a_4 \cdot r = \dfrac{2}{9} \cdot \dfrac{2}{3} = \dfrac{4}{27}$.

23. $3, 6i, -12, -24i, \ldots$. Since $\dfrac{6i}{3} = 2i$, $\dfrac{-12}{6i} = \dfrac{-2}{i} = \dfrac{-2i}{i^2} = 2i$, $\dfrac{-24i}{-12} = 2i$, this is a geometric sequence with common ratio $r = 2i$.

25. $a_6 = 17 = a + 5d$ and $a_4 = 11 = a + 3d$. Then, $a_6 - a_4 = 17 - 11 \;\Leftrightarrow\; (a + 5d) - (a + 3d) = 6 \;\Leftrightarrow\; 6 = 2d \;\Leftrightarrow\; d = 3$. Substituting into $11 = a + 3d$ gives $11 = a + 3 \cdot 3$, and so $a = 2$. Thus $a_2 = a + (2 - 1)d = 2 + 3 = 5$.

27. $a_3 = 9$ and $r = \dfrac{3}{2}$. Then $a_5 = a_3 \cdot r^2 = 9 \cdot \left(\dfrac{3}{2}\right)^2 = \dfrac{81}{4}$.

29. (a) $A_n = 32{,}000 \cdot 1.05^{n-1}$

(b) $A_1 = \$32{,}000$, $A_2 = 32{,}000 \cdot 1.05^1 = \$33{,}600$, $A_3 = 32{,}000 \cdot 1.05^2 = \$35{,}280$, $A_4 = 32{,}000 \cdot 1.05^3 = \$37{,}044$,

$A_5 = 32{,}000 \cdot 1.05^4 = \$38{,}896.20$, $A_6 = 32{,}000 \cdot 1.05^5 = \$40{,}841.01$, $A_7 = 32{,}000 \cdot 1.05^6 = \$42{,}883.06$,

$A_8 = 32{,}000 \cdot 1.05^7 = \$45{,}027.21$

31. Let a_n be the number of bacteria in the dish at the end of $5n$ seconds. So $a_0 = 3$, $a_1 = 3 \cdot 2$, $a_2 = 3 \cdot 2^2$, $a_3 = 3 \cdot 2^3$, ...

Then, clearly, a_n is a geometric sequence with $r = 2$ and $a = 3$. Thus at the end of $60 = 5\,(12)$ seconds, the number of

bacteria is $a_{12} = 3 \cdot 2^{12} = 12{,}288$.

33. Suppose that the common ratio in the sequence $a_1, a_2, a_3, \ldots$ is r. Also, suppose that the common ratio in the sequence

$b_1, b_2, b_3, \ldots$ is s. Then $a_n = a_1 r^{n-1}$ and $b_n = b_1 s^{n-1}$, $n = 1, 2, 3, \ldots$. Thus $a_n b_n = a_1 r^{n-1} \cdot b_1 s^{n-1} = (a_1 b_1)\,(rs)^{n-1}$.

So the sequence $a_1 b_1, a_2 b_2, a_3 b_3, \ldots$ is geometric with first term $a_1 b_1$ and common ratio rs.

35. (a) $6, x, 12, \ldots$ is arithmetic if $x - 6 = 12 - x$ $\Leftrightarrow$ $2x = 18$ $\Leftrightarrow$ $x = 9$.

(b) $6, x, 12, \ldots$ is geometric if $\dfrac{x}{6} = \dfrac{12}{x}$ $\Leftrightarrow$ $x^2 = 72$ $\Leftrightarrow$ $x = \pm 6\sqrt{2}$.

37. $\sum_{k=3}^{6} (k+1)^2 = (3+1)^2 + (4+1)^2 + (5+1)^2 + (6+1)^2 = 16 + 25 + 36 + 49 = 126$

39. $\sum_{k=1}^{6} (k+1)\,2^{k-1} = 2 \cdot 2^0 + 3 \cdot 2^1 + 4 \cdot 2^2 + 5 \cdot 2^3 + 6 \cdot 2^4 + 7 \cdot 2^5 = 2 + 6 + 16 + 40 + 96 + 224 = 384$

41. $\sum_{k=1}^{10} (k-1)^2 = 0^2 + 1^2 + 2^2 + 3^2 + 4^2 + 5^2 + 6^2 + 7^2 + 8^2 + 9^2$

43. $\sum_{k=1}^{50} \dfrac{3^k}{2^{k+1}} = \dfrac{3}{2^2} + \dfrac{3^2}{2^3} + \dfrac{3^3}{2^4} + \dfrac{3^4}{2^5} + \cdots + \dfrac{3^{49}}{2^{50}} + \dfrac{3^{50}}{2^{51}}$

45. $3 + 6 + 9 + 12 + \cdots + 99 = 3\,(1) + 3\,(2) + 3\,(3) + \cdots + 3\,(33) = \sum_{k=1}^{33} 3k$

47. $1 \cdot 2^3 + 2 \cdot 2^4 + 3 \cdot 2^5 + 4 \cdot 2^6 + \cdots + 100 \cdot 2^{102}$

$$= (1)\,2^{(1)+2} + (2)\,2^{(2)+2} + (3)\,2^{(3)+2} + (4)\,2^{(4)+2} + \cdots + (100)\,2^{(100)+2}$$

$$= \sum_{k=1}^{100} k \cdot 2^{k+2}$$

49. $1 + 0.9 + (0.9)^2 + \cdots + (0.9)^5$ is a geometric series with $a = 1$ and $r = \dfrac{0.9}{1} = 0.9$. Thus, the sum of the series is

$S_6 = \dfrac{1 - (0.9)^6}{1 - 0.9} = \dfrac{1 - 0.531441}{0.1} = 4.68559$.

51. $\sqrt{5} + 2\sqrt{5} + 3\sqrt{5} + \cdots + 100\sqrt{5}$ is an arithmetic series with $a = \sqrt{5}$ and $d = \sqrt{5}$. Then $100\sqrt{5} = a_n = \sqrt{5} + \sqrt{5}\,(n-1)$

$\Leftrightarrow$ $n = 100$. So the sum is $S_{100} = \dfrac{100}{2}\left(\sqrt{5} + 100\sqrt{5}\right) = 50\left(101\sqrt{5}\right) = 5050\sqrt{5}$.

53. $\sum_{n=0}^{6} 3 \cdot (-4)^n$ is a geometric series with $a = 3$, $r = -4$, and $n = 7$. Therefore, the sum of the series is

$S_7 = 3 \cdot \dfrac{1 - (-4)^7}{1 - (-4)} = \dfrac{3}{5}\left(1 + 4^7\right) = 9831$.

55. We have an arithmetic sequence with $a = 7$ and $d = 3$. Then

$S_n = 325 = \dfrac{n}{2}\,[2a + (n-1)\,d] = \dfrac{n}{2}\,[14 + 3\,(n-1)] = \dfrac{n}{2}\,(11 + 3n)$ $\Leftrightarrow$ $650 = 3n^2 + 11n$ $\Leftrightarrow$

$(3n + 50)\,(n - 13) = 0$ $\Leftrightarrow$ $n = 13$ (because $n = -\dfrac{50}{3}$ is inadmissible). Thus, 13 terms must be added.

57. This is a geometric sequence with $a = 2$ and $r = 2$. Then $S_{15} = 2 \cdot \dfrac{1 - 2^{15}}{1 - 2} = 2\left(2^{15} - 1\right) = 65{,}534$, and so the total

number of ancestors is 65,534.

59. $A = 10{,}000$, $i = 0.03$, and $n = 4$. Thus, $10{,}000 = R\dfrac{(1.03)^4 - 1}{0.03}$ $\Leftrightarrow$ $R = \dfrac{10{,}000 \cdot 0.03}{(1.03)^4 - 1} = \$2390.27$.

61. $1 - \dfrac{2}{5} + \dfrac{4}{25} - \dfrac{8}{125} + \cdots$ is a geometric series with $a = 1$ and $r = -\dfrac{2}{5}$. Therefore, the sum is $S = \dfrac{a}{1 - r} = \dfrac{1}{1 - \left(-\frac{2}{5}\right)} = \dfrac{5}{7}$.

63. $1 + \dfrac{1}{3^{1/2}} + \dfrac{1}{3} + \dfrac{1}{3^{3/2}} + \cdots$ is an infinite geometric series with $a = 1$ and $r = \dfrac{1}{\sqrt{3}}$. Thus, the sum is

$S = \dfrac{1}{1 - \frac{1}{\sqrt{3}}} = \dfrac{\sqrt{3}}{\sqrt{3} - 1} = \dfrac{1}{2}\left(3 + \sqrt{3}\right)$.

65. Let $P(n)$ denote the statement that $1 + 4 + 7 + \cdots + (3n - 2) = \dfrac{n(3n-1)}{2}$.

Step 1: $P(1)$ is the statement that $1 = \dfrac{1[3(1)-1]}{2} = \dfrac{1 \cdot 2}{2}$, which is true.

Step 2: Assume that $P(k)$ is true; that is, $1 + 4 + 7 + \cdots + (3k - 2) = \dfrac{k(3k-1)}{2}$. We want to use this to show that $P(k+1)$ is true. Now

$$
\begin{aligned}
1 + 4 + 7 + 10 + \cdots + (3k - 2) + [3(k+1) - 2] &= \frac{k(3k-1)}{2} + 3k + 1 \qquad \text{induction hypothesis} \\
&= \frac{k(3k-1)}{2} + \frac{6k+2}{2} = \frac{3k^2 - k + 6k + 2}{2} \\
&= \frac{3k^2 + 5k + 2}{2} = \frac{(k+1)(3k+2)}{2} \\
&= \frac{(k+1)[3(k+1)-1]}{2}
\end{aligned}
$$

Thus, $P(k+1)$ follows from $P(k)$. So by the Principle of Mathematical Induction, $P(n)$ is true for all n.

67. Let $P(n)$ denote the statement that $\left(1 + \frac{1}{1}\right)\left(1 + \frac{1}{2}\right)\left(1 + \frac{1}{3}\right) \cdots \cdots \left(1 + \frac{1}{n}\right) = n + 1$.

Step 1: $P(1)$ is the statement that $1 + \frac{1}{1} = 1 + 1$, which is clearly true.

Step 2: Assume that $P(k)$ is true; that is, $\left(1 + \frac{1}{1}\right)\left(1 + \frac{1}{2}\right)\left(1 + \frac{1}{3}\right) \cdots \cdots \left(1 + \frac{1}{k}\right) = k + 1$. We want to use this to show that $P(k+1)$ is true. Now

$$
\begin{aligned}
\left(1 + \tfrac{1}{1}\right)\left(1 + \tfrac{1}{2}\right)\left(1 + \tfrac{1}{3}\right) \cdots \cdots \left(1 + \tfrac{1}{k}\right)\left(1 + \tfrac{1}{k+1}\right) &= \left[\left(1 + \tfrac{1}{1}\right)\left(1 + \tfrac{1}{2}\right)\left(1 + \tfrac{1}{3}\right) \cdots \cdots \left(1 + \tfrac{1}{k}\right)\right]\left(1 + \tfrac{1}{k+1}\right) \\
&= (k+1)\left(1 + \tfrac{1}{k+1}\right) \qquad \text{induction hypothesis} \\
&= (k+1) + 1
\end{aligned}
$$

Thus, $P(k+1)$ follows from $P(k)$. So by the Principle of Mathematical Induction, $P(n)$ is true for all n.

69. $a_{n+1} = 3a_n + 4$ and $a_1 = 4$. Let $P(n)$ denote the statement that $a_n = 2 \cdot 3^n - 2$.

Step 1: $P(1)$ is the statement that $a_1 = 2 \cdot 3^1 - 2 = 4$, which is clearly true.

Step 2: Assume that $P(k)$ is true; that is, $a_k = 2 \cdot 3^k - 2$. We want to use this to show that $P(k+1)$ is true. Now

$$
\begin{aligned}
a_{k+1} &= 3a_k + 4 \qquad \text{definition of } a_{k+1} \\
&= 3\left(2 \cdot 3^k - 2\right) + 4 \qquad \text{induction hypothesis} \\
&= 2 \cdot 3^{k+1} - 6 + 4 = 2 \cdot 3^{k+1} - 2
\end{aligned}
$$

Thus $P(k+1)$ follows from $P(k)$. So by the Principle of Mathematical Induction, $P(n)$ is true for all n.

71. Let $P(n)$ denote the statement that $n! > 2^n$, for all natural numbers $n \geq 4$.

Step 1: $P(4)$ is the statement that $4! = 24 > 2^4 = 16$, which is true.

Step 2: Assume that $P(k)$ is true; that is, $k! > 2^k$. We want to use this to show that $P(k+1)$ is true. Now

$$
\begin{aligned}
(k+1)! &= (k+1)\,k! > (k+1) \cdot 2^k \qquad \text{induction hypothesis} \\
&> 2 \cdot 2^k \qquad \text{because } k + 1 > 2 \text{ for } k \geq 4 \\
&= 2^{k+1}
\end{aligned}
$$

Thus $P(k+1)$ follows from $P(k)$. So by the Principle of Mathematical Induction, $P(n)$ is true for all $n \geq 4$.

73. $\dbinom{10}{2} + \dbinom{10}{6} = \dfrac{10!}{2!\,8!} + \dfrac{10!}{6!\,4!} = \dfrac{10 \cdot 9}{2} + \dfrac{10 \cdot 9 \cdot 8 \cdot 7}{4 \cdot 3 \cdot 2} = 45 + 210 = 255.$

75. $\sum_{k=0}^{8} \binom{8}{k}\binom{8}{8-k} = 2\binom{8}{0}\binom{8}{8} + 2\binom{8}{1}\binom{8}{7} + 2\binom{8}{2}\binom{8}{6} + 2\binom{8}{3}\binom{8}{5} + \binom{8}{4}\binom{8}{4} = 2 + 2 \cdot 8^2 + 2 \cdot (28)^2 + 2 \cdot (56)^2 + (70)^2 = 12{,}870.$

77. $(2x + y)^4 = \binom{4}{0}(2x)^4 + \binom{4}{1}(2x)^3 y + \binom{4}{2}(2x)^2 y^2 + \binom{4}{3} \cdot 2xy^3 + \binom{4}{4} y^4 = 16x^4 + 32x^3 y + 24x^2 y^2 + 8xy^3 + y^4$

79. The first three terms in the expansion of $\left(b^{-2/3} + b^{1/3}\right)^{20}$ are $\binom{20}{0}\left(b^{-2/3}\right)^{20} = b^{-40/3}$, $\binom{20}{1}\left(b^{-2/3}\right)^{19}$

$\left(b^{1/3}\right) = 20b^{-37/3}$, and $\binom{20}{2}\left(b^{-2/3}\right)^{18}\left(b^{1/3}\right)^2 = 190b^{-34/3}$.

Chapter 12 Test

1. $a_1 = 1^2 - 1 = 0$, $a_2 = 2^2 - 1 = 3$, $a_3 = 3^2 - 1 = 8$, $a_4 = 4^2 - 1 = 15$, and $a_{10} = 10^2 - 1 = 99$.

2. $a_{n+2} = (a_n)^2 - a_{n+1}$, $a_1 = 1$ and $a_2 = 1$. Then $a_3 = a_1^2 - a_2 = 1^2 - 1 = 0$, $a_4 = a_2^2 - a_3 = 1^2 - 0 = 1$, and $a_5 = a_3^2 - a_4 = 0^2 - 1 = -1$.

3. (a) The common difference is $d = 5 - 2 = 3$.

   (b) $a_n = 2 + (n - 1)3$

   (c) $a_{35} = 2 + 3(35 - 1) = 104$

4. (a) The common ratio is $r = \frac{3}{12} = \frac{1}{4}$.

   (b) $a_n = a_1 r^{n-1} = 12\left(\frac{1}{4}\right)^{n-1}$

   (c) $a_{10} = 12\left(\frac{1}{4}\right)^{10-1} = \frac{3}{4^8} = \frac{3}{65,536}$

5. (a) $a_1 = 25$, $a_4 = \frac{1}{5}$. Then $r^3 = \frac{\frac{1}{5}}{25} = \frac{1}{125}$ $\Leftrightarrow$ $r = \frac{1}{5}$, so $a_5 = ra_4 = \frac{1}{25}$.

   (b) $S_8 = 25\dfrac{1 - \left(\frac{1}{5}\right)^8}{1 - \frac{1}{5}} = \dfrac{5^8 - 1}{12,500} = \dfrac{97,656}{3125}$

6. (a) $a_1 = 10$ and $a_{10} = 2$, so $9d = -8$ $\Leftrightarrow$ $d = -\frac{8}{9}$ and $a_{100} = a_1 + 99d = 10 - 88 = -78$.

   (b) $S_{10} = \frac{10}{2}\left[2a + (10 - 1)d\right] = \frac{10}{2}\left(2 \cdot 10 + 9\left(-\frac{8}{9}\right)\right) = 60$

7. Let the common ratio for the geometric series $a_1, a_2, a_3, \ldots$ be r, so that $a_n = a_1 r^{n-1}$, $n = 1, 2, 3, \ldots$. Then $a_n^2 = \left(a_1 r^{n-1}\right)^2 = \left(a_1^2\right)\left(r^2\right)^{n-1}$. Therefore, the sequence $a_1^2, a_2^2, a_3^2, \ldots$ is geometric with common ratio r^2.

8. (a) $\sum_{n=1}^5 \left(1 - n^2\right) = \left(1 - 1^2\right) + \left(1 - 2^2\right) + \left(1 - 3^2\right) + \left(1 - 4^2\right) + \left(1 - 5^2\right) = 0 - 3 - 8 - 15 - 24 = -50$

   (b) $\sum_{n=3}^6 (-1)^n 2^{n-2} = (-1)^3 2^{3-2} + (-1)^4 2^{4-2} + (-1)^5 2^{5-2} + (-1)^6 2^{6-2} = -2 + 4 - 8 + 16 = 10$

9. (a) The geometric sum $\frac{1}{3} + \frac{2}{3^2} + \frac{2^2}{3^3} + \frac{2^3}{3^4} + \cdots + \frac{2^9}{3^{10}}$ has $a = \frac{1}{3}$, $r = \frac{2}{3}$, and $n = 10$. So

   $S_{10} = \frac{1}{3} \cdot \frac{1 - (2/3)^{10}}{1 - (2/3)} = \frac{1}{3} \cdot 3\left(1 - \frac{1024}{59,049}\right) = \frac{58,025}{59,049}$.

   (b) The infinite geometric series $1 + \frac{1}{2^{1/2}} + \frac{1}{2} + \frac{1}{2^{3/2}} + \cdots$ has $a = 1$ and $r = 2^{-1/2} = \frac{1}{\sqrt{2}}$. Thus,

   $S = \frac{1}{1 - 1/\sqrt{2}} = \frac{\sqrt{2}}{\sqrt{2} - 1} = \frac{\sqrt{2}}{\sqrt{2} - 1} \cdot \frac{\sqrt{2} + 1}{\sqrt{2} + 1} = 2 + \sqrt{2}$.

10. Let $P(n)$ denote the statement that $1^2 + 2^2 + 3^2 + \cdots + n^2 = \dfrac{n(n+1)(2n+1)}{6}$.

Step 1: Show that $P(1)$ is true. But $P(1)$ says that $1^2 = \dfrac{1 \cdot 2 \cdot 3}{6}$, which is true.

Step 2: Assume that $P(k)$ is true; that is, $1^2 + 2^2 + 3^2 + \cdots + k^2 = \dfrac{k(k+1)(2k+1)}{6}$. We want to use this to show that $P(k+1)$ is true. Now

$$
\begin{aligned}
1^2 + 2^2 + 3^2 + \cdots + k^2 + (k+1)^2 &= \frac{k(k+1)(2k+1)}{6} + (k+1)^2 \qquad \text{induction hypothesis} \\[2mm]
&= \frac{k(k+1)(2k+1) + 6(k+1)^2}{6} = \frac{(k+1)(2k^2 + k) + (6k+6)(k+1)}{6} \\[2mm]
&= \frac{(k+1)(2k^2 + k + 6k + 6)}{6} = \frac{(k+1)(2k^2 + 7k + 6)}{6} \\[2mm]
&= \frac{(k+1)(k+2)(2k+3)}{6} = \frac{(k+1)[(k+1)+1][2(k+1)+1]}{6}
\end{aligned}
$$

Thus $P(k+1)$ follows from $P(k)$. So by the Principle of Mathematical Induction, $P(n)$ is true for all n.

11. $(2x + y^2)^5 = \binom{5}{0}(2x)^5 + \binom{5}{1}(2x)^4 y^2 + \binom{5}{2}(2x)^3 (y^2)^2 + \binom{5}{3}(2x)^2 (y^2)^3 + \binom{5}{4}(2x)(y^2)^4 + \binom{5}{5}(y^2)^5$

$\qquad = 32x^5 + 80x^4 y^2 + 80x^3 y^4 + 40x^2 y^6 + 10xy^8 + y^{10}$

12. $\binom{10}{3}(3x)^3 (-2)^7 = 120 \cdot 27x^3 (-128) = -414{,}720x^3$

13. **(a)** Each week he gains 24% in weight, that is, $0.24a_n$. Thus, $a_{n+1} = a_n + 0.24a_n = 1.24a_n$ for $n \geq 1$. a_0 is given to be 0.85 lb.

(b) $a_6 = 1.24a_5 = 1.24(1.24a_4) = \cdots = 1.24^6 a_0 = 1.24^6 (0.85) \approx 3.1$ lb

(c) The sequence $a_1, a_2, a_3, \ldots$ is geometric with common ratio 1.24.

Focus on Modeling: Modeling with Recursive Sequences

1. (a) Since there are 365 days in a year, the interest earned per day is $\dfrac{0.0365}{365} = 0.0001$. Thus the amount in the account at the end of the nth day is $A_n = 1.0001A_{n-1}$ with $A_0 = \$275{,}000$.

(b) $A_0 = \$275{,}000$, $A_1 = 1.0001A_0 = 1.0001 \cdot 275{,}000 = \$275{,}027.50$,
$A_2 = 1.0001A_1 = 1.0001\left(1.0001A_0\right) = 1.0001^2 A_0 = \$275{,}055.00$,
$A_3 = 1.0001A_2 = 1.0001\left(1.0001^2 A_0\right) = 1.0001^3 A_0 = \$275{,}082.51$,
$A_4 = 1.0001A_3 = 1.0001\left(1.0001^3 A_0\right) = 1.0001^4 A_0 = \$275{,}110.02$,
$A_5 = 1.0001A_4 = 1.0001\left(1.0001^4 A_0\right) = 1.0001^5 A_0 = \$275{,}137.53$,
$A_6 = 1.0001A_5 = 1.0001\left(1.0001^5 A_0\right) = 1.0001^6 A_0 = \$275{,}165.04$,
$A_7 = 1.0001A_6 = 1.0001\left(1.0001^6 A_0\right) = 1.0001^7 A_0 = \$275{,}192.56$

(c) $A_n = 1.0001^n \cdot 275{,}000$

3. (a) Since there are 12 months in a year, the interest earned per day is $\dfrac{0.03}{12} = 0.0025$. Thus the amount in the account at the end of the nth month is $A_n = 1.0025A_{n-1} + 100$ with $A_0 = \$100$.

(b) $A_0 = \$100$, $A_1 = 1.0025A_0 + 100 = 1.0025 \cdot 100 + 100 = \$200.25$, $A_2 = 1.0025A_1 + 100 = 1.0025\left(1.0025 \cdot 100 + 100\right) + 100 = 1.0025^2 \cdot 100 + 1.0025 \cdot 100 + 100 = \$300.75$, $A_3 = 1.0025A_2 + 100 = 1.0025\left(1.0025^2 \cdot 100 + 1.0025 \cdot 100 + 100\right) + 100 = 1.0025^3 \cdot 100 + 1.0025^2 \cdot 100 + 1.0025 \cdot 100 + 100 = \$401.50$, $A_4 = 1.0025A_3 + 100 = 1.0025\left(1.0025^3 \cdot 100 + 1.0025^2 \cdot 100 + 1.0025 \cdot 100 + 100\right) + 100 = 1.0025^4 \cdot 100 + 1.0025^3 \cdot 100 + 1.0025^2 \cdot 100 + 1.0025 \cdot 100 + 100 = \$502.51$

(c) $A_n = 1.0025^n \cdot 100 + \cdots + 1.0025^2 \cdot 100 + 1.0025 \cdot 100 + 100$, the partial sum of a geometric series, so
$$A_n = 100 \cdot \frac{1 - 1.0025^{n+1}}{1 - 1.0025} = 100 \cdot \frac{1.0025^{n+1} - 1}{0.0025}.$$

(d) Since 5 years is 60 months, we have $A_{60} = 100 \cdot \dfrac{1.0025^{61} - 1}{0.0025} \approx \$6{,}580.83$.

5. (a) The amount A_n of pollutants in the lake in the nth year is 30% of the amount from the preceding year ($0.30A_{n-1}$) plus the amount discharged that year (2400 tons). Thus $A_n = 0.30A_{n-1} + 2400$.

(b) $A_0 = 2400$, $A_1 = 0.30\left(2400\right) + 2400 = 3120$,
$A_2 = 0.30\left[0.30\left(2400\right) + 2400\right] + 2400 = 0.30^2\left(2400\right) + 2400\left(2400\right) + 2400 = 3336$,
$A_3 = 0.30\left[0.30^2\left(2400\right) + 2400\left(2400\right) + 2400\right] + 2400$
$\quad = 0.03^3\left(2400\right) + 0.30^2\left(2400\right) + 2400\left(2400\right) + 2400 = 3400.8$,

$A_4 = 0.30\left[0.03^3\left(2400\right) + 0.30^2\left(2400\right) + 2400\left(2400\right) + 2400\right] + 2400$
$\quad = 0.03^4\left(2400\right) + 0.03^3\left(2400\right) + 0.30^2\left(2400\right) + 2400\left(2400\right) + 2400 = 3420.2$

(c) A_n is the partial sum of a geometric series, so

$A_n = 2400 \cdot \dfrac{1 - 0.30^{n+1}}{1 - 0.30} = 2400 \cdot \dfrac{1 - 0.30^{n+1}}{0.70}$
$\quad \approx 3428.6\left(1 - 0.30^{n+1}\right)$

(d) $A_6 = 2400 \cdot \dfrac{1 - 0.30^7}{0.70} = 3427.8$ tons. The sum of a geometric

series, is $A = 2400 \cdot \dfrac{1}{0.70} = 3428.6$ tons.

(e)

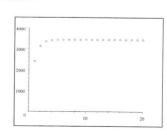

397

7. (a) In the nth year since Victoria's initial deposit the amount V_n in her CD is the amount from the preceding year (V_{n-1}), plus the 5% interest earned on that amount ($0.05V_{n-1}$), plus $500 times the number of years since her initial deposit ($500n$). Thus $V_n = 1.05V_{n-1} + 500n$.

(b) Ursula's savings surpass Victoria's savings in the 35th year.

9. (a) $R_1 = 104$, $R_2 = 108.0$, $R_3 = 112.0$, $R_4 = 115.9$, $R_5 = 119.8$, $R_6 = 123.7$, $R_7 = 127.4$

(b)

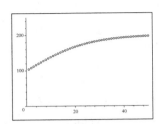

As n becomes large the raccoon population approaches 200.

13 Counting and Probability

13.1 Counting Principles

1. By the Fundamental Counting Principle, the number of possible single-scoop ice cream cones is

$$\begin{pmatrix} \text{number of ways to} \\ \text{choose the flavor} \end{pmatrix} \cdot \begin{pmatrix} \text{number of ways to} \\ \text{choose the type of cone} \end{pmatrix} = 4 \cdot 3 = 12.$$

3. By the Fundamental Counting Principle, the possible number of 3-letter words is

$$\begin{pmatrix} \text{number of ways to} \\ \text{choose the 1st letter} \end{pmatrix} \cdot \begin{pmatrix} \text{number of ways to} \\ \text{choose the 2nd letter} \end{pmatrix} \cdot \begin{pmatrix} \text{number of ways to} \\ \text{choose the 3rd letter} \end{pmatrix}.$$

   (a) Since repetitions are allowed, we have 4 choices for each letter. Thus there are $4 \cdot 4 \cdot 4 = 64$ words.

   (b) Since repetitions are not allowed, we have 4 choices for the 1st letter, 3 choices for the 2nd letter, and 2 choices for the 3rd letter. Thus there are $4 \cdot 3 \cdot 2 = 24$ words.

5. Since there are four choices for each of the five questions, by the Fundamental Counting Principle there are $4 \cdot 4 \cdot 4 \cdot 4 \cdot 4 = 1024$ different ways the test can be completed.

7. Since a runner can only finish once, there are no repetitions. And since we are assuming that there is no tie, the number of different finishes is $\begin{pmatrix} \text{number of ways to} \\ \text{choose the 1st runner} \end{pmatrix} \cdot \begin{pmatrix} \text{number of ways to} \\ \text{choose the 2nd runner} \end{pmatrix} \cdots \begin{pmatrix} \text{number of ways to} \\ \text{choose the 5th runner} \end{pmatrix} = 5 \cdot 4 \cdot 3 \cdot 2 \cdot 1 = 120.$

9. Since there are 4 main courses, there are 6 ways to choose a main course. Likewise, there are 5 drinks and 3 desserts so there are 5 ways to choose a drink and 3 ways to choose a dessert. So the number of different meals consisting of a main course, a drink, and a dessert is $\begin{pmatrix} \text{number of ways to} \\ \text{choose the main course} \end{pmatrix} \cdot \begin{pmatrix} \text{number of ways to} \\ \text{choose a drink} \end{pmatrix} \cdot \begin{pmatrix} \text{number of ways to} \\ \text{choose a dessert} \end{pmatrix} = (4)(5)(3) = 60.$

11. By the Fundamental Counting Principle, the number of different routes from town A to town D via towns B and C is

$$\begin{pmatrix} \text{number of routes} \\ \text{from A to B} \end{pmatrix} \cdot \begin{pmatrix} \text{number of routes} \\ \text{from B to C} \end{pmatrix} \cdot \begin{pmatrix} \text{number of routes} \\ \text{from C to D} \end{pmatrix} = (4)(5)(6) = 120.$$

13. The number of possible sequences of heads and tails when a coin is flipped 5 times is

$$\begin{pmatrix} \text{number of possible} \\ \text{outcomes on the 1st flip} \end{pmatrix} \cdot \begin{pmatrix} \text{number of possible} \\ \text{outcomes on the 2nd flip} \end{pmatrix} \cdots \begin{pmatrix} \text{number of possible} \\ \text{outcomes on the 5th flip} \end{pmatrix} = (2)(2)(2)(2)(2)$$
$$= 2^5 = 32$$

Here there are only two choices, heads or tails, for each flip.

15. Since there are six different faces on each die, the number of possible outcomes when a red die and a blue die and a white die are rolled is

$$\begin{pmatrix} \text{number of possible} \\ \text{outcomes on the red die} \end{pmatrix} \cdot \begin{pmatrix} \text{number of possible} \\ \text{outcomes on the blue die} \end{pmatrix} \cdot \begin{pmatrix} \text{number of possible} \\ \text{outcomes on the white die} \end{pmatrix} = (6)(6)(6) = 6^3 = 216.$$

17. The number of possible skirt-blouse-shoe outfits is

$$\left(\begin{array}{c}\text{number of ways} \\ \text{to choose a skirt}\end{array}\right) \cdot \left(\begin{array}{c}\text{number of ways} \\ \text{to choose a blouse}\end{array}\right) \cdot \left(\begin{array}{c}\text{number of ways} \\ \text{to choose shoes}\end{array}\right) = (5)\,(8)\,(12) = 480.$$

19. The number of possible ID numbers of one letter followed by two digits is

$$\left(\begin{array}{c}\text{number of ways} \\ \text{to choose a letter}\end{array}\right) \cdot \left(\begin{array}{c}\text{number of ways} \\ \text{to choose a digit}\end{array}\right) \cdot \left(\begin{array}{c}\text{number of ways} \\ \text{to choose a digit}\end{array}\right) = (26)\,(10)\,(10) = 2600.$$

Since $2600 < 2844$, it is not possible to give each of the company's employees a different ID number using this scheme.

21. The number of different California license plates possible is

$$\left(\begin{array}{c}\text{number of ways to} \\ \text{choose a nonzero digit}\end{array}\right) \cdot \left(\begin{array}{c}\text{number of ways} \\ \text{to choose 3 letters}\end{array}\right) \cdot \left(\begin{array}{c}\text{number of ways} \\ \text{to choose 3 digits}\end{array}\right) = (9)\,(26^3)\,(10^3) = 158{,}184{,}000.$$

23. There are two possible ways to answer each question on a true-false test. Thus the possible number of different ways to complete a true-false test containing 10 questions is

$$\left(\begin{array}{c}\text{number of ways} \\ \text{to answer question 1}\end{array}\right) \cdot \left(\begin{array}{c}\text{number of ways} \\ \text{to answer question 2}\end{array}\right) \cdots \left(\begin{array}{c}\text{number of ways} \\ \text{to answer question 10}\end{array}\right) = 2^{10} = 1024.$$

25. The number of different classifications possible is

$$\left(\begin{array}{c}\text{number of ways} \\ \text{to choose a major}\end{array}\right) \cdot \left(\begin{array}{c}\text{number of ways to} \\ \text{choose a minor}\end{array}\right) \cdot \left(\begin{array}{c}\text{number of ways} \\ \text{to choose a year}\end{array}\right) \cdot \left(\begin{array}{c}\text{number of ways} \\ \text{to choose a sex}\end{array}\right) = (32)\,(32)\,(4)\,(2) = 8192.$$

To see that there are 32 ways to choose a minor, we start with 32 fields, subtract the major (you cannot major and minor in the same subject), and then add one for the possibility of no minor.

27. The number of possible license plates of two letters followed by three digits is

$$\left(\begin{array}{c}\text{number of ways to} \\ \text{choose 2 letters}\end{array}\right) \cdot \left(\begin{array}{c}\text{number of ways} \\ \text{to choose 3 digits}\end{array}\right) = (26^2)\,(10^3) = 676{,}000.$$ Since $676{,}000 < 8{,}000{,}000$, there will not be

enough different license plates for the state's 8 million registered cars.

29. Since a student can hold only one office, the number of ways that a president, a vice-president and a secretary can be chosen from a class of 30 students is

$$\left(\begin{array}{c}\text{number of ways} \\ \text{to choose a president}\end{array}\right) \cdot \left(\begin{array}{c}\text{number of ways to} \\ \text{choose a vice-president}\end{array}\right) \cdot \left(\begin{array}{c}\text{number of ways} \\ \text{to choose a secretary}\end{array}\right) = (30)\,(29)\,(28) = 24{,}360.$$

31. The number of ways a chairman, vice-chairman, and a secretary can be chosen if the chairman must be a Democrat and the vice-chairman must be a Republican is

$$\left(\begin{array}{c}\text{number of ways to choose} \\ \text{a Democratic chairman} \\ \text{from the 10 Democrats}\end{array}\right) \left(\begin{array}{c}\text{number of ways to choose} \\ \text{a Republican vice-chairman} \\ \text{from the 10 Republicans}\end{array}\right) \cdot \left(\begin{array}{c}\text{number of ways to choose} \\ \text{a secretary from} \\ \text{the remaining 15 members}\end{array}\right) = (10)\,(7)\,(15) = 1050.$$

33. The possible number of 5-letter words formed using the letters A, B, C, D, E, F, and G is calculated as follows:

(a) If repetition of letters is allowed, then there are 7 ways to pick each letter in the word. Thus the number of 5-letter

words is $\left(\begin{array}{c}\text{number of ways to} \\ \text{choose the 1st letter}\end{array}\right) \cdot \left(\begin{array}{c}\text{number of ways to} \\ \text{choose the 2nd letter}\end{array}\right) \cdots \left(\begin{array}{c}\text{number of ways to} \\ \text{choose the 5th letter}\end{array}\right) = 7^5 = 16{,}807.$

(b) If no letter can be repeated in a word, then the number of 5-letter words is

$$\left(\begin{array}{c}\text{number of ways to} \\ \text{choose the 1st letter}\end{array}\right) \cdot \left(\begin{array}{c}\text{number of ways to} \\ \text{choose the 2nd letter}\end{array}\right) \cdots \left(\begin{array}{c}\text{number of ways to} \\ \text{choose the 5th letter}\end{array}\right) = (7)\,(6)\,(5)\,(4)\,(3) = 2520$$

(c) If each word must begin with the letter A, then the number of 5-letter words (letters can be repeated) is

$$\begin{pmatrix} \text{number of ways to} \\ \text{choose the letter A} \end{pmatrix} \cdot \begin{pmatrix} \text{number of ways to} \\ \text{choose the 2nd letter} \end{pmatrix} \cdots \cdot \begin{pmatrix} \text{number of ways to} \\ \text{choose the 5th letter} \end{pmatrix} = (1)\,(7)\,(7)\,(7)\,(7) = 7^4 = 2401.$$

(d) If the letter C must be in the middle, then the number of 5-letter words (letters can be repeated) is

$$\begin{pmatrix} \text{number of ways to} \\ \text{choose the 1st letter} \end{pmatrix} \cdot \begin{pmatrix} \text{number of ways to} \\ \text{choose the 2nd letter} \end{pmatrix} \cdot \begin{pmatrix} \text{number of ways to} \\ \text{choose the letter C} \end{pmatrix} \cdot \begin{pmatrix} \text{number of ways to} \\ \text{choose the 4th letter} \end{pmatrix} \cdot \begin{pmatrix} \text{number of ways to} \\ \text{choose the 5th letter} \end{pmatrix}$$

$$= (7)\,(7)\,(1)\,(7)\,(7) = 7^4 = 2401$$

(e) If the middle letter must be a vowel, then the number of 5-letter words is $(7)\,(7)\,(2)\,(7)\,(7) = 2 \cdot 7^4 = 4802.$

35. The number of possible variable names is $\begin{pmatrix} \text{number of ways to} \\ \text{choose a letter} \end{pmatrix} \cdot \begin{pmatrix} \text{number of ways to choose} \\ \text{a letter or a digit} \end{pmatrix} = (26)\,(36) = 936.$

37. The number of ways 4 men and 4 women may be seated in a row of 8 seats is calculated as follows:

(a) If the women are to be seated together and the men together, then the number of ways they can be seated is

$$\begin{pmatrix} \text{number of ways to seat the} \\ \text{women first and the men second} \end{pmatrix} + \begin{pmatrix} \text{number of ways to seat the} \\ \text{the men first and the women second} \end{pmatrix}$$

$$= (4)\,(3)\,(2)\,(1)\,(4)\,(3)\,(2)\,(1) + (4)\,(3)\,(2)\,(1)\,(4)\,(3)\,(2)\,(1) = (4!)\,(4!) + (4!)\,(4!) = 576 + 576 = 1152$$

(b) If they are to be seated alternately by gender, then the number of ways they can be seated is

$$\begin{pmatrix} \text{number of ways to seat them} \\ \text{alternately if a woman is first} \end{pmatrix} + \begin{pmatrix} \text{number of ways to seat them} \\ \text{alternately if a man is first} \end{pmatrix}$$

$$= (4)\,(4)\,(3)\,(3)\,(2)\,(2)\,(1)\,(1) + (4)\,(4)\,(3)\,(3)\,(2)\,(2)\,(1)\,(1) = 1152$$

39. The number of ways that 8 mathematics books and 3 chemistry books may be placed on a shelf if the math books are to be next to each other and the chemistry books next to each other is

$$\begin{pmatrix} \text{number of ways to place them} \\ \text{if the math books are first} \end{pmatrix} + \begin{pmatrix} \text{number of ways to place them} \\ \text{if the chemistry books are first} \end{pmatrix} = (8!)\,(3!) + (3!)\,(8!) = 483{,}840.$$

41. In order for a number to be odd, the last digit must be odd. In this case, it must be a 1. Thus, the other two digits must be chosen from the three digits 2, 4, and 6. So, the number of 3-digit odd numbers that can be formed using the digits 1, 2, 4, and 6, if no digit may be used more than once is

$$\begin{pmatrix} \text{number of ways to} \\ \text{choose a digit} \\ \text{from the 3 digits} \end{pmatrix} \cdot \begin{pmatrix} \text{number of ways to} \\ \text{choose a digit from} \\ \text{the 2 remaining digits} \end{pmatrix} \begin{pmatrix} \text{number of ways} \\ \text{to choose} \\ \text{an odd digits} \end{pmatrix} = (3)\,(2)\,(1) = 6.$$

43. (a) From Exercise 6, the number of seven-digit phone numbers is 8,000,000. Under the old rules, the possible number of telephone area codes consisting of 3 digits with the given conditions is $(8)\,(2)\,(10) = 160$. So the possible number of area code + telephone number combinations under the old rules is $160 \cdot 8{,}000{,}000 = 1{,}280{,}000{,}000.$

(b) Under the new rules, where the second digit can be any number, the number of area codes possible is $(8)\,(10)\,(10) = 800$. Thus the possible number of area code + telephone number combinations under the new rules is $800 \cdot 8{,}000{,}000 = 6{,}400{,}000{,}000.$

(c) Although over $1\frac{1}{4}$ billion telephone numbers may seem to be enough for a population of 300,000,000, so many individuals and businesses now have multiple lines for fax, Internet access, and cellular phones that this number has proven insufficient.

(d) There are now more "new" area codes than "old" area codes. Numerous answers are possible.

13.2   Permutations and Combinations

1. $P(8,3) = \dfrac{8!}{(8-3)!} = \dfrac{8!}{5!} = 8 \cdot 7 \cdot 6 = 336$

3. $P(11,4) = \dfrac{11!}{(11-4)!} = \dfrac{11!}{7!} = 11 \cdot 10 \cdot 9 \cdot 8 = 7920$

5. $P(100,1) = \dfrac{100!}{(100-1)!} = \dfrac{100!}{99!} = 100$

7. Here we have 6 letters, of which 3 are A, 2 are B, and 1 is C. Thus the number of distinguishable permutations is
$\dfrac{6!}{3!\,2!\,1!} = \dfrac{(6)(5)(4)(3!)}{(2)(3!)} = 60.$

9. Here we have 5 letters, of which 2 are A, 1 is B, 1 is C, and 1 is D. Thus the number of distinguishable permutations is
$\dfrac{5!}{2!\,1!\,1!\,1!} = \dfrac{(5)(4)(3)(2!)}{2!} = 60.$

11. Here we have 9 letters, of which 2 are X, 3 are Y, and 4 are Z. Thus the number of distinguishable permutations is
$\dfrac{9!}{2!\,3!\,4!} = \dfrac{9 \cdot 8 \cdot 7 \cdot 6 \cdot 5 \cdot 4!}{2 \cdot 3 \cdot 2 \cdot 4!} = 1260.$

13. $C(8,3) = \dfrac{8!}{3!\,(8-3)!} = \dfrac{8!}{3!\,5!} = \dfrac{8 \cdot 7 \cdot 6}{3 \cdot 2 \cdot 1} = 56$

15. $C(11,4) = \dfrac{11!}{4!\,7!} = \dfrac{11 \cdot 10 \cdot 9 \cdot 8}{4 \cdot 3 \cdot 2 \cdot 1} = 330$

17. $C(100,1) = \dfrac{100!}{1!\,99!} = \dfrac{100}{1} = 100$

19. We need the number of ways of selecting three students in order for the positions of president, vice president, and secretary from the 15 students in the class. Since a student cannot hold more than one position, this number is
$P(15,3) = \dfrac{15!}{(15-3)!} = \dfrac{15!}{12!} = 15 \cdot 14 \cdot 13 = 2730.$

21. Since a person cannot occupy more than one chair at a time, the number of ways of seating 6 people from 10 people in a row of 6 chairs is $P(10,6) = 10 \cdot 9 \cdot 8 \cdot 7 \cdot 6 \cdot 5 = 151{,}200.$

23. The number of ways of selecting 3 objects in order (a 3-letter word) from 6 distinct objects (the 6 letters) assuming that the letters cannot be repeated is $P(6,3) = 6 \cdot 5 \cdot 4 = 120.$

25. The number of ways of selecting 3 objects in order (a 3-digit number) from 4 distinct objects (the 4 digits) with no repetition of the digits is $P(4,3) = 4 \cdot 3 \cdot 2 = 24.$

27. The number of ways of ordering 9 distinct objects (the contestants) is $P(9,9) = 9! = 362{,}880.$ Here a runner cannot finish more than once, so no repetitions are allowed, and order is important.

29. The number of ways of ordering 1000 distinct objects (the contestants) taking 3 at a time is
$P(1000,3) = 1000 \cdot 999 \cdot 998 = 997{,}002{,}000.$ We are assuming that a person cannot win more than once, that is, there are no repetitions.

31. We first place Jack in the first seat, and then seat the remaining 4 students. Thus the number of these arrangements is
$\begin{pmatrix} \text{number of ways to} \\ \text{seat Jack in the 1st seat} \end{pmatrix} \cdot \begin{pmatrix} \text{number of ways to seat} \\ \text{the remaining 4 students} \end{pmatrix} = P(1,1) \cdot P(4,4) = 1!\,4! = 24.$

33. Here we have 6 objects, of which 2 are blue marbles and 4 are red marbles. Thus the number of distinguishable permutations is $\dfrac{6!}{2!\,4!} = \dfrac{6 \cdot 5 \cdot 4!}{2 \cdot 4!} = 15.$

35. The number of distinguishable permutations of 12 objects (the 12 coins), from like groups of size 4 (the pennies), of size 3 (the nickels), of size 2 (the dimes) and of size 3 (the quarters) is $\dfrac{12!}{4!\,3!\,2!\,3!} = 277{,}200.$

37. The number of distinguishable permutations of 12 objects (the 12 ice cream cones) from like groups of size 3 (the vanilla cones), of size 2 (the chocolate cones), of size 4 (the strawberry cones), and of size 5 (the butterscotch cones) is
$$\frac{14!}{3!\,2!\,4!\,5!} = 2{,}522{,}520.$$

39. The number of distinguishable permutations of 8 objects (the 8 cleaning tasks) from like groups of size 5, 2, and 1 workers, respectively is $\dfrac{8!}{5!\,2!\,1!} = 168.$

41. We want the number of ways of choosing a group of three from a group of six. This number is $C\,(6,3) = \dfrac{6!}{3!\,3!} = 20.$

43. We want the number of ways of choosing a group of six people from a group of ten people. The number of combinations of 10 objects (10 people) taken 6 at a time is $C\,(10,6) = \dfrac{10!}{6!\,4!} = 210.$

45. We want the number of ways of choosing a group (the 5-card hand) where order of selection is not important. The number of combinations of 52 objects (the 52 cards) taken 5 at a time is $C\,(52,5) = \dfrac{52!}{5!\,47!} = 2{,}598{,}960.$

47. The order of selection is not important, hence we must calculate the number of combinations of 10 objects (the 10 questions) taken 7 at a time, this gives $C\,(10,7) = \dfrac{10!}{7!\,3!} = 120.$

49. We assume that the order in which he plays the pieces in the recital is not important, so the number of combinations of 12 objects (the 12 pieces) taken 8 at a time is $C\,(12,8) = \dfrac{12!}{8!\,4!} = 495.$

51. We are just interested in the group of seven students taken from the class of 30 students, not the order in which they are picked. Thus the number is $C\,(30,7) = \dfrac{30!}{7!\,23!} = 2{,}035{,}800.$

53. We first take Jack out of the class of 30 students and select the 7 students from the remaining 29 students. Thus there are $C\,(29,7) = \dfrac{29!}{7!\,22!} = 1{,}560{,}780$ ways to pick the 7 students for the field trip.

55. We must count the number of different ways to choose the group of 6 numbers from the 53 numbers. There are $C\,(53,6) = \dfrac{53!}{6!\,47!} = 22{,}957{,}480$ possible tickets, so it would cost $\$22{,}957{,}480.$

57. (a) The number of ways to select 5 of the 8 objects is $C\,(8,5) = \dfrac{8!}{5!\,3!} = 56.$

(b) A set with 8 elements has $2^8 = 256$ subsets.

59. Each subset of toppings constitutes a different way a hamburger can be ordered. Since a set with 10 elements has $2^{10} = 1024$ subsets, there are 1024 different ways to order a hamburger.

61. We pick the two men from the group of ten men, and we pick the two women from the group of ten women. So the number of ways 2 men and 2 women can be chosen is
$$\begin{pmatrix}\text{number of ways to pick}\\ \text{2 of the 10 men}\end{pmatrix} \cdot \begin{pmatrix}\text{number of ways to pick}\\ \text{2 of the 10 women}\end{pmatrix} = C\,(10,2) \cdot C\,(10,2) = (45)\,(45) = 2025.$$

63. The leading and supporting roles are different (order counts), while the extra roles are not (order doesn't count). Also, the male roles must be filled by the male actors and the female roles filled by the female actresses. Thus, number of ways the actors and actresses can be chosen is
$$\begin{pmatrix}\text{number of ways the}\\ \text{leading and the supporting}\\ \text{actors can be chosen}\end{pmatrix} \cdot \begin{pmatrix}\text{number of ways the}\\ \text{leading and the supporting}\\ \text{actresses can be chosen}\end{pmatrix} \cdot \begin{pmatrix}\text{number of ways to}\\ \text{choose 5 of 8 male extras}\end{pmatrix} \cdot \begin{pmatrix}\text{number of ways to}\\ \text{choose 3 of 10 female extras}\end{pmatrix}$$
$$= P\,(10,2) \cdot P\,(12,2) \cdot C\,(8,5) \cdot C\,(10,3) = (90)\,(132)\,(56)\,(120) = 79{,}833{,}600.$$

65. To order a pizza, we must make several choices. First the size (4 choices), the type of crust (2 choices), and then the toppings. Since there are 14 toppings, the number of possible choices is the number of subsets of the 14 toppings, that is 2^{14} choices. So by the Fundamental Counting Principle, the number of possible pizzas is $(4)(2) \cdot 2^{14} = 131{,}072$.

67. We treat John and Jane as one object and Mike and Molly as one object, so we need the number of ways of permuting 8 objects. We then multiply this by the number of ways of arranging John and Jane within their group and arranging Mike and Molly within their group. Thus the number of possible arrangements is $P(8,8) \cdot P(2,2) \cdot P(2,2) = 8! \cdot 2! \cdot 2! = 161{,}280$.

69. (a) To find the number of ways the men and women can be seated we first select and place a man in the first seat and then arrange the other 7 people. Thus we get $\begin{pmatrix} \text{select 1 of} \\ \text{the 4 men} \end{pmatrix} \cdot \begin{pmatrix} \text{arrange the} \\ \text{remaining 7 people} \end{pmatrix} = C(4,1) \cdot P(7,7) = 4 \cdot 7! = 20{,}160$.

(b) To find the number of ways the men and women can be seated we first select and place a woman in the first and last seats and then arrange the other 6 people. Thus we get

$\begin{pmatrix} \text{arrange 2 of} \\ \text{the 4 women} \end{pmatrix} \cdot \begin{pmatrix} \text{arrange the} \\ \text{remaining 6 people} \end{pmatrix} = P(4,2) \cdot P(6,6) = 6 \cdot 6! = 8{,}640$.

71. The number of ways the top finalist can be chosen is

$\begin{pmatrix} \text{number of ways} \\ \text{to choose the 6} \\ \text{semifinalists from the 30} \end{pmatrix} \cdot \begin{pmatrix} \text{number of ways} \\ \text{to choose the 2} \\ \text{finalists from the 6} \end{pmatrix} \cdot \begin{pmatrix} \text{number of ways} \\ \text{to choose the winner} \\ \text{from the 2 finalists} \end{pmatrix} = C(30,6) \cdot C(6,2) \cdot C(2,1)$

$= (593{,}775)(15)(2) = 17{,}813{,}250$

73. We find the number of committees that can be formed and subtract those committees that contain the two picky people. There are $C(10,4)$ ways to select the committees without restrictions. The number of committees that contain the two picky people is $C(8,2)$ because once we place these two people on the committee, then we need only choose two more members from the remaining eight candidates to complete the committee. Thus the number of committees that do not contain these two people is $C(10,4) - C(8,2) = 210 - 28 = 182$.

75. We show two methods to solve this exercise.

Method 1: We consider the number of 5 member committees that can be formed from the 26 interested people and subtract off those that contain no teacher and those that contain no students. Thus the number of committees is $C(26,5) - C(12,5) - C(14,5) = 65{,}780 - 792 - 2002 = 62{,}986$.

Method 2: In the method we construct all the committees that are possible. Thus the number of committees is

$\begin{pmatrix} \text{committees with} \\ \text{1 student and 4 teachers} \end{pmatrix} + \begin{pmatrix} \text{committees with} \\ \text{2 students and 3 teachers} \end{pmatrix} + \begin{pmatrix} \text{committees with} \\ \text{3 students and 2 teachers} \end{pmatrix} + \begin{pmatrix} \text{committees with} \\ \text{4 students and 1 teacher} \end{pmatrix}$

$= C(14,1) \cdot C(12,4) + C(14,2) \cdot C(12,3) + C(14,3) \cdot C(12,2) + C(14,4) \cdot C(12,1)$

$= (14)(495) + (91)(220) + (364)(66) + (1001)(12)$

$= 6{,}930 + 20{,}020 + 24{,}024 + 12{,}012 = 62{,}986$

77. We are only interested in selecting a set of three marbles to give to Luke and a set of two marbles to give to Mark, not the order in which we hand out the marbles. Since both $C(10,3) \cdot C(7,2)$ and $C(10,2) \cdot C(8,3)$ count the number of ways this can be done, these numbers must be equal. (Calculating these values shows that they are indeed equal.) In general, if we wish to find two distinct sets of k and r objects selected from n objects ($k + r \le n$), then we can either first select the k objects from the n objects and then select the r objects from the $n - k$ remaining objects, or we can first select the r objects from the n objects and then the k objects from the $n - r$ remaining objects. Thus $\binom{n}{r} \cdot \binom{n-r}{k} = \binom{n}{k} \cdot \binom{n-k}{r}$.

13.3   Probability

1. Let H stand for head and T for tails.

   (a) The sample space is $S = \{HH, HT, TH, TT\}$.

   (b) Let E be the event of getting exactly two heads, so $E = \{HH\}$. Then $P(E) = \frac{n(E)}{n(S)} = \frac{1}{4}$.

   (c) Let F be the event of getting at least one head. Then $F = \{HH, HT, TH\}$, and $P(F) = \frac{n(F)}{n(S)} = \frac{3}{4}$.

   (d) Let G be the event of getting exactly one head, that is, $G = \{HT, TH\}$. Then $P(G) = \frac{n(G)}{n(S)} = \frac{2}{4} = \frac{1}{2}$.

3. (a) Let E be the event of rolling a six. Then $P(E) = \frac{n(E)}{n(S)} = \frac{1}{6}$.

   (b) Let F be the event of rolling an even number. Then $F = \{2, 4, 6\}$. So $P(F) = \frac{n(F)}{n(S)} = \frac{3}{6} = \frac{1}{2}$.

   (c) Let G be the event of rolling a number greater than 5. Since 6 is the only face greater than 5, $P(G) = \frac{n(G)}{n(S)} = \frac{1}{6}$.

5. (a) Let E be the event of choosing a king. Since a deck has 4 kings, $P(E) = \frac{n(E)}{n(S)} = \frac{4}{52} = \frac{1}{13}$.

   (b) Let F be the event of choosing a face card. Since there are 3 face cards per suit and 4 suits, $P(F) = \frac{n(F)}{n(S)} = \frac{12}{52} = \frac{3}{13}$.

   (c) Let F be the event of choosing a face card. Then $P(F') = 1 - P(F) = 1 - \frac{3}{13} = \frac{10}{13}$.

7. (a) Let E be the event of selecting a red ball. Since the jar contains 5 red balls, $P(E) = \frac{n(E)}{n(S)} = \frac{5}{8}$.

   (b) Let F be the event of selecting a yellow ball. Since there is only one yellow ball,
$$P(F') = 1 - P(F) = 1 - \frac{n(F)}{n(S)} = 1 - \frac{1}{8} = \frac{7}{8}.$$

   (c) Let G be the event of selecting a black ball. Since there are no black balls in the jar, $P(G) = \frac{n(G)}{n(S)} = \frac{0}{8} = 0$.

9. (a) Let E be the event of drawing a red sock. Since 3 pairs are red, the drawer contains 6 red socks, and so
$$P(E) = \frac{n(E)}{n(S)} = \frac{6}{18} = \frac{1}{3}.$$

   (b) Let F be the event of drawing another red sock. Since there are 17 socks left of which 5 are red, $P(F) = \frac{n(F)}{n(S)} = \frac{5}{17}$.

11. (a) Let E be the event of choosing a "T". Since 3 of the 16 letters are T's, $P(E) = \frac{3}{16}$.

   (b) Let F be the event of choosing a vowel. Since there are 6 vowels, $P(F) = \frac{6}{16} = \frac{3}{8}$.

   (c) Let F be the event of choosing a vowel. Then $P(F') = 1 - \frac{3}{8} = \frac{5}{8}$.

13. Let E be the event of choosing 5 cards of the same suit. Since there are 4 suits and 13 cards in each suit, $n(E) = 4 \cdot C(13, 5)$. Also, $n(S) = C(52, 5)$. Therefore, $P(E) = \frac{4 \cdot C(13,5)}{C(52,5)} = \frac{5,148}{2,598,960} \approx 0.00198$.

15. Let E be the event of dealing a royal flush (ace, king, queen, jack, and 10 of the same suit). Since there is only one such sequence for each suit, there are only 4 royal flushes, so $P(E) = \frac{4}{C(52,5)} = \frac{4}{2,598,960} \approx 1.53908 \times 10^{-6}$.

17. (a) Let B stand for "boy"and G stand for "girl". Then
$$S = \{BBBB, GBBB, BGBB, BBGB, BBBG, GGBB, GBGB, GBBG,$$
$$BGGB, BGBG, BBGG, BGGG, GBGG, GGBG, GGGB, GGGG\}$$

   (b) Let E be the event that the couple has only boys. Then $E = \{BBBB\}$ and $P(E) = \frac{1}{16}$.

   (c) Let F be the event that the couple has 2 boys and 2 girls. Then
$$F = \{GGBB, GBGB, GBBG, BGGB, BGBG, BBGG\}, \text{ so } P(F) = \frac{6}{16} = \frac{3}{8}.$$

   (d) Let G be the event that the couple has 4 children of the same sex. Then $G = \{BBBB, GGGG\}$, and
$$P(G) = \frac{2}{16} = \frac{1}{8}.$$

   (e) Let H be the event that the couple has at least 2 girls. Then H' is the event that the couple has fewer than two girls. Thus,
$$H' = \{BBBB, GBBB, BGBB, BBGB, BBBG\}, \text{ so } n(H') = 5, \text{ and } P(H) = 1 - P(H') = 1 - \frac{5}{16} = \frac{11}{16}.$$

19. Let E be the event that the ball lands in an odd numbered slot. Since there are 18 odd numbers between 1 and 36,
$$P(E) = \frac{18}{38} = \frac{9}{19}.$$

21. Let E be the event of picking the 6 winning numbers. Since there is only one way to pick these,
$$P(E) = \frac{1}{C(49,6)} = \frac{1}{13{,}983{,}816} \approx 7.15 \times 10^{-8}.$$

23. The sample space consist of all possible True-False combinations, so $n(S) = 2^{10}$.

(a) Let E be the event that the student answers all 10 questions correctly. Since there is only one way to answer all 10 questions correctly, $P(E) = \dfrac{1}{2^{10}} = \dfrac{1}{1024}$.

(b) Let F be the event that the student answers exactly 7 questions correctly. The number of ways to answer exactly 7 of the 10 questions correctly is the number of ways to choose 7 of the 10 questions, so $n(F) = C(10,7)$. Therefore,
$$P(E) = \frac{C(10,7)}{2^{10}} = \frac{120}{1024} = \frac{15}{128}.$$

25. (a) Let E be the event that the monkey types "Hamlet" as his first word. Since "Hamlet" contains 6 letters and there are 48 typewriter keys, $P(E) = \dfrac{1}{48^6} \approx 8.18 \times 10^{-11}$.

(b) Let F be the event that the monkey types "to be or not to be" as his first words. Since this phrase has 18 characters (including the blanks), $P(F) = \dfrac{1}{48^{18}} \approx 5.47 \times 10^{-31}$.

27. Let E be the event that the monkey will arrange the 11 blocks to spell PROBABILITY as his first word. The number of ways of arranging these blocks is the number of distinguishable permutations of 11 blocks. Since there are two blocks labeled B and two blocks labeled I, the number of distinguishable permutations is $\dfrac{11!}{2!\,2!}$. Only one of these arrangements spells the word PROBABILITY. Thus $P(E) = \dfrac{1}{\frac{11!}{2!\,2!}} = \dfrac{2!\,2!}{11!} \approx 1.00 \times 10^{-7}$.

29. (a) Let E be the event that the pea is tall. Since tall is dominant, $E = \{TT, Tt, tT\}$. So $P(E) = \frac{3}{4}$.

(b) E' is the event that the pea is short. So $P(E') = 1 - P(E) = 1 - \frac{3}{4} = \frac{1}{4}$.

31. (a) Yes, the events are mutually exclusive since a person cannot be both male and female.

(b) No, the events are not mutually exclusive since a person can be both tall and blond.

33. (a) Yes, the events are mutually exclusive since the number cannot be both even and odd. So
$$P(E \cup F) = P(E) + P(F) = \tfrac{3}{6} + \tfrac{3}{6} = 1.$$

(b) No, the events are not mutually exclusive since 6 is both even and greater than 4. So
$$P(E \cup F) = P(E) + P(F) - P(E \cap F) = \tfrac{3}{6} + \tfrac{2}{6} - \tfrac{1}{6} = \tfrac{2}{3}.$$

35. (a) No, the events E and F are not mutually exclusive since the Jack, Queen, and King of spades are both face cards and spades. So $P(E \cup F) = P(E) + P(F) - P(E \cap F) = \tfrac{13}{52} + \tfrac{12}{52} - \tfrac{3}{52} = \tfrac{11}{26}$.

(b) Yes, the events E and F are mutually exclusive since the card cannot be both a heart and a spade. So
$$P(E \cup F) = P(E) + P(F) = \tfrac{13}{52} + \tfrac{13}{52} = \tfrac{1}{2}.$$

37. (a) Let E be the event that the spinner stops on red. Since 12 of the regions are red, $P(E) = \tfrac{12}{16} = \tfrac{3}{4}$.

(b) Let F be the event that the spinner stops on an even number. Since 8 of the regions are even-numbered, $P(F) = \tfrac{8}{16} = \tfrac{1}{2}$.

(c) Since 4 of the even-numbered regions are red, $P(E \cup F) = P(E) + P(F) - P(E \cap F) = \tfrac{3}{4} + \tfrac{1}{2} - \tfrac{4}{16} = 1$.

39. Let E be the event that the ball lands in an odd numbered slot and F be the event that it lands in a slot with a number higher than 31. Since there are two odd-numbered slots with numbers greater than 31,
$$P(E \cup F) = P(E) + P(F) - P(E \cap F) = \tfrac{18}{38} + \tfrac{5}{38} - \tfrac{2}{38} = \tfrac{21}{38}.$$

41. Let E be the event that the committee is all male and F the event it is all female. The sample space is the set of all ways that 5 people can be chosen from the group of 14. These events are mutually exclusive, so
$$P(E \cup F) = P(E) + P(F) = \frac{C(6,5)}{C(14,5)} + \frac{C(8,5)}{C(14,5)} = \frac{6+56}{2002} = \frac{31}{1001}.$$

43. Let E be the event that the marble is red and F be the event that the number is odd-numbered. Then E' is the event that the marble is blue, and F' is the event that the marble is even-numbered.

(a) $P(E) = \frac{6}{16} = \frac{3}{8}$

(b) $P(F) = \frac{8}{16} = \frac{1}{2}$

(c) $P(E \cup F) = P(E) + P(F) - P(E \cap F) = \frac{6}{16} + \frac{8}{16} - \frac{3}{16} = \frac{11}{16}$

(d) $P(E' \cup F') = P(E') + P(F') - P(E' \cap F') = \frac{10}{16} + \frac{8}{16} - \frac{5}{16} = \frac{13}{16}$.

45. (a) Yes, the first roll does not influence the outcome of the second roll.

(b) The probability of getting a six on both rolls is $P(E \cap F) = P(E) \cdot P(F) = \left(\frac{1}{6}\right)\left(\frac{1}{6}\right) = \frac{1}{36}$.

47. (a) Let E_A and E_B be the event that the respective spinners stop on a purple region. Since these events are independent,
$$P(E_A \cap E_B) = P(E_A) \cdot P(E_B) = \frac{1}{4} \cdot \frac{2}{8} = \frac{1}{16}.$$

(b) Let F_A and F_B be the event that the respective spinners stop on a blue region. Since these events are independent,
$$P(F_A \cap F_B) = P(F_A) \cdot P(F_B) = \frac{1}{4} \cdot \frac{1}{8} = \frac{1}{32}.$$

49. Let E be the event of getting a 1 on the first roll, and let F be the event of getting an even number on the second roll. Since these events are independent, $P(E \cap F) = P(E) \cdot P(F) = \frac{1}{6} \cdot \frac{3}{6} = \frac{1}{12}$.

51. Let E be the event that the player wins on spin 1, and let F be the event that the player wins on spin 2. What happens on the first spin does not influence what happens on the second spin, so the events are independent. Thus,
$$P(E \cap F) = P(E) \cdot P(F) = \frac{1}{38} \cdot \frac{1}{38} = \frac{1}{1444}.$$

53. Let E, F and G denote the events of rolling two ones on the first, second, and third rolls, respectively, of a pair of dice. The events are independent, so $P(E \cap F \cap G) = P(E) \cdot P(F) \cdot P(G) = \frac{1}{36} \cdot \frac{1}{36} \cdot \frac{1}{36} = \frac{1}{36^3} \approx 2.14 \times 10^{-5}$.

55. The probability of getting 2 red balls by picking from jar B is $\left(\frac{5}{7}\right)\left(\frac{4}{6}\right) = \frac{10}{21}$. The probability of getting 2 red balls by picking one ball from each jar is $\left(\frac{3}{7}\right)\left(\frac{5}{7}\right) = \frac{15}{49}$. The probability of getting 2 red balls after putting all balls in one jar is $\left(\frac{8}{14}\right)\left(\frac{7}{13}\right) = \frac{4}{13}$. Hence, picking both balls from jar B gives the greatest probability.

57. Let E be the event that two of the students have the same birthday. It is easier to consider the complementary event E': no two students have the same birthday. Then

$$
\begin{aligned}
P(E') &= \frac{\text{number of ways to assign 8 different birthdays}}{\text{number of ways to assign 8 birthdays}} = \frac{P(365, 8)}{365^8} \\
&= \frac{365 \cdot 364 \cdot 363 \cdot 362 \cdot 361 \cdot 360 \cdot 359 \cdot 358}{365 \cdot 365 \cdot 365 \cdot 365 \cdot 365 \cdot 365 \cdot 365 \cdot 365} \approx 0.92566
\end{aligned}
$$

So $P(E) = 1 - P(E') \approx 0.07434$.

59. Let E be the event that she opens the lock within an hour. The number of combinations she can try in one hour is $10 \cdot 60 = 600$. The number of possible combinations is $P(40, 3)$ assuming that no number can be repeated. Thus
$$P(E) = \frac{600}{P(40, 3)} = \frac{600}{59{,}280} = \frac{5}{494} \approx 0.010.$$

61. Let E be the event that Paul stands next to Phyllis. To find $n(E)$ we treat Paul and Phyllis as one object and find the number of ways to arrange the 19 objects and then multiply the result by the number of ways to arrange Paul and Phyllis. So $n(E) = 19! \cdot 2!$. The sample space is all the ways that 20 people can be arranged. Thus $P(E) = \frac{19! \cdot 2!}{20!} = \frac{2}{20} = 0.10$.

63. The sample space for the genders of Mrs. Smith's children is $S_S = \{(b, g), (b, b)\}$, where we list the firstborn first. The sample space for the genders of Mrs. Jones' children is $S_J = \{(b, g), (g, b), (b, b)\}$. Thus, the probability that Mrs. Smith's other child is a boy is $\frac{1}{2}$, while the probability that Mrs. Jones' other child is a boy is $\frac{1}{3}$. The reason there is a difference is that we know that Mrs. Smith's firstborn is a boy.

13.4   Binomial Probability

1. $P\,(2 \text{ successes in } 5) = C\,(5,2) \cdot (0.7^2)\,(0.3^3) = 0.13230$

3. $P\,(0 \text{ success in } 5) = C\,(5,0) \cdot (0.7^0)\,(0.3^5) = 0.00243$

5. $P\,(1 \text{ success in } 5) = C\,(5,1) \cdot (0.7^1)\,(0.3^4) = 0.02835$

7. $P\,(\text{at least 4 successes}) = P\,(4 \text{ successes}) + P\,(5 \text{ successes}) = 0.36015 + 0.16807 = 0.52822$

9. $P\,(\text{at most 1 failure}) = P\,(0 \text{ failure}) + P\,(1 \text{ failure}) = P\,(5 \text{ successes}) + P\,(4 \text{ successes}) = 0.52822$

11. $P\,(\text{at least 2 successes})\quad = P\,(2 \text{ successes}) + P\,(3 \text{ successes}) + P\,(4 \text{ successes}) + P\,(5 \text{ successes})$
$$= 0.13230 + 0.30870 + 0.36015 + 0.16807 = 0.96922$$

13. Here "success" is "face is 4" and $P\,(\text{face is 4}) = \frac{1}{6}$. Then $P\,(2 \text{ successes in } 6) = C\,(6,2) \cdot \left(\frac{1}{6}\right)^2 \left(\frac{5}{6}\right)^4 = 0.20094$.

15. $P\,(4 \text{ successes in } 10) = C\,(10,4) \cdot (0.4^4)\,(0.6^6) \approx 0.25082$

17. (a) $P\,(5 \text{ in } 10) = C\,(10,5) \cdot (0.45^5)\,(0.55^5) \approx 0.23403$

   (b) $P\,(\text{at least 3})\quad = 1 - P\,(\text{at most 2}) = P\,(0 \text{ in } 10) + P\,(1 \text{ in } 10) + P\,(2 \text{ in } 10)$
$$= 1 - \left[C\,(10,0) \cdot (0.45^0)\,(0.55^{10}) + C\,(10,1) \cdot (0.45^1)\,(0.55^9) + C\,(10,2) \cdot (0.45^2)\,(0.55^8)\right]$$
$$\approx 0.90044$$

19. (a) The complement of at least 1 germinating is no seed germinating, so
$$P\,(\text{at least 1 germinates}) = 1 - P\,(0 \text{ germinates}) = 1 - C\,(4,0) \cdot (0.75^0)\,(0.25^4) \approx 0.99609.$$

   (b) $P\,(\text{at least 2 germinate})\quad = P\,(2 \text{ germinates}) + P\,(3 \text{ germinates}) + P\,(4 \text{ germinates})$
$$= C\,(4,2) \cdot (0.75^2)\,(0.25^2) + C\,(4,3) \cdot (0.75^3)\,(0.25^1) + C\,(4,4) \cdot (0.75^4)\,(0.25^0)$$
$$\approx 0.94922$$

   (c) $P\,(4 \text{ germinates}) = C\,(4,4) \cdot (0.75^4)\,(0.25^0) \approx 0.31641$

21. (a) $P\,(\text{all 10 are boys}) = C\,(10,10) \cdot (0.52^{10})\,(0.48^0) \approx 0.0014456.$

   (b) $P\,(\text{all 10 are girls}) = C\,(10,0) \cdot (0.52^0)\,(0.48^{10}) \approx 0.00064925.$

   (c) $P\,(5 \text{ in } 10 \text{ are boys}) = C\,(10,5) \cdot (0.52^5)\,(0.48^5) \approx 0.24413.$

23. (a) $P\,(3 \text{ in } 3) = C\,(3,3) \cdot (0.005^3)\,(0.995^0) \approx 0.000000125.$

   (b) The complement of "one or more bulbs is defective" is "none of the bulbs is defective." So
$$P\,(\text{at least 1 defective}) = 1 - P\,(\text{none is defective}) = 1 - C\,(3,0) \cdot (0.005^0)\,(0.995^3) \approx 0.014925.$$

25. The complement of "2 or more workers call in sick" is "0 or 1 worker calls in sick." So
$$P\,(2 \text{ or more})\quad = 1 - \left[P\,(0 \text{ in } 8) + P\,(1 \text{ in } 8)\right]$$
$$= 1 - \left[C\,(8,0) \cdot (0.04^0)\,(0.96^8) + C\,(8,1) \cdot (0.04^1)\,(0.96^7)\right] \approx 0.038147$$

27. (a) $P\,(6 \text{ in } 6) = C\,(6,6) \cdot (0.75^6)\,(0.25^0) \approx 0.17798$

   (b) $P\,(0 \text{ in } 6) = C\,(6,0) \cdot (0.75^0)\,(0.25^6) \approx 0.00024414$

   (c) $P\,(3 \text{ in } 6) = C\,(6,3) \cdot (0.75^3)\,(0.25^3) \approx 0.13184$

   (d) $P\,(\text{at least 2 seasick})\quad = 1 - P\,(\text{at most 1 seasick}) = 1 - \left[P\,(6 \text{ in } 6 \text{ OK}) + P\,(5 \text{ in } 6 \text{ OK})\right]$
$$= 1 - \left[C\,(6,6) \cdot (0.75^6)\,(0.25^0) + C\,(6,5) \cdot (0.75^5)\,(0.25^1)\right] \approx 0.46606$$

29. (a) The complement of "at least one gets the disease" is "none gets the disease." Then

$$P\left(\text{at least 1 gets the disease}\right) = 1 - P\left(0\text{ gets the disease}\right) = 1 - C\left(4, 0\right) \cdot \left(0.25^0\right)\left(0.75^4\right) \approx 0.68359.$$

(b) $P\left(\text{at least 3 get the disease}\right)$ $= P\left(3\text{ get the disease}\right) + P\left(4\text{ get the disease}\right)$

$$= C\left(4, 3\right) \cdot \left(0.25^3\right)\left(0.75^1\right) + C\left(4, 4\right) \cdot \left(0.25^4\right)\left(0.75^0\right) \approx 0.05078$$

31. Fred (a nonsmoker) is already in the room, so this exercise concerns the remaining 4 participants assigned to the room.

(a) $P\left(1\text{ in 4 is a smoker}\right) = C\left(4, 1\right) \cdot \left(0.3^1\right)\left(0.7^3\right) = 0.4116$

(b) $P\left(\text{at least 1 smoker}\right) = 1 - P\left(0\text{ in 4 is a smoker}\right) = 1 - C\left(4, 0\right) \cdot \left(0.3^0\right)\left(0.7^4\right) = 0.7599$

33. (a)

Number of heads	Probability
0	0.001953
1	0.017578
2	0.070313
3	0.164063
4	0.246094
5	0.246094
6	0.164063
7	0.070313
8	0.017578
9	0.001953

(b) Between 4 and 5 heads.

(c) If the coin is flipped 101 times, then 50 and 51 heads has the greatest probability of occurring. If the coin is flipped 100 times, then 50 heads has the greatest probability of occurring.

13.5   Expected Value

1. Mike gets \$2 with probability $\frac{1}{2}$ and \$1 with probability $\frac{1}{2}$. Thus, $E = \left(2\right)\left(\frac{1}{2}\right) + \left(1\right)\left(\frac{1}{2}\right) = 1.5$, and so his expected winnings are \$1.50 per game.

3. Since the probability of drawing the ace of spades is $\frac{1}{52}$, the expected value of this game is $E = \left(100\right)\left(\frac{1}{52}\right) + \left(-1\right)\left(\frac{51}{52}\right) = \frac{49}{52} \approx 0.94$. So your expected winnings are \$0.94 per game.

5. Since the probability that Carol rolls a six is $\frac{1}{6}$, the expected value of this game is $E = \left(3\right)\left(\frac{1}{6}\right) + \left(0.50\right)\left(\frac{5}{6}\right) = \frac{5.5}{6} \approx 0.9167$. So Carol expects to win \$0.92 per game.

7. Since the probability that the die shows an even number equals the probability that that die shows an odd number, the expected value of this game is $E = \left(2\right)\left(\frac{1}{2}\right) + \left(-2\right)\left(\frac{1}{2}\right) = 0$. So Tom should expect to break even after playing this game many times.

9. Since it costs \$0.50 to play, if you get a silver dollar, you win only $1 - 0.50 = \$0.50$. Thus the expected value of this game is $E = \left(0.50\right)\left(\frac{2}{10}\right) + \left(-0.50\right)\left(\frac{8}{10}\right) = -0.30$. So your expected winnings are $-\$0.30$ per game. In other words, you should expect to lose \$0.30 per game.

11. You can either win \$35 or lose \$1, so the expected value of this game is $E = \left(35\right)\left(\frac{1}{38}\right) + \left(-1\right)\left(\frac{37}{38}\right) = -\frac{2}{38} = -0.0526$. Thus the expected value is $-\$0.0526$ per game.

13. By the rules of the game, a player can win \$10 or \$5, break even, or lose \$100. Thus the expected value of this game is $E = \left(10\right)\left(\frac{10}{100}\right) + \left(5\right)\left(\frac{10}{100}\right) + \left(-100\right)\left(\frac{2}{100}\right) + \left(0\right)\left(\frac{78}{100}\right) = -0.50$. So the expected winnings per game are $-\$0.50$.

15. If the stock goes up to \$20, she expects to make $\$20 - \$5 = \$15$. And if the stock falls to \$1, then she has lost $\$5 - \$1 = \$4$. So the expected value of her profit is $E = \left(15\right)\left(0.1\right) - \left(4\right)\left(0.9\right) = -2.1$. Thus, her expected profit per share is $-\$2.10$, that is, she should expect to lose \$2.10 per share. She did not make a wise investment.

17. There are $C(49, 6)$ ways to select a group of six numbers from the group of 49 numbers, of which only one is a winning set. Thus the expected value of this game is $E = (10^6 - 1) \left(\dfrac{1}{C(49,6)} \right) + (-1) \left(1 - \dfrac{1}{C(49,6)} \right) \approx -\$0.93$.

19. Let x be the fair price to pay to play this game. Then the game is fair whenever $E = 0 \Leftrightarrow$
$(13 - x) \left(\frac{4}{52} \right) + (-x) \left(\frac{48}{52} \right) = 0 \Leftrightarrow 52 - 52x = 0 \Leftrightarrow x = 1$. Thus, a fair price to pay to play this game is $1.

Chapter 13 Review

1. The number of possible outcomes is

$\begin{pmatrix} \text{number of outcomes} \\ \text{when a coin is tossed} \end{pmatrix} \cdot \begin{pmatrix} \text{number of outcomes} \\ \text{a die is rolled} \end{pmatrix} \cdot \begin{pmatrix} \text{number of ways} \\ \text{to draw a card} \end{pmatrix} = (2)(6)(52) = 624.$

3. (a) Order is not important, and there are no repetitions, so the number of different two-element subsets is

$C(5, 2) = \dfrac{5!}{2!\,3!} = \dfrac{5 \cdot 4}{2} = 10.$

(b) Order is important, and there are no repetitions, so the number of different two-letter words is $P(5, 2) = \dfrac{5!}{3!} = 20$.

5. You earn a score of 70% by answering exactly 7 of the 10 questions correctly. The number of different ways to answer the questions correctly is $C(10, 7) = \dfrac{10!}{7!\,3!} = 120.$

7. You must choose two of the ten questions to omit, and the number of ways of choosing these two questions is

$C(10, 2) = \dfrac{10!}{2!\,8!} = 45.$

9. The maximum number of employees using this security system is

$\begin{pmatrix} \text{number of choices} \\ \text{for the first letter} \end{pmatrix} \cdot \begin{pmatrix} \text{number of choices} \\ \text{for the second letter} \end{pmatrix} \cdot \begin{pmatrix} \text{number of choices} \\ \text{for the third letter} \end{pmatrix} = (26)(26)(26) = 17,576.$

11. We could count the number of ways of choosing 7 of the flips to be heads; equivalently we could count the number of ways of choosing 3 of the flips to be tails. Thus, the number of different ways this can occur is $C(10, 7) = C(10, 3) = \dfrac{10!}{3!\,7!} = 120.$

13. Let x be the number of people in the group. Then $C(x, 2) = 10 \Leftrightarrow \dfrac{x!}{2!\,(x-2)!} = 10 \Leftrightarrow \dfrac{x!}{(x-2)!} = 20 \Leftrightarrow x(x-1) = 20$
$\Leftrightarrow x^2 - x - 20 = 0 \Leftrightarrow (x - 5)(x + 4) = 0 \Leftrightarrow x = 5$ or $x = -4$. So there are 5 people in this group.

15. A letter can be represented by a sequence of length 1, a sequence of length 2, or a sequence of length 3. Since each symbol is either a dot or a dash, the possible number of letters is

$\begin{pmatrix} \text{number of letters} \\ \text{using 3 symbols} \end{pmatrix} + \begin{pmatrix} \text{number of letters} \\ \text{using 2 symbols} \end{pmatrix} + \begin{pmatrix} \text{number of letters} \\ \text{using 1 symbol} \end{pmatrix} = 2^3 + 2^2 + 2 = 14.$

17. (a) Since we cannot choose a major and a minor in the same subject, the number of ways a student can select a major and a minor is $P(16, 2) = 16 \cdot 15 = 240.$

(b) Again, since we cannot have repetitions and the order of selection is important, the number of ways to select a major, a first minor, and a second minor is $P(16, 3) = 16 \cdot 15 \cdot 14 = 3360.$

(c) When we select a major and 2 minors, the order in which we choose the minors is not important. Thus the number of

ways to select a major and 2 minors is $\begin{pmatrix} \text{number of ways} \\ \text{to select a major} \end{pmatrix} \cdot \begin{pmatrix} \text{number of ways to} \\ \text{select two minors} \end{pmatrix} = 16 \cdot C(15, 2) = 16 \cdot 105 = 1680.$

19. Since the letters are distinct, the number of anagrams of the word TRIANGLE is $8! = 40,320.$

21. **(a)** The possible number of committees is $C(18,7) = 31{,}824$.

(b) Since we must select the 4 men from the group of 10 men and the 3 women from the group of 8 women, the possible

number of committees is $\left(\begin{array}{c}\text{number of ways to}\\\text{choose 4 of 10 men}\end{array}\right) \cdot \left(\begin{array}{c}\text{number of ways to}\\\text{choose 3 of 8 women}\end{array}\right) = C(10,4)\cdot C(8,3) = 210\cdot 56 = 11{,}760.$

(c) We remove Susie from the group of 18, so the possible number of committees is $C(17,7) = 19{,}448$.

(d) The possible number of committees is

$\left(\begin{array}{c}\text{possible number of}\\\text{committees with 5 women}\end{array}\right) + \left(\begin{array}{c}\text{possible number of}\\\text{committees with 6 women}\end{array}\right) + \left(\begin{array}{c}\text{possible number of}\\\text{committees with 7 women}\end{array}\right)$

$= C(8,5)\cdot C(10,2) + C(8,6)\cdot C(10,1) + C(8,7)\cdot C(10,0) = 56\cdot 45 + 28\cdot 10 + 8\cdot 1 = 2808$

(e) Since the committee is to have 7 members, "at most two men" is the same as "at least five women," which we found in part (d). So the number is also 2808.

(f) We select the specific offices first, then complete the committee from the remaining members of the group. So the number of possible committees is

$\left(\begin{array}{c}\text{number of ways to choose}\\\text{a chairman, a vice-chairman,}\\\text{and a secretary}\end{array}\right) \cdot \left(\begin{array}{c}\text{number of ways to choose}\\\text{4 other members}\end{array}\right) = P(18,3)\cdot C(15,4) = 4896\cdot 1365 = 6{,}683{,}040.$

23. Let R_n denote the event that the nth ball is red and let W_n denote the event that the nth ball is white.

(a) $P(\text{both balls are red}) = P(R_1 \cap R_2) = P(R_1)\cdot P(R_2 \mid R_1) = \frac{10}{15}\cdot\frac{9}{14} = \frac{3}{7}.$

(b) *Solution 1:* The probability that one is white and that the other is red is

$\dfrac{\text{number of ways to select one white and one red}}{\text{number of ways to select two balls}} = \dfrac{C(10,1)\cdot C(5,1)}{C(15,2)} = \dfrac{10}{21}.$

Solution 2:

$P(\text{one white and one red}) = P(W_1 \cap R_2) + P(R_1 \cap W_2) = P(W_1)\cdot P(R_2 \mid W_1) + P(R_1)\cdot P(W_2 \mid R_1)$
$= \frac{5}{15}\cdot\frac{10}{14} + \frac{10}{15}\cdot\frac{5}{14} = \frac{10}{21}$

(c) *Solution 1:* Let E be the event "at least one is red". Then E' is the event "both are white".
$P(E') = P(W_1 \cap W_2) = P(W_1)\cdot P(W_2 \mid W_1) = \frac{5}{15}\cdot\frac{4}{14} = \frac{2}{21}.$ Thus $P(E) = 1 - \frac{2}{21} = \frac{19}{21}.$
Solution 2: $P(\text{at least one is red}) = P(\text{one red and one white}) + P(\text{both red}) = \frac{10}{21} + \frac{9}{21} = \frac{19}{21}$ (from (a) and (b)).

(d) Since 5 of the 15 balls are both red and even-numbered, the probability that both balls are red and even-numbered is $\frac{5}{15}\cdot\frac{4}{14} = \frac{2}{21}.$

(e) Since 2 of the 15 balls are both white and odd-numbered, the probability that both are white is $\frac{2}{15}\cdot\frac{1}{14} = \frac{1}{105}.$

25. The probability that you select a mathematics book is $\dfrac{\text{number of ways to select a mathematics book}}{\text{number of ways to select a book}} = \dfrac{4}{10} = \dfrac{2}{5} = 0.4.$

27. **(a)** $P(\text{ace}) = \frac{4}{52} = \frac{1}{13}$

(b) Let E be the event the card chosen is an ace, and let F be the event the card chosen is a jack. Then
$P(E \cup F) = P(E) + P(F) = \frac{4}{52} + \frac{4}{52} = \frac{2}{13}.$

(c) Let E be the event the card chosen is an ace, and let F be the event the card chosen is a spade. Then
$P(E \cup F) = P(E) + P(F) - P(E \cap F) = \frac{4}{52} + \frac{13}{52} - \frac{1}{52} = \frac{4}{13}.$

(d) Let E be the event the card chosen is an ace, and let F be the event the card chosen is a red card. Then
$P(E \cap F) = \dfrac{n(E \cap F)}{n(S)} = \frac{2}{52} = \frac{1}{26}.$

29. (a) The probability the first die shows some number is 1, and the probability the second die shows the same number is $\frac{1}{6}$. So the probability each die shows the same number is $1 \cdot \frac{1}{6} = \frac{1}{6}$.

(b) By part (a), the event of showing the same number has probability of $\frac{1}{6}$, and the complement of this event is that the dice show different numbers. Thus the probability that the dice show different numbers is $1 - \frac{1}{6} = \frac{5}{6}$.

31. In the numbers game lottery, there are 1000 possible "winning" numbers.

(a) The probability that John wins $500 is $\frac{1}{1000}$.

(b) There are $P(3,3) = 6$ ways to arrange the digits 1, 5, 9. However, if John wins only $50, it means that his number 159 was not the winning number. Thus the probability is $\frac{5}{1000} = \frac{1}{200}$.

33. There are 36 possible outcomes in rolling two dice and 6 ways in which both dice show the same numbers, namely, $(1, 1)$, $(2, 2)$, $(3, 3)$, $(4, 4)$, $(5, 5)$, and $(6, 6)$. So the expected value of this game is $E = (5)\left(\frac{6}{36}\right) + (-1)\left(\frac{30}{36}\right) = 0$.

35. Since Mary makes a guess as to the order of ratification of the 13 original states, the number of such guesses is $P(13, 13) = 13!$, while the probability that she guesses the correct order is $\dfrac{1}{13!}$. Thus the expected value is

$$E = (1{,}000{,}000)\left(\frac{1}{13!}\right) + (0)\left(\frac{13! - 1}{13!}\right) = 0.00016.$$ So Mary's expected winnings are $0.00016.

37. (a) Since there are only two colors of socks, any 3 socks must contain a matching pair.

(b) *Method 1:* If the two socks drawn form a matching pair then they are either both red or both blue. So

$$P\left(\begin{array}{c}\text{choosing a}\\\text{matching pair}\end{array}\right) = P\left(\begin{array}{c}\text{both red or}\\\text{both blue}\end{array}\right) = P(\text{both red}) + P(\text{both blue}) = \frac{C(20,2)}{C(50,2)} + \frac{C(30,2)}{C(50,2)} \approx 0.51.$$

Method 2: The complement of choosing a matching pair is choosing one sock of each color. So

$$P(\text{choosing a matching pair}) = 1 - P(\text{different colors}) = 1 - \frac{C(20,1) \cdot C(30,1)}{C(50,2)} \approx 1 - 0.49 = 0.51.$$

39. (a) (number of codes) = (choices for 1st digit) · (choices for 2nd digit) · · · · · (choices for 5th digit)
$$= 10 \cdot 10 \cdot 10 \cdot 10 \cdot 10 = 10^5 = 100{,}000$$

(b) Since there are five numbers (0, 1, 6, 8 and 9) that can be read upside down, we have
(number of codes) = (choices for 1st digit) · (choices for 2nd digit) · · · · · (choices for 5th digit) $= 5^5 = 3125$.

(c) Let E be the event that a zip code can be read upside down. Then by parts (a) and (b), $P(E) = \dfrac{n(E)}{n(S)} = \dfrac{5^5}{10^5} = \dfrac{1}{32}$.

(d) Suppose a zip code is turned upside down. Then the middle digit remains the middle digit, so it must be a digit that reads the same when turned upside down, that is, a 0, 1 or 8. Also, the last digit becomes the first digit, and the next to last digit becomes the second digit. Thus, once the first two digits are chosen, the last two are determined. Therefore, the number of zip codes that read the same upside down as right side up is
(number of codes) = (choices for 1st digit) · (choices for 2nd digit) · · · · · (choices for 5th digit) $= 5 \cdot 5 \cdot 3 \cdot 1 \cdot 1 = 75$.

41. (a) Using the rule for the number of distinguishable combinations, the number of divisors of N is
$(7 + 1)(2 + 1)(5 + 1) = 144$.

(b) An even divisor of N must contain 2 as a factor. Thus we place a 2 as one of the factors and count the number of distinguishable combinations of $M = 2^6 3^2 5^5$. So using the rule for the number of distinguishable combinations, the number of even divisors of N is $(6 + 1)(2 + 1)(5 + 1) = 126$.

(c) A divisor is a multiple of 6 if 2 is a factor and 3 is a factor. Thus we place a 2 as one of the factors and a 3 as one of the factors. Then we count the number of distinguishable combinations of $K = 2^6 3^1 5^5$. So using the rule for the number of distinguishable combinations, the number of even divisors of N is $(6 + 1)(1 + 1)(5 + 1) = 84$.

(d) Let E be the event that the divisor is even. Then using parts (a) and (b), $P(E) = \frac{n(E)}{n(S)} = \frac{126}{144} = \frac{7}{8}$.

43. (a) $P\left(4\text{ sixes in 8 rolls}\right) = C\left(8, 4\right) \cdot \left(\frac{1}{6}\right)^4 \left(\frac{5}{6}\right)^4 \approx 0.026048$.

(b) There are three even numbers on a die and three odds numbers, so $P\left(\text{even}\right) = P\left(\text{odd}\right) = 0.5$. Thus

$$P\left(2\text{ or more evens in 8 rolls}\right) = 1 - P\left(\text{fewer than 2 evens}\right) = 1 - \left[P\left(0\text{ evens}\right) + P\left(1\text{ even}\right)\right]$$
$$= 1 - \left[C\left(8, 0\right) \cdot \left(0.5^0\right)\left(0.5^8\right) + C\left(8, 1\right) \cdot \left(0.5^1\right)\left(0.5^7\right)\right] \approx 0.96484$$

Chapter 13 Test

1. (a) If repetition is allowed, then each letter of the word can be chosen in 10 ways since there are 10 letters. Thus the number of possible five-letter words is $10 \cdot 10 \cdot 10 \cdot 10 \cdot 10 = 10^5 = 100{,}000$.

(b) If repetition is not allowed, then since order is important, we need the number of permutations of 10 objects (the 10 letters) taken 5 at a time. Therefore, the number of possible five-letter words is $P\left(10, 5\right) = 10 \cdot 9 \cdot 8 \cdot 7 \cdot 6 = 30{,}240$.

2. There are three choices to be made: one choice each of a main course, a dessert, and a drink. Since a main course can be chosen in one of five ways, a dessert in one of three ways, and a drink in one of four ways, there are $5 \cdot 3 \cdot 4 = 60$ possible ways that a customer could order a meal.

3. (a) Order is important in the arrangement, therefore the number of ways to arrange $P\left(30, 4\right) = 657{,}720$.

(b) Here we are interested in the group of books to be taken on vacation so order is not important, therefore the number of ways to choose these books is $C\left(30, 4\right) = 27{,}405$.

4. There are two choices to be made: choose a road to travel from Ajax to Barrie, and then choose a different road from Barrie to Ajax. Since there are 4 roads joining the two cities, we need the number of permutations of 4 objects (the roads) taken 2 at a time (the road there and the road back). This number is $P\left(4, 2\right) = 4 \cdot 3 = 12$.

5. A customer must choose a size of pizza and must make a choice of toppings. There are 4 sizes of pizza, and each choice of toppings from the 14 available corresponds to a subset of the 14 objects. Since a set with 14 objects has 2^{14} subsets, the number of different pizzas this parlor offers is $4 \cdot 2^{14} = 65{,}536$.

6. (a) We want the number of ways of arranging 4 distinct objects (the letters L, O, V, E). This is the number of permutations of 4 objects taken 4 at a time. Therefore, the number of anagrams of the word LOVE is $P\left(4, 4\right) = 4! = 24$.

(b) We want the number of distinguishable permutations of 6 objects (the letters K, I, S, S, E, S) consisting of three like groups of size 1 and a like group of size 3 (the S's). Therefore, the number of different anagrams of the word KISSES is $\dfrac{6!}{1!\,1!\,1!\,3!} = \dfrac{6!}{3!} = 120$.

7. We choose the officers first. Here order is important, because the officers are different. Thus there are $P\left(30, 3\right)$ ways to do this. Next we choose the other 5 members from the remaining 27 members. Here order is not important, so there are $C\left(27, 5\right)$ ways to do this. Therefore the number of ways that the board of directors can be chosen is

$$P\left(30, 3\right) \cdot C\left(27, 5\right) = 30 \cdot 29 \cdot 28 \cdot \frac{27!}{5!\,22!} = 1{,}966{,}582{,}800.$$

8. One card is drawn from a deck.

(a) Since there are 26 red cards, the probability that the card is red is $\frac{26}{52} = \frac{1}{2}$.

(b) Since there are 4 kings, the probability that the card is a king is $\frac{4}{52} = \frac{1}{13}$.

(c) Since there are 2 red kings, the probability that the card is a red king is $\frac{2}{52} = \frac{1}{26}$.

9. Let R be the event that the ball chosen is red. Let E be the event that the ball chosen is even-numbered.

(a) Since 5 of the 13 balls are red, $P(R) = \frac{5}{13} \approx 0.3846$.

(b) Since 6 of the 13 balls are even-numbered, $P(E) = \frac{6}{13} \approx .4615$.

(c) $P(R \text{ or } E) = P(R) + P(E) - P(R \cap E) = \frac{5}{13} + \frac{6}{13} - \frac{2}{13} = \frac{9}{13} \approx 0.6923$.

10. Let E be the event of choosing 3 men. Then $P(E) = \dfrac{n(E)}{n(S)} = \dfrac{\text{number of ways to choose 3 men}}{\text{number of ways to choose 3 people}} = \dfrac{C(5,3)}{C(15,3)} \approx 0.022$.

11. Two dice are rolled. Let E be the event of getting doubles. Since a double may occur in 6 ways, $P(E) = \dfrac{n(E)}{n(S)} = \dfrac{6}{36} = \dfrac{1}{6}$.

12. There are 4 students and 12 astrological signs. Let E be the event that at least 2 have the same astrological sign. Then E' is the event that no 2 have the same astrological sign. It is easier to find E'. So

$$P(E') = \frac{\text{number of ways to assign 4 different astrological signs}}{\text{number of ways to assign 4 astrological signs}} = \frac{P(12,4)}{12^4} = \frac{12 \cdot 11 \cdot 10 \cdot 9}{12 \cdot 12 \cdot 12 \cdot 12} = \frac{55}{96}.$$

Therefore, $P(E) = 1 - P(E') = 1 - \frac{55}{96} = \frac{41}{96} \approx 0.427$.

13. (a) $P(6 \text{ heads in } 10 \text{ tosses}) = C(10,6) \cdot (0.55^6)(0.45^4) = 0.23837$.

(b) "Fewer than 3 heads" is the same as "0, 1, or 2 heads." So

$$P(\text{fewer than 3 heads}) = P(0 \text{ head in } 10) + P(1 \text{ head in } 10) + P(2 \text{ heads in } 10)$$
$$= C(10,0) \cdot (0.45^0)(0.45^{10}) + C(10,1) \cdot (0.55^1)(0.45^9) + C(10,2) \cdot (0.55^2)(0.45^8)$$
$$= 0.02739$$

14. A deck of cards contains 4 aces, 12 face cards, and 36 other cards. So the probability of an ace is $\frac{4}{52} = \frac{1}{13}$, the probability of a face card is $\frac{12}{52} = \frac{3}{13}$, and the probability of a non-ace, non-face card is $\frac{36}{52} = \frac{9}{13}$. Thus the expected value of this game is $E = (10)\left(\frac{1}{13}\right) + (1)\left(\frac{3}{13}\right) + (-.5)\left(\frac{9}{13}\right) = \frac{8.5}{13} \approx 0.654$, that is, about \$0.65.

Focus on Modeling: The Monte Carlo Method

1. **(a)** You should find that with the switching strategy, you win about 90% of the time. The more games you play, the closer to 90% your winning ratio will be.

   (b) The probability that the contestant has selected the winning door to begin with is $\frac{1}{10}$, since there are ten doors and only one is a winner. So the probability that he has selected a losing door is $\frac{9}{10}$. If the contestant switches, he exchanges a losing door for a winning door (and vice versa), so the probability that he loses is now $\frac{1}{10}$, and the probability that he wins is now $\frac{9}{10}$.

3. **(a)** You should find that player A wins about $\frac{7}{8}$ of the time. That is, if you play this game 80 times, player A should win approximately 70 times.

   (b) The game will end when either player A gets one more head or player B gets three more tails. Each toss is independent, and both heads and tails have probability $\frac{1}{2}$, so we obtain the following probabilities.

Outcome	Probability
H	$\frac{1}{2}$
TH	$\frac{1}{2} \cdot \frac{1}{2} = \frac{1}{4}$
TTH	$\frac{1}{2} \cdot \frac{1}{2} \cdot \frac{1}{2} = \frac{1}{8}$
TTT	$\frac{1}{2} \cdot \frac{1}{2} \cdot \frac{1}{2} = \frac{1}{8}$

   Since Player A wins for any outcome that ends in heads, the probability that he wins is $\frac{1}{2} + \frac{1}{4} + \frac{1}{8} = \frac{7}{8}$.

5. With 1000 trials, you are likely to obtain an estimate for π that is between 3.1 and 3.2.

7. **(a)** We can use the following TI-83 program to model this experiment. It is a minor modification of the one given in Problem 5.

```
PROGRAM:PROB7
:0→P
:For(N,1,1000)
:rand→X:rand→Y
:P+((X+Y)<1)→P
:End
:Disp "PROBABILITY IS APPROX",P/1000
```

   You should find that the probability is very close to $\frac{1}{2}$.

   (b) Following the hint, the points in the square for which $x + y < 1$ are the ones that lie below the line $x + y = 1$. This triangle has area $\frac{1}{2}$ (it takes up half the square), so the probability that $x + y < 1$ is $\frac{1}{2}$.